T0190124

Organic Solutes, Oxidative Stress, and Antioxidant Enzymes Under Abiotic Stressors

Organic Solutes, Oxidative Stress, and Antioxidant Enzymes Under Abiotic Stressors

Edited by
Arafat Abdel Hamed Abdel Latef

CRC Press
Taylor & Francis Group
Boca Raton London New York

CRC Press is an imprint of the
Taylor & Francis Group, an **informa** business

First edition published 2021 by
CRC Press
6000 Broken Sound Parkway NW, Suite 300, Boca Raton, FL 33487-2742

and by CRC Press
2 Park Square, Milton Park, Abingdon, Oxon, OX14 4RN

© 2022 Taylor & Francis Group, LLC

CRC Press is an imprint of Taylor & Francis Group, LLC

Reasonable efforts have been made to publish reliable data and information, but the author and publisher cannot assume responsibility for the validity of all materials or the consequences of their use. The authors and publishers have attempted to trace the copyright holders of all material reproduced in this publication and apologize to copyright holders if permission to publish in this form has not been obtained. If any copyright material has not been acknowledged please write and let us know so we may rectify in any future reprint.

Except as permitted under U.S. Copyright Law, no part of this book may be reprinted, reproduced, transmitted, or utilized in any form by any electronic, mechanical, or other means, now known or hereafter invented, including photocopying, microfilming, and recording, or in any information storage or retrieval system, without written permission from the publishers.

For permission to photocopy or use material electronically from this work, access www.copyright.com or contact the Copyright Clearance Center, Inc. (CCC), 222 Rosewood Drive, Danvers, MA 01923, 978-750-8400. For works that are not available on CCC please contact mpkbookspermissions@tandf.co.uk

Trademark notice: Product or corporate names may be trademarks or registered trademarks and are used only for identification and explanation without intent to infringe.

Library of Congress Cataloging-in-Publication Data

Names: Latef, Arafat Abdel Hamed Abdel, editor.
Title: Organic solutes, oxidative stress and antioxidant enzymes under
abiotic stressors / Arafat Abdel Hamed Abdel Latef.
Description: First edition. | Boca Raton, FL : CRC Press, 2021. | Includes
bibliographical references and index. | Summary: "This book presents
evidence-based approaches and techniques used to diagnose and manage
organic solutes, oxidative stress, and antioxidant enzymes in crop
plants under abiotic stressors. It discusses strategies in abiotic
stress tolerance including osmoregulation, osmoprotectants, and the
regulation of compatible solutes and antioxidant enzymes in plants. With
contributions from 49 scholars worldwide, this authoritative guide is
educational for scientists working with plants and abiotic stressors"--
Provided by publisher.
Identifiers: LCCN 2021011174 (print) | LCCN 2021011175 (ebook) | ISBN
9780367901400 (hardback) | ISBN 9781032040523 (paperback) | ISBN
9781003022879 (ebook)
Subjects: LCSH: Crops--Effect of stress on. | Plants--Effect of stress on.
| Oxidative stress. | Antioxidants.
Classification: LCC SB112.5 .O74 2021 (print) | LCC SB112.5 (ebook) | DDC
632/.1--dc23
LC record available at https://lccn.loc.gov/2021011174
LC ebook record available at https://lccn.loc.gov/2021011175

ISBN: 978-0-367-90140-0 (hbk)
ISBN: 978-1-032-04052-3 (pbk)
ISBN: 978-1-003-02287-9 (ebk)

Typeset in Times
by Deanta Global Publishing Services, Chennai, India

Contents

Editor's Biography

Professor Arafat Abdel Hamed Abdel Latef is a professor of plant physiology at the Department of Botany and Microbiology, Faculty of Science, South Valley University, Egypt. He received his B.Sc in 1998, M.Sc. in 2003 and Ph.D. in 2005 from South Valley University, Egypt. He has published 60 research articles in peer reviewed journals and 6 book chapters in international books. He is the recipient of the ParOwn 1207 Post Doctor Fellowship 2007 granted by the Egyptian Ministry of Higher Education and Scientific Research, which was carried out in the Institute of Vegetables and Flowers, Chinese Academy of Agricultural Sciences, Beijing, China. He serves as an editorial member of several reputed journals such as Topic Editor of *Plants-Basel* and Review Editor of *Frontiers in Plant Science*. Furthermore, he is an expert reviewer for more than 80 international journals. Recently, he received Top Peer Reviewer Award 2019, powered by Publons. He has received several awards, such as the South Valley University Award of Scientific Publication for the years from 2010 to 2020. In addition, he has received the South Valley University Award of Excellence 2019. He is actively engaged in studying the physio-biochemical and molecular responses of different plants under environmental stresses and their tolerance strategies under these stressors.

List of Contributors

Farhan Ahmad
Department of Bioengineering
Integral University
Lucknow, India

Muhammad Shahedul Alam
Institute of Agronomy and Crop Physiology
Justus-Liebig-University
Giessen, Germany

Manivannan Alagarsamy
ICAR Central Institute for Cotton Research
Coimbatore, India

Dalia Z. Alomari
Leibniz Institute of Plant Genetics and Crop
   Plant Research (IPK)
Gatersleben, Germany

Ahmad M. Alqudah
Institute of Agricultural and Nutritional
   Sciences
Martin Luther University Halle-Wittenberg
Betty-Heimann, Halle (Saale), Germany

Md. Tahjib-Ul-Arif
Department of Biochemistry and Molecular
   Biology
Bangladesh Agricultural University
Mymensingh, Bangladesh

Md Ashrafuzzaman
Department of Genetic Engineering &
   Biotechnology
School of Life Sciences
Shahjalal University of Science & Technology
Sylhet, Bangladesh

Shyna Bhalla
Department of Botany
Panjab University
Chandigarh, India

Siamak Shirani Bidabadi
Department of Horticulture, College of
   Agriculture
Isfahan University of Technology
Isfahan, Iran

Aditi Bisht
Department of Botany
Panjab University
Chandigarh, India

Amandeep Cheema
Department of Botany
Panjab University
Chandigarh, India

Mona F. A. Dawood
Department of Botany & Microbiology
Faculty of Science
Assiut University
Assiut, Egypt

Murat Dikilitas
Department of Plant Protection
Faculty of Agriculture
Harran University
S. Urfa, Turkey

Titash Dutta
Department of Biochemistry and
   Bioinformatics
Institute of Science
GITAM University
Andhra Pradesh, India

Heikham Evelin
Department of Botany
Rajiv Gandhi University
Arunachal Pradesh, India

Abreeq Fatima
Ranjan Plant Physiology and Biochemistry
   Laboratory
Department of Botany
University of Allahabad
Prayagraj, India

Manu Pratap Gangola
The Flowr Group (Okanagan) Inc
Kelowna, British Columbia, Canada

Sourav Garai
Department of Agronomy
Bidhan Chandra Krishi Viswavidyalaya
West Bengal, India

Neera Garg
Department of Botany
Panjab University
Chandigarh, India

Md. Rezwanul Haque
Department of Biochemistry and Molecular
    Biology
Bangladesh Agricultural University
Mymensingh, Bangladesh

M. Afzal Hossain
Department of Biochemistry and Molecular
    Biology
Bangladesh Agricultural University
Mymensingh, Bangladesh
Visakhapatnam, India

Akbar Hossain
Bangladesh Wheat and Maize Research
    Institute (BWMRI)
Dinajpur, Bangladesh

Arshad Javaid
Institute of Agricultural Sciences
University of the Punjab
Lahore, Pakistan

Aisha Kamal
Department of Bioengineering
Integral University
Lucknow, India

Sema Karakas
Department of Soil Science and Plant
    Nutrition
Faculty of Agriculture,
Harran University
S. Urfa, Turkey

Mohammad Golam Kibria
Department of Soil Science
Bangladesh Agricultural University
Mymensingh, Bangladesh

Arafat Abdel Hamed Abdel Latef
Botany and Microbiology Department
Faculty of Science
South Valley University
Qena, Egypt

Seyedeh-Somayyeh Shafiei-Masouleh
Ornamental Plants Research Center (OPRC),
    Horticultural Sciences Research Institute
    (HSRI), Agricultural Research, Education
    and Extension Organization (AREEO)
Mahallat, Iran

Mousumi Mondal
Research Scholar, Department of Agronomy
Bidhan Chandra Krishi Viswavidyalaya
West Bengal, India

Amira M.I. Mourad
Department of Botany and Microbiology,
    Faculty of Science
Assiut University
Assiut, Egypt

Yasser S. Moursi
Department of Botany, Faculty of Science
University of Fayoum
Fayoum, Egypt

Nageswara Rao Reddy Neelapu
Department of Biochemistry and
    Bioinformatics
Institute of Science
GITAM University
Visakhapatnam, India

Mohammed Arif Sadik Polash
Plant Physiology Lab, Department of Crop
    Botany
Khulna Agricultural University
Khulna, Bangladesh

Sheo Mohan Prasad
Ranjan Plant Physiology and Biochemistry
    Laboratory, Department of Botany
University of Allahabad
Prayagraj, India

Bharathi R. Ramadoss
Bioriginal Food and Science Corporation
Saskatoon, Saskatchewan. Canada

Mohammad Saidur Rhaman
Department of Seed Science and Technology
Bangladesh Agricultural University
Mymensingh, Bangladesh

Ahmed Sallam
Department of Genetics, Faculty of Agriculture
Assiut University
Assiut, Egypt

Kiran Saroy
Department of Botany
Panjab University
Chandigarh, India

Vinay Shankar
Department of Botany
Gaya College, Magadh University
Bihar, India

Parisa Sharifi
Department of Agricultural Extension and
    Education
Higher Education Center Shahid Bakeri
    Miyandoab
Urmia University
Urmia, Iran

Amna Shoaib
Institute of Agricultural Sciences
University of the Punjab
Lahore, Pakistan

Shagata Islam Shorna
Plant Physiology Lab, Department of Crop
    Botany
Bangladesh Agricultural University
Mymensingh, Bangladesh

Eray Simsek
Department of Plant Protection,
Faculty of Agriculture,
Harran University
S. Urfa, Turkey

Madhulika Singh
Ganga River Ecology Res. Lab.
Environmental Sci.CAS in Botany Institute of
    Science Banaras
Hindu University Varanasi

Usharani Subramanian
Tamil Nadu Agricultural University
Coimbatore, India

Challa Surekha
Department of Biochemistry and
    Bioinformatics
Institute of Science, GITAM University
Visakhapatnam, India

Samar G. Thabet
Department of Botany, Faculty of Science
University of Fayoum
Fayoum, Egypt

Shabir H. Wani
Mountain Research Centre for Field Crops,
    Khudwani
Sher-e-Kashmir University of Agricultural
    Sciences and Technology
Kashmir, India

1 Abiotic Stresses and Their Interactions with Each Other on Plant Growth, Development and Defense Mechanisms

Murat Dikilitas, Eray Simsek, Sema Karakas and Arafat Abdel Hamed Abdel Latef

CONTENTS

1.1   INTRODUCTION

Abiotic stress factors, such as drought, temperature, cold, salt, heavy metal stress, environmental pollution and UV light, etc., largely influence plant development and crop production. Climate models forecast an increased frequency of droughts, extreme temperatures and floods, etc. (Gourdji et al. 2013; Verma and Deepti 2016; Abdel Latef et al. 2021; Dawood et al. 2020). Increases in global temperature along with drought would seriously jeopardize crop production and quality. The threat is real and now many scenarios have been proposed in all agricultural systems. Abiotic stress factors have become a very important issue to tackle in agriculture and most of the financial budget and energy have been spent there instead of developing better tasting and healthy crop plants. Therefore, the last century, from an agricultural point of view, has been spent focusing on the increase of crop productivity to meet the demands and requirements of the world's growing population (Krishna et al. 2019). In this century, the situation has not changed, yet extra complexities have arisen that need to be dealt with. Although great successes have been made in improving abiotic stress tolerance on crop plants via classical breeding and selection approaches, the success achieved so far has been low and slow as predicted. One of the main reasons behind this is the limited reliable knowledge of the hidden biochemistry in plant cells. The complexity and severity of abiotic stress factors as well as their interactions make it more difficult to generate stress-tolerant crop plants. In crops, abiotic stress tolerance is generally measured through crop yield capacity. However, to achieve this, we need to unravel the mechanisms of stress tolerance and need to evaluate the interactions among stress factors. The abiotic stress factors can affect the crop plants in two ways; simultaneous and sequential. In simultaneous cases, two or more stress factors come together at the same time, and the tolerance against stress factors must be generated for all stressors by the host plants. In sequential stress occurrences, stress factors may follow one another. The important point here is the order of stress. As the saying goes, "the one who comes first is served first", and plants prepare themselves for the first stress factor in terms of biochemical and molecular responses, and if the energy is left, then it is sacrificed for the forthcoming stress factors (Dikilitas et al. 2020). In some cases, the first stress may delay the responses of plants for the second stress or inhibit the responses to be made for the second stress factor. The signaling mechanism here plays a crucial role under these circumstances.

Since plants are immobile and depend on the environment for their growth, development and defense, they therefore must challenge the stressors throughout their life cycle via generating new strategies. There are many abiotic stressors such as heat, cold, drought, salinity, heavy metal stresses as well as air, water and soil pollution, etc. Unlike wild plants, agricultural crop plants have to be controlled by agronomical practices. Wild plants, due to small leaf structures or deep root formations, can tolerate the environmental and biotic stresses efficiently. Their life cycle is relatively short when compared to those of agricultural crop plants. As well as this, the life cycles of crop plants have to be longer to get better quality and higher crop yield. However, due to extreme weather conditions and various abiotic stress factors, the desired crop productivity is not always possible.

In recent decades, the world has experienced more severe and frequent abiotic stressors as compared to those of previous decades, and this trend is going upward. Increases in demand for food and food products along with increased population growth as well as decreases in agricultural lands and complex interactions among abiotic stressors exert new metabolic pathways in crop plants. Researchers are, therefore, under enormous pressure to develop multiple stress-tolerant agricultural crop plants. Initially, it was estimated that a 34% food increase would be needed by 2050 to meet the demand for the increasing global population (FAO 2009). However, estimations regarding requirements for food increases as well as world population have been updated, and now it is expected that the world's population will reach 9.7 billion by 2050 and will continue at a slower pace and peak at nearly 11 billion by 2100, according to United Nations (UN) Department of Economic and Social Affairs; therefore, more food will be needed (https://www.un.org/development/desa/en/news/population/world-population-prospects-2019.html). According to recent updates, agricultural production should now be increased by at least 60–110 %, and 70% more food is needed for an additional 2.3 billion people by 2050 (Tilman et al. 2011; Wani and Sah 2014).

Generally, resistance, tolerance and sensitivity terminologies have been used to describe the situation of crop plants. The term "tolerance" is often measured as "survival" in model plants such as *Arabidopsis*, however maintenance of "yield" and "productivity" is more suitable because of economic reasons (Dolferus 2014; El Moukhtari et al. 2020). Although most of the biochemical and molecular mechanisms have been evaluated and unraveled regarding one particular abiotic stress factor, the complexity of stressors through interactions or combinations lead to many mechanisms needing to be elucidated properly, and most of the defense or tolerance mechanisms still need to be evaluated under these circumstances to generate and breed more resistant or tolerant crop plants against single or multiple stress factors. The knowledge is still evolving and needs to be developed. Therefore, breeding in crop production has been slow and inefficient. So far, most of the achievements have been made through cultivation techniques and management practices such as cultural measures and irrigation regimes, etc., rather than breeding and genetic modification (Richards et al. 2014; Dolferus 2014). We hope that our knowledge and technical ability and advanced new technologies may provide us with the ability to develop new strategies against multiple stress factors.

In the next section, we evaluated individual abiotic stress factors and their effects on plant defense mechanisms and crop yield and quality. We also elucidated the interactions among them to help plant breeders evaluate breeding mechanisms of crop plants under complex stress conditions. It is of utmost importance to determine the stress-tolerant characteristics of a diverse range of genotypes of plant species and integrate them for crop improvement. Stress-tolerant characteristics should not only be identified via biochemical pathways but also be determined via genome-wide analysis through functional and genomics approaches. This way seems to be more practical and faster if are determined to act quickly to generate multiple stress-tolerant crop plants.

1.2 MAIN ABIOTIC STRESS TYPES

During plant growth and development, plants are exposed to various stress factors such as drought, heat, salinity, heavy metals, freezing, cold, flooding, air pollution and pesticide damage, etc. Among these stress factors, drought, heat, salinity, freezing, cold temperature and pollution are the major causes of crop loss (Zhou et al. 2017). Generally, plants can cope with a low or moderate level of stress efficiently without sacrificing more energy and metabolites. However, when the stress exceeds the threshold level, the biochemical, molecular and physiological mechanisms begin, and serious symptoms of abiotic stresses become visible and lead to the death of the organism. Plants, in general, have a strategy to adapt to extreme environmental conditions and therefore can be found everywhere around the globe including the Antarctic Peninsula and deserts (Bravo and Griffith 2005; Gechev et al. 2012). Adaptation strategies may follow different routes including biochemical, physical, molecular and metabolic pathways. Exploring the molecular mechanisms of drought, heat, heavy metal or salt tolerance and their signaling molecules help us develop resistant or tolerant crop plants. However, one should understand that we cannot transfer all the required traits to one particular crop plant. Transferring the entire gene responsible for all metabolic pathways including yield and quality traits is simply impossible at the moment and will not be easy in the future because the morphological and developmental traits would interfere with each other in one particular crop plant. Mechanisms for quality traits often cross with defensive traits, and therefore, it is very difficult at the moment to generate highly resistant crop plants having good quality. However, we could evaluate each stress factor individually and understand what stimulating or inhibition mechanisms could affect the desired characteristics of crop plants. Then, we could elucidate the response of crop plants under combined stress conditions to differentiate the symptoms and biochemical and molecular pathways from those of pathways in plants under single stress conditions. As the sequencing technologies and genomics are developed, the dependence on model plant systems such as *Arabidopsis* would be reduced, and whole genomes of cereals, crop plants and vegetables could be easily made, and their responses to individual and combined stresses could be cleared fast and efficiently.

Abiotic stresses, in general, can lead to a decrease in plant performances, such as seed viability, germination power, the weak establishment of seedlings, and growth (Amooaghaie and Nikzad 2013; Li et al. 2014; Krishna et al. 2019; Cohen et al. 2021). One of the earliest signals in many abiotic stresses involves reactive oxygen species (ROS) and reactive nitrogen species (RNS) which regulate enzymatic reactions and genes responsible for enzyme and metabolite synthesis (Das and Roychoudhury 2014). Almost all kinds of abiotic stresses lead to oxidative stresses in plants. Accumulation of ROS and RNS can damage macromolecules such as protein, DNA, lipid and carbohydrates and cause the death of the organism via leading to membrane damage. Additive or synergistic effects of stresses can cause more serious oxidative stress, therefore, control of ROS and RNS for homeostasis can be very difficult to enable multi-stress tolerance (Greco et al. 2012; Zandalinas et al. 2017; Balfagon et al. 2020).

Plants, on the other hand, defend themselves via physiological, biochemical and molecular approaches. The antioxidant enzymes and non-enzymatic metabolites produced during abiotic stress include catalase (CAT), ascorbate peroxidase (APX), dehydroascorbate reductase (DHAR), glutathione peroxidase (GPX), glutathione reductase (GR), monodehydroascorbate reductase (MDHAR), guaiacol peroxidase (GOPX), glutathione-S-transferase (GST) and superoxide dismutase (SOD), and non-enzymatic antioxidative components (ascorbic acid, ASH; glutathione, GSH) which act to regulate toxic effects of ROS (Gill and Tuteja 2010; Parmar et al. 2017; Zandalinas et al. 2017; Krishna et al. 2019; Abdel Latef et al. 2020 a, b, c). For example, plants maintain their osmotic homeostasis partially via the accumulation of osmoregulatory compounds such as free proline (Abdel Latef et al. 2017 a, b, c, 2018, 2019 a, b; Gujjar et al. 2018; Osman et al. 2020) and other amino acids, soluble proteins, soluble carbohydrates and ion balance (Abdel Latef and Chaoxing 2011 a, b, 2014; Krasensky and Jonak 2012; Golldack et al. 2014; Krishna et al. 2019). Osmoregulation is also necessary for normal plant physiology such as cell turgor pressure maintenance and stomatal movement and hence improves photosynthesis and growth.

Here, the general description and pathways for abiotic stresses were given rather than describing symptoms for each stress issue unless a specific description is needed. Almost all abiotic stresses interact with each other, and their combinations should be counted as another stress factor. Since they are interrelated, they might trigger similar cellular damage or their defense responses could be antagonistic. For example, drought, salinity and heat primarily result in osmotic stress and disruption of cellular homeostasis (Kumar and Verma 2018; Krishna et al. 2019; Zandalinas et al. 2020). Although antagonistic pathways are expected in combined stress conditions, especially with those of stressors e.g. between abiotic and biotic stressors, in combined abiotic stress conditions, we need to further elucidate the signaling pathways for antagonistic crosstalk due to the presence of inconclusive research outcomes.

Among abiotic stresses, drought is one of the first environmental stresses responsible for the decrease in crop production and crop quality worldwide. It is widespread and affects crop plants in terms of growth, yield, osmotic adjustment, pigment formation, water balance, photosynthesis, defense mechanisms and membrane integrity, etc. (Krishna et al. 2019; Imadi et al. 2016). Drought is a meteorological term and is defined as a period of precipitation below normal and seriously affects plant growth and development. It has become common in many parts of the world with the increase of global temperatures. Over-irrigation in agricultural areas, especially in places experiencing drought, has also increased the shortage of water and resulted in more severe drought stress symptoms in crop plants (Sehgal et al. 2017; Krishna et al. 2019). Briefly put, plants exposed to drought stress show increased transpiration and decreased relative water content (RWC), and because of this, they exhibit severe dehydration and osmotic stress symptoms. Drought affects virtually every phase of plant physiology and biochemistry. Unlike other stress responses, plants show first symptoms by closing their stomata to save water content, however, this could also reduce the photosynthesis efficacy through a reduced intake of CO_2 and thus, reduce crop production. In some studies, the benefit of drought stress was proposed to increase the contents of valuable organic compounds. For example, Paim et al. (2020) stated that lettuce plants exposed to moderate drought stress did not show any changes in firmness during storage, but exhibited an increase in the content of flavonoids in water stress. Several studies have shown that the application of moderate stresses could contribute to a better tolerance of the plant to subsequent stresses and improved the quality of crop plants at harvest and after storage. This strategy has been known as "priming" (Savvides et al. 2016; da Silva Leite et al. 2021). Chemical applications such as proline and sodium nitroprusside have been confirmed to increase the stomatal conductance, increase the availability of water and attenuate the water stress on crop plants (da Silva Leite et al. 2021). In some cases, pre-treatment of drought at early stages also contributed to increased drought tolerance at later stages in crop plants (Mendanha et al. 2020). Insufficient rainfall, irrigated agricultural practices and fast-growing urban areas increase water demand, and this puts more pressure on crops in that they have to grow more often in drought conditions. As is commonly known, signaling chemical abscisic acid (ABA) plays a crucial role. It accumulates in tissues subjected to water stress and regulates water movement by closing stomata and increasing stomatal conductance (Fraire-Velazquez et al. 2011). Drought-stressed plants undergo serious physiological, biochemical and molecular changes. They accumulate, to the best of their ability, osmotically active soluble compounds to increase water potential and to maintain turgor pressure. For example, ABA regulates the expression of many stress-responsive genes, including the late embryogenesis abundant (LEA) proteins, leading to a reinforcement of drought stress tolerance in plants (Aroca et al. 2008; Wang et al. 2020 b). One should remember and consider while generating drought stress-tolerant or other stress-tolerant crop plants, metabolic adjustments via accumulation of chemicals in response to adverse conditions are transient and totally depend on the genetic background and the severity of stress conditions. Therefore, we cannot make stress-sensitive plants tolerant by applying chemicals or priming with drought, salt or cold stress permanently. Water stress does not only modify the biochemical and molecular pathways but also structural systems could be damaged, e.g. thylakoid membranes (Yadav et al. 2020). Drought stress is, therefore, more complicated than most other abiotic stressors and it has the potential to affect crop plants directly (Yadav et al. 2020).

One of the major abiotic stresses is temperature stress that comes out as heat, cold or freezing stresses. Due to global warming and increased population growth, heat stress has become intensified and more common than in previous decades everywhere in the world. Along with heat stress, cold and freezing stresses have become common and intensified especially in orchards and fruiting stages of crop plants. Heat stress impacts the growth and development of a plant for a particular period; however, its effect is more serious and prolonged as compared to the effect of other stress factors. Heat stress might not constitute stress for some plants while some crop plants exhibit severe symptoms. Therefore, this trait could well be used for breeding heat-tolerant crop generations. Heat stress could have more drastic and devastating effects on crop plants in the future since the global temperature is estimated to further increase by 1.5–4.5 °C by 2050 (Masouleh and Sassine 2020). It is predicted that extreme climatic conditions such as heatwaves will also increase in the near future due to the increase of global warming. Even if the temperature increase is slow, the frequency and duration of heatwaves would be prevalent in several parts of the world (Krishna et al. 2019; Langworthy et al. 2020). Typical heat stress is caused by a 5–15 °C rise in temperature above the optimum temperature for plant growth and development for a certain period (Fragkostefanakis et al. 2015). Under these circumstances, frequent irrigation to ease the stress may not be a solution. For example, Langworthy et al. (2020) stated that heatwave conditions on perennial ryegrass growth were not mitigated by frequent irrigation. Adaptation and development of microbial plant pathogens under humid and hot conditions might be successful to penetrate the crop plants (Dikilitas et al. 2019 a). Although all abiotic stresses affect photosynthesis and respiration, temperature stress has an additional negative effect in a very short time on membrane integrity and leads to leakage of metabolites out of the cell. These injuries could be repaired via metabolic processes such as increased flavonoids, phenylpropanoids and phenolic compound accumulations, however, this could also reduce the yield and quality of the crops. One of the main effects of heat stress on crop plants comes out with the life cycle in that the heat stress reduces the process associated with carbon assimilation and shortens the phases of plant development (Masouleh and Sassine 2020). Under heat stress, plants tend to make more respiration for cooling the cell components and cell structures, however, this requires more carbon fixation and water demand. Heat stress does not come alone, it comes along with other stress factors such as drought and water stresses. So, the stress case under heat stress is very complex and interwoven. For example, high temperatures over 35 °C reduce Rubisco activity and reduce photosynthesis in many crop plants (Dikilitas et al. 2019 b; Masouleh and Sassine 2020). This affects whole defensive systems including membrane functions. Heat stress also reduces the efficiency of photosystem II in the chloroplasts, which are the most heat-sensitive parts of the photosynthesis process (Barnabás et al. 2008). Duration and severity of heat stress in air temperature may also contribute to increasing the soil temperature. If the increase in air temperature is accompanied by water stress, the stress level is further increased to the point at which even the tolerant cultivar might show symptoms of stress. At subcellular levels, the effect of heat stress might be expressed more profoundly. Anatomical structures such as drastic changes in chloroplast shape and shrinking of vacuoles and thylakoids reduce the capacity of photosynthesis (Lipiec et al. 2013). Similarly, Zhen et al. (2020) stated that root and steel diameters of rice under heat stress were decreased by 29 and 15%, respectively. Nazdar et al. (2019) reported that heat stress resulted in a significant increase in electrolyte leakage percentage in calendula (*Calendula officinalis* L.) cultivars. Leaf thickness and stomatal density were much higher in the heat-tolerant cultivar "Indian Prince" than the heat-sensitive cultivar under heat stress. They showed that leaf thickness originates from cuticular thickening as well as increased thickness in mesophyll and spongy parenchyma layers. They suggested that increased leaf thickness could be an avoidance mechanism to reduce evaporation. Biochemically, the main disturbances associated with heat stress are protein denaturation, inactivation of enzymes and disintegration of cell membranes, etc. At very high temperatures, these symptoms are expressed severely. These injuries have the potential to produce toxic compounds such as ROS and RNS and break off the cellular homeostasis and delay the growth and development of crop plants. If plants get exposed to high temperatures after a certain time, heat stress can cause permanent damage to the biological membranes and cytoskeleton, and also result in denaturation and aggregation of proteins through disruption of homeostasis. The high temperatures can also cause a huge alteration in gene expression, which results in higher amounts of molecular chaperones such as heat shock proteins (HSPs), scavenger proteins, ROS and other enzymes and leads to the accumulation of plant metabolites functioning in the protection of biological systems and cellular structures (Shah et al. 2013; Fragkostefanakis et al. 2015; Hassan et al. 2021).

Heat stress severely reduces fruit set, modifies the flower morphology and deteriorates the stress-protective metabolites. For example, many disorders such as sunscald, blossom end rot (BER), fruit cracking and unequal ripening of fruits and green shoulders were observed in tomatoes following heat stress (Toivonen et al. 2011; Tonhati et al. 2020). High-temperature stress can reduce the size, weight and appearance of the fruit and also affects components such as lycopene and sugar contents (Abdelmageed and Gruda 2009). Again, the flower set, the most sensitive stage, could be disrupted. The anthers could be heavily disrupted as compared to the female organs (Masouleh and Sassine 2020). Recently, the *TMS1* gene encoding a heat shock protein of Hsp40 was introduced as a thermotolerance gene in *Arabidopsis*-enabled pollen tube growing under stress conditions. However, any possible mutation in this gene might seriously affect seed production. The high temperature is beneficial to plants only when it is below the critical thresholds which increase the maturation and quality of fruits. Above threshold levels, enhanced transpiration rate to cool the entire system causes significant loss of water and reduces the photosynthesis rate. To ease the negative effects of heat stress, new promising approaches have been made. For example, Morrow and Tanguay (2012) stated that a small heat shock stimulant was found sufficient to induce the expression and synthesis of small HSPs, which in turn developed thermo-tolerance in plants. In plants, many types of small HSPs have been reported, and they are located in different cell organelles such as chloroplast, endoplasmic reticulum and mitochondria, cell membranes, and cytosol (Waters 2013).

Temperature stress could also include low and freezing temperatures. For example, cold stress results in poor germination, stunted growth, chlorosis, wilting and, in severe cases, death of the organism (Kong et al. 2020). Cold stress, unlike other stress factors, causes severe membrane damage and hampers the reproductive development of flower organs in plants (Yadav 2010). The damage is mainly due to acute dehydration during freezing or cold temperatures. Receptors at the cell membrane sites perceive the cold stress as they received signals from those of other stress factors, and the received signal is transduced to the cold-responsive genes to regulate the cold stress (Yadav 2010). Understanding the mechanisms of genes responsible for cold tolerance and the transcription factors involved in the cold stress signaling network are of crucial importance to generate cold-tolerant crop plants and to remediate the crop plants under combined stress that involves cold stress. Most crop plants exhibit tolerance as low as 0–15 °C without injury or damage, however, lower temperatures than this could disturb the biochemical and physiological mechanisms. Crop plants under cold stress adjust structures and modify the metabolism to tolerate the cold stress. As with heat stress, enzymatic and non-enzymatic metabolites and osmolytes are produced to reduce cellular dehydration. Osmolytes produced during cold stress are involved in restoring membrane lipid composition to reform the cellular structures (Yadav 2010). Cold stress symptoms depend on the genetic background and the severity and duration of cold stress. In general, the permanent effects of cold stress might appear after 48 to 72 h of stress exposure. Reproductive stages are far more sensitive to cold stress. Apart from the production capacity of osmolytes, saturated and unsaturated fatty acid levels in plasma membranes also play significant roles in determining the tolerance level of crop plants against cold stress. The high proportion of saturated fatty acid levels is characterized by cold-sensitive plants while the high proportion of unsaturated fatty acid levels is characterized by cold-resistant plants (Yadav 2010). Recently, genes involved in the regulation of cold-stress (*LeCOR413PM2*) were cloned from tomato leaves and then used to generate cold-tolerant transgenic tomato plants (Zhang et al. 2021). The authors showed that *LeCOR413PM2* localized in the plasma membrane of leaves was highly expressed when compared to those of other organs such as root, stem, flower and fruit. They suggested that overexpression of the gene not only reduced damage to the cell membrane and ROS accumulation but also maintained high activity of antioxidant enzymes and metabolites and increased cold tolerance of transgenic tomato plants. On the other hand, suppressed expression of the gene led to an increase in the sensitivity of tomato plants to cold stress. It is important to note that ice formation during cold stress is the real cause of plasma membrane damage. It leads to puncture and dehydration of these ruptures. Yadav (2010) stated that the formation of ice occurred in the apoplastic space of plant tissue due to having relatively lower solute concentration. Therefore, more punctures took place at those sites. This, of course, influences the cell metabolism and disintegrates the membrane from the cytoplasm, and this increases the rigidity of the cell wall.

Cold stress, unlike most of the other abiotic stresses, affects the quality of on-vine fruits as well as affecting the post-harvest stages of the fruits or crops. Although there are various methods to increase the cold tolerance of crop plants or reduce the effects of cold stress on crop plants, one of the most promising approaches, until fully developed cold-tolerant transgenic crop plants, is the use of microorganisms that increase the solubility in the plant cells exposed to cold stress. Apart from the use of cold-tolerant microorganisms, chemical applications have also given promising results. The important issue here is the selection of the right chemicals with appropriate doses. Kong et al. (2020) reported that the application of exogenous melatonin (100 µmol L^{-1}) on bell pepper during storage at 4 °C reduced cell structure damage and improved the chilling injury and membrane permeability. The membrane lipid content and the ratio of unsaturated to saturated fatty acids increased by reducing the enzymatic activity of lipoxygenase (LOX) through the regulation of gene transcription. Exposing plants to low but non-freezing temperatures also induces multigenic processes termed cold acclimation, which leads to an increased freezing tolerance (Fürtauer et al. 2019). In this case, plants undergo a series of transcriptomic, proteomic and metabolomic changes that affect the signaling pathways. Carbohydrates are thought to play a crucial role in this reprogramming.

Salt stress is considered as one of the major abiotic stresses significantly limiting agricultural crop production especially in semi-arid and arid regions (Flowers 2004; Abdel Latef 2010, 2011; Ahmad et al. 2016 b; Abdel Latef et al. 2020 a; Osman et al. 2020). Due to global warming, over-irrigation, water deficiency, rising sea levels and inefficient drainage systems in agricultural soils, crop plants have to be cultivated in saline conditions. These stressors with inappropriate irrigation regimes contribute to the increase in soil salinity. From this point onwards, soil salinity will have a much higher impact when combined with the above stressors. If the trend in saline areas goes up like this, almost half of cultivated land will be salinized by the middle of this century. Poor quality of irrigation water and poor drainage systems will further increase the problem of soil salinity. It affects the osmotic system and water balance in the plants, and as a result of that, necrosis and premature death of old leaves are inevitable (Julkowska and Testerink 2015). Salinity at high concentrations can cause toxicity in plant cells due to increased concentrations of Na^+ and Cl^- ions (Boscaiu and Fita 2020). They disrupt the membrane permeability and cause leakage of solutes out of the cell. Simply put, salinity stress leads to the accumulation of Na^+ which competes with K^+ to bind in proteins and triggers the inhibition of synthesis of metabolic enzymes and protein synthesis (Sarker and Oba 2020). High NaCl accumulation at the root zone of the crop plants reduces the water potential and restricts the availability of water and results in osmotic stress. High accumulation of salt decreases the conductivity of stomata and restricts the intake of CO_2 and creates the adverse chloroplast's CO_2/O_2 ratio (Remorini et al. 2009). Salinity also results in nutritional imbalance and inhibits the function of stomata, which leads to extreme water loss (Calero Hurtado et al. 2019). Recently, to follow up the stress progress, Razzaque et al. (2019) determined the gene expression level in rice plants exposed to salinity. They stated that gene expression decreased in the sensitive leaves while tolerant plants showed the opposite trend. They concluded that the tolerant cultivar of rice seedlings responded early while the sensitive ones showed much slower responses in terms of defense mechanisms. They interpreted that the delay in the activation of defense mechanisms in the sensitive cultivars prevented them from recovery during salt stress. Similarly, Nemati et al. (2011) investigated salt-tolerant (IR651) and salt-sensitive (IR29) rice (*Oryza sativa* L.) lines which were exposed to NaCl salinity (100 mmol L^{-1}) for three weeks. They showed that the tolerant genotype had mechanisms to prevent high Na$^+$ and Cl$^-$ ion accumulation. They thought that high total soluble sugars concentration in the shoot of IR651 was due to adjustment of osmotic potential and better water uptake under saline conditions. These mechanisms probably prevented tissues exposed to salinity from death and enabled them to continue their growth and development under saline conditions. Since salinity comes with drought or water stress, crop plants try to reduce their osmotic potential by increasing mineral ion concentrations and compatible solutes such as proline, carbohydrates and glycine betaine. Salinity has significant effects on Mn^{2+} and P content which remove toxic effects of oxidative stress due to involvement in antioxidant enzymes in plants (Nemati et al. 2011). Decreases in Mn^{2+} ion concentration under salinity result in a decrease of microelements' solubility in saline soils. Nemati et al. (2011) stated that tolerant plants with above salt tolerance mechanisms tolerated better salinity stress. These mechanisms have all been confirmed in *Chenopodium quinoa* Willd., a halophytic crop, which showed that not only antioxidant ability contributed to salt tolerance capacity, but also that leaf osmoregulation, K$^+$ retention, Na$^+$ exclusion and ion homeostasis played significant roles for salt tolerance capacity (Cai and Gao 2020).

During salt stress, the hormonal balance could also be affected. Stress hormones such as abscisic acid (ABA), salicylic acid (SA), jasmonic acid (JA), ethylene, gibberellic acid (GA), cytokinin, etc. hormones have all been found to play important roles in regulating salinity stress signals and controlling the balance between growth and stress responses (Yu et al. 2020). Under salt stress, crop plants have differing responses in terms of hormones. Some hormones positively regulate plant's salt tolerance while others have negative roles. Therefore, significant crosstalk occurs among different hormones. Although there are many inconsistent results, the case under the combined stress conditions is far more complex to solve. In the last few decades, great progress has been made, however many mechanisms under single or combined stress with that of salinity remain unsolved and yet to be elucidated. Further studies are needed to clarify the mechanisms of Na$^+$ and Cl$^-$ ions uptake.

As in the other stress factors, low or mild stress of salinity might also trigger the defense responses of plants and might increase the nutritional quality of fruits or vegetables by enhancing amino acids, glucose, fructose, titratable acid, vitamin C and total soluble solid contents (Nebauer et al. 2013). Similarly, Yuan et al. (2010) stated that five- to seven-day-old radish sprouts exposed to 100 mmol L^{-1} exhibited significantly high glucoraphasatin, total glucosinolate and total phenolic contents. They reported that the moderate level of salt stress could improve the nutritional value of radish sprouts as well as enhancing health-promoting compounds.

One of the abiotic stress factors posing a significant threat to crop plants and the environment since the start of the industrialization has been heavy metals and other environmental pollutants. Significant amounts of heavy metals have been added to agricultural soils due to natural and anthropogenic activities. Heavy metals are major environmental pollutants when they are present in low or moderate concentrations in the soil, unlike other stress factors, and show significant toxic effects on growth and development in plants. Uncontrolled discharge to nature of toxic chemicals that contain heavy metals exposes a serious threat to agricultural crop plants as well as to nature. The seed germination stage has been considered as the most sensitive stage to heavy metal toxicity (Solanki and Dhankhar 2011). Emerging roots face the toxic effects of heavy metals directly, and the toxic chemicals could not be sequestered due to the weak structure of roots. Tolerant plants could immobilize, exclude or reduce the transport rate of heavy metals depending on the genotype and the nature of the heavy metal. It is important to elucidate the toxicity response of a plant to heavy metals so that we can use appropriate plant species for the rehabilitation of contaminated areas. Heavy metals hamper the growth of plants by disturbing many biochemical, physiological and metabolic processes (Abdel Latef 2013; Abdel Latef and Abu Alhmad 2013; Ahmad et al. 2016 a; Abdel Latef et al. 2020 c). They trigger changes at the transcript level of numerous gene coding for proteins to have protective changes against damage caused by stress (Solanki and Dhankhar 2011; Singh et al. 2015). Based on chemical and physical properties, heavy metals lead to the production of ROS, and block the essential functional groups of biomolecules and compete with essential metal ions (Solanki and Dhankhar 2011). Early studies showed that increased production of ROS and other free radicals led to significant damage to cell membranes by binding to the sulphydryl groups of membrane proteins or by increasing rates of lipid peroxidation (Liu et al. 2004). This was associated with the damage to nucleic acids and membranes. For instance, Maheshwari & Dubey (2007) stated that nickel suppressed the hydrolysis of RNA and proteins by inhibiting the activity of RNase and protease respectively. Increased free radicals under heavy metal toxicity interfere with the enzymatic activities that lead to impairment of functional groups in the enzymes (Singh et al. 2015). An important mechanism of heavy metal toxicity is their ability to bind strongly to oxygen, nitrogen and sulfur atoms (Nieboer and Richardson 1980). A range of heavy metal cations have a high affinity for sulfides, and these products have very low solubility. Once entering the cell, heavy metal cations, especially those with high atomic numbers such as Hg^{2+}, Cd^{2+} and Ag^{2+}, tend to bind to $-SH$ groups and may inhibit the activity of sensitive enzymes (Nies 1999) and therefore, heavy metal can inactivate enzymes by binding to cysteine residues in the cell. For example, An et al. (2020) showed that Cd and Cu combination in soil resulted in lipid peroxidation in carrots, and the activity of SOD enzyme decreased gradually. Heavy metals can be accumulated in cells via uptake systems responsible for essential cations. Many enzymes contain heavy metals as co-factors or co-enzymes, and the displacement of one particular metal by another metal will inhibit or lead to loss of enzymatic activity. Divalent cations such as Ni^{2+}, Zn^{2+} and Co^{2+} were found to displace Mg^{2+} in ribulose-1,5–biphosphate carboxylase/oxygenase and resulted in the loss of activity (Van Assche and Clijsters 1986; Ghori et al. 2019). Heavy metal generally causes mutations or cell death in plants and severe health hazards in human beings through the food chain. Recently, Dutta et al. (2018) reported that heavy metals not only caused toxicity and oxidative stress in plants but also disturbed the chemical or physical structures of DNA and induced both cytotoxic and genotoxic stresses. They stated that these types of stresses led to genome instability and severely affected plant health and crop yield.

Sugars and the amylase enzymes are considered important elements in plant metabolism not only because they are involved in photosynthesis, but also because they are part of the respiratory system. Sugars play several roles against wounds and infection as well as in the detoxification of foreign substances (Kaur et al. 2000). Amylase activity, on the other hand, plays a crucial role in the hydrolysis of carbohydrates, and its low activity, meaning lower sugar content, might lead to increased concentrations of heavy metals (Devi et al. 2007; Gopal et al. 2008). For example, Kuriakose & Prasad (2008) reported that cadmium toxicity greatly impaired not only the breakdown of soluble sugars but also the translocation of soluble sugars to the growing embryonic axis. Therefore, heavy metal toxicity has great potential to retard the growth and development of crop plants.

As is the case with other stresses, low concentrations of heavy metal might have also priming effects. For example, (Solanki and Dhankhar 2011) stated that a low concentration of copper treatment led to an increase in the total carbohydrates in seedlings. The mobilized carbohydrates to the embryonic axis supported the growth and provided energy for the development of crop plants (Müntz et al. 2001; Shahid et al. 2017). Heavy metal stress likely induces senescence through enhancement of degradation of chlorophyll, protein and RNA molecules (Khudsar et al. 2004). It is also likely that heavy metals induced lipid peroxidation in plants, and fragmentation of proteins due to the toxic effects of ROS might also lead to reduced protein content. A decline in protein level may reduce the activity of nitrate reductase, which has a crucial role in the detoxification of heavy metal toxicity. For example, (Singh Sengar et al. 2008) observed that the nitrate reductase enzyme was more sensitive than other components such as protein and organic nitrogen to heavy metal pollution. The decrease in nitrate reductase activity in heavy metal treated plants could also reflect a decrease in photosynthesis because sugars are essential for nitrate reductase expression (Kaiser et al. 1999). Recently, similar findings were made by Shakeel et al. (2020) who stated that seeds of beetroot (*Beta vulgaris* L.) grown in 15% fly ash-amended soil showed significantly higher plant growth as well as pigment content as compared to the control. The authors stated that improving nitrate reductase activity enabled plants to increase the contents of total protein and carbohydrate as well as increasing the proline and the total antioxidant activity. They suggested that the increased levels of proline, antioxidant activity and the enzymes might be responsible for scavenging the oxidative stress. Similarly, increased carotenoid and SOD levels led to increased tolerance against high concentrations of Pb and Cd in aquatic plant *Hydrilla verticillate* (Singh et al. 2013). Alteration in the activity of acid and alkali phosphatase enzymes in plants has also been noted under heavy metal stress (Sharma and Dubey 2005). Heavy metals are not easily degradable substances and can move around food chains. Since contamination of soil and water by heavy metals poses significant threats to the environment, agricultural lands and emerges as a major hazard to public health, their interactions with other stress factors could have far more devastating effects on crops and food chains. As a consequence, crop plants are now under more pressure than before in terms of heavy metal toxicity and its combined stress. Therefore, strategies to resolve heavy problems have been focused on detoxification, transport and/or sequestration (Cong et al. 2019). Until a firm solution is produced to tackle the abiotic stresses, exogenous application of various organic or inorganic compounds and their ameliorative effects on heavy metal-induced toxicity in plants remains as a promising solution. For example, Tiwari & Lata (2018) reported that the use of microbes could improve the heavy metal tolerance in crop plants and could be employed as an alternate technology. The use of beneficial microorganisms could be used as a promising method for safe crop-management practices. The authors stated that plant-associated microbes decreased metal accumulation in plant tissues and helped to reduce metal bioavailability in soils. To increase the efficacy of microorganisms, genetically transformed bacteria could be a better approach to increase the remediation of heavy metals and stress tolerance in plants. Recently, heavy metal-transporting P-type ATPases (HMAs) were also shown to have a significant role in the uptake and translocation of heavy metals in rice plants (Cong et al. 2019). The authors stated that rice HMA genes in plants were upregulated and the transgenerational memory via changes in gene regulation carried on even after the removal of heavy metals.

Anthropologic activities not only cause pollution in soils but also result in atmospheric pollution through cycles of gases and heavy metals. Therefore, environmental pollution, heavy metals, drought or temperature stresses not only threaten the agricultural soils but also cause significant pollution in the air. In particular, increases in atmospheric gases, such as CO_2 and O_3, and an increase in ultraviolet radiation cause significant acidification and lead to increased acidic rains in agricultural soils (Oksanen 2010; Barnes et al. 2019). Acidification is the biggest threat for plant ecosystems in developing countries along with pollution of other atmospheric gases. The most harmful effects of ozone on crop plants are loss of membrane integrity, inefficient photosynthesis, accelerated leaf senescence and reduced crop loss. As in the other stresses, plants tend to close their stomata to reduce ozone uptake, however this could affect all metabolic processes as recounted above. Defense mechanisms against ozone in crop plants are similar to those of responses exerted against other abiotic stress conditions. However, in high concentrations, the detoxification capacity of plants could be limited and the ultrastructural disorders, as well as foliar injuries leading to programmed cell death, could be observed. Atmospheric heavy metals may be absorbed via foliar organs of plants. Although root metal transfer has largely been studied, very little is known about heavy metal uptake by plant leaves from the atmosphere (Shahid et al. 2017). Metal deposited on plant leaves followed by atmospheric pollution might enter the plants through penetration of the cuticle and stomata and then is translocated within plants. If heavy metals are not excluded from plant tissues, they are then stored in the cellular parts of plants. Storage in the edible parts is of significance to health. Other abiotic stresses such as temperature, drought or excess humidity might increase the impact of atmospheric pollutants. Therefore, its interactions with other abiotic stresses should also be considered. Defense mechanisms of crop plants followed by atmospheric pollution and its interactions with other abiotic stresses remain elusive.

Abiotic stresses not only threaten the vegetative and pre-harvest stages of crop plants but also significantly deteriorate the postharvest fruits and vegetables. Although many abiotic stress factors

TABLE 1.1
Main Abiotic Stress Factors and Their Effects on Crop Plants

Abiotic stress	Plant Responses and Expected Symptoms
Drought	Plants cannot uptake water and transport it upwards. Due to the closure of stomata as defense response, photosynthesis activity is reduced. Accumulation of solutes may prevent desiccation. Leaf rolling, decreased leaf area and thickness are expected.
Heat	Due to high evaporation, heavy water loss is expected, and leaf roll, chlorosis and wilting are characteristic symptoms. Antioxidant enzymes are decreased while phenolic and other flavonoid-related substances can be expected to increase.
Cold, chilling and freezing	Metabolism, in general, slows down. Due to the formation of ice in the apoplastic tissues, metabolites could be lost upon cessation of freezing temperatures. Photosynthesis, especially photosystem II, is heavily affected. Flower and fruiting stages are the most sensitive stages. Osmolytes are expected to increase while ethylene is synthesis is inhibited.
Hypoxia	Reduced oxygen supply, anaerobic respiration takes place. Elasticity in roots may increase, and roots lose their uniform structure and cannot function properly. Necrosis and death may be inevitable.
Salinity	Ionic imbalance, nutritional competitivity and toxicity are the main issues. Water potential decreases in soils and plants. Chlorosis and leaf roll, necrosis and wilting are the main symptoms.
Heavy metal	Toxicity and ionic homeostasis are disrupted. Photosynthesis is the prime target. Reduced protein and enzyme synthesis, reduced nutrient uptake are expected. Membrane leakage is observed.
Atmospheric pollution	Water potential in leaves is expected to drop. Defense metabolites increase while the quality of crops is expected to decrease.
Pesticide and high concentrations of fertilizers	Acute pesticide toxicity leads to necrosis in leaves and stems and reduces the water potential in soils. Leaf roll, chlorosis and necrosis are common symptoms. Reduced photosynthetic activity is expected.
Light	Excess light can lead to increase in photooxidation and increase ROS concentration. Photosynthesis is inhibited.

affect postharvest fruits and vegetables, climatic changes are of particular importance as they determine the limit of crop production. Therefore, elucidating the effects of abiotic stresses and the defense responses of fruits would help us develop more stress-tolerant fruits and vegetables, and we could increase the storage and shelf life of fruits and vegetables (Toivonen et al. 2011). Since we have been losing crop yields due to abiotic and biotic stresses through cultivation processes, we should not lose the great majority of the production after harvesting during storage and shelf-life, if the aim is to maintain sustainability. It is not clear that breeding for stress tolerance in the field will be carried on through postharvest stages although some promising results have been obtained.

Abiotic stresses and their effects on crop plants have been summarized in Table 1.1. The illustrated mechanism and pathways under the combined abiotic stress issues are given in the next section.

1.3 POSSIBLE INTERACTIONS BETWEEN STRESSORS

Plants may show resistance or tolerance to the effects of individual stresses; however, it would be challenging for a plant to overcome the combined stress factors. Plants have to adapt to severe stress conditions created by the combined stress, or they should be supported firmly in terms of biochemical and molecular pathways to compensate for the negative effects of stressors. The combined stress comes out in two different ways; simultaneous and sequential (consecutive). Abiotic stress can impose limitations on crop productivity and limit the use of agricultural lands for farming. As sessile organisms, plants cannot physically run away from environmental stresses. Therefore, they have had to evolve strategies to cope with abiotic and biotic stresses. Recent studies have revealed that single abiotic stress could be coordinated by plants via stress-specific responses such as transcriptions and metabolites (Zandalinas et al. 2020). However, in nature, plants are continuously exposed to a combination of more than one abiotic stress factor, each potentially triggering its stress-specific systemic response. It is important to determine if plants have the capacity to integrate two different systemic signals at the same time during stress combinations. Many findings showed that plants actually can integrate two or more different systemic signals at the same time during stress combinations. Plants sense the different stresses. However, this depends on the stress source, duration and genetic background of the plants. In fact, the ability to respond and adapt to abiotic stress conditions may be a driving force in speciation (Lexer and Fay 2005). Crosstalk between the abiotic-stresses was also explored and a high degree of similarity was found in many stress interactions. Specifically, drought and salt stress were found to induce mostly similar responses. Plant response to drought and salt stress involved similar pathways, such as ionic and osmotic homeostasis, which were disrupted and the signaling pathways were found to be similar. Both drought and salt stress results led to an increase in ABA expression, which plays a significant role in the tolerance of osmotic stress and regulation of plant water balance through regulation of guard cells. Under drought conditions, these plants rapidly wilt and die if the stress prolongs. For example, Attia et al. (2019) reported that the combined stress of drought and salt significantly restricted the growth of pea plants (*Pisum sativum* L.). Restriction in leaf biomass was more pronounced in cv. Douce de Provence than cv. Lincoln in which the cv Douce de Provence had a shorter breeding cycle (45 days) than Lincoln (60 days). Similarly, episodes of prolonged drought coupled with heat waves had devastating effects on agricultural production and crop quality (Cohen et al. 2020). Combined stresses not only deteriorate the quality and production of the crops but also shorten the life cycle of crops. Therefore, even crop plants become resistant to abiotic stresses; the effect of the combination of stresses would give less time to recover and limit the production of stress-related organic metabolites. Zhou et al. (2017) tried to differentiate the drought and heat stress in combination, and they found that drought stress had a predominant effect on tomato plants over heat stress. They concluded that in the generation of tolerant cultivars to both heat and drought stress conditions, drought had prime importance. Although there are many interactions among abiotic stresses, a recent review mentioned boron and other abiotic stresses in combination with those of salinity, light and CO_2 etc. (García-Sánchez et al. 2020).

In recent studies, apart from the mechanisms of the combination of abiotic stresses on crop plants, the remediation of crop plants under the combined stresses has been elucidated, and possible promising approaches have been made. For example, Lakaew et al. (2020) showed that ozone (O_3) and heat stress combination induced oxidative damage in *Oryza sativa* L. The authors tried for the first time to apply calcium acetate, spraying on foliar leaves, and they showed that pre-treatment of rice with 5 mmol L^{-1} calcium acetate stimulated activity and expressions of ascorbate peroxidase, and lowered lipid peroxidation level. In this way, promising results have been obtained with one chemical that reduced the negative effects of the multiple stress in rice plants. Again, Zhou et al. (2020) reported the crosstalk of miRNAs and their target genes in tomato plants under simultaneous drought and heat stress conditions in fields. They have provided promising results on how crops could adapt to combined drought and heat stress by regulating miRNA and mRNAs. They found that the miRNAs-mRNAs formed complex regulation networks under combined stress conditions.

We outlined the general concept of combined abiotic stresses and their mechanisms in brief, and we mentioned the recent promising strategies to remediate the crop plants under abiotic stress conditions. From now on, instead of discussing and elucidating each stress factor with each other, we will summarize the interactions between stressors (Table 1.2) and illustrate the stress pathways in single and multiple stress conditions due to limited space and time availability (Figure 1.1 and Figure 1.2).

1.4 BREEDING CROP PLANTS AGAINST MULTIPLE STRESS FACTORS

To cope with abiotic stress, plants respond with several pathways including molecular, cellular and physiological modifications to adapt to such stress conditions. Better elucidation of the plants to abiotic stress helps us breed more tolerant or resistant crop plants. Studies on wild plant species with high-stress tolerance would also help us generating stress tolerant crop plants and contribute to our

TABLE 1.2

Abiotic Stress Factors, Mode of Action on Crop Plants, Possible Treatments and Their Interactions with Each Other

Abiotic stressors	Plant Species	Stress Interactions	Mode of Action	Reference
Drought	*Cupressus arizonica* G.		Antioxidant enzymes along with the hydrogen peroxide, malondialdehyde and proline increased as a result of water stress.	Aalipour et al. (2020)
Drought	*Anthyllis cytisoides* and *Artemisia herba–alba*		Reduced crop yield.	Verwijmeren et al. (2019)
Drought	*Salvia dolomitica* Codd		Moderate and severe drought stress led to modulate the expression of some genes involved in a volatile organic compound and essential oil biosynthesis.	Caser et al. (2019)
Drought	*Triticum aestivum* L.		Drought stress caused a reduction in leaf water potential, net photosynthesis and contents of chlorophylls and carotenoids.	Abid et al. (2018)
Heat	*Nicotiana tabacum* cv. Bright-Yellow 2		Heat stress-induced cell death accompanied by oxidative stress.	Malerba and Cerana (2018)
Heat	*Triticum aestivum* L.		Heat stress reduced grain yield (approx. 40%) due to grain abortion and reduced grain filling.	Chavan et al. (2019)
Heat	*Zea mays* L. (heat-resistant HZS and heat-sensitive L9K)		Malondialdehyde contents increased significantly, the level of SOD, CAT and POD enzyme activities were found in resistant lines.	Wu et al. (2020)
Heat	*Medicago sativa*		Overexpression of genes, (40 °C). Transgenic plants showed high water potential and increased non-enzymatic antioxidant contents.	Matthews et al. (2019)
Salinity	*Triticum aestivum* L.		Proline content and soluble sugar and SOD expressions were evident. Biochar application mitigated the negative effects of salinity.	Kanwal et al. (2018)
Salinity	*Zea mays*		Nano-silica primed seeds showed a higher germination rate. Priming encouraged the germination and increased the antioxidant enzymes.	Naguib and Abdalla (2019)
Salinity	*Phaseolus vulgaris* L.		Maize grain extract treatments protected plants against the deleterious effects of salinity and improved growth and yield components.	Rady et al. (2019)
Salinity	*Triticum aestivum* L.		Treatment with 12.5 mmol L^{-1} of NaCl significantly maximized β-carotene, phenolic acid, flavonoid and vitamin C levels.	Islam et al. (2019)
Heavy Metal	*Spinacea oleracea* L.		Decreased plant growth and development. Affected plants cannot be efficiently used in phytoremediation studies.	Shilev et al. (2019)
Heavy Metal	*Triticum aestivum* L.		Cd and Pb caused oxidative damage, with a higher level of hydrogen peroxide and malondialdehyde contents.	Huang et al. (2019)

(Continued)

TABLE 1.2 (CONTINUED)

Abiotic Stress Factors, Mode of Action on Crop Plants, Possible Treatments and Their Interactions with Each Other

Abiotic stressors	Plant Species	Stress Interactions	Mode of Action	Reference
Heavy Metal	*Oryza sativa* L. and *Arabidopsis thaliana* L.		Essential heavy metal overaccumulation was found toxic to plant cells and inhibited their growth.	Li et al. (2020 b)
Heavy Metal	*Arabidopsis thaliana* L.		The phytotoxic damage of heavy metals was manifested through reduced root length and chlorophyll content, as well as increased toxic metabolites.	Wu et al. (2019)
Heavy Metal	*Lycopersicon esculentum* L.		Cd toxicity led to a modification in morphological parameters, plant pigments and gaseous exchange parameters. Levels of phenolic compounds and osmoprotectants were stimulated.	Khanna et al. (2019)
Chilling	*Arabidopsis thaliana* L.		Low temperatures induced cold regulating genes and reached a maximum level just after 4 h of chilling stress.	Meng and Sui (2019)
Chilling	*Oryza sativa* L.		Chilling stress significantly reduced the rice seed germination and seedling growth as well as the respiration rate and ATP synthesis.	Nie et al. (2020)
Chilling	*Brassica campestris* L.		Chilling stress disrupted cellular structures and damaged membrane functions. Photosynthetic apparatus were negatively affected. CO_2 assimilation and activity of PSII were affected.	Wang et al. (2020 a)
Chilling	*Citrullus lanatus*		Overproduction of ROS was evident. Accumulation of proline was detected.	Jiao et al. (2021)
Chilling	*Zea mays* L.		Chilling temperature (0–15 °C) reduced growth and development.	Aydinoglu (2020)
Frost Damage	*Camellia sinensis* (L.) Kuntze		Water potential is greatly reduced and metabolic activities are ceased.	Samarina et al. (2020)
Frost Damage	*Persea americana* Mill, cv. "Hass"		Stomatal conductance is disrupted and rapid formation of ROS was observed.	Joshi et al. (2020)
Frost Damage	*Sorghum bicolor* L.		Flavonoid metabolites have positive correlations with protective traits in diverse crops under frost stress. Luteolinidin increased whereas total flavonoids decreased.	Cloutier et al. (2020)
Frost Damage	*Triticum aestivum* L.		Plant height decreased gradually with an increase in freezing degree. After freeze stress, sensitivity to stress was observed, growth and yield were decreased.	Xie et al. (2020)
Frost Damage	*Triticum aestivum* L.		Early frost and low temperatures caused irreparable damage to vital organs and the electron transport systems, which could lead to increased ROS concentrations. This processes influenced the membrane lipids and damaged to proteins and nucleic acids.	Bao et al. (2020)

(Continued)

TABLE 1.2 (CONTINUED)

Abiotic Stress Factors, Mode of Action on Crop Plants, Possible Treatments and Their Interactions with Each Other

Abiotic stressors	Plant Species	Stress Interactions	Mode of Action	Reference
Heavy metal and salinity	*Oryza sativa* L. and *Arabidopsis thaliana* L.	Positive interactions	Combined salt and heavy metals (Cd, Pb, Cu) synergistically increased gene expressions. However, responses of monocotyledon and dicotyledon plants responded differently to combined stresses.	Kim and Kang (2018)
Heavy metal and salinity	*Cucumis sativus* L.	Positive interactions	Salinity decreased Zn uptake and increased another heavy metal uptake by plants.	Taghipour and Jalali (2019)
Heavy metal and salinity stress	*Oryza sativa* L.	Positive interactions	Severe symptoms were observed when compared to those of each stress factors.	Mekawy et al. (2018)
Heavy metal and salinity	*Mesembryanthemum crystallinum*	Positive interactions	Stressor-induced alterations in genes were observed.	Nosek et al. (2020)
Heavy metal and drought	*Oryza sativa* L.	Positive interactions	Biochemical and physiological analyses revealed that OsGSTU30 overexpression lines have improved tolerance against both stresses as compared to wild-type plants.	Srivastava et al. (2019)
Heavy metal and drought	*Amaranthus retroflexus* and *A. tricolor*	Positive interactions	Heavy metal and drought induced a greater stress intensity.	Wang et al. (2020 b)
Drought x Salinity	*Oryza sativa* L.	Positive interactions	OsMYB6-overexpressing transgenic plants showed increased tolerance to drought and salt stress compared with wild-type plants via higher proline content, CAT and SOD activities and lower malondialdehyde levels.	Tang et al. (2019)
Drought x Salinity	*Sorghum bicolor* L. Moench.	Positive interactions	Plant growth, biomass and leaf chlorophyll were significantly reduced. A significant increase in hydrogen peroxide content was evident.	Nxele et al. (2017)
Drought x Salinity	*Agropyron cristatum* L.	Positive interactions	Drought and salinity stress increased the malondialdehyde and hydrogen peroxide contents.	Sheikh-Mohamadi et al. (2017)
Drought x Heat	Different cultivars of *Solanum lycopersicum*	Positive interactions	The cultivars differing in heat sensitivity did not show the difference in the combined stress sensitivity.	Zhou et al. (2017)
Drought x Heat	*Catharanthus roseus*	Positive interactions	Plant height and weight were significantly decreased and the effects were more pronounced in combined stress conditions. Drought and/or heat stress triggered the accumulation of osmolytes with maximum accumulation in response to the combined stress.	Alhaithloul et al. (2019)
Drought x Heat	*Eucalyptus globulus*	Positive interactions	Drought-stressed plants subject to a heat shock exhibited a sharp decrease in gas exchange.	(Correia et al. (2018)

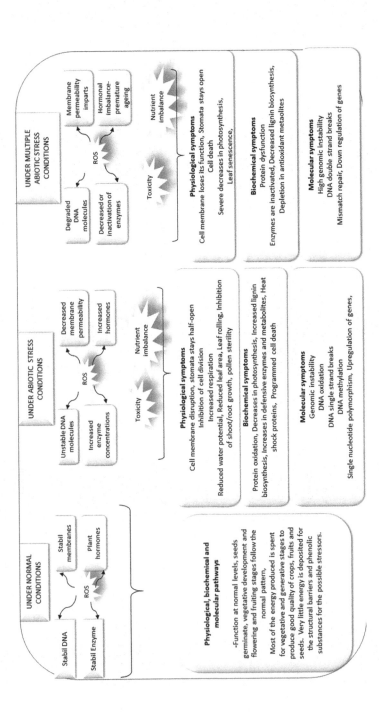

FIGURE 1.1   A simple illustration for physiological, biochemical and molecular mechanisms of a plant cell under normal, single and multiple abiotic stress conditions. Single abiotic stress conditions were considered as one abiotic stress in general. If abiotic stressors are heavy metal and high concentrations of salt, then ionic competition and toxicity of ions can be observed that directly damage nuclear components. Multiple stress conditions involve the combination of various abiotic stress conditions and their observed symptoms and metabolic mechanisms. Color from green to red indicates the severity of stress and difficulties in metabolic pathways. Minimal ROS production is necessary for sustaining cell metabolism. ROS act as signaling molecules at low levels and keep the cells alert against abiotic/biotic stressors. ROS produced at high concentrations under stress conditions damage cell components. Under multiple stress conditions, the concentration of ROS would tend to further increase, however, this depends on the genetic background, stress severity and duration. ROS at high concentrations seriously damage cell membranes and result in breaks in DNA molecules, modify the hormonal and ionic homeostasis and lead to enzymatic dysfunctions and lipid peroxidation, and destroy the pigments and other antioxidant molecules. Under single abiotic stress, a great amount of energy is sacrificed for defensive purposes instead of flowering and fruiting formation. Fruiting is accelerated to produce seeds to deliver the genetic information for the next generation. However, the formed fruits are very small and less as compared to those of healthy plants. Under multiple stress conditions, crop plants could reach the fruiting or seeding stages in most cases, and severe symptoms of the stressors could be observed in every stage of the crop plants. Cell metabolism under no stress conditions. (▨) Cell metabolism under abiotic stress conditions. ( ) Cell metabolism under multiple abiotic stress conditions (▩).

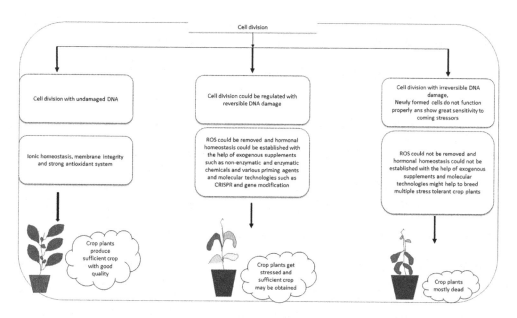

FIGURE 1.2 Status of cell division and plant responses under varying stress conditions. Cell metabolism under no stress conditions. (■) Cell metabolism under abiotic stress conditions. ( ) Cell metabolism under multiple abiotic stress conditions: (■)

understanding of stress tolerance and avoidance. With the advancement of molecular techniques in biology and genomics, advancements have been possible in crop breeding although the common application of the results has not been observed yet (Huang et al. 2013). Understanding the mechanisms for stress tolerance in these plants can be directly or indirectly applied in agricultural areas.

The main principle to reduce the impact of abiotic stresses could be enabled through stress mitigation which depends on the reduced accumulation of ROS and RNS, increased activity of antioxidant enzymes and scavenging of toxic compounds. Plants activate their defense signaling cascades to overcome the negative effects of stress. However, it is very difficult to judge whether plants activate their defense signaling pathways simultaneously or sequentially or whether a single cascade controls a multitude of stress responses (Dikilitas et al. 2019 a). Due to their sessile nature, plants are exposed to single or multiple abiotic stressors continuously, either sequentially or simultaneously. To breed tolerant crop plants against abiotic stress factors, signaling molecules, genes, proteins, metabolites and hormones under the combined stress conditions should be elucidated in detail both under single and multiple stress conditions. Advances in genomics technology and computational biology made these complex interactions possible to elucidate them effectively. Next-generation sequencing, proteomics, microarrays and systems biology mark-up language have all been used to understand systems biology. To start up the breeding program, we should discover and find out signaling molecules. A few of them activated during abiotic stress conditions are ROS, RNS, Ca^{2+}, etc. (Hasanuzzaman et al. 2020). These signal molecules act as secondary messengers, modulate gene expression and activate the enzymes involved in cell defense systems. Coordination between these signal molecules helps maintain cellular homeostasis and plays an effective role in maintaining stress tolerance. For example, abscisic acid (ABA) controls drought and osmotic tolerance through ABA-mediated stomatal closure (Dar et al. 2017). On the other hand, ethylene interacts with ABA during these stresses. Several other biomolecules and hormones also have crucial roles in a coordinated manner to protect and maintain homeostasis in the cellular system. It is very difficult for plants to continue the defense process via synthesizing defense-related metabolites, hormones and enzymes or activate/stop the defense process upon exposure to intermittent stresses. Defense strategies depending on signaling mechanisms are affected according to stress combination, duration, stress severity and plant genetic background. Simultaneous and consecutive stresses differ in terms of responses. Plants can only sense the stress upon exposure and decide which one would be tackled first. Here, it is thought that the order of the stress makes the difference for the activation of defense responses. Plants that are unable to adapt and fail to show proper stress responses undergo death. The strategies for genetic engineering for abiotic stress tolerance depend on the expression of genes involved in signaling, regulatory pathways encoding proteins that confer stress tolerance, or enzymes present for structural defense (Latha et al. 2017). For example, many genes and gene products have been characterized following abiotic stress in crop plants. Transferring these genes into crop plants on target may increase the tolerance to those stressors. However, newly inserted genes and gene products should not interfere with other metabolic processes in crop plants such as flowering and fruiting. Although a lot of work has been done to generate abiotic stress-tolerant crop plants, there is still a need to work in detail to generate tolerant crop plants. Interactions among stress factors make the generation of stress-tolerant crops more difficult due to cross and antagonistic signaling pathways and cross inhibition of metabolites during multiple stress occurrences. Despite the discovery of many stress tolerant genes, a big effort is needed to meet the demand for more tolerant crop plants.

Wild crop species have stress tolerance genes; however, it is very difficult to transfer them into crop plants at the moment due to high genetic distance and crossing barriers (Krishna et al. 2019). Along with the advancement in transgenic technology, gene transfer across species would be much easier. Many plants have been developed and many more are on the way, and these stress-tolerant plants have been tested in terms of producing compatible solutes such as betaines, sugars, amino acids, etc., and toxic-compound scavenging capacities (Krishna et al. 2019).

The sessile nature of plants means they are continuously exposed to vagaries of environmental conditions and, therefore, they have adopted a cellular mechanism to respond to extreme environmental conditions by altering their growth, productivity and quality. Several studies have been carried out to understand this mechanism of the plant's response to various stresses such as drought, heat, salinity, cold, heavy metals, oxidative stress and hypoxia due to waterlogging (Karkute et al. 2019). To avoid these abiotic stresses, plants have evolved convoluted perception and signaling network pathways which perceive signals and modulate the expression of gene/s that further lead to biochemical, physiological and morphological changes to make the plants adapt to the stress condition. Although significant achievements have been made to understand these pathways in *Arabidopsis* as a model plant system, a few studies and promising results have been reported that characterized these pathways in crop plants as well.

Traditional plant breeding has been enormously used in agricultural crop improvement, for fruit quality and disease resistance. However, it is limited by several factors such as time to produce crosses and backcrosses, and fulfillment of the genetic pools that can be utilized for improvement, and that almost a decade is required to release a new variety. Further, conventional breeding may not be helpful to transfer the target trait from the wild gene pool to developed varieties (Gould 1992). The alternative to traditional breeding that overcomes all these limitations is a recombinant-DNA technology or genetic engineering which allows transplanting of genetic distinctiveness across the species. Transgenic technology is based on the identification and characterization of genes responsible for the desired trait. These genes can be then utilized to transform and manipulate any existing crop variety.

One of the ways to improve abiotic stress tolerance is to increase the level of osmolytes in plants under stress. It has been observed that the level of glycine betaine, which is a quaternary ammonium compound in many organisms, increases under salt or other osmotic stresses. Transgenic rice overexpressing choline oxidase A (coda) gene showed an increased synthesis of glycine betaine and thereby enhanced salinity tolerance (Mohanty et al. 2002). More studies about the physiological and molecular pathways of abiotic stresses have been made in "omic" studies to identify the genes involved in defense protein synthesis and their expression patterns during stress perception and responses.

Due to space and time limitations, we summarized genetics and biochemical approaches in Tables 1.3 and 1.4 for plants under single or multiple stress conditions. Instead of evaluating and characterizing each improvement strategy, we outlined how crop plants were improved through biochemical and genetic traits.

TABLE 1.3
Genetic Engineering of Functional Genes in Abiotic Stress Tolerance

Gene	Function	Response	Reference
*abi*3	Abscisic Acid-induced protein	Marked increase in expression of low temperature-induced freezing tolerance.	Saijo et al. (2000)
adc	Arginine decarboxylase (putrescine biosynthesis)	Minimal chrorophyll loss under salt and cold stress.	Singla-Pareek et al. (2003), Kou et al. (2018), Espasandin et al. (2018)
p5cs	Δ^1-pyrroline-5-carboxylate synthase	Enhanced biomass under salt stress.	Wang et al. (2004), Chen et al. (2013)
Cnb1	Calcineurin B1	Transformants showed substantial NaCl tolerance by co-expression of catalytic and the regulatory subunits.	Mittler (2002)
Os cdpk7	Calcium dependent	Overexpression induction of some stress responsive genes in response to salinity/ drought and cold.	Apel and Hirt (2004)
Atgs k1	*Arabidopsis* homologue of GSK3/shaggy like kinase	Transformants showed induced expression of NaCl stress responsive genes.	Babu et al. (2004)
Hsp 101	Heat shock protein 101	Transformants showed tolerance to sudden shifts to extreme temperature.	Nagamiya et al. (2007)
Apx3	Ascorbate peroxidase	Transformants showed increased protection against oxidative stress.	Agarwal and Jha (2010)
glyI and *glyII*	Glyoxalase pathway genes	Improved salinity tolerance in tobacco.	Vannini et al. (2007)
GmNAC109	regulates biotic and abiotic stress responses	Increased lateral root formation in transgenic *Arabidopsis* lines in drought stress tolerance.	Yang et al. (2019)
SlbZIP1	bZIP transcription factor6	Plays crucial role as regulator in ABA-mediated stress response in plants	Zhu et al. (2018)
NRT1.1	nitrate transporter 1.1	NRT1.1 regulates NO_3^- allocation to roots by coordinating Cd^{2+} accumulation in root vacuoles, which facilitates Cd^{2+} detoxification.	Jian et al. (2019)
Amt	ammonium transporter1	In the roots of winter wheat in response to soil drought under conditions of limited nitrogen (N).	Duan et al. (2016), Breria et al. (2020)
Zat6	putative c2h2 zinc finger transcription factor mRNA	Improved seed germination under salt stress via ZAT6's effects on the MAPK cascade.	Han et al. (2020)
OSISAP1	zinc finger A20 and AN1 domain-containing stress-associated protein 1	Encoding a zinc-finger protein is induced after different types of stresses such as cold, desiccation, salt, submergence and heavy metals.	Hu et al. (2020)
MdSAP15	zinc finger AN1 domain-containing stress-associated protein 15-like	Enhances drought tolerance in *Arabidopsis* plants	Dong et al. (2018)
COP1	Constitutively photomorphogenic 1 E3 ubiquitin ligase	Plays a fundamental role in the regulation of stomatal movements in response to dehydration.	Moazzam-Jazi et al. (2018)

(Continued)

TABLE 1.3 (CONTINUED)
Genetic Engineering of Functional Genes in Abiotic Stress Tolerance

Gene	Function	Response	Reference
Kin2/Cor6.6	stress-responsive protein (KIN2) / stress-induced protein (KIN2) / cold-responsive protein (COR6.6) / cold-regulated protein (COR6.6)	Directly involved in response to abiotic stresses, such as drought and cold	Zhang et al. (2018)
COR413IM1	Cold regulated 413 inner membrane 1	Chloroplastic protein and low-temperature induced protein.	(Woldesemayat and Ntwasa (2018)
AtMYB12	myb domain protein 12	Regulated the expression of key enzyme genes involved in flavonoid biosynthesis.	Wang et al. (2016)
VlWRKY3	Vitis WRKY DNA-binding protein 3	VlWRKY3 plays important roles in responses to both abiotic and biotic stress.	Guo et al. (2018)
WRKY	WRKY transcription factor	The expression of GmWRKY47 and GmWRKY 58 genes was decreased upon dehydration.	Song et al. (2016)
DWF4	DWF4 encoding a C-22 hydroxylase that catalyzes a rate-determining step in BR biosynthesis	AtDWF4 over expression enhances drought tolerance	Sahni et al. (2016)
HKT1	high-affinity K+ transporter 1	Expression of the gene HKT1,2, which encode a plasmalemma Na+/H+ antiporter, showed the greatest change in expression as a response to salt stress.	Ali et al. (2019)
SlSOS1	plasmalemma Na+/H+ antiporter	Downregulation of SlSOS1 ultimately reduces the plant tolerance to salt stress.	Waqas et al. (2020)
TaNHX2	TaNHX2 has significant sequence homology to NHX sodium exchangers	TaNHX2 gene in transgenic sunflower conferred improved salinity stress tolerance and growth performance.	Mushke et al. (2019)
TaAVP1	Inorganic H pyrophosphatase family protein	AVP1 and MIOX4 genes improved stress tolerance under water-limited and salt-stress conditions.	Nepal et al. (2020)
NHX	Plant Na+/H+ exchangers (NHXs)	Na+/H+antiporters (NHXs) are coupled with V-ATPase for generating a driving force to maintain ion homeostasis.	Yarra (2019)
LEA	Late embryogenesis abundant (LEA) proteins	Is involved in the plant response to salt and drought stress as a reactive oxygen species scavenger.	Zheng et al. (2019)
LEA	Late embryogenesis abundant (LEA) proteins	LEA genes involved at improving the stress tolerance of wheat.	Liu et al. (2019)

1.5 CONCLUSIONS AND PERSPECTIVES

Domestication of wild plants into agricultural crop plants to feed the increasing world population had affected our lives significantly. Many crop plants suffer from increasing abiotic stress conditions. During cross-breeding or gene modification, some of the genes have been lost or modified as compared to natural conditions. For example, a comparison of the wild rice (*Oryza rufipogon*) and cultivated rice (*O. sativa japonica*) genome sequences revealed that genes were lost in the cultivated species (Richards et al. 2014). Dolferus (2014) stated that some gene deletions, mutations and variations in gene expression level

TABLE 1.4
Improvement Strategy of Plants under Abiotic Stress Conditions

Plant Species	Stress Type	Improving Strategy	Reference
Arabidopsis thaliana	Drought stress	CRISPR/dCas9 fusion	Roca Paixão et al. (2019)
Arabidopsis thaliana	Drought stress	Plant growth-promoting rhizobacteria (PGPR)	Liu et al. (2020)
Oryza sativa L.	Drought stress	Genetic engineering	Latha et al. (2019)
Oryza sativa L.	Drought and Salinity Stress	Regulation of gene expression (*OsMYB6*)	Tang et al. (2019)
Triticum aestivum	Salinity stress	PGPR + Halophyte plant use	Orhan (2016)
Oryza sativa L. "Aiswarya"	NaCl, PEG	UV-B priming	Dhanya Thomas et al. (2020)
Zea mays	Heavy metal stress (arsenic toxicity)	Foliage salicylic acid supplementation	Kaya et al. (2020)
Brassica napus L.	Moderate and short-term drought exposure	Exogenous Si	Hasanuzzaman et al. (2018)
Phoenix dactylifera L.	Heat Stress	Silicon and Gibberellin treatments	Khan et al. (2020)
Solanum lycopersicum L.	Cold, heat, salinity and drought	epigallocatechin-3-gallate (EGCG, bioactive flavonoids)	Li et al. (2019)
Arabidopsis thaliana	NaCl and drought stresses	Use of gene	Li et al. (2020 a)
Phaseolus vulgaris L.	Water Stress Conditions	Ascorbic Acid	Gaafar et al. (2020)
Oryza sativa L.	Salinity stress	CRISPR/Cas9-targeted mutagenesis of the *OsRR22* gene	Zhang et al. (2019)
Solanum lycopersicum L.	Salinity and Osmotic Stress	CRISPR/Cas9 induced *SlARF4* mutagenesis	Bouzroud et al. (2020)
Oryza sativa L.	Drought stress	CRISPR/Cas9-Induced Mutagenesis of Semi-Rolled Leaf	Liao et al. (2019)
Oryza sativa L.	Drought and salinity stresses	CRISPR-Cas9 mediated genome editing	Santosh Kumar et al. (2020)
Hordeum vulgare	Salinity stress	The desired mutant has been produced by CRISPR/Cas9	Vlčko and Ohnoutková (2020)
Medicago truncatula	Cold stress	Hydrogen peroxide and nitric oxide play a crucial role as an elicitor	Arfan et al. (2019)
Stevia rebaudiana Bertoni	Salinity stress	Chitosan as an elucitor	Gerami et al. (2020)
Zea mays L.	Salinity stress	Endogenous Selenium	Benincasa et al. (2020)
Oryza sativa L.	Chilling and drought stress	A consortium of two rhizobacteria *Bacillus amyloliquefaciens* Bk7 and *Brevibacillus laterosporus* B4, and biochemical elicitors salicylic acid and β-aminobutyric acid	Kakar et al. (2016)
Arabidopsis thaliana	Drought and biotrophic pathogen stresses	Acibenzolar S-methyl (ASM) β-Aminobutyric acid (BABA)	Venegas-Molina et al. (2020)
Coelogyne ovalis Lindl.	Salicylic acid, methyl jasmonate, abscisic acid, chitosan and yeast extract	*CoCHS* gene expression	Singh and Kumaria (2020)

(Continued)

TABLE 1.4 (CONTINUED)
Improvement Strategy of Plants under Abiotic Stress Conditions

Plant Species	Stress Type	Improving Strategy	Reference
Citrullus lanatus	Vanadium stress	Melatonin pretreatment	Nawaz et al. (2018)
Trithicum aestivum L.	Drought stress	Abscissic acid. Improved shoot under mild drought conditions.	Valluru et al. (2016)
Oryza sativa L.	Salt stress	Polyamines-spermidine and spermine	Paul and Roychoudhury (2017)
Zea mays	Salt stress	*Pseudomonas fluorescens*. Promoted root growth	Zerrouk et al. (2016)

directly or indirectly affected abiotic stress tolerance. As stated before, wild species are more tolerant to abiotic stresses. When domestication is made to increase the quality and other related characteristics, some of the traits related to stress tolerance could be lost. However, these genes could be re-introduced in domesticated crops (Prasanna 2012; Atwell et al. 2014). Transgenic technology has the potential to contribute to sustainable agriculture, food and nutritional security. Considering the drastic changes in temperature and rainfall patterns, heavy metal, drought, salinity and other abiotic stress-tolerant crop plants have to be developed. It has been reported that manipulation of single or multiple genes linked to abiotic stress response provides plants with a tolerance to such stresses. In the future, plants with a high tolerance to single abiotic stresses will not be a solution for sustainable agriculture, because the abiotic stresses are interrelated. For example, salt stress often results in water deficit in plants, or heat stress results in drought stress in plants (Krishna et al. 2019). Therefore, plants with multiple stress tolerance are required to not only survive in harsh environmental conditions but also to give enough yields. A deeper understanding of the mechanism of stress tolerance through metabolomics is needed to develop stress-tolerant crop plants. Modification of metabolic pathways is quite complicated, as the majority of the proteins of a pathway interact with several other proteins. Thus, efficient metabolic engineering will only be achieved by controlling multiple genes on the same or interlinked pathways. Although transgenic crops have been successfully cultivated for almost two decades in the United States and more than a decade in India without any known detrimental impact on humans or the environment, the move to produce new transgenic crop plants is still surrounded by debate, and this should be the subject for political and social affairs. It is important to discuss the safety of transgenic crops for public health. Until generating multiple stress-tolerant crop plants, we have to apply biochemical and physical priming agents that trigger the defense response-related pathways under multiple stress conditions.

Plants continuously adjust growth and development according to their exposure to the threat. This growth adjustment mechanism is quite remarkable but poorly understood (Dolferus 2014). Field crop plants are continuously exposed to a combination of stress factors to increase their tolerance capacity.

ACKNOWLEDGEMENT

There is no conflict of interest among the authors. We apologize for not citing all of the relevant references due to space limitations.

REFERENCES

Aalipour, H., A. Nikbakht, N. Etemadi, F. Rejali, and M. Soleimani. 2020. Biochemical Response and Interactions between Arbuscular Mycorrhizal Fungi and Plant Growth Promoting Rhizobacteria during Establishment and Stimulating Growth of Arizona Cypress (*Cupressus arizonica* G.) under Drought Stress. *Scientia Horticulturae* 261 (February). doi:10.1016/j.scienta.2019.108923. http://www.ncbi.nlm.nih.gov/pubmed/108923.

Abdel Latef, A., M.F. Abu Alhmad, and S.A. Hammad. 2017a. Foliar Application of Fresh Moringa Leaf Extract Overcomes Salt Stress in Fenugreek (*Trigonella foenum-Graecum*). *Plants* 57(1): 157–79.

Abdel Latef, A.A. 2010. Changes of Antioxidative Enzymes in Salinity Tolerance among Different Wheat Cultivars. *Cereal Research Communications* 38(1): 43–55. doi:10.1556/CRC.38.2010.1.5.

Abdel Latef, A.A.H. 2011. Ameliorative Effect of Calcium Chloride on Growth, Antioxidant Enzymes, Protein Patterns and Some Metabolic Activities of Canola (Brassica napus L.) under Seawater Stress. *Journal of Plant Nutrition* 34(9): 1303–20. doi:10.1080/01904167.2011.580817.

Abdel Latef, A.A.H. 2013. Growth and Some Physiological Activities of Pepper (Capsicum annuum L.) in Response to Cadmium Stress and Mycorrhizal Symbiosis. *Journal of Agricultural Science and Technology* 15(Suppl): 1437–48.

Abdel Latef, A.A.H., and M.F. Abu Alhmad. 2013. Strategies of Copper Tolerance in Root and Shoot of Broad Bean (Vicia Faba L.). *Pakistan Journal of Agricultural Sciences* 50(3): 223–328.

Abdel Latef, A.A.H., M.F. Abu Alhmad, and K.E. Abdelfattah. 2017b. The Possible Roles of Priming with ZnO Nanoparticles in Mitigation of Salinity Stress in Lupine (Lupinus Termis) Plants. *Journal of Plant Growth Regulation* 36(1): 60–70. doi:10.1007/s00344-016-9618-x.

Abdel Latef, A.A.H., M.F. Abu Alhmad, M. Kordrostami, A.-B.A.-E. Abo–Baker, and A. Zakir. 2020a. Inoculation with Azospirillum lipoferum or Azotobacter chroococcum Reinforces Maize Growth by Improving Physiological Activities Under Saline Conditions. *Journal of Plant Growth Regulation* 39(3): 1293–306. doi:10.1007/s00344-020-10065-9.

Abdel Latef, A.A.H., and H. Chaoxing. 2011a. Effect of Arbuscular Mycorrhizal Fungi on Growth, Mineral Nutrition, Antioxidant Enzymes Activity and Fruit Yield of Tomato Grown under Salinity Stress. *Scientia Horticulturae* 127(3): 228–33. doi:10.1016/j.scienta.2010.09.020.

Abdel Latef, A.A.H., and H. Chaoxing. 2011b. Arbuscular Mycorrhizal Influence on Growth, Photosynthetic Pigments, Osmotic Adjustment and Oxidative Stress in Tomato Plants Subjected to Low Temperature Stress. *Acta Physiologiae Plantarum* 33(4): 1217–25. doi:10.1007/s11738-010-0650-3.

Abdel Latef, A.A.H., and H. Chaoxing. 2014. Does Inoculation with Glomus Mosseae Improve Salt Tolerance in Pepper Plants? *Journal of Plant Growth Regulation* 33(3): 644–53. doi:10.1007/s00344-014-9414-4.

Abdel Latef, A.A.H., M.F.A. Dawood, H. Hassanpour, M. Rezayian, and N.A. Younes. 2020b. Impact of the Static Magnetic Field on Growth, Pigments, Osmolytes, Nitric Oxide, Hydrogen Sulfide, Phenylalanine Ammonia-Lyase Activity, Antioxidant Defense System, and Yield in Lettuce. *Biology* 9(7): 172. doi:10.3390/biology9070172.

Abdel Latef, A.A.H., M. Kordrostami, A. Zakir, H. Zaki, and O. Saleh. 2019a. Eustress with H2O2 Facilitates Plant Growth by Improving Tolerance to Salt Stress in Two Wheat Cultivars. *Plants* 8(9): 303. doi:10.3390/plants8090303.

Abdel Latef, A.A.H., M.G. Mostofa, M.M. Rahman, I.B. Abdel-Farid, and L. P. Tran. 2019b. Extracts from Yeast and Carrot Roots Enhance Maize Performance under Seawater-Induced Salt Stress by Altering Physio-Biochemical Characteristics of Stressed Plants. *Journal of Plant Growth Regulation* 38(3): 966–79. doi:10.1007/s00344-018-9906-8.

Abdel Latef, A.A.H., A.M. Omer, A.A. Badawy, M.S. Osman, and M.M. Ragaey. 2021. Strategy of Salt Tolerance and Interactive Impact of *Azotobacter chroococcum* and/or *Alcaligenes faecalis* Inoculation on Canola (*Brassica napus* L.) Plants Grown in Saline Soil. *Plants* 10(1): 110.

Abdel Latef, A.A.H., A.K. Srivastava, M.S.A. El-sadek, M. Kordrostami, and L.-S.P. Tran. 2018. Titanium Dioxide Nanoparticles Improve Growth and Enhance Tolerance of Broad Bean Plants under Saline Soil Conditions. *Land Degradation and Development* 29(4): 1065–73. doi:10.1002/ldr.2780.

Abdel Latef, A.A.H., A.K. Srivastava, H. Saber, E.A. Alwaleed, and L.-S.P. Tran. 2017c. Sargassum muticum and Jania Rubens Regulate Amino Acid Metabolism to Improve Growth and Alleviate Salinity in Chickpea. *Scientific Reports* 7(1): 10537. doi:10.1038/s41598-017-07692-w.

Abdel Latef, A.A.H., A. Zaid, A.-B.A.-E. Abo-Baker, W. Salem, and M.F. Abu Alhmad. 2020c. Mitigation of Copper Stress in Maize by Inoculation with Paenibacillus polymyxa and Bacillus circulans. *Plants* 9(11): 1513. doi:10.3390/plants9111513.

Abdelmageed, A.H.A., and N. Gruda. 2009. Influence of High Temperatures on Gas Exchange Rate and Growth of Eight Tomato Cultivars under Controlled Heat Stress Conditions. *European Journal of Horticultural Science* 74(4): 152–59.

Abid, M., A. Hakeem, Y. Shao, et al. 2018. Seed Osmopriming Invokes Stress Memory against Post-Germinative Drought Stress in Wheat (*Triticum aestivum* L.). *Environmental and Experimental Botany* 145(January): 12–20. doi:10.1016/j.envexpbot.2017.10.002.

Agarwal, P.K., and B. Jha. 2010. Transcription Factors in Plants and ABA Dependent and Independent Abiotic Stress Signalling. *Biologia Plantarum* 54(2): 201–12. doi:10.1007/s10535-010-0038-7.

Ahmad, P., A.A. Abdel Latef, E.F. Abd Allah, et al. 2016a. Calcium and Potassium Supplementation Enhanced Growth, Osmolyte Secondary Metabolite Production, and Enzymatic Antioxidant Machinery in Cadmium-Exposed Chickpea (Cicer arietinum L.). *Frontiers in Plant Science* 7(Apr 2016): 513. doi:10.3389/fpls.2016.00513.

Ahmad, P., A.A. Abdel Latef, A. Hashem, et al. 2016b. Nitric Oxide Mitigates Salt Stress by Regulating Levels of Osmolytes and Antioxidant Enzymes in Chickpea. *Frontiers in Plant Science* 7(Mar 2016): 347. doi:10.3389/fpls.2016.00347.

Alhaithloul, H.A., M.H. Soliman, K.L. Ameta, M.A. El-Esawi, and A. Elkelish. 2019. Changes in Ecophysiology, Osmolytes, and Secondary Metabolites of the Medicinal Plants of Mentha piperita and *Catharanthus roseus* Subjected to Drought and Heat Stress. *Biomolecules* 10(1): 43. doi:10.3390/biom10010043.

Ali, A., A. Maggio, R. Bressan, and D.-J. Yun. 2019. Role and Functional Differences of HKT1-Type Transporters in Plants under Salt Stress. *International Journal of Molecular Sciences* 20(5): 1059. doi:10.3390/ijms20051059.

Amooaghaie, R., and K. Nikzad. 2013. The Role of Nitric Oxide in Priming-Induced Low-Temperature Tolerance in Two Genotypes of Tomato. *Seed Science Research* 23(2): 123–31. doi:10.1017/S0960258513000068.

An, Q., X. He, N. Zheng, et al. 2020. Physiological and Genetic Effects of Cadmium and Copper Mixtures on Carrot under Greenhouse Cultivation. *Ecotoxicology and Environmental Safety* 206(December): 111363. doi:10.1016/j.ecoenv.2020.111363.

Apel, K., and H. Hirt. 2004. Reactive Oxygen Species: Metabolism, Oxidative Stress, and Signal Transduction. *Annual Review of Plant Biology* 55(1): 373–99. doi:10.1146/annurev.arplant.55.031903.141701.

Arfan, M., D.-W. Zhang, L.-J. Zou, et al. 2019. Hydrogen Peroxide and Nitric Oxide Crosstalk Mediates Brassinosteroids Induced Cold Stress Tolerance in Medicago truncatula. *International Journal of Molecular Sciences* 20(1): 144. doi:10.3390/ijms20010144.

Aroca, R., P. Vernieri, and J.M. Ruiz-Lozano. 2008. Mycorrhizal and Non-Mycorrhizal Lactuca sativa Plants exHibit Contrasting Responses to Exogenous ABA during Drought Stress and Recovery. *Journal of Experimental Botany* 59(8): 2029–41. doi:10.1093/jxb/ern057.

Assche, F. Van, and H. Clijsters. 1986. Inhibition of Photosynthesis in Phaseolus vulgaris by Treatment with Toxic Concentration of Zinc: Effect on Ribulose-1,5-Bisphosphate Carboxylase/ Oxygenase. *Journal of Plant Physiology* 125(3–4): 355–60. doi:10.1016/S0176-1617(86)80157-2.

Attia, H., K.H. Alamer, C. Ouhibi, S. Oueslati, and M. Lachaal. 2019. Interaction Between Salt Stress and Drought Stress on Some Physiological Parameters in Two Pea Cultivars. *International Journal of Botany* 16(1): 1–8. doi:10.3923/ijb.2020.1.8.

Atwell, B.J., H. Wang, and A.P. Scafaro. 2014. Could Abiotic Stress Tolerance in Wild Relatives of Rice Be Used to Improve Oryza sativa? *Plant Science: An International Journal of Experimental Plant Biology* 215–216: 48–58. doi:10.1016/j.plantsci.2013.10.007.

Aydinoglu, F. 2020. Elucidating the Regulatory Roles of MicroRNAs in Maize (Zea mays L.) Leaf Growth Response to Chilling Stress. *Planta* 251(2): 38. doi:10.1007/s00425-019-03331-y.

Balfagon, D., S.I. Zandalinas, R. Mittler, and A. Gómez-Cadenas. 2020. High Temperatures Modify Plant Responses to Abiotic Stress Conditions. *Physiologia Plantarum* 170(3): 335–44. doi:10.1111/ppl.13151.

Bao, Y., N. Yang, J. Meng, et al. 2020. Adaptability of Winter Wheat Dongnongdongmai 1 (Triticum Aestivum L.) to Overwintering in Alpine Regions. *Plant Biology*. doi:10.1111/plb.13200.

Barnabás, B., K. Jäger, and A. Fehér. 2008. The Effect of Drought and Heat Stress on Reproductive Processes in Cereals. *Plant, Cell and Environment* 31(1): 11–38. doi:10.1111/j.1365-3040.2007.01727.x.

Barnes, P.W., C.E. Williamson, R.M. Lucas, et al. 2019. Ozone Depletion, Ultraviolet Radiation, Climate Change and Prospects for a Sustainable Future. *Nature Sustainability* 2(7): 569–79. doi:10.1038/s41893-019-0314-2.

Benincasa, P., R. D'Amato, B. Falcinelli, et al. 2020. Grain Endogenous Selenium and Moderate Salt Stress Work as Synergic Elicitors in the Enrichment of Bioactive Compounds in Maize Sprouts. *Agronomy* 10(5): 735. doi:10.3390/agronomy10050735.

Boscaiu, M., and A. Fita. 2020. Physiological and Molecular Characterization of Crop Resistance to Abiotic Stresses. *Agronomy* 10(9): 1308. doi:10.3390/agronomy10091308.

Bouzroud, S., K. Gasparini, G. Hu, et al. 2020. Down Regulation and Loss of Auxin Response Factor 4 Function Using CRISPR/Cas9 Alters Plant Growth, Stomatal Function and Improves Tomato Tolerance to Salinity and Osmotic Stress. *Genes* 11(3): 272. doi:10.3390/genes11030272.

Bravo, L.A., and M. Griffith. 2005. Characterization of Antifreeze Activity in Antarctic Plants. *Journal of Experimental Botany* 56(414): 1189–96. doi:10.1093/jxb/eri112.

Breria, C.M., C.-H. Hsieh, T.-B. Yen, et al. 2020. A SNP-Based Genome-Wide Association Study to Mine Genetic Loci Associated to Salinity Tolerance in Mungbean (Vigna radiata L.). *Genes* 11(7): 759. doi:10.3390/genes11070759.

Cai, Z.Q., and Q. Gao. 2020. Comparative Physiological and Biochemical Mechanisms of Salt Tolerance in Five Contrasting Highland Quinoa Cultivars. *BMC Plant Biology* 20(1): 1–15. doi:10.1186/s12870-020-2279-8.

Calero Hurtado, A.C., D. Aparecida Chiconato, R. de Mello Prado, G. da Silveira Sousa Junior, and G. Felisberto. 2019. Silicon Attenuates Sodium Toxicity by Improving Nutritional Efficiency in Sorghum and Sunflower Plants. *Plant Physiology and Biochemistry: PPB* 142(September): 224–33. doi:10.1016/j.plaphy.2019.07.010.

Caser, M., W. Chitarra, F. D'Angiolillo, et al. 2019. Drought Stress Adaptation Modulates Plant Secondary Metabolite Production in Salvia Dolomitica Codd. *Industrial Crops and Products* 129(March): 85–96. doi:10.1016/j.indcrop.2018.11.068.

Chandra Babu, R.C., J. Zhang, A. Blum, et al. 2004. HVA1, a LEA Gene from Barley Confers Dehydration Tolerance in Transgenic Rice (Oryza sativa L.) via Cell Membrane Protection. *Plant Science* 166(4): 855–62. doi:10.1016/j.plantsci.2003.11.023.

Chavan, S.G., R.A. Duursma, M. Tausz, and O. Ghannoum. 2019. Elevated CO_2 Alleviates the Negative Impact of Heat Stress on Wheat Physiology but Not on Grain Yield. *Journal of Experimental Botany* 70(21): 6447–59. doi:10.1093/jxb/erz386.

Chen, J.B., J.W. Yang, Z.Y. Zhang, X.F. Feng, and S.M. Wang. 2013. Two *P5CS* Genes from Common Bean Exhibiting Different Tolerance to Salt Stress in Transgenic *Arabidopsis*. *Journal of Genetics* 92(3): 461–69. doi:10.1007/s12041-013-0292-5.

Cloutier, M., D. Chatterjee, D. Elango, et al. 2020. Sorghum Root Flavonoid Chemistry, Cultivar, and Frost Stress Effects on Rhizosphere Bacteria and Fungi. *Phytobiomes Journal*, August. doi: 10.1094/PBIOMES-01-20-0013-FI.

Cohen, I., S.I. Zandalinas, C. Huck, F.B. Fritschi, and R. Mittler. 2021. Meta-Analysis of Drought and Heat Stress Combination Impact on Crop Yield and Yield Components. *Physiologia Plantarum*: 1–11. doi:10.1111/ppl.13203.

Cong, W., Y. Miao, L. Xu, et al. 2019. Transgenerational Memory of Gene Expression Changes Induced by Heavy Metal Stress in Rice (Oryza sativa L.). *BMC Plant Biology* 19(1): 282. doi:10.1186/s12870-019-1887-7.

Correia, B., R.D. Hancock, J. Amaral, et al. 2018. Combined Drought and Heat Activates Protective Responses in Eucalyptus Globulus That Are Not Activated When Subjected to Drought or Heat Stress Alone. *Frontiers in Plant Science* 9(June): 819. doi:10.3389/fpls.2018.00819.

Dar, N.A., I. Amin, W. Wani, et al. 2017. Abscisic Acid: A Key Regulator of Abiotic Stress Tolerance in Plants. *Plant Gene* 11(September): 106–11. doi:10.1016/j.plgene.2017.07.003.

Das, K., and A. Roychoudhury. 2014. Reactive Oxygen Species (ROS) and Response of Antioxidants as ROS-Scavengers during Environmental Stress in Plants. *Frontiers in Environmental Science*. doi:10.3389/fenvs.2014.00053.

Dawood, M.F.A., M. Tahjib-Ul-Arif, A.A.M. Sohag, A.A.H.A. Latef, and M.M. Ragaey. 2020. Mechanistic Insight of Allantoin in Protecting Tomato Plants against Ultraviolet C Stress. *Plants* 10(1): 11.

Devi, P.U., S. Murugan, S. Akilapriyadharasini, S. Suja, and P. Chinnaswamy. 2007. Effect of Mercury and Efluents on Seed Germination, Root-Shoot Length, Amylase Activity and Phenolic Compounds in Vigna unguiculata. *Nature Environment and Pollution Technology* 6(3): 457–62.

Dhanya Thomas, T.T., C. Dinakar, and J.T. Puthur. 2020. Effect of UV-B Priming on the Abiotic Stress Tolerance of Stress-Sensitive Rice Seedlings: Priming Imprints and Cross-Tolerance. *Plant Physiology and Biochemistry : PPB* 147(February): 21–30. doi:10.1016/j.plaphy.2019.12.002.

Dikilitas, M., E. Simsek, and S. Karakas. 2019a. Stress Responsive Signaling Molecules and Genes Under Stressful Environments in Plants. In: *Plant Signaling Molecules*, edited by M.I.R. Khan, P.S. Reddy, A. Ferrante, and N.A. Khan, 19–42. Elsevier. doi:10.1016/B978-0-12-816451-8.00002-2.

Dikilitas, M., E. Simsek, S. Karakas, and P. Ahmad. 2019b. High-Temperature Stress and Photosynthesis Under Pathological Impact. In: *Photosynthesis, Productivity and Environmental Stress*, 39–64. Wiley. doi:10.1002/9781119501800.ch3.

Dikilitas, M., E. Simsek, and A. Roychoudhury. 2020. Role of Proline and Glycine Betaine in Overcoming Abiotic Stresses. In: *Protective Chemical Agents in the Amelioration of Plant Abiotic Stress*, 1–23. doi:10.1002/9781119552154.ch1.

Dolferus, R. 2014. To Grow or Not to Grow: A Stressful Decision for Plants. *Plant Science* 229: 247–61. doi:10.1016/j.plantsci.2014.10.002.

Dong, Q., D. Duan, S. Zhao, et al. 2018. Genome-Wide Analysis and Cloning of the Apple Stress-Associated Protein Gene Family Reveals MdSAP15, Which Confers Tolerance to Drought and Osmotic Stresses in Transgenic Arabidopsis. *International Journal of Molecular Sciences* 19(9): 2478. doi:10.3390/ijms19092478.

Duan, J., H. Tian, and Y. Gao. 2016. Expression of Nitrogen Transporter Genes in Roots of Winter Wheat (Triticum aestivum L.) in Response to Soil Drought with Contrasting Nitrogen Supplies. *Crop and Pasture Science* 67(2): 128–36. doi:10.1071/CP15152.

Dutta, S., M. Mitra, P. Agarwal, et al. 2018. Oxidative and Genotoxic Damages in Plants in Response to Heavy Metal Stress and Maintenance of Genome Stability. *Plant Signaling and Behavior.* doi:10.1080/15592324.2018.1460048.

El Moukhtari, A., C. Cabassa-Hourton, M. Farissi, and A. Savouré. 2020. How Does Proline Treatment Promote Salt Stress Tolerance During Crop Plant Development? *Frontiers in Plant Science* 11(July): 1127. doi:10.3389/fpls.2020.01127.

Espasandin, F.D., P.I. Calzadilla, S.J. Maiale, O.A. Ruiz, and P.A. Sansberro. 2018. Overexpression of the Arginine Decarboxylase Gene Improves Tolerance to Salt Stress in Lotus tenuis Plants. *Journal of Plant Growth Regulation* 37(1): 156–65. doi:10.1007/s00344-017-9713-7.

FAO. 2009. How to Feed the World in 2050. http://www.fao.org/fileadmin/templates/wsfs/docs/expert_paper/How:to_Feed_the_World_in_2050.pdf.

Flowers, T.J. 2004. Improving Crop Salt Tolerance. *Journal of Experimental Botany* 55(396): 307–19. doi:10.1093/jxb/erh003.

Fragkostefanakis, S., S. Simm, P. Paul, et al. 2015. Chaperone Network Composition in Solanum Lycopersicum Explored by Transcriptome Profiling and Microarray Meta-Analysis. *Plant, Cell and Environment* 38(4): 693–709. doi:10.1111/pce.12426.

Fraire-Velazquez, S., R. Rodriguez-Guerra, and L. Sanchez-Caldero. 2011. Abiotic and Biotic Stress Response Crosstalk in Plants. In: *Abiotic Stress Response in Plants - Physiological, Biochemical and Genetic Perspectives*, edited by A. Shanker and B. Venkateswarlu. IntechOpen. doi:10.5772/23217.

Fürtauer, L., J. Weiszmann, W. Weckwerth, and T. Nägele. 2019. Dynamics of Plant Metabolism during Cold Acclimation. *International Journal of Molecular Sciences* 20(21): 5411. doi:10.3390/ijms20215411.

Gaafar, A.A., S.I. Ali, M.A. El-Shawadfy, et al. 2020. Ascorbic Acid Induces the Increase of Secondary Metabolites, Antioxidant Activity, Growth, and Productivity of the Common Bean under Water Stress Conditions. *Plants* 9(5): 627. doi:10.3390/plants9050627.

García-Sánchez, F., S. Simón-Grao, J.J. Martínez-Nicolás, et al. 2020. Multiple Stresses Occurring with Boron Toxicity and Deficiency in Plants. *Journal of Hazardous Materials* 397(February): 122713. doi:10.1016/j.jhazmat.2020.122713.

Gechev, T.S., C. Dinakar, M. Benina, V. Toneva, and D. Bartels. 2012. Molecular Mechanisms of Desiccation Tolerance in Resurrection Plants. *Cellular and Molecular Life Sciences: CMLS* 69(19): 3175–86. doi:10.1007/s00018-012-1088-0.

Gerami, M., P. Majidian, A. Ghorbanpour, and Z. Alipour. 2020. Stevia Rebaudiana Bertoni Responses to Salt Stress and Chitosan Elicitor. *Physiology and Molecular Biology of Plants: An International Journal of Functional Plant Biology* 26(5): 965–74. doi:10.1007/s12298-020-00788-0.

Ghori, N.-H., T. Ghori, M.Q. Hayat, et al. 2019. Heavy Metal Stress and Responses in Plants. *International Journal of Environmental Science and Technology* 16(3): 1807–28. doi:10.1007/s13762-019-02215-8.

Gill, S.S., and N. Tuteja. 2010. Reactive Oxygen Species and Antioxidant Machinery in Abiotic Stress Tolerance in Crop Plants. *Plant Physiology and Biochemistry* 48(12): 909–30. doi:10.1016/j.plaphy.2010.08.016.

Gopal, R., V. Giri, and N. Nautiyal. 2008. Excess Copper and Manganese Alters the Growth and Vigour of Maize Seedlings in Solution Culture. *Indian Journal of Plant Physiology* 13(1): 44–9.

Golldack, D., C. Li, H. Mohan, and N. Probst. 2014. Tolerance to Drought and Salt Stress in Plants: Unraveling the Signaling Networks. *Frontiers in Plant Science* 5: 151. doi:10.3389/fpls.2014.00151.

Gould, W.A. 1992. *Tomato Production, Processing and Technology. Tomato Production, Processing and Technology* (3rd ed). CTI Publications Inc, Baltimore, MD. doi:10.1533/9781845696146.

Gourdji, S.M., A.M. Sibley, and D.B. Lobell. 2013. Global Crop Exposure to Critical High Temperatures in the Reproductive Period: Historical Trends and Future Projections. *Environmental Research Letters* 8(2): 24041. doi:10.1088/1748-9326/8/2/024041.

Greco, M., A. Chiappetta, L. Bruno, and M.B. Bitonti. 2012. In Posidonia Oceanica Cadmium Induces Changes in DNA Methylation and Chromatin Patterning. *Journal of Experimental Botany* 63(2): 695–709. doi:10.1093/jxb/err313.

Gujjar, R.S., S.G. Karkute, A. Rai, M. Singh, and B. Singh. 2018. Proline-Rich Proteins May Regulate Free Cellular Proline Levels during Drought Stress in Tomato. *Current Science* 114(4): 915–20. doi: 10.18520/cs/v114/i04/915-920.

Guo, R., H. Qiao, J. Zhao, et al. 2018. The Grape VLWRKY3 Gene Promotes Abiotic and Biotic Stress Tolerance in Transgenic Arabidopsis thaliana. *Frontiers in Plant Science* 9(April): 545. doi:10.3389/fpls.2018.00545.

Han, G., C. Lu, J. Guo, et al. 2020. C2H2 Zinc Finger Proteins: Master Regulators of Abiotic Stress Responses in Plants. *Frontiers in Plant Science.* doi:10.3389/fpls.2020.00115.

Hasanuzzaman, M., M.H.M.B. Bhuyan, F. Zulfiqar, et al. 2020. Reactive Oxygen Species and Antioxidant Defense in Plants under Abiotic Stress: Revisiting the Crucial Role of a Universal Defense Regulator. *Antioxidants* 9(8): 681. doi:10.3390/antiox9080681.

Hasanuzzaman, M., K. Nahar, T.I. Anee, M.I.R. Khan, and M. Fujita. 2018. Silicon-Mediated Regulation of Antioxidant Defense and Glyoxalase Systems Confers Drought Stress Tolerance in Brassica napus L. *South African Journal of Botany* 115(March): 50–7. doi:10.1016/j.sajb.2017.12.006.

Hassan, M.U., M.U. Chattha, I. Khan et al. 2021. Heat Stress in Cultivated Plants: Nature, Impact, Mechanisms, and Mitigation Strategies—A Review. *Plant Biosystems—An International Journal Dealing with all Aspects of Plant Biology* 155(2): 211–34. doi:10.1080/11263504.2020.1727987.

Hu, S., J. Liu, Y. Zhang, et al. 2020. Analysis of Resistance to Alkaline Salt Stress in Arabidopsis thaliana with Overexpression of OsiSAP1 Gene. *Genomics and Applied Biology* 2: 1–10. doi:10.5376/gab.2020.11.0002.

Huang, H., M. Rizwan, M. Li, et al. 2019. Comparative Efficacy of Organic and Inorganic Silicon Fertilizers on Antioxidant Response, Cd/Pb Accumulation and Health Risk Assessment in Wheat (Triticum aestivum L.). *Environmental Pollution* 255(December): 113146. doi:10.1016/j.envpol.2019.113146.

Huang, J., A. Levine, and Z. Wang. 2013. Plant Abiotic Stress. *The Scientific World Journal.* doi:10.1155/2013/432836.

Imadi, S.R., A. Gul, M. Dikilitas, et al. 2016. Water Stress: Types, Causes, and Impact on Plant Growth and Development. In: *Water Stress and Crop Plants: a Sustainable Approach*, Vol. 2–2, 343–55. Chichester, UK: John Wiley & Sons, Ltd. doi:10.1002/9781119054450.ch21.

Islam, M.Z., B.J. Park, and Y.T. Lee. 2019. Effect of Salinity Stress on Bioactive Compounds and Antioxidant Activity of Wheat Microgreen Extract under Organic Cultivation Conditions. *International Journal of Biological Macromolecules* 140(November): 631–36. doi:10.1016/j.ijbiomac.2019.08.090.

Jian, S., J. Luo, Q. Liao, et al. 2019. NRT1.1 Regulates Nitrate Allocation and Cadmium Tolerance in Arabidopsis. *Frontiers in Plant Science* 10(March): 384. doi:10.3389/fpls.2019.00384.

Jiao, C., G. Lan, Y. Sun, G. Wang, and Y. Sun. 2021. Dopamine Alleviates Chilling Stress in Watermelon Seedlings via Modulation of Proline Content, Antioxidant Enzyme Activity, and Polyamine Metabolism. *Journal of Plant Growth Regulation*: 1–16. doi:10.1007/s00344-020-10096-2.

Joshi, N.C., D. Yadav, K. Ratner, et al. 2020. Sodium Hydrosulfide Priming Improves the Response of Photosynthesis to Overnight Frost and Day High Light in Avocado (Persea Americana Mill, Cv. 'Hass') (Persea Americana Mill, Cv. "Hass"). *Physiologia Plantarum* 168(2): 394–405. doi:10.1111/ppl.13023.

Julkowska, M.M., and C. Testerink. 2015. Tuning Plant Signaling and Growth to Survive Salt. *Trends in Plant Science.* doi:10.1016/j.tplants.2015.06.008.

Kaiser, W.M., H. Weiner, and S.C. Huber. 1999. Nitrate Reductase in Higher Plants: A Case Study for Transduction of Environmental Stimuli into Control of Catalytic Activity. *Physiologia Plantarum* 105(2): 384–89. doi:10.1034/j.1399-3054.1999.105225.x.

Kakar, K.U., X. -l. Ren, Z. Nawaz, et al. 2016. A Consortium of Rhizobacterial Strains and Biochemical Growth Elicitors Improve Cold and Drought Stress Tolerance in Rice (*Oryza sativa* L.). *Plant Biology* 18(3): 471–83. doi:10.1111/plb.12427.

Kanwal, S., N. Ilyas, S. Shabir, et al. 2018. Application of Biochar in Mitigation of Negative Effects of Salinity Stress in Wheat (*Triticum aestivum* L.). *Journal of Plant Nutrition* 41(4): 526–38. doi:10.1080/01904167.2017.1392568.

Karkute, S.G., R. Krishna, W.A. Ansari, et al. 2019. Heterologous Expression of the AtDREB1A Gene in Tomato Confers Tolerance to Chilling Stress. *Biologia Plantarum* 63(1): 268–77. doi:10.32615/bp.2019.031.

Kaur, S., A.K. Gupta, and N. Kaur. 2000. Effect of GA3, Kinetin and Indole Acetic Acid on Carbohydrate Metabolism in Chickpea Seedlings Germinating under Water Stress. *Plant Growth Regulation* 30(1): 61–70. doi:10.1023/A:1006371219048.

Kaya, C., M. Ashraf, M.N. Alyemeni, F.J. Corpas, and P. Ahmad. 2020. Salicylic Acid-Induced Nitric Oxide Enhances Arsenic Toxicity Tolerance in Maize Plants by Upregulating the Ascorbate-Glutathione Cycle and Glyoxalase System. *Journal of Hazardous Materials* 399(May): 123020. doi:10.1016/j.jhazmat.2020.123020.

Khan, A., S. Bilal, A.L. Khan, et al. 2020. Silicon and Gibberellins: Synergistic Function in Harnessing ABA Signaling and Heat Stress Tolerance in Date Palm (Phoenix dactylifera L.). *Plants* 9(5): 620. doi:10.3390/plants9050620.

Khanna, K., S.K. Kohli, P. Ohri, et al. 2019. Microbial Fortification Improved Photosynthetic Efficiency and Secondary Metabolism in Lycopersicon esculentum Plants Under Cd Stress. *Biomolecules* 9(10): 581. doi:10.3390/biom9100581.

Khudsar, T., M. Mahmooduzzafar, M. Iqbal, and R.K. Sairam. 2004. Zinc-Induced Changes in Morpho-Physiological and Biochemical Parameters in Artemisia annua. *Biologia Plantarum* 48(2): 255–60. doi:10.1023/B:BIOP.0000033453.24705.f5.

Kim, Y.O., and H. Kang. 2018. Comparative Expression Analysis of Genes Encoding Metallothioneins in Response to Heavy Metals and Abiotic Stresses in Rice (Oryza sativa) and Arabidopsis thaliana. *Bioscience, Biotechnology, and Biochemistry* 82(9): 1656–65. doi:10.1080/09168451.2018.1486177.

Kong, X., W. Ge, B. Wei, et al. 2020. Melatonin Ameliorates Chilling Injury in Green Bell Peppers during Storage by Regulating Membrane Lipid Metabolism and Antioxidant Capacity. *Postharvest Biology and Technology* 170(December). doi:10.1016/j.postharvbio.2020.111315. http://www.ncbi.nlm.nih.gov/pubmed/111315.

Kou, S., L. Chen, W. Tu, et al. 2018. The Arginine Decarboxylase Gene *ADC1* , Associated to the Putrescine Pathway, Plays an Important Role in Potato Cold-Acclimated Freezing Tolerance as Revealed by Transcriptome and Metabolome Analyses. *The Plant Journal : For Cell and Molecular Biology* 96(6): 1283–98. doi:10.1111/tpj.14126.

Krasensky, J., and C. Jonak. 2012. Drought, Salt, and Temperature Stress-Induced Metabolic Rearrangements and Regulatory Networks. *Journal of Experimental Botany* 63(4): 1593–1608. doi:10.1093/jxb/err460.

Krishna, R., S.G. Karkute, W.A. Ansari, et al. 2019. Transgenic Tomatoes for Abiotic Stress Tolerance: Status and Way Ahead. *3 Biotech* 9(4): 143. doi:10.1007/s13205-019-1665-0.

Kumar, A., and J.P. Verma. 2018. Does Plant-Microbe Interaction Confer Stress Tolerance in Plants: A Review? Microbiological Research 207(March): 41–52. doi:10.1016/j.micres.2017.11.004.

Kuriakose, S.V., and M.N.V. Prasad. 2008. Cadmium Stress Affects Seed Germination and Seedling Growth in Sorghum bicolor (L.) Moench by Changing the Activities of Hydrolyzing Enzymes. *Plant Growth Regulation* 54(2): 143–56. doi:10.1007/s10725-007-9237-4.

Lakaew, K., S. Akeprathumchai, and P. Thiravetyan. 2020. Foliar Spraying of Calcium Acetate Alleviates Yield Loss in Rice (Oryza sativa L.) by Induced Anti-Oxidative Defence System under Ozone and Heat Stresses. *Annals of Applied Biology* (April): 1–13. doi:10.1111/aab.12653.

Langworthy, A.D., R.P. Rawnsley, M.J. Freeman, et al. 2020. Can Irrigating More Frequently Mitigate Detrimental Heat Wave Effects on Perennial Ryegrass Growth and Persistence? *Agricultural and Forest Meteorology* 291(September). doi:10.1016/j.agrformet.2020.108074. http://www.ncbi.nlm.nih.gov/pubmed/108074.

Latha, G., T. Mohapatra, A.S. Geetanjali, and K.R.S.S. Rao. 2017. Engineering Rice for Abiotic Stress Tolerance: A Review. *Current Trends in Biotechnology and Pharmacy* 11(4): 396–413. http://abap.co.in/files/CTBP_11_4.pdf%0Ahttps://www.cabdirect.org/cabdirect/abstract/20193302559.

Latha, G.M., K.V. Raman, J.M. Lima, et al. 2019. Genetic Engineering of Indica Rice with AtDREB1A Gene for Enhanced Abiotic Stress Tolerance. *Plant Cell, Tissue and Organ Culture* 136(1): 173–88. doi:10.1007/s11240-018-1505-7.

Lexer, C., and M.F. Fay. 2005. Adaptation to Environmental Stress: A Rare or Frequent Driver of Speciation? *Journal of Evolutionary Biology* 18(4): 893–900. doi:10.1111/j.1420-9101.2005.00901.x.

Li, J., M. Li, S. Yao, G. Cai, and X. Wang. 2020a. Patatin-Related Phospholipase PPLAIIIγ Involved in Osmotic and Salt Tolerance in Arabidopsis. *Plants* 9(5): 650. doi:10.3390/plants9050650.

Li, J., M. Zhang, J. Sun, et al. 2020b. Heavy Metal Stress-Associated Proteins in Rice and Arabidopsis: Genome-Wide Identification, Phylogenetics, Duplication, and Expression Profiles Analysis. *Frontiers in Genetics* 11(May): 477. doi:10.3389/fgene.2020.00477.

Li, X., Y. Li, G.J. Ahammed, et al. 2019. RBOH1-Dependent Apoplastic H2O2 Mediates Epigallocatechin-3-Gallate-Induced Abiotic Stress Tolerance in Solanum Lycopersicum L. *Environmental and Experimental Botany* 161: 357–66. doi:10.1016/j.envexpbot.2018.11.013.

Li, Z., Y. Peng, X.Q. Zhang, et al. 2014. Exogenous Spermidine Improves Water Stress Tolerance of White Clover (Trifolium repens L.) Involved in Antioxidant Defence, Gene Expression and Proline Metabolism. *Plant OMICS* 7(6): 517–26.

Liao, S., X. Qin, L. Luo, et al. 2019. CRISPR/Cas9-Induced Mutagenesis of Semi-Rolled Leaf1,2 Confers Curled Leaf Phenotype and Drought Tolerance by Influencing Protein Expression Patterns and ROS Scavenging in Rice (Oryza sativa L.). *Agronomy* 9(11): 728. doi:10.3390/agronomy9110728.

Lipiec, J., C. Doussan, A. Nosalewicz, and K. Kondracka. 2013. Effect of Drought and Heat Stresses on Plant Growth and Yield: A Review. *International Agrophysics*. doi:10.2478/intag-2013-0017.

Liu, H., M. Xing, W. Yang, et al. 2019. Genome-Wide Identification of and Functional Insights into the Late Embryogenesis Abundant (LEA) Gene Family in Bread Wheat (Triticum aestivum). *Scientific Reports* 9(1): 1–11. doi:10.1038/s41598-019-49759-w.

Liu, J., Z. Xiong, T. Li, and H. Huang. 2004. Bioaccumulation and Ecophysiological Responses to Copper Stress in Two Populations of Rumex dentatus L. from Cu Contaminated and Non-Contaminated Sites. *Environmental and Experimental Botany* 52(1): 43–51. doi:10.1016/j.envexpbot.2004.01.005.

Liu, W., E. Sikora, and S.W. Park. 2020. Plant Growth-Promoting Rhizobacterium, Paenibacillus Polymyxa CR1, Upregulates Dehydration-Responsive Genes, RD29A and RD29B, during Priming Drought Tolerance in Arabidopsis. *Plant Physiology and Biochemistry: PPB* 156(November): 146–54. doi:10.1016/j.plaphy.2020.08.049.

Maheshwari, R., and R.S. Dubey. 2007. Nickel Toxicity Inhibits Ribonuclease and Protease Activities in Rice Seedlings: Protective Effects of Proline. *Plant Growth Regulation* 51(3): 231–43. doi:10.1007/s10725-006-9163-x.

Malerba, M., and R. Cerana. 2018. Effect of Selenium on the Responses Induced by Heat Stress in Plant Cell Cultures. *Plants* 7(3): 64. doi:10.3390/plants7030064.

Masouleh, S.S.S., and Y.N. Sassine. 2020. Molecular and Biochemical Responses of Horticultural Plants and Crops to Heat Stress. *Ornamental Horticulture* 26(2): 148–58. doi:10.1590/2447-536X.v26i2.2134.

Matthews, C., M. Arshad, and A. Hannoufa. 2019. Alfalfa Response to Heat Stress Is Modulated by MicroRNA156. *Physiologia Plantarum* 165(4): 830–42. doi:10.1111/ppl.12787.

Mekawy, A.M.M., D.V.M. Assaha, R. Munehiro, et al. 2018. Characterization of Type 3 Metallothionein-Like Gene (OsMT-3a) from Rice, Revealed Its Ability to Confer Tolerance to Salinity and Heavy Metal Stresses. *Environmental and Experimental Botany* 147(March): 157–66. doi:10.1016/j.envexpbot.2017.12.002.

Mendanha, T., E. Rosenqvist, B. Nordentoft Hyldgaard, J.H. Doonan, and C. Ottosen. 2020. Drought Priming Effects on Alleviating the Photosynthetic Limitations of Wheat Cultivars (*Triticum aestivum* L.) with Contrasting Tolerance to Abiotic Stresses. *Journal of Agronomy and Crop Science* 206(6): 651–64. doi:10.1111/jac.12404.

Meng, C., and N. Sui. 2019. Overexpression of Maize MYB-IF35 Increases Chilling Tolerance in Arabidopsis. *Plant Physiology and Biochemistry: PPB* 135(February): 167–73. doi:10.1016/j.plaphy.2018.11.038.

Mittler, R. 2002. Oxidative Stress, Antioxidants and Stress Tolerance. *Trends in Plant Science*. doi:10.1016/S1360-1385(02)02312-9.

Moazzam-Jazi, M., S. Ghasemi, S.M. Seyedi, and V. Niknam. 2018. COP1 Plays a Prominent Role in Drought Stress Tolerance in Arabidopsis and Pea. *Plant Physiology and Biochemistry: PPB* 130(September): 678–91. doi:10.1016/j.plaphy.2018.08.015.

Mohanty, A., H. Kathuria, A. Ferjani, et al. 2002. Transgenics of an Elite Indica Rice Variety Pusa Basmati 1 Harbouring the Coda Gene Are Highly Tolerant to Salt Stress. *TAG. Theoretical and Applied Genetics. Theoretische und Angewandte Genetik* 106(1): 51–7. doi:10.1007/s00122-002-1063-5.

Morrow, G., and R.M. Tanguay. 2012. Small Heat Shock Protein Expression and Functions during Development. *International Journal of Biochemistry and Cell Biology* 44(10): 1613–21. doi:10.1016/j.biocel.2012.03.009.

Müntz, K., M.A. Belozersky, Y.E. Dunaevsky, A. Schlereth, and J. Tiedemann. 2001. Stored Proteinases and the Initiation of Storage Protein Mobilization in Seeds during Germination and Seedling Growth. *Journal of Experimental Botany* 52(362): 1741–52. doi:10.1093/jexbot/52.362.1741.

Mushke, R., R. Yarra, and P.B. Kirti. 2019. Improved Salinity Tolerance and Growth Performance in Transgenic Sunflower Plants via Ectopic Expression of a Wheat Antiporter Gene (TaNHX2). *Molecular Biology Reports* 46(6): 5941–53. doi:10.1007/s11033-019-05028-7.

Nagamiya, K., T. Motohashi, K. Nakao, et al. 2007. Enhancement of Salt Tolerance in Transgenic Rice Expressing an Escherichia coli Catalase Gene, KatE. *Plant Biotechnology Reports* 1(1): 49–55. doi:10.1007/s11816-007-0007-6.

Naguib, D.M., and H. Abdalla. 2019. Metabolic Status during Germination of Nano Silica Primed Zea mays Seeds under Salinity Stress. *Journal of Crop Science and Biotechnology* 22(5): 415–23. doi:10.1007/s12892-019-0168-0.

Nawaz, M.A., Y. Jiao, C. Chen, et al. 2018. Melatonin Pretreatment Improves Vanadium Stress Tolerance of Watermelon Seedlings by Reducing Vanadium Concentration in the Leaves and Regulating Melatonin Biosynthesis and Antioxidant-Related Gene Expression. *Journal of Plant Physiology* 220(January): 115–27. doi:10.1016/j.jplph.2017.11.003.

Nazdar, T., A. Tehranifar, A. Nezami, H. Nemati, and L. Samiei. 2019. Physiological and Anatomical Responses of Calendula (*Calendula officinalis* L.) Cultivars to Heat-Stress Duration. *The Journal of Horticultural Science and Biotechnology* 94(3): 400–11. doi:10.1080/14620316.2018.1532324.

Nebauer, S.G., M. Sánchez, L. Martínez, et al. 2013. Differences in Photosynthetic Performance and Its Correlation with Growth among Tomato Cultivars in Response to Different Salts. *Plant Physiology and Biochemistry: PPB* 63(February): 61–69. doi:10.1016/j.plaphy.2012.11.006.

Nemati, I., F. Moradi, S. Gholizadeh, M.A. Esmaeili, and M.R. Bihamta. 2011. The Effect of Salinity Stress on Ions and Soluble Sugars Distribution in Leaves, Leaf Sheaths and Roots of Rice (Oryza sativa L.) Seedlings. *Plant, Soil and Environment* 57(1): 26–33. doi: 10.17221/71/2010-PSE.

Nepal, N., J.P. Yactayo-Chang, R. Gable, et al. 2020. Phenotypic Characterization of *Arabidopsis thaliana* Lines Overexpressing *AVP1* and *MIOX4* in Response to Abiotic Stresses. *Applications in Plant Sciences* 8(8). doi:10.1002/aps3.11384.

Nie, L., H. Liu, L. Zhang, and W. Wang. 2020. Enhancement in Rice Seed Germination via Improved Respiratory Metabolism under Chilling Stress. *Food and Energy Security* 9(4): e234. doi:10.1002/fes3.234.

Nieboer, E., and D.H.S. Richardson. 1980. The Replacement of the Nondescript Term "Heavy Metals" by a Biologically and Chemically Significant Classification of Metal Ions. *Environmental Pollution Series B, Chemical and Physical* 1(1): 3–26. doi:10.1016/0143-148X(80)90017-8.

Nies, D.H. 1999. Microbial Heavy-Metal Resistance. *Applied Microbiology and Biotechnology.* doi:10.1007/s002530051457.

Nosek, M., A. Kaczmarczyk, R.J. Jędrzejczyk, et al. 2020. Expression of Genes Involved in Heavy Metal Trafficking in Plants Exposed to Salinity Stress and Elevated Cd Concentrations. *Plants* 9(4): 475. doi:10.3390/plants9040475.

Nxele, X., A. Klein, and B.K. Ndimba. 2017. Drought and Salinity Stress Alters ROS Accumulation, Water Retention, and Osmolyte Content in Sorghum Plants. *South African Journal of Botany* 108(January): 261–66. doi:10.1016/j.sajb.2016.11.003.

Oksanen, E.J. 2010. Environmental Pollution and Function of Plant Leaves. *Medical and Health Sciences* 5: 217.

Orhan, F. 2016. Alleviation of Salt Stress by Halotolerant and Halophilic Plant Growth-Promoting Bacteria in Wheat (Triticum aestivum). *Brazilian Journal of Microbiology : [Publication of the Brazilian Society for Microbiology]* 47(3): 621–27. doi:10.1016/j.bjm.2016.04.001.

Osman, M.S., A.A. Badawy, A.I. Osman, and A.A.H. Abdel Latef. 2020. Ameliorative Impact of an Extract of the Halophyte Arthrocnemum Macrostachyum on Growth and Biochemical Parameters of Soybean Under Salinity Stress. *Journal of Plant Growth Regulation* July: 1–12. doi:10.1007/s00344-020-10185-2.

Paim, B.T., R.L. Crizel, S.J. Tatiane, et al. 2020. Mild Drought Stress Has Potential to Improve Lettuce Yield and Quality. *Scientia Horticulturae* 272(October): 109578. doi:10.1016/j.scienta.2020.109578.

Parmar, N., K.H. Singh, D. Sharma, et al. 2017. Genetic Engineering Strategies for Biotic and Abiotic Stress Tolerance and Quality Enhancement in Horticultural Crops: A Comprehensive Review. *3 Biotech* 7(4): 239. doi:10.1007/s13205-017-0870-y.

Paul, S., and A. Roychoudhury. 2017. Seed Priming with Spermine and Spermidine Regulates the Expression of Diverse Groups of Abiotic Stress-Responsive Genes during Salinity Stress in the Seedlings of Indica Rice Varieties. *Plant Gene* 11(September): 124–32. doi:10.1016/j.plgene.2017.04.004.

Prasanna, B.M. 2012. Diversity in Global Maize Germplasm: Characterization and Utilization. *Journal of Biosciences* 37(5): 843–55. doi:10.1007/s12038-012-9227-1.

Rady, M.M., N.B. Talaat, M.T. Abdelhamid, B.T. Shawky, and E.-S.M. Desoky. 2019. Maize (*Zea mays* L.) Grains Extract Mitigates the Deleterious Effects of Salt Stress on Common Bean (*Phaseolus vulgaris* L.) Growth and Physiology. *The Journal of Horticultural Science and Biotechnology* 94(6): 777–89. doi:10.1080/14620316.2019.1626773.

Razzaque, S., S.M. Elias, T. Haque, et al. 2019. Gene Expression Analysis Associated with Salt Stress in a Reciprocally Crossed Rice Population. *Scientific Reports* 9(1): 1–17. doi:10.1038/s41598-019-44757-4.

Remorini, D., J.C. Melgar, L. Guidi, et al. 2009. Interaction Effects of Root-Zone Salinity and Solar Irradiance on the Physiology and Biochemistry of Olea europaea. *Environmental and Experimental Botany* 65(2–3): 210–19. doi:10.1016/j.envexpbot.2008.12.004.

Richards, R.A., J.R. Hunt, J.A. Kirkegaard, and J.B. Passioura. 2014. Yield Improvement and Adaptation of Wheat to Water-Limited Environments in Australia - A Case Study. *Crop and Pasture Science* 65(7): 676–89. doi:10.1071/CP13426.

Roca Paixão, J.F., F.X. Gillet, T.P. Ribeiro, et al. 2019. Improved Drought Stress Tolerance in Arabidopsis by CRISPR/DCas9 Fusion with a Histone Acetyltransferase. *Scientific Reports* 9(1): 1–9. doi:10.1038/s41598-019-44571-y.

Sahni, S., B.D. Prasad, Q. Liu, et al. 2016. Overexpression of the Brassinosteroid Biosynthetic Gene DWF4 in Brassica napus Simultaneously Increases Seed Yield and Stress Tolerance. *Scientific Reports* 6(1): 28298. doi:10.1038/srep28298.

Saijo, Y., S. Hata, J. Kyozuka, K. Shimamoto, and K. Izui. 2000. Over-Expression of a Single Ca2+-Dependent Protein Kinase Confers Both Cold and Salt/Drought Tolerance on Rice Plants. *Plant Journal: For Cell and Molecular Biology* 23(3): 319–27. doi:10.1046/j.1365-313X.2000.00787.x.

Samarina, L.S., L.S. Malyukova, A.M. Efremov, et al. 2020. Physiological, Biochemical and Genetic Responses of Caucasian Tea (Camellia sinensis (L.) Kuntze) Genotypes under Cold and Frost Stress. *PeerJ* 8(August): e9787. doi:10.7717/peerj.9787.

Santosh Kumar, V.V., R.K. Verma, S.K. Yadav, et al. 2020. CRISPR-Cas9 Mediated Genome Editing of Drought and Salt Tolerance (OsDST) Gene in Indica Mega Rice Cultivar MTU1010. *Physiology and Molecular Biology of Plants : An International Journal of Functional Plant Biology* 26(6): 1099–110. doi:10.1007/s12298-020-00819-w.

Sarker, U., and S. Oba. 2020. The Response of Salinity Stress-Induced A. Tricolor to Growth, Anatomy, Physiology, Non-Enzymatic and Enzymatic Antioxidants. *Frontiers in Plant Science* 11(October): 1–14. doi:10.3389/fpls.2020.559876.

Savvides, A., S. Ali, M. Tester, and V. Fotopoulos. 2016. Chemical Priming of Plants Against Multiple Abiotic Stresses: Mission Possible? *Trends in Plant Science* 21(4): 329–40. doi:10.1016/j.tplants.2015.11.003.

Sehgal, A., K. Sita, J. Kumar, et al. 2017. Effects of Drought, Heat and Their Interaction on the Growth, Yield and Photosynthetic Function of Lentil (Lens Culinaris Medikus) Genotypes Varying in Heat and Drought Sensitivity. *Frontiers in Plant Science* 8: 1776. doi:10.3389/fpls.2017.01776.

Singh, M., J. Kumar, S. Singh, V.P. Singh, S.M. Prasad, and MPVVB, Singh. 2015. Adaptation Strategies of Plants against Heavy Metal Toxicity: A Short Review. *Biochemistry and Pharmacology: Open Access (Los Angel)* 4: 161. doi:10.4172/2167-0501.1000161.

Singh Sengar, R., M. Gautam, S. Kumar Garg, R. Chaudhary, and K. Sengar. 2008. Effect of Lead on Seed Germination, Seedling Growth, Chlorophyll Content and Nitrate Reductase Activity in Mung Bean (Vigna radiata). *Research Journal of Phytochemistry* 2(2): 61–68. doi:10.3923/rjphyto.2008.61.68.

Shah, K., M. Singh, and A.C. Rai. 2013. Effect of Heat-Shock Induced Oxidative Stress Is Suppressed in BcZAT12 Expressing Drought Tolerant Tomato. *Phytochemistry* 95 (November): 109–17. doi:10.1016/j.phytochem.2013.07.026.

Shahid, M., C. Dumat, S. Khalid, et al. 2017. Foliar Heavy Metal Uptake, Toxicity and Detoxification in Plants: A Comparison of Foliar and Root Metal Uptake. *Journal of Hazardous Materials*. doi:10.1016/j.jhazmat.2016.11.063.

Shakeel, A., A.A. Khan, and K.R. Hakeem. 2020. Growth, Biochemical, and Antioxidant Response of Beetroot (Beta vulgaris L.) Grown in Fly Ash-Amended Soil. *SN Applied Sciences* 2(8): 1–9. doi:10.1007/s42452-020-3191-4.

Sharma, P., and R.S. Dubey. 2005. Lead Toxicity in Plants. *Brazilian Journal of Plant Physiology*. doi: 10.1590/S1677-04202005000100004.

Sheikh-Mohamadi, M.H., N. Etemadi, A. Nikbakht, et al. 2017. Antioxidant Defence System and Physiological Responses of Iranian Crested Wheatgrass (Agropyron cristatum L.) to Drought and Salinity Stress. *Acta Physiologiae Plantarum* 39(11): 245. doi:10.1007/s11738-017-2543-1.

Shilev, S., I. Babrikova, and T. Babrikov. 2019. Consortium of Plant Growth-promoting Bacteria Improves Spinach (Spinacea Oleracea L.) Growth under Heavy Metal Stress Conditions. Journal of Chemical Technology & Biotechnology 95(4): jctb.6077. doi:10.1002/jctb.6077.

da Silva Leite, R., M.N. do Nascimento, A.L. da Silva, and R. de Jesus Santos. 2021. Chemical Priming Agents Controlling Drought Stress in Physalis Angulata Plants. *Scientia Horticulturae* 275(January): 109670. doi:10.1016/j.scienta.2020.109670.

Singh, A., C.S. Kumar, and A. Agarwal. 2013. Effect of Lead and Cadmium on Aquatic Plant Hydrilla verticillata. *Journal of Environmental Biology* 34(6): 1027–31. https://pubmed.ncbi.nlm.nih.gov/24555332/.

Singh, N., and S. Kumaria. 2020. Deciphering the Role of Stress Elicitors on the Differential Modulation of Chalcone Synthase Gene and Subsequent Production of Secondary Metabolites in Micropropagated Coelogyne ovalis Lindl., a Therapeutically Important Medicinal Orchid. *South African Journal of Botany* July. doi:10.1016/j.sajb.2020.06.019.

Singla-Pareek, S.L., M.K. Reddy, and S.K. Sopory. 2003. Genetic Engineering of the Glyoxalase Pathway in Tobacco Leads to Enhanced Salinity Tolerance. *Proceedings of the National Academy of Sciences of the United States of America* 100(25): 14672–77. doi:10.1073/pnas.2034667100.

Solanki, R., and R. Dhankhar. 2011. Biochemical Changes and Adaptive Strategies of Plants under Heavy Metal Stress. *Biologia*. doi:10.2478/s11756-011-0005-6.

Song, H., P. Wang, L. Hou, et al. 2016. Global Analysis of WRKY Genes and Their Response to Dehydration and Salt Stress in Soybean. *Frontiers in Plant Science* 7(Feb 2016): 9. doi:10.3389/fpls.2016. 00009.

Srivastava, D., G. Verma, A.S. Chauhan, V. Pande, and D. Chakrabarty. 2019. Rice (Oryza sativa L.) Tau Class Glutathione S-Transferase (OsGSTU30) Overexpression in Arabidopsis thaliana Modulates a Regulatory Network Leading to Heavy Metal and Drought Stress Tolerance. *Metallomics: Integrated Biometal Science* 11(2): 375–89. doi:10.1039/c8mt00204e.

Taghipour, M., and M. Jalali. 2019. Impact of Some Industrial Solid Wastes on the Growth and Heavy Metal Uptake of Cucumber (Cucumis sativus L.) under Salinity Stress. *Ecotoxicology and Environmental Safety* 182(October): 109347. doi:10.1016/j.ecoenv.2019.06.030.

Tang, Y., X. Bao, Y. Zhi, et al. 2019. Overexpression of a MYB Family Gene, OsMYB6, Increases Drought and Salinity Stress Tolerance in Transgenic Rice. *Frontiers in Plant Science* 10(February): 168. doi:10.3389/ fpls.2019.00168.

Tilman, D., C. Balzer, J. Hill, and B.L. Befort. 2011. Global Food Demand and the Sustainable Intensification of Agriculture. *Proceedings of the National Academy of Sciences of the United States of America* 108(50): 20260–64. doi:10.1073/pnas.1116437108.

Tiwari, S., and C. Lata. 2018. Heavy Metal Stress, Signaling, and Tolerance Due to Plant-Associated Microbes: An Overview. *Frontiers in Plant Science*. doi:10.3389/fpls.2018.00452.

Toivonen, P.M., and D.M. Hodges. 2011. Abiotic Stress in Harvested Fruits and Vegetables. *Abiotic Stress in Plants - Mechanisms and Adaptations* September. doi:10.5772/22524.

Tonhati, R., S.C. Mello, P. Momesso, and R.M. Pedroso. 2020. L-Proline Alleviates Heat Stress of Tomato Plants Grown under Protected Environment. *Scientia Horticulturae* 268(June). doi:10.1016/j.scienta.2020.109370. http://www.ncbi.nlm.nih.gov/pubmed/109370.

Valluru, R., W.J. Davies, M.P. Reynolds, and I.C. Dodd. 2016. Foliar Abscisic Acid-to-Ethylene Accumulation and Response Regulate Shoot Growth Sensitivity to Mild Drought in Wheat. *Frontiers in Plant Science* 7(Apr 2016): 461. doi:10.3389/fpls.2016.00461.

Vannini, C., M. Campa, M. Iriti, et al. 2007. Evaluation of Transgenic Tomato Plants Ectopically Expressing the Rice Osmyb4 Gene. *Plant Science* 173(2): 231–39. doi:10.1016/j.plantsci.2007.05.007.

Venegas-Molina, J., S. Proietti, J. Pollier, et al. 2020. Induced Tolerance to Abiotic and Biotic Stresses of Broccoli and Arabidopsis after Treatment with Elicitor Molecules. *Scientific Reports* 10(1): 1–17. doi:10.1038/s41598-020-67074-7.

Verma, AK, and S. Deepti. 2016. Abiotic Stress and Crop Improvement: Current Scenario. *Advances in Plants & Agriculture Research* 4(4): 345–46. doi:10.15406/apar.2016.04.00149.

Verwijmeren, M., C. Smit, S. Bautista, M.J. Wassen, and M. Rietkerk. 2019. Combined Grazing and Drought Stress Alter the Outcome of Nurse: Beneficiary Interactions in a Semi-Arid Ecosystem. *Ecosystems* 22(6): 1295–307. doi:10.1007/s10021-019-00336-2.

Vlčko, T., and L. Ohnoutková. 2020. Allelic Variants of CRISPR/Cas9 Induced Mutation in an Inositol Trisphosphate 5/6 Kinase Gene Manifest Different Phenotypes in Barley. *Plants* 9(2): 195. doi:10.3390/ plants9020195.

Wang, F., W. Kong, G. Wong, et al. 2016. AtMYB12 Regulates Flavonoids Accumulation and Abiotic Stress Tolerance in Transgenic Arabidopsis thaliana. *Molecular Genetics and Genomics: MGG* 291(4): 1545–59. doi:10.1007/s00438-016-1203-2.

Wang, J., R. Fang, L. Yuan, et al. 2020a. Response of Photosynthetic Capacity and Antioxidative System of Chloroplast in Two Wucai (*Brassica campestris* L.) Genotypes against Chilling Stress. *Physiology and Molecular Biology of Plants : An International Journal of Functional Plant Biology* 26(2): 219–32. doi:10.1007/s12298-019-00743-8.

Wang, S., M. Wei, H. Cheng, et al. 2020b. Indigenous Plant Species and Invasive Alien Species Tend to Diverge Functionally under Heavy Metal Pollution and Drought Stress. *Ecotoxicology and Environmental Safety* 205(December):111160. doi:10.1016/j.ecoenv.2020.111160.

Wang, W., B. Vinocur, O. Shoseyov, and A. Altman. 2004. Role of Plant Heat-Shock Proteins and Molecular Chaperones in the Abiotic Stress Response. *Trends in Plant Science*. doi:10.1016/j.tplants.2004.03.006.

Wani, S.H., and S.K. Kumar Sah. 2014. Biotechnology and Abiotic Stress Tolerance in Rice. *Rice Research: Open Access* 2(2). doi:10.4172/jrr.1000e105.

Waqas, M., L. Shahid, K. Shoukat, et al. 2020. Role of DNA-Binding with One Finger (Dof) Transcription Factors for Abiotic Stress Tolerance in Plants. *Transcription Factors for Abiotic Stress Tolerance in Plants*, 1–14. doi:10.1016/b978-0-12-819334-1.00001-0.

Waters, E.R. 2013. The Evolution, Function, Structure, and Expression of the Plant SHSPs. *Journal of Experimental Botany*. doi:10.1093/jxb/ers355.

Woldesemayat, A.A., and M. Ntwasa. 2018. Pathways and Network Based Analysis of Candidate Genes to Reveal Cross-Talk and Specificity in the Sorghum (Sorghum bicolor (L.) Moench) Responses to Drought and It's Co-Occurring Stresses. *Frontiers in Genetics* 9(November): 557. doi:10.3389/fgene.2018.00557.

Wu, C., F. Li, H. Xu, et al. 2019. The Potential Role of Brassinosteroids (BRs) in Alleviating Antimony (Sb) Stress in Arabidopsis thaliana. *Plant Physiology and Biochemistry: PPB* 141(August): 51–9. doi:10.1016/j.plaphy.2019.05.011.

Wu, D.C., J.F. Zhu, Z.Z. Shu, et al. 2020. Physiological and Transcriptional Response to Heat Stress in Heat-Resistant and Heat-Sensitive Maize (Zea mays L.) Inbred Lines at Seedling Stage. *Protoplasma* 257(6): 1615–37. doi:10.1007/s00709-020-01538-5.

Xie, Y., C. Wang, W. Yang, et al. 2020. Canopy Hyperspectral Characteristics and Yield Estimation of Winter Wheat (Triticum aestivum) under Low Temperature Injury. *Scientific Reports* 10(1): 1–10. doi:10.1038/s41598-019-57100-8.

Yadav, S., P. Modi, A. Dave, et al. 2020. Effect of Abiotic Stress on Crops. In: *Sustainable Crop Production*, edited by M. Hasanuzzaman, M. Fujita, M.C.M.T. Filho, and T.A.R. Nogueira. IntechOpen. doi:10.5772/intechopen.88434.

Yadav, S.K. 2010. Cold Stress Tolerance Mechanisms in Plants. A Review. *Agronomy for Sustainable Development. (EDP Sciences)*. doi:10.1051/agro/2009050.

Yang, X., M.Y. Kim, J. Ha, and S.H. Lee. 2019. Overexpression of the Soybean NAC Gene GmNAC109 Increases Lateral Root Formation and Abiotic Stress Tolerance in Transgenic Arabidopsis Plants. *Frontiers in Plant Science* 10(August): 1036. doi:10.3389/fpls.2019.01036.

Yarra, R. 2019. The Wheat NHX Gene Family: Potential Role in Improving Salinity Stress Tolerance of Plants. *Plant Gene*. doi:10.1016/j.plgene.2019.100178.

Yu, Z., X. Duan, L. Luo, et al. 2020. How Plant Hormones Mediate Salt Stress Responses. *Trends in Plant Science* 25(11): 1117–30. doi:10.1016/j.tplants.2020.06.008.

Yuan, G., X. Wang, R. Guo, and Q. Wang. 2010. Effect of Salt Stress on Phenolic Compounds, Glucosinolates, Myrosinase and Antioxidant Activity in Radish Sprouts. *Food Chemistry* 121(4): 1014–19. doi:10.1016/j.foodchem.2010.01.040.

Zandalinas, S.I., D. Balfagón, V. Arbona, and A. Gómez-Cadenas. 2017. Modulation of Antioxidant Defense System Is Associated with Combined Drought and Heat Stress Tolerance in Citrus. *Frontiers in Plant Science* 8: 953. doi:10.3389/fpls.2017.00953.

Zandalinas, S.I., Y. Fichman, A.R. Devireddy, et al. 2020. Systemic Signaling during Abiotic Stress Combination in Plants. *Proceedings of the National Academy of Sciences of the United States of America* 117(24): 13810–20. doi:10.1073/pnas.2005077117.

Zerrouk, I.Z., M. Benchabane, L. Khelifi, et al. 2016. A Pseudomonas strain Isolated from Date-Palm Rhizospheres Improves Root Growth and Promotes Root Formation in Maize Exposed to Salt and Aluminum Stress. *Journal of Plant Physiology* 191(February): 111–19. doi:10.1016/j.jplph.2015.12.009.

Zhang, A., Y. Liu, F. Wang, et al. 2019. Enhanced Rice Salinity Tolerance via CRISPR/Cas9-Targeted Mutagenesis of the OsRR22 Gene. *Molecular Breeding : New Strategies in Plant Improvement* 39(3): 1–10. doi:10.1007/s11032-019-0954-y.

Zhang, L., J. Cheng, X. Sun, et al. 2018. Overexpression of VaWRKY14 Increases Drought Tolerance in Arabidopsis by Modulating the Expression of Stress-Related Genes. *Plant Cell Reports* 37(8): 1159–72. doi:10.1007/s00299-018-2302-9.

Zhang, L., X. Guo, Z. Zhang, A. Wang, and J. Zhu. 2021. Cold-Regulated Gene LeCOR413PM2 Confers Cold Stress Tolerance in Tomato Plants. *Gene* 764(January): 145097. doi:10.1016/j.gene.2020.145097.

Zhen, B., H. Li, Q. Niu, et al. 2020. Effects of Combined High Temperature and Waterlogging Stress at Booting Stage on Root Anatomy of Rice (Oryza sativa L.). *Water* 12(9): 2524. doi:10.3390/w12092524.

Zheng, J., H. Su, R. Lin, et al. 2019. Isolation and Characterization of an Atypical LEA Gene (IpLEA) from Ipomoea pes-caprae Conferring Salt/Drought and Oxidative Stress Tolerance. *Scientific Reports* 9(1): 1–21. doi:10.1038/s41598-019-50813-w.

Zhou, R., X. Yu, C.O. Ottosen, et al. 2017. Drought Stress Had a Predominant Effect over Heat Stress on Three Tomato Cultivars Subjected to Combined Stress. *BMC Plant Biology* 17(1): 1–13. doi:10.1186/s12870-017-0974-x.

Zhou, R., X. Yu, C.O. Ottosen, et al. 2020. Unique MiRNAs and Their Targets in Tomato Leaf Responding to Combined Drought and Heat Stress. *BMC Plant Biology* 20(1): 1–10. doi:10.1186/s12870-020-2313-x.

Zhu, M., X. Meng, J. Cai, et al. 2018. Basic Leucine Zipper Transcription Factor SlbZIP1 Mediates Salt and Drought Stress Tolerance in Tomato. *BMC Plant Biology* 18(1): 1–14. doi:10.1186/s12870-018-1299-0.

2 Contribution of Organic Solutes in Amelioration of Salt Stress in Plants

Vinay Shankar and Heikham Evelin

CONTENTS

2.1 INTRODUCTION

Salts constitute a fundamental component of soil. Salts may be deposited naturally due to the weathering of rocks or may be introduced by irrigation with brackish water (Munns and Gilliham 2015). Salinization of land, has, however, become more widespread due to extensive irrigation, improper agricultural practices and land clearing (FAO 2015; Munns and Gilliham 2015). Major salts that contribute to salinity are chlorides and sulfates of calcium, magnesium and sodium (Parihar et al. 2015). Salinity can also alter the physical and chemical properties of soil by destabilizing soil structure, decreasing porosity and hydraulic conductivity (Amini et al. 2016). Excess salt levels in the soil cause (i) osmotic stress, (ii) ionic toxicity and (iii) oxidative stress in plants. While osmotic stress is a result of water paucity in the plant owing to reduced water potential in the rhizosphere, ionic toxicity arises due to increased uptake of Na^+ and Cl^- ions resulting in disruption of ionic homeostasis in the plant. As a result, the production of reactive oxygen species (ROS) shoots up in the plants causing oxidative stress. These salt-induced effects affect the growth, reproduction and productivity of plants. It is well known that salt is the most lethal substance to impede the growth of plants.

Salt-induced osmotic stress occurs when the water potential of the soil at the root surface decreases thereby reducing water uptake by plants (Yang and Guo 2018; Abdel Latef et al. 2020). Osmotic stress causes loss of cellular turgor affecting its normal functioning, subsequently hampering plant growth and development. Therefore, osmotic adjustment is of prime importance to prevent cellular dehydration (Somero and Yancey 2010; Osman et al. 2020). Osmotic adjustment occurs through the accumulation of organic solutes and inorganic ions in the cell to maintain turgor and reduce the detrimental effects of salt stress on plants (Giri 2011). Organic solutes can also restrict the entry of Na^+ ions in the cytoplasm as well as promote their compartmentalization in vacuoles to shield plants from ionic toxicity (Slama et al. 2015; Abdel Latef et al. 2019 a).

Plants are dynamic systems and can adjust their osmotic potential with respect to soil that contains salt. As a primary adaptive strategy, they activate osmotic stress pathways to induce the biosynthesis and buildup of highly soluble, low molecular weight organic solutes known as osmolytes (Yang and Guo 2018). The fundamental role of these solutes is osmoregulation and stabilization of native protein structure in plants under salt stress (Somero and Yancey 2010; Abdel Latef et al. 2019 b). The mechanism of osmoregulation in plants growing under salt stress conditions is an intensely investigated topic. This knowledge has been applied in enhancing plant salt tolerance. In this regard, the chapter will discuss osmolytes, their classification and mechanisms of action. The most commonly accumulated osmolytes are also described. The approaches to increase their concentration in plants have also been discussed.

2.2 OSMOLYTES

An osmolyte is defined as "any inorganic or organic species whose concentration is regulated during the process of volume regulation in parallel with osmotic stress imposed on the cell" (Somero and Yancey 2010). Initially, inorganic osmolytes accumulate to respond to osmotic stress, which later on are replaced by organic osmolytes, as high concentrations of inorganic species can disrupt cellular processes.

Organic osmolytes can be categorized based on their chemical nature as well as their effect on protein stabilization. Chemically, organic osmolytes are classified into four major groups: (i) sugar and polyhydric alcohols (glucose, mannitol, methyl-inositols, myo-inositol, sorbitol and sucrose); (ii) amino acids and derivatives (proline, lysine, arginine, histidine, etc.); (iii) methylated ammonium and sulfonium solutes (choline-o-sulfate, β-alanine betaine, glycine-betaine, proline betaine and Dimethylsufoniopropionate); and (iv) urea (Somero and Yancey 2010). Out of these, urea is not accumulated in plants (hence, not discussed here). The osmolytes are predominantly accumulated in the cytoplasm (Slama et al. 2015). However, the type of organic solutes synthesized is dependent on the plants, its tissues and the type of stress, indicating their functional diversity (Table 2.1) (Yang and Guo 2018).

Based on their effect on protein stability, organic osmolytes are grouped into two categories –protecting and denaturing osmolytes. Protecting osmolytes make protein more stable while denaturing osmolytes decrease protein stability (discussed below) (Street et al. 2006; Burg and Ferraris 2008; Hu et al. 2009).

2.2.1 MECHANISM OF ACTION OF ORGANIC OSMOLYTES

Organic osmolytes protect plant cells under salt stress via two complementary mechanisms (Burg and Ferraris 2008). One, as salt stress induces the outward movement of water from the cells, the cell volume decreases. Under such conditions, the level of all cellular components, including inorganic salts are raised, followed by osmotic uptake of water to rapidly restore the cell volume. The elevated inorganic salt concentration is disadvantageous as it can perturb cellular processes. Accumulation of organic osmolytes ensues later, replacing and decreasing the intracellular concentration of

TABLE 2.1

Representative Osmolytes Involved in Osmoregulation in Plants under Salt Stress

Osmolyte	Osmolytes	Probable Function	Species	Tissue-Specific Accumulation	Reference
Proline	Charged metabolites	Osmotic adjustment; scavenger of ROS; stabilization of protein, membrane and subcellular structure	A variety of plant species, such as *Arabidopsis thaliana*, *Distichlis spicata*, *Helianthus annuus* and *Oryza sativa*	Seedling, seed	Vanrensburg et al. (1993); reviewed by Kishor et al. (2005); reviewed by Mansour and Ali (2017)
Glycine betaine		Osmotic adjustment; decrease of ROS concentration and lipid peroxidation; stabilization of membrane and macromolecules	A variety of plant species, such as *Beta vulgaris*, cyanobacterium and *Nicotiana tabacum*	Young leaf, root, suspension cell	Makela et al. (2000); Banu et al. (2009); Tsutsumi et al. (2015); Yamada et al. (2015)
Choline-O-sulfate, Beta-alanine betaine, hydroxyproline		Osmotic adjustment; sulfate detoxification	Family *Plumbaginaceae*	Leaf, shoot	Reviewed by Rhodes and Hanson (1993); Hanson et al. (1994)
Dimethylsulfonium propionate		Osmotic adjustment	Marine algae	Marine alga cell	Giordano et al. (2005); Ito et al. (2011)
Polyamine		Osmotic adjustment; decrease of membrane leakage; modulation of activity of ion channel; activation of antioxidant enzyme	A variety of plant species, such as *Arabidopsis thaliana*, *Oryza sativa* and *Spinacia oleracea*	Shoot, root	Reviewed by Groppa and Benavides (2008); Zapata et al. (2017)
Mannitol	Polyols	Osmotic adjustment; stabilization of macromolecular structure; scavenger of ROS; protection of photosynthetic apparatus	A variety of plant species, such as marine algae and *Arabidopsis thaliana*	Marine alga cell, seedling	Reviewed by Bohnert and Jensen (1996); Shen et al. (1997); Chan et al. (2011); reviewed by Slama et al. (2015)
Glucosylglycerol		Osmotic adjustment; sustainability of cell division.	Marine cyanobacteria	Cyanobacterium cell	Borowitzka et al. (1980); Ferjani et al. (2003)
Glycerol		Osmotic adjustment	*Dunaliella parva*	*Dunaliella parva* cell	Ben-Amotz and Avron (1973)
Sorbitol		Osmotic adjustment; stabilization of protein	*Plantago coronopus*, *Prunus cerasus*	Leaf, phloem, seedling	Ahmad et al. (1979); Smekens and Tienderen (2001); Pommerrenig et al. (2007)

(Continued)

TABLE 2.1 (CONTINUED)
Representative Osmolytes Involved in Osmoregulation in Plants under Salt Stress

Osmolyte	Probable Function	Osmolytes	Species	Tissue-Specific Accumulation	Reference
Sucrose	Osmotic adjustment; affecting redox homeostasis; protection of macromolecular structure; signal molecule	Soluble sugars	*Plantago major*, *Solanum tuberosum*, etc.	Cyanobacteria cell, leaf, phloem	Reviewed by Gupta and Kaur (2005); Eggert et al. (2016)
Glucose, fructose	Osmotic adjustment; carbon energy reserve; signal molecule; protecting lipid bilayer		*Oryza sativa*, *Solanum lycopersicum*, etc.	Root, leaf, fruit	Arbona et al. (2005); reviewed by Gupta and Kaur (2005); Yin et al. (2010); Boriboonkaset et al. (2013)
Raffinose	Osmotic adjustment	Complex sugars	*Coffea arabica*	Leaf	dos Santos et al. (2011)
Trehalose	Osmotic adjustment; stabilization of membrane and protein		*Oryza sativa*, *Phaseolus vulgaris*, *Zea mays*, etc.	Leaf, root	Crowe et al. (1984); Fougéré et al. (1991); Garcia et al. (1997); Henry et al. (2015)

inorganic salts. Organic osmolytes are non-perturbing even at relatively high intracellular concentrations (Burg and Ferraris 2008). Hence, these osmolytes are also known as compatible solutes (Somero and Yancey 2010).

Second, organic osmolytes participate in the stabilization of protein structure (Street et al. 2006; Burg and Ferraris 2008). Depending on their effect on protein stability, osmolytes can be protecting or denaturing types. Protecting osmolytes increase protein stability, by pushing the equilibrium of protein folding toward the native form, and denaturing osmolytes shift the reversible process of folding towards the unfolded state (Rossky 2008). Stabilization or denaturation of protein is determined by the configuration of the protein backbone (Street et al. 2006). Proteins are stabilized when osmolytes show osmophobic effect, that is, they are strongly excluded from the protein surface. On the other hand, when osmolyte favorably interacts with the protein backbone, that is, they are attached to the protein surface, protein gets denatured (Hu et al. 2009). Hence, a high concentration of protecting osmolytes stabilize protein structure (Burg and Ferraris 2008).

2.2.2 Functional Diversity of Organic Osmolytes

The primary function of osmolytes is osmotic adjustment. Besides this, the osmolytes play diverse roles to overcome stress. They are involved in: i) stabilization of membrane integrity and protein; ii) prevention of oxidative damage by enhancing antioxidant defense system through the mediation of gene up-regulation; iii) providing energy and carbon during recovery from stress; iv) as signaling molecules; v) protection of photosynthetic machinery; and vi) preventing NaCl-induced K efflux from the cytoplasm (Lokhande and Suprasanna 2012) (Figure 2.1).

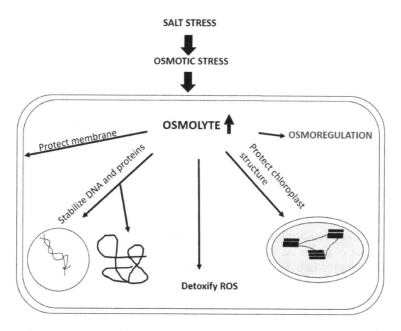

FIGURE 2.1 Protective mechanism of organic osmolytes in cells exposed to saline condition. In response to salt induced osmotic stress, cells enhanced synthesis and accumulation of organic osmolytes. The primary role of osmolytes is osmoregulation. In addition, they also protect membranes, stabilize DNA and proteins, detoxify reactive oxygen species (ROS) and protect chloroplast structure.

2.3   TYPES OF OSMOLYTES

2.3.1   SUGAR AND POLYOLS

Salt stress induces the accumulation of sugars like fructose, fructans, glucose, sucrose and trehalose in several plant species belonging to the families Amaranthaceae, Brassicaceae, Cyperaceae, Juncaceae, Plumbaginaceae, Poaceae, etc. (Slama et al. 2015; Jawahar et al. 2019). Sugars are the dominant solutes in many plants and are extremely sensitive to changing environmental cues. They are highly soluble in water and hence act as compatible osmolytes. The possible roles of sugars in plants under osmotic stress are: (i) osmotic adjustment; (ii) stabilizing membranes upon water stress; (iii) act as a reservoir of energy for sustaining the growth of sink tissues; (iv) influence sugar-mediated signaling to mediate expression of genes involved in the regulation of photosynthesis, respiration and the metabolism of starch and sucrose (Lokhande and Suprasanna 2012).

Polyols or sugar alcohols are reduced forms of ketose or aldose sugars. More than 13 polyols have been isolated until now, but the most commonly reported polyols are mannitol, sorbitol and galactitol (Williamson et al. 2002). Under salt stress, plants accumulate polyols which serve as osmoprotectants. When the osmotic potential is low, polyols protect macromolecules and prevent their inactivation by forming an artificial sphere of hydration owing to the water-like hydroxyl groups in them (Williamson et al. 2002; Slama et al. 2015). In addition, polyols also support redox control (Slama et al. 2015).

2.3.2   AMINO ACIDS

Free amino acids are recognized as important organic osmolytes in many organisms including algae, insects, marine invertebrate phyla, methanogens, protozoa and vascular plants (Somero and Yancey 2010). The most commonly accumulated amino acids for osmoprotection are the small, polar, zwitterionic molecules, such as alanine, glycine, glutamine and proline. It is also reported that negatively charged amino acids occasionally function as osmolytes while positively charged amino acids, such as lysine and arginine, do not take part in osmoregulation. In vascular plants, proline is the most frequently accumulated osmolyte amongst the amino acids.

2.3.3   METHYLATED AMMONIUM AND SULPHONIUM COMPOUNDS

Zwitterions with charged methylated nitrogen or sulfur atoms generally referred to as methylamines and sulfonium compounds, respectively, constitute the third type of osmolytes. They are also known as quaternary ammonium compounds (QACs) or tertiary sulfonium compounds (TSCs), respectively.

In plants, frequently accumulated QACs include β-alanine betaine, choline-O-sulphate, glycine betaine and proline betaine. Glycine betaine is accumulated in many organisms, such as algae, bacteria, cyanobacteria, fungi, plants and animals (Mansour and Ali 2017 a). In plants exposed to abiotic stresses, glycine betaine (GB) is the most commonly accumulated QACs (Lokhande and Suprasanna 2012). β-alanine betaine is accumulated by most members of the highly stress-tolerant plumbago family (*Plumbaginaceae*) instead of GB (Hanson et al. 1991, 1994). Under saline hypoxic conditions, synthesis of β-alanine betaine is advantageous evolutionarily over GB synthesis as it does not require O_2 during its synthesis from the ubiquitous primary metabolite, β-alanine. On the other hand, GB synthesis cannot oxidize choline, the first step of GB synthesis in the absence of molecular oxygen (Hanson et al. 1991, 1994). β-alanine betaine is synthesized by *N*-methylation of β-alanine catalyzed by S-adenosyl-*L*-methionine (SAM) enzyme via N-methyl b-alanine and N,N-dimethyl b-alanine (Raman and Rathinasabapathi 2003). Four pathways – aspartate, polyamine, propionate and uracil – have been identified for the synthesis of β-alanine betaine.

Proline betaine, also known as stachydrine is a dimethyl proline found in non-halophytic *Citrus* and *Medicago* species (Trinchant et al. 2004). It is an analogue of GB, but has been found to accumulate in a few halophytic members of Asteraceae, Capparidaceae, Labiatae, Leguminosae, Myrtaceae, Plumbaginaceae and Rutaceae, (Carter et al. 2006). Proline betaine is as effective as GB and more efficient than proline (Amin et al. 1995).

Choline-O-sulfate is accumulated by the members of Plumbaginaceae and some marine algae (Hanson et al. 1994). It is synthesized from choline by the catalytic action 3'-phosphoadenosine-5'-phosphosulfate-dependent choline sulfotransferase. It is involved in osmoprotection and detoxification of sulfate, abundantly present in saline environments (Hanson et al. 1994).

Tertiary sulphonium compounds contain a full methyl-substituted sulfur atom and are prevalent in many algae. In flowering plants, it is restricted to a few species of *Spartina*, a salt marsh grass, *Saccharum* spp. (Poaceae) and *Wollastonia biflora* (Asteraceae) (Otte et al. 2004). Dimethylsufoniopropionate (DMSP) is an important part of sulfur metabolism in photosynthetic organisms (Haworth et al. 2017). It has been proposed to be involved in osmoregulation, detoxification of excess sulfur and detoxification of ROS (Otte et al. 2004; Haworth et al. 2017). DMSP is synthesized from methionine in both algae and higher plants; however, the pathways of synthesis differ among them (Hanson et al. 1994; Kocsis and Hanson 2000). The methionine (MET) pathway is involved in plant stress responses and is linked to the production of antioxidants, glutathione and the stress hormone, ethylene. Three pathways have been identified – algal, *W. biflora* and *S. alternifolia* (McNeil et al. 1999).

In algae, DMSP is synthesized in four steps: i) transamination of Met to give the corresponding 2-oxo acid, ii) reduction of 2-oxo acid to a 2-hydroxy acid, iii) then *S*-methylation and iv) oxidative decarboxylation (Gage et al. 1997).

In flowering plants, *W. biflora* and *S. alterniflora*, conversion of methionine to S-methyl methionine (SMM), catalyzed by cytosolic MetS-methyltransferase, is the first step, and the last is the oxidation of DMSP-aldehyde, catalyzed by betaine-aldehyde dehydrogenase (BADH) (Trossat et al. 1996, 1997; Kocsis et al. 1998). In *W. biflora*, the conversion of SMM to DMSP-aldehyde occurs in the chloroplast in a single step via an unusual coupled transamination-decarboxylation reaction (Trossat et al. 1996). Conversely, the conversion of SMM to DMSP-aldehyde in *S. alterniflora* occurs in two steps via the intermediate DMSP-amine (Kocsis and Hanson 2000).

2.4 OSMOLYTES COMMONLY ACCUMULATED IN PLANTS UNDER SALINE CONDITIONS

Some of the commonly accumulated osmolytes are discussed in this section. Organic solutes not included in the basic classification of osmolytes but proposed to contribute to osmoregulation are also discussed.

2.4.1 TREHALOSE

Trehalose (α-D-lucopyranosyl-1, 1-α-D-glucopyranoside) is a non-reducing disaccharide found in diverse organisms like bacteria, fungi, nematodes and plants (Lunn et al. 2014). It is one of the most studied osmoprotectants and also contributes to the structural and functional stabilization of proteins, enzymes, membranes and guard biological structures during desiccation (Elbein et al. 2003). It can protect plants from the damaging salt effects by maintaining a favorable K^+/Na^+ ratio, preventing ROS accumulation and increasing the concentration of soluble sugars in them (Redillas et al. 2012; Chang et al. 2014).

In plants, trehalose is biosynthesized from glucose-6-phosphate and uridine diphosphate glucose by the catalytic action of two key enzymes: trehalose-6-phosphate synthase (TPS) and trehalose-6-phosphate phosphatase (TPP) (Figure 2.2).

Studies have shown that transgenic lines overexpressing *OsTPS1* improved salt tolerance of rice seedlings by enhancing trehalose and proline levels accompanied by regulation of the expression of stress-related genes (Table 2.2) (Li et al. 2011). Salt-tolerant *Brassica juncea* species had higher TPS

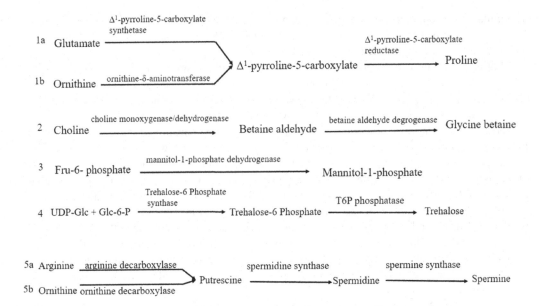

FIGURE 2.2 Biosynthetic pathways of commonly studied organic osmolytes – proline (1a and 1b), glycin betaine (2), mannitol (3), trehalose (4) and polyamines (5a and 5b).

activity accompanied by higher trehalose accumulation and better salt tolerance (Chakraborty and Sairam 2016). According to Redillas et al. (2012), transgenic rice plants expressing a bifunctional fusion enzyme TPSP resulted in the elevated production of trehalose and increased soluble sugar content to enhance tolerance to salt stress.

Exogenous treatment of plants with trehalose has been shown to relieve salt-induced damages in *Arabidopsis*, *Brassica*, *Oryza sariva* and *Triticum aestivum* (Table 2.3) (Zeid 2009; Yang et al. 2014; Abdallah et al. 2016; Shahbaz et al. 2017; Nemat Alla et al. 2019). Under salt stress, trehalose pretreatment reduces Na^+ accumulation, protects root integrity, prevents chlorophyll degradation and maintains optimum net photosynthetic rate, induces a strong antioxidant defense system and increases the synthesis of GB and total soluble proteins (Zeid 2009; Yang et al. 2014; Abdallah et al. 2016; Shahbaz et al. 2017; Nemat Alla et al. 2019). In *Arabidopsis*, exogenous trehalose (0.5, 1, and 5 mM) treatment ameliorated salt-induced ionic imbalance, ROS spurt and programmed cell death (Yang et al. 2014). Trehalose (25 mM) treated rice seedlings showed higher solute concentration and antioxidant enzyme activity contributing to osmotic adjustment and quenching of ROS.

2.4.2 RAFFINOSE

Raffinose family oligosaccharides are non-structural carbohydrates which are extensively found in higher plants. They constitute the second most abundant sugars after sucrose and are being increasingly recognized as crucial molecules during plant's response to stress (dos Santos et al. 2011; Sengupta et al. 2015). Studies report the accumulation of raffinose family of oligosaccharides (RFOs) as osmolytes in abiotic as well as biotic stresses (dos Santos et al. 2011). RFOs are known to stabilize membrane, scavenge ROS and protect macromolecules by forming a vitreous state. They are also implicated in signaling and stabilizing photosystem II, water retention in the cytosol and compartmentalization of Na^+ in cytosol or apoplast (Foyer and Shigeoka 2011; Knaupp et al. 2011; Valluru and Van den Ende 2011). RFOs are sucrose molecules with extensions of α-1, 6-galactosyl moieties donated by galactinol (Gol; 1-*O*-α-D-galactopyranosyl-L-*myo*-inositol) (Sengupta et al. 2015). Thus, synthesis of galactinol is a prerequisite for RFO synthesis and begins with the activity of galactinol synthase (GolS; EC 2.4.1.123). GolS belongs to GT8 family glycosyltransferase and

TABLE 2.2

Some Recent Examples where Transgenic Plants Overexpressing Genes Involved Osmolyte Biosynthesis and Increased Salt Tolerance in Host Plants

Transgenic Plant	Salt level (mM NaCl)	Gene Incorporated	Osmolyte	Effects on Transgenics	Reference
Sorghum bicolor	100	Δ¹ Pyrroline-5-carboxylate synthetase P5CS	Proline	More proline; maintenance of PSII activity and protection of antioxidant enzyme activities.	Surender Reddy et al. (2015)
Glycine max	200	P5CS		Transgenics have longer plant, greater leaf area, higher relative chlorophyll content and more number of fresh pods than wild type plants; more proline in transgenics and low lipid peroxidation in membranes	Zhang et al. (2015)
Arabidopsis thaliana	250	P5CS gene family (P5CS1, P5CS2 and P5CS3)		LrP5CS2 lines accumulated less proline than others, LrP5CS1 lines grew better in root elongation. LrP5CS3 lines grew better than all others under unstressed condition and osmotic stress	Wei et al. (2016)
Panicum virgatum	0, 150, 250 and 350	P5CS		Transgenic had more tillers, broader leaf blade, bigger internode diameter, longer internode and panicle length, early flowering and higher expression of flowering related genes; higher expression levels of ROS scavenger genes (PvSOD, PvPOD, PvCAT and PvGST); lower electrolyte leakage; higher water content; higher K⁺/Na⁺ ratio than control plants	Guan et al. (2018)
Populus alba × Populus glandulosa	200	bacterial choline oxidase (codA) gene	Glycinebetaine	Increased efficiency of photosystem II activity	Ke et al. (2016)
Triticum aestivum	200	betaine aldehyde dehydrogenase, BADH		Transgenics had more chlorophyll and carotenoid, higher Hill reaction and Ca²⁺-ATPase activities than wild type. Enhanced protection of thylakoid and better photosynthesis	Tian et al. (2017)
Solanum lycopersicum cv. "Moneymaker"	200	codA		Glycine betaine enhances salt tolerance in tomato plants. Regulation of ion channels and alleviation of K⁺ efflux, maintain Na⁺-K⁺ homeostasis and enhance antioxidant capacity	Wei et al. (2016)
Eucalyptus camaldulensis	70	coda		Alleviated salt-induced photosynthetic disorder	Tran et al. (2018)

(Continued)

TABLE 2.2 (CONTINUED)

Some Recent Examples where Transgenic Plants Overexpressing Genes Involved Osmolyte Biosynthesis and Increased Salt Tolerance in Host Plants

Transgenic Plant	Salt level (mM NaCl)	Gene Incorporated	Osmolyte	Effects on Transgenics	Reference
Arachis hypogaea	0, 50, 100, 150 and 200	Mannitol dehydrogenase gene (*mtlD*) gene	Mannitol	Mannitol shielded salt stress in transgenics. Transgenics showed better growth and physio-biochemical parameters like mannitol content, proline levels, total chlorophyll content, osmotic potential, electrolytic leakage and relative water content than wild type	Bhauso et al. (2014)
Triticum aestivum	85	*mtlD*		Transgenics showed better growth, more chlorophylls, proline, total soluble sugars, soluble sugar fractions and mannitol, better antioxidant system, high yield and better yield quality	El Yazal et al. (2016)
*Arachis hypogaea	1, 2 and 3	*mtlD* gene		Transgenics showed more mannitol, low oxidative stress and better antioxidative responses	Patel et al. (2017)
Arabidopsis thaliana	200	S6PDH (D-Sorbitol-6-phosphate 2-dehydrogenase)	Sorbitol	Transgenics accumulate more sorbitol, had more germination rate and root elongation capacity	Gao et al. (2018)
Oryza sativa	150	TPP (Trehalose-6-Phosphatase)	Trehalose	Transgenics showed salt tolerance, activated stress-inducible genes	Ge et al. (2008)
Solanum lycopersicum cv. Joyful	150	TPSP (trehalose-6-phosphate synthase/phosphatase fusion gene)		Transgenic plants had more trehalose in their leaves and enhanced salt tolerance and photosynthetic rates than wild-type plants.	Lyu et al. (2013)
Arabidopsis thaliana	0–200	*TsGolS* (*Galactinol synthase*)	Raffinose	Transgenics had more galactinol, raffinose and a-ketoglutaric acid than WT plants Transgenic plants had a higher germination rate, photosynthesis ability and seedling growth	Sun et al. (2013)

(Continued)

TABLE 2.2 (CONTINUED)
Some Recent Examples where Transgenic Plants Overexpressing Genes Involved Osmolyte Biosynthesis and Increased Salt Tolerance in Host Plants

Transgenic Plant	Salt level (mM NaCl)	Gene Incorporated	Osmolyte	Effects on Transgenics	Reference
Oryza sativa	150	S-adenosylmethionine decarboxylase (*SAMDC*)	Polyamines	Transgenics had more spermidine and spermine, increased seedling growth	Roy and Wu 2002)
Pyrus communis "Ballad"	150	spermidine synthase (*SPDS*)		Transgenics showed a substantial increase in salt tolerance with low electrolyte leakage and thiobarbituric acid concentration	Wen et al. (2008)
Nicotiana tabacum	200	S-adenosylmethionine synthetase (*SAMS2*)		Transgenics were more tolerant to salt stress and maintained higher photosynthetic rate and biomass	Qi et al. (2010)

* Salt level expressed as dS m[-1]

TABLE 2.3
Some Recent Examples where Exogenous Supplementation of Osmolytes Enhances Salt Tolerance in Plants

Plant	*Salt Level (mM NaCl)	Exogenous Osmolyte Concentration (mM)	Effect on Osmolyte Treated Plants	Reference
Proline				
Oryza sativa MR 220 and MR 253 cultivars	0 and 150	5 and 10	Treated plants were taller, had more number of roots, root NO_3^- content, root nitrate reductase and root glutamine synthetase activities, higher nitrite reductase (NiR) and glutamate synthase (NADH–GOGAT) activities	Teh et al. (2016)
Lupinus termis L Giza 1 and Giza 2 varieties	*6.35–6.45 dS m^{-1}	3, 6 or 9	Increased growth and yield, maintenance of leaf photosynthetic pigments, decreased concentration of alkaloids and improved stem anatomical traits	Rady et al. (2018)
Oryza sativa (salt sensitive BRRI dhan29 and moderately salt-tolerant BRRI dhan47)	25, 50 and 100	0, 25 and 50	Increased growth, chlorophyll, intracellular proline, ascorbate contents, K$^+$/Na$^+$ ratio and activities of antioxidant enzymes in salt-sensitive rice; no significant response in salt-tolerant rice	Bhusan et al. (2016)
Zea mays	80	30	Reduced Na$^+$ and Cl$^-$ accumulation, increased K$^+$ content, better antioxidant systems	De Freitas et al. (2018)
Glycine max	0 and 15	0 and 25	increased biological nitrogen fixation and specific nodule activity under stress conditions	El Sabagh et al. (2017)
Sorghum bicolor	75	30	Improved ionic homeostasis, reduced Na$^+$ and Cl$^-$ ions and increased K$^+$ and Ca^{2+} ions, increased compatible solutes, reduced membrane damage, punctual modulation in proline synthesis (down-regulation of Δ^1-pyrroline-5-carboxylate synthetase activity) and degradation (up-regulation of proline dehydrogenase activity) enzymes	De Freitas et al. (2018)
Lactuca sativa cv. Teide and cv. *Impulsion*	0–15	0–15**	Exogenous application of foliar proline (up to 5 µM) increased the yield under control condition and enhanced the plant response to salinity.	Orsini et al. (2018)
Glycine betaine				
Glycine max	150	0, 5, 25 and 50	Decrease in proline, MDA and Na$^+$ content; increase in CAT and SOD activity	Malakzadeh et al. (2015)
				(Continued)

TABLE 2.3 (CONTINUED)
Some Recent Examples where Exogenous Supplementation of Osmolytes Enhances Salt Tolerance in Plants

Plant	*Salt Level (mM NaCl)	Exogenous Osmolyte Concentration (mM)	Effect on Osmolyte Treated Plants	Reference
Lactuca sativa	0 and 100	0, 5, 10 and 25	Increase in gibberellic acid, salicylic acid and indole acetic acid concentrations; reduce membrane permeability, malondialdehyde and H_2O_2 concentration, reduced Na^+ accumulation but significantly increased other element contents	Yildrim et al. (2015)
Lactuca sativa	0 and 100	0, 5, 10 and 25	Increased in total antioxidants and total phenolics, regulating antioxidant enzyme activity and altering the contents of organic acids and amino acids.	Shams et. al. (2016)
Allium cepa	*4.80	0, 25 and 50	Increased yield and water-use efficiency, antioxidant capacity increased, enhanced salt tolerance	Rady et al. (2018)
Gossypium hirsutum	150 mM	5.0 mM	Promoted gas exchange and fluorescence increased chlorophyll pigments, and stimulated the antioxidant enzyme activity	Hamani et al. (2020)
Mannitol				
Zea mays L. cv., DK 647 F1	0 or 100	15 and 30	Checked salt-induced shoot growth inhibition, reduced the activity of antioxidant enzymes (CAT, SOD, POD and PPO), increased K^+, Ca^{2+} and P, but decreased that of Na^+	Kaya et al. (2013)
Triticum aestivum	100	100	Alleviated salt-induced oxidative damage by enhancing antioxidant enzyme activities in the roots	Dinler et al. (2008)
Sorbitol				
Oryza sativa Salt sensitive (KDML105) Salt tolerant (PK)	175	5 and 10	Enhanced growth and reduced H_2O_2 and MDA in salt sensitive (KDML105) No effect on PK	Theerakulpisut and Gunnula (2012)
Solanum melongena	1.5, 3, 6 dSm^{-1}	5 and 10 g L^{-1}	Sorbitol treatment reduced the adverse effects of salinity on eggplants by increasing water content and fruit size.	Issa et al. (2020)
Trehalose				
Arabidopsis thaliana	0, 150 and 250	0.5, 1 and 5	Retained K^+ and K^+/Na^+ ratio in plant tissues to improve salt tolerance, antagonized salt-induced damages in redox state and cell death	Yang et al. (2014

(Continued)

Organic Solutes, Oxidative Stress, Antioxidant Enzymes

TABLE 2.3 (CONTINUED)
Some Recent Examples where Exogenous Supplementation of Osmolytes Enhances Salt Tolerance in Plants

Plant	*Salt Level (mM NaCl)	Exogenous Osmolyte Concentration (mM)	Effect on Osmolyte Treated Plants	Reference
Oryza sativa varieties (Giza 177 and Giza 178)	0, 30 and 60	25	Better effect on Giza 178, improved osmotic adjustment and antioxidant capacity	Abdallah et al. (2016)
Triticum aestivum	75, 150 and 225	10	Decreased H_2O_2 and lipid peroxides, increased reducing power, sugars, K^+, K^+/Na^+ ratio, phenolics, proline and Na as well as the expression of AOX, NHX1 and SOS1.	Nemat Alla et al. (2019)
Fragaria × ananassa Duch. cv. "Gaviota"	50	10 and 30	Improvement of carotenoids, flavonoids and anthocyanins compounds in leaves resulting in normal photochemical functioning, the activation of the enzymatic antioxidants and the compartmentalization of Na for better growth under salt stress.	Samadi et al. (2019)
Polyamines				
Zoysia japonica Steud sensitive (cv. Z081) or tolerant (cv. Z057)	200	0.15, 0.30, 0.45, 0.6	Increased Spd and Spm contents and ornithine decarboxylase, S-adenosylmethionine decarboxylase and diamine oxidase activities in both cultivars, induced antioxidant enzyme activities, reduced H_2O_2 and MDA levels and improved the tolerance	Li et al. (2016)
Cucumis sativus L., cv. *Jingyou No. 4*)	75	0.3 Spermine	Increased the activities of antioxidant enzymes which counteracted the adverse effects of salinity on the structure of the photosynthetic apparatus	Shu et al. (2013)
Pistacia vera	#25, 50, 100, 150	putrescine, spermidine and spermine (0.1 and 1 mM)	Increased SOD, CAT activity, decreased H_2O_2 balance ion exchange, had lower $Na^+:K^+$ ratio and Cl^- in leaves and better $Na^+:K^+$ discrimination under salt stress condition	Kamiab et al. (2014)

* Salt level expressed as dS m^{-1}.

** osmolytes concentration expressed in μM.

\# Mixture of salts including NaCl, $CaCl_2$ and $MgCl_2$.

catalyzes the galactosylation of *myo*-inositol to produce galactinol. Three stress-responsive GolS genes (*AtGolS1*, *AtGolS2* and *AtGolS3*) have been identified in *Arabidopsis thaliana* out of seven GolS genes. Raffinose and the subsequent higher molecular weight RFOs (Stachyose, Verbascose and Ajugose) are synthesized from sucrose by subsequent addition of activated galactose moieties donated by galactinol. It is noteworthy that GolS are functional only in angiosperms despite ROFs being widely distributed in higher plants (dos Santos et al. 2011; Sengupta et al. 2015). Within a plant, RFOs are found in the endosperm, flowers, leaves, shoots, roots and pericarp of fruits (dos Santos et al. 2011). In *Coffea arabica*, the mRNA expression of GolS (*CaGolS1*, *CaGolS2* and *CaGolS3*) is up-regulated under salt stress and was associated with extreme elevation of raffinose levels in them and subsequently increased tolerance to salt stress (dos Santos et al. 2011).

Overexpression of *Thellungiella salsuginea GolS2* in *A. thaliana* resulted in increased accumulation of galactinol, raffinose and a-ketoglutaric acid. This resulted in improved salt tolerance in the transgenic plants by increasing germination rate, ensuring better growth of seedlings and enhancing photosynthetic efficiency (Sun et al. 2013).

2.4.3 Mannitol

Mannitol is one of the most extensively distributed soluble carbohydrates in biological systems (Patel and Williamson 2016). It is a polyol and serves as a compatible solute in plants. Besides this, it also participates in the quenching of ROS, stabilizing the structure of membrane and protein, protecting photosynthetic apparatus and acts as a low-molecular-weight chaperone in plants under salt stress (Patel and Williamson 2016).

Mannitol is generally synthesized as a primary product of photosynthesis and can be metabolized by some plant species. Plants synthesize mannitol-1-phosphate from Fru-6- phosphate in a reversible reaction catalyzed by mannitol-1-phosphate dehydrogenase encoded by *MtlD* gene (Figure 2.2) (Abebe et al. 2003). Thus, accumulation of mannitol in plants is modulated by *MtlD*, and therefore, its complex regulation is central to the balanced integration of mannitol metabolism (Patel and Williamson 2016). However, mannitol is not synthesized in many important agronomical crops such as wheat and tobacco (Seckin et al. 2009).

Plant salt tolerance has been enhanced by increasing mannitol concentration via overexpression of *MtlD* gene (Table 2.2). Genetic engineering has been successful to incorporate *MtlD* gene even in mannitol non-accumulating plants. Abebe et al. (2003) produced transgenic wheat plants that overexpressed *MtlD* gene obtained from *Escherichia coli* and observed that the transgenic plants improved growth under salt stress. The better growth in transgenic plants was attributed in part to the role of mannitol as osmolyte, besides other stress-protective functions (Abebe et al. 2003). Transgenic peanut plants overexpressing *MtlD* gene showed improved salt tolerance by accumulating more mannitol; moderating antioxidative reactions; regulating osmotic potential; maintaining better water status; increasing concentration of proline and total chlorophyll; and decreasing electrolytic leakage over wild type (Bhauso et al. 2014; Patel et al. 2017). Similar findings were reported in wheat genotypes overexpressing *MtlD* gene (El Yazal et al. 2016).

Mannitol has also been exogenously applied in plants to improve salt tolerance (Table 2.3). Exogenous mannitol application enhanced antioxidant enzyme activities to shield wheat plants from oxidative stress caused by salt stress (Seckin et al. 2009). Foliar-applied mannitol (15 and 30 mM) increased the contents of K^+, Ca^{2+} and P, but decreased that of Na^+ in the salt-stressed maize plants concerning those of the salt-stressed plants not supplemented with mannitol (Kaya et al. 2013), suggesting that mannitol application can modulate ionic concentrations to improve salt tolerance.

2.4.4 Sorbitol

Sorbitol is one of the many osmolytes accumulated in plants under salt stress (Singh et al. 2015) Parihar et al. 2015). Sorbitol is synthesized in leaves by successive conversion of glucose-6-phosphate

(G-6-P) to sorbitol-6-phosphate to sorbitol by aldose-6-phosphate reductase and sorbitol-6-phosphatase, respectively. Sorbitol catabolism is catalyzed by NAD^+-dependent sorbitol-6-phosphate 2-dehydrogenase (S6PDH). Also, S6PDH is known to catalyze NADH-dependent conversion of D-fructose 6-phosphate to D-sorbitol 6-phosphate (Gao et al. 2018)

Sorbitol accumulation was reported in three *Plantago* species: *P. major* (salt-sensitive), P. *crassifolia* and *P. coronopus* (halophytes) subjected to 100, 200, 400, 600 and 800 mM NaCl (Al Hasan et al. 2016). However, the elevation in the sorbitol level in salt-stressed plants was relatively low as compared to the corresponding control plants suggesting that strong stress is unnecessary for the induction of sorbitol synthesis in these plants (Al Hasan et al. 2016). Deguchi et al. (2002) reported that in Japanese pear, sorbitol is the main osmoprotectant accumulated to counter salt-induced osmotic stress, and not glucose, fructose or sucrose, endorsing that accumulation of osmolyte is dependent on plant species.

Arabidopsis thaliana transgenics overexpressing D-Sorbitol-6-phosphate 2-dehydrogenase from *Haloarcula marismortui* (HmS6PDH) showed significantly higher accumulation of sorbitol than wild type upon exposure to 200 mM NaCl (Gao et al. 2018). HmS6PDH expression increased seed germination and root elongation capacities of transgenic *A. thaliana* plants. *Haloarcula marismortui* is an extremophile found in the Dead Sea which accumulates salt above saturating concentrations. Therefore, *HmS6PDH* gene can be a potential candidate to generate salt-tolerant transgenic plants.

Theerakulpisut and Gunnula (2012) studied the effect of exogenous sorbitol in salt-sensitive (KDML05) and salt-tolerant (PK) varieties of rice under salt stress. They observed that sorbitol imparted a beneficial effect on salt-sensitive rice by reducing H_2O_2 and MDA content, but not in salt-tolerant PK variety, suggesting that accumulation of osmolyte is dependent on the salt-sensitiveness of the plant. Recently, Issa et al. (2020) showed that sorbitol treatment reduced the adverse effects of salinity on eggplants by increasing water content and fruit size.

2.4.5 PROLINE

Accumulation of proline is one of the most frequently described salt tolerance mechanisms in plants (Kishor et al. 2005; Wang et al. 2015; Kumar et al. 2010; Surender Reddy et al. 2015; Mansour and Ali 2017 b). Proline protects salt-induced osmotic stress in plants via osmoregulation in cells. It also participates in a wide range of cellular processes, including quenching of ROS, stabilizing DNA, proteins and membranes, and decreasing salt-induced damage to photosynthetic apparatus (Christgen and Becker 2019). In addition, it can also serve as a carbon and nitrogen source to support growth and meet energy demands, thereby aiding in recovery from stress (Tanner 2008). Proline helps in the prevention of acidosis in the cytoplasm and the maintenance of favorable $NADP^+$/ NADPH ratios for metabolism (Hare and Cress 1997). Therefore, under stress conditions, proline serves as a reservoir for excess NAD^+ and $NADP^+$ to be consumed during respiration and photosynthesis. High $NADP^+$ concentration is essential to regenerate NADPH in the pentose phosphate pathway and the synthesis of purines for the supply of ribose-5-phosphate (Kishor et al. 2005).

Under salt stress conditions, proline accumulates in plant cells owing to enhanced biosynthesis coupled with decreased degradation (Sharma et al. 2011). Proline is synthesized mainly in the photosynthetic tissues, particularly in chloroplast and upregulated by light (Kaur and Asthir 2015; Signorelli 2016). Biosynthesis of proline in higher plants occurs via two pathways: the glutamate (Glu) and ornithine (Orn) pathway. Glu pathway operates in chloroplasts and cytosol and predominantly occurs under stress (Figure 2.2). In this pathway, P5CS (Δ^1-pyrroline-5-carboxylate synthetase) reduces Glu to GSA (glutamate-semialdehyde), which gets converted to P5C (Δ^1-pyrroline-5-carboxylate). P5CR (Δ^1-pyrroline-5-carboxylate reductase) reduces P5C to proline (Hu et al. 1992; Roosens et al. 1998). Orn pathway occurs in mitochondria and is active in seedling development. In this pathway, δ-OAT (ornithine-δ-aminotransferase) transaminates Orn to P5C, which is then relayed to the cytosol to be reduced to proline by P5CR (Armengaud et al. 2004).

Proline gets degraded in mitochondria where proline dehydrogenase (PDH) converts proline to P5C, which is then converted to Glu by Δ^1-pyrroline-5-carboxylate dehydrogenase (P5CDH). Proline accumulation has also been found to be influenced by phytohormones. Gibberellic acid and cytokinins promoted proline accumulation while indole acetic acid did not.

P5CS is considered as the key enzyme in the biosynthesis of proline and has garnered utmost attention, being frequently studied to decipher proline metabolism in plants under salt stress (Table 2.2). Transgenic moth bean plants overexpressing *P5CS* gene have been shown to enhance the accumulation of proline (Su and Wu 2004). In transgenic *indica* rice (cv KJT-3), overexpression of a mutagenized variant of *P5CS* gene, *P5CSF129A*, led to increased proline accumulation, better growth and biomass and lower lipid peroxidation when subjected to salt stress (Kumar et al. 2010). Surender Reddy et al. (2015) also observed that overexpression of *P5CSF129A* gene increased proline accumulation in leaves of transgenic sorghum plants subjected to 100 mM NaCl. Enhanced proline accumulation was credited for decreased negative salt effects on photosynthetic parameters, water use efficiencies and antioxidant status (superoxide dismutase, catalase, glutathione reductase). Proline concentration is dependent on duration of stress and salt concentration as observed in roots, leaves and stems of *Kosteletzkya virginica* seedlings (Wang et al. 2015).

Alternatively, exogenous application of proline has also been used to improve plant tolerance to salt as well as plant recovery from salt stress (Table 2.3). Kumar et al. (2015) found that foliar proline application (25 mM) enhanced rice plants' (salt-sensitive BRRI dhan 29 and moderately salt-tolerant Binadhan 8) tolerance to salt. Proline foliar spray improved growth and yield by increasing chlorophyll, intracellular proline and ascorbic acid contents, maintaining a higher K^+/Na^+ ratio and activation of antioxidant enzymes. According to Nounjan et al. (2012), exogenous application of proline (10 mM) reduced Na^+/K^+ ratio, increased endogenous proline and transcript levels of *P5CS* and *P5CR* in Thai aromatic rice seedlings and promoted a stronger ability of plants to recover from stress. However, dose of exogenous proline and the plant species are major players in determining the ameliorative effects of proline application (El Moukhtar et al. 2020).

2.4.6 GLYCINE BETAINE (N, N, N-TRIMETHYL GLYCINE)

GB is an amphoteric compound that is commonly accumulated as an osmolyte in many plants. A buildup of GB in plants has been correlated with improved plant salt stress tolerance (Chakraborty and Sairam 2016; Nawaz and Ashraf 2010; Hasanuzzaman et al. 2012). It is exceptionally water soluble and known to function as a capable compatible solute (Le Rudulier et al. 1984). It has a stabilizing effect on the structure and functions of enzymes and protein complexes, as it can interact with both hydrophilic and hydrophobic domains of these macromolecules (Sakamato and Murata 2000). It also allays the detrimental effects of salt stress on the integrity of membranes (Gorham 1995).

In plants, GB is synthesized from its precursor, choline, by the successive action of choline monooxygenase (CMO) and betaine aldehyde dehydrogenase (BADH), which convert choline to betaine aldehyde and GB, respectively. In animals and bacteria, CMO is absent and choline is transformed into betaine aldehyde by choline dehydrogenases (CDH) (Figure 2.2).

Accumulation of GB in plants can be enhanced by overexpressing the genes encoding GB biosynthetic enzymes, CMO, CDH and BADH (Table 2.2). Transgenic cotton plants containing *AhCMO* (*CMO* cloned from *Atriplex hortensis*) have been found to accumulate 26 and 131% more GB than non-transgenic plants under normal and salt stress conditions, respectively (Zhang et al. 2009). More GB in transgenic cotton enhanced salt tolerance, as it provided greater protection of cell membranes and photosynthetic capacity (Zhang et al. 2009). Transgenic wheat plants containing *betA* gene that encodes *CDH* from *Escherichia coli* were found to have lower Na^+/K^+ and solute potential than wild type plants when subjected to 200 mM NaCl. This was attributed to high GB concentration that protected cell membranes and photosynthetic apparatus and imparted vigorous growth (He et al. 2010). According to Fan et al. (2012), transgenic sweet potato overexpressing BADH gene from *Spinacea oleracea* (*SoBADH*) resulted in a subsequent increase in BADH activity and GB accumulation. Enhanced GB concentration in plant tissues protected cell damage by preserving cell membrane integrity, maintaining optimum photosynthetic activity, reducing the generation of ROS and stimulating induction of antioxidant enzymes and proline accumulation. Similar observations were reported in *Brassica juncea* species under salt stress (Chakraborty and Sairam 2016).

Another approach to enhance plant salt tolerance via GB is its application to the plant exogenously (Table 2.3). Foliar spray of GB has been found to improve salt resistance in maize (Nawaz and Ashraf 2010), rice (Hasanuzzaman et al. 2014) and okra (Habib et al. 2012) as compared to non-treated plants. GB treated plants had better water use efficiency, higher photosynthetic activity, strong antioxidant systems and regulation of ionic balance versus non-treated counterparts (Nawaz and Ashraf 2010; Hasanuzzaman et al. 2014; Habib et al. 2012). This subsequently translated into better growth, and higher biomass and yield of plants in GB-treated plants.

2.4.7 POLYAMINES

Polyamines are aliphatic, low molecular weight and polycationic nitrogen compounds found in all living cells. Plants exhibit enhanced polyamine biosynthesis and accumulate as compatible solute under salt stress to contribute to cellular osmoregulation (Minocha et al. 2014; Singh et al. 2015; Yang and Guo 2018; Zapata et al. 2017). Additionally, polyamines contribute to plant protection under salt stress by providing stability to membranes, scavenging ROS, modulating cellular pH and photosynthesis (Pang et al. 2007).

In plants, the precursor of polyamines is arginine or ornithine. Arginine decarboxylase (ADC) or ornithine decarboxylase (ODC) catalyzes the conversion of arginine or ornithine, respectively, to putrescine (Put). Thereafter, putrescine undergoes two sequential additions of aminopropyl groups to synthesize spermidine and spermine, catalyzed by spermidine synthase (SPDS) and spermine synthase (SPMS), respectively (Figure 2.2) (Slocum 1991).

Polyamine concentration in plants has been successfully enhanced via genetic engineering to improve salt tolerance in plants (Table 2.2). Transgenic rice plants overexpressing arginine decarboxylase (*ADC*) from oat (*Avena sativa* L.) led to stress-induced up-regulation of ADC activity and polyamine levels. Second-generation transgenic rice plants showed higher biomass than wild type under saline stress (Roy and Wu 2002). Transgenic tobacco plants overexpressing *ODC* resulted in increased Put synthesis and enhanced plant tolerance to salinity (Kumria and Rajam 2002).

Salt stress in plants can also be alleviated by exogenous application of polyamines. Exogenous application of Put (0.1 mmol/L) improved growth of seedlings under salinity (50–250 mmol/L NaCl) by preventing peroxidation of membranes and biomolecule denaturation brought about by effective sequestration of ROS through induction of antioxidants (Verma and Mishra 2005). Shu et al. (2012) observed that exogenous Spd (1 mmol l^{-1}) application induced biosynthesis of endogenous Spd and protected plants under salt stress by preventing damage in photosystems and stabilizing xanthophyll components in cucumber. On the other hand, the application of spermine increased the efficiency of photochemical reactions accompanied by modulation of the chloroplast antioxidant system in salt-stressed cucumber leaves (Shu et al. 2013).

2.4.8 ORGANIC ACIDS

Under salt stress, plants are also known to accumulate organic acids. These acids are emerging as important molecules in providing tolerance to salt stress (Guo et al. 2010). They are considered to participate in osmotic adjustment, balancing excess cation and regulation of pH. Besides this, they deliver CO_2 to the Calvin cycle and promote sugar synthesis in C_4 plants and prevent the cellular buildup of toxic Cl^-; (Guo et al. 2010). Liu and Shi (2010) reported the accumulation of formate, acetate and glycolate in sunflower plants exposed to salt stress. However, the concentration of succinate decreased in treated plants as compared to control.

Salt stress is known to severely affect the tricarboxylic acid (TCA) cycle by decreasing the concentration of participating metabolites and carbon catabolism (Richter et al. 2015). Torre-González et al. (2017) found that under salt stress, the concentrations of organic acid and the activity of the corresponding enzymes increased in salt-resistant tomato cv. Grand Brix while it decreased in salt-sensitive cv. Marmanade RAF. They suggested that oxaloacetic acid could complex with Na$^+$ thereby preventing plants from Na$^+$ toxicity and enhancing plant stress tolerance.

According to Sun et al. (2013), salt stress resulted in the accumulation of α-ketoglutaric acid in transgenic *A. thaliana* plants overexpressing *GolS2*, a stress-inducible galactinol synthase gene. It is a crucial molecule involved in the synthesis of amino acids, nitrogen transport and oxidation reactions. It also serves as an antioxidant. Therefore, elevated α-ketoglutaric acid may impart salinity tolerance by being a source of biomolecules and an antioxidant.

2.5 CONCLUSIONS

The contribution of organic solutes in allaying the detrimental effects of salt in plants and improving their tolerance is mirrored in the literature cited above. Osmoregulation seems to be the major role of organic solutes in the amelioration of salt stress. However, in addition to osmoregulation, these solutes also protect plants under salt stress in various other ways as described above. The type of osmolytes accumulated is dependent on the nature of stress and the plant. For example, fructans are mostly accumulated in plants under chilling stress.

One of the emerging molecules in the alleviation of salt-induced osmotic stress are organic acids. However, there have been few studies to decipher its osmoregulatory role. The inclusion of organic acids in osmotic stress research can reveal their osmoregulatory mechanisms.

Biotechnological approaches are successful in producing transgenic plants that overexpress key genes involved in the metabolism of organic solute to impart salinity tolerance and improved growth. In the alternative approach, these solutes have been applied to plants exogenously to enhance salt tolerance. However, the findings are too random to be conclusive. For example, the range of concentration of the solute applied exogenously is quite big for the same plant. Hence, in this case, standardization of the concentration of organic solute is critical as well as the time and duration of application. Besides this, the exact mechanism of how exogenous application improves salinity tolerance remains to be unraveled. Another method to enhance the concentration of organic solutes is the application of biofertilizers, such as arbuscular mycorrhizal fungi (AMF). AMF has been consistently shown to improve the salt tolerance level of plants, and the accumulation of organic solutes has been one of the strategies employed. As such, this approach may also be considered to increase osmolyte accumulation in plants under salt stress.

REFERENCES

Abdallah, M.M.-S., Abdelgawad, Z.A. and El-Bassiouny, H.M.S. 2016. Alleviation of the adverse effects of salinity stress using trehalose in two rice varieties. *South African Journal of Botany* 103: 275–282.

Abdel Latef, A.A., Kordrostami, M., Zakir, A., Zaki, H. and Saleh, O.M. 2019a. Eustress with H$_2$O$_2$ facilitates plant growth by improving tolerance to salt stress in two wheat cultivars. *Plants* 8(9): 303.

Abdel Latef, A.A., Mostofa, M.G., Rahman, M.M., Abdel-Farid, I.B. and Tran, L.P. 2019b. Extracts from yeast and carrot roots enhance maize performance under seawater-induced salt stress by altering physio-biochemical characteristics of stressed plants. *Journal of Plant Growth Regulation* 38(3): 966–979.

Abdel Latef, A.A.H., Abu Alhmad, M.F., Kordrostami, M., Abo–Baker, A.A. and Zakir, A. 2020. Inoculation with azospirillum lipoferum or azotobacter chroococcum reinforces maize growth by improving physiological activities under saline conditions. *Journal of Plant Growth Regulation* 39(3): 1293–1306.

Abebe, T., Guenzi, A.C., Martin, B. and Cushman, J.C. 2003. Tolerance of mannitol-accumulating transgenic wheat to water stress and salinity. *Plant Physiology* 131(4): 1748–1755.

Ahmad, I., Larher, F. and Stewart, G.R. 1979. Sorbitol, a compatible osmotic solute in *Plantago maritima*. *New Phytologist* 82: 671–678.

Al Hassan, M., Pacurar, A., López-Gresa, M.P., Donat-Torres, M.P., Llinares, J.V., Boscaiu, M. and Vicente, O. 2016. Effects of salt stress on three ecologically distinct Plantago species. *PLOS ONE* 11(8).

Amin, U.S., Lash, T.D. and Wilkinson, B.J. 1995. Proline betaine is a highly effective osmoprotectant for Staphylococcus aureus. *Archives of Microbiology* 163(2): 138–142.

Amini, S., Ghadiri, H., Chen, C. and Marschner, P. 2016. Salt-affected soils, reclamation, carbon dynamics, and biochar: A review. *Journal of Soils and Sediments* 16(3): 939–953.

Arbona, V., Marco, A.J., Iglesias, D.J., López-Climent, M.F., Talon, M. and Gómez-Cedenas, A. 2005. Carbohydrate depletion in roots and leaves of salt-stressed potted *Citrus clementina* L. *Journal of Plant Growth Regulation* 46: 153–160.

Armengaud, P., Thiery, L., Buhot, N., Grenier-De March, G.G. and Savouré, A. 2004. Transcriptional regulation of proline biosynthesis in *Medicago truncatula* reveals developmental and environmental specific features. *Physiologia Plantarum* 120(3): 442–450.

Banu, N.A., Hoque, A., Watanabesugimoto, M., Matsuoka, K., Nakamura, Y., Shimoishi, Y. and Murata, Y. 2009. Proline and glycinebetaine induce antioxidant defense gene expression and suppress cell death in cultured tobacco cells under salt stress. *Journal of Plant Physiology* 166: 146.

Ben-Amotz, A. and Avron, M. 1973. The role of glycerol in the osmotic regulation of the halophilic alga *Dunaliella parva. Plant Physiology* 51: 875–878.

Bhauso, T.D., Thankappan, R., Kumar, A., Mishra, G.P., Dobaria, J.R. and Rajam, M.V. 2014. Over-expression of bacterial mtlD gene confers enhanced tolerance to salt-stress and waterdeficit stress in transgenic peanut (*Arachis hypogaea*) through accumulation of mannitol. *Australian Journal of Crop Science* 8(3): 413–421.

Bhusan, D., Das, D.K., Hossain, M., Murata, Y. and Hoque, M.A. 2016. Improvement of salt tolerance in rice (*Oryza sativa* L.) by increasing antioxidant defense systems using exogenous application of proline. *Australian Journal of Crop Science* 10(1): 50–56.

Bohnert, H.J. and Jensen, R.G. 1996. Strategies for engineering water-stress tolerance in plants. *Trends in Biotechnology* 14: 89–97.

Boriboonkaset, T., Theerawitaya, C., Yamada, N., Pichakum, A., Supaibulwatana, K., Cha-Um, S., Takabe, T. and Kirdmanee, C. 2013. Regulation of some carbohydrate metabolism-related genes, starch and soluble sugar contents, photosynthetic activities and yield attributes of two contrasting rice genotypes subjected to salt stress. *Protoplasma* 250: 1157–1167.

Borowitzka, L.J., Demmerle, S., Mackay, M.A. and Norton, R.S. 1980. Carbon-13 nuclear magnetic resonance study of osmoregulation in a blue-green alga. *Science* 210: 650–651.

Burg, M.B. and Ferraris, J.D. 2008. Intracellular organic osmolytes: Function and regulation. *The Journal of Biological Chemistry* 283(12): 7309–7313.

Carter, J.L., Colmer, T.D. and Veneklaas, E.J. 2006. Variable tolerance of wetland tree species to combined salinity and waterlogging is related to regulation of ion uptake and production of organic solutes. *New Phytologist* 169(1): 123–133.

Chakraborty, K. and Sairam, R.K. 2016. Induced-expression of osmolyte biosynthesis pathway genes improves salt and oxidative stress tolerance in *Brassica* species. *Indian Journal of Experimental Botany* 55(10): 711–721.

Chang, B., Yang, L., Cong, W., Zu, Y. and Tang, Z. 2014. The improved resistance to high salinity induced by trehalose is associated with ionic regulation and osmotic adjustment in *Catharanthus roseus. Plant Physiology and Biochemistry: PPB* 77: 140–148.

Christgen, S.L. and Becker, D.F. 2019. Role of proline in pathogen and host interactions. *Antioxidants and Redox Signaling* 30(4): 683–709.

Crowe, J.H., Crowe, L.M. and Chapman, D. 1984. Preservation of membranes in anhydrobiotic organisms: The role of trehalose. *Science* 223: 701–703.

de Freitas, P.A.F., de Souza Miranda, R.S., Marques, E.C., Prisco, J.T. and Gomes-Filho, E. 2018. Salt tolerance induced by exogenous proline in maize is related to low oxidative damage and favorable ionic homeostasis. *Journal of Plant Growth Regulation* 37(3): 911–924.

Deguchi, M., Watanabe, M. and Kanayama, Y. 2002. Increase in sorbitol biosynthesis in stressed Japanese pear leaves. In: International Symposium on Asian Pears, Commemorating the 100th Anniversary of Nijisseiki Pear 587 (pp 511–517).

de la Torre-González, A., Albacete, A., Sánchez, E., Blasco, B. and Ruiz, J.M. 2017. Comparative study of the toxic effect of salinity in different genotypes of tomato plants: Carboxylates metabolism. *Scientia Horticulturae* 217: 173–178.

Dinler, S.B., Sekmen, A.H. and Turkan, I. 2008. An enhancing effect of exogenous mannitol on the antioxidant enzyme activities in roots of wheat under salt stress. *Journal of Plant Growth Regulation* 28(1): 12–20.

dos Santos, T.B., Budzinski, I.G., Marur, C.J., Petkowicz, C.L., Pereira, L.F. and Vieira, L.G. 2011. Expression of three galactinol synthase isoforms in *Coffea arabica* L. and accumulation of raffinose and stachyose in response to abiotic stresses. *Plant Physiology and Biochemistry: PPB* 49(4): 441–448.

Eggert, E., Obata, T., Gerstenberger, A., Gier, K., Brandt, T., Fernie, A.R., Schulze, W. and Kúhn C. 2016. A sucrose transporter-interacting protein disulphide isomerise affects redox homeostasis and links sucrose partitioning with abiotic stress tolerance. *Plant, Cell and Environment* 39: 1366–1380.

Elbein, A.D., Pan, Y.T., Pastuszak, I. and Carroll, D. 2003. New insights on trehalose: A multifunctional molecule. *Glycobiology* 13(4): 17R–27R.

El Moukhtari, A., Cabassa-Hourton, C., Farissi, M. and Savouré, A. 2020. How does proline treatment promote salt stress tolerance during crop plant development? *Frontiers in Plant Science* 11: 1127.

El-Yazal, M.A.S., Eissa, H.F., Ahmed, S.M.A.E., Howla dar, S.M., Zaki, S.S. and Rady, M.M. 2016. The *mtlD* gene-overexpressed transgenic wheat tolerates salt stress through accumulation of mannitol and sugars. *Planta* 4(6): 78–90.

Fan, W., Zhang, M., Zhang, H. and Zhang, P. 2012. Improved tolerance to various abiotic stresses in transgenic sweet potato (*Ipomoea batatas*) expressing spinach betaine aldehyde dehydrogenase. *PLOS ONE* 7(5): e37344.

Ferjani, A., Mustardy, L., Sulpice, R., Marin, K., Suzuki, I., Hagemann, M. and Murata, N. 2003. Glucosylglycerol, a compatible solute, sustains cell division under salt stress. *Plant Physiology* 131: 1628–1637.

Food and Agriculture Organization [FAO] 2015. *Status of the Worlds's Soil Resources (SWSR)—Main Report.* United Nations.

Fougéré, F., Le Rudulier, D. and Streeter, J.G. 1991. Effects of salt stress on amino acid, organic acid, and carbohydrate composition of roots, bacteroids, and cytosol of alfalfa (*Medicago sativa* L.). *Plant Physiology* 96: 1228–1236.

Foyer, C.H. and Shigeoka, S. 2011. Understanding oxidative stress and antioxidant functions to enhance photosynthesis. *Plant Physiology* 155(1): 93–100.

Gage, D.A., Rhodes, D., Nolte, K.D., Hicks, W.A., Leustek, T., Cooper, A.J. and Hanson, A.D. 1997. A new route for synthesis of dimethylsulphoniopropionate in marine algae. *Nature* 387(6636): 891–894.

Gao, J.J., Sun, Y.R., Zhu, B., Peng, R.H., Wang, B., Wang, L.J., Li, Z., Chen, L. and Yao, Q.H. 2018. Ectopic expression of sorbitol-6-phosphate 2-dehydrogenase gene from *Haloarcula marismortui* enhances salt tolerance in transgenic *Arabidopsis thaliana*. *Acta Physiologiae Plantarum* 40(6): 108.

Garcia, B., de Almeida Engler, J., Iyer, S., Gerats, T., Van Montagu, M. and Caplan, A.B. 1997. Effects of osmoprotectants upon NaCl stress in rice. *Plant Physiology* 115: 159–169.

Ge, L., Chao, D., Shi, M., Zhu, M.Z., Gao, J.P. and Lin, H.X. 2008. Overexpression of the trehalose-6-phosphate phosphatase gene *OsTPP1* confers stress tolerance in rice and results in the activation of stress responsive genes. *Planta* 228(1): 191–201.

Giordano, M., Norici, A. and Hell, R. 2005. Sulfur and phytoplankton: Acquisition, metabolism and impact on the environment. *New Phytologist* 166: 371–382.

Giri, J. 2011. Glycinebetaine and abiotic stress tolerance in plants. *Plant Signaling and Behavior* 6(11): 1746–1751.

Gorham, J. 1995. Betaines in higher plants—Biosynthesis and role in stress metabolism. In: *Amino Acids and Their Derivatives in Higher Plants* (ed. R.M. Wallsgrove), 171–203. Cambridge University Press, Cambridge.

Groppa, M.D. and Benavides, M.P. 2008. Polyamines and abiotic stress: Recent advances. *Amino Acids* 34: 35–45.

Guan, C., Huang, Y., Cui, X., Liu, S.J., Zhou, Y.Z. and Zhang, Y.W. 2018. Overexpression of gene encoding the key enzyme involved in proline-biosynthesis (*PuP5CS*) to improve salt tolerance in switchgrass (*Panicum virgatum* L.). *Plant Cell Reports* 37(8): 1187–1199.

Guo, L.Q., Shi, D.C. and Wang, D.L. 2010. The key physiological response to alkali stress by the alkali-resistant halophyte *Puccinellia tenuiflora* is the accumulation of large quantities of organic acids and into the rhyzosphere. *Journal of Agronomy and Crop Science* 196(2): 123–135.

Gupta, A.K. and Kaur, N. 2005. Sugar signaling and gene expression in relation to carbohydrate metabolism under abiotic stresses in plants. *Journal of Biosciences* 30: 101–116.

Habib, N., Ashraf, M., Ali, Q. and Perveen, R. 2012. Response of salt stressed okra (*Abelmoschus esculentus* Moench) plants to foliar-applied glycine betaine and glycine betaine containing sugarbeet extract. *South African Journal of Botany* 83: 151–158.

Hamani, A.K.M., Wang, G., Soothar, M.K., Shen, X., Gao, Y., Qiu, R. and Mehmood, F. 2020. Responses of leaf gas exchange attributes, photosynthetic pigments and antioxidant enzymes in NaCl-stressed cotton (*Gossypium hirsutum* L.) seedlings to exogenous glycine betaine and salicylic acid. *BMC Plant Biology* 20(1): 434.

Hanson, A.D., Rathinasabapathi, B., Chamberlin, B. and Gage, D.A. 1991. Comparative physiological evidence that β-alanine betaine and choline-O-sulfate act as compatible osmolytes in halophytic Limonium species. *Plant Physiology* 97(3): 1199–1205.

Hanson, A.D., Rathinasabapathi, B., Rivoal, J., Burnet, M., Dillon, M.O. and Gage, D.A. 1994. Osmoprotective compounds in the Plumbaginaceae: A natural experiment in metabolic engineering of stress tolerance. *Proceedings of the National Academy of Sciences of the United States of America* 91(1): 306–310.

Hare, P.D. and Cress, W.A. 1997. Metabolic implications of stress-induced proline accumulation in plants. *Plant Growth Regulation* 21(2): 79–102.

Hasanuzzaman, M., Alam, M.M., Rahman, A., Hasanuzzaman, M., Nahar, K. and Fujita, M. 2014. Exogenous proline and glycine betaine mediated upregulation of antioxidant defense and glyoxalase systems provides better protection against salt-induced oxidative stress in two rice (*Oryza sativa* L.) varieties. *BioMed Research International*: 757219.

Haworth, M., Catola, S., Marino, G., Brunetti, C., Michelozzi, M., Riggi, E., Avola, G., Cosentino, S.L., Loreto, F. and Centritto, M. 2017. Moderate drought stress induces increased foliar dimethylsulfoniopropionate (DMSP) concentration and isoprene emission in two contrasting ecotypes of Arundo donax. *Frontiers in Plant Science* 8: 1016.

He, C., Yang, A., Zhang, W., Gao, Q. and Zhang, J. 2010. Improved salt tolerance of transgenic wheat by introducing betA gene for glycine betaine synthesis. *Plant Cell, Tissue and Organ Culture* 101(1): 65–78.

Henry, C., Bledsoe, S.W., Griffiths, C.A., Kollman, A., Paul, M.J., Sakr, S. and Lagrimini, L.M. 2015. Differential role for trehalose metabolism in salt-stressed maize. *Plant Physiology* 169: 1072–1089.

Hu, C.A., Delauney, A.J. and Verma, D.P. 1992. A bifunctional enzyme (delta 1-pyrroline-5-carboxylate synthetase) catalyzes the first two steps in proline biosynthesis in plants. *Proceedings of the National Academy of Sciences* 89(19): 9354–9358.

Hu, C.Y., Pettitt, B.M. and Roesgen, J. 2009. Osmolyte solutions and protein folding. *F1000 Biology Reports* 1: 41.

Issa, D.B., Alturki, S.M., Sajyan, T.K. and Sassine, Y.N. 2020. Sorbitol and lithovit-guano25 mitigates the adverse effects of salinity on eggplant grown in pot experiment.

Ito, T., Asano, Y., Tanaka, Y. and Takabe, T. 2011. Regulation of biosynthesis of dimethylsulfoniopropionate and its uptake in sterile mutant of *Ulva pertusa* (chlorophyta). *Journal of Phycology* 47: 517–523.

Jawahar, G., Rajasheker, G., Maheshwari, P., Punita, D.L., Jalaja, N., Kumari, P.H., Kumar, S.A., Afreen, R., Karumanchi, A.R., Rathnagiri, P. and Sreenivasulu, N. 2019. Osmolyte diversity, distribution, and their biosynthetic pathways. In: *Plant Signaling Molecules*, 449–458. Woodhead Publishing.

Kamiab, F., Talaie, A., Khezri, M. and Javanshah, A. 2014. Exogenous application of free polyamines enhance salt tolerance of pistachio (*Pistacia vera* L.) seedlings. *Plant Growth Regulation* 72(3): 257–268.

Kaur, G. and Asthir, B. 2015. Proline: A key player in plant abiotic stress tolerance. *Biologia Plantarum* 59(4): 609–619.

Kaya, C., Sonmez, O., Aydemir, S., Ashraf, M. and Dikilitas, M. 2013. Exogenous application of mannitol and thiourea regulates plant growth and oxidative stress responses in salt-stressed maize (*Zea mays* L.). *Journal of Plant Interactions* 8(2): 34–241.

Ke, Q., Wang, Z., Ji, C.Y., Jeong, J.C., Lee, H.S., Li, H., Xu, B., Deng, X. and Kwak, S.S. 2016. Transgenic poplar expressing *codA* exhibits enhanced growth and abiotic stress tolerance. *Plant Physiology and Biochemistry: PPB* 100: 75–84.

Kishor, P.B.K., Sangam, S., Amrutha, R.N., Laxmi, P.S., Naidu, K.R., Rao, K.R.S.S., Rao, S., Reddy, K.J., Theriappan, P. and Sreenivasulu, N. 2005. Regulation of proline biosynthesis, degradation, uptake and transport in higher plants: Its implications in plant growth and abiotic stress tolerance. *Current Science* 88: 424–438.

Knaupp, M., Mishra, K.B., Nedbal, L. and Heyer, A.G. 2011. Evidence for a role of raffinose in stabilizing photosystem II during freeze–thaw cycles. *Planta* 234(3): 477–486.

Kocsis, M.G. and Hanson, A.D. 2000. Biochemical evidence for two novel enzymes in the biosynthesis of 3-dimethylsulfoniopropionate in Spartina alterniflora. *Plant Physiology* 123(3): 1153–1161.

Kocsis, M.G., Nolte, K.D., Rhodes, D., Shen, T.L., Gage, D.A. and Hanson, A.D. 1998. Dimethylsulfoniopropionate biosynthesis in Spartina alterniflora1. Evidence that S-methylmethionine and dimethylsulfoniopropylamine are intermediates. *Plant Physiology* 117(1): 273–281.

Kumar, D.P., Murata, Y., Hoque, M.A. and Ali, M.A. 2015. Effect of soil salinity and exogenous proline application on rice growth, yield, biochemical and antioxidant enzyme activities. *EC Agriculture* 2: 229–240.

Kumar, V., Shriram, V., Kavi Kishor, P.B.K., Jawali, N. and Shitole, M.G. 2010. Enhanced proline accumulation and salt stress tolerance of transgenic indica rice by over-expressing P5CSF129A gene. *Plant Biotechnology Reports* 4(1): 37–48.

Kumria, R. and Rajam, M.V. 2002. Ornithine decarboxylase transgene in tobacco affects polyamines, in *in vitro*-morphogenesis and response to salt stress. *Journal of Plant Physiology* 159(9): 983–990.

Le Rudulier, D., Strom, A.R., Dandekar, A.M., Smith, L.T. and Valentine, R.C. 1984. Molecular biology of osmoregulation. *Science* 224(4653): 1064–1068.

Li, H.W., Zang, B.S., Deng, X.W. and Wang, X.P. 2011. Overexpression of the trehalose-6-phosphate synthase gene OsTPS1 enhances abiotic stress tolerance in rice. *Planta* 234(5): 1007–1018.

Li, S., Jin, H. and Zhang, Q. 2016. The effect of exogenous spermidine concentration on polyamine metabolism and salt tolerance in zoysia grass (Zoysia *japonica* Steud) subjected to short-term salinity stress. *Frontiers in Plant Science.* 7: 1221–1233.

Liu, J. and Shi, D.C. 2010. Photosynthesis, chlorophyll fluorescence, inorganic ion and organic acid accumulations of sunflower in responses to salt and salt-alkaline mixed stress. *Photosynthetica* 48(1): 127–134.

Lokhande, V.H. and Suprasanna, P. 2012. Prospects of halophytes in understanding and managing abiotic stress tolerance. In: *Environmental Adaptations and Stress Tolerance of Plants in the Era of Climate Change* (eds P. Ahmad, Prasad M.). Springer, New York, NY.

Lunn, J.E., Delorge, I., Figueroa, C.M., Van Dijck, P. and Stitt, M. 2014. Trehalose metabolism in plants. *Plant Journal: For Cell and Molecular Biology* 79(4): 544–567.

Lyu, J.I., Min, S.R., Lee, J.H., Lim, Y.H., Kim, J., Bae, C. and Liu, J.R. 2013. Overexpression of a trehalose-6-phosphate synthase/phosphatase fusion gene enhances tolerance and photosynthesis during drought and salt stress without growth aberrations in tomato. *Plant Cell, Tissue and Organ Culture* 112(2): 257–262.

Makela, P., Karkkainen, J. and Somersalo, S. 2000. Effect of glycine betaine on chloroplast ultrastructure, chlorophyll and protein content and RUBPCO activities in tomato grown under drought or salinity. *Biologia Plantarum* 43: 471–475.

Malekzadeh, P. 2015. Influence of exogenous application of glycinebetaine on antioxidative system and growth of salt-stressed soybean seedlings (*Glycine max* L.). *Physiology and Molecular Biology of Plants* 21: 225–232

Mansour, M.M.F. and Ali, E. 2017a. Glycinebetaine in saline conditions: An assessment of the current state of knowledge. *Acta Physiologiae Plantarum* 39: 1–17.

Mansour, M.M.F. and Ali, E.F. 2017b. Evaluation of proline functions in saline conditions. *Phytochemistry* 140: 52.

McNeil, S.D., Nuccio, M.L. and Hanson, A.D. 1999. Betaines and related osmoprotectants. Targets for metabolic engineering of stress resistance. *Plant Physiology* 120(4): 945–950.

Minocha, R., Majumdar, R. and Minocha, S.C. 2014. Polyamines and abiotic stress in plants: A complex relationship. *Frontiers in Plant Science* 5: 175.

Munns, R. and Gilliham, M. 2015. Salinity tolerance of crops—what is the cost? *New Phytologist* 208(3): 668–673.

Nawaz, K. and Ashraf, M. 2010. Exogenous application of glycinebetaine modulates activities of antioxidants in maize plants subjected to salt stress. *Journal of Agronomy and Crop Science* 196(1): 28–37.

Nemat Alla, M., Badran, E. and Mohammed, F. 2019. Exogenous trehalose alleviates the adverse effects of salinity stress in wheat. *Turkish Journal of Botany* 43(1): 48–57.

Nounjan, N., Nghia, P.T. and Theerakulpisut, P. 2012. Exogenous proline and trehalose promote recovery of rice seedlings from salt-stress and differentially modulate antioxidant enzymes and expression of related genes. *Journal of Plant Physiology* 169(6): 596–604.

Orsini, F., Pennisi, G., Mancarella, S.Al., Al Nayef, M., Sanoubar, R., Nicola, S. and Gianquinto, G. 2018. Hydroponic lettuce yields are improved under salt stress by utilizing white plastic film and exogenous applications of proline Nayef, M. Sanoubar, R. Nicola, S. and Gianquinto, G. *Scientia Horticulturae* 233: 283–293.

Osman, M.S., Badawy, A.A., Osman, A.I. and Abdel Latef, A.A.H. 2020. Ameliorative impact of an extract of the halophyte *Arthrocnemum macrostachyum* on growth and biochemical parameters of soybean under salinity stress. *Journal of Plant Growth Regulation.* https://doi.org/10.1007/s00344-020-10185-2.

Otte, M.L., Wilson, G., Morris, J.T. and Moran, B.M. 2004. Dimethylsulphoniopropionate (DMSP) and related compounds in higher plants. *Journal of Experimental Botany* 55(404): 1919–1925.

Pang, X.M., Zhang, Z.Y., Wen, X.P., Ban, Y. and Moriguchi, T. 2007. Polyamines, all-purpose players in response to environment stresses in plants. *Plant Stress* 1(2): 173–188.

Parihar, P., Singh, S., Singh, R., Singh, V.P. and Prasad, S.M. 2015. Effect of salinity stress on plants and its tolerance strategies: A review. *Environmental Science and Pollution Research International* 22(6): 4056–4075.

Patel, K.G., Thankappan, R., Mishra, G.P., Mandaliya, V.B., Kumar, A. and Dobaria, J.R. 2017. Transgenic Peanut (Arachis hypogaea L.) Overexpressing mtlD gene showed improved photosynthetic, physio-bio-chemical, and yield-parameters under soil-moisture deficit stress in lysimeter system. *Frontiers in Plant Science* 8: 1881.

Patel, T.K. and Williamson, J.D. 2016. Mannitol in plants, fungi, and plant-fungal interactions. *Trends in Plant Science* 21(6): 486–497.

Pommerrenig, B., Papini-Terzi, F.S. and Sauer, N. 2007. Differential regulation of sorbitol and sucrose loading into the phloem of *Plantago* major in response to salt stress. *Plant Physiology* 144: 1029–1038.

Qi, Y., Wang, F., Zhang, H. and Liu, W. 2010. Overexpression of suadea salsa *S*-adenosylmethionine synthetase gene promotes salt tolerance in transgenic tobacco. *Acta Physiologiae Plantarum* 32(2): 263–269.

Rady, M.O.A., Semida, W.M., Abd El-Mageed, T.A., Hemida, K.A. and Rady, M.M. 2018. Up-regulation of antioxidative defense systems by glycine betaine foliar application in onion plants confer tolerance to salinity stress. *Scientia Horticulturae* 240: 614–622.

Raman, S.B. and Rathinasabapathi, B. 2003. β-alanine N-methyltransferase of *Limonium latifolium*. cDNA cloning and functional expression of a novel N-methyltransferase implicated in the synthesis of the osmoprotectant β-alanine betaine. *Plant Physiology* 132(3): 1642–1651.

Redillas, M.C.F.R., Park, S., Lee, J.W., Kim, Y.S., Jeong, J.S., Jung, H., Bang, S.W., Hahn, T. and Kim, J. 2012. Accumulation of trehalose increases soluble sugar contents in rice plants conferring tolerance to drought and salt stress. *Plant Biotechnology Reports* 6(1): 89–96.

Rhodes, A.D. and Hanson, A.D. 1993. Quaternary ammonium and tertiary sulfonium compounds in higher plants. *Annual Review of Plant Biology* 44: 357–384.

Richter, J.A., Erban, A., Kopka, J. and Zörb, C. 2015. Metabolic contribution to salt stress in two maize hybrids with contrasting resistance. *Plant Science: An International Journal of Experimental Plant Biology* 233: 107–115.

Roosens, N.H.C.J., Thu, T.T., Iskandar, H.M. and Jacobs, M. 1998. Isolation of the ornithine-δ-aminotransferase cDNA and effect of salt stress on its expression in *Arabidopsis thaliana*. *Plant Physiology* 117(1): 263–271.

Rossky, P.J. 2008. Protein denaturation by urea: Slash and bond. *Proceedings of the National Academy of Sciences of the United States of America* 105(44): 16825–16826.

Roy, M. and Wu, R. 2002. Overexpression of S-adenosylmethionine decarboxylase gene in rice increases polyamine level and enhances sodium chloride-stress tolerance. *Plant Science* 163(5): 987–992.

Sabagh, El, Sorour, A., Ragab, S., Saneoka, A.H. and Islam, M.S. 2017. The effect of exogenous application of proline and glycine betaine on the nodule activity of soybean under saline condition. *Journal of Agriculture Biotechnology* 2(1): 1–5.

Sakamoto, A. and Murata, N. 2000. Genetic engineering of glycinebetaine synthesis in plants: Current status and implications for enhancement of stress tolerance. *Journal of Experimental Botany* 51(342): 81–88.

Samadi, S., Habibi, G. and Vaziri, A. 2019. Exogenous trehalose alleviates the inhibitory effects of salt stress in strawberry plants. *Acta Physiologiae Plantarum* 41(7): 112.

Seckin, B., Sekmen, A.H. and Turkan, İ. 2009. An enhancing effect of exogenous mannitol on the antioxidant enzyme activities in roots of wheat under salt stress. *Journal of Plant Growth Regulation* 28(1): 12–20.

Sengupta, S., Mukherjee, S., Basak, P. and Majumder, A.L. 2015. Significance of galactinol and raffinose family oligosaccharide synthesis in plants. *Frontiers in Plant Science* 6: 656.

Shahbaz, M., Abid, A., Masood, A. and Waraich, E.A. 2017. Foliar-applied trehalose modulates growth, mineral nutrition, photosynthetic ability, and oxidative defense system of rice (*Oryza sativa* L.) under saline stress. *Journal of Plant Nutrition* 40(4): 584–599.

Shams, M., Yildirim, E., Ekinci, M., et al. 2016. Exogenously applied glycine betaine regulates some chemical characteristics and antioxidative defence systems in lettuce under salt stress. *Horticulture Environment and Biotechnology* 57: 225–231.

Sharma, S., Villamor, J.G. and Verslues, P.E. 2011. Essential role of tissue-specific proline synthesis and catabolism in growth and redox balance at low water potential. *Plant Physiology* 157(1): 292–304.

Shen, B., Jensen, R.C. and Bohnert, H. 1997. Mannitol protects against oxidation by hydroxyl radicals. *Plant Physiology* 115: 527–532.

Shu, S., Yuan, L., Guo, S., Sun, J. and Yuan, Y. 2013. Effects of exogenous spermine on chlorophyll fluorescence, antioxidant system and ultrastructure of chloroplasts in *Cucumis sativus* L. under salt stress. *Plant Physiology and Biochemistry: PPB* 63: 209–216.

Shu, S., Yuan, L.Y., Guo, S.R., Sun, J. and Liu, C.J. 2012. Effects of exogenous spermidine on photosynthesis, xanthophyll cycle and endogenous polyamines in cucumber seedlings exposed to salinity. *African Journal of Biotechnology* 11(22): 6064–6074.

Signorelli, S. 2016. The fermentation analogy: A point of view for understanding the intriguing role of proline accumulation in stressed plants. *Frontiers in Plant Science* 7: 1339.

Singh, M., Kumar, J., Singh, S., Singh, V.P. and Prasad, S.M. 2015. Roles of osmoprotectants in improving salinity and drought tolerance in plants: A review. *Reviews in Environmental Science and Bio/Technology* 14(3): 407–426.

Slama, I., Abdelly, C., Bouchereau, A., Flowers, T. and Savouré, A. 2015. Diversity, distribution and roles of osmoprotective compounds accumulated in halophytes under abiotic stress. *Annals of Botany* 115(3): 433–447.

Slocum, R.D. 1991. Polyamine biosynthesis in plants. *Biochemistry and Physiology of Polyamines in Plants*: 23–40.

Smekens, M.J. and Tienderen, P.H.V. 2001. Genetic variation and plasticity of *Plantago coronopus* under saline conditions. *Acta Oecologica* 22: 187–200.

Somero, G.N. and Yancey, P.H. 2010. Osmolytes and cell-volume regulation: Physiological and evolutionary principles. *Comprehensive Physiology*: 441–484.

Street, T.O., Bolen, D.W. and Rose, G.D. (2006). *Proceedings of the National Academy of Sciences of the United States of America*. 103(38): 13997–14002.

Su, J. and Wu, R. 2004. Stress-inducible synthesis of proline in transgenic rice confers faster growth under stress conditions than that with constitutive synthesis. *Plant Science* 166(4): 941–948.

Sun, Z., Qi, X., Wang, Z., Li, P., Wu, C., Zhang, H. and Zhao, Y. 2013. Overexpression of TsGOLS2, a galactinol synthase, in Arabidopsis thaliana enhances tolerance to high salinity and osmotic stresses. *Plant Physiology and Biochemistry: PPB* 69: 82–89.

Surender Reddy, P., Jogeswar, G., Rasineni, G.K., Maheswari, M., Reddy, A.R., Varshney, R.K. and Kavi Kishor, P.B. 2015.Proline over-accumulation alleviates salt stress and protects photosynthetic and antioxidant enzyme activities in transgenic sorghum [*Sorghum bicolor* (L.) Moench]. *Plant Physiology and Biochemistry: PPB* 94: 104–113.

Tanner, J.J. 2008. Structural biology of proline catabolism. *Amino Acids* 35(4): 719–730.

Teh, C., Shaharuddin, N.A., Ho, C. and Mahmood, M. 2016. Exogenous proline significantly affects the plant growth and nitrogen assimilation enzymes activities in rice (*Oryza sativa*) under salt stress. *Acta Physiologiae Plantarum* 38(6): 151.

Theerakulp, P. and Gunnula, W. 2012. Exogenous sorbitol and trehalose mitigated salt stress damage in salt-sensitive but not salt-tolerant rice seedlings. *Asian Journal of Crop Science* 4(4): 165–170.

Tian, F., Wang, W., Liang, C., Wang, X., Wang, G. and Wang, W. 2017. Over accumulation of glycine betaine makes the function of the thylakoid membrane better in wheat under salt stress. *The Crop Journal* 5(1): 73–82.

Tran, N.T., Oguchi, T., Matsunaga, E., Kawaoka, A., Watanabe, K.N. and Kikuchi, A. 2018. Transcriptional enhancement of a bacterial *choline oxidase A* gene by an *HSP* terminator improves the glycine betaine production and salinity stress tolerance of *Eucalyptus camaldulensis* trees. *Plant Biotechnology (Tokyo, Japan)* 35(3): 215–224.

Trinchant, J.C., Boscari, A., Spennato, G., Van de Sype, G. and Le Rudulier, D. 2004. Proline betaine accumulation and metabolism in alfalfa plants under sodium chloride stress. Exploring its compartmentalization in nodules. *Plant Physiology* 135(3): 1583–1594.

Trossat, C., Nolte, K.D. and Hanson, A.D. 1996. Evidence that the pathway of dimethylsulfoniopropionate biosynthesis begins in the cytosol and ends in the chloroplast. *Plant Physiology* 111(4): 965–973.

Trossat, C., Rathinasabapathi, B. and Hanson, A.D. 1997. Transgenically expressed betaine aldehyde dehydrogenase efficiently catalyzes oxidation of dimethylsulfoniopropionaldehyde and [omega]-aminoaldehydes. *Plant Physiology* 113(4): 1457–1461.

Tsutsumi, K., Yamada, N., Cha-Um, S., Tanaka, Y. and Takabe, T. 2015. Differential accumulation of glycinebetaine and choline monooxygenase in bladder hairs and lamina leaves of *Atriplex gmelini* under high salinity. *Journal of Plant Physiology* 176: 101–107.

Valluru, R. and Van den Ende, W. 2011. Myo-inositol and beyond--emerging networks under stress. *Plant Science: An International Journal of Experimental Plant Biology* 181(4): 387–400.

Vanrensburg, L., Kruger, G.H.J. and Kruger, R.H. 1993. Proline accumulation as drought tolerance selection criterion: Its relationship to membrane integrity and chloroplast ultrastructure in *Nicotiana tabacum* L. *Journal of Plant Physiology* 141: 188–194.

Verma, S. and Mishra, S.N. 2005. Putrescine alleviation of growth in salt stressed *Brassica juncea* by inducing antioxidative defense system. *Journal of Plant Physiology* 162(6): 669–677.

Wang, H., Tang, X., Wang, H. and Shao, H. 2015. Proline accumulation and metabolism-related genes expression profiles in *Kosteletzkya virginica* seedlings under salt stress. *Frontiers in Plant Science* 6: 792.

Wei, C., Cui, Q. Zhang, Zhang, X.Q., Zhao, Y. and Jia, G.X. 2016. Three P5CS genes including a novel one from *Lilium regale* play distinct roles in osmotic, drought and salt stress tolerance. *Journal of Plant Biology* 59(5): 456–466.

Wen, X., Pang, X., Matsuda, N., Kita, M., Inoue, H., Hao, Y.J., Honda, C. and Moriguchi, T. 2008. Overexpression of the apple *spermidine synthase* gene in pear confers multiple abiotic stress tolerance by altering polyamine titers. *Transgenic Research* 17(2): 251–263.

Williamson, J.D., Jennings, D.B., Guo, W.W., Pharr, D.M. and Ehrenshaft, M. 2002. Sugar alcohols, salt stress, and fungal resistance: Polyols—Multifunctional plant protection? *Journal of the American Society for Horticultural Science* 127(4): 467–473.

Yamada, N., Takahashi, H., Kitou, K., Sahashi, K., Tamagake, H., Tanaka, Y. and Takabe, T. 2015. Suppressed expression of choline monooxygenase in sugar beet on the accumulation of glycine betaine. *Plant Physiology and Biochemistry* 96: 217–221.

Yang, L., Zhao, X., Zhu, H., Paul, M., Zu, Y. and Tang, Z. 2014. Exogenous trehalose largely alleviates ionic unbalance, ROS burst, and PCD occurrence induced by high salinity in *Arabidopsis* seedlings. *Frontiers in Plant Science* 5: 570.

Yang, Y. and Guo, Y. 2018. Elucidating the molecular mechanisms mediating plant salt-stress responses. *New Phytologist* 217(2): 523–539.

Yin, Y.G., Kobayashi, Y., Sanuki, A., Kondo, S., Fukuda, N., Ezura, H., Sugaya, S. and Matsukura, C. 2010. Salinity induces carbohydrate accumulation and sugar regulated starch biosynthetic genes in tomato (*Solanum lycopersicum* L. cv. 'Micro- Tom') fruits in an ABA- and osmotic stress-independent manner. *Journal of Experimental Botany* 61: 563–574.

Zapata, P.J., Serrano, M., García-Legaz, M.F., Pretel, M.T. and Botella, M.A. 2017. Short term effect of salt shock on ethylene and polyamines depends on plant salt sensitivity. *Frontiers in Plant Science* 8: 855.

Zeid, I.M. 2009. Trehalose as osmoprotectant for maize under salinity-induced stress. *Research Journal of Agriculture and Biological Sciences* 5(5): 613–622.

Zhang, G., Zhu, W., Gai, J., Zhu, Y. and Yang, L. 2015. Enhanced salt tolerance of transgenic vegetable soybeans resulting from overexpression of a novel Δ1-pyrroline-5-carboxylate synthetase gene from Solanum torvum Swartz. *Horticulture, Environment, and Biotechnology* 56(1): 94–104.

Zhang, H., Dong, H., Li, W., Sun, Y., Chen, S. and Kong, X. 2009. Increased glycine betaine synthesis and salinity tolerance in AhCMO transgenic cotton lines. *Molecular Breeding* 23(2): 289–298.

3 Significance of Organic Solutes in Modulating Plant Inherent Responses under Heavy Metal Stress

Aditi Bisht and Neera Garg

CONTENTS

3.1 INTRODUCTION

Heavy metals (HMs) are elements having an atomic number >20, a density 5 times greater than water or more than 5 g cm^{-3} (Khan et al. 2015). HMs belong to a non-biodegradable group of elements that have genotoxic, cytotoxic and mutagenic effects on plants, humans and animals. Anthropogenic events, including mining, use of pesticides, disposal of munitions and nuclear wastes, burning of fossil fuel and waste from industries as well as natural processes (continental dust and volcanic discharge), add a high level of metal contaminants into the atmosphere. HMs are classified into nonessential (Hg, Cd, As, Ag, Sb, Pb, Co and Cr) and essential (Fe, Mg, Mn, Mo, Cu, Ni and Zn) micronutrients (Maleki et al. 2017). Among these, Co, Cu, Ni, Mn, Mo and Zn are required in small quantities by plants but induce detrimental effects when present in abundance. The HM concentration in agricultural soils is in the range of 0.1–200 mg kg^{-1} (O'Neill 1995), while in garden soils the range varies from 3.06–579.84 mg kg^{-1} (Szolnoki et al. 2013).

In the present day, soil contamination with HMs is of major worry because of their deleterious effects on food safety as well as on soil ecosystems, primarily on plant growth and microbial biomass. The accessibility of metals in soil depends on soil pH and texture, organic matter (OM) content, metals and ions concentration, cation exchange capacity (CEC), and moisture and redox potential (Wang et al. 2013). The disturbance in soil microbial activity by the presence of HMs reduces soil fertility, which can affect the quality and productivity of plants (Tecon and Or 2017). HM ions can be taken up by plant roots from the soil when present in excessive concentration, and the ions can be translocated by the roots to their above-ground parts (shoots), which disturbs the growth and various metabolic processes. In addition to soil uptake, HMs can enter the plants through foliar tissue (e.g. leaf cuticle, stomata and ectodesmata) (Karmakar and Padhy 2019). HMs cause toxicity in plants through the replacement of essential metal ions present in various enzymes or pigments and, thus, interrupt their natural function. Excessive accumulation of HMs in plants causes necrosis, chlorosis, imbalance in nutrient uptake and water metabolism, variation in enzymatic activity and changes in plant's morphology, which ultimately lead to damaged cell structures. In addition, they impair many metabolic processes including the formation of bonds between HMs and sulfhydryl (-SH) groups of protein, which creates altered or dysfunctional protein structure and disruption of functionality of necessary metal ions present in diverse biomolecules (enzymes or pigments), has a negative impact on the integrity of the biomembrane and hinders functional groups of essential cellular molecules, thus, results in inhibition of diverse vital processes such as respiration, photosynthesis and enzymatic activities in plants (Wani et al. 2018). The higher concentration of HMs causes a rise in the level of ROS-reactive oxygen species which include hydroxyl free radicals- OH^-, superoxide free radicals- O_2^-, non-free radicals such as $^1O_2$ (singlet oxygen) and H_2O_2 (hydrogen peroxide), and enhances the level of cytotoxic compounds (methylglyoxal-MG), which results in oxidative burden to the plants and disturbs the cellular homeostasis between pro-oxidant and antioxidant (Abdel Latef 2013; Berni et al. 2019; Abdel Latef et al. 2020).

To attain ionic homeostasis, plants must constantly keep the physiological concentrations of metal ions (essential and nonessential). Moreover, this homeostasis in a plant can be maintained at the cell, tissue, organ as well as genotype-specific manner and to some extent, controlled by environmental conditions (Ovečka and Takáč 2014). The main approaches used by plants to alleviate the negative impact of HMs are escaping from exposure, reducing their uptake and intracellular sequestering. HM induces osmotic stress in plants which leads to upregulation of various cellular adaptive responses including induction of stress proteins and hastening of ROS scavenging systems by generating various antioxidants (enzymatic and non-enzymatic) and accretion of compatible solutes that are also known as organic solutes/osmolytes (Handa et al. 2018 a). Compatible solutes are low molecular weight-LMW, non-toxic, greatly soluble electroneutral compounds that have a major role in turgor pressure maintenance as well as build up gradient for water, regulation of photosynthesis and respiration, redox equilibrium and cellular homeostasis. Usually, they defend the plants from abiotic stresses through different strategies like maintaining membrane integrity, enzymes/proteins stabilization, ROS detoxification and contribution toward cellular osmotic adjustment (Nishizawa et al. 2008; Peshev et al. 2013; Mostofa et al. 2015; Handa et al. 2018 a). Moreover, many of these can also shield cellular constituents from injury caused during dehydration, therefore, they are generally mentioned as osmoprotectants. Generally, organic solutes include sugar molecules (trehalose, sucrose, fructose and polysaccharides), amino acids-AAs (proline, histidine, etc.), glycine betaine-GB, polyamines-PAs and sugar alcohols (polyols) (Fariduddin et al. 2013 a; Slama et al. 2015; Handa et al. 2018 a).

Recently, at the molecular level, biotechnological approaches emerge as a considerable tool against abiotic stress tolerance (especially HM induced stress) in plants. The transgenic strategy has emerged as an important means to acclimate crops under hastily altering environmental conditions (Chater et al. 2019). Improvement of crop plants that are tolerant of environmental constraints is considered a promising method which can fulfill the growing food demands of the under-developed and developing nations. Use of transgenic plants, having improved metal tolerance, is one of the most commonly used strategies to introduce and upregulate the genes intricate in the uptake, translocation and sequestration of HMs (Koźmińska et al. 2018). Before continuing with transgenic, the crucial step is the recognition of particular genes that serve as key controllers of diverse metabolic pathways including antioxidant (enzymatic and non-enzymatic) defense system, ion homeostasis, osmolyte synthesis and various additional forefront defense pathways. After selecting a particular gene of interest, integration of that genes into the desired species can be done with the help of genome editing tools. Hence, this chapter focuses on updated information about the bioavailability of HMs in soil, their impact on uptake and toxicity and tolerance mechanisms in plants with special reference to the role of organic solutes in metal-induced stress alleviation.

3.2   BIOAVAILABILITY OF HMS IN SOIL

Soil plays an essential role in supporting primary productivity, modulating nutrient flow and promoting the microbial biomass in the underground system (Xian et al. 2015). The excessive accretion of HMs in agrarian soils has led to increased HM uptake which affects the quality as well as productivity of the crop plants and may likely be dangerous to humans by arriving in the food chain. The total concentration of HM in the soil doesn't deliver any precise information about the metal availability and agility, therefore, it remains compulsory to help in finding the specific bio-availability, mobility as well as reactivity of HMs and how plants or other soil microflora uptake these metal ions. The bioavailability of the HM can be demarcated as the total amount of HMs present in a soil that can likely be taken up by plant roots. Factors that affect the bioavailability and accumulation of HMs in the soil are physical and chemical properties of the soils, presence of microbial biomass, plant genotypes and their root exudates (Oves et al. 2016). Physical factors include soil structure and penetrability while chemical characteristics are soil pH, particle-size distribution, SOM-soil organic matter content, CEC-cation exchange capacity and redox potential (Qian et al. 1996). The mobility and bioavailability of HMs increase inversely with soil's pH and decreases inversely with SOM contents as well as CEC. Besides this, soil proteins including functional and detrital proteins are known to play vital roles in the persistence of SOM contents (Rillig et al. 2007).

Soil microbial diversity can act as a vital pool of plant nutrients and is frequently highly interrelated with the SOM contents. Metal availability and transport in the rhizosphere are influenced by microbial diversity such as AMF-arbuscular mycorrhizal fungi, soil bacteria, plant growth-promoting bacteria (PGPB), etc., which are recommended as biological indicators of soil quality. These microbes can modify the availability of HMs by the release of oxygen, protons and organic compounds such as LMW organic acids, enzymes and carbohydrates (Patel et al. 2008). Availability of HM alters the activities of soil enzymes (arylsulfatase, acid and alkaline phosphatase, dehydrogenase, β-glucosidase, cellulase, invertase, protease and urease) in polluted sites. These soil enzymes are needed for the cycling of nutrients (C, P, S and N) in the soils (Xian et al. 2015). The diverse types and amount of root exudates like organic acids, e.g. citric acids, oxalic acid, malic acid, succinic acid, under metal contaminated soil, can alter the rhizospheric environment by improving soil enzyme activity, nutrient availability and microbial community structure (Dotaniya and Meena 2015). These root exudates help in the formation of stable metal chelates by complexing with metal ions present in the soil, and alter their mobility and bioavailability, thus, prevent their entry into the neighboring plant and can reduce the negative impact of HMs on plants (Zhang et al. 2019).

3.3   PATHWAY OF HM UPTAKE IN PLANTS

Plants absorb the majority of HMs accessible in the soil system principally via the root system. HMs are co-transported in the roots through the plasma membrane (PM) via various transporter familys, having an extensive range of substrate specificity. In addition to root uptake, the aerial tissues of plants like leaves, fruits and flowers can also take up HMs, implying that plant aerial parts are also efficacious absorbing structures for metal uptake (Bondada et al. 2004). Below, we will discuss the pathway for HM uptake inside the plant cell.

3.3.1 From Soil

In plants, HMs are taken up from the soil, through the root's cortical tissues because of their resemblance with some essential micronutrients, and reach the xylem tissue by adopting a symplastic or apoplastic pathway (Oves et al. 2016). Passive uptake of HMs (e.g. Cu, Mn, Fe, Zn and Ni) in plant roots occur through the electrochemical gradients or via ion channels that are intricated in the transport of specific essential metal ions. The movement of HM ions toward the root surface relies on the diffusion of ions along the concentration gradient, root interventions where soil volume is dispelled by volume of the root because of root growth and bulk transport from soil solution along the water potential gradient. The majority of HMs are taken up commonly in their cationic forms; the exception is Mo, which is taken up as molybdate anion. After reaching the apoplast, metals are further transported through the PM into the cytoplasm.

Plants have various families or classes of PM transporters which are significantly involved in the uptake of HMs. Such transporters include heavy metal P1B-ATPase (CPx-type), the NRAMP-natural resistance-associated macrophage proteins, the ZIP families-Zinc/Iron Regulated Transporter and the CDF-cation diffusion facilitator (Clemens and Ma 2016). Some of the HM ions are highly moveable and have analogous transporter such as other essential elements. Cd is readily taken up via natural resistance-associated macrophage proteins (NRAMP) (Sasaki et al. 2012) and Zinc Regulated Transporter/Iron Regulated Transporter (ZIP/IRT) family of the transporters (Wu et al. 2016) which are especially for the uptake of Fe, Ca, Zn and Mn (Clemens and Ma 2016). Arsenic (As) is generally accessible to the plants in two forms (arsenate-AsV and arsenite-AsIII), where As V enters through phosphorous (P) transporters (Pht1) (Wu et al. 2011) while As III enters via aquaporin channels (Clemens and Ma 2016). Metal ions such as Zn, Cu, Pb and Cd can be transported across the cell membrane through the heavy metal P-type ATPases, which is also known as CPx -type ATPase, by consuming ATP (Eren and Arguello 2004). Hg can enter into the plant cells by contending with other metals like Cd, Fe, Zn and Cu. Chromium (Cr) exists in two stable forms (Cr VI and Cr III). Cr VI enters into the plants via active transport through sulphur transporters (SULTR) while Cr III enters passively (Oliveira 2012). The copper transporter family (CTR) displays a high affinity to Cu ions, located on PM of various organisms, including plants, and plays an important role in Cu transport (Puig and Thiele 2002). Ca^{2+}/Cation Antiporter (CaCA) transporter family makes use of Ca^{2+} via effluxing the Ca^{2+} across PM against the concentration gradient with the use of K^+, H^+ or Na^+ as counter ions (Emery et al. 2012). After crossing through all epidermal (root surface and root cortex) and apoplasmic barriers (Casparian strips) of roots, HM ions enter the symplastic pathway and are then transported to the stele as well as xylem tissues. After entering the xylem, heavy metals are transported to all the plant parts. Also, HMs can translocate from source to sink via phloem tissue through the heavy metal ATPases (HMA) family of transporters, the Multidrug and Toxic Compound Extrusion or Multi-Antimicrobial Extrusion (MATE) family of efflux proteins and the oligopeptide transporters family (Manara 2012).

3.3.2 Through Foliar Deposition

Contamination of the environment through metal-enriched particulate matter is becoming a major concern for plants and human health (Shahid et al. 2019). The World Health Organization has recognized the health risk of HMs in atmospheric air (WHO 2007). Plants growing near smelting, mining and metropolitan zones exhibit enhanced foliar concentrations of HMs (Xiong et al. 2014). Various HMs (As, Cd, Pb, Cu, Cr, Sb, Sn, etc.) discharged from industrial zones are transported up to a distance of a few miles away from the sources by dry and wet deposition. Through the air, particulate matter deposits on the leaf surfaces of plants which causes accretion of HMs inside the leaves by foliar transfer. Apart from root metal uptake, very few reports are available on HM uptake through plant foliar tissue from the atmosphere (Shahid et al. 2014 a). HM uptake by foliar tissues occurs via cuticular cracks, stomatal aperture, lenticels, aqueous pores and ectodesmata (Fernández et al. 2013). Uptake of HMs through foliar tissue is observed as a surface phenomenon, however, adaxial features of cuticular surface are key constituents in assisting high HM uptake (Bondada et al. 2004). However, foliar uptake of HM critically depends on various factors including the physicochemical features of the cuticle and adsorbed HMs, exposure time, the morphology, surface area and texture (pubescence and roughness) of the leaves, gas exchangeability, plant habitus (evergreen or deciduous) and environmental situations. Particulate matter adsorbed on the external surface of plant leaves is majorly retained by trichomes as well as cuticular waxes, however, some of the metal contents can enter inside the leaf tissue. There are many reports which indicate that plants growing near industrial sites show several-fold increases in HM concentration in leaves as well as inside the plants. A direct relation has been reported in *Pteris vittata* between foliar-applied As and its uptake by the fronds (Bondada et al. 2004; Shahid et al. 2018). After foliar uptake, plants can translocate metal ions towards the root tissues. In *Shorea robusta*, foliar transfer of some HMs (Cd, Pb, Cr, Fe, Cu, Zn, Mn and Ni) has been reported when metals were present in the form of suspended particulate matter in the air (Karmakar and Padhy 2019). Xiong et al. (2014) reported that when cabbage (*Brassica oleracea*) and spinach (*Spinacia oleracea*) plants were exposed to exogenous particulate matters or mono-metallic oxide particles containing (Cd, Sb, Zn, Pb), both the plants displayed high quantities of metals in the plant leaves. Schreck et al. (2012) reported enhanced foliar uptake of various metal ions (As, Cd, Cu, Sn, Sb, Zn and especially Pb) in *Lactuca sativa*, *Lolium perenne* and *Petroselinum crispum* grown near a battery-recycling factory.

Heavy metals accumulation within the plant cells (from soil or through foliar uptake) leads to alteration in various metabolic processes which ultimately affect the plant growth and yield. To counteract the negative impact of HMs, plants have established a network of various defense strategies including synthesis of diverse cellular biomolecules (phytochelatins-PCs, glutathione, metallothioneins-MTs), production of antioxidants and especially the accumulation of organic solutes which play major roles in maintaining osmotic balance or homeostasis (Figure 3.1).

3.4   ROLE OF ORGANIC SOLUTES IN HM DETOXIFICATION IN PLANTS

In plants, HM toxicity induces oxidative burden and upregulation of stress-related proteins. Toxic effects of HMs lead to chlorosis, decrease in biomass production as well as growth inhibition, metal accumulation, altered water as well as mineral balance, which ultimately cause senescence and plant death (Wani et al. 2018). In the last few years, efforts have been made by various researchers to study the physiological, biochemical and molecular processes which are involved in HM uptake and translocation from the soil to different plant parts. Plants adopt diverse approaches to overcome metal-induced toxicity, e.g. upregulation of numerous tolerance mechanisms including immobilization, chelation, sequestration (mainly in vacuole), exclusion and the expression of stress-inducible proteins. Vacuolar sequestration of HMs takes place with the help of phytochelatins, i.e. in the form of PCs-HM complex, which is one of the main processes of metal sequestration (Anjum et al. 2015). Moreover, metal transporters on PM and vacuolar membrane (tonoplast) are considerably involved in the HM uptake and homeostasis. After reaching inside the cells, HMs sequestration in the vacuole takes place by tonoplast transporters including cation diffusion facilitator (CDF) transporters, ATP-binding cassette transporters (ABC transporters), CaCA Transporters, HMA Transporters and NRAMP Transporters (Manara 2012).

Besides this, to overcome or tolerate metal ions-induced toxicity, biosynthesis of different cellular biomolecules is also one of the well-known strategies. This includes the generation of LMW protein metallo-chaperones or chelators such as organic acids, nicotianamine, mugineic acids, metallothioneins or cellular exudes such as flavonoid as well as phenolic compounds, heat shock proteins (HSPs) and phytohormones like ethylene, jasmonic acid and salicylic acid (Emamverdian

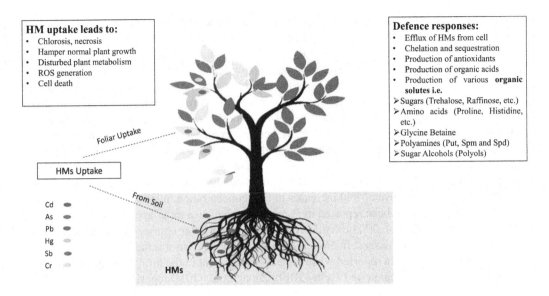

HM uptake leads to:
- Chlorosis, necrosis
- Hamper normal plant growth
- Disturbed plant metabolism
- ROS generation
- Cell death

Foliar Uptake

HMs Uptake

From Soil

Cd
As
Pb
Hg
Sb
Cr

HMs

Defence responses:
- Efflux of HMs from cell
- Chelation and sequestration
- Production of antioxidants
- Production of organic acids
- Production of various **organic solutes** i.e.
  - ➤ Sugars (Trehalose, Raffinose, etc.)
  - ➤ Amino acids (Proline, Histidine, etc.)
  - ➤ Glycine Betaine
  - ➤ Polyamines (Put, Spm and Spd)
  - ➤ Sugar Alcohols (Polyols)

FIGURE 3.1 Diagrammatic presentation of heavy metal (HM) uptake, toxicity and defense responses in plant.

et al. 2015). Moreover, the toxic effects of metals are countered by non-enzymatic (glutathione reductase-GR, ascorbate-AsA, carotenoids, tocopherol, phenolic compounds, etc.) as well as enzymatic (catalase-CAT, ascorbate peroxidase-APOX, superoxide dismutase-SOD, etc.) antioxidants.

In addition to the non-enzymatic and enzymatic defense mechanism, one of the strategies to combat metal-induced stresses is the accumulation and upregulation of osmolytes. Organic solutes/compatible (solutes/osmoprotectants) are small, LMW compounds, which accumulate at a concentration of 5–50 μmolg^{-1} fresh weight in plants (Handa et al. 2018 a). Enhanced biosynthesis and accretion of osmolytes in plants exposed to various stresses, is one of the major processes that evolved to sustain their cellular integrity and give genuineness of endurance (Wani et al. 2013). Organic solutes are extremely soluble and do not obstruct other normal metabolic processes because they are non-toxic even when accumulating at higher concentrations in cells. They are capable of forming hydrophilic complexes, therefore, help in reducing the effects of stress inducers (Slama et al. 2015). The major organic solutes present in plants are sugars (trehalose, raffinose, fructans, etc.), amino acids (proline, histidine, etc.), glycine betaine-GB, polyamines (spermidine, spermine and putrescine) and sugar alcohol (polyols) (Handa et al. 2018 a). Their primary function is to protect the plants from abiotic stresses via various mechanisms like maintaining vital processes (photosynthesis and respiration) and redox environment in the cells, adjusting cellular homeostasis, detoxification of ROS, stabilization of proteins or enzymes and by upkeeping membrane integrity (Hasanuzzaman et al. 2019). In the following section, the roles of various organic solutes in plants have been discussed for HM stress alleviation (Table 3.1).

3.4.1 Sugars

Plants are autotrophic photosynthetic organisms that can synthesize as well as consume sugar in diverse forms. Besides their imperious role in sustaining growth in plants, they also contribute to stress management. Sugar signaling causes down or up-regulation of multiple genes in plants which are particularly involved in photosynthesis, oxidative phosphorylation and sugar metabolism (Slama et al. 2015). Non-reducing sugars-NRS (raffinose, sucrose and trehalose), reducing sugars like fructans and polysaccharides are major classes of soluble sugars present in plants and are broadly described for their participation in stress tolerance (Peshev et al. 2013). These sugar molecules

TABLE 3.1

Role of Diverse Organic Solute in Heavy Metal Induced Stress Alleviation in Various Plant Species

S. No.	Organic Solutes	Heavy Metals	Plant Species	References
1.	Sugars	Cd	*Arabidopsis thaliana*	Sun et al. (2010)Abu-Muriefah (2015)Garg
		Pb	*Vicia faba*	and Singh (2018 b)Duman et al. (2011)Saha
		Cu	*Cajanus cajan*	et al. (2016)Aly and Mohamed (2012)Bali et
		Cr	*Lemna gibba* L.	al. (2016)Handa et al. (2018 b)Garg and
		As	*Cicer arietinum*	Singla (2012)
			Zea mays	
			Brassica juncea	
			Brassica juncea	
			Pisum sativum	
2.	**Amino acid**	Se	*Phaseolus vulgaris* L.	Aggarwal et al. (2011)Dai et al. (2017)Zouari
	Proline	Cd	*Avicennia marina, Kandelia obovate*	et al. (2016 a, b)Siddique and Dubey
		Cu	*Olea europaea* L.	(2017)Garg and Chandel (2012)Garg and
		Zn	*Triticum aestivum* L.	Singh (2018 a)Rasheed et al. (2014)Singh et
		Ni	*Caja`nus cajan* L. Milllsp	al. (2016)Noreen et al. (2018)Fariduddin et
		Co	*Cajanus cajan* L. Milllsp	al. (2013 a)Wen et al. (2013)Li et al. (2013
		Pb	*Triticum aestivum* L.	a, b)
		Al	*Triticum aestivum* L.	Kotapati et al. (2017)Balal et al.
		Cr	*Triticum aestivum* L.	(2016)Mishra and Dubey (2013)Rizwan et
		As	*Cucumis sativus* L.	al. (2017)Saif and Khan (2018)Imtiyaz et al.
		Hg	*Zea mays*	(2014)Imtiyaz et al. (2014)Amooaghaie and
			Triticum aestivum L.	Enteshari (2017)Gohari et al. (2012)Ashraf
			Eleusine coracana L.	and Tang (2017)Surapu et al. (2014)Yu et al.
			Lolium perenne	(2017)Ruscitti et al. (2011)Garg and Singla
			Oryza sativa	(2012)Wang et al. (2009)
			Cicer arietinum	
			Glycine max L.	
			Glycine max L.	
			Sesamum indicum	
			Brassica napus L.	
			Oryza sativa	
			Lycopersicum esculentum	
			Oryza sativa L.	
			Capsicum annuum L.	
			Pisum sativum	
			Oryza sativa	
	Histidine	Ni	*Thlaspi caerulescens*	Richau et al. (2009)Mozafari et al. (2016)Salt
		Zn	*Solanum lycopersicum*	et al. (1999)
			Thlaspi caerulescens	
	Other amino acids	As	*Oryza sativa* L.	Dixit et al. (2015 b)Sun et al.
		Cd	*Arabidopsis thaliana*	(2010)Zemanová et al. (2017)Zhu et al.
		Cr	*Noccaea caerulescens*	(2018)Dubey et al. (2010)Bhatia et al.
		Ni	*Noccaea praecox*	(2005)Zafari et al. (2016)
		Pb	*Crassocephalum crepidioides*	
			Oryza sativa L.	
			Stackhousia tryonii	
			Prosopis farcta	

(Continued)

TABLE 3.1 (CONTINUED)

Role of Diverse Organic Solute in Heavy Metal Induced Stress Alleviation in Various Plant Species

S. No.	Organic Solutes	Heavy Metals	Plant Species	References
3.	Glycine Betaine	Cd As Zn, Cd Al	*Spinacia oleracea* *Gossypium sp.* *Oryza sativa* *Lemna gibba* L. *Triticum aestivum* L. *Vigna radiata* *Pisum sativum* *Cajanus cajan* *Oryza sativa*	Aamer et al. (2018)Farooq et al. (2016)Cao et al. (2013)Duman et al. (2011)Rasheed et al. (2014)Hossain et al. (2010)Garg and Singla (2012)Kaur and Garg (2017)Sharma and Dubey (2005)
4.	Polyamines Putrescine	Cd Cr Cu Hg Ni	*Vigna radiata* *Triticum aestivum* L. *Potamogeton crispus* *Chenopodium quinoa* *Actinidia deliciosa* *Raphanus sativus* *Raphanus sativus* *Raphanus sativus* *Sagittaria sagittifolia* *Nymphoides peltatum* *Hydrocharis dubia* *Cajanus cajan*	Nahar et al. (2016 a)Rady and Hemida (2015)Yang et al. (2010)Scoccianti et al. (2016)Scoccianti et al. (2013)Choudhary et al. (2012 a)Choudhary et al. (2012 b)Choudhary et al. (2010)Xu et al. (2011)Wang et al. (2004)Zhao et al. (2008)Garg and Saroy (2020)
	Spermidine	Cd Cr Cu Hg Ni	*Vigna radiata* *Hydrocharis dubia* *Raphanus sativus* *Helianthus annuus* *Eichornia crassipes* *Cajanus cajan*	Nahar et al. (2016 a)Yang et al. (2013)Choudhary et al. (2010)Groppa et al. (2008)Ding et al. (2010)Garg and Saroy (2020)
	Spermine	Cd Cu Ni	*Vigna radiata* *Hydrocharis dubia* *Helianthus annus* *Oryza sativa* *Raphanus sativus* *Sagittaria sagittifolia* *Cajanus cajan*	Nahar et al. (2016 a, b)Yang et al. (2013)Groppa et al. (2008)Roychoudhury et al. (2012)Choudhary et al. (2010) Xu et al. (2011) Garg and Saroy (2020)

can act as antioxidants and osmoregulators that play the main role in ROS detoxification and stabilization of cell membrane as well as turgor pressure maintenance respectively. Enhanced sugars accessibility at endogenous level results in corroboration of oxidative pentose phosphate pathway which subsequently leads to enhanced scavenging of ROS. Under oxidative burden, these sugar molecules have been observed to scavenge OH$^-$ radicals better than O_2^-, therefore, are reflected as critical components of the plant defense responses (Bolouri-Moghaddam et al. 2010). Sugar has also been reported to refurbish the amount of nicotinamide adenine dinucleotide phosphate hydrogen (NADPH) which is obligatory as a stimulus of many enzymes such as monodehydroascorbate reductase-MDHAR and glutathione reductase-GR (Nishikawa et al. 2005). Moreover, under stressful conditions, sugar accretion results in increased efficiency of respiration along with boosting ascorbate synthesis (Nishikawa et al. 2005). Oligosaccharides have been reported to elicit antioxidative defense in roots of *Medicago sativa*, suggested to be the active contribution of sugar signaling in plant tolerance (Camejo et al. 2012). In a HM polluted habitat, roots of *Populus nigra* L. have displayed the altered activity of sugar synthesis and alteration in oxidative pentose phosphate pathway and glycolysis cycle (Stobrawa and Lorenc-Plucinska 2007). Trehalose has been known to stabilize proteins and membranes under stress conditions and participates in ROS scavenging (Fernandez et al. 2010). Trehalose has been reported to scavenge superoxide anion and hydrogen peroxide in a concentration-dependent manner (Morgutti et al. 2019). Mostofa et al. (2015) reported that exogenous application of trehalose to Cu treated rice seedling displayed a reduction in Cu-induced oxidative harm with lowered malondialdehyde (MDA) content as well as ROS concentration. However, significant improvement in ascorbate (AsA) content and activities of antioxidant enzymes including CAT, guiacol peroxidase (GPOX), Glutathione S-transferase (GST) and SOD were also reported by them. Garg and Singh (2018b) reported that under Zn and Cd toxicity, increased tolerance was observed in *Cajanus cajan* because of enhanced trehalose turnover and phytochelatin production which reinforced the antioxidant defense. In addition, galactinol and raffinose family of oligosaccharides (RFOs) have also been reported to increase the defense responses against oxidative stress in plants (Kumar et al. 2018). The oxidized RFOs resulted in the production of AsA as well as some other antioxidant-like flavonoids. Nishizawa et al. (2008) reported that the level of RFOs was enhanced under elevated oxidative stressful conditions which resulted in lowering the peroxidation of lipid in membranes of wild type *Arabidopsis* and provides tolerance. Various antioxidant defense mechanisms have been elicited by fructans, therefore, they can act as strong ROS scavengers (Hasanuzzaman et al. 2019). Besides this, fructans have altered the activity of non-enzymatic antioxidants, i.e. AsA and GSH, which help in stress tolerance (Rigui et al. 2019). Guangqiu et al. (2007) reported that mangrove (*Aegiceras corniculatum*) plants under the influence of different HMs have displayed decreased fresh biomass as well as starch content and increased soluble sugars, total tannin and phenolic content. Under metal ions stress (Co, Fe, Mn), enhanced production of different metabolites including carbohydrate have been reported by Jahangir et al. (2008).

3.4.2 Amino Acids (AAs)

Amino acids are also effective compatible/organic solutes that behave as a potent oxidative protectant. They maintain homeostasis in the cell by equilibrating the ratio of NADPH/NADP$^+$ therefore, regulate the diverse vital processes inside the plant cells like respiration and photosynthesis (Slama et al. 2015). Some of the AAs which can act as osmolytes are proline, histidine, alanine, glycine, etc. and their role under metal stress has been discussed below.

3.4.2.1 Proline-Pro

This is a proteinogenic 5C (five-carbon) α-amino or so-called imino acid that can act as a compatible as well as a metabolic organic solute, free radical scavenger (antioxidant) and macromolecules stabilizer (Szepesi and Szőllősi 2018). In higher plants, under various abiotic stress including HM stress, the production of Pro increases, which is an emblematic non-enzymatic response. Also, Pro plays diverse roles in plants to combat various abiotic stresses via adaptation, recovery and signaling (Fidalgo et al. 2013). Different researchers have observed numerous mechanisms by which Pro boosts the resistance against HMs in plants. Clemens (2006) investigated that HM-induced Pro accretion in plants is not directly emanated from HM stress, but due to water imbalance, which happens as a result of excessive metal uptake. Detoxification of ROS (i.e. quenching singlet oxygen and hydroxyl radicals) has been done by overproduction of Pro during HM toxicity. Also, enhancement in antioxidant enzyme activities and their protection, protein stabilization, regulation of intracellular pH as well as cellular homeostasis, chelation of HMs and biosynthesis of chlorophyll are associated with the activity of Pro (Rastgoo et al. 2011). The organ-specific amassing of Pro where roots have more Pro than shoots has been reported in various plant species including hybrid poplar

(*Populus trichocarpa* × *P. deltoides*) under Cd toxicity (Nikolic et al. 2008), *Cymbopogon flexuosus* exposed to Hg and Cd stress (Handique and Handique 2009), *Brassica juncea* L. subjected to Pb and Cd stress (John et al. 2012) and *Solanum nigrum* to Cu toxicity (Fidalgo et al. 2013). In contrast to the above-mentioned reports, some studies observed that Pro tends to gather more in shoots than roots in HM-stressed plants (Singh et al. 2016; Saif and Khan 2018; Zhou et al. 2019). The increased concentration of Pro was reported under Cd, Hg and Zn treated *Brassica oleracea* seedlings (Theriappan et al. 2011), Cd, Pb and As exposed *Shorea robusta* seedlings (Pant et al. 2011), Cu and Zn subjected wheat seedlings (Vinod et al. 2012), Cd, Co, Pb and Ag stressed *Aeluropus littoralis* (Rastgoo et al. 2011), which indicated that HMs are stronger evoker of proline in shoots. The above-mentioned studies indicated that the ability of a particular HM to induce Pro accumulation in plants depends upon the plant species, their toxicity threshold, concentration and specificity of HMs. To overcome the negative impacts of HMs on the plants, exogenous foliar application of Pro is an effective strategy to decrease the toxicity of metals. Exogenously applied Pro on Cd stressed chickpea (Hayat et al. 2013), Ni treated pea (Shahid et al. 2014 b) and Hg exposed rice (Wang et al. 2009) displayed improved stress tolerance by reducing the effects of HMs through acting as a compatible solute. For HM detoxification, exogenously applied Pro plays a major role, therefore, more attention should be paid to this practice to cope with HM induced toxicity. Furthermore, seeds priming with Pro for better tolerance to the plants under metal-induced toxicity has increased the merit of investigation (Emamverdian et al. 2015).

3.4.2.2 Histidine-His

His is one of the proteinogenic α-amino acids which can be *de novo* synthesized by plants and has a great affinity for metal binding when present as free AA as well as metal-coordinating residues in proteins (Kulis-Horn et al. 2014). The Ni-hyperaccumulation peculiarity of *Alyssum* sp. (Brassicaceae) has been confirmed because of its especially linked capability for free His production. Ni-hyperaccumulator *Thlaspi goesingense* and *A. lesbiacum* have shown a large and proportionate increase in His concentration under Ni toxicity than non-accumulators i.e. *T. arvense* and *Brassica juncea* respectively (Persans et al. 1999; Kerkeb and Krämer 2003). Besides Ni, in the roots of *T. caerulescens*, Zn-His complexes have been reported which provided Zn tolerance to given plant species (Salt et al. 1999). Kuroda et al. (2001) reported that in contrast to the parental strain, mutant yeast strain which was cell surface-engineered for His displayed enhanced His oligopeptide (hexa-His) that adsorbed three to eight times more Cu ions as well as being more resistant to Cu. Zemanova et al. (2014) reported that His played an important role under Cd stress alleviation in *Noccaea caerulescens* and *Arabidopsis halleri*. Under Ni toxicity, histidine in combination with calcium alleviated the Ni stress by inhibiting its uptake and translocation in tomato plants (Mozafari et al. 2016). Exogenous application of histidine, citrate and malate in *Mesembryanthemum crystallinum* and *Brassica juncea* enhanced Ni-chelation and sequestration in different organs, thus, provided metal tolerance (Amari et al. 2016).

3.4.2.3 Other Amino Acids

When plants are exposed to different HMs, not only proline and histidine but also some other amino acids synthesized within the plants which play a major role in stress alleviation. Arsenic exposed rice (*Oryza sativa*) plants displayed increased concentration of essential and non-essential AAs namely valine, methionine, leucine, alanine, his, pro, glutamate, cystine (cys) and γ-glutamyl cys (Tripathi et al. 2013 b; Dixit et al. 2015 a). Zemanova et al. (2014) investigated the role of methionine and tryptophan under Cd stressed *Noccaea caerulescens* and *Arabidopsis halleri*, and reported their importance in metal stress amelioration. In *Stackhousia tryonii* i.e. one of the Ni hyperaccumulators, under Ni toxicity, alteration in AA contents of xylem sap has been reported (Bhatia et al. 2005). Cys is essential for met and GSH/ PC synthesis in plants, therefore, it acts as a major metabolite for HM homeostasis as well in antioxidant defense. *Arabidopsis* plants exposed to Cd displayed enhanced levels of AAs (alanine, β-alanine, proline and serine) as well as of other metabolites (Domínguez-Solís et al. 2004). Zhu et al. (2018) reported that under Cd toxicity, higher tolerance in *Crassocephalum crepidioides* was related to greater accretion of amino acids, especially for glutamine and asparagine. Alteration in amino acid content has been reported under Pb toxicity in *Prosopis farcta* (Zafari et al. 2016). The alterations in amino acids concentrations under metal stress concluded that they play a very important role in stress alleviation.

3.4.3 GLYCINE BETAINE-GB

GB belongs to the class of organic compounds which have great solubility in water and are non-harmful at elevated levels or concentrations for the plants (Fariduddin et al. 2013b). The synthesis of GB induced during various stresses and *in vivo* its concentration varies from 40 to 400 μ mol g^{-1} dry weight depending upon plant species. Synthesis of GB involves two different pathways and including two substrates i.e. choline and glycine. In higher plants, synthesis of GB majorly takes place in the chloroplast, and the reactions are catalyzed by NAD$^+$-dependent betaine aldehyde dehydrogenase (BADH) and choline monooxygenase (CMO) enzymes, both of which are present in the stroma of chloroplasts (Allakhverdieva et al. 2001). GB has an important role in antagonizing the effects of diverse type of stresses by providing osmotic adjustments between environment and plant cells, protecting thylakoid membrane system to regulate various photosynthetic attributes and proteins stabilization, protecting various sub-cellular structures and is also involved in the refolding of enzymes by behaving as a molecular chaperone (Wani et al. 2013). In addition to endogenous GB, exogenous application of the same also provides stress tolerance against various metal ions. *Triticum aestivum* plants when exogenously treated with GB under HM stress, have shown enhancement of ROS scavenging, upregulation of cellular homeostasis and provision of integrity and stability for biomembranes (Ali et al. 2015). Stepien et al. (2016) reported reduced toxicity of aluminum (Al) through exogenous foliar application of GB on the leaves of *Cucumis sativus* by regulating ETC-electron transport chain and photosynthesis. Bharwana et al. (2014) reported that in *Gossypium* plants, the contrary effects of Pb were overcome by the exogenous application of GB. GB improved plant growth by lowering the levels of MDA and H$_2$O$_2$ and membrane permeability and enhanced the production of antioxidant enzymes. In addition, Jabeen et al. (2016) have reported that GB alleviated the Cr toxicity in mung bean (*Vigna radiata*) by decreasing the uptake of Cr and upregulating the antioxidative system. Varshney (2011) reported that in Cu-stressed *Brassica juncea* plants, the negative impact of oxidative stress was counteracted through exogenous application of GB which led to the elevated level of Pro and antioxidant enzymes (POX, CAT and SOD), and ultimately improved plant growth and yield. Vezza et al. (2018) reported that under As toxicity, *Glycine max* L. displayed enhanced accretion of osmolytes including GB.

3.4.4 POLYAMINES-PAs

Polyamines are LMW, positively charged, omnipresent organic molecules having a high tendency to bind with negatively charged molecules such as proteins, RNA, DNA and biomembrane, hence, play a vital role in the stabilization of membranes and functions via altering the gene expression (Pathak et al. 2014). In plants, there are three major types of PAs, i.e putrescine- Put (di-amine) which is the precursor for two other PAs, spermine-Spm (tetra-amine) and spermidine-Spd (tri-amine). Besides this, cadaverine is one of the PAs that has been reported in various plant species. PAs can regulate various metabolic processes in plants including scavenging of free radicals as well as of ROS (as an antioxidant) and are, therefore, suggested by many researchers to be a class of plant growth regulator (like phytohormone) and secondary messengers. The physiological consequences of PAs in plants are specified by their contribution to several stress responses (Pál et al. 2017). To overcome the phytotoxic effects of HMs, PAs exogenous application has been testified as another approach for sustainable agricultural management. Exogenous application of PAs has alleviated Cd, Cu and Ni induced toxicity in *Malus hupehensis*, *Nymphoides peltatum* and *Cajanus cajan* respectively

(Wang et al. 2007; Zhao and Yang 2008; Garg and Saroy 2020). Many studies also recommended the defensive role of PAs (Put, Spd and Spm) under Cd stress in various plant species including *Boehmeria nivea*, *Hydrocharis dubia* and *Triticum aestivum* (Yang et al. 2013; Rady and Hemida 2015; Gong et al. 2016).

3.4.5 POLYHYDROXYLATED SUGAR ALCOHOLS-POLYOLS

Polyols are the molecules that are a reduced form of keto or aldo sugars/carbohydrates which can be of two types (acyclic or cyclic). The acyclic sugar form of sugar alcohols is also called glycitols or alditols or true polyols, and in higher plants mannitol, galactitol and sorbitol are found most commonly (Dutta et al. 2019). Cyclic polyols comprise of myo-inositol (MI) and the byproducts are galactinol, ononitol and pinitol. Commonly, a single type of sugar alcohol presents in a plant species, and it is also family-specific (Williamson et al. 2002). Their accretion in plants takes place during stressed environmental circumstances specifically under osmotic stress. They play various roles like maintenance of cellular osmostasis, have high affinity for membranes, enzymes or protein complexes and protect them from oxidative stress by scavenging ROS. HM induces the accumulation of polyols in plants by altering sugar metabolism. Various polyols have been found to accumulate under HM induced stresses e.g. mannitol accumulation under metal stressed (Co, Cd, Ni and Cr) *Salvinia* plants (Dhir et al. 2012), sorbitol and mannitol in Cd exposed *Brassica juncea* (brown mustard) (Bali et al. 2016), galactinol and inositol in Cd stressed *Populus tremula* (poplar plants) (Kieffer et al. 2009) and various sugar alcohols in Pb treated *Chrysopogon zizanioides* (Pidatala et al. 2016), revealing the involvement of polyols in stress responses. Keunen et al. (2013) have reported that in *Nicotiana tobacum*, mannitol possessed strong antioxidant activity, therefore, helps in ROS scavenging as well as protecting the functions of various antioxidants (thioredoxins, GSH and ferredoxin). In addition, Cuin and Shabala (2007) found that mannitol, myo-inositol and other polyols have strong antioxidative features that contributed toward oxidative stress management in roots of *Arabidopsis* by reducing the hydroxyl radical and stimulating the efflux of K^+. Exogenous foliar spraying of mannitol to wheat seedlings and maize plants displayed enhanced growth as well as biomass under Cr toxicity (Adrees et al. 2015; Habiba et al. 2019). The higher concentration of polyols under diverse environmental circumstances (including HM toxicity) scavenge ROS more efficiently and more precisely, therefore, they can behave as effective antioxidants against oxidative stress.

3.5 BIOENGINEERING OF PLANTS FOR OSMOLYTE SYNTHESIS

In plant biology, to improve plant tolerance toward abiotic stresses including HM stress, the transgenic approach has arisen as an essential tool to acclimatize crops under rapidly altering environmental circumstances for identification and good understanding of numerous key steps involved at the molecular level (Dixit et al. 2015 b). Increased tolerance in plants to HM induced stresses by overexpressing metal-binding peptides-PCs and MTs genes has been broadly reported (Shukla et al. 2013). Also, various scientists have underlined the genetic engineering efforts for improved osmolyte synthesis and avoidance of stress damage (Wani et al. 2013). Progress in biological tools (both cellular as well as molecular) has made it conceivable to clone essential biosynthetic genes, including those for osmolyte synthesis, and transfer them into plants. The crucial step before continuing with transgenic is the recognition of a particular and specific gene that serves as a key controller for osmolyte synthesis and in its metabolic pathways. Exogenous application as well as genetically engineered biosynthesis of GB increase the plant tolerance to various abiotic stresses (Vinocur and Altman 2005). In bioengineered crop plants, the upregulation of polyamine biosynthetic genes leads to enhancement in stress tolerance (Wen et al. 2008; Soudek et al. 2016). From many plant species, numerous gene coding for the enzymes required for different steps during PAs synthesis and their metabolism have been discovered and cloned. Under high concentration of HMs (Cu, Cd, Pb and

Zn), transgenic *Pyrus communis*-European pear which has overexpressed SPDS-spermidine synthase gene of apple (*MdSPDS1*), resulted in improved stress tolerance by increasing the endogenous PAs level and displayed the increased activity of antioxidative enzymes in transgenics plants (Wen et al. 2008). Many studies have clearly shown the direct correlation between polyamines and PC metabolism during HM stress (Paul et al. 2018). Wycisk et al. (2004) reported the useful role of high His levels in *Arabidopsis thaliana* transgenic plants which expressed a *Salmonella typhimurium* ATP phosphoribosyltransferase enzyme (*StHisG*). Under Ni toxicity, the transgenic plants amassed about two times higher His levels than wild types which ultimately led to higher biomass production (almost ten-fold). Under favorable conditions, mannitol dehydrogenase-MTD breaks down the mannitol for acclimatization and growth. During a stressful environment and under carbohydrate suppression, the activity of enzyme MTD was lessened which led to heightened accretion of mannitol (Williamson et al. 2002). Shen et al. (1997) reported that *MTD* gene-containing transgenic plants displayed over-accumulation of mannitol, however, accumulated mannitol concentration was low i.e. 5–100 mM, which was not sufficient for osmotic homeostasis but was effective in enhancing stress tolerance, indicating the positive effect of mannitol under HM induced responses and upregulation of antioxidant defense machinery. Inositols (polyol) are cyclic hexitols that perform diverse physiological functions in plants, and among the nine forms of inositol isomers, MI is most common (Saxena et al. 2013). Biosynthesis of MI takes place in two steps. The first step comprises of conversion of D-glucose 6-phosphate to L-myo-inositol 1-phosphate which is catalyzed by L-myo-inositol 1-phosphate synthase-MIPS, i.e. a rate-limiting enzyme. In various plant species, to resist environmental stresses (abiotic and biotic), induction of *MIPS* gene is one of the major factors (Wang et al. 2011; Zhai et al. 2016). Overexpression of transgene *MfMIPS1* from *Medicago falcata* in *Nicotiana tobaccum* resulted in enhanced *MIPS* activity which revealed improved stress tolerance (Tan et al. 2013). Another study in transgenic *Arabidopsis* plants revealed that the constitutive overexpression of gene *GsMIPS2* taken from *Glycine soja* displayed improved salt tolerance (Nisa et al. 2016). Upregulation of *MIPS* gene under As toxicity in *Cicer arietinum*, suggested its role in As stress tolerance (Tripathi et al. 2013 a). GolS is an essential enzyme for biosynthesis of raffinose and galactinol. In *Arabidopsis*, the expression of *GolS* genes is controlled by *HsfA*, i.e. heat shock transcription factor-A (Schramm et al. 2006). Under various abiotic stresses, the level of *HsfA2* increases resulted in the accumulation of galactose and raffinose, which help to overcome the negative impact of stress inducers (Nishizawa et al. 2006). Various researchers observed that both sugars, i.e. raffinose and galactose, exhibited an *in vitro* hydroxyl radical scavenging activity, therefore, suggesting they play a key role under oxidative stress (Nishizawa et al. 2008; Sun et al. 2010; Contreras et al. 2018). Wang et al. (2016) investigated the function of GolS under HM stress and found that exposure of wheat to HMs like Cu and Zn induced expression of *TcGolS3* gene. Under Zn exposure, overexpression of wheat *TaGolS3* gene in transgenic rice and *Arabidopsis* ensued improved growth, proline content, antioxidant enzyme activities and ROS scavenging and reduced membrane permeability as well as lipid peroxidation than wild type plants (Wang et al. 2016). These studies indicated that transgenic plants having an osmolyte synthesizing gene could be used as an effective strategy for abiotic stress tolerance.

3.6 CONCLUSIONS AND FUTURE PROSPECTS

Among the abiotic stresses, contamination of agricultural soil by HMs poses a serious threat to the environment as well as human health. Exposure of plants to hazardous HMs (As, Cd, Cu, Hg, Pb and Al) causes cellular damage via the production of ROS, disrupting physiological, cellular and molecular processes. Plants can overcome this damage by modulating plant inherent responses such as the production of stress-related peptides (phytochelatins and metallothioneins), sequestration of HMs inside the vacuole, production of antioxidants (enzymatic and non-enzymatic) and osmolyte synthesis. In this chapter, we gave updated information about HM stress, their bioavailability and uptake, the various detrimental consequences on plants and alleviation of toxicity with the help of endogenous as well as exogenous osmolyte application. Genetic engineering programs have confirmed the beneficial effects of osmolyte production under stressful environments. Transgenic crops with altered osmolyte synthetic genes have a high tolerance capacity as compared to the non-transgenic ones, indicating that osmolyte production can be more beneficial against HM induced stress. However, further research is required in the field of genetic engineering to identify osmolyte synthesizing genes in various crops in response to HM stresses to understand their role in reducing stress-induced damages in plants.

ACKNOWLEDGMENTS

The authors are grateful to Council of Scientific & Industrial Research (CSIR), New Delhi and Department of Biotechnology (DBT), Ministry of Science and Technology, Govt. of India for financial assistance in carrying out related research.

CONFLICT OF INTEREST

There is no conflict of interest between the authors.

REFERENCES

Aamer M, Muhammad UH, Li Z, Abid A, Su Q, Liu Y, Adnan R, Muhammad AU, Tahir AK, Huang G (2018) Foliar application of glycinebetaine (GB) alleviates the cadmium (Cd) toxicity in spinach through reducing cd uptake and improving the activity of anti-oxidant system. *Appl Ecol Env Res* 16(6):7575–7583

Abdel Latef AA (2013) Growth and some physiological activities of pepper (*Capsicum annuum* L.) in response to cadmium stress and mycorrhizal symbiosis. *J Agri Sci Tech* 15:1437–1448

Abdel Latef AA, Zaid A, Abo-Baker AA, Salem W, Abu Alhmad MF (2020) Mitigation of copper stress in maize by inoculation with *Paenibacillus polymyxa* and *Bacillus circulans*. *Plants* 9(11):1513

Abu-Muriefah SS (2015) Effects of silicon on faba bean (*Vicia faba* L.) plants grown under heavy metal stress conditions. *Afr J Agric Sci Technol* 3:255–268

Adrees M, Ali S, Iqbal M, Aslam Bharwana SA, Siddiqi Z, Farid M, Ali Q, Saeed R, Rizwan M (2015) Mannitol alleviates chromium toxicity in wheat plants in relation to growth, yield, stimulation of antioxidative enzymes, oxidative stress and Cr uptake in sand and soil media. *Ecotoxicol Environ Saf* 122:1–8

Aggarwal M, Sharma S, Kaur N, Pathania D, Bhandhari K, Kaushal N, Kaur R, Singh K, Srivastava A, Nayyar H (2011) Exogenous proline application reduces phytotoxic effects of selenium by minimising oxidative stress and improves growth in bean (*Phaseolus vulgaris* L.) seedlings. *Biol Trace Elem Res* 140(3):354–367

Ali S, Chaudhary A, Rizwan M, Anwar HT, Adrees M, Farid M, Irshad MK, Hayat T, Anjum SA (2015) Alleviation of chromium toxicity by glycine betaine is related to elevated antioxidant enzymes and suppressed chromium uptake and oxidative stress in wheat (*Triticum aestivum* L.). *Environ Sci Pollut Res* 22(14):10669–10678

Allakhverdieva MY, Mamedov DM, Gasanov RA (2001) The effect of glycine betaine on the heat stability of photosynthetic reactions in thylakoid membranes. *Turk J Bot* 25:11–17

Aly AA, Mohamed AA (2012) The impact of copper ion on growth, thiol compounds and lipid peroxidation in two maize cultivars (*Zea mays* L.) grown in vitro. *Aust J Crop Sci* 6:541–549

Amari T, Lutts S, Taamali M, Lucchini G, Sacchi GA, Abdelly C, Ghnaya T (2016) Implication of citrate, malate and histidine in the accumulation and transport of nickel in *Mesembryanthemum crystallinum* and *Brassica juncea*. *Ecotoxicol Environ Saf* 126:122–128

Amooaghaie R, Zangene-Madar F, Enteshari S (2017) Role of two-sided crosstalk between NO and H_2S on improvement of mineral homeostasis and antioxidative defense in *Sesamum indicum* under lead stress. *Ecotoxicol Environ Saf* 139:210–218

Anjum NA, Hasanuzzaman M, Hossain MA, Thangavel P, Roychoudhury A, Gill SS, Rodrigo MA, Adam V, Fujita M, Kizek R, Duarte AC, Pereira E, Ahmad I (2015) Jacks of metal/metalloid chelation trade in plants—an overview. *Front Plant Sci* 6:192. https://doi.org/10.3389/fpls.2015.00192

Ashraf U, Tang X (2017) Yield and quality responses, plant metabolism and metal distribution pattern in aromatic rice under lead (Pb) toxicity. *Chemosphere* 176:141–155

Balal RM, Shahid MA, Javaid MM, Anjum MA, Ali HH, Mattson NS, Garcia-Sanchez F (2016) Foliar treatment with *Lolium perenne* (Poaceae) leaf extract alleviates salinity and nickel-induced growth inhibition in pea. *Braz J Bot* 39(2):453–463

Bali S, Kohli SK, Poonam Kaur H, Bhardwaj R (2016) Improvement in photosynthetic efficiency of *Brassica juncea* under copper stress by plant steroid hormone. *J Chem Pharm Res* 8:464–470

Berni R, Luyckx M, Xu X, Legay S, Sergeant K, Hausman JF, Lutts S, Cai G, Guerriero G (2019) Reactive oxygen species and heavy metal stress in plants: impact on the cell wall and secondary metabolism. *Env Exp Bot* 161: 98–106

Bharwana SA, Ali S, Farooq MA, Iqbal N, Hameed A, Abbas F, Ahmad MSA (2014) Glycine betaine-induced lead toxicity tolerance related to elevated photosynthesis, antioxidant enzymes suppressed lead uptake and oxidative stress in cotton. *Turk J Bot* 38:281–292

Bhatia NP, Nkang AE, Walsh KB, Baker AJ, Ashwath N, Midmore DJ (2005) Successful seed germination of the nickel hyperaccumulator *Stackhousia tryonii. Ann Bot* 96(1):159–163

Bolouri-Moghaddam MR, Le Roy K, Xiang L, Rolland F, Van den Ende W (2010) Sugar signalling and anti-oxidant network connections in plant cells. *FEBS J* 277(9):2022–2037

Bondada BR, Tu S, Ma LQ (2004) Absorption of foliar-applied arsenic by the arsenic hyperaccumulating fern (*Pteris vittata* L.). *Sci Total Env* 332(1–3):61–70

Camejo D, Martí MC, Olmos E, Torres W, Sevilla F, Jiménez A (2012) Oligogalacturonides stimulate antioxidant system in alfalfa roots. *Biologia Plant* 56(3):537–544

Cao F, Liu L, Ibrahim W, Cai Y, Wu F (2013) Alleviating effects of exogenous glutathione, glycinebetaine, brassinosteroids and salicylic acid on cadmium toxicity in rice seedlings (*Oryza sativa*). *Agrotechnology* 02(1):107–112

Chater CC, Covarrubias AA, Acosta-Maspons A (2019) Crop biotechnology for improving drought tolerance: targets, approaches, and outcomes. *Ann Plant Rev*:1–39. https://doi.org/10.1002/9781119312994. apr0669

Choudhary SP, Bhardwaj R, Gupta BD, Dutt P, Gupta RK, Kanwar M, Dutt P (2010) Changes induced by Cu^{2+} and Cr^{6+} metal stress in polyamines, auxins, abscisic acid titers and antioxidative enzymes activities of radish seedlings. *Braz J Plant Physiol* 22(4):263–270

Choudhary SP, Kanwar M, Bhardwaj R, Yu JQ, Tran LSP (2012a) Chromium stress mitigation by polyamine-brassinosteroid application involves phytohormonal and physiological strategies in *Raphanus sativus* L. *PLOS ONE* 7(3):e33210. https://doi.org/10.1371/journal.pone.0033210

Choudhary SP, Oral HV, Bhardwaj R, Yu J-Q, Tran L-SP (2012b) Interaction of brassinosteroids and polyamines enhances copper stress tolerance in raphanus sativus. *J Exp Bot* 63(15):5659–5675

Clemens S (2006) Toxic metal accumulation, responses to exposure and mechanisms of tolerance in plants. *Biochimie* 88(11):1707–1719

Clemens S, Ma JF (2016) Toxic heavy metal and metalloid accumulation in crop plants and foods. *Annu Rev Plant Biol* 67:489–512

Contreras RA, Pizarro M, Köhler H, Sáez CA, Zúñiga GE (2018) Copper stress induces antioxidant responses and accumulation of sugars and phytochelatins in Antarctic *Colobanthus quitensis* (Kunth) Bartl. *Biol Res* 51(1):1–10

Cuin TA, Shabala S (2007) Compatible solutes reduce ROS-induced potassium efflux in *Arabidopsis* roots. *Plant Cell Environ* 30(7):875–885

Dai M, Lu H, Liu W, Jia H, Hong H, Liu J, Yan C (2017) Phosphorus mediation of cadmium stress in two mangrove seedlings *Avicennia marina* and *Kandelia obovata* differing in cadmium accumulation. *Ecotoxicol Environ Saf* 139:272–279

Dhir B, Nasim SA, Samantary S, Srivastava S (2012) Assessment of osmolyte accumulation in heavy metal exposed *Salvinia natans. Int J Bot* 8(3):153–158

Ding C, Shi G, Xu X, Yang H, Xu Y (2010) Effect of exogenous spermidine on polyamine metabolism in water hyacinth leaves under mercury stress. *Plant Growth Regul* 60(1):61–67

Dixit G, Singh AP, Kumar A, Dwivedi S, Deeba F, Kumar S, Suman S, Adhikari B, Shukla Y, Trivedi PK, Pandey V, Tripathi RD (2015) Sulfur alleviates arsenic toxicity by reducing its accumulation and modulating proteome, amino acids and thiol metabolism in rice leaves. *Sci Rep* 5:16205

Dixit R, Wasiullah, Malaviya D, Pandiyan K, Singh U, Sahu A, Shukla R, Singh B, Rai J, Sharma P, Lade H, Paul D (2015b) Bioremediation of heavy metals from soil and aquatic environment: an overview of principles and criteria of fundamental processes. *Sustainability* 7(2):2189–2212

Domínguez-Solís JR, López-Martín MC, Ager FJ, Ynsa MD, Romero LC, Gotor C (2004) Increased cysteine availability is essential for cadmium tolerance and accumulation in *Arabidopsis thaliana. Plant Biotech J* 2(6):469–476

Dotaniya ML, Meena VD (2015) Rhizosphere effect on nutrient availability in soil and its uptake by plants: a review. *Proc Natl Acad Sci India Sect B Biol Sci* 85(1): 1–12

Dubey S, Misra P, Dwivedi S, Chatterjee S, Bag SK, Mantri S, Asif MH, Rai A, Kumar S, Shri M, Tripathi P, Tripathi RD, Trivedi PK, Chakrabarty D, Tuli R (2010) Transcriptomic and metabolomic shifts in rice roots in response to Cr (VI) stress. *BMC Genom* 11:648. https ://doi.org/10.1186/1471-2164-11-648

Duman F, Aksoy A, Aydin Z, Temizgul R (2011) Effects of exogenous glycinebetaine and trehalose on cadmium accumulation and biological responses of an aquatic plant (*Lemna gibba* L.). *Water Air Soil Pollut* 217(1–4):545–556

Dutta T, Neelapu NR, Wani SH, Surekha C (2019) Role and regulation of osmolytes as signaling molecules to abiotic stress tolerance. In: Khan M, Iqbal R, Reddy PS, Ferrante A, Khan NA (eds) *Plant signaling molecules.* Cambridge, UK: Woodhead Publishing, pp 459–477

Emamverdian A, Ding Y, Mokhberdoran F, Xie Y (2015) Heavy metal stress and some mechanisms of plant defense response. *Sci World J* 2015. http://doi.org/10.1155/2015/756120

Emery L, Whelan S, Hirschi KD, Pittman JK (2012) Protein phylogenetic analysis of Ca(2+)/cation Antiporters and Insights into their evolution in plants. *Front Plant Sci* 3:1. https://doi.org/10.3389/fpls.2012.00001

Eren E, Argüello JM (2004) Arabidopsis HMA2, a divalent heavy metal-transporting P(IB)-type ATPase, is involved in cytoplasmic Zn2+ homeostasis. *Plant Physiol* 136(3):3712–3723

Fariduddin Q, Khalil RR, Mir BA, Yusuf M, Ahmad A (2013a) 24-Epibrassinolide regulates photosynthesis, antioxidant enzyme activities and proline content of *Cucumis sativus* under salt and/or copper stress. *Environ Monit Assess* 185(9):7845–7856

Fariduddin Q, Varshney P, Yusuf M, Ali A, Ahmad A (2013b) Dissecting the role of glycine betaine in plants under abiotic stress. *Plant Stress* 7:8–18

Farooq MA, Ali S, Hameed A, Bharwana SA, Rizwan M, Ishaque W, Farid M, Mahmood K, Iqbal Z (2016) Cadmium stress in cotton seedlings: physiological, photosynthesis and oxidative damages alleviated by glycinebetaine. *S Afr J Bot* 104:61–68

Fernandez O, Béthencourt L, Quero A, Sangwan RS, Clément C (2010) Trehalose and plant stress responses: friend or foe? *Trends Plant Sci* 15(7):409–417

Fernández V, Sotiropoulos T, Brown PH (2013) Foliar fertilization: scientific principles and field pratices. International Fertilizer Industry Association. https://www.researchgate.net/publication/264810852

Fidalgo F, Azenha M, Silva AF, de Sousa A, Santiago A, Ferraz P, Teixeira J (2013) Copper-induced stress in *Solanum nigrum* L. and antioxidant defense system responses. *Food Energy Secur* 2(1):70–80

Garg N, Chandel S (2012) Role of arbuscular mycorrhizal (AM) fungi on growth, cadmium uptake, osmolyte, and phytochelatin synthesis in *Cajanus cajan* (L.) Millsp. under NaCl and Cd stresses. *J Plant Growth Regul* 31(3):292–308

Garg N, Saroy K (2020) Interactive effects of polyamines and arbuscular mycorrhiza in modulating plant biomass, N2 fixation, ureide, and trehalose metabolism in Cajanus cajan (L.) Millsp. genotypes under nickel stress. *Environ Sci Pollut Res Int* 27(3):3043–3064

Garg N, Singh S (2018a) Arbuscular mycorrhiza *Rhizophagus irregularis* and silicon modulate growth, proline biosynthesis and yield in *Cajanus cajan* L. Millsp. (pigeonpea) genotypes under cadmium and zinc stress. *J Plant Growth Regul* 37(1):46–63

Garg N, Singh S (2018b) Mycorrhizal inoculations and silicon fortifications improve rhizobial symbiosis, antioxidant defense, trehalose turnover in pigeon pea genotypes under cadmium and zinc stress. *Plant Growth Regul* 86(1):105–119

Garg N, Singla P (2012) The role of *Glomus mosseae* on key physiological and biochemical parameters of pea plants grown in arsenic contaminated soil. *Sci Hort* 143:92–101

Gohari M, Habib-Zadeh AR, Khayat M (2012) Assessing the intensity of tolerance to lead and its effect on amount of protein and proline in root and aerial parts of two varieties of rape seed (*Brassica napus* L.). *JBASR* 2(1):935–938

Gong X, Liu Y, Huang D, Zeng G, Liu S, Tang H, Zhou L, Hu X, Zhou Y, Tan X (2016) Effects of exogenous calcium and spermidine on cadmium stress moderation and metal accumulation in *Boehmeria nivea* (L.) Gaudich. *Environ Sci Pollut Res* 23(9):8699–8708

Groppa MD, Zawoznik MS, Tomaro ML, Benavides MP (2008) Inhibition of root growth and polyamine metabolism in sunflower (*Helianthus annuus*) seedlings under cadmium and copper stress. *Biol Trace Elem Res* 126(1–3):246. https://doi.org/10.1007/s12011-008-8191-y

Guangqiu Q, Chongling Y, Haoliang L (2007) Influence of heavy metals on the carbohydrate and phenolics in mangrove, *Aegiceras corniculatum* L., seedlings. *Bull Environ Contam Toxicol* 78(6):440–444

Habiba U, Ali S, Rizwan M, Ibrahim M, Hussain A, Shahid MR, Alamri SA, Alyemeni MN, Ahmad P (2019) Alleviative role of exogenously applied mannitol in maize cultivars differing in chromium stress tolerance. *Environ Sci Pollut Res* 26(5):5111–5121

Handa N, Kohli SK, Kaur R, Sharma A, Kumar V, Thukral AK, Arora S, Bhardwaj R (2018a) Role of compatible solutes in enhancing antioxidative defense in plants exposed to metal toxicity. In: Hasanuzzaman M, Nahar K, Fujita M (eds) *Plants under metal and metalloid stress*. Springer, Singapore, pp 207–228

Handa N, Kohli SK, Thukral AK, Bhardwaj R, Alyemeni MN, Wijaya L, Ahmad P (2018b) Protective role of selenium against chromium stress involving metabolites and essential elements in *Brassica juncea* L. seedlings. *3 Biotech* 8(1):66. https://doi.org/10.1007/s13205-018-1087-4

Handique GK, Handique AK (2009) Proline accumulation in lemongrass (*Cymbopogon flexuosus* Stapf.) due to heavy metal stress. *J Environ Biol* 30(2):299–302

Hasanuzzaman M, Anee TI, Bhuiyan TF, Nahar K, Fujita M (2019) Emerging role of osmolytes in enhancing abiotic stress tolerance in rice. In: Hasanuzzaman M, Fujita M, Nahar K, Biswas JK (eds) *Advances in rice research for abiotic stress tolerance*. Woodhead Publishing, pp 677–708. https://doi.org/10.1016/b978-0-12-814332-2.00033-2

Hayat S, Hayat Q, Alyemeni MN, Ahmad A (2013) Proline enhances antioxidative enzyme activity, photosynthesis and yield of *Cicer arietinum* L. exposed to cadmium stress. *Acta Botanica Croatica* 72(2):323–335

Hossain MA, Hasanuzzaman M, Fujita M (2010) Up-regulation of antioxidant and glyoxalase systems by exogenous glycinebetaine and proline in mung bean confer tolerance to cadmium stress. *Physiol Mol Biol Plants* 16(3):259–272

Imtiyaz S, Agnihotri RK, Ganie SA, Sharma R (2014) Biochemical response of *Glycine max* (L.) Merr. to cobalt and lead stress. *J Stress Physiol Biochem* 10(3):259–272

Jabeen N, Abbas Z, Iqbal M, Rizwan M, Jabbar A, Farid M, Ali S, Ibrahim M, Abbas F (2016) Glycine betaine mediates chromium tolerance in mung bean through lowering of Cr uptake and improved antioxidant system. *Arch Agron Soil Sci* 62(5):648–662

Jahangir M, Abdel-Farid IB, Choi YH, Verpoorte R (2008) Metal ion-inducing metabolite accumulation in *Brassica rapa*. *J Plant Physiol* 165(14):1429–1437

John R, Ahmad P, Gadgil K, Sharma S (2012) Heavy metal toxicity: effect on plant growth, biochemical parameters and metal accumulation by *Brassica juncea* L. *Int J Plant Prod* 3(3):65–76

Karmakar D, Padhy PK (2019) Metals uptake from particulate matter through foliar transfer and their impact on antioxidant enzymes activity of *S. robusta* in a tropical forest, West Bengal, India. *Arch Environ Contam Toxicol* 76(4):605–616

Kaur H, Garg N (2017) Zinc-arbuscular mycorrhizal interactions: effect on nutrient pool, enzymatic antioxidants, and osmolyte synthesis in pigeonpea nodules subjected to Cd stress. *Commun Soil Sci Plant Anal* 48(14):1684–1700

Kerkeb L, Krämer U (2003) The role of free histidine in xylem loading of nickel in *Alyssum lesbiacum* and *Brassica juncea*. *Plant Physiol* 131(2):716–724

Keunen ELS, Peshev D, Vangronsveld J, Van Den Ende WVD, Cuypers ANN (2013) Plant sugars are crucial players in the oxidative challenge during abiotic stress: extending the traditional concept. *Plant Cell Environ* 36(7):1242–1255

Khan A, Khan S, Khan MA, Qamar Z, Waqas M (2015) The uptake and bioaccumulation of heavy metals by food plants, their effects on plants nutrients, and associated health risk: a review. *Environ Sci Pollut Res* 22(18):13772–13799

Kieffer P, Planchon S, Oufir M, Ziebel J, Dommes J, Hoffmann L, Hausman JF, Renaut J (2009) Combining proteomics and metabolite analyses to unravel cadmium stress-response in poplar leaves. *J Proteome Res* 8(1):400–417

Kotapati KV, Palaka BK, Ampasala DR (2017) Alleviation of nickel toxicity in finger millet (*Eleusine coracana* L.) germinating seedlings by exogenous application of salicylic acid and nitric oxide. *Crop J* 5(3):240–250

Koźmińska A, Wiszniewska A, Hanus-Fajerska E, Muszyńska E (2018) Recent strategies of increasing metal tolerance and phytoremediation potential using genetic transformation of plants. *Plant Biotechnol Rep* 12(1):1–14

Kulis-Horn RK, Persicke M, Kalinowski J (2014) Histidine biosynthesis, its regulation and biotechnological application in *Corynebacterium glutamicum*. *Microb Biotechnol* 7(1):5–25

Kumar V, Khare T, Shaikh S, Wani SH (2018) Compatible solutes and abiotic stress tolerance in plants. In: Ramakrishna A, Gill SS (eds) *Metabolic adaptations in plants during abiotic stress*. Taylor & Francis (CRC Press), USA, pp 213–220

Kuroda K, Shibasaki S, Ueda M, Tanaka A (2001) Cell surface-engineered yeast displaying a histidine oligopeptide (hexa-His) has enhanced adsorption of and tolerance to heavy metal ions. *Appl Microbiol Biotechnol* 57(5–6):697–701

Li X, Yang Y, Jia L, Chen H, Wei X (2013b) Zinc-induced oxidative damage, antioxidant enzyme response and proline metabolism in roots and leaves of wheat plants. *Ecotoxicol Environ Saf* 89:150–157

Li ZG, Ding XJ, Du PF (2013a) Hydrogen sulfide donor sodium hydrosulfide-improved heat tolerance in maize and involvement of proline. *J Plant Physiol* 170(8):741–747

Maleki M, Ghorbanpour M, Kariman K (2017) Physiological and antioxidative responses of medicinal plants exposed to heavy metals stress. *Plant Gene* 11:247–254

Manara A (2012) Plant responses to heavy metal toxicity. In: Furini A (ed) *Plants and Heavy Metals*. Springer, Dordrecht, pp 27–53

Mishra P, Dubey RS (2013) Excess nickel modulates activities of carbohydrate metabolizing enzymes and induces accumulation of sugars by upregulating acid invertase and sucrose synthase in rice seedlings. *Biometals* 26(1):97–111

Morgutti S, Negrini N, Pucciariello C, Sacchi GA (2019) Role of trehalose and regulation of its levels as a signal molecule to abiotic stresses in plants. In: Khan Iqbal MR, Reddy AS, Ferrante A, Khan NA (eds) *Plant signaling molecules*. Cambridge, UK: Woodhead Publishing, pp 235–255

Mostofa MG, Hossain MA, Fujita M, Tran LS (2015) Physiological and biochemical mechanisms associated with trehalose-induced copper-stress tolerance in rice. *Sci Rep* 5:11433. https://www.nature.com/articles/srep11433

Mozafari H, Asrar Z, Yaghoobi MM, Salari H, Mozafari M (2016) Calcium and L-histidine interaction on nutrients accumulation in three tomato cultivars under nickel stress. *J Plant Nutr* 39(5):628–642

Nahar K, Hasanuzzaman M, Alam MM, Rahman A, Suzuki T, Fujita M (2016a) Polyamine and nitric oxide crosstalk: antagonistic effects on cadmium toxicity in mung bean plants through upregulating the metal detoxification, antioxidant defense and methylglyoxal detoxification systems. *Ecotoxicol Environ Saf* 126:245–255

Nahar K, Rahman M, Hasanuzzaman M, Alam MM, Rahman A, Suzuki T, Fujita M (2016b) Physiological and biochemical mechanisms of spermine-induced cadmium stress tolerance in mung bean (*Vigna radiata* L.) seedlings. *Environ Sci Pollut Res* 23(21):21206–21218

Natasha Shahid M, Dumat C, Khalid S, Rabbani F, Farooq ABU, Amjad M, Abbas G, Niazi NK (2019) Foliar uptake of arsenic nanoparticles by spinach: an assessment of physiological and human health risk implications. *Environ Sci Pollut Res Int* 26(20):20121–20131

Nikolić NA, Kojić DA, Pilipović AN, Pajević SL, Krstić BO, Borišev MI, Orlović SA (2008) Responses of hybrid poplar to cadmium stress: photosynthetic characteristics, cadmium and proline accumulation, and antioxidant enzyme activity. *Acta Biol Cracov Bot* 50(2):95–103

Nisa Z, Chen C, Yu Y, Chen C, Mallano AI, Xiang-Bo D, Xiao-Li S, Yan-Ming Z (2016) Constitutive overexpression of myo-inositol-1-phosphate synthase gene (*GsMIPS2*) from *Glycine soja* confers enhanced salt tolerance at various growth stages in *Arabidopsis*. *J Northeast Agric Univ* 23(2):28–44

Nishikawa F, Kato M, Hyodo H, Ikoma Y, Sugiura M, Yano M (2005) Effect of sucrose on ascorbate level and expression of genes involved in the ascorbate biosynthesis and recycling pathway in harvested broccoli florets. *J Exp Bot* 56(409):65–72

Nishizawa A, Yabuta Y, Shigeoka S (2008) Galactinol and raffinose constitute a novel function to protect plants from oxidative damage. *Plant Physiol* 147(3):1251–1263

Nishizawa A, Yabuta Y, Yoshida E, Maruta T, Yoshimura K, Shigeoka S (2006) *Arabidopsis* heat shock transcription factor A2 as a key regulator in response to several types of environmental stress. *Plant J* 48(4):535–547

Noreen S, Akhter MS, Yaamin T, Arfan M (2018) The ameliorative effects of exogenously applied proline on physiological and biochemical parameters of wheat (*Triticum aestivum* L.) crop under copper stress condition. *J Plant Interact* 13(1):221–230

O'Neill (1995) Arsenic. In: Alloway BJ (ed) *Heavy metals in soils*. Blackie Academic and Professional, London, pp 105–121

Oliveira H (2012) Chromium as an environmental pollutant: insights on induced plant toxicity. *J Bot* 2012. http://doi.org/10.1155/2012/375843 http://www.ncbi.nlm.nih.gov/pubmed/375843

Ovečka M, Takáč T (2014) Managing heavy metal toxicity stress in plants: biological and biotechnological tools. *Biotechnol Adv* 32(1):73–86

Oves M, Saghir Khan M, Huda Qari A, Nadeen Felemban M, Almeelbi T (2016) Heavy metals: biological importance and detoxification strategies. *J Bioremediat Biodegrad* 7(2):1–5

Pál M, Csávás G, Szalai G, Oláh T, Khalil R, Yordanova R, Gell G, Birinyi Z, Németh E, Janda T (2017) Polyamines may influence phytochelatin synthesis during Cd stress in rice. *J Hazard Mater* 340:272–280

Pant PP, Tripathi AK, Dwivedi V (2011) Effect of heavy metals on some biochemical parameters of sal (*Shorea robusta*) seedling at nursery level, Doon Valley, India. *J Agric Sci* 2(1):45–51

Patel DK, Archana G, Kumar GN (2008) Variation in the nature of organic acid secretion and mineral phosphate solubilization by Citrobacter sp. DHRSS in the presence of different sugars. *Curr Microbiol* 56(2):168–174

Pathak MR, Teixeira da Silva JA, Wani SH (2014) Polyamines in response to abiotic stress tolerance through transgenic approaches. *GM Crops Food* 5(2):87–96

Paul S, Banerjee A, Roychoudhury A (2018) Role of polyamines in mediating antioxidant defense and epigenetic regulation in plants exposed to heavy metal toxicity. In: Hasanuzzaman M, Nahar K, Fujita M (eds) *Plants under metal and metalloid stress*. Springer, Singapore, pp 229–247

Persans MW, Yan X, Patnoe JM, Krämer U, Salt DE (1999) Molecular dissection of the role of histidine in nickel hyperaccumulation in *Thlaspi goesingense* (Hálácsy). *Plant Physiol* 121(4):1117–1126

Peshev D, Vergauwen R, Moglia A, Hideg E, Van den Ende WVD (2013) Towards understanding vacuolar antioxidant mechanisms: a role for fructans? *J Exp Bot* 64(4):1025–1038

Pidatala VR, Li K, Sarkar D, Ramakrishna W, Datta R (2016) Identification of biochemical pathways associated with lead tolerance and detoxification in *Chrysopogon zizanioides* L. Nash (vetiver) by metabolic profiling. *Environ Sci Technol* 50(5):2530–2537

Puig S, Thiele DJ (2002) Molecular mechanisms of copper uptake and distribution. *Curr Opin Chem Biol* 6(2):171–180

Qian JI, Shan XQ, Wang ZJ, Tu Q (1996) Distribution and plant availability of heavy metals in different particle-size fractions of soil. *Sci Total Environ* 187(2):131–141

Rady MM, Hemida KA (2015) Modulation of cadmium toxicity and enhancing cadmium-tolerance in wheat seedlings by exogenous application of polyamines. *Ecotoxicol Environ Saf* 119:178–185

Rasheed R, Ashraf MA, Hussain I, Haider MZ, Kanwal U, Iqbal M (2014) Exogenous proline and glycinebetaine mitigate cadmium stress in two genetically different spring wheat (*Triticum aestivum* L.) cultivars. *Braz J Bot* 37(4):399–406

Rastgoo L, Alemzadeh A, Afsharifar A (2011) Isolation of two novel isoforms encoding zinc-and copper-transporting P1B-ATPase from Gouan (*Aeluropus littoralis*). *Plant OMICS* 4(7):377–383

Richau KH, Kozhevnikova AD, Seregin IV, Vooijs R, Koevoets PLM, Smith JAC, Ivanov VB, Schat H (2009) Chelation by histidine inhibits the vacuolar sequestration of nickel in roots of the hyperaccumulator *Thlaspi caerulescens*. *New Phytol* 183(1):106–116

Rigui AP, Carvalho V, Wendt Dos Santos AL, Morvan-Bertrand A, Prud'homme MP, Machado de Carvalho MA, Gaspar M (2019) Fructan and antioxidant metabolisms in plants of Lolium perenne under drought are modulated by exogenous nitric oxide, Prud'homme MP, de Carvalho MA, Gaspar M. *Plant Physiol Biochem* 145:205–215

Rillig MC, Caldwell BA, Wösten HA, Sollins P (2007) Role of proteins in soil carbon and nitrogen storage: controls on persistence. *Biogeochemistry* 85(1):25–44

Rizwan M, Imtiaz M, Dai Z, Mehmood S, Adeel M, Liu J, Tu S (2017) Nickel stressed responses of rice in Ni subcellular distribution, antioxidant production, and osmolyte accumulation. *Environ Sci Pollut Res* 24(25):20587–20598

Roychoudhury A, Basu S, Sengupta DN (2012) Antioxidants and stress-related metabolites in the seedlings of two indica rice varieties exposed to cadmium chloride toxicity. *Acta Physiol Plant* 34(3):835–847

Ruscitti M, Arango M, Ronco M, Beltrano J (2011) Inoculation with mycorrhizal fungi modifies proline metabolism and increases chromium tolerance in pepper plants (*Capsicum annuum* L.). *Braz J Plant Physiol* 23(1):15–25

Saha M, Sarkar S, Sarkar B, Sharma BK, Bhattacharjee S, Tribedi P (2016) Microbial siderophores and their potential applications: a review. *Environ Sci Pollut Res* 23(5):3984–3999

Saif S, Khan MS (2018) Assessment of toxic impact of metals on proline, antioxidant enzymes, and biological characteristics of *Pseudomonas aeruginosa* inoculated *Cicer arietinum* grown in chromium and nickel-stressed sandy clay loam soils. *Environ Monit Assess* 190(5):290. https://doi.org/10.1007/s10661-018-6652-0

Salt DE, Prince RC, Baker AJ, Raskin I, Pickering IJ (1999) Zinc ligands in the metal hyperaccumulator thlaspi caerulescens as determined using X-ray absorption spectroscopy. *Environ Sci Technol* 33(5):713–717

Sasaki A, Yamaji N, Yokosho K, Ma JF (2012) Nramp5 is a major transporter responsible for manganese and cadmium uptake in rice. *Plant Cell* 24(5):2155–2167

Saxena SC, Kaur H, Verma P, Petla BP, Andugula VR, Majee M (2013) Osmoprotectants: potential for crop improvement under adverse conditions. In: Tuteja N, Singh Gill S (eds) *Plant acclimation to environmental stress*. Springer, New York, pp 197–232

Schramm F, Ganguli A, Kiehlmann E, Englich G, Walch D, von Koskull-Döring P (2006) The heat stress transcription factor HsfA2 serves as a regulatory amplifier of a subset of genes in the heat stress response in *Arabidopsis*. *Plant Mol Biol* 60(5):759–772

Schreck E, Foucault Y, Sarret G, Sobanska S, Cécillon L, Castrec-Rouelle M, Uzu G, Dumat C (2012) Metal and metalloid foliar uptake by various plant species exposed to atmospheric industrial fallout: mechanisms involved for lead. *Sci Total Environ* 427–428:253–262

Scoccianti V, Bucchini AE, Iacobucci M, Ruiz KB, Biondi S (2016) Oxidative stress and antioxidant responses to increasing concentrations of trivalent chromium in the Andean crop species *Chenopodium quinoa* willd. *Ecotoxicol Environ Saf* 133:25–35

Scoccianti V, Iacobucci M, Speranza A, Antognoni F (2013) Over-accumulation of putrescine induced by cyclohexylamine interferes with chromium accumulation and partially restores pollen tube growth in *Actinidia deliciosa*. *Plant Physiol Biochem* 70:424–432

Shahid M, Austruy A, Echevarria G, Arshad M, Sanaullah M, Aslam M, Nadeem M, Nasim W, Dumat C (2014a) EDTA-enhanced phytoremediation of heavy metals: a review. *Soil Sediment Contam Int J* 23(4):389–416

Shahid MA, Balal RM, Pervez MA, Abbas T, Aqeel MA, Javaid MM, Garcia-Sanchez F (2014b) Exogenous proline and proline-enriched *Lolium perenne* leaf extract protects against phytotoxic effects of nickel and salinity in *Pisum sativum* by altering polyamine metabolism in leaves. *Turk J Bot* 38(5):914–926

Sharma P, Dubey RS (2005) Modulation of nitrate reductase activity in rice seedlings under aluminium toxicity and water stress: role of osmolytes as enzyme protectant. *J Plant Physiol* 162(8):854–864

Shen BO, Jensen RG, Bohnert HJ (1997) Increased resistance to oxidative stress in transgenic plants by targeting mannitol biosynthesis to chloroplasts. *Plant Physiol* 113(4):1177–1183

Shukla D, Kesari R, Tiwari M, Dwivedi S, Tripathi RD, Nath P, Trivedi PK (2013) Expression of *Ceratophyllum demersum* phytochelatin synthase, *CdPCS1*, in *Escherichia coli* and *Arabidopsis* enhances heavy metal (loid) s accumulation. *Protoplasma* 250(6):1263–1272

Siddique A, Dubey AP (2017) Phyto-toxic effect of heavy metal (CdCl$_2$) on seed germination, seedling growth and antioxidant defence metabolism in wheat (*Triticum aestivum* L.) variety HUW-234. *Int J Bio-Resource Environ Agric* 8(2):261–267

Singh V, Tripathi BN, Sharma V (2016) Interaction of Mg with heavy metals (Cu, Cd) in *T. aestivum* with special reference to oxidative and proline metabolism. *J Plant Res* 129(3):487–497

Slama I, Abdelly C, Bouchereau A, Flowers T, Savouré A (2015) Diversity, distribution and roles of osmoprotective compounds accumulated in halophytes under abiotic stress. *Ann Bot* 115(3):433–447

Soudek P, Ursu M, Petrová Š, Vaněk T (2016) Improving crop tolerance to heavy metal stress by polyamine application. *Food Chem* 213:223–229

Stepien P, Gediga K, Piszcz U, Karmowska K (2016) Effects of the exogenous glycinebetaine on photosynthetic apparatus in cucumber leaves challenging Al stress. In: Proceedings of the 18th International Conference on Heavy Metals in the Environment

Stobrawa K, Lorenc-Plucińska G (2007) Changes in antioxidant enzyme activity in the fine roots of black poplar (*Populus nigra* L.) and cottonwood (*Populus deltoides* Bartr. ex Marsch) in a heavy-metal-polluted environment. *Plant Soil* 298(1–2):57–68

Sun X, Zhang J, Zhang H, Ni Y, Zhang Q, Chen J, Guan Y (2010) The responses of *Arabidopsis thaliana* to cadmium exposure explored via metabolite profiling. *Chemosphere* 78(7):840–845

Surapu V, Ediga A, Meriga B (2014) Salicylic acid alleviates aluminum toxicity in tomato seedlings (*Lycopersicum esculentum* Mill.) through activation of antioxidant defense system and proline biosynthesis. *Adv Biosci Biotechnol* 05(9):777. https://doi.org/10.4236/abb.2014.59091

Szepesi Á, Szőllősi R (2018) Mechanism of proline biosynthesis and role of proline metabolism enzymes under environmental stress in plants. In: Ahmad P, Ahanger MA, Singh VP, Tripathi DK, Alam P, Alyemeni MN (eds) *Plant metabolites and regulation under environmental stress*. Elsevier, Academic Press, pp 337–353

Szolnoki ZS, Farsang A, Puskás I (2013) Cumulative impacts of human activities on urban garden soils: origin and accumulation of metals. *Environ Pollut* 177:106–115

Tan JL, Wang CY, Xiang B, Han R, Guo Z (2013) Hydrogen peroxide and nitric oxide mediated cold- and dehydration-induced myo-inositol phosphate synthase that confers multiple resistances to abiotic stresses. *Plant Cell Environ* 36(2):288–299

Tecon R, Or D (2017) Biophysical processes supporting the diversity of microbial life in soil. *FEMS Microbiol Rev* 41(5):599–623

Theriappan P, Gupta AK, Dhasarrathan P (2011) Accumulation of proline under salinity and heavy metal stress in cauliflower seedlings. *J Appl Sci Environ Manag* 15(2). http://doi.org/10.4314/jasem.v15i2.68497

Tripathi P, Singh PC, Mishra A, Chaudhry V, Mishra S, Tripathi RD, Nautiyal CS (2013a) *Trichoderma* inoculation ameliorates arsenic induced phytotoxic changes in gene expression and stem anatomy of chickpea (*Cicer arietinum*). *Ecotoxicol Environ Saf* 89:8–14

Tripathi P, Tripathi RD, Singh RP, Dwivedi S, Chakrabarty D, Trivedi PK, Adhikari B (2013b) Arsenite tolerance in rice (Oryza sativa L.) involves coordinated role of metabolic pathways of thiols and amino acids. *Environ Sci Pollut Res* 20(2):884–896

Varshney P (2011) *Effect of 24-Epibrassinolide and Glycine Betaine Application on Brassica juncea under Copper Stress* (Doctoral dissertation, Aligarh Muslim University)

Vezza ME, Llanes A, Travaglia C, Agostini E, Talano MA (2018) Arsenic stress effects on root water absorption in soybean plants: physiological and morphological aspects. *Plant Physiol Biochem* 123:8–17

Vinocur B, Altman A (2005) Recent advances in engineering plant tolerance to abiotic stress: achievements and limitations. *Curr Opin Biotechnol* 16(2):123–132

Vinod K, Awasthi G, Chauchan PK (2012) Cu and Zn tolerance and responses of the biochemical and physio-chemical system of wheat. *J Stress Physiol Biochem* 8(3):203–213

Wang C, Yang Z, Yuan X, Browne P, Chen L, Ji J (2013) The influences of soil properties on Cu and Zn availability in soil and their transfer to wheat (*Triticum aestivum* L.) in the Yangtze River delta region, China. *Geoderma* 193–194:131–139

Wang F, Zeng B, Sun Z, Zhu C (2009) Relationship between proline and Hg2+-induced oxidative stress in a tolerant rice mutant. *Arch Environ Contam Toxicol* 56(4):723–731

Wang X, Shi G, Xu Q, Hu J (2007) Exogenous polyamines enhance copper tolerance of *Nymphoides peltatum*. *J Plant Physiol* 164(8):1062–1070

Wang X, Shi G-X, Ma G-Y, Xu Q-S, Khaled R, Hu J-Z (2004) Effects of exogenous spermidine on resistance of *Nymphoides peltatum* to Hg^{2+} stress. *Zhi Wu Sheng Li Yu Fen Zi Sheng Wu Xue Xue Bao* 30(1):69–74

Wang Y, Huang J, Gou CB, Dai X, Chen F, Wei W (2011) Cloning and characterization of a differentially expressed cDNA encoding myo-inositol-1-phosphate synthase involved in response to abiotic stress in *Jatropha curcas*. *Plant Cell Tissue Organ Cult* 106(2):269–277

Wang Y, Liu H, Wang S, Li H, Xin Q (2016) Overexpression of a common wheat gene galactinol synthase3 enhances tolerance to zinc in *Arabidopsis* and rice through the modulation of reactive oxygen species production. *Plant Mol Biol Rep* 34:794–806

Wani SH, Singh NB, Haribhushan A, Mir J (2013) Compatible solute engineering in plants for abiotic stress tolerance—role of glycine betaine. *Curr Genomics* 14(3):157–165

Wani W, Masoodi KZ, Zaid A, Wani SH, Shah F, Meena VS, Wani SA, Mosa KA (2018) Engineering plants for heavy metal stress tolerance. *Rend Fis Acc Lincei* 29(3):709–723

Wen JF, Gong M, Liu Y, Hu JL, Deng MH (2013) Effect of hydrogen peroxide on growth and activity of some enzymes involved in proline metabolism of sweet corn seedlings under copper stress. *Sci Hort* 164:366–371

Wen XP, Pang XM, Matsuda N, Kita M, Inoue H, Hao YJ, Honda C, Moriguchi T (2008) Over-expression of the apple spermidine synthase gene in pear confers multiple abiotic stress tolerance by altering polyamine titers. *Transgen Res* 17(2):251–263

WHO (2007) *Health risks of heavy metals from longrange transboundary air pollution*. WHO Regional Office for Europe, Copenhagen, Denmark

Williamson JD, Jennings DB, Guo WW, Pharr DM, Ehrenshaft M (2002) Sugar alcohols, salt stress, and fungal resistance: polyols—multifunctional plant protection? *J Am Soc Hortic Sci* 127(4):467–473

Wu D, Yamaji N, Yamane M, Kashino-Fujii M, Sato K, Feng Ma J (2016) The HvNramp5 transporter mediates uptake of cadmium and manganese, but not iron. *Plant Physiol* 172(3):1899–1910

Wu Z, Ren H, McGrath SP, Wu P, Zhao F-J (2011) Investigating the contribution of the phosphate transport pathway to arsenic accumulation in rice. *Plant Physiol* 157(1):498–508

Wycisk K, Kim EJ, Schroeder JI, Krämer U (2004) Enhancing the first enzymatic step in the histidine biosynthesis pathway increases the free histidine pool and nickel tolerance in *Arabidopsis thaliana*. *FEBS Lett* 578(1–2):128–134

Xian Y, Wang M, Chen W (2015) Quantitative assessment on soil enzyme activities of heavy metal contaminated soils with various soil properties. *Chemosphere* 139:604–608

Xiong TT, Leveque T, Austruy A, Goix S, Schreck E, Dappe V, Sobanska S, Foucault Y, Dumat C (2014) Foliar uptake and metal (loid) bioaccessibility in vegetables exposed to particulate matter. *Environ Geochem Health* 36(5):897–909

Xu X, Shi G, Jia R (2011) Changes of polyamine levels in roots of *Sagittaria sagittifolia* L. under copper stress. *Environ Sci Pollut Res* 19(7):2973–2982

Yang H, Shi G, Wang H, Xu Q (2010) Involvement of polyamines in adaptation of *Potamogeton crispus* L. to cadmium stress. *Aquat Toxicol* 100(3):282–288

Yang HY, Shi GX, Li WL, Wu WL (2013) Exogenous spermidine enhances *Hydrocharis dubia* cadmium tolerance. *Russ J Plant Physiol* 60(6):770–775

Yu XZ, Lin YJ, Fan WJ, Lu MR (2017) The role of exogenous proline in amelioration of lipid peroxidation in rice seedlings exposed to Cr (VI). *Int Biodeterior Biodegrad* 123:106–112

Zafari S, Sharifi M, Ahmadian Chashmi NA, Mur LA (2016) Modulation of Pb-induced stress in *Prosopis* shoots through an interconnected network of signaling molecules, phenolic compounds and amino acids. *Plant Physiol Biochem* 99:11–20

Zemanová V, Pavlík M, Pavlíková D (2017) Cadmium toxicity induced contrasting patterns of concentrations of free sarcosine, specific amino acids and selected microelements in two *Noccaea* species. *PLOS ONE* 12(5):e0177963. https://doi.org/10.1371/journal.pone.0177963

Zemanová V, Pavlík M, Pavlíková D, Tlustoš P (2014) The significance of methionine, histidine and tryptophan in plant responses and adaptation to cadmium stress. *Plant Soil Environ* 60(9):426–432

Zhai H, Wang F, Si Z, Huo J, Xing L, An Y, He S, Liu Q (2016) A myo-inositol-1-phosphate synthase gene, *IbMIPS1*, enhances salt and drought tolerance and stem nematode resistance in transgenic sweet potato. *Plant Biotechnol J* 14(2):592–602

Zhang Q, Yu R, Fu S, Wu Z, Chen HY, Liu H (2019) Spatial heterogeneity of heavy metal contamination in soils and plants in Hefei, China. *Sci Rep* 9(1):1049. https://www.nature.com/articles/s41598-018-36582-y

Zhao H, Yang H (2008) Exogenous polyamines alleviate the lipid peroxidation induced by cadmium chloride stress in *Malus hupehensis* Rehd. *Sci Hort* 116(4):442–447

Zhao J, Shi G, Yuan Q (2008) Polyamines content and physiological and biochemical responses to ladder concentration of nickel stress in *Hydrocharis dubia* (Bl.) backer leaves. *Biometals* 21(6):665–674

Zhou MX, Renard ME, Quinet M, Lutts S (2019) Effect of NaCl on proline and glycinebetaine metabolism in *Kosteletzkya pentacarpos* exposed to Cd and Zn toxicities. *Plant Soil* 441(1–2):525–542

Zhu G, Xiao H, Guo Q, Zhang Z, Zhao J, Yang D (2018) Effects of cadmium stress on growth and amino acid metabolism in two compositae plants. *Ecotoxicol Environ Saf* 158:300–308

Zouari M, Ben Ahmed C, Elloumi N, Bellassoued K, Delmail D, Labrousse P, Ben Abdallah F, Ben Rouina B (2016) Impact of proline application on cadmium accumulation, mineral nutrition and enzymatic antioxidant defense system of Olea europaea L. cv Chemlali exposed to cadmium stress. *Ecotoxicol Environ Saf* 128:195–205

Zouari M, Elloumi N, Ahmed CB, Delmail D, Rouina BB, Abdallah FB, Labrousse P (2016b) Exogenous proline enhances growth, mineral uptake, antioxidant defense, and reduces cadmium-induced oxidative damage in young date palm (*Phoenix dactylifera* L.). *Ecol Eng* 86:202–209

4 Role of Organic Solutes on Mineral Stress Tolerance

Md. Tahjib-Ul-Arif, Shagata Islam Shorna, Mohammed Arif Sadik Polash, Md Rezwanul Haque, Mohammad Golam Kibria, Mohammad Saidur Rhaman and M. Afzal Hossain

CONTENTS

4.1 INTRODUCTION

Organic osmolytes, also known as compatible solutes, are uncharged, polar and water-soluble compounds in nature that do not interfere with the cellular metabolism even at higher concentrations and act upon reducing osmotic imbalance (Jadhav et al. 2018; Saxena et al. 2019). Under stress conditions, the biosynthesis and accumulation of organic osmolytes are varied diversely. The well documented compatible solutes found in plants during different stress conditions are proline (Meloni et al. 2001; Yoshiba et al. 1997), glycinebetaine (Gadallah 1999; Yang and Lu 2005), trehalose (Ibrahim and Abdellatif 2016; Kosar et al. 2020), polyols (Almaghamsi et al. 2021; Skliros et al. 2018) and soluble sugars (Salehi et al. 2016; Skliros et al. 2018). The biosynthesis and accumulation of different organic osmolytes in plant cells are mainly associated with the soil osmotic potential, and therefore the key functions of those organic osmolytes are to defend the cell structure and photosynthetic apparatus from water desiccation (Cha-um et al. 2010; Saxena et al. 2019). They also protect plants under stress by reactive oxygen species (ROS) scavenging, singlet oxygen ($^1O_2$) quenching and maintaining osmotic balance through continuous water influx (Suprasanna et al. 2016).

Pollution of soils, air and ground water with an excessive amount of toxic metals is one of the major problems in recent times. Minerals are non-biodegradable and are entering the human food chain, thereby leading to enhanced risks for human health (Sharma and Dietz 2009). Minerals can be categorized as essential or non-essential for plants. Essential minerals such as copper (Cu), iron (Fe), nickel (Ni) and zinc (Zn) act as cofactors in over 1500 proteins that are crucial for the normal metabolic activities in plants (Hänsch and Mendel 2009). Plants need those essential minerals at an optimal level, and both deficiency and excessiveness are problematic for plants. However, even low concentrations of non-essential metals such as cadmium (Cd), cobalt (Co), aluminum (Al), arsenic

(As), chromium (Cr), lead (Pb) and mercury (Hg) disturb biochemical and physiological processes and decrease plant growth and productivity (Asati et al. 2016).

So far, it has been well documented from plenty of researches that the growth and developmental processes of plants are distressed due to the extensive absorption of metals. For instance, unwarranted levels of Cd in plant cells considerably decrease the production of biomass, reduce photosynthetically active pigments and hamper plant's photosynthetic ability and related chemical routes (Akhtar et al. 2017; Liu et al. 2018). Exposure to excessive concentrations of Hg interferes the mitochondrial activity and induces oxidative stress by triggering the generation of ROS (Elbaz et al. 2010; Tamás and Zelinová 2017; Zhou et al. 2008). Moreover, Cr instigates harmful consequences on plant physiological courses such as photosynthesis, water relations and mineral nutrition. Metabolic alterations by Cr exposure have also been described in plants by a direct effect on enzymes and metabolites or by the generation of ROS (Pandey et al. 2009; Trinh et al. 2014). High levels of Pb cause inhibition of enzyme activities, water imbalance and alterations in membrane permeability, and disturbed mineral nutrition induces oxidative stress by increasing the production of ROS in plants (Kaur et al. 2015; Zhong et al. 2020). Excess of Co restricted the concentration of Fe, chlorophyll, protein and catalase (CAT) activity in leaves, and affected the translocation of P, S, Mn, Zn and Cu from roots to aerial parts (Chatterjee and Chatterjee 2000). Though Cu and Zn are essential minerals, they are toxic at higher concentrations, and as revealed in many researches, Cu and Zn inhibit plant growth by hampering different biochemical processes (Assareh et al. 2008; Chen et al. 2000; Doncheva and Stoyanova 2007; Yurekli and Porgali 2006).

Higher concentrations of metals in growing media can cause osmotic stress directly or indirectly in plants (Rucińska-Sobkowiak 2016). Metal stress causes an imbalance between the generation and detoxification of ROS such as superoxide ($O_2^{\bullet-}$), hydrogen peroxide (H_2O_2) and the hydroxyl radical ($^{\bullet}OH$) (Ghori et al. 2019; Sharma and Dietz 2009). In view of the different chemical properties of metals, two modes of action can be distinguished. Under physiological conditions, redox-active metals such as Cu and Fe exist in different oxidation states ($Cu^{+/2+}$ and $Fe^{2+/3+}$) which enables both metals to directly participate in the Fenton and Haber-Weiss reactions, finally leading to the formation of highly toxic $^{\bullet}OH$ radicals from H_2O_2 (Schutzendubel and Polle 2002; Sharma and Dietz 2009). On the other hand, physiologically non-redox-active metals such as Cd, Hg and Zn only indirectly contribute to increased ROS production, for example by depleting or inhibiting cellular antioxidants (Schutzendubel and Polle 2002; Sharma and Dietz 2009). Plants try to alleviate the adversities of mineral stress-induced osmotic and oxidative stress by accumulating organic solutes into the cells.

In the present chapter, we describe how different mineral stress affects levels of organic osmolytes in plants and how the exogenous application of organic osmolytes can mitigate the adverse effects of mineral stress. Moreover, we discuss the role of endogenous organic osmolytes in mineral homeostasis in plants under different stress conditions.

4.2 HOW MINERAL STRESS AFFECTS THE ORGANIC OSMOLYTES IN PLANTS

Aluminum stress of 50 µM in the root of wheat (*Triticum aestivum*) increased the soluble sugars content (Hossain et al. 2006). Proline and soluble sugars content increased in the leaves of *Fagopyrum sp.* plants when exposed to Al stress (Pirzadah et al. 2019). Moreover, glycinebetaine content increased in the leaves of cucumber (*Cucumis sativus*) when treated with 0.1 or 1.0 mM Al stress (Stepien 2016). The proline content remained unchanged upon 600 µM Al stress exposure in sorghum (*Sorghum bicolor* L.) (Galvez et al. 1991). On the other hand, Bhamburdekar and Chavan (2011) showed a decreasing trend of proline accumulation in pigeon pea (*Cajanas cajan*) seed when treated with 100 ppm Al stress (Table 4.1).

An increase of proline content was shown by several researchers at different levels of As stress in several plants. For example, in eggplant (*Solanum melongena*) at 5 and 25 µM (Singh et al. 2015), rice (*Oryza sativa*) seedlings at 25 and 50 µM (Mishra and Dubey 2006), coriander (*Coriandrum sativum*) seeds at 100 and 200 µM (Karam et al. 2016), black gram (*Vigna mungo*) seeds at 100

TABLE 4.1
Effect of Mineral Stresses on Different Organic Osmolytes in Plants

Minerals	Concentrations	Plant Species	Organs/Tissues	Endogenous Osmolytes and Levels	References
Al	50 μM	*Triticum aestivum*	Root	Sugars (↑)	Hossain et al. (2006)
Al	100, 200 and 300 μM	*Fagopyrum* sp.	Leaf	Proline (↑), Sugars (↑)	Pirzadah et al. (2019)
Al	0.1 and 1 mM	*Cucumis sativus*	Leaf	Glycinebetaine (↑)	Stepien (2016)
Al	300, 450 and 600 μM	*Sorghum bicolor*	Seedling	Proline (↑↓)	Galvez et al. (1991)
Al	10, 50 and 100 ppm	*Cajanas cajan*	Seed	Proline (↓)	Bhamburdekar and Chavan (2011)
As	30, 60 and 90 mg kg⁻¹	*Pisum sativum*	Root, shoot	Sugars (↑), Proline (↑), Glycinebetaine (↑)	Garg and Singla (2012)
As	5 and 25 μM	*Solanum melongena*	Seedling	Proline (↑)	Singh et al. (2015)
As	200 μM	*Zea mays*	Leaf	Proline (↑), Sugars (↑)	Anjum et al. (2016)
As	25 and 50 μM	*Oryza sativa*	Seedling	Proline (↑)	Mishra and Dubey (2006)
As	100 and 200 μM	*Coriandrum sativum*	Seed	Proline (↑)	Karam et al. (2016)
As	100 and 200 μM	*Vigna mungo*	Seed	Proline (↑)	Srivastava and Sharma (2013)
As	1.28 and 10.8 mg L⁻¹	*Trapa natans*	Plant	Proline (↑)	Baruah et al. (2014)
As	50–400μM	*Triticum aestivum*	Seedling	Proline (↑)	Zengin (2015)
Cd	1 and 5 μM	*Gossypium hirsutum*	Seedling	Glycinebetaine (↑)	Farooq et al. (2016)
Cd	1mM	*Nicotiana tabacum*	Leaf cell	Proline (↑), Glycinebetaine (↑)	Islam et al. (2009 a)
Cd	30 and 60 mgL⁻¹	*Cyperus alopecuroides*	Leaf	Proline (↑), Sugars (↑)	Batool et al. (2014)
Cd	0.5, 1 and 3 mM	*Lemna gibba*	Plant	Proline (↑), Trehalose (↓), Glycinebetaine (↓)	Duman (2011)
Cd	10⁻⁶ and 10⁻⁴ M	*Helianthus annuus*	Seedling	Proline (↑)	Kastori et al. (1992)
Cd	150 mg L⁻¹	*Solanum lycopersicum*	Seedling	Proline (↑), Glycinebetaine (↑)	Alyemeni et al. (2018)
Cd	150 mg kg⁻¹	*Mentha arvensis*	Seedling	Proline (↑), Glycinebetaine (↑)	Zaid and Mohammad (2018)
Cd	10 and 25 μM	*Cicer arietinum*	Seed	Proline (↑)	Tantrey and Agnihotri (2010)
Cd	0.05, 0.06 and 0.08 mM	*Phaseolus vulgaris*	Seedling	Proline (↑)	Zengin and Munzuroglu (2005)
Cd	50, 100, 200, 350 and 500 mg kg⁻¹	*Cymbopogon flexuosus*	Leaf	Proline (↑)	Handique and Handique (2009)
Cr	2, 10, 25 and 50 ppm	*Vigna sinensis*	Seeds	Sugars (↓)	Nath et al. (2008)
Cr	1 and 2 mg L⁻¹	*Salvinia minima*	-	Sugars (↑)	Nichols et al. (2000)

(Continued)

TABLE 4.1 (CONTINUED)
Effect of Mineral Stresses on Different Organic Osmolytes in Plants

Minerals	Concentrations	Plant Species	Organs/ Tissues	Endogenous Osmolytes and Levels	References
Cr	10, 20, 50 and 100 μM	Ocimum tenuiflorum	Leaf	Proline (↑)	Rai et al. (2004)
Cr	0.25 and 0.5 mM	Triticum aestivum	Seed	Glycinebetaine (↑)	Ali et al. (2015)
Cr	50 mL	Oryza sativa	Seedling	Proline (↑)	Yu and Lu (2016)
Cr	250 and 500 μM	Vigna radiata	Leaf	Glycinebetaine (↑)	Jabeen et al. (2016)
Cr	30, 60, 90, 120 and 150 μmol L^{-1}	Zea mays	Leaf	Proline (↑), Soluble sugars (↑)	Anjum et al. (2017)
Cu	100 μM	Oryza sativa	Seedling	Trehalose (↑)	Mostofa et al. (2015)
Cu	10 ppm	Triticum aestivum	Shoot, leaf	Proline (↑)	Bassi and Sharma (1993 a)
Cu	5 ppm	Lemna minor	Leaf	Proline (↓)	Bassi and Sharma (1993 b)
Cu	10^{-6} and 10^{-4} M	Helianthus annuus	Seedling	Proline (↑)	Kastori et al. (1992)
Cu	10 mL	Cajanus cajan	Seed	Proline (↑), Glycinebetaine (↑), Sugars (↑)	Sirhindi et al. (2015)
Cu	1, 5 and 10 mM	Cajanus cajan	Seedling	Proline (↑), Glycinebetaine (↑), Sugars (↑)	Sharma et al. (2017)
Cu	0.1, 0.2 and 0.3 mM	Phaseolus vulgaris	Seedling	Proline (↑)	Zengin and Munzuroglu (2005)
Hg	15 and 30 μM	Cicer arietinum	Seedling	Proline (↑), Glycinebetaine (↑)	Ahmad et al. (2018)
Hg	20 μg L^{-1}	Suaeda salsa	Shoot and root	Glycinebetaine (↓), Sugars (↑)	Wu et al. (2012)
Hg	10 and 25 μM	Cicer arietinum	Seed	Proline (↑)	Tantrey and Agnihotri (2010)
Hg	25, 50 and 75 μM	Fagopyrum tataricum	Shoot and root	Proline (↓), Soluble sugars (↑)	Pirzadah et al. (2018)
Hg	5, 10, 20, 50, 100, 200, 500 and 1000 mg kg^{-1}	Cyrtomium macrophyllum	Seedling	Proline (↑)	Xun et al. (2017)
Hg	0.02, 0.04 and 0.06 mM	Phaseolus vulgaris	Seedling	Proline (↑)	Zengin and Munzuroglu (2005)
Hg	50, 100, 200, 350 and 500 mg kg^{-1}	Cymbopogon flexuosus	Leaf	Proline (↑)	Handique and Handique (2009)
Ni	50, 100 and 200 μmol	Zea mays	Seed	Sugars (↓)	Fatemeh et al. (2012)
Ni	30 and 60 mgL^{-1}	Cyperus alopecuroides	Leaf	Proline (↑), Soluble sugars (↑)	Batool et al. (2014)
Ni	20, 50, 100, 250 and 500 μM	Zea mays	Root, Shoot	Sugars (↑)	Baccouch et al. (1998)
Ni	2 mM	Glycine max	Seedling	Proline (↑), Glycinebetaine (↑), Sugars (↑)	Sirhindi et al. (2016)

(Continued)

TABLE 4.1 (CONTINUED)
Effect of Mineral Stresses on Different Organic Osmolytes in Plants

Minerals	Concentrations	Plant Species	Organs/ Tissues	Endogenous Osmolytes and Levels	References
Ni	10, 50 and 100 μM	*Vigna mungo*	Seedling	Proline (↑)	Singh et al. (2012)
Pb	1.5, 2.0 and 2.5 mM	*Phaseolus vulgaris*	Seedling	Proline (↑)	Zengin and Munzuroglu (2005)
Pb	50, 100, 200, 350 and 500 mg kg⁻¹	*Cymbopogon flexuosus*	Leaf	Proline (↑)	Handique and Handique (2009)
Pb	100–400 μM	*Brassica napus*	Root	Proline (↑)	Gohari et al. (2012)
Pb	500 and 1000 μM	*Oryza sativa*	Root	Proline (↑)	Sharma and Dubey (2005)
Pb	10⁻⁶ and 10⁻⁴ M	*Helianthus annuus*	Seedling	Proline (↑)	Kastori et al. (1992)
Pb	5, 10, and 25 μM	*Lemna gibba*	Plant	Glycinebetaine (↑), Trehalose (↑)	Duman (2011)
Pb	0.01, 0.1, 0.5, 1.0, 2.5 and 5.0 mM	*Cajanus cajan, Vigna mungo, Triticum aestivum*	Seed	Proline (↑)	Saradhi (1991)
Pb	0, 250, 500, 1000 and 2000 mg L⁻¹	*Robinia pseudoacacia*	Leaf	Proline (↑↓)	Dezhban et al. (2015)
Pb	1, 2 and 4 mM	*Tritium aestivum*	Seed and leaf	Proline (↑)	Yang et al. (2011)
Pb	10, 50 and 100 μM	*Vigna mungo*	Seedling	Proline (↑)	Singh et al. (2012)
Zn	10 ppm	*Triticum aestivum*	Shoot, leaf	Proline (↑)	Bassi and Sharma (1993 a)
Zn	600 and 1000 mg kg⁻¹	*Cajanus cajan*	Root	Trehalose (↑)	Garg and Singh (2018)
Zn	0.5, 1 and 3 mM	*Triticum aestivum*	Leaf and root	Proline (↑), Sugars (↑)	Li et al. (2013)
Zn	10 ppm	*Lemna minor*	Leaf	Proline (↓)	Bassi and Sharma (1993 b)
Zn	20.6 ppm	*Salvia offcinalis*	Leaf	Proline (↑)	Hendawy and Khalid (2005)
Zn	10⁻⁶ and 10⁻⁴ M	*Helianthus annuus*	Seedling	Proline (↑)	Kastori et al. (1992)

↑, increase; ↓, decrease.

and 200 μM (Srivastava and Sharma 2013), *Trapa natans* at 1.28 and 10.80 mg L⁻¹ (Baruah et al. 2014) and wheat at 50 to 400 μM (Zengin 2015) of As stress, the proline accumulation enhanced significantly. Proline and soluble sugar increased in maize (*Zea mays*) leaf at 200 μM (Anjum et al. 2016). Furthermore, in addition to glycinebetaine, proline and soluble sugars increased in garden pea (*Pisum sativum*) at 30, 60 and 90 mg kg⁻¹ of As stress (Garg and Singla 2012) (Table 4.1).

Cadmium stress increased proline content in different plants such as in sunflower (*Helianthus annuus*) (Kastori et al. 1992), chickpea (*Cicer arietinum*) (Tantrey and Agnihotri 2010), french bean (*Phaseolus vulgaris*) seeds (Zengin and Munzuroglu 2005) and *Cymbopogon flexuosus* leaves (Handique and Handique 2009). Glycinebetaine content increased in cotton (*Gossypium hirsutum*) seedlings when exposed to Cd stress (Farooq et al. 2016). Proline and glycinebetaine increased in tobacco (*Nicotiana tabacum*) leaf cells (Islam et al. 2009 a) and tomato (*Solanum lycopersicum*) seedlings (Alyemeni et al. 2018). Again proline and soluble sugar increased in *Cyperus alopecuroides* leaves due to Cd stress (Batool et al. 2014). In *Lemna gibba*, Cd stress increased proline but decreased trehalose and glycinebetaine contents (Duman 2011) (Table 4.1).

Chromium stress increased the proline content of basil (*Ocimum tenuiflorum*) leaves up to 100 μM (Rai et al. 2004) and in rice seedlings up to 50 mM (Yu and Lu 2016). Glycinebetaine increased in wheat seeds at 0.25 and 0.5 mM (Ali et al. 2015) and in mungbean (*Vigna radiata*) leaves at 250 and 500 μM (Jabeen et al. 2016) of Cr stress. Due to Cr stress, soluble sugars decreased in cowpea (Vigna sinensis) seeds (Nath et al. 2008) but increased in *Salvinia minima* (Nichols et al. 2000). Finally, Anjum et al. (2017) showed proline and soluble sugars increased in maize leaves at 30, 60, 90, 120 and 150 μmol L^{-1} of Cr stress (Table 4.1).

Copper stress increased proline content in wheat at 10 ppm (Bassi and Sharma 1993 a), sunflower seedlings at 10^{-6} and 10^{-4} M (Kastori et al. 1992) and in french bean seedlings at 0.1, 0.2 and 0.3 mM (Zengin and Munzuroglu 2005) but decreased in *Lemna minor* leaves at 5 ppm of Cu stress. Trehalose increased in rice seedlings at 100 μM of Cu stress (Mostofa et al. 2015). Proline, glycinebetaine and soluble sugars increased in pigeon pea up to 10 mM of Cu stress (Sharma et al. 2017; Sirhindi et al. 2015) (Table 4.1).

Mercury stress increased proline in chickpea seeds (Tantrey and Agnihotri 2010), *Cyrtomium macrophyllum* seedlings (Xun et al. 2017), french bean seedlings (Zengin and Munzuroglu 2005) and *Cymbopogon flexuosus* leaves (Handique and Handique 2009). Conversely, Hg stress decreased proline and increased soluble sugars in *Fagopyrum tataricum* (Pirzadah et al. 2018). Mercury stress in chickpea seedlings increased proline and glycinebetaine (Ahmad et al. 2018). Again, in *Suaeda salsa* shoots and roots, mercury stress increased glycinebetaine and soluble sugar contents (Wu et al. 2012) (Table 4.1).

Nickel stress increased soluble sugars in maize shoots and roots (Baccouch et al. 1998) but decreased in maize seeds (Fatemeh et al. 2012). Increase in proline showed in black gram (*Vigna mungo*) seedlings due to Ni stress (Singh et al. 2012). Proline and soluble sugars increased in *Cyperus alopecuroides* leaves (Batool et al. 2014). Similar increase in glycinebetaine was observed in soybean (*Glycine max*) seedlings (Sirhindi et al. 2016) (Table 4.1).

Increasing in proline was shown in several plants due to Pb stress, for instance, french bean (Zengin and Munzuroglu 2005), *Cymbopogon flexuosus* leaves (Handique and Handique 2009), rapeseed (*Brassica napus*) roots (Gohari et al. 2012), rice roots (Sharma and Dubey 2005), sunflower (Kastori et al. 1992), pigeon pea, black gram and wheat seeds (Alia and Saradhi 1991), wheat (Yang et al. 2011) and black gram (Singh et al. 2012). Proline remains unchanged in *Robinia pseudoacacia* leaves due to Pb stress (Dezhban et al. 2015). Glycinebetaine and trehalose increased in *Lemna gibba* at 5, 10 and 25 μM of Pb stress (Duman 2011) (Table 4.1).

Trehalose increased in pigeon pea exposed at 600 and 1000 mg kg^{-1} of Zn stress (Garg and Singh 2018). Proline and sugar contents increased in wheat up to 3 mM of Zn stress (Li et al. 2013), where only proline increased in garden sage leaves at 20.6 ppm (Hendawy and Khalid 2005) and sunflower seedlings at 10^{-6} and 10^{-4} M (Kastori et al. 1992) of Zn stress. 10 ppm of Zn stress increased proline in wheat and decreased proline in *Lemna minor* (Bassi and Sharma 1993 b) (Table 4.1).

So far research has shown that mineral stresses generally increase the organic solutes contents including proline, glycinebetaine, trehalose and soluble sugars (Table 4.1), though some decrease and unchanged level of osmolytes were also found by different researchers. It has also been found that a higher concentration of metal toxicity increases the osmolytes content compared to control, though in some cases same concentrations of metal toxicity either increase the osmolytes or decrease the osmolytes secretion, which varies among plants. Thus mineral stress affects the osmolytes in several plants at different concentration rates.

4.3   ROLES OF ORGANIC SOLUTES IN ALLEVIATING MINERAL STRESS

Generalized functions of osmolytes in mineral stress alleviation are described in Figure 4.1. The detailed mechanism of mineral stress alleviation are described below.

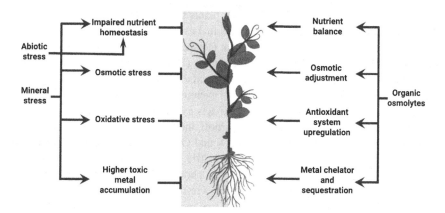

FIGURE 4.1 Simplified mechanism of mineral stress-induced growth inhibition and organic solutes mediated stress tolerance in plants.

4.3.1 Proline in Metal Toxicity Alleviation

Metals such as Cd, Cu, Pb, Mo and Zn are the most common environmental pollutants that can create severe unfavorable conditions resulting in serious physiological and structural disturbances in plants (Hayat et al. 2012 a). Among metal pollutants, Cd is very toxic to plants because it can alter the uptake of essential minerals, the activity of functional enzymes and structures of proteins (Garg and Bhandari 2014). In the contrast, plenty of studies have shown that the application of exogenous proline potentially alleviated the Cd-phototoxicity (Islam et al. 2009 a; Konotop et al. 2017; Rasheed et al. 2014; Tian et al. 2016; Zouari et al. 2016 a, b). Pretreatment with proline reduced the cellular ROS accumulation and also shielded the integrity of cell membrane date palm (*Phoenix dactylifera* L.) and olive (*Olea europaea* L. cv. Chemlali) plants subjected to Cd stress, thereby improving the Cd-stress tolerance (Zouari et al. 2016 a, b). Importantly, exogenous proline application decreased the Cd content in roots and leaves of Cd-treated date palm, olive and wheat plants (Rasheed et al. 2014; Zouari et al. 2016 a, b). Similar findings were also reported in many plant species including in cell-cultured tobacco treated with Cd in presence of proline (Islam et al. 2009 a, b).

Copper as a plant nutrient is involved in the maintenance of a number of enzymes as a cofactor (Emamverdian et al. 2015). However, its excessive concentration in root zones causes oxidative stress and reduces the efficacy of physiological and biochemical processes as well as photosynthesis and thus hampers the growth of plants (Adrees et al. 2015; Emamverdian et al. 2015; Peñarrubia et al. 2015). Foliar spraying of proline enhanced the cell division in the apical meristem of wheat plants, and that eventually contributed to higher plant height (Noreen et al. 2018). Similar findings were reported in rice seedlings treated with Cu stress (Chen et al. 2004). Higher amount of proline accumulation in cells in response to mineral stresses has been reported in lower plant species, for example *Chlorella vulgaris* (Mehta and Gaur 1999).

Molybdenum is also an essential micronutrient that is required at a minimum level by plants for metabolism, and high concentrations cause damage to plant cells (Huamain et al. 1999; Zimmer and Mendel 1999). It has been reported that more plants accumulate higher amounts of endogenous proline in cells to encounter Mo stress and to maintain the plant's ability to tolerate stress-induced adverse conditions (Hadi et al. 2016). In addition, exogenous application of proline improved cabbage seedlings' tolerance against Mo stress and recovered their growth by scavenging ROS as well as upregulating anthocyanin genes' expression and antioxidants enzymes' activity (Kumchai et al. 2013).

Enhanced endogenous content of proline was observed in response to several metals such as Cd, Cu and Zn in *Solanum nigrum* L. (Al Khateeb and Al-Qwasemeh 2014); Pb and Ni in *Vigna mungo* L. (Singh et al. 2012); and Pb, Cd, Cu and Zn in sunflower (Kastori et al. 1992). Therefore, it can be presumed that the accumulation of a higher amount of proline under mineral stress conditions may be a stress-stimulated defensive reaction that may help plants to adapt against these stresses (Theriappan et al. 2011). Plenty of studies have already confirmed that exogenous use of proline can mitigate several mineral toxicities such as Cd stress in spring wheat (Rasheed et al. 2014), Cr stress in rice (Yu et al. 2017) and Cd, Ni and Mn stress in *Scenedesmus armatus* (El-Enany and Issa 2001). Hayat et al. (2012 b) demonstrated that foliar application of proline triggered the accumulation of endogenous proline that counteracted the selenium-induced phytotoxicity and eventually improved the growth traits of plants.

Besides acting as an osmoregulator, proline also acts as a ROS quencher (Kaur and Asthir 2015). Proline can effectively detoxify •OH in plants (Smirnoff and Cumbes 1989). Previously there has been a lot of controversy about the role of proline on $^1O_2$ and $O_2^{•-}$ scavenging, where it was reported that proline does not scavenge those ROS (Signorelli et al. 2013, 2016). However, a recent study has proved that proline is an effective quencher of $^1O_2$ and $O_2^{•-}$ in photosynthetic systems and isolated thylakoids of plants (Ur Rehman et al. 2020). Additional searches testified that both free and polypeptide-bound proline reacts with H_2O_2 and •OH and makes stable free radical adducts of proline and hydroxyproline derivatives (Kaul et al. 2008). Nevertheless, mechanistic evidence was collected by using density functional theory coupled with a polarizable continuum model to sustain the role for proline as a protective OH• scavenger under stress in plants. Proposed was a proline-proline cycle operation in which proline captures the first OH• by H-abstraction followed by the second H-abstraction which also captures another OH• yielding pyrroline-5-carboxylate, which is then recycled back to proline by the action of the pyrroline-5-carboxylate reductase/NADPH enzymatic system (Signorelli et al. 2014).

Several studies have attributed exogenous proline application to an enhancement of the activity of antioxidative enzymes in cultured tobacco cells (Islam et al. 2009 b), date palm (Zouari et al. 2016 b), wheat (Rasheed et al. 2014) under Cd stress; in faba bean under Se stress (Aggarwal et al. 2011); in *Solanum melongena* L. seedlings under As stress (Singh et al. 2015); in rice under Cr stress (Yu et al. 2017); in wheat under Cu stress (Noreen et al. 2018); and in pea under Ni stress (Shahid et al. 2014). Likewise, Xu et al. (2009) revealed that exogenous proline can detoxify the $^1O_2$ radicals by enhancing the SOD activity in Cd-stressed *Solanum nigrum* L. plants. The activities of enzymes of ascorbate-glutathione cycle such as APX, monodehydroascorbate reductase, glutathione reductase and dehydro-ascorbate reductase are considerably improved in mung bean plants due to the application of proline under Cd stress (Hossain et al. 2010). Moreover, Aggarwal et al. (2011) observed higher contents of ascorbic acid and glutathione, and activities of APX, SOD and CAT together with a decrease in H_2O_2 and malondialdehyde (MDA) content in Se-stressed faba bean plants treated with exogenous proline.

Proline also acts as a metal cheater, thereby alleviating metal-induced stress (Hayat et al. 2012 b; Mehta and Gaur 1999). It can mediate the phytochelatin formation process and chelate with metal ions and as a result decrease the metal-induced phytotoxicity (Singh et al. 2010; Sun et al. 2007). It has been reported that exogenous proline can protect the normal activity of nitrate reductase and glucose-6-P dehydrogenase enzymes against inhibition by Zn and Cd (Sharma et al. 1998). This protection was due to the formation of a proline-metal complex. On the other hand, proline supplementation seems to constitute a barrier by which the influx of Cd in roots was controlled. This mechanism appeared to be very effective in limiting Cd content and avoiding the appearance of Cd toxicity symptoms (Zouari et al. 2016 a, b).

4.3.2 GLYCINEBETAINE IN METAL TOXICITY ALLEVIATION

Glycinebetaine is an organic osmolyte that accumulates in a variety of plant species in response to mineral toxicity (Ashraf and Foolad 2007; Lou et al. 2015). Exogenous glycinebetaine application can induce the expression of certain stress-responsive genes, including those for enzymes that scavenge ROS and methylglyoxal, and decrease accumulation and detoxification of ROS, thereby restoring photosynthesis and reducing oxidative stress (Hossain et al. 2010, 2016). It takes part in stabilizing membranes and macromolecules. It is also involved in the stabilization and protection of photosynthetic components, quaternary enzyme and protein complex structures under environmental stresses including mineral toxicity (Farooq et al. 2016; Papageorgiou and Murata 1995).

Lead stress decreased gas exchange characteristics and the performance of antioxidant enzymes (Duman 2011), and an increase in electrolyte leakage and oxidative stress markers, such as hydrogen peroxide and MDA, was observed despite the elevation of antioxidant enzymes activities as well as non-enzymatic antioxidants (Zouari et al. 2018). Exogenous application of glycinebetaine significantly alleviated the Pb-induced toxicity by improving the growth, biomass, photosynthetic parameters (chlorophyll synthesis) and antioxidant enzyme activities and by lowering the electrolytic leakage, MDA and H_2O_2 levels in cotton plants (Bharwana et al. 2014). Similar findings were reported in *Lemna gibba* (Duman 2011) and *O. europaea* trees (Zouari et al. 2018).

Cadmium toxicity caused a significant decrease in plant height, root length, number of leaves per plant, fresh and dry weights of leaf, stem, photosynthetic pigments and gas exchange characteristics in leaves. Cadmium toxicity increased the concentration of ROS as indicated by the increased production of MDA; H_2O_2 and electrolyte leakage; SOD, CAT and POD activity; and oxalic and tartaric acid content in both leaves and roots (Farooq et al. 2016; Lou et al. 2015). Exogenous application of glycinebetaine resulted in a decrease in lipid peroxidation and an increase in CAT activity with reducing Cd accumulation (Islam et al. 2009 b). Similar findings were reported by Lou et al. (2015) on perennial ryegrass. Many studies showed that it alleviated the oxidative damage as evidenced by the decreased production of electrolyte leakage, H_2O_2 and MDA contents and by increasing the performance of the antioxidant enzymatic system in cotton (Farooq et al. 2016), in *T. aestivum* cultivars (Rasheed et al. 2014) and in mung bean (Hossain et al. 2010). It also ameliorated the Cd-induced damage to the leaf/root ultrastructure in maize plants (Li et al. 2016). Exogenous application of glycine betaine significantly restored the membrane integrity and increased the activities of ascorbate-glutathione cycle enzymes under Cd stress (Islam et al. 2009 b). Foliar application of glycinebetaine alleviates the Cd toxicity in spinach (*Spinacia oleracea* L.) through the reduction in Cd uptake and improvement in the antioxidant defensive system under Cd stress (Aamer et al. 2018) and in the aquatic plant *Lemna gibba* L. (Duman 2011).

Many studies revealed that Cr reduces plant growth and development by influencing photosynthetic performance and antioxidant enzyme activities (Ahmad et al. 2020 b; Jabeen et al. 2016). Exogenous application of glycinebetaine can diminish or lessen the lethal consequences produced by Cr stress in many crop species, for example in cauliflower (*Brassica oleracea* L.) (Ahmad et al. 2020 a), wheat (Ali et al. 2015) and sorghum (Kumar et al. 2019). Parallel results were stated by Jabeen et al. (2016) in mung bean plants, where increased plant growth and photosynthetic pigments under Cr stress were observed. It has also been reported that glycinebetaine could decrease oxidative stress and improve oxidant scavenging enzymes' activity and play a positive role in maintaining optimal plant morphology and photosynthetic attributes of cauliflower under Cr-stressed conditions (Ahmad et al. 2020 a). Also, in wheat plants, the enhanced activity of antioxidant enzymes in both shoots and roots was observed (Ali et al. 2015).

Moreover, excessive Ni in growth media hampered plant growth and plasma membrane stability of *Pennisetum typhoideum* plant, which was highly linked with the cellular accumulated Ni and ROS. On the other hand, the application of exogenous glycinebetaine served as a protective tool to impart Ni-stress tolerance through the upregulation of antioxidant systems and thereby helped plants to mitigate Ni-toxicity (Xalxo et al. 2017). Furthermore, exposure to Al stress induced quick stomatal closure that caused CO_2 entry limitation. Therefore, Al accumulation directly resulted in the impairment of CO_2 assimilation in plants. Conversely, foliar exogenous glycinebetaine application can significantly alleviate the adverse effects of Al stress through the protection of the photosynthetic apparatus, resulting in an improved electron transport chain, gas exchange and enzymatic CO_2 fixation (Stepien 2016).

4.3.3 TREHALOSE IN METAL TOXICITY ALLEVIATION

Trehalose, a non-reducing disaccharide, is produced in large amounts when plants encounter various environmental stresses (Iordachescu and Imai 2008). Several researchers proved the protective roles of trehalose in metabolism, physiology and development processes of plants (Fernandez et al. 2010; Garg et al. 2002). Trehalose is found to be a direct and indirect scavenger of ROS, and the protective effect of trehalose seems to be primarily caused by its function as a protectant against damage to lipids and membranes by physical interaction (Stolker 2010). Treatment with exogenous trehalose altered levels of transcripts and modified cell wall, nitrogen metabolism and several genes related to stress-related genes (Bae et al. 2005).

Many studies have revealed the toxic effects of excessive Cu on photosynthetic pigments and plant growth-related parameters (Mostofa et al. 2015; Tahjib-Ul-Arif et al. 2020). The improved tolerance induced by trehalose could be attributed to its ability to reduce Cu uptake and decrease Cu-induced oxidative damage by lowering the accumulation of ROS and MDA in Cu-stressed plants (Mostofa et al. 2015). Trehalose counteracted the Cu-induced increase in proline and glutathione content, but significantly improved ascorbic acid content and redox status. The activities of major antioxidant enzymes were largely stimulated by trehalose pretreatment in rice plants exposed to excessive Cu (Mostofa et al. 2015). Trehalose pretreatment improved Cu tolerance in rice plants by inhibiting Cu uptake and regulating the antioxidant and glyoxalase systems and thereby demonstrated the important role of trehalose in mitigating heavy metal toxicity (Mostofa et al. 2015). Several studies revealed that Cd causes a remarkable decrease in pigments (chlorophyll *a* and *b)* content (Wang et al. 2020). Trehalose significantly protected plants from lipid peroxidation, and improved photosynthetic activity, proline content and antioxidant enzyme activities (Duman et al. 2011). Moreover, a study revealed that engineered bacteria produce trehalose and found that they then reduced 1 mM Cr (VI) to Cr (III), whereas wild-type cells were only able to reduce half that amount and reduced Cr toxicity (Frederick et al. 2013).

4.3.4 POLYOLS IN METAL TOXICITY ALLEVIATION

Mineral toxicity alters sugar metabolism in plants and induces the accumulation of sugar alcohols (also known as polyols) (Oikonomou et al. 2019). The most commonly occurring polyols in higher plants are myo-inositol, mannitol and sorbitol (Noiraud et al. 2001), which are thought to protect plants from oxidative damage (Yildizli et al. 2018).

Hindrance in nutrient uptake by plants due to Cd and Pb stress might have resulted in plant biomass reduction (Tauqeer et al. 2016). Chromium toxicity destroys chlorophyll and carotenoids contents in maize (Habiba et al. 2019), sunflower (Singh et al. 2013), wheat and mung bean (Jabeen et al. 2016). Mannitol is an important polyol that significantly alleviated the Cd/Pb-induced deterioration in growth and biomass (Tauqeer et al. 2016). Exogenous application of mannitol reduces the Cr uptake through roots, plays a protective role in cell membrane stability and improves photosynthetic pigments (Habiba et al. 2019; Kaya et al. 2013). Mannitol significantly decreased the ROS production under metal stress by enhancing the antioxidant defense system (Habiba et al. 2019).

Sorbitol is also found in higher plants, but it is a polyol that is less frequently tested on stressed conditions (Lambers et al. 1981). Salvucci (2000) reported that sorbitol protects cytoplasmic proteins and cell membranes from desiccation in stressed conditions. Exogenous spraying of sorbitol reduces cell electrolyte leakage. Sorbitol showed significant effects when applied in low concentrations and had adverse effects when applied in high concentrations (Issa et al. 2020).

4.4 ROLE OF ORGANIC SOLUTES IN CONFERRING MINERAL HOMEOSTASIS UNDER STRESS

The natural environment for plants is composed of a complex set of abiotic stresses and biotic stresses. The use of osmolytes may be a strategy to overcome the adverse effect of these abiotic stresses. However, researchers showed that the use of different concentrations of osmolytes can help plants to maintain mineral homeostasis under abiotic stress conditions (Table 4.2).

TABLE 4.2

Role of Exogenous Organic Solutes in Maintaining Mineral Homeostasis under Stress

Stress	Concentration	Plant Species	Exogenous Osmolytes	Mineral Contents	References
Salinity	1.84, 6.03 or 8.97 dSm^{-1}	*Phaseolus vulgaris*	Proline (5 mM)	K$^+$ ↑, Na$^+$ ↓	Abdelhamid et al. (2013)
	200 mM NaCl	*Viburnum lucidum Callistemon citrinus*	Glycinebetaine (2.5 mM), Proline (5.0 mM)	Na$^+$ ↓	Cirillo et al. (2016)
	50, 100 and 150 mM NaCl	*Oryza sativa*	Trehalose (10 and 20 mM)	Na$^+$ ↓, Ca$^+$ ↓	Shahbaz et al. (2017)
	100 mM NaCl	*Olea europaea L. cv. Chemlali*	Proline (25 mM)	Na$^+$ ↓, K$^+$ ↑, Ca$^+$ ↑	Ahmed et al. (2011)
	40 and 160 mM NaCl	*Atriplex halimus*	Proline (1 mM) Glycinebetaine (1 mM)	Na$^+$ =, K$^+$ =	Hassine et al. (2008)
	200 mM NaCl	*Eurya emarginata*	Proline (10 mM)	K$^+$ ↑, Na$^+$ ↓	Zheng et al. (2015)
	50, 100 and 150 mM NaCl	*Carthamus tinctorius*	Glycinebetaine (10, 30 and 60 mM)	K$^+$ ↑, Na$^+$ ↓	Alasvandyari et al. (2017)
	135 mM NaCl	*Lycopersicon esculentum*	Proline (1 and 10 mM), Glycinebetaine (1 and 5 mM)	K$^+$↑, Na$^+$ ↓	Heuer (2003)
	100 mM NaCl	*Oryza sativa*	Proline (30 mM)	K$^+$↓, Na$^+$ ↓	Roy et al. (1993)
	-0.2 MPa	*Zea mays*	Trehalose (2–20 mM)	K$^+$/Na$^+$ ↑	Zeid (2009)
	150 and 250 mM NaCl	*Arabidopsis sp.*	Trehalose (0.5, 1 and 5 mM)	Na$^+$ ↓, K$^+$ ↑, Mg^{2+} ↓, Ca^{2+} ↓	Yang et al. (2014)
	50 and 100 mM NaCl	*Phaseolus vulgaris*	Glycinebetaine (25 and 50 mM)	Na$^+$ ↓, K$^+$ ↑	Sofy et al. (2020)
	2.84, 15 dSm^{-1}	*Triticum aestivum*	Glycinebetaine (50 and 100 mM)	Na$^+$ ↓, K$^+$ ↑, Ca^{2+}↑	Raza et al. (2007)
	75, 150 and 225 mM NaCl	*Triticum aestivum*	Trehalose (10 mM)	Na$^+$ ↓, K$^+$ ↑	Nemat Alla et al. (2019)
	11 and 17 dSm^{-1}	*Gossypium hirsutum*	Trehalose (5 and 50 mM)	Na$^+$ =, K$^+$ =	Shahzad et al. (2020)
	0.70, 3.15 and 6.30 dSm^{-1}	*Vigna unguiculata*	Trehalose (1 gL^{-1})	K$^+$↑	Eisa and Ibrahim (2016)
	150 mM NaCl	*Pisum sativum*	Glycinebetaine (5 and 10 mM)	Na$^+$ ↓, Ca^{2+} ↑	Nusrat et al. (2014)
	250 mM NaCl	*Catharanthus roseus*	Trehalose (10 mM)	Na$^+$ ↓, K$^+$ ↑	Chang et al. (2014)
Drought	60% FC	*Triticum aestivum*	Glycinebetaine (50 and 100 mM)	K$^+$ ↑	Shahbaz et al. (2012)
	60% FC	*Triticum aestivum*	Glycinebetaine (10 mM)	K$^+$ ↓, Na$^+$ ↓, Cl$^-$ ↓, Mg^{2+} ↓, Ca^{2+} ↓	Aldesuquy et al. (2013)
	85 ± 5% and 65 ± 5% FC	*Zea mays*	Glycinebetaine (20, 40, 60 and 60 ml)	K$^+$ ↑	LiXin et al. (2009)
	-	*Triticum aestivum*	Glycinebetaine (50, 100 and 150 mM)	K$^+$ ↓, Ca^{2+} ↑, Na$^+$ ↓	Raza et al. (2015)
	60% FC	*Zea mays*	Proline (30 and 60 mmolL^{-1})	K$^+$ ↑, Ca^{2+} ↑, Mg^{2+} ↑	Ali et al. (2008)
	60% FC	*Raphanus sativus*	Trehalose (25 mM)	K$^+$ ↑, Ca^{2+} ↑	Akram et al. (2016)
	60% FC	*Raphanus sativus*	Trehalose (25 and 50 mM)	K$^+$ =, Ca^{2+} ↑	Akram et al. (2015)

=, unchanged; ↑, increase; ↓, decrease; FC, field capacity.

Under salinity stress, different rates of proline applications decreased the Na^+ and increased the K^+ in several plants. For example, 5 mM proline application in french bean (Abdelhamid et al. 2013), 25 mM proline in olive (*Olea europaea cv. Chemlali*) (Ahmed et al. 2011), 10 mM proline in *Eurya emarginata* (Zheng et al. 2015) and tomato (Heuer 2003) decreased the Na^+ and increased the K^+ accumulation. Similar decreasing of Na^+ was found in *Viburnum lucidum* L. and Crimson Bottlebrush (*Callistemon citrinus*) (Cirillo et al. 2016). However, both Na^+ and K^+ decreased in rice under salinity at 30 mM of exogenous proline (Roy et al. 1993). Conversely, Na^+ and K^+ remained unchanged in Mediterranean saltbush (*Atriplex halimus*) at 1 mM exogenous proline application (Hassine et al. 2008). Exogenous proline of 25 mM in olive increased Ca^{2+} content during salinity stress (Ahmed et al. 2011).

Similar to proline, exogenous glycinebetaine application during salinity stress increased K^+ and decreased Na^+. For instance, glycinebetaine up to 60 mM in safflower (*Carthamus tinctorius*) (Alasvandyari et al. 2017), glycinebetaine up to 5 mM in tomato (Heuer 2003) and glycinebetaine up to 50 mM in french bean (Sofy et al. 2020) increased K^+ and decreased Na^+ under salinity stress. However, both Na^+ and K^+ remain unchanged in Mediterranean saltbush at 1 mM exogenous glycinebetaine application during salinity stress (Hassine et al. 2008). Exogenous glycinebetaine up to 100 mM decreased Na^+ and increased both K^+ and Ca^{2+} contents in wheat under salinity stress (Raza et al. 2007). Again, Na^+ decreasing and Ca^{2+} increasing was found in pea (*Pisum sativum*) up to 10 mM exogenous glycinebetaine application during salinity (Nusrat et al. 2014). Furthermore, in *Viburnum lucidum* and crimson bottlebrush, 2.5 mM glycinebetaine foliar application decreased Na^+ (Cirillo et al. 2016).

Na^+ decreasing and K^+ increasing were found at 10 mM exogenous trehalose application in wheat under salinity stress (Nemat Alla et al. 2019) and Madagascar periwinkle (*Catharanthus roseus*) (Chang et al. 2014). Similar results with decreasing Mg^{2+} and Ca^{2+} were found in *Arabidopsis spp.* under salinity up to 5 mM exogenous trehalose applications (Yang et al. 2014). Exogenous trehalose application in cowpea increased K^+ during salinity stress (Eisa and Ibrahim 2016). Again, trehalose application in rice under salinity decreased Na^+ and Ca^{2+} (Shahbaz et al. 2017). The ratio of K^+ and Na^+ increased in exogenous trehalose application in maize (Zeid 2009). Conversely, both Na^+ and K^+ remained unchanged in cotton under salinity stress (Shahzad et al. 2020).

Exogenous glycinebetaine applications increased K^+ in maize under drought stress (LiXin et al. 2009; Shahbaz et al. 2012). Decrease of K^+, Na^+, Cl^-, Mg^{2+} and Ca^{2+} was shown in wheat at 10 mM exogenous glycinebetaine application (Aldesuquy et al. 2013). Again, exogenous glycinebetaine up to 150 mM in wheat increased Ca^{2+}, and decreased K^+ and Na^+ (Raza et al. 2015). During drought stress, 30 and 60 mmol L^{-1} exogenous proline applications increased K^+, Ca^{2+} and Mg^{2+} in maize (Ali et al. 2008). Exogenous trehalose application at 25 mM increased K^+ and Ca^{2+} in radish (*Raphanus sativus*) (Akram et al. 2016). On the other hand, exogenous trehalose application up to 50 mM increased Ca^{2+}, but K^+ remained unchanged in radish (Akram et al. 2015).

From analyzing the above, it can be said that using different concentrations of osmolytes affects the plant minerals under different abiotic stress.

4.5   L-CARNITINE: AN UNREVEALED OSMOLYTE IN PLANTS

L-carnitine is a well-known organic compound which has been frequently studied in animal systems. All living beings depend on these small compatible solutes that accumulate in cells without altering any cellular functions, and these compounds usually protect cells from the adverse effects of osmotic stress (Yancey 2005). L-carnitine is a quaternary ammonium compound that has osmolyte properties (Peluso et al. 2000). Under salt- and chilling-induced osmotic stress conditions, L-carnitine accumulated in higher levels and functioned as osmotic adjustment, which eventually reduced the adverse effects of such stress. The higher accumulation of intracellular L-carnitine *Escherichia coli* in under salt stress occurred due to the transformation of crotonobetaine into L-carnitine (Cánovas et al. 2007). Moreover, under chilling stress, the transportation of exogenous

L-carnitine enhanced from the growth medium to cytosol in *Listeria monocytogenes* (Angelidis and Smith 2003). In animals, L-carnitine accumulation in the epididymal lumen is facilitated by the Na^{2+}/L-carnitine plasma membrane symporter *OCTN2*, which is activated by osmotic stress (Cotton et al. 2010). In plants, quaternary ammonium compounds, such as glycine betaine, act as osmolytes (Ashraf and Foolad 2007). Recently it has been reported that L-carnitine supplementation in *Arabidopsis thaliana* acts as a compatible solute and helps to counteract against salt-induced osmotic stress, similarly to proline (Charrier et al. 2012). Hence, L-carnitine could function as a compatible solute due to its chemical closeness with glycinebetaine. Detailed and compartment-specific accumulation of L-carnitine has not been studied yet. Therefore, the role of L-carnitine as a compatible solute in plants remains hypothetical. On the other hand, in bacteria and animals the L-carnitine transporters *OpuC* (Angelidis et al. 2002) and *OCTN2* (Tamai et al. 2001) are associated with osmoregulation, respectively. In *A. thaliana*, the homologueous protein of *OCTN2* is *OCT1* (Lelandais-Brière et al. 2007). Moreover, the *OCT1* protein localizes onto the vacuolar membrane that helps plants to accumulate Na^+ under salt stress. Therefore, it is possible that *OCT1* could be an L-carnitine/Na^+ symporter, and it can be presumed that L-carnitine plays a part in osmotic adjustment through ionic homeostasis under mineral stress conditions in plant cells.

4.6 CONCLUSIONS

In this era of industrialization, metal pollution is increasing day by day and therefore plants are facing higher metal stresses. Metal stress factors considerably affect plant growth and development and eventually yield of crops. Plants employ several defensive mechanisms, such as metabolic changes, gene expression and osmolytes synthesis, to counteract metal toxicity. Several organic solutes like proline, glycinebetaine, trehalose and polyols show protective roles against metal stress, which has been proved by many researchers. Usually, those osmolytes adjust metal-induced osmotic imbalance and protect plants from oxidative damages as well as act as metal chelators. Therefore, manipulation of crop plants for stress-related pathways of osmolytes via genetic engineering to ensure higher organic solutes production and accumulation could be an effective means to develop metal stress-tolerant plants.

REFERENCES

Aamer, M., Muhammad, U., Li, Z., Abid, A., Su, Q., Liu, Y., Adnan, R., Muhammad, A., Tahir, A., Huang, G., 2018. Foliar application of glycinebetaine (gb) alleviates the cadmium (Cd) toxicity in spinach through reducing cd uptake and improving the activity of anti-oxidant system. Applied Ecology and Environmental Research 16, 7575–7583.

Abdelhamid, M.T., Rady, M.M., Osman, A.S., Abdalla, M.A., 2013. Exogenous application of proline alleviates salt-induced oxidative stress in *Phaseolus vulgaris* L. plants. *The Journal of Horticultural Science and Biotechnology* 88(4), 439–446.

Adrees, M., Ali, S., Rizwan, M., Ibrahim, M., Abbas, F., Farid, M., Zia-ur-Rehman, M., Irshad, M.K., Bharwana, S.A., 2015. The effect of excess copper on growth and physiology of important food crops: a review. *Environmental Science and Pollution Research International* 22(11), 8148–8162.

Aggarwal, M., Sharma, S., Kaur, N., Pathania, D., Bhandhari, K., Kaushal, N., Kaur, R., Singh, K., Srivastava, A., Nayyar, H., 2011. Exogenous proline application reduces phytotoxic effects of selenium by minimising oxidative stress and improves growth in bean (*Phaseolus vulgaris* L.) seedlings. *Biological Trace Element Research* 140(3), 354–367.

Ahmad, P., Ahanger, M.A., Egamberdieva, D., Alam, P., Alyemeni, M.N., Ashraf, M., 2018. Modification of osmolytes and antioxidant enzymes by 24-epibrassinolide in chickpea seedlings under mercury (Hg) toxicity. *Journal of Plant Growth Regulation* 37(1), 309–322.

Ahmad, R., Ali, S., Abid, M., Rizwan, M., Ali, B., Tanveer, A., Ahmad, I., Azam, M., Ghani, M.A., 2020a. Glycinebetaine alleviates the chromium toxicity in *Brassica oleracea* L. by suppressing oxidative stress and modulating the plant morphology and photosynthetic attributes. *Environmental Science and Pollution Research International* 27(1), 1101–1111.

Ahmad, R., Ali, S., Rizwan, M., Dawood, M., Farid, M., Hussain, A., Wijaya, L., Alyemeni, M.N., Ahmad, P., 2020b. Hydrogen sulfide alleviates chromium stress on cauliflower by restricting its uptake and enhancing antioxidative system. *Physiologia Plantarum* 168(2), 289–300.

Akhtar, T., Zia-ur-Rehman, M., Naeem, A., Nawaz, R., Ali, S., Murtaza, G., Maqsood, M.A., Azhar, M., Khalid, H., Rizwan, M., 2017. Photosynthesis and growth response of maize (*Zea mays* L.) hybrids exposed to cadmium stress. *Environmental Science and Pollution Research International* 24(6), 5521–5529.

Akram, N.A., Noreen, S., Noreen, T., Ashraf, M., 2015. Exogenous application of trehalose alters growth, physiology and nutrient composition in radish (*Raphanus sativus* L.) plants under water-deficit conditions. *Brazilian Journal of Botany* 38(3), 431–439.

Akram, N.A., Waseem, M., Ameen, R., Ashraf, M., 2016. Trehalose pretreatment induces drought tolerance in radish (*Raphanus sativus* L.) plants: some key physio-biochemical traits. *Acta Physiologiae Plantarum* 38(1), 3.

Al Khateeb, W., Al-Qwasemeh, H., 2014. Cadmium, copper and zinc toxicity effects on growth, proline content and genetic stability of *Solanum nigrum* L., a crop wild relative for tomato; comparative study. *Physiology and Molecular Biology of Plants* 20(1), 31–39.

Alasvandyari, F., Mahdavi, B., Madah, H., 2017. Glycine betaine affects the antioxidant system and ion accumulation and reduces salinity-induced damage in safflower seedlings. *Archives of Biological Sciences* 69(1), 139–147.

Aldesuquy, H.S., Abbas, M.A., Abo-Hamed, S.A., Elhakem, A.H., 2013. Does glycine betaine and salicylic acid ameliorate the negative effect of drought on wheat by regulating osmotic adjustment through solutes accumulation? *Journal of Stress Physiology and Biochemistry* 9, 5–22.

Ali, Q., Ashraf, M., Shahbaz, M., Humera, H., 2008. Ameliorating effect of foliar applied proline on nutrient uptake in water stressed maize (*Zea mays* L.) plants. *Pakistan Journal of Botany* 40, 211–219.

Ali, S., Chaudhary, A., Rizwan, M., Anwar, H.T., Adrees, M., Farid, M., Irshad, M.K., Hayat, T., Anjum, S.A., 2015. Alleviation of chromium toxicity by glycinebetaine is related to elevated antioxidant enzymes and suppressed chromium uptake and oxidative stress in wheat (*Triticum aestivum* L.). *Environmental Science and Pollution Research International* 22(14), 10669–10678.

Alia, P.P., Saradhi, P.P., 1991. Proline accumulation under heavy metal stress. *Journal of Plant Physiology* 138(5), 554–558.

Almaghamsi, A., Nosarzewski, M., Kanayama, Y., Archbold, D.D., 2021. Effects of abiotic stresses on sorbitol biosynthesis and metabolism in tomato (*Solanum lycopersicum*). *Functional Plant Biology* 48(3), 286–297.

Alyemeni, M.N., Ahanger, M.A., Wijaya, L., Alam, P., Bhardwaj, R., Ahmad, P., 2018. Selenium mitigates cadmium-induced oxidative stress in tomato (*Solanum lycopersicum* L.) plants by modulating chlorophyll fluorescence, osmolyte accumulation, and antioxidant system. *Protoplasma* 255(2), 459–469.

Angelidis, A.S., Smith, G.M., 2003. Role of the glycine betaine and carnitine transporters in adaptation of Listeria monocytogenes to chill stress in defined medium. *Applied and Environmental Microbiology* 69(12), 7492–7498.

Angelidis, A.S., Smith, L.T., Hoffman, L.M., Smith, G.M., 2002. Identification of OpuC as a chill-activated and osmotically activated carnitine transporter in *Listeria monocytogenes*. *Applied and Environmental Microbiology* 68(6), 2644–2650.

Anjum, S.A., Ashraf, U., Khan, I., Tanveer, M., Shahid, M., Shakoor, A., Wang, L., 2017. Phyto-toxicity of chromium in maize: oxidative damage, osmolyte accumulation, anti-oxidative defense and chromium uptake. *Pedosphere* 27(2), 262–273.

Anjum, S.A., Tanveer, M., Hussain, S., Shahzad, B., Ashraf, U., Fahad, S., Hassan, W., Jan, S., Khan, I., Saleem, M.F., Bajwa, A.A., Wang, L., Mahmood, A., Samad, R.A., Tung, S.A., 2016. Osmoregulation and antioxidant production in maize under combined cadmium and arsenic stress. *Environmental Science and Pollution Research International* 23(12), 11864–11875.

Asati, A., Pichhode, M., Nikhil, K., 2016. Effect of heavy metals on plants: an overview. *International Journal of Application or Innovation in Engineering & Management* 5, 2319–4847.

Ashraf, M., Foolad, M.R., 2007. Roles of glycine betaine and proline in improving plant abiotic stress resistance. *Environmental and Experimental Botany* 59(2), 206–216.

Assareh, M., Shariat, A., Ghamari-Zare, A., 2008. Seedling response of three Eucalyptus species to copper and zinc toxic concentrations. *Caspian Journal of Environmental Sciences* 6, 97–103.

Baccouch, S., Chaoui, A., Ferjani, E.E., 1998. Nickel toxicity: effects on growth and metabolism of maize. *Journal of Plant Nutrition* 21(3), 577–588.

Bae, Hanhong, Herman, E., Bailey, B., Bae, Hyeun-Jong, Sicher, R., 2005. Exogenous trehalose alters Arabidopsis transcripts involved in cell wall modification, abiotic stress, nitrogen metabolism, and plant defense. *Physiologia Plantarum* 125(1), 114–126.

Baruah, S., Borgohain, J., Sarma, K., 2014. Phytoremediation of arsenic by *Trapa natans* in a hydroponic system. *Water Environment Research* 86(5), 422–432.

Bassi, R., Sharma, S.S., 1993a. Proline accumulation in wheat seedlings exposed to zinc and copper. *Phytochemistry* 33(6), 1339–1342.

Bassi, R., Sharma, S.S., 1993b. Changes in proline content accompanying the uptake of zinc and copper by *Lemna minor. Annals of Botany* 72(2), 151–154.

Batool, R., Hameed, M., Ashraf, M., Fatima, S., Nawaz, T., Ahmad, M.S.A., 2014. Structural and functional response to metal toxicity in aquatic *Cyperus alopecuroides* Rottb. *Limnologica* 48, 46–56.

Ben Ahmed, Ch, Magdich, S., Ben Rouina, B., Sensoy, S., Boukhris, M., Ben Abdullah, F., 2011. Exogenous proline effects on water relations and ions contents in leaves and roots of young olive. *Amino Acids* 40(2), 565–573.

Ben Hassine, A.B., Ghanem, M.E., Bouzid, S., Lutts, S., 2008. An inland and a coastal population of the Mediterranean xero-halophyte species *Atriplex halimus* L. differ in their ability to accumulate proline and glycinebetaine in response to salinity and water stress. *Journal of Experimental Botany* 59(6), 1315–1326.

Bhamburdekar, S., Chavan, P., 2011. Effect of some stresses on free proline content during pigeonpea (*Cajanas cajan*) seed germination. *Journal of Stress Physiology and Biochemistry* 7(3), 235–241.

Bharwana, S.A., Ali, S., Farooq, M.A., Iqbal, N., Hameed, A., Abbas, F., Ahmad, M.S.A., 2014. Glycine beta-ine-induced lead toxicity tolerance related to elevated photosynthesis, antioxidant enzymes suppressed lead uptake and oxidative stress in cotton. *Turkish Journal of Botany* 38, 281–292.

Cánovas, M., Bernal, V., Sevilla, A., Torroglosa, T., Iborra, J.L., 2007. Salt stress effects on the central and carnitine metabolisms of Escherichia coli. *Biotechnology and Bioengineering* 96(4), 722–737.

Charrier, A., Rippa, S., Yu, A., Nguyen, P.-J., Renou, J.-P., Perrin, Y., 2012. The effect of carnitine on Arabidopsis development and recovery in salt stress conditions. *Planta* 235(1), 123–135.

Chang, B., Yang, L., Cong, W., Zu, Y., Tang, Z., 2014. The improved resistance to high salinity induced by trehalose is associated with ionic regulation and osmotic adjustment in *Catharanthus roseus. Plant Physiology and Biochemistry* 77, 140–148.

Chatterjee, J., Chatterjee, C., 2000. Phytotoxicity of cobalt, chromium and copper in cauliflower. *Environmental Pollution* 109(1), 69–74.

Cha-um, S., Takabe, T., Kirdmanee, C., 2010. Osmotic potential, photosynthetic abilities and growth charac-ters of oil palm (*Elaeis guineensis* Jacq.) seedlings in responses to polyethylene glycol-induced water deficit. *African Journal of Biotechnology* 9, 6509–6516.

Chen, C.-T., Chen, T.-H., Lo, K.-F., Chiu, C.-Y., 2004. Effects of proline on copper transport in rice seedlings under excess copper stress. *Plant Science* 166(1), 103–111.

Chen, L.-M., Lin, C.C., Kao, C.H., 2000. Copper toxicity in rice seedlings: changes in antioxidative enzyme activities, H2O2 level, and cell wall peroxidase activity in roots. *Botanical Bulletin of Academia Sinica* 41(2), 99–103.

Cirillo, C., Rouphael, Y., Caputo, R., Raimondi, G., Sifola, M., De Pascale, S., 2016. Effects of high salinity and the exogenous application of an osmolyte on growth, photosynthesis, and mineral composition in two ornamental shrubs. *The Journal of Horticultural Science and Biotechnology* 91(1), 14–22.

Cotton, L.M., Rodriguez, C.M., Suzuki, K., Orgebin-Crist, M., Hinton, B.T., 2010. Organic cation/carni-tine transporter, OCTN2, transcriptional activity is regulated by osmotic stress in epididymal cells. *Molecular Reproduction and Development: Incorporating Gamete Research* 77(2), 114–125.

Dezhban, A., Shirvany, A., Attarod, P., Delshad, M., Matinizadeh, M., Khoshnevis, M., 2015. Cadmium and lead effects on chlorophyll fluorescence, chlorophyll pigments and proline of *Robinia pseudoacacia. Journal of Forestry Research* 26(2), 323–329.

Doncheva, S., Stoyanova, Z., 2007. Plant response to copper and zinc hydroxidesulphate and hydroxidecar-bonate used as an alternative copper and zinc sources in mineral nutrition. *Romanian Agricultural Research* 7, 15–23.

Duman, F., 2011. Effects of exogenous glycinebetaine and trehalose on lead accumulation in an aquatic plant (*Lemna gibba* L.). *International Journal of Phytoremediation* 13(5), 492–497.

Duman, F., Aksoy, A., Aydin, Z., Temizgul, R., 2011. Effects of exogenous glycinebetaine and trehalose on cadmium accumulation and biological responses of an aquatic plant (*Lemna gibba* L.). *Water, Air, and Soil Pollution* 217(1–4), 545–556.

Eisa, G., Ibrahim, S.A., 2016. Mitigation the harmful effects of diluted sea water salinity on physiological and anatomical traits by foliar spray with trehalose and sodium nitroprusside of cowpea plants. *Middle East Journal of Agriculture Research* 5, 672–686.

Elbaz, A., Wei, Y.Y., Meng, Q., Zheng, Q., Yang, Z.M., 2010. Mercury-induced oxidative stress and impact on antioxidant enzymes in Chlamydomonas reinhardtii. *Ecotoxicology* 19(7), 1285–1293.

El-Enany, A., Issa, A., 2001. Proline alleviates heavy metal stress in *Scenedesmus armatus*. *Folia Microbiologica* 46(3), 227–230.

Emamverdian, A., Ding, Y., Mokhberdoran, F., Xie, Y., 2015. Heavy metal stress and some mechanisms of plant defense response. *The Scientific World Journal* Vol. 2015, ID 756120.

Farooq, M., Ali, S., Hameed, A., Bharwana, S., Rizwan, M., Ishaque, W., Farid, M., Mahmood, K., Iqbal, Z., 2016. Cadmium stress in cotton seedlings: physiological, photosynthesis and oxidative damages alleviated by glycinebetaine. *South African Journal of Botany* 104, 61–68.

Fatemeh, G., Reza, H., Rashid, J., Latifeh, P., 2012. Effects of Ni^{2+} toxicity on Hill reaction and membrane functionality in maize. *Journal of Stress Physiology and Biochemistry* 8(4):55–61.

Fernandez, O., Béthencourt, L., Quero, A., Sangwan, R.S., Clément, C., 2010. Trehalose and plant stress responses: friend or foe? *Trends in Plant Science* 15(7), 409–417.

Frederick, T.M., Taylor, E.A., Willis, J.L., Shultz, M.S., Woodruff, P.J., 2013. Chromate reduction is expedited by bacteria engineered to produce the compatible solute trehalose. *Biotechnology Letters* 35(8), 1291–1296.

Gadallah, M., 1999. Effects of proline and glycinebetaine on *Vicia faba* responses to salt stress. *Biologia Plantarum* 42(2), 249–257.

Galvez, L., Clark, R., Klepper, L., Hansen, L., 1991. Organic acid and free proline accumulation and nitrate reductase activity in sorghum (*Sorghum bicolor*) genotypes differing in aluminium tolerance. In: Wright R.J., Baligar V.C., Murrmann R.P. (eds), *Plant-Soil Interactions at Low Ph*. Dordrecht: Springer, 859–867.

Garg, A.K., Kim, J.-K., Owens, T.G., Ranwala, A.P., Do Choi, Y., Kochian, L.V., Wu, R.J., 2002. Trehalose accumulation in rice plants confers high tolerance levels to different abiotic stresses. *Proceedings of the National Academy of Sciences of the United States of America* 99(25), 15898–15903.

Garg, N., Bhandari, P., 2014. Cadmium toxicity in crop plants and its alleviation by arbuscular mycorrhizal (AM) fungi: an overview. *Plant Biosystems - an International Journal Dealing with All Aspects of Plant Biology* 148(4), 609–621.

Garg, N., Singh, S., 2018. Mycorrhizal inoculations and silicon fortifications improve rhizobial symbiosis, antioxidant defense, trehalose turnover in pigeon pea genotypes under cadmium and zinc stress. *Plant Growth Regulation* 86(1), 105–119.

Garg, N., Singla, P., 2012. The role of *Glomus mosseae* on key physiological and biochemical parameters of pea plants grown in arsenic contaminated soil. *Scientia Horticulturae* 143, 92–101.

Ghori, N.-H., Ghori, T., Hayat, M., Imadi, S., Gul, A., Altay, V., Ozturk, M., 2019. Heavy metal stress and responses in plants. *International Journal of Environmental Science and Technology* 16(3), 1807–1828.

Gohari, M., Habib-Zadeh, A., Khayat, M., 2012. Assessing the intensity of tolerance to lead and its effect on amount of protein and proline in root and aerial parts of two varieties of rape seed (*Brassica napus* L.). *Journal of Basic and Applied Scientific Research* 2, 935–938.

Habiba, U., Ali, S., Rizwan, M., Ibrahim, M., Hussain, A., Shahid, M.R., Alamri, S.A., Alyemeni, M.N., Ahmad, P., 2019. Alleviative role of exogenously applied mannitol in maize cultivars differing in chromium stress tolerance. *Environmental Science and Pollution Research International* 26(5), 5111–5121.

Hadi, F., Ali, N., Fuller, M.P., 2016. Molybdenum (Mo) increases endogenous phenolics, proline and photosynthetic pigments and the phytoremediation potential of the industrially important plant *Ricinus communis* L. for removal of cadmium from contaminated soil. *Environmental Science and Pollution Research International* 23(20), 20408–20430.

Handique, G., Handique, A., 2009. Proline accumulation in lemongrass (*Cymbopogon flexuosus* Stapf.) due to heavy metal stress. *Journal of Environmental Biology* 30(2), 299–302.

Hänsch, R., Mendel, R.R., 2009. Physiological functions of mineral micronutrients (Cu, Zn, Mn, Fe, Ni, Mo, B, Cl). *Current Opinion in Plant Biology* 12(3), 259–266.

Hayat, S., Alyemeni, M.N., Hasan, S.A., 2012a. Foliar spray of brassinosteroid enhances yield and quality of *Solanum lycopersicum* under cadmium stress. *Saudi Journal of Biological Sciences* 19(3), 325–335.

Hayat, S., Hayat, Q., Alyemeni, M.N., Wani, A.S., Pichtel, J., Ahmad, A., 2012b. Role of proline under changing environments: a review. *Plant Signaling and Behavior* 7(11), 1456–1466.

Hendawy, S., Khalid, K.A., 2005. Response of sage (*Salvia officinalis* L.) plants to zinc application under different salinity levels. *Journal of Applied Sciences Research* 1, 147–155.

Heuer, B., 2003. Influence of exogenous application of proline and glycinebetaine on growth of salt-stressed tomato plants. *Plant Science* 165(4), 693–699.

Hossain, M.A., Burritt, D.J., Fujita, M., 2016. *Proline and Glycine Betaine Modulate Cadmium-Induced Oxidative Stress Tolerance in Plants: Possible Biochemical and Molecular Mechanisms. Plant-Environment Interaction: Responses and Approaches to Mitigate Stress.* Wiley, Chichester, 97–123.

Hossain, M.A., Hasanuzzaman, M., Fujita, M., 2010. Up-regulation of antioxidant and glyoxalase systems by exogenous glycinebetaine and proline in mung bean confer tolerance to cadmium stress. *Physiology and Molecular Biology of Plants* 16(3), 259–272.

Huamain, C., Chunrong, Z., Cong, T., Yongguan, Z., 1999. Heavy metal pollution in soils in China: status and countermeasures. *Ambio—A Journal of the Human Environment* 28, 130–134.

Ibrahim, H.A., Abdellatif, Y.M., 2016. Effect of maltose and trehalose on growth, yield and some biochemical components of wheat plant under water stress. *Annals of Agricultural Sciences* 61(2), 267–274.

Iordachescu, M., Imai, R., 2008. Trehalose biosynthesis in response to abiotic stresses. *Journal of Integrative Plant Biology* 50(10), 1223–1229.

Islam, M.M., Hoque, M.A., Okuma, E., Banu, M.N.A., Shimoishi, Y., Nakamura, Y., Murata, Y., 2009a. Exogenous proline and glycinebetaine increase antioxidant enzyme activities and confer tolerance to cadmium stress in cultured tobacco cells. *Journal of Plant Physiology* 166(15), 1587–1597.

Issa, D., Alturki, S., Sajyan, T., Sassine, Y., 2020. Sorbitol and lithovit-guano25 mitigates the adverse effects of salinity on eggplant grown in pot experiment. *Agronomy Research* 18(1), 113–126.

Jabeen, N., Abbas, Z., Iqbal, M., Rizwan, M., Jabbar, A., Farid, M., Ali, S., Ibrahim, M., Abbas, F., 2016. Glycinebetaine mediates chromium tolerance in mung bean through lowering of Cr uptake and improved antioxidant system. *Archives of Agronomy and Soil Science* 62(5), 648–662.

Jadhav, K., Kushwah, B., Jadhav, I., 2018. Insight into compatible solutes from halophiles: exploring significant applications in biotechnology. In: Singh J., Sharma D., Kumar G., Sharma N. (eds), *Microbial Bioprospecting for Sustainable Development.* Singapore: Springer, 291–307.

Karam, E.A., Keramat, B., Asrar, Z., Mozafari, H., 2016. Triacontanol-induced changes in growth, oxidative defense system in Coriander (*Coriandrum sativum*) under arsenic toxicity. *Indian Journal of Plant Physiology* 21(2), 137–142.

Kastori, R., Petrović, M., Petrović, N., 1992. Effect of excess lead, cadmium, copper, and zinc on water relations in sunflower. *Journal of Plant Nutrition* 15(11), 2427–2439.

Kaul, S., Sharma, S., Mehta, I., 2008. Free radical scavenging potential of L-proline: evidence from in vitro assays. *Amino Acids* 34(2), 315–320.

Kaur, G., Asthir, B., 2015. Proline: a key player in plant abiotic stress tolerance. *Biologia Plantarum* 59(4), 609–619.

Kaur, G., Singh, H.P., Batish, D.R., Mahajan, P., Kohli, R.K., Rishi, V., 2015. Exogenous nitric oxide (NO) interferes with lead (Pb)-induced toxicity by detoxifying reactive oxygen species in hydroponically grown wheat (*Triticum aestivum*) roots. *PLOS ONE* 10(9), e0138713.

Kaya, C., Sonmez, O., Aydemir, S., Ashraf, M., Dikilitas, M., 2013. Exogenous application of mannitol and thiourea regulates plant growth and oxidative stress responses in salt-stressed maize (*Zea mays* L.). *Journal of Plant Interactions* 8(3), 234–241.

Konotop, Y., Kovalenko, M., Matušíková, I., Batsmanova, L., Taran, N., 2017. Proline application triggers temporal redox imbalance, but alleviates cadmium stress in wheat seedlings. *Pakistan Journal of Botany* 49, 2145–2151.

Kosar, F., Akram, N.A., Ashraf, M., Ahmad, A., Alyemeni, M.N., Ahmad, P., 2020. Impact of exogenously applied trehalose on leaf biochemistry, achene yield and oil composition of sunflower under drought stress. *Physiologia Plantarum.* https://doi.org/10.1111/ppl.13155.

Kumar, P., Tokas, J., Singal, H., 2019. Amelioration of chromium VI toxicity in sorghum (*Sorghum bicolor* L.) using glycine betaine. *Scientific Reports* 9(1), 1–15.

Kumchai, J., Huang, J., Lee, C., Chen, F., Chin, S., 2013. Proline partially overcomes excess molybdenum toxicity in cabbage seedlings grown in vitro. *Genetics and Molecular Research* 12(4), 5589–5601.

Lambers, H., Blacquière, T., Stuiver, B., 1981. Interactions between osmoregulation and the alternative respiratory pathway in *Plantago coronopus* as affected by salinity. *Physiologia Plantarum* 51(1), 63–68.

Lelandais-Brière, C., Jovanovic, M., Torres, G.A., Perrin, Y., Lemoine, R., Corre-Menguy, F., Hartmann, C., 2007. Disruption of AtOCT1, an organic cation transporter gene, affects root development and carnitine-related responses in Arabidopsis. *The Plant Journal: For Cell and Molecular Biology* 51(2), 154–164.

Li, M., Wang, G., Li, J., Cao, F., 2016. Foliar application of betaine alleviates cadmium toxicity in maize seedlings. *Acta Physiologiae Plantarum* 38(4), 95.

Li, X., Yang, Y., Jia, L., Chen, H., Wei, X., 2013. Zinc-induced oxidative damage, antioxidant enzyme response and proline metabolism in roots and leaves of wheat plants. *Ecotoxicology and Environmental Safety* 89, 150–157.

Liu, L., Shang, Y., Li, L., Chen, Y., Qin, Z., Zhou, L., Yuan, M., Ding, C., Liu, J., Huang, Y., Yang, R.W., Zhou, Y.H., Liao, J.Q., 2018. Cadmium stress in Dongying wild soybean seedlings: growth, Cd accumulation, and photosynthesis. *Photosynthetica* 56(4), 1346–1352.

LiXin, Z., ShengXiu, L., ZongSuo, L., 2009. Differential plant growth and osmotic effects of two maize (*Zea mays* L.) cultivars to exogenous glycinebetaine application under drought stress. *Plant Growth Regulation* 58(3), 297–305.

Lou, Y., Yang, Y., Hu, L., Liu, H., Xu, Q., 2015. Exogenous glycinebetaine alleviates the detrimental effect of Cd stress on perennial ryegrass. *Ecotoxicology* 24(6), 1330–1340.

Mehta, S., Gaur, J., 1999. Heavy-metal-induced proline accumulation and its role in ameliorating metal toxicity in *Chlorella vulgaris*. *The New Phytologist* 143(2), 253–259.

Meloni, D.A., Oliva, M.A., Ruiz, H.A., Martinez, C.A., 2001. Contribution of proline and inorganic solutes to osmotic adjustment in cotton under salt stress. *Journal of Plant Nutrition* 24(3), 599–612.

Mishra, S., Dubey, R.S., 2006. Inhibition of ribonuclease and protease activities in arsenic exposed rice seedlings: role of proline as enzyme protectant. *Journal of Plant Physiology* 163(9), 927–936.

Mostofa, M.G., Hossain, M.A., Fujita, M., Tran, L.S., 2015. Physiological and biochemical mechanisms associated with trehalose-induced copper-stress tolerance in rice. *Scientific Reports* 5, 11433.

Nath, K., Singh, D., Shyam, S., Sharma, Y.K., 2008. Effect of chromium and tannery effluent toxicity on metabolism and growth in cowpea (*Vigna sinensis* L. Saviex Hassk) seedling. *Research in Environment and Life Sciences* 1, 91–94.

Nemat Alla, M., Badran, E., Mohammed, F., 2019. Exogenous trehalose alleviates the adverse effects of salinity stress in wheat. *Turkish Journal of Botany* 43(1), 48–57.

Nichols, P.B., Couch, J.D., Al-Hamdani, S.H., 2000. Selected physiological responses of *Salvinia minima* to different chromium concentrations. *Aquatic Botany* 68(4), 313–319.

Noiraud, N., Maurousset, L., Lemoine, R., 2001. Transport of polyols in higher plants. *Plant Physiology and Biochemistry* 39(9), 717–728.

Noreen, S., Akhter, M.S., Yaamin, T., Arfan, M., 2018. The ameliorative effects of exogenously applied proline on physiological and biochemical parameters of wheat (*Triticum aestivum* L.) crop under copper stress condition. *Journal of Plant Interactions* 13(1), 221–230.

Nusrat, N., Shahbaz, M., Perveen, S., 2014. Modulation in growth, photosynthetic efficiency, activity of antioxidants and mineral ions by foliar application of glycinebetaine on pea (*Pisum sativum* L.) under salt stress. *Acta Physiologiae Plantarum* 36(11), 2985–2998.

Oikonomou, A., Ladikou, E.-V., Chatziperou, G., Margaritopoulou, T., Landi, M., Sotiropoulos, T., Araniti, F., Papadakis, I.E., 2019. Boron excess imbalances root/shoot allometry, photosynthetic and chlorophyll fluorescence parameters and sugar metabolism in apple plants. *Agronomy* 9(11), 731.

Pandey, V., Dixit, V., Shyam, R., 2009. Chromium effect on ROS generation and detoxification in pea (*Pisum sativum*) leaf chloroplasts. *Protoplasma* 236(1–4), 85–95.

Papageorgiou, G.C., Murata, N., 1995. The unusually strong stabilizing effects of glycine betaine on the structure and function of the oxygen-evolving photosystem II complex. *Photosynthesis Research* 44(3), 243–252.

Peluso, G., Barbarisi, A., Savica, V., Reda, E., Nicolai, R., Benatti, P., Calvani, M., 2000. Carnitine: an osmolyte that plays a metabolic role. *Journal of Cellular Biochemistry* 80(1), 1–10.

Peñarrubia, L., Romero, P., Carrió-Seguí, A., Andrés-Bordería, A., Moreno, J., Sanz, A., 2015. Temporal aspects of copper homeostasis and its crosstalk with hormones. *Frontiers in Plant Science* 6, 255.

Pirzadah, T.B., Malik, B., Tahir, I., Irfan, Q.M., Rehman, R.U., 2018. Characterization of mercury-induced stress biomarkers in *Fagopyrum tataricum* plants. *International Journal of Phytoremediation* 20(3), 225–236.

Pirzadah, T.B., Malik, B., Tahir, I., Rehman, R.U., Hakeem, K.R., Alharby, H.F., 2019. Aluminium stress modulates the osmolytes and enzyme defense system in Fagopyrum species. *Plant Physiology and Biochemistry* 144, 178–186.

Rai, V., Vajpayee, P., Singh, S.N., Mehrotra, S., 2004. Effect of chromium accumulation on photosynthetic pigments, oxidative stress defense system, nitrate reduction, proline level and eugenol content of *Ocimum tenuiflorum* L. *Plant Science* 167(5), 1159–1169.

Rasheed, R., Ashraf, M.A., Hussain, I., Haider, M.Z., Kanwal, U., Iqbal, M., 2014. Exogenous proline and glycinebetaine mitigate cadmium stress in two genetically different spring wheat (*Triticum aestivum* L.) cultivars. *Brazilian Journal of Botany* 37(4), 399–406.

Raza, M.A.S., Saleem, M.F., Khan, I.H., 2015. Combined application of glycinebetaine and potassium on the nutrient uptake performance of wheat under drought stress. *Pakistan Journal of Agriculture Science* 52, 19–26.

Raza, S.H., Athar, H.R., Ashraf, M., Hameed, A., 2007. Glycinebetaine-induced modulation of antioxidant enzymes activities and ion accumulation in two wheat cultivars differing in salt tolerance. *Environmental and Experimental Botany* 60(3), 368–376.

Rehman, A., Bashir, F., Ayaydin, F., Kóta, Z., Páli, T., Vass, I., 2020. Proline is a quencher of singlet oxygen and superoxide both in in vitro systems and isolated thylakoids. *Physiologia Plantarum*. https://doi.org/10.1111/ppl.13265.

Roy, D., Basu, N., Bhunia, A., Banerjee, S., 1993. Counteraction of exogenous L-proline with NaCl in salt-sensitive cultivar of rice. *Biologia Plantarum* 35(1), 69.

Rucińska-Sobkowiak, R., 2016. Water relations in plants subjected to heavy metal stresses. *Acta Physiologiae Plantarum* 38(11), 257.

Salehi, A., Tasdighi, H., Gholamhoseini, M., 2016. Evaluation of proline, chlorophyll, soluble sugar content and uptake of nutrients in the German chamomile (*Matricaria chamomilla* L.) under drought stress and organic fertilizer treatments. *Asian Pacific Journal of Tropical Biomedicine* 6(10), 886–891.

Salvucci, M.E., 2000. Sorbitol accumulation in whiteflies: evidence for a role in protecting proteins during heat stress. *Journal of Thermal Biology* 25(5), 353–361.

Saxena, R., Kumar, M., Tomar, R.S., 2019. Plant responses and resilience towards drought and salinity stress. *Plant Archives* 19, 50–58.

Schützendübel, A., Polle, A., 2002. Plant responses to abiotic stresses: heavy metal-induced oxidative stress and protection by mycorrhization. *Journal of Experimental Botany* 53(372), 1351–1365.

Shahbaz, M., Abid, A., Masood, A., Waraich, E.A., 2017. Foliar-applied trehalose modulates growth, mineral nutrition, photosynthetic ability, and oxidative defense system of rice (*Oryza sativa* L.) under saline stress. *Journal of Plant Nutrition* 40(4), 584–599.

Shahbaz, M., Masood, Y., Perveen, S., Ashraf, M., 2012. Is foliar-applied glycinebetaine effective in mitigating the adverse effects of drought stress on wheat (*Triticum aestivum* L.)? *Journal of Applied Botany and Food Quality* 84, 192.

Shahid, M.A., Balal, R.M., Pervez, M.A., Abbas, T., Aqeel, M.A., Javaid, M.M., Garcia-Sanchez, F., 2014. Exogenous proline and proline-enriched *Lolium perenne* leaf extract protects against phytotoxic effects of nickel and salinity in Pisum sativum by altering polyamine metabolism in leaves. *Turkish Journal of Botany* 38, 914–926.

Shahzad, A.N., Qureshi, M.K., Ullah, S., Latif, M., Ahmad, S., Bukhari, S.A.H., 2020. Exogenous trehalose improves cotton growth by modulating antioxidant defense under salinity-induced osmotic stress. *Pakistan Journal of Agricultural Research* 33(2), 270–279.

Sharma, P., Dubey, R.S., 2005. Lead toxicity in plants. *Brazilian Journal of Plant Physiology* 17(1), 35–52.

Sharma, P., Sirhindi, G., Singh, A.K., Kaur, H., Mushtaq, R., 2017. Consequences of copper treatment on pigeon pea photosynthesis, osmolytes and antioxidants defense. *Physiology and Molecular Biology of Plants* 23(4), 809–816.

Sharma, S.S., Dietz, K.-J., 2009. The relationship between metal toxicity and cellular redox imbalance. *Trends in Plant Science* 14(1), 43–50.

Sharma, S.S., Schat, H., Vooijs, R., 1998. In vitro alleviation of heavy metal-induced enzyme inhibition by proline. *Phytochemistry* 49(6), 1531–1535.

Signorelli, S., Arellano, J.B., Melø, T.B., Borsani, O., Monza, J., 2013. Proline does not quench singlet oxygen: evidence to reconsider its protective role in plants. *Plant Physiology and Biochemistry* 64, 80–83.

Signorelli, S., Coitiño, E.L., Borsani, O., Monza, J., 2014. Molecular mechanisms for the reaction between (˙) OH radicals and proline: insights on the role as reactive oxygen species scavenger in plant stress. *The Journal of Physical Chemistry. B* 118(1), 37–47.

Signorelli, S., Imparatta, C., Rodríguez-Ruiz, M., Borsani, O., Corpas, F.J., Monza, J., 2016. In vivo and in vitro approaches demonstrate proline is not directly involved in the protection against superoxide, nitric oxide, nitrogen dioxide and peroxynitrite. *Functional Plant Biology* 43(9), 870–879.

Singh, H.P., Mahajan, P., Kaur, S., Batish, D.R., Kohli, R.K., 2013. Chromium toxicity and tolerance in plants. *Environmental Chemistry Letters* 11(3), 229–254.

Singh, M., Pratap Singh, V., Dubey, G., Mohan Prasad, S., 2015. Exogenous proline application ameliorates toxic effects of arsenate in *Solanum melongena* L. seedlings. *Ecotoxicology and Environmental Safety* 117, 164–173.

Singh, V., Bhatt, I., Aggarwal, A., Tripathi, B.N., Munjal, A.K., Sharma, V., 2010. Proline improves copper tolerance in chickpea (*Cicer arietinum*). *Protoplasma* 245(1–4), 173–181.

Singh, G., Agnihotri, R.K., Reshma, R.S., Ahmad, M., 2012. Effect of lead and nickel toxicity on chlorophyll and proline content of Urd (*Vigna mungo* L.) seedlings. *International Journal of Plant Physiology and Biochemistry*. 4(6), 136–41.

Sirhindi, G., Mir, M.A., Abd-Allah, E.F., Ahmad, P., Gucel, S., 2016. Jasmonic acid modulates the physio-biochemical attributes, antioxidant enzyme activity, and gene expression in *Glycine max* under nickel toxicity. *Frontiers in Plant Science* 7, 591.

Sirhindi, G., Sharma, P., Singh, A., Kaur, H., Mir, M., 2015. Alteration in photosynthetic pigments, osmolytes and antioxidants in imparting copper stress tolerance by exogenous jasmonic acid treatment in *Cajanus cajan*. *International Journal of Plant Physiology and Biochemistry* 7, 30–39.

Skliros, D., Kalloniati, C., Karalias, G., Skaracis, G.N., Rennenberg, H., Flemetakis, E., 2018. Global metabolomics analysis reveals distinctive tolerance mechanisms in different plant organs of lentil (*Lens culinaris*) upon salinity stress. *Plant and Soil* 429(1–2), 451–468.

Smirnoff, N., Cumbes, Q.J., 1989. Hydroxyl radical scavenging activity of compatible solutes. *Phytochemistry* 28(4), 1057–1060.

Sofy, M.R., Elhawat, N., Tarek Alshaal, T., 2020. Glycine betaine counters salinity stress by maintaining high K^+/Na^+ ratio and antioxidant defense via limiting Na^+ uptake in common bean (*Phaseolus vulgaris* L.). *Ecotoxicology and Environmental Safety* 200. http://www.ncbi.nlm.nih.gov/pubmed/110732.

Srivastava, S., Sharma, Y.K., 2013. Impact of arsenic toxicity on black gram and its amelioration using phosphate. *ISRN Toxicology* 2013.

Stepien, P., 2016. Effects of the exogenous glycinebetaine on photosynthetic apparatus in cucumber leaves challenging Al stress. In: Proceedings of the 18th International Conference on Heavy Metals in the Environment.

Stolker, R., 2010. *Combating Abiotic Stress Using Trehalose*. M.Sc. Wageningen University & Research Centre.

Sun, R.-L., Zhou, Q.-X., Sun, F.-H., Jin, C.-X., 2007. Antioxidative defense and proline/phytochelatin accumulation in a newly discovered Cd-hyperaccumulator, *Solanum nigrum* L. *Environmental and Experimental Botany* 60(3), 468–476.

Suprasanna, P., Nikalje, G., Rai, A., 2016. Osmolyte accumulation and implications in plant abiotic stress tolerance. In: Iqbal N., Nazar R., A. Khan N. (eds), *Osmolytes and Plants Acclimation to Changing Environment: Emerging Omics Technologies*. New Delhi: Springer, 1–12.

Tahjib-Ul-Arif, M., Sohag, A.A.M., Mostofa, M.G., Polash, M.A.S., Mahamud, A.G.M.S.U., Afrin, S., Hossain, M.A., Hossain, M.A., Murata, Y., Tran, L.P., 2020. Comparative effects of ascobin and glutathione on copper homeostasis and oxidative stress metabolism in mitigation of copper toxicity in rice. *Plant Biology*. https://doi.org/10.1111/plb.13222.

Tamai, I., China, K., Sai, Y., Kobayashi, D., Nezu, J., Kawahara, E., Tsuji, A., 2001 *1512*. Na^+-coupled transport of L-carnitine via high-affinity carnitine transporter OCTN2 and its subcellular localization in kidney. *Biochimica et Biophysica Acta (BBA)-Biomembranes* 1512(2), 273–284.

Tamás, L., Zelinová, V., 2017. Mitochondrial complex II-derived superoxide is the primary source of mercury toxicity in barley root tip. *Journal of Plant Physiology* 209, 68–75.

Tantrey, M.S., Agnihotri, R., 2010. Chlorophyll and proline content of gram (*Cicer arietinum* L.) under cadmium and mercury treatments. *Agricultural Science Research Journal* 1, 119–122.

Tauqeer, H.M., Ali, S., Rizwan, M., Ali, Q., Saeed, R., Iftikhar, U., Ahmad, R., Farid, M., Abbasi, G.H., 2016. Phytoremediation of heavy metals by *Alternanthera bettzickiana*: growth and physiological response. *Ecotoxicology and Environmental Safety* 126, 138–146.

Theriappan, P., Gupta, A.K., Dhasarrathan, P., 2011. Accumulation of proline under salinity and heavy metal stress in cauliflower seedlings. *Journal of Applied Sciences and Environmental Management* 15(2), 251–255.

Tian, B., Qiao, Z., Zhang, L., Li, H., Pei, Y., 2016. Hydrogen sulfide and proline cooperate to alleviate cadmium stress in foxtail millet seedlings. *Plant Physiology and Biochemistry* 109, 293–299.

Trinh, N., Huang, T., Chi, W., Fu, S., Chen, C., Huang, H., 2014. Chromium stress response effect on signal transduction and expression of signaling genes in rice. *Physiologia Plantarum* 150(2), 205–224.

Wang, S., Wei, M., Wu, B., Cheng, H., Wang, C., 2020. Combined nitrogen deposition and Cd stress antagonistically affect the allelopathy of invasive alien species Canada goldenrod on the cultivated crop lettuce. *Scientia Horticulturae* 261. http://www.ncbi.nlm.nih.gov/pubmed/108955.

Wu, H., Liu, X., Zhao, J., Yu, J., 2012. Toxicological responses in halophyte *Suaeda salsa* to mercury under environmentally relevant salinity. *Ecotoxicology and Environmental Safety* 85, 64–71.

Xalxo, R., Yadu, B., Chakraborty, P., Chandrakar, V., Keshavkant, S., 2017. Modulation of nickel toxicity by glycinebetaine and aspirin in *Pennisetum typhoideum*. *Acta Biologica Szegediensis* 61, 163–171.

Xu, J., Yin, H., Li, X., 2009. Protective effects of proline against cadmium toxicity in micropropagated hyper-accumulator, *Solanum nigrum* L. *Plant Cell Reports* 28(2), 325–333.

Xun, Y., Feng, L., Li, Y., Dong, H., 2017. Mercury accumulation plant *Cyrtomium macrophyllum* and its potential for phytoremediation of mercury polluted sites. *Chemosphere* 189, 161–170.

Yancey, P.H., 2005. Organic osmolytes as compatible, metabolic and counteracting cytoprotectants in high osmolarity and other stresses. *Journal of Experimental Biology* 208(15), 2819–2830.

Yang, L., Zhao, X., Zhu, H., Paul, M., Zu, Y., Tang, Z., 2014. Exogenous trehalose largely alleviates ionic unbalance, ROS burst, and PCD occurrence induced by high salinity in Arabidopsis seedlings. *Frontiers in Plant Science* 5, 570.

Yang, X., Lu, C., 2005. Photosynthesis is improved by exogenous glycinebetaine in salt-stressed maize plants. *Physiologia Plantarum* 124(3), 343–352.

Yang, Y., Zhang, Y., Wei, X., You, J., Wang, W., Lu, J., Shi, R., 2011. Comparative antioxidative responses and proline metabolism in two wheat cultivars under short term lead stress. *Ecotoxicology and Environmental Safety* 74(4), 733–740.

Yildizli, A., Çevik, S., Ünyayar, S., 2018. Effects of exogenous myo-inositol on leaf water status and oxidative stress of *Capsicum annuum* under drought stress. *Acta Physiologiae Plantarum* 40(6), 122.

Yoshiba, Y., Kiyosue, T., Nakashima, K., Yamaguchi-Shinozaki, K., Shinozaki, K., 1997. Regulation of levels of proline as an osmolyte in plants under water stress. *Plant and Cell Physiology* 38(10), 1095–1102.

Yu, X., Lu, M., 2016. Responses of endogenous proline in rice seedlings under chromium exposure. *Global Journal of Environmental Science and Management* 2, 319–326.

Yu, X.-Z., Lin, Y.-J., Fan, W.-J., Lu, M.-R., 2017. The role of exogenous proline in amelioration of lipid per-oxidation in rice seedlings exposed to Cr (VI). *International Biodeterioration and Biodegradation* 123, 106–112.

Yurekli, F., Porgali, Zb., 2006. The effects of excessive exposure to copper in bean plants. *Acta Biologica Cracoviensia Series Botanica* 48, 7–13.

Zaid, A., Mohammad, F., 2018. Methyl jasmonate and nitrogen interact to alleviate cadmium stress in Mentha arvensis by regulating physio-biochemical damages and ROS detoxification. *Journal of Plant Growth Regulation* 37(4), 1331–1348.

Zakir Hossain, A.K., Koyama, H., Hara, T., 2006. Growth and cell wall properties of two wheat cultivars dif-fering in their sensitivity to aluminum stress. *Journal of Plant Physiology* 163(1), 39–47.

Zeid, I., 2009. Trehalose as osmoprotectant for maize under salinity-induced stress. *Research Journal of Agriculture and Biological Sciences* 5, 613–622.

Zengin, F., 2015. Effects of exogenous salicylic acid on growth characteristics and biochemical content of wheat seeds under arsenic stress. *Journal of Environmental Biology* 36(1), 249.

Zengin, F.K., Munzuroglu, O., 2005. Effects of some heavy metals on content of chlorophyll, proline and some antioxidant chemicals in bean (*Phaseolus vulgaris* L.) seedlings. *Acta Biologica Cracoviensia Series Botanica* 47, 157–164.

Zheng, J.-L., Zhao, L.-Y., Wu, C.-W., Shen, B., Zhu, A.-Y., 2015. Exogenous proline reduces NaCl-induced damage by mediating ionic and osmotic adjustment and enhancing antioxidant defense in *Eurya emar-ginata*. *Acta Physiologiae Plantarum* 37(9), 181.

Zhong, W., Xie, C., Hu, D., Pu, S., Xiong, X., Ma, J., Sun, L., Huang, Z., Jiang, M., Li, X., 2020. Effect of 24-epibrassinolide on reactive oxygen species and antioxidative defense systems in tall fescue plants under lead stress. *Ecotoxicology and Environmental Safety* 187, 109831.

Zhou, Z.S., Wang, S.J., Yang, Z.M., 2008. Biological detection and analysis of mercury toxicity to alfalfa (*Medicago sativa*) plants. *Chemosphere* 70(8), 1500–1509.

Zimmer, W., Mendel, R., 1999. Molybdenum metabolism in plants. *Plant Biology* 1(2), 160–168.

Zouari, M., Ben Ahmed, C., Elloumi, N., Bellassoued, K., Delmail, D., Labrousse, P., Ben Abdallah, F., Ben Rouina, B., 2016a. Impact of proline application on cadmium accumulation, mineral nutrition and enzymatic antioxidant defense system of *Olea* europaea L. cv Chemlali exposed to cadmium stress. *Ecotoxicology and Environmental Safety* 128, 195–205.

Zouari, M., Ben Ahmed, Ch, Zorrig, W., Elloumi, N., Rabhi, M., Delmail, D., Ben Rouina, B., Labrousse, P., Ben Abdallah, F., 2016b. Exogenous proline mediates alleviation of cadmium stress by promoting photosynthetic activity, water status and antioxidative enzymes activities of young date palm (*Phoenix dactylifera* L.). *Ecotoxicology and Environmental Safety* 128, 100–108.

Zouari, M., Elloumi, N., Labrousse, P., Ben Rouina, B., Ben Abdallah, F., Ben Ahmed, C., 2018. Olive trees response to lead stress: exogenous proline provided better tolerance than glycine betaine. *South African Journal of Botany* 118, 158–165.

5 Contents of Organic Solutes in Horticultural Plants and Crops under Environmental Pollutants

Seyedeh-Somayyeh Shafiei-Masouleh

CONTENTS

5.1   INTRODUCTION

Plants are effective constituents of the environment because they convert elements from abiotic into biotic forms. The main sources of elements include air, water and soil. The most important elements are arsenic (As), mercury (Hg), cadmium (Cd) and lead (Pb), which attention needs to be paid to in terms of pollutants in the food chain. Besides this, some micronutrients, including copper (Cu), zinc (Zn), chromium (Cr) and nickel (Ni), also have toxicity features for plants and animals in high concentrations. Factors which affect the absorption or uptake of elements by plants include active/ non-active transfer paths, sequestration and speciation, redox conditions, root characteristics and how plants respond to elements in seasonal cycles (Chojnacka et al. 2005). Metal ions have different effects on the cells based on their chemical structures. Symplastic uptake* of heavy metals in plants can cause phytotoxicity in most parts of the plant, particularly the cytosol and chloroplast (stroma). Either the instant and direct impairment of metabolic pathways or signaling induced by

---

* The inner side of the plasma membrane in which water and low molecular-weight solutes (small molecules, including sugars, amino acids and ions) can have freely diffusion.

metals can alter plant development. The important mechanism for metal detoxification is the transport processes. Exposure to metals often causes plants to synthesize a set of different metabolites in millimolar amounts, including amino acids, peptides and amines. Therefore, plants respond to heavy metals by this main metabolism i.e. nitrogen metabolism (Sharma and Dietz 2006). Crops are frequently vulnerable to industrial and natural toxins, including pesticides, allelochemicals, organic contaminants and heavy metals.

Heavy metals are one of the major environmental pollutants, and their toxicity must be increasingly considered for nutritional, environmental, evolutionary and ecological reasons. Every metal element with comparatively high density that is toxic even in low amounts is known as a heavy metal element. These elements have an atomic density of more than four g/cm³, or five times or above more than H_2O. Nevertheless, the chemical characteristics of heavy metals are more significant than their density. Elemental contaminants encompass radionuclides and toxic heavy metals (Meagher 2000). The elements known as heavy metals include lead, cadmium, nickel, cobalt, iron, zinc, chromium, iron, arsenic, silver and the platinum group elements. Any substance in the environment that causes a decrease in life quality and may ultimately cause death is defined as a pollutant (Nagajyoti et al. 2010).

The sensitivity of plants to these elements is different based on interactions of molecular networks and aspects of life in plants, which include: 1) absorbing and accumulating of those metals by binding with extracellular secretions and parts of the cell wall; 2) flowing heavy metals from the cytoplasm to outer parts (vacuoles); 3) forming the complexes of heavy metal ions in the cell; 4) activating antioxidative enzymes and accumulating the osmolytes and osmoprotectants; and 5) changing the metabolism of the plant to allow the adequate function of metabolic pathways and fast fixing of injured structures of the cell (John et al. 2012). In plants, the damaging effects of metals are linked to the impairment of plant functions, for example, mineral nutrition, photosynthesis, water relations and overproduction of the reactive oxygen species (ROS). Crop protection activities (chemical and crop management) have been practiced to reduce the toxicity of heavy metals. Furthermore, exogenous usage of osmolytes, including polyamines, proline and glycine betaine, has been an alternate program to make plants able to tolerate the harmful condition of toxicity caused by heavy metals. Proline is an amino acid that has several functions, and acts as a signaling molecule and starts a series of signaling pathways. Regarding proline functions, we can control the osmotic pressure of the cells, prevent the protein denature, decrease the toxic ROS, stabilize the enzymes and preserve the integrity of the membrane. Based on the evidence and backed by research, exogenous usage of proline may promote the health of plants that are exposed to cadmium and selenium. It seems that proline has a major role in enhancing the tolerance of plants to stress conditions and helping to minimize negative effects caused by heavy metals. Therefore, heavy metals negatively influence the entire plant and restrict the production of important molecules. For that reason, it is necessary to investigate the toxicity mechanisms of heavy metals to understand how to reduce the risks and regulate their accumulation into crops (Singh et al. 2015).

There are various resistance and tolerance techniques for protecting the plants from the toxic impacts of metals – tolerance allows plants to be alive during high accumulation of metals, and avoidance reduces the cellular accumulation of metals (Elloumi et al. 2017).

Under different stresses, plants accumulate the solutes – which protect membranes or encourage cellular osmotic regulation – that protect them against the stress (Mukherjee et al. 2019). Because plants are exposed to different sources of pollutants, including soil, air and water, in this chapter all types of pollution will be discussed (Figure 5.1). We introduce one of the defense mechanisms in plants (horticultural and crops), i.e. osmolytes, for the best understanding of the behavior of plants in their reactions during pollution stresses.

5.2 SOIL OR LAND POLLUTION

We can find the dispersed-form heavy metals in rock formations largely. Industry and urbanization have been causing an increase in heavy metals in the biosphere. Soil and aquatic ecosystems have

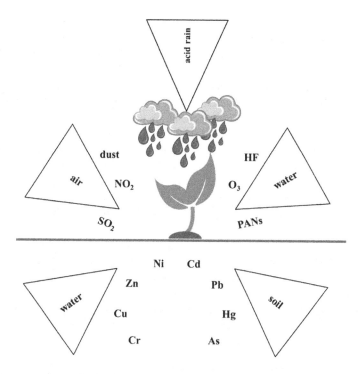

FIGURE 5.1 Various sources of pollutants that induce the stress conditions in plants; the water-sourced pollutants may affect plants through both root and shoot, acid rain may affect plants through a combination of water and CO_2 or gaseous pollutants. The gases, including nitrogen oxide (NO_2), sulfur dioxide (SO_2), ozone (O_3) and hydrogen fluoride (HF), and peroxy acyl nitrates (PANs) are toxic for plants.

the greatest amount of these metals and there are smaller parts in the air as vapors or harmful dust. These metals can be toxic for plants depending on metal type, plant species, their concentrations and structures or compositions, and soil acidity. Nevertheless, many heavy metals are essential to the growth and development of plants. For example, Cu and Zn are the cofactors and activators in enzyme reactions and have the catalytic feature (as a prosthetic group in metalloproteins). These metals as micronutrients are involved in electron transfer, nucleic acid biosynthesis and redox reactions. However, heavy metals such as As, Cd and Hg are seriously toxic for metal-sensitive enzymes and lead to inhibit the growth and finally cause the death of living plants. Plants are often affected by both the deficiency and oversupply of some heavy metal ions that are micronutrients (Nagajyoti et al. 2010; Abdel Latef 2013; Abdel Latef and Abu Alhmad 2013; Ahmad et al. 2016; Abdel Latef et al. 2017, 2020). Heavy metals may present a risk to soil microorganisms and agro-ecosystems as well as plants. However, using a tolerant strain of microorganisms can reduce or remove the harmful effects of heavy metals. Based on a report, the tolerant strain of *Pseudomonas aeruginosa* in the growing media of chickpea exposed to Cr and Ni reduced the level of stress markers, including proline, superoxide dismutase (SOD), ascorbate peroxidase (APX), glutathione reductase (GR) and catalase (CAT), and decreased the uptake of Cr and Ni by plant roots (Saif and Khan 2018). Moreover, Shavalikohshori et al. (2020), using inoculation of wheat seeds with this bacteria, observed that the harmful effects of salinity and Cd were diminished and the genes of proline pathway were more expressed as well as decreasing Na^+/K^+ ratio and Cd^{+2} in plant tissues.

Heavy metals like Cd may damage food safety as well as plant yields. Nowadays, using phytoremediation technology can overcome these harmful effects. Based on a study, the application of molybdenum as a foliar spray on *Ricinus communis* plants increased plant growth, Cd accumulation and the levels of phenolic compounds and proline, therefore the soils would be without Cd and suitable for cultivation of safe crops (Hadi et al. 2016).

Heavy metal toxicity in plants causes imbalances in the redox metabolism and this leads to oxidative damage by enhanced production of ROS. Antioxidative defense activity of plants includes the enzymatic scavengers such as CAT, SOD, APX, GR and guaiacol peroxidase (GPX) and non-enzymatic components like ascorbic acid (AA), α-tocopherol, carotenoids, flavonoids and proline (Ishtiyaq et al. 2018). Nickel stress in rice seedlings according to the reports by Rizwan et al. (2017) causes some symptoms as follows: 1) a decrease in fresh and dry weight, and root and shoot lengths as well as depletion of photosynthetic pigments, 2) more accumulation of Ni in shoots and roots, in this order, 3) damage in oxidative metabolism and increase in the level of hydrogen peroxide and malondialdehyde content, 4) increase of the activities of peroxidase, catalase and the content of glutathione and ascorbic acid, 5) decrease of the activity of superoxide dismutase, soluble protein and soluble sugars and 6) triggering of the rate of proline accumulation. Ahmad et al. (2018) stated that the application of nitric oxide in tomato can modulate antioxidants, osmolyte metabolism and enzymes of the ascorbate-glutathione cycle, and improves growth under Cd stress.

Furthermore, many synthetic organic substances, as well as heavy metals in urban and industrial areas, affect plant growth. These compounds could include textile dyes and antibiotics that are remarkable pollutants in soils and water. These persistent pollutants can have a negative influence on plant growth and development, and the levels of secondary metabolites. In wheat seedlings, contaminant solutions with a high concentration of textile dyes and antibiotics resulted in a reduction in flavonoid content while low doses of antibiotics reduced the content of flavonoids. Therefore, the chronic pollution of synthetic organic contaminants can importantly change the antioxidative capacity of plants (Copaciu et al. 2016).

5.3   WATER POLLUTION

An increase in urbanization, industrialization and population has released high amounts of wastewater that is continuously used as an irrigation resource in urban and semi-urban areas. Large amounts of heavy metals in wastewater are entering food products because of irrigation with wastewater and its accumulation in the soil and uptake by plants. Pollution caused by heavy metals could prevent the growth and yield of plants largely, and promote exposure to danger in the lives of humans and animals through biomagnification* (Singh et al. 2015).

Over the world, we can observe that heavy metals pollute water, which is considered to be of great concern. One of the critical environmental contaminants is cadmium, which is found in wastewater generated from industries that manufacture, amongst others, color stabilizers, rechargeable nickel-cadmium batteries, pigment production, electronic compounds and alloys. Cadmium is not important for plants and other organisms and thus is toxic to them. Aquatic macrophytes can easily uptake and accumulate Cd. For example, *Lemna gibba* can be used as phytoremediation for removing contaminants, caused by metals, from the water. Nevertheless, cadmium builds up in plants and leads to biochemical and physiological modifications. Cadmium is toxic to plants because of ROS production. Organisms can upregulate their antioxidation mechanisms, and can adapt those mechanisms to increase ROS production. The antioxidant techniques comprise different kinds of antioxidant enzymes, including ascorbate peroxidase, superoxide dismutase and catalase. Nevertheless, through very serious stressful conditions, the organism's antioxidant ability could be enough to stop the adverse impact of the stress producer. In the mentioned conditions, the osmotic potential of plants is decreased by osmolyte accumulation, including glycine betaine and proline. The exogenous usage of glycine betaine enhances tolerance to stress in various species of plants (the accumulators and non-accumulators of glycine betaine). Glycine betaine is environmentally friendly, non-toxic and water-soluble. Glycine betaine is readily synthesized from sugar beets as a kind of low-cost by-product. The exogenous usage of glycine betaine improves stress tolerance

---

* Biomagnification is the increasing concentration of toxic material in the tissue of tolerant organisms as the material progresses through the food chain, thereby becoming very dangerous. This is also known as bioamplification.

by decreasing metal accumulation, lipid peroxidation, stopping photo-inhibition and up-regulating antioxidant enzymes (Duman et al. 2011).

5.4   AIR POLLUTION

Air pollution is considered a global issue. It is important to test plants for their sensitivity/tolerance level to air pollutants because sensitive plants can act as vital indicators. Also, in industrial and urban regions, pollution-resistant plants can withstand air contamination (Patidar et al. 2016). Air pollution as well as some atmospheric gases at the extra level can affect plant health due to changing plant physiology, biochemistry and morphology. Anthropogenic emissions* into the atmosphere and then into the biosphere through modification, transformation and reaction lead to a diversity of severe and persistent diseases at regional, global and local scales. It must be noted that plant communities are more sensitive to air pollution compared to other organisms. The state of the environment can be understood by the symptoms in plants or effects on plant physiology, anatomy or biochemistry, due to contaminants. The continuous exchange of gases in the plant leaves means that any modification in the atmosphere can be observed in the plants' physiology. Nevertheless, plants also play a significant role in monitoring and maintaining environmental balance since they actively take part in recycling nutrients and gases, which includes oxygen and carbon dioxide. These plants usually have large leaves to accumulate and absorb the air contaminants (Saxena and Kulshrestha 2016). Plants are the main receptors for pollutants of the atmosphere (gases and particles) because they are constantly exposed to air. In terrestrial plant species, the canopy of plants acts as a natural sink for contaminants especially particles (Rai 2016). The main physiological processes that are affected by air pollutants are carbon allocation, stomatal function, respiration and photosynthesis. The sensitivity of plant photosynthesis varies depending on air contaminants such as nitrogen oxide, sulfur dioxide, ozone (O_3) and hydrogen fluoride (HF) and is based on species and cultivars. The variation of plant response to pollutants is related to genetic factors (Darrall 1989). On a large scale, dust, smoke, gases and particulates are among the main causes of air pollution, which are dangerous to humans, crops and horticultural plants. The leaves absorb atmospheric air pollutants by stomata, while particulates precipitate on surfaces of leaves (Patidar et al. 2016).

When plants are exposed to pollutants, plants may uptake, absorb, accumulate or integrate them into their body and show specific symptoms. In general, sensitive species exhibit quick injury symptoms compared to tolerant plants. For this reason, early warning symptoms of pollution is seen in sensitive species. However, air pollutants are scavenged by the tolerant species and the total pollution is reduced. Gases such as NO_2, SO_2, HF, peroxy acyl nitrates (PANs) and O_3 are highly toxic for plants. The destruction of higher plants (vascular plants) can occur drastically and quickly when affected by these gases. For instance, various plant species are severely sensitive upon exposure to hydrogen fluoride, for example the species of stone fruit (like peaches, apricots, and plums), natural crops (as *Picea abies* L. and *Hypericum perforatum* L.), monocotyledonous ornamental plants (like gladioli and tulips) and other plants (such as maize). A clear example is the accumulation of F- ion in the tips and edges of plant leaves which leads to leaf tissue necrosis with a distinct red-brown boundary zone (Saxena and Kulshrestha 2016). Acute and chronic injuries of particles are in two major types: 1) clear symptoms on the foliage, often as necrotic lesions that are caused by a high concentration of gas at a relatively short period of time; detection of this type of injury is simple and 2) chronic injury that is more easy and obvious to detect. These are observed because of long exposure to low concentrations of gas by changing growth and/or reducing yield without obvious symptoms. The reasons for leaf injury, premature senescence, stomatal destruction, decreased growth and yield, reduced photosynthetic activity and membrane permeability in sensitive plant species can be pollutants. Long term and low concentration of contamination can have detrimental

---

* Any emissions (emissions of greenhouse gases and aerosols) associated with human activities are known as anthropogenic emission.

effects on plant leaves without obvious injuries (Rai 2016). The air pollutants injure leaf cuticles and stomatal conductance. Moreover, air pollutants immediately affect leaf longevity, carbon allocation patterns and photosynthetic systems. Regionally, the interaction between contaminants and environmental agents can change the relations between the environment and plant. Air contaminants influence the growth and development of plants either by obvious leaf damage (necrosis and chlorosis between or in the veins) or by affecting the growth and reproduction of the plant. There are no specific symptoms produced by nitrogen dioxide (NO_2), chlorine (Cl_2), ammonia (NH_3) or hydrochloric acid (HCl) but they increase leaf chlorosis and necrosis and reduce plant's growth (Saxena and Kulshrestha 2016). Accordingly, for detecting, recognizing and monitoring cases, higher plants can be used as indicators of air pollutants. Among the important physiological processes that are affected by air pollution are respiration, photosynthesis, carbon allocation and stomatal function. The levels of biochemical indices such as protein, chlorophyll, ascorbic acid, soluble sugar, peroxidase and superoxide dismutase may depend on pollution.

Plants are considered as principal parts of the ecosystem, hence regard for their sensitivity to air contamination is more important than standards of air pollution. Nitrogen oxides (NO_x), sulfur dioxide (SO_2), troposphere ozone, carbon monoxide (CO), heavy metals and suspended particulate matter are among the most dangerous industrial contaminants (Assadi et al. 2011; Mukherjee et al. 2019). Other air contaminants are known as phytotoxic agents. The toxicities in plants due to gases have been identified as SO_2 for about a century, the effects of O_3 for more than 30 years, acid deposition for almost 20 years, and effects of high levels of nitrogen, nitrogen oxides (NOx) and ammonia (NH_3) in the last decade. Besides this, the influences of other contaminants such as fluorides, PAN or heavy metals have also been recognized (Assadi et al. 2011). Exposure of plants to harmful types of environmental pollution prompts them to interact through appropriate anatomical, morphological, biochemical and physiological modifications to survive under difficult conditions. Uptake of fluoride in plants is either through the cuticle of the leaf or by stomata that are transferred to the apoplast by transpiration (Elloumi et al. 2017).

Pollutants have detrimental effects on photosynthetic pigments, cell membranes, photosynthetic parameters and foliar water status. Plants that are exposed to fluoride induce biochemical responses, including activation of the antioxidant defense system and osmolytes. These biochemical responses induced by plants greatly assist the important ability of antioxidant enzymes in loquat trees to tolerate fluoride contamination in the air (Elloumi et al. 2017). Nevertheless, modifications in plant physiology may exist before the occurrence of apparent morphological damage. Physiological changes of plants can properly be used as early indicators of the harmful effects of air pollution in forests. Changes in the contents of chlorophylls and carotenoids take place when plants absorb pollutants, thus yield and growth of plants are affected. Leaf proline, soluble sugars, the contents of chlorophyll and carotenoids can reveal the plant's physiological status (Assadi et al. 2011).

The tolerant plants are those that can preserve their chlorophyll under contaminated conditions. The reduction of chlorophyll content affects the growth of the plant. The net photosynthetic rate is usually an indicator of increased air pollutants in the development of a plant. Air pollution causes closure of stomata that consequently decreases CO_2 for leaves and inhibits the carbon fixation. Air pollutants that are absorbed by plant leaves, including sulfur dioxides, nitrogen dioxides, CO_2 and suspended particulate matter, cause a reduction in the contents of photosynthetic pigments (chlorophyll and carotenoids) that directly affect the growth and development of a plant. Acidic substances can convert chlorophyll to pheophytin. In this process, replacement of Mg^{2+} by two atoms of hydrogen occurs in the chlorophyll molecule followed by altering properties of the light spectrum of chlorophyll particles (Saxena and Kulshrestha 2016). Reducing the leaf surface area and petiole length minimize the environmental contaminants (mainly air pollutants) and enhance the resistance of plants to contamination. A decrease in leaf area may occur for different plant species growing near heavy contaminants. The reduction in morphology and anatomy of plant species may be due to the occurrence of some hidden infections or physiological disorders. The length of the leaf is one of the traits that shows the plant's ability to protect against stress (Assadi et al. 2011).

Carotenoids and chlorophyll are the key parts in the production of energy in green plants, and their amounts are largely altered by environmental impacts on plant metabolism. Plant species responses against the same stress are often not the same. High levels of SO_2 on three *Eucalyptus* species have different effects depending on genotypes. SO_2 and NO_2 are the major contaminants that produce acid rain when they react with oxygen and water vapor. Materials that enter plants by acid rain cause the conversion of chlorophyll to pheophytin, because Mg^{+2} can be replaced by H^+ in the chlorophyll molecule. There is an increase in the amounts of total chlorophyll, chlorophyll *a*, chlorophyll *b*, soluble sugar, carotenoids and proline in *Eucalyptus camaldulensis* leaves in polluted regions. Furthermore, the morphological features of *E. camaldulensis* leaves will decrease in a polluted area compared to a clean area. Generally, *E. camaldulensis* shows high resistance to air contamination, and municipalities can use it to filter air contaminants (Assadi et al. 2011).

The damage to the morphology is generally visible through lesions on the leaves, flowers and fruits while biochemical and physiological changes, which are invisible, can be assayed and quantified. The symptoms can be indicators of air pollution stress for its early diagnosis and as markers of a particular physiological disorder (Saxena and Kulshrestha 2016).

Dust pollution is a problem for plant yield, especially in hot and dry regions. It affects plants by reducing leaf number, shoot length, dry weights of root and shoot as well as reducing relative water content, membrane stability and increasing proline, malondialdehyde (MDA) and H_2O_2 (Karami et al. 2017). Cement dust is introduced as a major particulate air pollutant and can cause negative impacts on plant growth and development (Abu-Romman and Alzubi 2016).

5.5 ACID RAIN

Progressively, acid rain is the main contaminant that can influence the production of the plant. Acid rain, as its name implies, has low pH and causes damage and death of plant tissue. However, plant species' susceptibility to acid rain varies and some plants are not sensitive to it. This could be related to the ability of plant tissue to store acid before any major physiological or physical injury (Craker and Bernstein 1984). Increasing the acid input to the soil may increase the leaching of calcium and other nutrient elements. Even though these losses will be small for environments and may cause short-term damage to fertile areas, they represent additional stresses to the environment. Rainfall that reaches forestland is less acidic than others (Likens and Bormann 1974). When carbon dioxide dissolves in the water, it forms carbonic acid, and the pH value of rainwater will be approximately 5.5, although over the past 20 years, most reports indicated that rainwater contains much greater acidity. It is generally viewed that the substances that lead to high acidity are formed from industrial contaminants discharged into the atmosphere by strong acids. Sulfuric acid, hydrofluoric and nitric acids cause the acidity. Sulfuric acid is formed from dissolving and oxidizing sulfur dioxide. Acid rain is an essential agent that increases the mobility and activity of heavy metals in soils. Leaching of Cd and Cu in soils is influenced by acid rain, hence, as the acidity of acid rain increases, the released amounts of Cd and Cu increase (Liao et al. 2005).

Studies on acid rain have concentrated on leaf damage and shoot growth since leaves and shoots are directly exposed to the precipitation, however, root-based symbiotic associations and root growth may be more sensitive. Acid rain mostly affects root crops, in which atmospheric pollutants can reduce the development of the roots. *Rhizobium* species or other coexistence with legumes may also be affected by acidity (Stroo and Alexander 1985). Modifications in the physiology and cellular biochemistry of plants are induced by acid rain. Acid deposition on the plant has various and complex biological impacts, including visible effects of damage (necrosis and/or chlorosis) and invisible symptoms like loss of nutrients from leaves, reduced photosynthesis, a variation of several enzyme activities and modified water balance. For example, the photochemical activity of bean plants and photosynthetic CO_2 fixation is reduced after acid precipitation (pH 2.2–1.8). There is a relationship between the plant's ability to overcome the acid rain impact and maintaining the production and purification of stress-induced oxygen species, like $O_2^{\bullet-}$ (superoxide radical), H_2O_2 (hydrogen peroxide) and $\bullet OH$ (hydroxyl radical) (Velikova et al. 2000).

Acid rain is one of the serious environmental issues causing morphological and physiological changes in plants. With the increasing levels of acidity of simulated acid rain, according to Debnath et al. (2018), plant growth, antioxidant activity, photosynthetic pigment contents, the contents of soluble protein and sugars in leaves of tomato cultivars decreased. However, the contents of H_2O_2, MDA and proline increased, as well as phenolic and flavonoid compounds.

5.6   ORGANIC SOLUTES

The addition of small quantities of antioxidants to materials leads to a rapid reaction with the free-radical molecules of an auto-oxidation chain and finally scavenges them. The antioxidant system involves essential components, which are amino acids and amines, alkaloids, carotenoids, other phenols, polyamines, phenolic acids, glutathione, ascorbate, vitamin E (α-tocopherols), flavonoids, miscellaneous compounds and chlorophyll derivatives (Figure 5.2) (Saxena and Kulshrestha 2016).

5.6.1   AMINO ACIDS

Antioxidant activities are reported for specific amino acids, including threonine, lysine, cysteine, methionine, tryptophan, arginine and histidine. The antioxidant potential is exhibited by amino acids under certain statuses of heat, oxygen or pH; however, these amino acids do not influence the

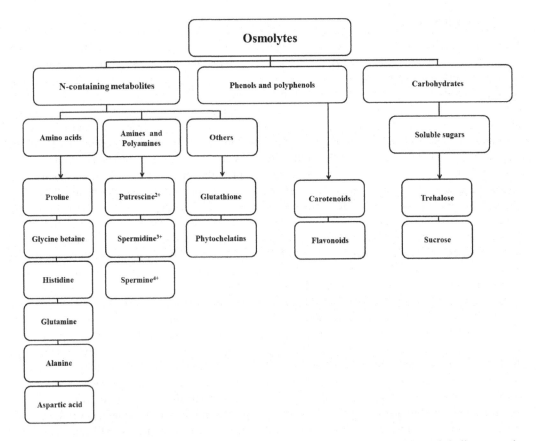

FIGURE 5.2   Classification of osmolytes in plants based on the most important compounds in literature; the range of color in the class of amino acids shows importance value.

enhancement of oxidation in other conditions. For instance, it has been reported that linoleic acid oxidation is enhanced by histidine and alanine at pH 7.5 and prevented at pH 9.5 (Ferenbaugh 1976; Saxena and Kulshrestha 2016).

The important effects caused by air contaminants on the life of the plant are through biochemical parameters, including the content of proline, chlorophyll and different actions of enzymes that work as bio-indicators to control the life of plants. Furthermore, SO_2 and O^3 are among the air contaminants that greatly influence the metabolism of plants and reduce their development. During plant health analysis, both physiological and morphological symptoms are observable (Saxena and Kulshrestha 2016). The effects of unfavorable abiotic stress in plants lead to a building up of soluble sugars, polyphenols and proline, which are amongst the global multifunctional stress-protective substances. *E. japonica* is one of the plants where soluble sugars, polyphenols and proline build up in its leaves at significantly high concentrations in the contaminated areas (Elloumi et al. 2017).

5.6.1.1 Proline

Plant pollutants, including gases and particulates, greatly increase due to the expansion of human population, transportation and various kinds of industries. The gradual damage of chlorophyll and yellowing of the leaves are among the most common effects of air pollution. Consequently, the plant's photosynthetic rate will decrease. In addition, gaseous emissions, such as SO_2, and CO, from vehicles cause an increase in the proline content of plants (Patidar et al. 2016). Proline is a protein-based amino acid acting as a radical scavenger, an osmolyte,* an electron sink, a cell wall component and a stabilizer of macromolecules (Sharma and Dietz 2006). Many plants accumulate the proline in response to heavy metals. The water balance of plants is unfavorably affected by heavy metals, mainly Cd. The functional importance of proline accumulation in plants exposed to stress caused by heavy metals seems to be due to the scavenging of hydroxyl radicals and water balance maintenance (Schat et al. 1997). This compatible solute[†] is accumulated in different plants under several abiotic stresses, including low temperature, high temperature, water deficit and salinity. The capacity of metals to induce proline is not the same (Sharma and Dietz 2006; Shafiei Masouleh et al. 2019). The strong proline inducers in plant species are Cu and Cd (Sharma and Dietz 2006). Based on a report by Alyemeni et al. (2018), treatment with selenium in cadmium-stressed plants (Cd + Se) further enhanced the activity of SOD, CAT, APX, and glutathione reductase. Moreover, osmolytes such as proline and glycine betaine increase with selenium application, and this shows their role in improving the osmotic stability of *Solanum lycopersicum* under cadmium stress as well as reducing cadmium uptake.

Exogenous application of proline inhibits the detrimental impacts of cadmium in crops and causes the improvement of growth and photosynthesis. Simultaneously, heavy metal stress reduces the regulation of leaf water potential which is recovered through the exogenous usage of proline and improves the water potential in the leaves thus keeping the membranes protected from damages caused by metals (Singh et al. 2015). Under different biotic and abiotic stresses, proline may build up in the cytosol of plants. It was noticed that the exogenous application of proline might improve the internal level of proline in stress caused by heavy metals. This action will protect enzymes, reducing the risk of peroxidation of proteins and lipids and 3-D structure of proteins and vital organelles such as the cell membranes, and finally helps in maintaining intracellular redox homeostasis potential. Additionally, proline may regulate water potential, decrease metal uptake, maintain osmotic amendment by cellular homeostasis and improve crop tolerance for clogging metals in the cytoplasm. In *Anacystis nidulans*, the minimal K ions flow after proline application upon copper

---

* Osmolytes are substances affecting osmosis. They are soluble in the solution within a cell, or the surrounding fluid, for example plasma. They have a role in maintaining the cell volume and fluid balance. For instance, when a cell swells due to external osmotic pressure, membrane channels open and allow efflux of osmolytes which carry water with them to recover the normal cell volume.

† Osmoprotectants or compatible solutes are small molecules that act as osmolytes and help organisms to continue to live in severe osmotic stress.

stress refers to the protecting function of proline in which the plasma membrane is protected from toxicity caused by Cu (Singh et al. 2015).

The compatible solutes buildup causes an action for plant adaptation to unfavorable environmental conditions. These adaptations occur to scavenge excess free radicals, maintain the osmotic balance, sustain the photosystem II, adjust the cellular redox potential and stabilize the cell membrane structure. Hence, proline always accumulates under stress conditions (Elloumi et al. 2017).

Proline builds up in a large variety of living things ranging from microorganisms to higher plants when exposed to environmental stresses. Under abiotic stress, plants will normally build up proline. Proline accumulation could regulate osmotic conditions at cell level and protect the enzyme. Hence, this can strengthen the structure of macromolecules and organelles (John et al. 2012). Proline may be involved in heavy metal stressed plants through various processes such as metal chelation, free radicals scavenging and osmo- and redox-regulation. Also, proline protects glucose-6- nitrate reductase and phosphate dehydrogenase against Cd and Zn-induced prohibition. Additionally, proline acts as an antioxidant: when plants are exposed to active oxidation and reduction together, for instance, copper, mercury, zinc and cadmium, this leads to the production of stimulated free radicals that cause oxidative stress. Moreover, it seems that the main reasons for toxicity are especially due to redox metals. For reducing oxidative stress, cells are supplied with a well-established connection of low molecular weight metabolites and antioxidative enzymes (Sharma and Dietz 2006). Exogenous usage of proline decreases the impacts of selenium on plants by improving growth in *Phaseolus* and reducing oxidative stress (Singh et al. 2015).

Based on a report, in *Chlorella vulgaris* pretreatment by exogenous spraying of proline prevents peroxidation of lipids as well as significant K^+ flow after exposure to the heavy metals of Cr, Zn, Cu and Ni. It was evident that the accumulation of proline in *C. vulgaris* was common because of the exposure to the metal. Identically, the use of proline in a medium available with nutrient minimized the inhibitory effect of copper on the development of cyanobacterium *Anacystis nidulans*. That said, the impact is much more noticeable upon the use of proline before the treatment of copper. Proline preserves the membrane by stopping K^+ outflowing toward copper. The *Chlorella* cells, which contain high concentrations of proline, are responsible for reducing Cu (Sharma and Dietz 2006).

Adverse impacts of vehicular pollutants are made through the production of ROS in plants that cause the peroxidizing destruction of cellular components. Proline protects the plants from injury caused by oxidative stress. Proline has an essential function in saving plants from different types of stresses. The accumulation of proline by plants is considered as a physiological response of plants when exposed to osmotic stress (Saxena and Kulshrestha 2016; Patidar et al. 2016). A negative relationship exists between proline content and the contents of chlorophyll (Chl) a, b and total Chl. During air pollution, Chl content will be decreased due to damaging chloroplast by active oxygen species. Plants with their proline contents (as the most common compatible osmolytes) survive under stress. When proline content increases this can be an indicator of vehicular pollution (Patidar et al. 2016). Proline is an element of several enzymes and proteins. It has an essential role in agricultural plants as the origin of osmoprotectant and energy in unfavorable statuses, including UV light and heavy metal pollution. The accumulation of proline in environmental stress decreases the degradation of other proteins. The buildup of proline in plant cells may be due to an increase in proline synthesis, decrease in proline degradation and hydrolysis of proteins (Assadi et al. 2011; Saxena and Kulshrestha 2016). Proline buildup is associated with increased tolerance against water deficit and salinity stresses in many plants (like rice and wheat). In a study, proline accumulation exhibited good resistance of *E. camaldulensis* against air pollution in the contaminated areas (Assadi et al. 2011).

Under environmental stresses, proline builds up as an organic solute in several plants (Duman et al. 2011). In abiotic stress, the accumulation of proline decreases other proteins' degeneration.

Based on another report (Zhou et al. 2019), the presence of cadmium but not zinc in the nutrient solution of the halophyte *Kosteletzkya pentacarpos* increased the contents of proline and glycine betaine of leaves. However, salinity reduced the content of proline in Cd-treated plants but increased it in Cd + Zn-exposed plants. Proline is synthesized through both glutamate and ornithine pathways while proline dehydrogenase is inhibited under plant exposure to heavy metals. The expression of the enzyme gene depends on the metal and plant organ, which is more in leaves compared to roots (Zhou et al. 2019). Stress exposure causes the expression of amino acids such as proline that acts as an osmoregulator, scavenger of radicals and stabilizer of macromolecules. Moreover, quaternary ammonium compounds, especially glycine betaine, are widely known as osmoregulators (Elloumi et al. 2018). That said, the application of exogenous proline in young olive plants could affect an increase in antioxidant enzyme activities (catalase, superoxide dismutase, glutathione peroxidase), proline content, relatively high amounts of hydrogen peroxide, thiobarbituric acid reactive substances and electrolyte leakage. Moreover, this treatment decreased Cd accumulation- induced oxidative damage due to increasing antioxidant enzyme activities. On the other side, nutritional status, photosynthetic activity, plant growth and oil content of olive fruit increased. In addition, cadmium decreased the uptake of essential elements (Ca, Mg and K) (Zouari et al. 2016).

Zouari et al. (2018) reported that Pb stress-induced oxidative damage in olive plants could be reduced through treating plants with the application of proline and glycine betaine in irrigation water and plants produced a better level of biomass, and proline acted better than glycine betaine. Konotop et al. (2017) reported that the treatment of cadmium-stressed wheat with proline increased the intracellular proline in leaves and water content and removed cadmium-caused dehydration.

In *Laurus nobilis* L., air pollution could lead to a remarkable increase in proline and protein content and activity of the nitrate reductase. Thus, *Laurus nobilis* L. as an evergreen and broadleaf plant could be a good choice for polluted urban areas (Sanaeirad et al. 2017). Cement dust as air contamination can affect the biochemical content of plant tissues. For example, in *Vinga mungo* L., it decreased the contents of amino acid, proline and sugars, but increasing the levels of cement dust caused an enhancement to the content of amino acid and proline (Rajasubramaniyam et al. 2018).

5.6.1.2 Histidine

About three-quarters of plants that hyperaccumulate metals are considered as accumulators of Ni. In *Alyssum* species (Brassicaceae), the accumulation of Ni is particularly related to the producing capacity of histidine. While Ni exposure leads to an increase in histidine concentration in a great and consistent manner in the xylem sap of *A. lesbiacum* (Ni-hyperaccumulator), this increase was not present in *A. montanum* (non-accumulator) (Sharma and Dietz 2006). The X-ray analysis of xylem sap detected the presence of Ni with histidine in *A. lesbiacum*. Hence, the hyperaccumulation of Ni depends on the displacement of Ni from root to shoot. Moreover, the supplementation of *A. montanum* by exogenous histidine (with equimolar concentrations) leads to the transmission of more Ni to the shoot due to the flow of Ni into the xylem. This reveals the tolerance to Ni concentrations. The same outcomes are observed with *Brassica juncea*, which is a Ni non-accumulator plant (Kerkeb and Krämer 2003). That said, in *A. lesbiacum* roots, the content of histidine was 4.4 times greater than that in *B. juncea*. In *A. lesbiacum* and *B. juncea*, Kerkeb and Krämer (2003) investigated the role of free histidine in xylem loading. Resultant data revealed a model in which Ni absorption is not related to the intake of histidine. The improved Ni emission inside the xylem is related to the synchronous emission of histidine from increasing free histidine of roots. Furthermore, *Thlaspi goesingense* is considered as a hyperaccumulator of Ni, which shows the raised histidine concentrations in roots in comparison to *T. arvense*.

5.6.2 Other Amino Acids

In *Deschampsia cespitosa*, whether tolerant or not to metals, toxicity caused by zinc differed with nitrogen origin in the medium of development. Zn is always not toxic when N is applied as ammonium instead of nitrate (Smirnoff and Stewart 1987). Smirnoff and Stewart (1987) found that ammonium non-tolerant clones build up asparagine inside the roots and that the produced zinc-asparagine compound may decrease the toxicity of zinc. A particular characteristic of some species is the accumulation of certain amino acids and overaccumulation of metals in some organs. Amino acids were identified by some researchers in species of Philippine Ni-hyperaccumulators, mainly in three species, which are *Dechampetalum geloniodes, Phyllanthus palwanensis* and *Walsura monophylla*. Proline was the very abundant amino acid in all cases. As a result, the correlation between metals and amino acids could be essential because Ni-amino acid complexes are steadier than complexes with carboxylic acids. Analysis of the modification in amino acids amounts in the xylem sap of Ni-hyperaccumulator *Stackhousia tryonii* was done. Glutamine amount decreased from 48% to 22%, and the amounts of total amino acid reduced a little bit from 22 to 18 mM while aspartic acid, glutamine and alanine increased. Alanine and aspartic acid can participate in producing the complex of Ni in xylem (Sharma and Dietz 2006).

Zinc has greatly influenced the concentrations of amino acids in soybean and tomato. A study of the balance of metal compounds in xylem sap indicates that the accumulation of Ni and Cu was associated with histidine and asparagine. The cysteine is a major metabolite in metal sequestration and antioxidant defense, which is required for glutathione/phytochelatin and methionine synthesis. It has been shown that the increased genetic ability to synthesize metal-induced cysteine maintains the *Arabidopsis* survival under severe cadmium stress. It is difficult to assess the importance of these modifications in the formation of amino acids for metal binding in vivo. However, it is important to remember that amino acids are distributed unsymmetrically between secretory and plasmatic compartments (Sharma and Dietz 2006).

5.6.2.1 Polyamines

Polyamines are present everywhere in all living organisms. They are a class of plant growth regulators and act as second messengers, which affect growth and development processes in plants. The polyamines, i.e. putrescine^{2+}, spermidine^{3+} and spermine^{4+}, are the cations due to protonation at cytoplasmic pH. This explains their binding ability to nucleic acids and phospholipid bonds in plasma membranes. The levels of polyamines and their biosynthetic enzymatic activity in plants increase under abiotic stresses. Polyamines are produced from citrulline, arginine or ornithine, following decarboxylation to produce putrescine. By adding one or two aminopropyl groups from S-adenosyl methionine, spermidine and spermine are produced. Mutant and transgenic plants with varied polyamine levels show strong deviations from growth. Moreover, polyamines have anti-oxidative properties through suppressing $O_2^{\bullet-}$ accumulations by inhibiting NADPH oxidase (Sharma and Dietz 2006).

Exposure to heavy metals causes an alteration to the polyamine contents. Cd treatment of oat seedlings and detached leaves leads to an increase up to ten times in putrescine contents, with the rise of spermine and spermidine content (Weinstein et al. 1986). Difluoromethyl arginine, which is an arginine decarboxylase inhibitor, blocks the Cd-dependent increase in putrescine and links this enzyme with Cd-induced putrescine biosynthesis. That said, reactions of polyamines to cadmium stress greatly differ in *Phaseolus vulgaris* in a particular way. Spermidine decreases in leaves, increases in hypocotyl and does not change in roots, while putrescine increases in hypocotyl, epicotyls and roots. Conversely, the level of spermine reduces in all parts of the seedling. The modifications promoted by Al in putrescine of *Catharanthus roseus* were affected by cellular age. Based on a report (Sharma and Dietz 2006), the addition of Al to the suspended growth medium caused the rise of the content of putrescine within one day in newly transmitted cells, however, it decreased the same at the latest phases (from two to seven days). Another report found that Hg treatment induced an increase in 1,3-diamino propane and putrescine in green alga *Chlorogonium elongatum*.

Polyamines such as spermine and spermidine play important physiological roles in the development and growth of plants. They are known as inhibitors of lipid peroxidation and strong ROS scavengers. Moreover, the exogenous application of polyamines protects the plants from different stress conditions, including wilting, contamination, salinity and cold. The property of free radical scavenging of polyamines is responsible for crop protection from ozone injury through an exogenous application (Saxena and Kulshrestha 2016).

Exogenous use of polyamines (spermidine and spermine) before acid rain can protect the bean plants in the stress condition. Polyamines can prepare the cell to meet and fight the stress by stabilizing the membranes and forming potential for increased buffering and antioxidant capacity. Its longer chains and a greater number of positive charges can explain the obvious protective effects of spermine compared to spermidine, which allows neutralizing and stabilizing the membrane (Velikova et al. 2000).

5.6.2.2 Phytochelatins and Glutathione

The importance of metal-induced phytochelatins[*] and glutathione[†] for tolerance to metals are broadly summed up in good surveys. Glutathione depletion seems to be the main reason for heavy metal toxicity in a low period. According to this assumption, a perfect association between the amounts of glutathione and tolerance index was observed through ten genetic patterns of pea with different sensitivities to cadmium. There is a great correlation between the concentrations of O-acetyl-L-serine, cysteine and glutathione in shoot tissue with the capacity of Ni-hyperaccumulation of different species of hyperaccumulator *Thlaspi* (Freeman et al. 2004). In hyperaccumulator *T. goesingense*, the increased glutathione concentrations synchronize with the great activity of serine acetyltransferase, in which serine acetyltransferase stimulates the acetylation of L-seine to O-acetyl-L-serine. O-acetyl-L-serine has a role in providing a carbon skeleton for cysteine biosynthesis (Sharma and Dietz 2006).

5.6.3 OTHER N-CONTAINING METABOLITES

When higher plants are exposed to metal stress, they synthesize the N-containing amino acid-derived metabolites. These include mugineic acid, nicotianamine and betaines. Mugineic acid and nicotianamine have a particular role in the homeostasis of metals and defense against metal stress, whereas betaines appear to be as general stress metabolites. Glycine betaine (trimethylglycine) is common as an osmolyte and rises in plants under drought and salinity conditions. Its composition is based on decarboxylation from serine to ethanolamine, which is methylated using S-adenosylmethionine as a methyl donor (Wani et al. 2013). Ecotypes of *Armeria maritima* are gathered from salt marshes, sandy pastures or sites containing zinc. When exposed to water deficit stress, plants grown in sites containing Zn accumulate more proline and betaine compared to crops developed in the absence of metal pollution (Sharma and Dietz 2006).

Nicotianamine is found in every crop and is produced by nicotianamine synthase (NAS) from three molecules of methionine. The genome of *Arabidopsis thaliana* includes four NAS genes (Sharma and Dietz 2006). Nicotianamine is a plant-derived chelator of metals. It plays a critical role in metal detoxification and phytoremediation (Kim et al. 2005).

5.6.4 SOLUBLE SUGARS

Soluble sugars are introduced as cryoprotectant and osmoprotectant, which are essential for the plasma membrane. Besides this, they are indispensable parts in the plant structure and as a power source in all living things. The contents of soluble sugars are signals that enable plant functions. Moreover, it was mentioned that soluble sugars act in a protective role against stress. A report found that the contents of soluble sugars increased in the plants grown in polluted sites and which showed the symptoms of stress (Koziol and Jordan 1978).

Soluble sugars are essential components of plants and act as an energy source. Sugars produced by plants in photosynthesis are broken down through the respiration process. Soluble sugars concentration expresses the functions of the plant, which control the sensitivity of plants toward

---

[*] Phytochelatins are oligomers of glutathione that are produced by the enzyme phytochelatin synthase. They are found in plants, fungi, nematodes and all groups of algae, including cyanobacteria. The roles of phytochelatins are as chelators and detoxifiers of heavy metals.

[†] Glutathione is an antioxidant in animals, plants, fungi and some bacteria, and archaea. It is capable of protecting cells from ROS.

contamination of the air. The low amount of soluble sugar in contaminated areas may be referred to as the decreased CO_2 fixation and increased respiration rate due to the deterioration of chlorophyll. Gases like H_2S, NO_2 and SO_2 lead to the soluble sugars being consumed in plant leaves under stressful conditions in a contaminated environment. In plants exposed to SO_2, the reaction of sulfite with ketones of carbohydrates and aldehydes leads to a decrease in the carbohydrate content in the plant. Hence, the SO_2-exposed plant shows rising contents of soluble sugars. SO_2 fumigation (0.34 and 0.51 ppm) of *Pinus banksiana* causes an increase in the amounts of the reducing sugars and a decrease in the contents of non-reducing sugars. The increase is due to the breaking down of polysaccharides to reducing sugars. The exposure of *Phaseolus vulgaris* seedlings to SO_2 reduces the content of starch in this plant (Koziol and Jordan 1978). Also, the nonstructural total carbohydrates are reduced after SO_2 (Saxena and Kulshrestha 2016).

Trehalose is a non-reducing disaccharide of glucose. It can be found in fungi, plants and yeasts. There is a correlation between exposure to trehalose and tolerance to various stresses. Hence, it is characterized by the ability to protect biomolecules from abiotic stresses. The protective effect of trehalose is shown in two methods, either by membrane stabilizing or by biological macromolecules stabilizing like protein. The exogenous application of trehalose increases the levels of intracellular trehalose. Similarly, *Arabidopsis thaliana* and yeasts can absorb exogenous trehalose and accumulate it. The oxidative stress resulting from metals is the peroxidation of lipids. Malondialdehyde is a common product of lipid peroxidation that is used to determine the oxidative stress level of plant cells. Generally, the primary sites of metal damage are the cell membranes of the plant. Trehalose plays a role in maintaining the stability of proteins and membranes and reducing the aggregation of denatured proteins. For instance, the exposure of *Lemna* plants to Cd causes a reduction in the total chlorophyll content. It was reported that Cd stress leads to a decrease in pigment concentration, whereas the application of glycine betaine and trehalose could influence the total chlorophyll content positively. As a result, both trehalose and glycine betaine are effective in reducing the harmful effects of cadmium stress on the photosynthetic activity of *L. gibba* (Duman et al. 2011).

Sugars play a role in stress management as well as plant growth through signaling and then up-regulation or down-regulation of a variety of genes of photosynthesis, respiration and sugar metabolism. Trehalose and sucrose are the water-soluble sugars that accumulate in higher plants under environmental stresses and are critical constituents of the plant defense response to oxidative stress (Elloumi et al. 2018).

5.7 MOLECULAR REGULATION

Some genes may express in plants that are exposed to heavy metal stress; this activates the specific enzymes to respond to negative impacts. The direct impact of heavy metals may appear as over generation of ROS in a plant that is exposed to stress, and afterward causes damage to important macromolecules, including proteins, lipids and nucleic acids, and therefore, will restrict crop yield. Thus, to engineer more metal tolerant plants, it is important to find out the key parts of the metal tolerance network in plants. These parts include different genes, transcription factors, proteins and metabolites (osmolytes, phytohormones, lipids, etc.) which may be applied to engineer plants for their increased metal tolerance. Despite the availability of many kinds of literature, understanding the interaction of metal stress and defense systems of plants (antioxidant and non-antioxidant) at the physiological, biochemical as well as molecular levels is still a scientific issue (Singh et al. 2015).

5.8 CONCLUSION AND PROSPECTS

Environmental pollution, including air, land or soil, water, and acid rains could affect plant growth and development. These kinds of pollution include heavy metals and gases that originate from human activities in the environment such as industry, commercial and transportation activities. Some heavy metals are essential elements for plants as micronutrients, including chromium (Cr), copper (Cu), nickel (Ni) and zinc (Zn). However, their high concentrations are toxic. The most relevant heavy metals that are toxic for plants and food chains are cadmium (Cd), arsenic (As), lead (Pb) and mercury (Hg). Plants show some morphological and physio-biochemical symptoms against heavy metals and other environment pollutants, including gases of SO_2, NO_2, HF, O_3, PANs and acid rain, due to the combination of water and these gases, especially CO_2, and consequently respond to them through enzymatic, antioxidative and molecular regulations. Organic solutes are one of the mechanisms in plants to respond to environmental pollutants. They can divide into three classes, including nitrogen-containing metabolites, phenols and polyphenols, and carbohydrates. In many reports, proline is introduced as scavenging ROS, is synthesized by oxidative stress and is more important than other organic solutes and regulative mechanisms even in other stresses generally. However, other organic solutes and regulative mechanisms against environmental stresses are specific to each plant.

With regard to future prospects, it is proposed to further research the ability to assay the molecular basis of organic solutes in plants exposed to pollutants, and identify which direct or indirect pathway of biosynthesis are more important in response to pollutant stress.

REFERENCES

Abdel Latef, A. A. 2013. Growth and some physiological activities of pepper (*Capsicum annuum* L.) in response to cadmium stress and mycorrhizal symbiosis. *Journal of Agricultural Science and Technology* 15: 1437–1448.

Abdel Latef, A. A., A. Zaid, A.-B.-E. Abo-Baker, W. Salem, and M. F. Abu Alhmad. 2020. Mitigation of copper stress in maize by inoculation with *Paenibacillus polymyxa* and *Bacillus circulans*. *Plants* 9(11): 1513.

Abdel Latef, A. A., and M. F. Abu Alhmad. 2013. Strategies of copper tolerance in root and shoot of broad bean (*Vicia faba* L.). *Pakistan Journal of Agricultural Sciences* 50: 323–328.

Abdel Latef, A. A., M. F. Abu Alhmad, and K. E. Abdelfattah. 2017. The possible roles of priming with ZnO nanoparticles in mitigation of salinity stress in Lupine (*Lupinus termis*) plants. *Journal of Plant Growth Regulation* 36(1): 60–70.

Abu-Romman, S., and J. Alzubi. 2016. Transcriptome analysis of Arabidopsis thaliana in response to cement dust. *Genes and Genomics* 38(9): 865–878.

Ahmad, P., A.A. Abdel Latef, E.F. Abd_Allah, A. Hashem, M. Sarwat, N.A. Anjum, and S. Gucel. 2016. Calcium and potassium supplementation enhanced growth, osmolyte secondary metabolite production, and enzymatic antioxidant machinery in cadmium-exposed chickpea (Cicer arietinum L.). *Frontiers in Plant Science* 7: 513.

Ahmad, P., M. A. Ahanger, M. N. Alyemeni, L. Wijaya, and P. Alam. 2018. Exogenous application of nitric oxide modulates osmolyte metabolism, antioxidants, enzymes of ascorbate-glutathione cycle and promotes growth under cadmium stress in tomato. *Protoplasma* 255(1): 79–93.

Alyemeni, M. N., M. A. Ahanger, L. Wijaya, P. Alam, R. Bhardwaj, and P. Ahmad. 2018. Selenium mitigates cadmium-induced oxidative stress in tomato (*Solanum lycopersicum* L.) plants by modulating chlorophyll fluorescence, osmolyte accumulation, and antioxidant system. *Protoplasma* 255(2): 459–469.

Assadi, A., A. G. Pirbalouti, F. Malekpoor, N. Teimori, and L. Assadi. 2011. Impact of air pollution on physiological and morphological characteristics of Eucalyptus camaldulensis. *Journal of Food, Agriculture and Environment* 9(2): 676–679.

Chojnacka, K., A. Chojnacki, H. Górecka, and H. Górecki. 2005. Bioavailability of heavy metals from polluted soils to plants. *Science of the Total Environment* 337(1–3): 175–182.

Copaciu, F., O. Opriş, Ü. Niinemets, and L. Copolovici. 2016. Toxic influence of key organic soil pollutants on the total flavonoid content in wheat leaves. *Water, Air, and Soil Pollution* 227(6): 196.

Craker, L. E., and D. Bernstein. 1984. Buffering of acid rain by leaf tissue of selected crop plants. *Environmental Pollution Series A, Ecological and Biological* 36(4): 375–381.

Darrall, N. M. 1989. The effect of air pollutants on physiological processes in plants. *Plant, Cell and Environment* 12(1): 1–30.

Debnath, B., M. Irshad, S. Mitra, M. Li, H. M. Rizwan, S. Liu, T. Pan, and D. Qiu. 2018. Acid rain deposition modulates photosynthesis, enzymatic and non-enzymatic antioxidant activities in tomato. *International Journal of Environmental Research* 12(2): 203–214.

Duman, F., A. Aksoy, Z. Aydin, and R. Temizgul. 2011. Effects of exogenous glycinebetaine and trehalose on cadmium accumulation and biological responses of an aquatic plant (*Lemna gibba* L.). *Water, Air, and Soil Pollution* 217(1–4): 545–556.

Elloumi, N., I. Mezghani, B. Ben Rouina, F. Ben Abdallah, and C. Ben Ahmed. 2018. Physiological evaluation of apricot (*Prunus armeniaca* L.) leaves to air pollution for biomonitoring of atmospheric quality. *Pollution* 4(4): 563–570.

Elloumi, N., M. Zouari, I. Mezghani, F. B. Ben Abdallah, S. Woodward, and M. Kallel. 2017. Adaptive biochemical and physiological responses of *Eriobotrya japonica* to fluoride air pollution. *Ecotoxicology* 26(7): 991–1001.

Ferenbaugh, R. W. 1976. Effects of simulated acid rain on *Phaseolus vulgaris* L.(Fabaceae). *American Journal of Botany* 63(3): 283–288.

Freeman, J. L., M. W. Persans, K. Nieman, C. Albrecht, W. Peer, I. J. Pickering, and D. E. Salt. 2004. Increased glutathione biosynthesis plays a role in nickel tolerance in *Thlaspi* nickel hyperaccumulators. *The Plant Cell* 16(8): 2176–2191.

Hadi, F., N. Ali, and M. P. Fuller. 2016. Molybdenum (Mo) increases endogenous phenolics, proline and photosynthetic pigments and the phytoremediation potential of the industrially important plant *Ricinus communis* L. for removal of cadmium from contaminated soil. *Environmental Science and Pollution Research International* 23(20): 20408–20430.

Ishtiyaq, S., H. Kumar, M. Varun, B. Kumar, and M. S. Paul. 2018. Heavy metal toxicity and antioxidative response in plants: An overview. In: Hasanuzzaman M., Fujita M., Nahar K. (eds), *Plants under Metal and Metalloid Stress* (pp. 77–106). Springer, Singapore.

John, R., P. Ahmad, K. Gadgil, and S. Sharma. 2012. Heavy metal toxicity: Effect on plant growth, biochemical parameters and metal accumulation by *Brassica juncea* L. *International Journal of Plant Production* 3(3): 65–76.

Karami, L., N. Ghaderi, and T. Javadi. 2017. Morphological and physiological responses of grapevine (*Vitis vinifera* L.) to drought stress and dust pollution. *Folia Horticulturae* 29(2): 231–240.

Kerkeb, L., and U. Krämer. 2003. The role of free histidine in xylem loading of nickel in *Alyssum lesbiacum* and *Brassica juncea*. *Plant Physiology* 131(2): 716–724.

Kim, S., M. Takahashi, K. Higuchi, K. Tsunoda, H. Nakanishi, E. Yoshimura, S. Mori, and N. K. Nishizawa. 2005. Increased nicotianamine biosynthesis confers enhanced tolerance of high levels of metals, in particular nickel, to plants. *Plant and Cell Physiology* 46(11): 1809–1818.

Konotop, Y. E. V. H. E. N. I. I. A., M. A. R. I. I. A. Kovalenko, I. L. D. I. K. O. Matušíková, L. Batsmanova, and N. Taran. 2017. Proline application triggers temporal redox imbalance, but alleviates cadmium stress in wheat seedlings. *Pakistan Journal of Botany* 49(6): 2145–2151.

Koziol, M. J., and C. F. Jordan. 1978. Changes in carbohydrate levels in red kidney bean (*Phaseolus vulgaris* L.) exposed to sulphur dioxide. *Journal of Experimental Botany* 29(5): 1037–1043.

Liao, B. H., H. Y. Liu, Q. R. Zeng, P. Z. Yu, A. Probst, and J. L. Probst. 2005. Complex toxic effects of Cd^{2+}, Zn^{2+}, and acid rain on growth of kidney bean (*Phaseolus vulgaris* L.). *Environment International* 31(6): 891–895.

Likens, G. E., and F. H. Bormann. 1974. Acid rain: A serious regional environmental problem. *Science* 184(4142): 1176–1179.

Masouleh, S. S., N. Jamal Aldine, and Y. N. Sassine. 2019. The role of organic solutes in the osmotic adjustment of chilling-stressed plants (vegetable, ornamental and crop plants). *Ornamental Horticulture* 25(4): 434–442.

Meagher, R. B. 2000. Phytoremediation of toxic elemental and organic pollutants. *Current Opinion in Plant Biology* 3(2): 153–162.

Mukherjee, S., A. Chakraborty, S. Mondal, S. Saha, A. Haque, and S. Paul. 2019. Assessment of common plant parameters as biomarkers of air pollution. *Environmental Monitoring and Assessment* 191(6): 400.

Nagajyoti, P. C., K. D. Lee, and T. V. M. Sreekanth. 2010. Heavy metals, occurrence and toxicity for plants: A review. *Environmental Chemistry Letters* 8(3): 199–216.

Patidar, S., A. Bafna, A. R. Batham, and K. Panwar. 2016. Impact of urban air pollution on photosynthetic pigment and proline content of plants growing along the A,B road Indore City, India. *International Journal of Current Microbiology and Applied Sciences* 5(3): 107–113.

Rai, P. K. 2016. Impacts of particulate matter pollution on plants: Implications for environmental biomonitoring. *Ecotoxicology and Environmental Safety* 129: 120–136.

Rajasubramaniyam, K., K. R. Hanumappa, N. Harish, and K. Narendra. 2018. Estimation of amino acid, protein, proline and sugars in *Vigna mungo* L. exposed to cement dust pollution. *International Journal of Advanced Research in Biological Sciences* 5(7): 173–186.

Rizwan, M., M. Imtiaz, Z. Dai, S. Mehmood, M. Adeel, J. Liu, and S., Tu. 2017. Nickel stressed responses of rice in Ni subcellular distribution, antioxidant production, and osmolyte accumulation. *Environmental Science and Pollution Research International* 24(25): 20587–20598.

Saif, S., and M. S. Khan. 2018. Assessment of toxic impact of metals on proline, antioxidant enzymes, and biological characteristics of *Pseudomonas aeruginosa* inoculated *Cicer arietinum* grown in chromium and nickel-stressed sandy clay loam soils. *Environmental Monitoring and Assessment* 190(5): 290.

Sanaeirad, H., A. Majd, H. Abbaspour, and M. Peyvandi. 2017. The effect of air pollution on proline and protein content and activity of nitrate reductase enzyme in *Laurus nobilis* L. plants. *Journal of Molecular Biology Research* 7(1): 99–105.

Saxena, P., and U. Kulshrestha. 2016. Biochemical effects of air pollutants on plants. In: Kalshrestha U. and Saxena P. (eds), *Plant Responses to Air Pollution* (pp. 59–70). Springer, Singapore.

Schat, H., S. S. Sharma, and R. Vooijs. 1997. Heavy metal-induced accumulation of free proline in a metal-tolerant and a non-tolerant ecotype of *Silene vulgaris*. *Physiologia Plantarum* 101(3): 477–482.

Sharma, S. S., and K. J. Dietz. 2006. The significance of amino acids and amino acid-derived molecules in plant responses and adaptation to heavy metal stress. *Journal of Experimental Botany* 57(4): 711–726.

Shavalikohshori, O., R. Zalaghi, K. Sorkheh, and N. Enaytizamir. 2020. The expression of proline production/degradation genes under salinity and cadmium stresses in Triticum aestivum inoculated with Pseudomonas sp. *International Journal of Environmental Science and Technology* 17(4): 2233–2242.

Singh, M., J. Kumar, S. Singh, V. P. Singh, S. M. Prasad, and M. P. V. V. B. Singh. 2015. Adaptation strategies of plants against heavy metal toxicity: A short review. *Biochemistry and Pharmacology* 4(161): 2167–0501.

Smirnoff, N., and G. R. Stewart. 1987. Nitrogen assimilation and zinc toxicity to zinc-tolerant and non-tolerant clones of *Deschampsia cespitosa* (L.) Beauv. *New Phytologist* 107(4): 671–680.

Stroo, H. F., and M. Alexander. 1985. Effect of simulated acid rain on mycorrhizal infection of *Pinus strobus* L. *Water, Air, and Soil Pollution* 25(1): 107–114.

Velikova, V., I. Yordanov, and A. Edreva. 2000. Oxidative stress and some antioxidant systems in acid rain-treated bean plants: Protective role of exogenous polyamines. *Plant Science* 151(1): 59–66.

Wani, S. H., N. B. Singh, A. Haribhushan, and J. I. Mir. 2013. Compatible solute engineering in plants for abiotic stress tolerance - role of glycine betaine. *Current Genomics* 14(3): 157–165.

Weinstein, L. H., R. Kaur-Sawhney, M. V. Rajam, S. H. Wettlaufer, and A. W. Galston. 1986. Cadmium-induced accumulation of putrescine in oat and bean leaves. *Plant Physiology* 82(3): 641–645.

Zhou, M. X., M. E. Renard, M. Quinet, and S. Lutts. 2019. Effect of NaCl on proline and glycinebetaine metabolism in *Kosteletzkya pentacarpos* exposed to Cd and Zn toxicities. *Plant and Soil* 441(1–2): 525–542.

Zouari, M., C. Ben Ahmed, N. Elloumi, K. Bellassoued, D. Delmail, P. Labrousse, F. Ben Abdallah, and B. Ben Rouina. 2016. Impact of proline application on cadmium accumulation, mineral nutrition and enzymatic antioxidant defense system of *Olea europaea* L. cv Chemlali exposed to cadmium stress. *Ecotoxicology and Environmental Safety* 128: 195–205.

Zouari, M., N. Elloumi, P. Labrousse, B. Ben Rouina, F. Ben Abdallah, and C. Ben Ahmed. 2018. Olive trees response to lead stress: Exogenous proline provided better tolerance than glycine betaine. *South African Journal of Botany* 118: 158–165.

6 Oxidative Stress Tolerance in Response to Salinity in Plants

Nageswara Rao Reddy Neelapu, Titash Dutta,
Shabir H. Wani and Challa Surekha

CONTENTS

6.1 INTRODUCTION

Plants, due to their sessile habitat, are more frequently exposed to abiotic stresses (e.g. drought, salt stress and extreme temperature). These environmental stresses are detrimental for plant growth and development causing widespread loss of crop productivity. Abiotic stress, predominantly salinity and drought, account for roughly 70% reduction in crop yield (Wild 2003). The situation is further aggravated by climate change, which acts as a catalyst for these abiotic stresses by escalating their frequency and severity. Due to climate change, there has been a gradual shift in rainfall patterns, and this has led to decreased precipitation across the globe (Wani et al. 2017; Yadav et al. 2020).

Salt stress is the bane of the agricultural sector, as more than 6% of irrigated land and 20% of the total agricultural landscape is grappling under its influence (Hussain et al. 2019; Wani et al. 2020). This gradual loss of cultivable land to salinity is a threat to sustainable crop productivity and poses obstacles in upholding the food demands of the population. Exposure to salinity triggers stomatal closure and leads to a drastic decrease in the carbon dioxide/oxygen ratio. This sudden decline in carbon dioxide/oxygen ratio results in rapid production and cellular accumulation of reactive oxygen species (ROS), leading to a condition called "oxidative burst". The major ROS molecules that are generated range from singlet oxygen ($^1O_2$), superoxide radical (O_2^-), hydrogen peroxide (H_2O_2) and hydroxyl radical (HO^-) (Dar et al. 2017; Singh et al. 2015). These toxic molecules are responsible for oxidative damage to DNA and functional proteins, lipid peroxidation and breakdown of photosynthetic pigment, and hamper enzyme activity. Disruption to all these essential organelles and biological functions triggers secondary stress conditions such as oxidative and ionic stress in plants, which has a deleterious impact on plant growth and survival (Munns and Tester 2008; Wani et al. 2013; Das and Roychoudhury 2014; Dutta et al. 2019).

In plants, cellular organelles (chloroplasts, mitochondria and peroxisomes) under normal growth conditions produce low ROS levels. Though ROS play an integral role in stress signaling and response pathway, prolonged exposure to salt stress causes excessive ROS production, which are not effectively scavenged or neutralized leading to ROS accumulation (Mittler et al. 2004; Bhattacharjee 2019). It is essential to scavenge these ROS predominantly during stress conditions, and in this regard, plants are equipped with an intricate ROS detoxification machinery that comprises of non-enzymatic antioxidant molecules: ascorbic acid (AsA) and a-tocopheroland glutathione (GSH); and ROS-scavenging enzymes: superoxide dismutase (SOD), ascorbate peroxidase (APX), catalase (CAT), glutathione peroxidase (GPX) and peroxiredoxin (PrxR) (Dietz et al. 2006; Kadioglu et al. 2011; Bose et al. 2014).

TABLE 6.1
Candidate Genes Involved in Enhancement of Oxidative Salt Tolerance

Name of the Gene	Source	Recipient	Application	Reference
SOD2	Schizossccharomycespombe	O. sativa L.	Enhanced SOD accumulation, drought and salt tolerance	Zhao et al. (2006)
katE	E.coli	O. sativa L.	Enhanced production of catalase thereby increased salt tolerance	Prodhan et al. (2008)
katE	E.coli	O. sativa japonica	Enhanced production of catalase thereby increased salt tolerance	Kaku et al. (2000)
BcGR1	Chinese cabbage	Nicotiana Tabacum	Reduces oxidative stress and enhance drought and salt tolerance	Lee and Jo (2004)
Aldehyde dehydrogenase family7 member (GmPP55)	Glycine max L.	A. Thaliana; Solanum tuberosum L	Improved tolerance to H_2O_2 , salt and drought stress	Rodrigues et al. (2006)
Ascorbate peroxidase (APX)	A. thaliana	Nicotiana Tabacum	Enhanced H_2O_2 metabolism and reduced oxidative stress	Badawi et al. (2004)
Mn-SOD	A. thaliana	A. thaliana	Enhanced production of Mn-SOD, Cu/Zn-SOD, Fe-SOD, CAT and peroxidase (POD) thereby increased salt tolerance	Wang et al. (2005)
SOS2 gene	Populustrichocarpa	hybrid poplar clone T89 (Populustremula× Populustremuloides Michx clone T89)	Enhanced SOD activity, improved growth attributes and salt tolerance	Zhou et al. (2014)
CuZnSODAPX	A. thaliana	Ipomoea batatas	CuZnSOD and APX levels in transgenic plants increased by 13.3 and 7.8 fold respectively. Enhanced salt tolerance	Yan et al. (2016)
cytosolic APX (cAPX)	Panax ginseng	A. thaliana	Reduces oxidative stress and enhance and salt tolerance	Sukweenadhi et al. (2017)
cytosolic APX (cAPX)	Populustomentosa	Nicotianatabacum	Decreased H_2O_2 levels, increased AsAcontent and APX activity. Enhanced salt tolerance	Cao et al. (2017)
glutathione peroxidase 5 (RcGPX5)	Rhodiolacrenulata	Salvia miltiorrhiza	Reduces oxidative stress and enhance and salt tolerance	Zhang et al. (2018)

Modified from Wani et al. 2017.

Though the majority of research has been focused on the deleterious effects of ROS and oxidative damage on plants, recent scientific developments have highlighted the positive impacts of ROS. Some findings suggest that ROS are involved in signal transduction during salt stress exposure, and its inter-play with auxin triggers the stress response in plants (Krishnamurthy and Rathinasabapathi 2013). Moreover, they also assist in apoptosis and initiate a cellular response to pathogenic infections. This dual role of ROS has been a hot prospect of research in the scientific community and several efforts are being made to understand its dual function of signal transduction and as well as its toxic implications. Understanding and elucidation of cellular ROS signaling pathways can help improve the salt tolerance capacity of plants and, in turn, reduce loss in crop productivity. In this chapter, we focus on unraveling the salt stress response mechanism and ROS's dual role, and we discuss the various ROS scavenging mechanisms that can be employed to enhance crop saline tolerance (Table 6.1).

6.2 SALINITY TRIGGERS THE BIPHASIC RESPONSE IN PLANTS

Saline water encompasses nearly 70% of the earth's surface. Salinity is defined as excessive salt accumulation in water as well as soil. Saline soil is characterized by high electrical conductivity of above 4 dS/m (approximately 50 mM NaCl). The ions that predominantly account for salinity are Na^+, Ca^{2+}, Mg^{2+}, K^+, Cl^-, SO_4^{2-}, HCO_3^-, CO_3^{2-} and NO_3^{2-} (Flowers et al. 2000). Other agents like mineral weathering, precipitation, movement of salt in land surface and groundwater, use of saline water for irrigation, use of potassium fertilizer and aquaculture also contributes to salinity.

6.2.1 BIPHASIC RESPONSE MECHANISM

Exposure to saline environments induces various changes expressed by a cascade of morphological, physiological, metabolic and molecular alterations. Prolonged exposure to salinity triggers the onset of secondary stress conditions such as osmotic and oxidative stress. This phenomenon is termed as the biphasic response and was coined by Munns and Termaat (1986). The first phase induces disrup-tion of cellular homeostasis, while the second phase is linked with disruption of ion distribution and oxidative damage. The biphasic response is linked to a decrease in plant growth and yield attributes under salt stress (Bhattachrjee 2005) (Figure 6.1).

The onset of the first (osmotic) phase is initiated by a high level of salt accumulation in the soil sur-rounding the plants. This increases the extracellular Na^+ and Cl^- ion levels while the intracellular Na^+ and Cl^- ion concentrations are unaltered, thereby diminishing the plant's water uptake capacity. The hallmark of the osmotic phase reduction in leaf surface and stomatal closure is due to water deficit.

The duration of the second (ionic/oxidative) phase of the biphasic response is critical for the plants' survival. During this phase, the toxic effects of the above ions overpower the plant machinery. As the exposure time increases, translocation of salts from roots to the shoots becomes frequent. They start accumulating in the older leaves, resulting in excessive intercellular Na^+ and Cl^- ions. The sudden spike in these ions causes significant oxidative damage to cellular organelles and membranes by disrupting their structural integrity by generating various ROS; diminishes photosynthetic activity and rate of tran-spiration; and inhibits essential enzyme activity (Gilroy et al. 2014). Moreover, excessive ROS accumu-lation drastically lowers the photosynthetic pigment contents, Hill activity and photosystem II (PSII) activity. As the photosynthetic efficiency diminishes, the rate of new leaf formation is stalled, and the plants succumb due to a depletion of nutrients that are essential for survival. (Kalaji et al. 2011).

The biphasic response has been observed and validated in major crops such as *Zea mays* L. (Fortmeier and Schubert 1995), wheat (Wakabayashi et al. 1997), tomato (Maggio et al. 2007) and rice (Negrao et al. 2011). The biphasic response in plants due to saline stress unravels the cross-talks between other environmental stress factors in relation to changes in gene expression targeting abiotic stress response. To ameliorate the deleterious effects of these stress conditions, plants have evolved three fundamental strategies to overcome the biphasic growth inhibition and promote growth under salinity. The strate-gies employed to regulate low cytosolic sodium concentration are ion exclusion, compartmentalization and osmoprotection. The ion exclusion strategy revolves around the exclusion of the toxic accumulated

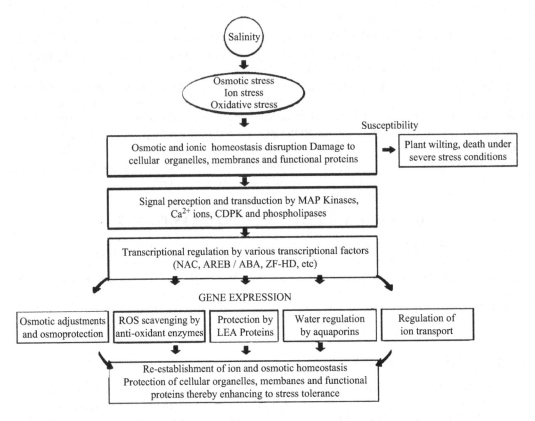

FIGURE 6.1   Salt stress responses in plants. (Modified from Wani et al. 2017.)

ions via ion exchangers (NHX). Compartmentalization of Na^+ ions into vacuoles facilitated by ion channels (V-PPases, V-ATPases, Na^+/K^+ antiporters) improves tissue tolerance. The third strategy of cellular osmoprotection involves tissue specific expression of potential genes to enhance the synthesis of osmolytes, aquaporins and antioxidant enzymes to improve salinity tolerance. (Roychoudhury and Chakraborty 2013; Sah et al. 2016; Joshi et al. 2016; Wani et al. 2018).

6.3   ROS: DUAL ROLE IN SALT-STRESSED PLANTS

Plants exposed to abiotic stress, such as salinity, tend to accumulate significantly high levels of ROS. These reactive oxygen intermediates (ROIs) are the reduced forms of atmospheric oxygen (O_2) and have serious toxic implications on plants if their cytosolic levels are not regulated efficiently (Sewelam et al. 2016; Jithesh et al. 2006). However, ROS at low levels also play an integral role in signal perception and subsequent stress response of the plants. In this segment the two facets of cellular ROS will be discussed and unravel their role as acclamatory signal molecules as well as toxic effects they induce in plants concerning salt stress exposure (Figure 6.2).

6.3.1   ROS: ROLE IN STRESS SIGNAL PERCEPTION AND TRANSDUCTION

Recent studies highlight the importance of ROS as signaling molecules in the regulation of abiotic stress responses and promoting plant growth and development (Foyer and Noctor 2005; Amir et al. 2019). Low cellular levels of ROS are a prerequisite for the plant to counter abiotic stress conditions and initiate mandatory regulatory pathways to curb the deleterious stress implications on plant growth and survival. ROS and abscisic acid (ABA) regulation has been associated with the induction of stomatal closure. As plants are exposed to salinity, they tend to accumulate ABA, which in

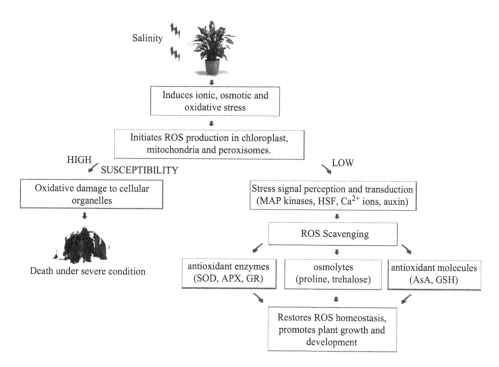

FIGURE 6.2 Dual role of ROS in salt stressed plants.

turn triggers stomatal closure signaling cascade to limit the loss of water by transpiration and also to regulate the uptake of CO_2 for photosynthesis (Kwak et al. 2003; Wani and Kumar 2015; Wani et al. 2016). H_2O_2, a key ROS, is involved in activating the calcium-permeable channels which are the first step in this signaling cascade (Miller et al. 2007, 2010).

Apart from ABA signal transduction, ROS have also been linked to activation of heat-stress transcription factors (HSFs), auxin homeostasis and MAPK (mitogen-activated protein kinase) cascades in response to salinity (Yang and Guo 2018). These ROS-dependent redox sensors are activated once the cellular ROS levels increase, and they assist the quenching of these ROIs via their respective regulatory pathway. Miller and Mittler (2006) reported that in *Arabidopsis thailiana* under an oxidative stress environment, the elevated levels of cellular ROS are perceived by HsfA4a transcription factor which then upregulates the oxidative stress-response genes in *A. thailiana*. Similarly, a 46 kDa phospho-p38-like MAPK (p46-MAPK) was activated in response to cytosolic ROS levels in *Triticum turgidum* and *Arabidopsis*. This activated p46-MAPK was responsible for initiating microtubule disruption and assemblage of tubulin polymers directed towards regulating ROS homeostasis. Low cellular ROS levels trigger MAPK cascades, which are a key regulatory detoxification-signaling pathway. They play an integral role in regulating ROS homeostasis by inducing antioxidant enzyme scavenging in response to salt/oxidative stress. For example, the MEKK1-MKK1/MKK2-MPK4 cascade is involved in ROS and salicylic acid (SA)-induced stress signal transduction in *Arabidopsis* (Pitzschke et al. 2009).

Auxin, a vital plant hormone, is also associated with salt stress regulation in plants. Studies have suggested that auxin homeostasis in response to salinity is triggered by elevated levels of cellular H_2O_2. Both auxin homeostasis and H2O2 levels are involved in the regulation of root gravitropism and growth parameters (such as pollen tube) to prevent susceptibility of plants in saline conditions (Krishnamurthy and Rathinasabapathi 2013)

6.3.2 ROS: Toxic Effects on Plants

Though low cytosolic ROS levels are reported to benefit the plants in terms of stress signal perception and transduction, high levels of ROS accumulation due to salinity are detrimental for the survival of the plants (Waszczak et al. 2018). As the cellular ROS levels increase they become toxic and cause a series of deleterious impacts in plants. They lead to oxidative damage of cellular membrane and organelles, and disruption of the structural integrity of DNA, functional proteins and enzymes present in chloroplasts and mitochondria (Allen 1995). The chloroplast is the major production unit of ROS during normal growth conditions as well as when plants encounter stress environments. The generated ROS cause considerable damage to photosystem II due to their highly unstable and reactive nature. Damage to photosystem II reduces the rate of photosynthesis in plants due to the impairment of photosynthetic enzymes and reduced CO_2 uptake (Mittler 2002; Gill and Tuteja 2010). As the rate of photosynthesis diminishes, the rate of nutrient assimilation is also hampered. This leads to a shortage of nutrients and drastically affects plant growth and development. ROS detoxification becomes essential under such circumstances to prevent the plant from entering into apoptosis and subsequent death of the plant. ROS detoxification is an intricate pathway which is carried out by antioxidant enzymes and non-enzymatic scavengers specializing in quenching these ROIs to maintain cellular ROS homeostasis during stress (Li et al. 2017).

6.4 REGULATION OF ROS IN CELLULAR ORGANELLES

The cellular organelles that contribute to low levels of ROS production are chloroplasts, mitochondria and peroxisomes during optimal growth conditions (Miller et al. 2010; Hossain and Dietz 2016). However, as abiotic stress such as salinity overpowers the normal growing conditions, the rate at which ROS is produced increases drastically, which needs to be regulated for the plants to survive. In this regard, various intricate ROS-quenching pathways have been elucidated in plants, such as the water-water cycle in chloroplasts, the ascorbate–glutathione cycle in chloroplasts, cytosol, mitochondria and peroxisomes, and glutathione peroxidase and CAT mediated scavenging in peroxisomes (Suo et al. 2017; Tanveer and Shabala 2018). Among these, the ascorbate-glutathione cycle occurs in all cellular compartments and is considered to be crucial for maintaining cellular ROS homeostasis.

6.4.1 CHLOROPLAST

The photosystem I (PSI) and photosystem II (PSII) machinery present in the thylakoids of chloroplast serves as a major ROS production site. Under salt stress, high levels of ROS are generated as the CO_2 levels diminish due to stomatal closure (Asada 1999). Since the photon intensity is in excess of that essential for CO assimilation, the excessive photons get involved in electron transfer to an oxygen molecule, resulting in the generation of superoxide radical (O_2^-) in PSI by what is commonly known as the Mehler reaction (Asada 2006). Quenching of these superoxide radical (O_2^-) is mandatory to prevent oxidative damage of PSI, and in this regard, a membrane-bound copper/zinc superoxide (Cu/ZnSOD) is located in PSI and is responsible for the conversion of the superoxide radicals into H_2O_2. This is followed by conversion of the H_2O_2 intermediate into the water by a membrane-bound thylakoid ascorbate peroxidase (tylAPX). This cycle is also known as the water-water cycle is a continuous process and continues until all the superoxide radicals are quenched.

In the light-harvesting complex of PSII, another ROS known as singlet oxygen ($^1O_2$) is generated by excited triplet-state chlorophyll as the ETC is reduced. Some of the intermediate H_2O_2 from the water-water cycle is directed towards quenching the formation of singlet oxygen by promoting quinone A (QA) oxidation, the primary plastoquinone (PQ) electron acceptor. This subsequently enhances the photosynthetic electron transport flow and invariably slows down generation of singlet oxygen intermediates. This regulation of ROS homeostasis in terms of controlled production and its subsequent quenching in the chloroplast was found to be beneficial for enhancing tolerance to salinity in many transgenic salt-tolerant cultivars (Van Camp et al. 1996; Mittler and Berkowitz 2001; Tseng et al. 2007). In contrast, inadequate ROS-scavenging in chloroplast has been linked to increased stress sensitivity and oxidative damage of PSI and PSII (Wang et al. 2005).

6.4.2 Mitochondria

Mitochondria is also a potential generator of ROS, though the levels of ROS generated are low in comparison to chloroplasts and peroxisomes (Rhoads et al. 2006; Amor et al. 2019). The major sites of ROI production in mitochondria include complex I and complex III of theelectron transport chain (mtETC) (Rhoads et al. 2006; Møller et al. 2007; Calderon et al. 2018). The ubi-semiquinone (UQ) intermediate produced in complex I and III directly transfers electrons to molecular oxygen, converting it to superoxide radical (O_2^-) and finally reducing it to H_2O_2 (Raha and Robinson 2000; Rhoads et al. 2006). To curb the levels of these ROS generated, two key enzymes, 1 alternative oxidase (AOX1) and manganese SOD (Mn-SOD), are involved in their detoxification pathways and also prevent the cells from entering the programmed cell death (PCD) pathway under salt stress. (Foyer and Noctor 2005; Sánchez-Guerrero et al. 2019). AOX functions by preventing the reduction of the UQ complex and blocks the electron transfer flow thereby lowering mitochondrial ROS production. The mode of action of Mn-SOD is somewhat complex, as it initially assists the conversion of superoxide radical to H_2O_2 and then reduces it to water via the water-water cycle.

Giraud et al. (2008) reported that *Arabidopsis* plants with mutant mitochondrial AOX1a are susceptible to drought and salinity stress, and they exhibited significantly lower antioxidant scavenging capacity in comparison to its wild type. Similarly, in a salt-tolerant tomato cultivar remarkably elevated levels of mitochondrial Mn-SOD were observed in comparison to its salt-sensitive counterpart under 100mM NaCl stress (Mittova et al. 2003).

Further scavenging via ascorbate–glutathione cycle is also common in mitochondria. Ascorbic acid (AsA) and glutathione (GSH) molecules have been reported to possess significant potential in combating oxidative stress, and these molecules are deemed crucial for plant defense mechanisms. The importance of these antioxidants is also validated by their high concentrations in mitochondria and other cellular components (5–20 mM ascorbic acid and 1–5 mM glutathione). It has been demonstrated by Foyer and Noctor (2005) that a high ascorbic acid/glutathione ratio is a prerequisite for effective scavenging of mitochondrial ROS to maintain an equilibrium between ROS production and quenching. This high ratio is maintained by three enzymes, namely, glutathione reductase (GR), monodehydro ascorbate reductase (MDAR) and dehydro ascorbate reductase (DHAR) utilizing NADPH as reducing equivalent.

6.4.3 Peroxisomes

Abiotic stresses, predominantly salinity and water-deficit stress, are responsible for inhibiting the antioxidant molecules present in peroxisomes and subsequently increase ROS production (Mittova et al. 2003; Nyathi and Baker 2006). Mittova et al. (2003) reported that in tomato cultivars, as low as 50 mM NaCl stress significantly lowered the AsA and GSH levels and induced lipid peroxidation in peroxisomes.

Several metabolic activities result in a rapid generation of ROIs (O_2^- and H_2O_2) in peroxisomes. Exposure to prolonged saline stress reduces water uptake rate, triggering stomatal closure, which in turn decreases the mesophyll CO_2/O_2 ratio. This alteration in the CO_2/O_2 ratio increases the rate of photorespiration and glycolate production in chloroplasts. These glycolate molecules are translocated to peroxisomes for their oxidation by glycolate-oxidase and account for the vast majority of H_2O_2 produced (Karpinski et al. 2003). The other metabolic activities that contribute to H_2O_2 generation include fatty acid β- oxidation, flavin oxidase pathway and dismutation of superoxide radicals (Palma et al. 2009). ROS regulation in peroxisomes is carried out primarily by CAT enzymes. They partake in H_2O_2 detoxification under elevated photorespiration conditions due to oxidative stress (Vandenabeele et al. 2004). APX and the AsA–GSH cycle also account for H_2O_2 quenching in peroxisomes. The superoxide radicals generated by xanthine oxidase (XOD) in the matrix of leaf peroxisomes are oxidized to the stable molecular oxygen thereby restoring ROS homeostasis (Corpas et al. 2001).

6.5 CONCLUSION

ROS signaling and regulation form an integral component of plant response to salt stress, and they often tend to exhibit dual response. On one hand, onset of salinity induces low ROS generation which contributes towards signal transduction of the stress in plants, whereas prolonged exposure to salinity significantly beefed up cytosolic ROS accumulation which is detrimental to the survival of plants. These two facets of ROS are closely interrelated, and a slight alteration in ROS homeostasis can bring about drastic changes to plant life cycle. Since a low cytosolic ROS level was able to activate other signaling pathways in stress response, this attribute is being exploited by various researchers to develop transgenic varieties to improve tolerance towards salinity. However, in this regard, extensive knowledge is still required because plants counter multiple abiotic and biotic stresses at the same time. So, it is imperative to understand the cross talks among these multiple stresses in relation to ROS metabolism and regulation to successfully develop transgenic varieties of plants with improved stress tolerance.

ACKNOWLEDGMENT

The authors are grateful to GITAM (Deemed to be University) for providing the necessary facilities to carry out the research work and for extending constant support in writing this review. TD is thankful for financial support in the form of the DST Inspire Fellowship (IF 160964), Department of Science and Technology, New Delhi.

REFERENCES

Amir R, Munir F, Kubra G, Nauman I, Noor N (2019) Role of signaling pathways in improving salt stress in plants. In: Akhtar M. (ed.), *Salt Stress, Microbes, and Plant Interactions: Mechanisms and Molecular Approaches*. Springer, Singapore, pp 183–211

Asada K (2006) Production and scavenging of reactive oxygen species in chloroplasts and their functions. *Plant Physiol* 141:391–396

Amor N, Jimenez A, Boudabbous M, Sevilla F, Abdelly C (2019) Implication of peroxisomes and mitochondria in the halophyte Cakile maritima tolerance to salinity stress. *Biologia Plant* 63(1):113–121

Asada K (1999) The water-water cycle in chloroplasts: Scavenging of active oxygens and dissipation of excess photons. *Annu Rev Plant Physiol Plant Mol Biol* 50(1):601–639

Badawi GH, Kawano N, Yamauchi Y, Shimada E, Sasaki R, Kubo A, Tanaka K (2004) Over-expression of ascorbate peroxidase in tobacco chloroplasts enhances the tolerance to salt stress and water deficit. *Physiol Plant* 121(2):231–238

Bhattacharjee S (2005) Reactive oxygen species and oxidative burst: Roles in stress, senescence and signal transducation in plants. *Curr Sci* 89(7):1113–1121

Bhattacharjee S (2019) ROS and antioxidants: Relationship in green cells. In: Bhattacharjee S. (ed.), *Reactive Oxygen Species in Plant Biology*. Springer, New Delhi, pp 33–63

Bose J, Rodrigo-Moreno A, Shabala S (2014) ROS homeostasis in halophytes in the context of salinity stress tolerance. *J Exp Bot* 65(5):1241–1257

Calderón A, Sánchez-Guerrero A, Ortiz-Espín A, Martínez-Alcalá I, Camejo D, Jiménez A, Sevilla F (2018) Lack of mitochondrial thioredoxin o1 is compensated by antioxidant components under salinity in *Arabidopsis thaliana* plants. *Physiol Plant* 164(3):251–267

Corpas FJ, Barroso JB, del Río LA (2001) Peroxisomes as a source of reactive oxygen species and nitric oxide signal molecules in plant cells, del Río LA. *Trends Plant Sci* 6(4):145–150

Dar MI, Naikoo MI, Khan FA, Rehman F, Green ID, Naushin F, Ansari AA (2017) An introduction to reactive oxygen species metabolism under changing climate in plants. In: *Reactive Oxygen Species and Antioxidant Systems in Plants: Role and Regulation under Abiotic Stress*. Springer, Singapore, pp 25–52

Das K, Roychoudhury A (2014) Reactive oxygen species (ROS) and response of antioxidants as ROS-scavengers during environmental stress in plants. *Front Environ Sci* 2:53

Dietz KJ, Jacob S, Oelze ML, Laxa M, Tognetti V, de Miranda SM, Baier M, Finkemeier I (2006) The function of peroxiredoxins in plant organelle redox metabolism. *J Exp Bot* 57(8):1697–1709

Dutta T, Neelapu NRR, Wani SH, Challa S (2019) Role and regulation of osmolytes as signaling molecules to abiotic stresses. In: MIR K, Palakolanu SR, Ferrante A, Khan NA (eds) *Plant Signaling Molecules: Role and Regulation under Stressful Environments*. Woodhead Publishing, Duxford, pp 459–477

Flowers TJ, Koyama ML, Flowers SA, Sudhakar C, Singh KP, Yeo AR (2000) QTL: Their place in engineering tolerance of rice to salinity. *J Exp Bot* 51(342):99–106

Fortmeier R, Schubert S (1995) Salt tolerance of maize (*Zea mays* L.): The role of sodium exclusion. *Plant Cell Environ* 18(9):1041–1047

Foyer CH, Noctor G (2005) Oxidant and antioxidant signalling in plants: A re-evaluation of the concept of oxidative stress in a physiological context. *Plant Cell Environ* 28(8):1056–1071

Gill SS, Tuteja N (2010) Reactive oxygen species and antioxidant machinery in abiotic stress tolerance in crop plants. *Plant Physiol Biochem* 48(12):909–930

Gilroy S, Suzuki N, Miller G, Choi WG, Toyota M, Devireddy AR, Mittler R (2014) A tidal wave of signals: Calcium and ROS at the forefront of rapid systemic signaling. *Trends Plant Sci* 19(10):623–630

Giraud E, Ho LH, Clifton R, Carroll A, Estavillo G, Tan YF, Howell KA, Ivanova A, Pogson BJ, Millar AH, Whelan J (2008) The absence of alternative oxidase1a in *Arabidopsis* results in acute sensitivity to combined light and drought stress. *Plant Physiol* 147(2):595–610

Hossain MS, Dietz KJ (2016) Tuning of redox regulatory mechanisms, reactive oxygen species and redox homeostasis under salinity stress. *Front Plant Sci* 7:548

Hussain S, Shaukat M, Ashraf M, Zhu C, Jin Q, Zhang J (2019) Salinity stress in arid and semi-arid climates: Effects and management in field crops. In: *Clim Change Agric Intechopen* DOI: 10.5772/intechopen.87982

Jithesh MN, Prashanth SR, Sivaprakash KR, Parida AK (2006) Antioxidative response mechanisms in halophytes: Their role in stress defence. *J Genet* 85(3):237

Joshi R, Wani SH, Singh B, Bohra A, Dar ZA, Lone AA, Pareek A, Singla-Pareek SL (2016) Transcription factors and plants response to drought stress: Current understanding and future directions. *Front Plant Sci* 7:1029

Kadioglu A., Saruhan N., Sağlam A. et al. (2011) Exogenous salicylic acid alleviates effects of long term drought stress and delays leaf rolling by inducing antioxidant system. *Plant Growth Regul* 64, 27–37.

Kalaji HM, Govindjee, Bosa K, Kościelniak J, Żuk-Gołaszewska K (2011) Effects of salt stress on photosystem II efficiency and CO_2 assimilation of two Syrian barley landraces. *Environ Exp Bot* 73:64–72

Karpinska B, Wingsle G, Karpinski S (2000) Antagonistic effects of hydrogen peroxide and glutathione on acclimation to excess excitation energy in *Arabidopsis*. *IUBMB Life* 50(1):21–26

Krishnamurthy A, Rathinasabapathi B (2013) Oxidative stress tolerance in plants: Novel interplay between auxin and reactive oxygen species signaling. *Plant Signal Behav* 8(10):e25761

Kwak JM, Mori IC, Pei ZM, Leonhardt N, Torres MA, Dangl JL, Bloom RE, Bodde S, Jones JD, Schroeder JI (2003) NADPH oxidase AtrbohD and AtrbohF genes function in ROS-dependent ABA signaling in Arabidopsis. *EMBO J* 22(11):2623–2633

Li G, Li J, Hao R, Guo Y (2017) Activation of catalase activity by a peroxisome-localized small heat shock protein Hsp17. 6CII. *J Genet Genomics* 44(8):395–404

Maggio A, Raimondi G, Martino A, De Pascale S (2007) Salt stress response in tomato beyond the salinity tolerance threshold. *Environ Exp Bot* 59(3):276–282

Miller G, Mittler R (2006) Could heat shock transcription factors function as hydrogen peroxide sensors in plants? *Ann Bot* 98(2):279–288

Miller G, Suzuki N, Ciftci-Yilmaz S, Mittler R (2010) Reactive oxygen species homeostasis and signalling during drought and salinity stresses. *Plant Cell Environ* 33(4):453–467

Miller G, Suzuki N, Rizhsky L, Hegie A, Koussevitzky S, Mittler R (2007) Double mutants deficient in cytosolic and thylakoid ascorbate peroxidase reveal a complex mode of interaction between reactive oxygen species, plant development, and response to abiotic stresses. *Plant Physiol* 144(4):1777–1785

Mittler R (2002) Oxidative stress, antioxidants and stress tolerance. *Trendsplant Sci* 7(9):405–410

Mittler R, Berkowitz G (2001) Hydrogen peroxide, a messenger with too many roles? *Redox Rep* 6(2):69–72

Mittler R, Vanderauwera S, Gollery M, Van Breusegem F (2004) Reactive oxygen gene network of plants. *Trends Plant Sci* 9(10):490–498

Møller IM, Jensen PE, Hansson A (2007) Oxidative modifications to cellular components in plants. *Annu Rev Plant Biol* 58:459–481

Munns R, Termaat A (1986) Whole-plant responses to salinity. *Functional Plant Biol* 13(1):143–160

Munns R, Tester M (2008) Mechanisms of salinity tolerance. *Ann Rev Plant Biol* 59:651–681

Negrão S, Courtois B, Ahmadi N, Abreu I, Saibo N, Oliveira MM (2011) Recent updates on salinity stress in rice: From physiological to molecular responses. *Crit Rev Plant Sci* 30(4):329–377

Nyathi Y, Baker A (2006) Plant peroxisomes as a source of signalling molecules. *Biochim Biophys Acta* 1763(12):1478–1495

Pitzschke A, Djamei A, Bitton F, Hirt H (2009) A major role of the MEKK1–MKK1/2–MPK4 pathway in ROS signalling. *Molplant* 2(1):120–137

Rhoads DM, Umbach AL, Subbaiah CC, Siedow JN (2006) Mitochondrial reactive oxygen species. Contribution to oxidative stress and interorganellar signaling. *Plant Physiol* 141(2):357–366

Rodrigues SM, Andrade MO, Gomes APS, DaMatta FM, Baracat-Pereira MC, Fontes EP (2006) Arabidopsis and tobacco plants ectopically expressing the soybean antiquitin-like ALDH7 gene display enhanced tolerance to drought, salinity, and oxidative stress. *J Exp Bot* 57(9):1909–1918

Roychoudhury A, Chakraborty M (2013) Biochemical and molecular basis of varietal difference in plant salt tolerance. *Annu Res Rev Biol*: 422–454

Sah SK, Kaur G, Wani SH (2016) Metabolic engineering of compatible solute trehalose for abiotic stress tolerance in plants. In: Iqbal N, Nazar R, Khan AN (eds), *Osmolytes and Plants Acclimation to Changing Environment: Emerging Omics Technologies*. Springer, New Delhi, pp 83–96

Sánchez-Guerrero A, Fernández del-Saz NF, Florez-Sarasa I, Ribas-Carbó M, Fernie AR, Jiménez A, Sevilla F (2019) Coordinated responses of mitochondrial antioxidative enzymes, respiratory pathways and metabolism in *Arabidopsis thaliana* thioredoxin trxo1 mutants under salinity. *Environ Exp Bot* 162:212–222

Sewelam N, Kazan K, Schenk PM (2016) Global plant stress signaling: Reactive oxygen species at the crossroad. *Front Plant Sci* 7:187

Singh M, Kumar J, Singh S, Singh VP, Prasad SM (2015) Roles of osmoprotectants in improving salinity and drought tolerance in plants: A review. *Rev Environ Sci Bio* 14(3):407–426

Suo J, Zhao Q, David L, Chen S, Dai S (2017) Salinity response in chloroplasts: Insights from gene characterization. *Int J Mol Sci* 18(5):1011

Tanveer M, Shabala S (2018) Targeting redox regulatory mechanisms for salinity stress tolerance in crops. In: Kumar V, Wani S, Suprasanna P, Tran LS (eds), *Salinity Responses and Tolerance in Plants*, Springer, Cham, pp 213–234

Tseng MJ, Liu CW, Yiu JC (2007) Enhanced tolerance to sulfur dioxide and salt stress of transgenic Chinese cabbage plants expressing both superoxide dismutase and catalase in chloroplasts. *Plant Physiol Biochem* 45(10–11):822–833

Van Camp W, Capiau K, Van Montagu M, Inzé D, Slooten L (1996) Enhancement of oxidative stress tolerance in transgenic tobacco plants overproducing Fe-superoxide dismutase in chloroplasts. *Plant Physiol* 112(4):1703–1714

Vandenabeele S, Vanderauwera S, Vuylsteke M, Rombauts S, Langebartels C, Seidlitz HK, Zabeau M, Van Montagu M, Inzé D, Van Breusegem F (2004) Catalase deficiency drastically affects gene expression induced by high light in *Arabidopsis thaliana*. *Plant J* 39(1):45–58

Wakabayashi K, Hoson T, Kamisaka S (1997) Osmotic stress suppresses cell wall stiffening and the increase in cell wall-bound ferulic and diferulic acids in wheat coleoptiles. *Plant Physiol* 113(3):967–973

Wang FZ, Wang QB, Kwon SY, Kwak SS, Su WA (2005) Enhanced drought tolerance of transgenic rice plants expressing a pea manganese superoxide dismutase. *J Plant Physiol* 162(4):465–472

Wani SH, Dutta T, Neelapu NRR, Surekha C (2017) Transgenic approaches to enhance salt and drought tolerance in plants. *Plant Gene* 11:219–231

Wani SH, Kumar V (2015) Plant stress tolerance: Engineering ABA: A Potent phytohormone. *Transcriptomics* 3(2):1000113

Wani SH, Kumar V, Khare T, Guddimalli R, Parveda M, Solymosi K, Suprasanna P, Kavi Kishor PB (2020) Engineering salinity tolerance in plants: Progress and prospects. *Planta* 251(4):76

Wani SH, Kumar V, Shriram V, Sah SK (2016) Phytohormones and their metabolic engineering for abiotic stress tolerance in crop plants. *Crop J* 4(3):162–176

Wani SH, Singh NB, Haribhushan A, Mir JI (2013) Compatible solute engineering in plants for abiotic stress tolerance - role of glycine betaine. *Curr Genomics* 14(3):157

Wani SH, Tripathi P, Zaid A, Challa GS, Kumar A, Kumar V, Upadhyay J, Joshi R, Bhatt M (2018) Transcriptional regulation of osmotic stress tolerance in wheat (Triticum aestivum L.). *Plant Mol Biol* 97(6):469–487

Waszczak C, Carmody M, Kangasjärvi J (2018) Reactive oxygen species in plant signaling. *Annu Rev Plant Biol* 69:209–236

Wild A (2003) *Soils, Land and Food: Managing the Land During the Twenty-First Century*. Cambridge University Press

Yadav A, Singh J, Ranjan K, Kumar P, Khanna S, Gupta M, Kumar V, Wani SH, Sirohi A (2020) Heat shock proteins: Master players for heat-stress tolerance in plants during climate change. In: Wani SH and Kumar V (eds), *Heat Stress Tolerance in Plants: Physiological, Molecular and Genetic Perspectives*: 189–211

Yang Y, Guo Y (2018) Unraveling salt stress signaling in plants. *J Integr Plant Biol* 60(9):796–804

Zhang L, Wu M, Teng Y, Jia S, Yu D, Wei T, Chen C, Song W (2018) Overexpression of the glutathione peroxidase 5 (RcGPX5) gene from rhodiola crenulata increases drought tolerance in Salvia miltiorrhiza. *Front Plant Sci* 9:1950

Zhou J, Wang J, Bi Y, Wang L, Tang L, Yu X, Ohtani M, Demura T, Zhuge Q (2014) Overexpression of PtSOS2 enhances salt tolerance in transgenic poplars plant. *Mol Biol Rep* 32(1):185–197

7 Oxidative Stress in Plants Exposed to Heavy Metals

Amna Shoaib and Arshad Javaid

CONTENTS

7.1   INTRODUCTION

The term heavy metals is used to describe a group of metals and metalloids exhibiting metallic properties with atomic masses of >20 and specific gravities of >5 g cm^3 (Edelstein and Ben-Hur 2018). These metals are of significant environmental concern because of their toxicity, diverse sources, accumulative behaviors and non-biodegradable properties (Wang and Chen 2009). Almost all heavy metals at high concentrations wreak havoc with every living being on our planet. The Industrial Revolution transformed the course of the earth's history into the Anthropocene era and aggravated the issue. Earth's envelope (atmosphere) and Earth's fuel (soil) are the ultimate source of heavy metal accumulation from anthropogenic or natural sources, and it is a prerequisite for the Earth's engine (plants) to modify systems organization and outputs to adapt to heavy metal pollution (Vardhan et al. 2019). Various heavy metals like nickel (Ni), cobalt (Co), manganese (Mn), copper (Cu), molybdenum (Mo), iron (Fe) and zinc (Zn) are ranked as essential in trace amount for the basic needs of plants, though become toxic beyond their permissible limits. Others like mercury (Hg), chromium (Cr), cadmium (Cd), arsenic (As) and lead (Pb) are classified as a "main threat", since they are harmful to plant growth even at very low concentrations (Table 7.1) (Chibuike and Obiora 2014; Page and Feller 2015; Vardhan et al. 2019).

In general, the physiological stressful concentration of the heavy metals directly or indirectly accumulated inside the plant through atmosphere or soil and accumulated metal has diverse sites of action inside the plant. A nominal quantity of heavy metals is bearable for both sensitive or tolerant plant species, whereas the toxic level induces a rather frequent and deteriorating oxidative stress at the cellular level due to over-accumulation of reactive oxygen species (ROS). ROS are synthesized by heavy metals directly through the Fenton and Haber-Weiss reactions, or indirectly

TABLE 7.1

Heavy Metals: Micronutrients or Pollutants?

Importance	Heavy Metal	Function of Heavy Metal	Reference
Micro nutrient	Fe	Required for electron transport chain, oxidative stress tolerance, nitrogen fixation, hormone biosynthesis and organelle integrity.	Welch and Shuman (1995), Campbell (1999), Hänsch and Mendel (2009), Ravet et al. (2009), Connorton et al. (2017)
	Mn	Essential for the oxygen evolution in photosystem II and for a series of enzymatic reactions (e.g. phosphoenolpyruvate carboxykinase and superoxide dismutase).	Welch (1995), Hänsch and Mendel (2009), Filiz and Tombuloglu (2015), Alejandro et al. (2020)
	Cu	Present in the plastidial plastocyanin, mitochondrial cytochrome c oxidase, Cu-Zn superoxide dismutase and in a series of other proteins.	Welch and Shuman (1995), Hänsch and Mendel (2009), Yruela (2013), Łukowski and Dec (2018)
	Zn	Required in electron transport and antioxidant metabolism, but it is also an integral component of many transcription factors.	Welch and Shuman (1995), Campbell (1999), Hänsch and Mendel (2009), Yruela (2013), Łukowski and Dec (2018)
	Ni	Involved in urease activity, essential for efficient nitrogen metabolism.	Welch and Shuman (1995), Campbell (1999), Hänsch and Mendel (2009), Freitas et al. (2018)
	Mo	Present in the soil solution as molybdate anion, it is part of the molybdenum cofactor, which is required for nitrate reductase or xanthine dehydrogenase activity.	Welch and Shuman (1995), Campbell (1999), Hänsch and Mendel (2009), Manuel et al. (2018)
	Co	Not required by the higher plant itself, but is essential for the microorganisms involved in symbiotic nitrogen fixation.	Welch and Shuman (1995), Jayakuma et al. (2008), Gad et al. (2019)
Pollutant	As, Cd, Cr, Hg and Pb	Not required by plant	Liu et al. (2009), Cenkci et al. (2010), Pinho and Ladeiro (2012), Rahman and Singh (2019)

by overwhelming the intrinsic antioxidant defenses (Jalmi et al. 2018). To cope with heavy metal-induced stress, plants adsorb them either onto the chelating molecules, such as phytochelatins and metallothionines, or by confiscation into the vacuoles. However, metal-induced oxidative stress leads to crosstalk among the several signal transduction units, which perceive the signal from upstream receptors and transmit them into the nucleus for biological response. The major signaling networks working in metal stresses are Mitogen-activated protein kinase (MAPK), and calcium and hormone signalings. Calcium (Ca^{2+}) is a versatile signal involved in the control of functions and processes of cells. It modulates processes of generation and clearance of ROS and thus shifts the redox state to either a more oxidized or reduced state by employing a multitude of calcium-sensing proteins that bind to Ca^{2+} and trigger different downstream signaling pathways (Zhou et al. 2020). MAPK cascades are connected to each other by phosphorylation and can lead to various kinds of responses including cell division and differentiation, expression of various stress-related genes or can inhibit the action of others (Jalmi et al. 2018). Different plant hormones play roles in metal stress response in the case of hormone signaling (Bücker-Neto et al. 2017; Jalmi et al. 2018).

Crosstalk of complex signaling pathways through metal-induced oxidative stress modulates the redox signaling pathways, changing their stability, integrity and functionality, thereby causing a wide range of physiological and metabolic alterations in plants. Hence, the combination of ROS with the oxidative stress leads to negative consequences for biomolecules (DNA, RNA, proteins and lipids) and associated processes during photosynthesis, tricarboxylic acid cycle and Calvin cycle (Rascio and Navari-Izzo 2011; Hasan et al. 2017; Mhamdi and Breusegem 2018). For instance, inactivation and denaturation of enzymes and proteins, blocking of functional groups of metabolically important molecules, displacement/substitution of essential metal ions from biomolecules and functional cellular units, conformational modifications, and disruption of membrane integrity alter plant metabolism with widespread visual evidence. Some of the morphological evidence includes a reduction in seed germination rate and plant growth, crippled photosynthetic apparatus, homeostatic events including water uptake, transport, transpiration and nutrient metabolism, all of which are often correlated with progressing senescence processes or with plant death Dutta et al. 2018).

This chapter is about the accumulation of heavy metal in plants, main reactive oxygen species and their generation, ROS-mediated cellular signaling and oxidative damages to macromolecules under metal toxicity.

7.2 HEAVY METAL ACCUMULATION IN PLANTS

7.2.1 Accumulation through the Atmosphere

Atmospheric heavy metals find their entry through the leaf (stomata, aqueous pores, ectodesmata, lenticels and cuticular cracks), whereas leaves' morpho-physiological attributes (foliage layout, leaf size, epidermis structure and cuticle composition) and metal speciation contribute significantly to metal adsorption (Shahid et al. 2017; Natasha et al. 2020). The detailed mechanism of metal allocation in aerial plant parts is still ambiguous and need to be explored. However, recent advances have highlighted that heavy metals are retained by cuticular waxes (pores in cuticle), penetrate through stomata (ectocythodes), desorb in the apoplast and absorb by the subjacent cells via active transport through the symplastic or apoplastic pathway. Finally, absorbed metal experiences phloem loading for transport to different "sinks" in a plant (Figure 7.1) (Shahid et al. 2017; Natasha et al. 2020). Generally, most of the absorbed metals are stored in leaf tissue (mostly in parenchyma), and only <1% is transported to the root (Shahid et al. 2017).

7.2.2 Accumulation through the Lithosphere

Regardless of the total concentration of heavy metals in the soil, only a small fraction can access the plants due to the strong affinity of heavy metals with the soil components (Shahid et al. 2015). Generally speaking, many factors including soil pH, organic matter, metal concentration, etc. determine the availability of metals to plants. Heavy metals in soil particles are initially adsorbed on root cells, subsequently bind with carboxyl groups of uronic acid in the root vicinity or directly enter to root epidermis. Root exudates release protons (H^+) and enzymes, which aid in acidification and electron transfer in the rhizosphere and lead to improved bioavailability of the metal. Metal enters the root cell wall and then moves across the cell membrane of the epidermis through the apoplastic pathway, driven by ATP-dependent proton pumps. After passing through all the hurdles from the root surface over the root cortex, finally metal ions enter the symplast and are transported to the xylem elements with the transpiration stream then to transpiring shoot parts (e.g. photosynthesizing leaves) (Figure 7.1) (Shahid et al. 2015). Several efflux transporters are located in plasma or intracellular membranes to monitor the symplastic transport of heavy metals. Mature leaves are destination houses of most of the metals, whereas young leaves accumulate only minor concentrations. Furthermore, metal transport in the xylem is usually presented as free hydration cation (Cao et al. 2020), while in the phloem, the metal was transported in either an upward or downward direction by the osmotic pressure gradient. The phloem companion cells can also accumulate metal from apoplastic fluid to cytosol by utilizing cell membrane metal influx transporters (Deng et al. 2016).

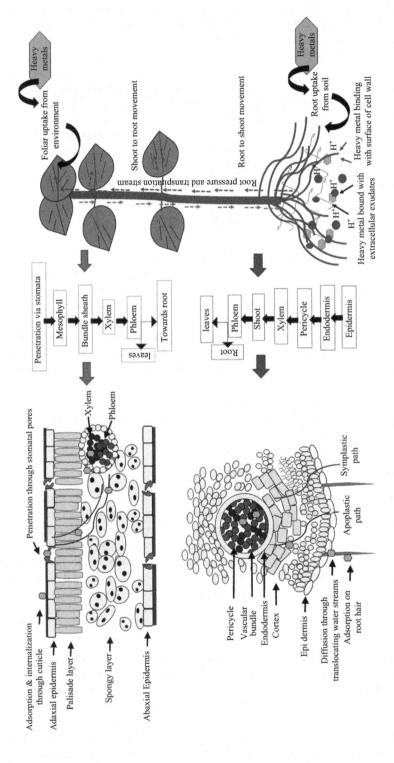

FIGURE 7.1   Schematic diagram of the uptake and translocation pathways of heavy metals in plants.

Besides this, metal can also be distributed from fully developed leaves to the unfolded and expanding leaves, emerging organs and maturing fruits and seeds (Hu et al. 2019). Moreover, plasmodesmata facilitate the transfer of metal-ligand complexes from companion cells to phloem vessels (Cao et al. 2020).

7.3 OXIDANTS

ROS is an umbrella term often used to describe a myriad of small "double-faced" molecules, which on one side contribute to signal transduction for homeostasis and redox regulation, while also acting as destructive species for cellular compounds (Sinegovskaya et al. 2020). ROS certainly are inseparable companions of atmospheric oxygen (O_2) since the oxygenation of the Earth's atmosphere about 2.4–3.8 billion years ago (Mittler 2017) and are continuously generated under natural conditions in different cellular compartments (mitochondria, chloroplasts, peroxisome and apoplast as a first place, while cell wall, endoplasmic reticulum and plasma membrane as other places) during energy transfer reactions from the reduction of O_2 into H_2O (Figure 7.2). For example, mitochondrial electron transport chain ($O_2^{\bullet-}$, H_2O_2) at the respiratory complexes (I, II and III); photosynthesis ($^1O_2$, $O_2^{\bullet-}$ and H_2O_2) in the chloroplast; oxidation pathways (H_2O_2) in peroxisomes; and plasma membrane-located respiratory burst oxidase homolog (RboH or NADPH oxidases) proteins ($O_2^{\bullet-}$) in the apoplast are engines of ROS signaling. Furthermore, each cellular compartment establishes and controls its own ROS homeostasis. However, ROS in specific subcellular compartments may lead to ROS accumulation in other compartments (Liu and He 2017).

These partially reduced or activated forms of O_2 (e.g. $O_2^{\bullet-}$, $OH^\bullet$, H_2O_2, HOO, $^1O_2$) are more reactive, short-lived and unstable than an oxygen molecule itself. So far, ROS stands for both oxygen radicals as well as non-radicals. Radicals are usually more reactive compared to non-radicals, as electrons are more stable in the paired form in the orbitals, but when an electron occupies an orbital by itself it has two possible directions of spin. However, H_2O_2 and $^1O_2$, themselves, can be quite toxic to cells although they are non-radicals (Halliwell 1991). Each type of ROS (reactive: $O_2^{\bullet-}$, highly stable: H_2O_2 and most reactive: $OH^\bullet$) has unique and distinct chemical properties (Mittler 2017).

Anion superoxide ($O_2^{\bullet-}$) is a byproduct of electron transport chains of respiration and photosynthesis and is formed by NADPH oxidases and cell wall peroxidases. It is an unreactive, short-lived molecule, unable to diffuse through membranes and has poor signaling molecules because of negative charge. Additionally, $O_2^{\bullet-}$ is a reductant of the transition metal ions in the Haber-Weiss reaction to produce $OH^\bullet$ from H_2O_2. In addition to that, $O_2^{\bullet-}$ oxidizes the quinols prepared via disproportionation of $OH^\bullet$ and thiols to give thiyl radicals, which can initiate radical chain reactions (Sewelam et al. 2016; Mittler 2017).

Hydrogen peroxide (H_2O_2) is a poor oxidant, mostly produced during the processes of photosynthesis, photorespiration and, to a lesser extent, cellular respiration. H_2O_2 is diffusible through the biological membranes through aquaporins and can inactivate cell molecules even at a very low concentration. Its selective reactivity, stability and diffusability make it fit to work as a signaling molecule during oxidation in peroxisomes. It changes thiols to the sulfenic acids, reacts with cysteine slowly in proteins and also suppresses CO_2 fixation in chloroplasts (Sewelam et al. 2016; Jalmi et al. 2018).

Hydroxyl radical ($OH^\bullet$) does not diffuse across a long distance and exhibits high indiscriminate reactivity for organic and inorganic molecules close to its site of production (Štolfa et al. 2015).

Singlet oxygen ($^1O_2$) can move only a very small distance in cells and reacts quickly with amino acids, unsaturated lipids and other components of the cell. Consequently, $^1O_2$ can directly react with only those molecules close to its site of production, i.e. in the chloroplast (Triantaphylidès and Havaux 2009). Besides this, ROS-mediated alterations in the cellular redox homeostasis could set very specific signaling roles for ROS. Various pathways could sense and weigh the modifications in cellular redox balance resulting from the change of intracellular ROS concentration, then translate

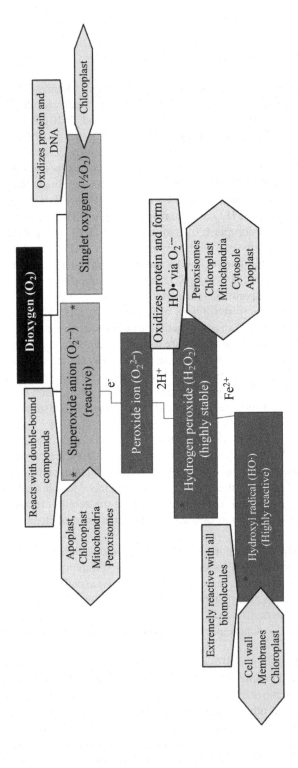

FIGURE 7.2   Reactive oxygen species, their location and mode of action in plant cell.

these modifications into very specific cellular signals, which direct the cell to fabricate a pertinent adaptive response (Sewelam et al. 2016).

7.3.1 Heavy Metal-Induced Oxidants

Productions of ROS are increased under the exposure of unfavorable environmental cues; which alter redox homeostasis and impart oxidative stress in the cell (Bhattacharjee 2019). However, the toxicity of heavy metals resides in their direct generation of ROS. Redox-active metals (Fe, Cu and Cr) directly generate ROS via Fenton-type reactions and Haber-Weiss cycling, yet non-redox metals (Al, Cd, Hg, Mn, Ni, Pb and Zn) indirectly generate ROS through stimulating NADPH oxidases (ROS enzymes) or displacing crucial metals from the essential binding sites in biomolecules (Jalmi et al. 2018). An increase in chloroplastic as well as mitochondrial ROS production has been suggested to contribute to signaling during metal stress (Keunen et al. 2013). The increase in ROS production under Cd and Cu toxicity may be related to improved photorespiration, leading to the production of H_2O_2 by glycolate oxidase (Del Río and López-Huertas 2016). Reduction in availability of CO_2 due to closure of stomata can increase ROS production in the plasma membrane, mitochondria, apoplast, peroxisomes, chloroplasts and glyoxysomes, which can initiate retrograde and anterograde signaling in response to abiotic stress and heavy metal toxicity (Choudhury et al. 2017).

Plenty of literature has highlighted that high heavy metal concentrations severely disturb redox homeostasis of the plant cell and inflict oxidative damage by upsetting photosynthetic electron flow (Bhattacharjee 2019). In *Oryzae sativa*, Cu stress caused high production of ROS (H_2O_2 and $O_2^{\bullet-}$) and a key reduction in the levels of antioxidants (Thounaojam et al. 2014). Cr can generate ROS either by direct electron transfer or by suppressing metabolic reactions (Panda et al. 2016). An excessive amount of $O_2^{\bullet-}$, H_2O_2 and $OH^{\bullet-}$ in *Zea mays* and *Helianthus annulus* has been recorded on Cr exposure (Kharbech et al. 2017; Farid et al. 2018). For example, accumulation of H_2O_2 in *Z. mays* on exposure to Cu (Liu et al. 2018), $O_2^{\bullet-}$ and H_2O_2 in *Brassica juice* under Zn and Cd toxicity (Małecka et al. 2019) and $O_2^{\bullet-}$ in *Brassica napus* under Cr toxicity have been documented (Gill et al. 2015). However, generation of $OH^{\bullet}$ radicals by $O_2^{\bullet-}$ and H_2O_2 in the presence of heavy metals is the main reason for producing toxic effects in plants (Jalmi et al. 2018).

7.3.2 Oxidant-Mediated Cellular Signaling

Like any other living organism, the plant also activates its defense system to resist oxidative stress either by chelating heavy metals with phytochelatin and metallothionines (chelating molecule) as well as by sequestration within the vacuoles (Jalmi et al. 2018). When metal-ligands bind to a specific cellular receptor in the cell, the ROS act as important secondary messengers to broadcast the initial signal (the "first message"), which leads to the transmission of extracellular stimuli to complex signaling pathways for the intracellular response. Receptor-like protein flagellin sensitive 2 (FLS2), kinases (RLKs), ethylene resistance1/2 (ETR1/2), EF-Tu receptor (EFR), ERECTA (ER) and salt intolerance 1 (SIT1) etc. are among the receptors, which have been known to perceive the signals from the external environment to the nucleus (Jalmi et al. 2018). Henceforward, crosstalk among the major signaling networks like calcium signaling, MAPK signaling and hormone signaling activates stress-responsive genes by transcription to relieve plant from stress (Figure 7.3) (Jalmi et al. 2018).

7.3.2.1 Calcium Signaling Steers

Calcium (Ca^{2+}) has a dual role as a structural component of cell walls and membranes and as an intracellular second messenger (cytosolic Ca^{2+}) in dynamic signaling systems. Ca^{2+} channels at the surface, and intracellular organelles of the cell are mostly activated by H_2O_2 and $OH^{\bullet-}$. The inwardly-rectifying ROS-activated ion channels mediate Ca^{2+}-influx for growth and development in pollen tubes and roots. On the other hand, the outwardly-rectifying group facilitates K^+ efflux

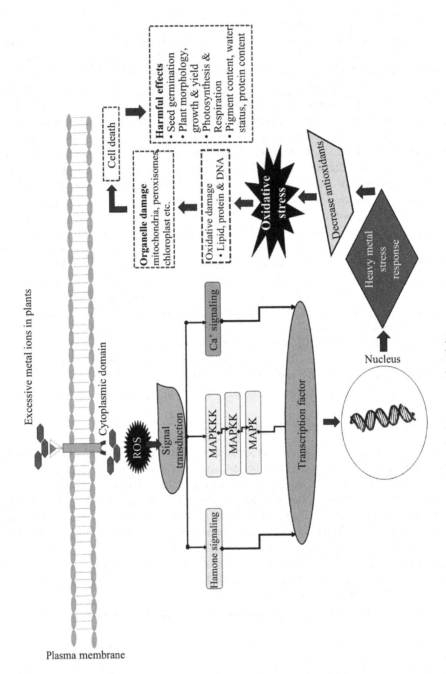

FIGURE 7.3 Consequences of oxidative stress generated under heavy metal toxicity.

for osmotic pressure regulation in guard cells, induction of programmed cell death and autophagy in roots. Both Ca^{2+} influx and K^+ efflux are mediated by a voltage-independent group (Demidchik 2018). Calcium sensing proteins [Calmodulins (CaMs), CaM like proteins (CMLs), Calcineurin B-like proteins (CBLs) and Ca^{2+}-dependent protein kinases (CDPKs)] are responsible for mediating ROS induced chemical signals into a biological response (Jalmi et al. 2018; Ghori et al. 2019).

Ca^{2+} channels have been regarded as a major route of entry of metal ions into cells (Chen et al. 2018), and these channels can be affected directly by oxidation to change Ca^{2+} signaling (Hampel and Trebak 2017). Cd, by blocking root hair Ca^{2+} influx channels or Cd outcompeting Ca^{2+} for that influx pathway, can inhibit root hair growth in *Arabidopsis* (Fan et al. 2011). Cd^{2+} and Pb^{2+} mimic Ca^{2+} at their specific sites, bind proteins (e.g. CaMs, protein kinase and synaptic proteins) and antagonize Ca^{2+} influx through voltage-dependent Ca^{2+} channels (Marchetti 2013). Likewise, Cd^{2+}, by participating in Ca^{2+} dependent pathways, can affect cytoskeletal dynamics and apoptotic cell death (Choong et al. 2014). Imbalance in Ca^{2+} transport from the plasma membrane into cytoplasm Cr was found to inhibit cell division in plants (Emamverdian et al. 2015). Cu has been known to activate as well as block Ca^{2+} channels through ROS production (Wilkins et al. 2016). CaM has been regarded as an important cellular target in Cd, Ni and Pb stress. Transgenic *Nicotia natabacum* overexpressed calmodulin-binding protein 4 (NtCBP4) in Pb uptake across the plasma membrane (Zeng et al. 2015). Numerous researchers have demonstrated that the Ca-permeable channel is the major pathway by which Pb enters the root (Kumar et al. 2017). The irreversible collapse of the electric potential across the plasma membrane and stable loss of cell turgor has been associated with an increase in fluxes of heavy metals (e.g. Cu) (Demidchik 2018).

It was also found that Ca^{2+} provides tolerance against heavy metal stress. The increase in cytoplasmic Ca^{2+} shows an early response to various biotic and abiotic stresses, including oxidative stresses (Dodd et al. 2010). Cr stress was alleviated in foxtail millet (*Setariaitalica*) by the improved activity of antioxidant enzymes, where hydrogen sulfide interacts with Ca^{2+} signaling (Fang et al. 2014). CDPK-like protein activity increased in rice seedlings, reflecting the role of Ca^{2+} signaling in alleviating the Cd-induced stress responses (Huang et al. 2014). Ahmad et al. (2015) showed that exogenous application of Ca^{2+} in *Brassica juice* can modulate the physiological and biochemical responses in plants against Cd stress. They also stated the involvement of Ca^{2+}/CaM system in Pb^{2+} and Ni^{2+} mediated stress. Zhao et al. (2014) revealed Ca^{2+} alleviates the toxic effects of Cd in *Arabidopsis* seedlings by retaining auxin homeostasis, representing crosstalk between signaling pathways to fight heavy metal stress. High concentrations of Ca^{2+} enhanced the cation exchange strength, lessened the fracture degree of fibrous roots, lowered the vascular bundle atrophy, protected the cell wall and decreased the oxidization of alkynyl acetylene bonds in *Eichhorniacrassipe* exposed to Pb (Zhou et al. 2020).

It is also known that in response to heavy metal, CDPK collaborates with MAPK to transmit signals to adapt against the changing environment (Opdenakker et al. 2012). It was observed that under Cu and Cd, CDPK and phosphatidylinositol are needed to activate MAPK (Thapa et al. 2012). Huang and Huang (2008) suggested that Ca^{2+} is involved in Pb^{2+}-mediated cell death by increasing the generation of ROS and triggering MAPK activity through the CDPK pathway. The interplay between the MAPK signaling pathway and CaMs has been reported to respond against metal-induced oxidative stress (Jalmi et al. 2018). The literature highlighted the substantiate role of Ca^{2+} and Ca^{2+}-dependent signaling pathways to impart heavy metal tolerance stresses in plants, though mechanisms by which these responses are regulated are still insufficient and need further elaboration (Jalmi et al. 2018).

7.3.2.2 Mitogen-Activated Kinases

MAPKs cascade comprises a family of serine/threonine protein kinases, which are a universal module and major pathway of signal transduction linking upstream signaling to the nucleus against diverse responses (developmental, hormonal, biotic and abiotic stresses) (Lie and He 2017; Jalmi et al. 2018). Three tiers of MAPK cascade (MAPKKKs, MAPKKs and MAPKs) mediating phosphorylation reactions occur in the presence of protein-tyrosine kinase, G-protein receptors and two-component histidine kinases from upstream receptor to downstream target from cytoplasm to nucleus (Jalmi et al. 2018).

MAPK signaling plays a key role in plant responses to metal stress. Exposure of plants to heavy metal leads to activation of MAPKs, which are activated through the accumulation of ROS or by perceiving specific metal-ligand and which convert the perception of metals to intracellular signals to the nucleus, where appropriate responses are initiated (Jalmi et al. 2018). Furthermore, single stress conditions lead to the interplay between different pathways, while the interplay between the MAPK pathways and ROS generation is known as inducing MAPK signaling and is common to other abiotic and biotic stress responses (Opdenakker et al. 2012). It is also known that ROS not only activate MAPK but also disturb MAPK cascades, which in turn can modulate ROS production and responses (Liu and He 2017).

Plenty of literature is available on the activation of MAPKs in response to heavy metals (Cd, Cu and As), but the mechanism behind the specific activation of MAPK cascade by ROS is still unclear (Jalmi et al. 2018; Liu et al. 2018). Generally, mRNA, as well as activity levels, are improved rapidly after exposure to the metal, ranging from 5 min to 1 h, and activation of MAPKs is transient (Opdenakker et al. 2012). The induction of OsMPK3 and OsMPK6 genes in *O. sativa* after an increase in ROS has been observed by both Cd^{2+} and Cu^{2+} (Yet et al. 2007). Involvement of OsMPK3, OsMPK4 and OsMKK4 has been shown in As(III) mediated in *O.sativa* (Rao et al. 2011). In *Arabidopsis thaliana*, MPK3 and MPK6, while in *O. sativa* transcript OsMSRMK2 (OsMPK3 homolog), OsMSRMK3 (OsMPK7 homolog) and OsWJUMK1 (OsMPK20-4 homolog), have been best characterized after Cd and Cu induced oxidative stress (Sethi et al. 2014). Likewise, activation of MEKK, MKKK2 and WRKY25 cascade in *Arabidopsis thaliana* under Fe stress (Smeets et al. 2013), MAPK cascade in *Loliumperenne* under Cu toxicity (Dong et al. 2014), and MPK3/6 in *A. thaliana* under Cd stress have been documented through increased production of ROS (Schellingen et al. 2015). Furthermore, Liu et al. (2018) reported oxidative stress in *Z. mays* through increase H_2O_2 accumulation and ZmMPK3 activation on exposure to excess Cu. It is suggested that more attention be paid to MAPK signaling in other biotic and abiotic stresses and its interplay with other signaling pathways, to generate a framework where the participation of MAPK signaling in metal stress should be studied (Opdenakker et al. 2012).

7.3.2.3 Hormonal Signaling

Numerous plant hormones produce ROS as part of the mechanism that regulates the growth and development of plants (Xia et al. 2015). Biosynthesis, transportation and accumulation of auxin, abscisic acid, ethylene, gibberellins, ethylene, brassinosteroid and cytokinin (hormones) as well as the interaction of salicylic acid and jasmonic acid (signaling molecules) increase the signaling pathways, activate various antioxidant gene expressions and stimulate the production of osmoprotectants like soluble sugars, proline and amino acids against heavy metals stress (Jalmi et al. 2015; Bücker-Neto et al. 2017; Shukla et al. 2017; Sharma et al. 2020). In plants, the reaction of the hormone to heavy stress relies on the kind of heavy metal, concentration of it and interactions of the complex with other stresses. The literature revealed that heavy metals (e.g. Cu, Cd, As and Hg) boost ROS, calcium and MAPK signaling, and increase abscisic acid biosynthetic genes, which lead to enhanced tolerance in plants against metal toxicity (Bücker-Neto et al. 2017). Enhanced H_2O_2 in Cu-treated seedlings has no role in Cu-regulated auxin redistribution for the inhibition of primary root elongation (Yuan et al. 2013).

Detrimental consequences of As (arsenic) have been documented on the level of auxins in *Brassica juncea* and *Hordeumvulgare* (Srivastava et al. 2013; Zelinová et al. 2015). Increased ethylene synthesis following Cd toxicity is a characteristic feature of the alarm situation during severe stress (Schellingen et al. 2014). The activity of ethylene associated gene (ACS) has been reported to increase in *Brassica juncea* plants exposed to Zn or Ni (Khan and Khan 2014), Cd (Asgher et al. 2015) and in Cd-exposed plants of *Triticum aestivum* (Khan et al. 2015).

Alteration in root systems has been reported to link through signal transduction of crosstalk in indole acetic acid (IAA), ethylene and ROS under metal stress. For example, pinformed1 (PIN1) protein participates in the distribution of IAA under heavy metal stress conditions (Sytar et al. 2019). Likewise, IAA distribution is changed under boron (B) starvation, which resulted in lower regulation of the PIN1 protein and restriction of root elongation (Li et al. 2015). Gibberellins and cytokinins have also been shown to play important roles in improved plant protection against heavy metal stress (Bucker-Neto et al. 2017). Reduction in cytokinin production, as well as intensive involvement of cytokinin with other hormones under heavy metal stress, has been reported (Sytar et al. 2019). Salicylic acid concentration has been documented to increase during metal stress, as it regulates RO, interacts with other plant hormones (e.g. auxins, abscisic acid, gibberellin) and stimulates antioxidant compounds, thereby can alleviate the heavy metal stress in the plant (Sharma et al. 2020).

Phytohormones can be involved in phytochelatin biosynthesis, while a clear link among heavy metals, hormonal pathways and metal-binding ligands in plants still needs to be elaborated on (Bucker-Neto et al. 2017).

Deregulated ROS would modify redox reactions and associated signal transduction pathways with finally appropriate cellular responses. An unfortunate consequence of heavy metal toxicity in oxidative stress in cells would result in oxidative modification of all the classes of biologically important macromolecules (lipids, proteins and DNA), which induce subtle effects on plant's growth, morphology, physiology and yield (Jalmi et al. 2018; Sharma et al. 2020; Sinegovskaya et al. 2020). Quantity and site of ROS are equally important as oxidation rate and abundance of target moieties for macromolecular oxidation. In addition, the reaction with target molecules, such as DNA, proteins, lipids and other ROS, depends on the redox environment of the cell. The relatively high reactivity of some oxidants restricts their diffusion and role as true signaling molecules (Hempel and Trebak 2017).

7.3.3 Oxidative Modifications in Biomolecules

7.3.3.1 Lipid Oxidation

Plasma membrane injury is taken as the main event of heavy metal-induced oxidative stress, where ROS are accumulated either due to impaired function of oxidoreductase enzymes, leakage of the electron transport chain or plasma membrane-bound NADPH oxidase. Via apoplast to symplast, heavy metal ions bind to different molecules concurrently in the cytoplasm as well as in cell organelles, which results in ROS generation in the vicinity of cellular and organellar membranes. For example, in single-membrane organelles (peroxisomes and glyoxysomes), H_2O_2 is formed through glycolate oxidase, whereas xanthine oxidase, urate oxidase and NADPH oxidases generate $O_2^{\bullet-}$. ROS are mainly produced via electron transport chains of chloroplasts and mitochondria, and through peroxisomal activity, while chloroplasts are an essential source of ROS in plants under the metal-induced oxidative stress (Štolfa et al. 2015).

The concentration of ROS above threshold level directly modifies plasma membrane, and cellular and organellar membranes by abstracting hydrogen from membrane lipids (unsaturated fatty acid). Thus, the creation of several double bonds in the lipid structure also participates in the formation of peroxide radicals (Sharma et al. 2012). Single oxidation leads to side by side chemical reactions for the propagation of other lipids, amplification in lipid peroxides and generation of truncated or fragmented oxidized lipids. Furthermore, cellular nucleophiles including protein side chains and the polar head of some lipids (e.g. phosphatidylethanolamine) can react with some of these oxidized products, further increasing the diversity and complexity of the species (e.g. lipid epoxides, aldehydes, lipid alkoxyl radicals, alkanes and alcohols) (Ali et al. 2019).

An increase in the level of ROS within sub-cellular compartments under high Cu levels is thought to be linked with inhibition in Mehler reaction on the PSI and Hill reaction on PSII, while leading to an alteration in the carbon metabolism, cellular responses and signal transduction in the plant cell (Rochaix 2011). Heavy metal-induced photo-oxidative damage to the chloroplast membrane changes its functions and components of the photosynthetic electron transport chain, thus impairing the light phase of photosynthesis (Ventrella et al. 2009). Ni treatment (100 mg kg^{-1}) caused an enhanced lipid oxidation level in *Brassica juncea*, which decreased after the external application of a known antioxidant sulfosalicylic acid. Reduction in photosynthetic rate, transpiration rate, stomatal conductivity, plant growth and yield have been documented after exposure to heavy metals (Yusuf et al. 2012). Moreover, As can affect primary photosynthetic reactions by reacting with chlorophyll and carotenoid integrity, thus impairing the activity of photosystems' antenna complexes and the electron transport chain, which results in declined ATP and NADPH syntheses and/or enhanced energy dissipation through fluorescence or as heat (Gusman et al. 2013). Impairment of cell membrane and enzymatic imbalance have been found associated with Ni-induced oxidative stress in both agronomic and non-agronomic plants (Emamverdian et al. 2015). Excessive Cu, Zn, Pb and Cd strongly generate oxidative damage, leading to loss of photosynthesis and water uptake, disturbance of nutrient uptake, degradation of chlorophyll and disturbance of enzyme activities (Panday et al. 2016).

Oves et al. (2016) have shown that metal ions cause peroxidation of lipids of the plasma and chloroplast membranes. Uptake of Cu by the plant tissues reduced the content of photosynthetic pigments, and stimulated lipid peroxidation and permeability of the membrane. Also, membrane disruption by free radicals generated under metal stress modulates the activities of membrane-bound ATPase, which reduces the transport activities of the root plasma membrane and ultimately limits nutrient uptake by the roots (Mroczek-Zdyrska and Wojcik 2012; Oves et al. 2016). The changes in membrane lipids also lead to abnormal cell structures and alterations in the cell membrane of organelles such as mitochondria, peroxisomes and chloroplasts (Amari et al. 2017).

Malook et al. (2017) revealed that heavy metals contaminated soil significantly and affected the photosynthetic system and total soluble proteins by increasing lipid peroxidation in *Spinacia oleracea*. Likewise, Georgiadou et al. (2018) acknowledged that a high concentration of Cu or Fe caused lipid peroxidation through the highest accumulation of free radicals. Abbas et al. (2018) and Sharma et al. (2020) also explored the ROS ameliorated membrane lipid peroxidation reaction products, along with electrolyte leakage due to As and Cr toxicity in the plant cell, respectively. Impairment of the photosynthetic process was also corroborated by lower sugar levels, stomatal closure, reduction in efficiency of PSII and the more severe growth reduction in As-sensitive *Nicotiana sylvestris* (Kofronová et al. 2020).

The risen level of malondialdehyde (MDA) has been widely used to judge the degree of damage to the cell membrane. . Several reports exist supporting this correlation between high levels of heavy metal and increased MDA content (Georgiadou et al. 2018). Ali et al. (2015) explored the link between the deterioration of membrane permeability and MDA in *Triticumestivum*. Pospíšil and Yamamoto (2017) recognized that MDA accumulation is associated with the peroxidation of the phospholipid layer of the cell membrane under metal stress. El-Amier et al.'s (2019) study reflected the lipid peroxidation as a consequence of ROS production, which also increased the content of MDA in *Pisum sativum* treated with Cd and Ni.

It is clear that lipid-derived radicals worsen oxidative stress, especially of the cellular membranes causing them to leak, and the radicals themselves react with proteins and DNA, thus affect normal cellular functioning (Sharma et al. 2012; Ghori et al. 2019).

7.3.3.2   Protein Oxidation

Proteins are a key target of heavy metals. Oxygen radicals and other ROS species are created due to heavy metal toxicity, modified peptides and proteins (folded and unfolded). Direct modification occurs through nitrosylation, carbonylation, disulfide bond formation and glutathionylation, and indirectly by coalescence with bi-products of lipid peroxidation (Dutta et al. 2018; Ali et al. 2019). Additionally, thiol and sulfur-containing amino acid residues from protein and peptides are preferred by ROS, as H$^+$ or OH$^-$ abstraction leads to the formation of more radicals (peroxyl radical, oxygen- or sulfur-centered radicals), while disulfide bridge formation has also been observed due to protein and peptide cross-linking. Methionine sulphoxide derivatives may also be formed after the addition of oxygen to methionine, several metals have been shown to cause changes in protein-bound thiol groups, and protein peroxides can oxidize both proteins and other targets (Sharma et al. 2012).

Ghasemi et al. (2012) found that excess Ni perniciously influenced photosynthetic protein complexes and decreased the rate of Hill reaction in maize. Cd binding to cysteine residues in its active sites resulted in inhibition in the activity of thiol transferase, which led to oxidative damage in *Brassica juncea* (D'Alessandro et al. 2013). Cd-induced oxidative stress can alter activity proteins involved in monitoring the redox state of storage proteins by oxidizing Trx isoforms in *P. sativum* seeds (Smiri et al. 2013). As-induced ROS have been shown to reduce biomass and total protein content in *Brassica juncea* and *Oryza sativa* (Kanwar et al. 2015; Upadhyay et al. 2016). As has also been reported to react with sulfhydryl (–SH) groups of proteins and enzymes, culminating in a decline in cellular function and eventually leading to cell death (Abbas et al. 2019).

The Cd-induced modifications may also disrupt the stabilizing interactions connected with alteration in the tertiary structure and result in the loss of important functions of that protein. Fallout dysfunction of protein stimulates the danger of protein aggregation (Hasan et al. 2017). Alterations in cellular proteomes due to the degradation of a number of proteins with increasing concentrations of Cu and Zn have been reported (Georgiadou et al. 2018). Metal-induced ROS also has destructive effects on some enzymes responsible for protein synthesis machinery in *P. sativum* on exposure to both Cd and Ni (El-Amier et al. 2019). Protein oxidation through the formation of carbonyl derivate has been associated with its degradation and malfunctioning (Abbas et al. 2019).

Cr(VI)-induced ROS accumulation in various plant species has been noticed to suppress enzymatic antioxidant system, which also harmed cellular and subcellular membranes; induced ultrastructural changes in cell organelles such as mitochondria, plastids and thylakoids; inhibited protein and enzymes at the transcriptional or post-transcriptional level; as well as degraded various enzymes and proteins and damaged DNA (Wakeel et al. 2020).

Heavy metal toxicity can also increase the sensitivity of cellular protein homeostasis by interfering with the folding process. Misfolding or aggregation in 3D structure of native protein leads to inappropriate association with other cellular components, which results in the malfunctioning of these quality-control systems (Chen et al. 2018). Hasan et al. (2017) described how under extreme conditions, metal ions stimulate the aggregation of nascent or non-native proteins, resulting in endoplasmic reticulum stress and reduced viability of the cell. The increase in the vulnerability of plants to diseases has been associated with deterioration in protein structure under heavy metal stress (Emamverdian et al. 2015).

7.3.3.3 DNA Oxidation

Heavy metal-mediated oxidative stress in plants leads to DNA (nuclear, mitochondrial and chloroplastic) damage through exposing it to reactive species, and adduction more often results in malfunctioning or thorough deactivation of the encoded proteins (Sharma et al. 2012; Shahid et al. 2014; Ali et al. 2019). Interaction of Cd, Cr, Cu, Hg, Pb and Zn with sulfhydryl groups and the phosphate backbone has been reported to cause DNA damage (Aravind and Prasad 2005; Sheng et al.2008; Igiri et al. 2018). Nucleic acid impairments, chromosomal misjoining and cell division deregulation are among the important genotoxic effect of oxidative stress (Srivastava et al. 2017; Dutta et al. 2018). Base modification, base-free sites, damage to pyrimidine dimers, DNA-protein cross-links and strand breaks are among the variegated amount of modifications in DNA structure, induced by metal-mediated oxidative stress (Dutta et al. 2018; Srivastava et al. 2017; Ali et al. 2019). The most frequently occurring event has been documented as base deletion, which might be responsible for a frameshift mutation in DNA (Payus et al. 2016; Srivastava et al. 2017).

Altered DNA occurring over a cell cycle under heavy metal toxicity results in the sticky chromosome (poised chromosomes) that pave the way for cell death. High levels of Zn (100 mg·L^{-1}) in cells can result in abnormal chromosomes, followed by sticky metaphase and premature separation of chromosomes in *Vigna subterranean* (Oladele et al. 2013). Likewise, Truta et al. (2013) noted that the rate of ana-telophase aberrations was two to three times higher than the control treatment when *Hordeum vulgare* seedling was treated with 250 to $500\,\mu M$ Zn^{2+}. It has been found that Cd-induced oxidative stress increased chromosomal aberrations in mitotic cells of root tips in *Capsicum annum*. Henceforth, the interaction of Cd with chromatin proteins steered incorrect coding of some non-histone proteins responsible for chromosome organization to negatively alter the number of mitotic cells in root tips, and root length declined (Aslam et al. 2014). Abnormalities in chromosome generated altered nucleotide sequences of alleles that may result in the production of slightly different types of amino acids or variant forms of proteins. These proteins code for the development of varied morphological, anatomical and physiological characteristics in the organism (Bhat et al. 2012).

DNA damage, chromosomal aberrations, micronuclei and chromosomal fragmentation and bridging in *Allium cepa* (Patnaik et al. 2013), and, likewise, micronucleus and chromosomal fragmentation and bridging, and an increase in the percentage tail DNA, tail moment and tail length in *Vicia faba* have been documented on exposure to Cr-induced oxidative stress (Loubna et al. 2015). Toxicity of Al has also been reported to cause a reduction in root respiration and disturbances in the enzymatic regulation of sugar phosphorylation probably due to the hindrance of cell division in roots by Al ions binding to DNA. Likewise, excess of Zn is found to have genotoxic effects on plants, resulting in genetic-related disorders and damages to plants (Emamverdian et al. 2015).

Moreover, Abubacker and Sathya's (2017) study has proved that Cr, Cu, Pb and Zn induced sticky and broken chromosomes, polyploid prophase, C-mitosis multipolar metaphase and multipolar anaphase, which may be the cause of inhibition of DNA synthesis, since these abnormalities are related to nuclear abnormalities and cell death. Besides this, modification in chromatin structure (histone complexes) may alter gene expression and accessibility of transcription factors under various heavy metal stress (Dutta et al. 2018). Variation in chromatin structure and transgene silencing has been observed with an increased level of demethylation in histone (H3K4, H3K9), which resulted in expanded cell cycle duration and growth inhibition after exposure to Ni (Sun et al. 2013; Zhao et al. 2014).

Besides this, oxidation stress induced by heavy metals contributed to the propagation of mutated DNA (Hirano and Tamae 2010), and 8-oxoguanine (oxidized base) is the most abundant oxidation product of guanine in DNA (Hirano and Tamae 2010), which may result in malignant transformation with DNA (Bal and Kasprzak 2002). Furthermore, ROS generated during heavy metal (Cd, Pb, Cr and As) stress also caused the formation of pro-mutagenic adduct 7,8-dihydro-8-oxoguanine (8-OxoG), which can induce DNA lesions (Markkanen et al. 2011). DNA lesions such as DNA single-strand breaks (SSBs) and double-strand breaks (DSBs) frequently lead to structural malformation in chromosomes, which induce errors in DNA replication and result in cell death (Dutta et al. 2018).

7.4 CONCLUSION

Higher concentrations of heavy metal accumulation in plants, through leaf and root, lead to oxidative stress in plants, which contributes to damaging biomolecules as well as disrupting signaling pathways. This in turn leads to disorders associated with plant growth and yield.

REFERENCES

Abbas G, Murtaza B, Bibi I, Shahid M, Niazi N, Khan M, Amjad M, Hussain M & Natasha. 2018. Arsenic uptake, toxicity, detoxification, and speciation in plants: Physiological, biochemical, and molecular aspects. *Int J Environ Res Pub Health*, 15(1): 59.

Abubacker MN & Sathya C. 2017. Genotoxic effect of heavy metals Cr, Cu, Pb and Zn using *Allium cepa* L. *Bio Sci Bio Tecnol Res Asia*, 14: 1181–1186.

Ahmad A, Hadi F & Ali N. 2015. Effective phytoextraction of cadmium, Cd with increasing concentration of total phenolics and free proline in *Cannabis sativa* (L) plant under various treatments of fertilizers, plant growth regulators and sodium salt. *Int J Phytoremed*, 17: 56–65.

Alejandro S, Höller S, Meier B & Peiter E. 2020. Manganese in plants: From acquisition to subcellular allocation. *Front Plant Sci*, 11: 300.

Ali MA, Fahad S, Haider I, Ahmed N, Ahmad S, Hussain S & Arshad M. 2019. Plant Abiotic Stress Tolerance. In: M. Hasanuzzaman, V. Fotopoulos, K. Nahar, M. Fujita (Eds). Springer, Cham. .

Ali S, Bharwana SA, Rizwan Farid M, Kanwal S, Ali Q, Ibrahim M, Gill RA & Khan MD. 2015. Fulvic acid mediates chromium (Cr) tolerance in wheat (*Triticumaestivum* L.) through lowering of Cr uptake and improved antioxidant defense system. *Environ Scipollut Res*, 22: 10601–10609.

Amari T, Ghnaya T & Abdelly C. 2017. Nickel, cadmium and lead phytotoxicity and potential of halophytic plants in heavy metal extraction. *South African Journal of Botany*, 111: 99–110.

Aravind P & Prasad MN. 2005. Cadmium-zinc interactions in a hydroponic system using *Ceratophyllumdemersum* L.: Adaptive ecophysiology, biochemistry and molecular toxicology. *Braz J Plant Physiol*, 17(1): 3–20.

Asgher M, Khan MIR, Anjum NA & Khan NA. 2015. Minimising toxicity of cadmium in plants—Role of plant growth regulators. *Protoplasma*, 252(2): 399–413.

Aslam R, Ansari MY, Choudhary S, Bhat TM & Jahan N. 2014. Genotoxic effects of heavy metal cadmium on growth, biochemical, cyto-physiological parameters and detection of DNA polymorphism by RAPD in *Capsicum annuum* L. an important spice crop of India. *Saudi J Biol Sci*, 21(5): 465–472.

Bal W & Kasprzak KS. 2002. Induction of oxidative DNA damage by carcinogenic metals. *Toxicollett*, 127(1–3): 55–62.

Bhat TM, Ansari MYK, Alka R & Aslam R. 2012. Sodium azide (NaN3) induced genetic variation of Psoraleacorylifolia L. and analysis of variants using RAPD markers. *Nucleus*, 55(3): 149–154.

Bhattacharjee S. 2019. *ROS and Oxidative Stress: Origin and Implication. Reactive Oxygen Species in Plant Biology*. (pp: 1–31). Springer.

Bücker-Neto L, Paiva ALS, Machado RD, Arenhart RA & Margis-Pinheiro M. 2017. Interactions between plant hormones and heavy metals responses. *Genet Mol Biol*, 40(1): 373–386.

Campbell WH. 1999. Nitrate reductase structure, function and regulation: Bridging the gap between biochemistry and physiology. *Annu Rev Plant Physiol Plant Mol Biol*, 50: 277–303.

Cao Y, Ma C, Chen H, Zhang J, White JC, Chen G & Xing, B. 2020. Xylem-based long-distance transport and phloem remobilization of copper in *Salix integra* Thunb. *J Hazard Mater*, 15: 392.

Cenkci S, Cigerci İH, Yıldız M, Özay C, Bozdağ A & Terzi H. 2010. Lead contamination reduces chlorophyll biosynthesis and genomic template stability in *Brassica rapa* L. *Environ Exp Bot*, 67(3): 467–473.

Chen X, Ouyang Y, Fan Y, Qiu B, Zhang G & Zeng F. 2018. The pathway of transmembrane cadmium influx via calcium-permeable channels and its spatial characteristics along rice root. *J Exp Bot*, 69(21): 5279–5291.

Chen Z, Zhu D, Wu J, Cheng Z, Yan X, Deng X & Yan Y. 2018. Identification of differentially accumulated proteins involved in regulating independent and combined osmosis and cadmium stress response in *Brachypodium* seedling roots. *Sci Rep*, 8: 7790.

Chibuike GU & Obiora SC. 2014. Heavy metal polluted soils: Effect on plants and bioremediation methods. *Appl Envi Soil Sci*: 752708.

Choong G, Liu Y & Templeton DM. 2014. Interplay of calcium and cadmium in mediating cadmium toxicity. *Chem Biol Interact*, 211: 54–65.

Choudhury FK, Rivero RM, Blumwald E & Mittler R. 2017. Reactive oxygen species, abiotic stress and stress combination. *Plant J*, 90(5): 856–867.

D'Alessandro A, Taamalli M, Gevi F, Timperio AM, Zolla L & Ghnaya T. 2013. Cadmium stress responses in *Brassica juncea*: Hints from proteomics and metabolomics. *J Proteome Res*, 12(11): 4979–4997.

Del Río LA & López-Huertas E. 2016. ROS generation in peroxisomes and its role in cell signaling. *Plant Cell Physiol*, 57(7): 1364–1376.

Demidchik V. 2018. ROS-activated ion channels in plants: Biophysical characteristics, physiological functions and molecular nature. *Int J Mol Sci*, 19(4): 1263.

Deng THB, van der Ent A, Tang YT, Sterckeman T, Echevarria G, Morel J-L & Qiu R-L. 2016. Nickel translocation via the phloem in the hyperaccumulator *Noccaea caerulescens* (Brassicaceae). *Plant Soil*, 404: 35–45.

Dodd AN, Kudla J & Sanders D. 2010. The language of calcium signaling. *Annu Rev Plant Biol*, 61: 593–620.

Dong Y, Xu L, Wang Q, Fan Z, Kong J & Bai X. 2014. Effects of exogenous nitric oxide on photosynthesis, antioxidative ability, and mineral element contents of perennial ryegrass under copper stress. *J Plant Interact*, 9(1): 402–411.

Dutta S, Mitra M, Agarwal P, Mahapatra K, De S, Sett U & Roy S. Oxidative and genotoxic damages in plants in response to heavy metal stress and maintenance of genome stability. *Plant Signal Behav*, 13(8): e1460048.

Edelstein M & Ben-Hur M. 2018. Heavy metals and metalloids: Sources, risks and strategies to reduce their accumulation in horticultural crops. *Scihortic*, 234: 431–444.

El-Amier Y, Elhindi K, El-Hendawy S, Al-Rashed S & Abd-ElGawad A. 2019. Antioxidant system and bio-molecules alteration in *Pisum sativum* under heavy metal stress and possible alleviation by 5-aminolev-ulinic acid. *Molecules*, 24(22): 4194.

Emamverdian A, Ding Y, Mokhberdoran F & Xie Y. 2015. Heavy metal stress and some mechanisms of plant defense response. *Sci World J*: 2015: 1–18.

Fan J-L, Wei X-Z, Wan L-C, Zhang L-Y, Zhao X-Q, Liu W-Z, Hao HQ & Zhang HY. 2011. Disarrangement of actin filaments and Ca^{2+} gradient by $CdCl_2$ alters cell wall construction in Arabidopsis thaliana root hairs by inhibiting vesicular trafficking. *J Plant Physiol*, 168(11): 1157–1167.

Fang Y, Xie K & Xiong L. 2014. Conserved miR164-targeted NAC genes negatively regulate drought resis-tance in rice. *J Exp Bot*, 65(8): 2119–2135.

Farid M, Ali S, Rizwan M, Ali Q, Saeed R, Nasir T, Abbasi GH, Rehmani MIA, Ata-Ul-Karim ST, Bukhari SAH & Ahmad T. 2018. Phyto-management of chromium contaminated soils through sunflower under exogenously applied 5-aminolevulinic acid-Ul-Karim ST & Bukhari SAH. *Ecotoxicol Environ Saf*, 151: 255–265.

Fİlİz E & Tombuloğlu H. Genome-wide distribution of superoxide dismutase (SOD) gene families in Sorghum bicolor. *Turk J Biol*, 39: 49–59.

Freitas SD, Rodak WB & Reis RDA. 2018. Hidden Nickel Deficiency? nickel fertilization via soil improves nitrogen metabolism and grain yield in soybean genotypes. Reis dBF, Carvalho SDT, Schulze J, Carneiro CMA & Guilherme GLR. *Front Plant Sci*, 9: 614.

Gad N, Sekara A & Abdelhamid MT. 2019. The potential role of cobalt and/or organic fertilizers in improving the growth, yield, and nutritional composition of Moringaoleifera. *Agronomy*, 9(12): 862.

Georgiadou EC, Kowalska E, Patla K, Kulbat K, Smolińska B, Leszczyńska J & Fotopoulos V. 2018. Influence of heavy metals (Ni, Cu, and Zn) on nitro-oxidative stress responses, proteome regulation and allergen production in basil (*Ocimumbasilicum* L.) plants. *Front Plant Sci*, 9: 862.

Ghasemi F, Heidari R, Jameii R & Purakbar L. 2012. Effects of Ni^{2+} toxicity on Hill reaction and membrane functionality in maize. *J Stress PhiolBiochem*, 8: 55–61.

Ghori N-H, Ghori T, Hayat MQ, Imadi SR, Gul A, Altay V & Ozturk M. 2019. Heavy metal stress and responses in plants. *Int J Environ Sci Technol*, 16(3): 1807–1828.

Gill RA, Zang LL, Ali B, Farooq MA, Cui P, Yang S, Ali S & Zhou WJ. 2015. Chromium-induced physio-chemical and ultrastructural changes in four cultivars of *Brassica napus* L. *Chemosphere*, 120: 154–164.

Gusman GS, Oliveira JA, Farnese FS & Cambraia J. 2013. Arsenate and arsenite: The toxic efects on photo-synthesis and growth of lettuce plants. *Acta Physiol Plant*, 35(4): 1201–1209.

Halliwell B. 1991. Reactive oxygen species in living systems: Source, biochemistry, and role in human disease. *Am J Med*, 91(3C): 14S–22S.

Hänsch R & Mendel RR. 2009. Physiological functions of mineral micronutrients (Cu, Zn, Mn, Fe, Ni, Mo, B, Cl). *Curr Opin Plant Biol*, 12(3): 259–266.

Hasan M, Cheng Y, Kanwar MK, Chu XY, Ahammed GJ & Qi ZY. 2017. Responses of plant proteins to heavy metal stress—A review. *Front Plant Sci*, 8: 1492.

Hempel N & Trebak M. 2017.Crosstalk between calcium and reactive oxygen species signaling in cancer. *Cell Cal.*, 63: 70–96.

Hirano T & Tamae K. 2010. Heavy metal-induced oxidative DNA damage in earthworms: A review. *Appl Environ Soil Sci*: 726946.

Hu Y, Tian SK, Foyer CH, Hou DD, Wang HX, Zhou WW, Liu T, Ge J, Lu LL & Lin XY. 2019. Efficient phloem transport significantly remobilizes cadmium from old to young organs in a hyperaccumula-torSedum alfredii. *J Hazard Mater*, 365: 421–429.

Huang TL, Huang LY, Fu SF, Trinh NN & Huang HJ. 2014. Genomic profiling of rice roots with short-and long-term chromium stress. *Plant Mol Biol*, 86(1–2): 157–170.

Huang T-L & Huang H-J. 2008. ROS and CDPK-like kinase-mediated activation of MAP kinase in rice roots exposed to lead. *Chemosphere*, 71(7): 1377–1385.

Igiri BE, Okoduwa SI, Idoko GO, Akabuogu EP, Adeyi AO & Ejiogu IK. 2018. Toxicity and bioremediation of heavy metals contaminated ecosystem from tannery wastewater: A review. *J Toxicol*, 2018.

Jalmi SK, Bhagat PK, Verma D, Noryang S, Tayyeba S, Singh K, Sharma D & Sinha AK. 2018. Traversing the links between heavy metal stress and plant signaling. *Front Plant Sci*, 9: 12. doi: 10.3389/fpls.2018.00012.

Jayakumar K, Vijayarengan P, Changxing Z, Gomathinayagam M & Jaleel CA. 2008. Soil applied cobalt alters the nodulation, leg-haemoglobin content and antioxidant status of *Glycine max* (L.) Merr. *Colloids Surf B Biointerfaces*, 67(2): 272–275.

Kanwar MK, Poonam R & Bhardwaj R. 2015. Arsenic induced modulation of antioxidative defense system and brassinosteroids in *Brassica juncea* L. *Ecotoxicol Environ Saf*, 115: 119–125.

Keunen EL, Peshev D, Vangronsveld J, Van Den Ende WI & Cuypers AN. 2013. Plant sugars are crucial players in the oxidative challenge during abiotic stress: Extending the traditional concept. *Plant Cell Environ*, 36(7): 1242–1255.

Khan MIR, Nazir F, Asgher M, Per TS & Khan NA. 2015. Selenium and sulfur influence ethylene formation and alleviate cadmium-induced oxidative stress by improving proline and glutathione production in wheat. *J Plant Physiol*, 173: 9–18.

Khan MIR & Khan NA. 2014. Ethylene reverses photosynthetic inhibition by nickel and zinc in mustard through changes in PS II activity, photosynthetic nitrogen use efficiency, and antioxidant metabolism. *Protoplasma*, 251(5): 1007–1019.

Kharbech O, Houmani H, Chaoui A & Corpas FJ. 2017. Alleviation of Cr(VI)-induced oxidative stress in maize (*Zea mays* L.) seedlings by NO and H$_2$S donors through differential organ-dependent regulation of ROS and NADPH-recycling metabolisms. *J Plant Physiol*, 219: 71–80.

Kofronová M, Hrdinová A, Mašková P, Tremlová J, Soudek P, Petrová Š, Pinkas D & Lipavská H. 2020. Multi-component antioxidative system and robust carbohydrate status, the essence of plant arsenic tolerance. *Antioxidants*, 9(4): 283.

Kumar B, Smita K & Cumbal Flores LC. 2017. Plant mediated detoxification of mercury and lead. *Arab J Chem*, 10: S2335–S2342.

Li K, Kamiya T & Fujiwara T. 2015. Differential roles of PIN1 and PIN2 in root meristem maintenance under low-B conditions in *Arabidopsis thaliana*. *Plant Cell Physiol*, 56(6): 1205–1214.

Liu J, Wang J, Lee S & Wen R. 2018. Copper-caused oxidative stress triggers the activation of antioxidant enzymes via ZmMPK3 in maize leaves. *PLOS ONE*, 13(9): e0203612.

Liu W, Yang YS, Li PJ, Zhou QX, Xie LJ & Han YP. 2009. Risk assessment of cadmium-contaminated soil on plant DNA damage using RAPD and physiological indices. *J Hazard Mater*, 161(2–3): 878–883.

Liu Y & He C. 2017. A review of redox signaling and the control of MAP kinase pathway in plants. *Redox Biol*, 11: 192–204.

Loubna EF, Hafidi M, Silvestre J, Kallerhoff J, Merlina G & Pinelli E. 2015. Efficiency of co-composting process to remove genotoxicity from sewage sludge contaminated with hexavalent chromium. *Ecol. Eng*, 82: 355–360.

Łukowski A & Dec D. 2018. Influence of Zn, Cd, and Cu fractions on enzymatic activity of arable soils. *Environ Monit Assess*, 190(5): 278.

Małecka A, Konkolewska A, Hanć A, Barałkiewicz D, Ciszewska L, Ratajczak E, Staszak AM, Kmita H & Jarmuszkiewicz W. 2019. Insight into the phytoremediation capability of *Brassica juncea* (v. Malopolska): Metal accumulation and antioxidant enzyme activity. *Int J Mol Sci*, 20(18): 4355.

Malook I, Rehman SU, Khan MD, El-Hendawy SE, Al-Suhaibani NA, Aslam MM & Jamil M. 2017. Heavy metals induced lipid peroxidation in spinach mediated with microbes. *Pak J Bot*, 49: 2301–2308.

Manuel TJ, Alejandro CA, Angel L, Aurora G & Emilio F. 2018. *Roles of Molybdenum in Plants and Improvement of Its Acquisition and Use Efficiency in Plant Micronutrient Use Efficiency*. In: Hossain MA, Kamiya T, Burritt DJ, Tran LP and Fujiwara T (ed.). Academic Press. London, UK (pp 137–159).

Marchetti C. 2013. Role of calcium channels in heavy metal toxicity. *ISRN Toxicol*. doi: 10.1155/2013/184360.

Markkanen EE, van Loon B, Ferrari E, Parsons JL, Dianov GL & Hübscher U. Regulation of oxidative DNA damage repair by DNA polymerase λ and MutYH by cross-talk of phosphorylation and ubiquitination. *Proc Natl Acad Sci U S A*, 109(2): 437–442.

Mhamdi A & Van Breusegem F. 2018. Reactive oxygen species in plant development. *Development*, 145(15): dev164376.

Mittler R. 2017. ROS are good. *Trends Plant Sci*, 22(1): 11–19.

Mroczek-Zdyrska M & Wójcik M. 2012. The influence of selenium on root growth and oxidative stress induced by lead in *Viciafaba* L. minor plants. *Biol Trace Elem Res*, 147(1–3): 320–328.

Natasha, Shahid M & Khalid S. 2020. Foliar application of lead and arsenic solutions to Spinacia oleracea: Biophysiochemical analysis and risk assessment. *Environ Sci Pollut Res*, 27(32): 39763–39773.

Oladele EO, Odeigah PGC & Taiwo IA. 2013. The genotoxic effect of lead and zinc on Bambara groundnut (*Vigna subterranean*). *Afr J Environ Sci Technol*, 7: 9–13.

Opdenakker K, Remans T, Vangronsveld J & Cuypers A. 2012. Mitogen-activated protein (MAP) kinases in plant metal stress: regulation and responses in comparison to other biotic and abiotic stresses. *Int J Mol Sci*, 13(6): 7828–7853.

Oves M, Saghir Khan M, Huda Qari A, NadeenFelemban M & Almeelbi T. 2016. Heavy Metals: Biological importance and detoxification strategies. *J Bioremed Biodeg*, 7: 334.

Page, V & Feller U. 2015. Heavy Metals in Crop Plants: Transport and Redistribution Processes on the Whole Plant Level. *Agronomy*, 5: 447–463.

Panda SK, Choudhury S & Patra HK. 2016. Chapter 11. Abiotic stress response in plants. Book. In: N Tuteja, SS Gill (Eds). *Heavy-Metal-Induced Oxidative Stress in Plants: Physiological and Molecular Perspectives*.Wiley, Hoboken. (pp 221–236).

Pandey B, Suthar S & Singh V. 2016. Accumulation and health risk of heavy metals in sugarcane irrigated with indusial effluent in some rural areas of Uttarakhand, India. *Process Saf Environ*, 102: 655–666.

Patnaik AR, Achary VMM & Panda BB. 2013. Chromium (VI)-induced hormesis and genotoxicity are mediated through oxidative stress in root cells of Allium cepa L *Plant Growth Regul*, 71(2): 157–170.

Payus C, Ying TS & Nyet Kui W. 2016. Effect of heavy metal contamination on the DNA mutation on nepenthes plant from abandoned mine. *Research J of Environmental Toxicology*, 10(4): 193–204.

Pinho S & Ladeiro B. 2012. Phytotoxity by lead as heavy metal focus on oxidative stress. *J Bot*, 2012: 1–10.

Pospíšil P & Yamamoto Y. 2017. Damage to photosystem II by lipid peroxidation products. *Biochim Biophys Acta Gen Subj*, 1861(2): 457–466.

Rahman Z & Singh VP. 2019. The relative impact of toxic heavy metals (THMs)(arsenic (As), cadmium (Cd), chromium (Cr)(VI), mercury (Hg), and lead (Pb)) on the total environment: An overview. *Environ Monit Assess*, 191(7): 419.

Rao KP, Vani G, Kumar K, Wankhede DP, Misra M, Gupta M & Sinha AK. 2011. Arsenic stress activates MAP kinase in rice roots and leaves. *Arch Biochembiophys*, 506(1): 73–82.

Rascio N & Navari-Izzo F. 2011. Heavy metal hyperaccumulating plants: How and why do they do it? And what makes them so interesting? And what makes them so interesting? *Plant Sci*, 180(2): 169–181.

Ravet K, Touraine B, Boucherez J, Briat JF, Gaymard F & Cellier F. 2009. Ferritins control interaction between iron homeostasis and oxidative stress in Arabidopsis. *Plant J*, 57(3): 400–412.

Rochaix JD. 2011. Regulation of photosynthetic electron transport. *Biochim Biophys Acta*, 1807(3): 375–383.

Schellingen K, Straeten D Van Der, Vandenbussche F, Prinsen E, Remans T, Vangronsveld J & Cuypers A. 2014. Cadmium-induced ethylene production and responses in *Arabidopsis thaliana* rely on ACS2 and ACS6 gene expression. *BMC Plant Biol*, 14: 214–214.

Schellingen K, Van Der Straeten D, Remans T, Loix C, Vangronsveld J & Cuypers A. 2015. Ethylene biosynthesis is involved in the early oxidative challenge induced by moderate Cd exposure in *Arabidopsis thaliana*. *Environ Exp Bot*, 117: 1–11.

Sethi V, Raghuram B, Sinha AK & Chattopadhyay S. 2014. A mitogen-activated protein kinase cascade module, MKK3-MPK6 and MYC2, is involved in blue light-mediated seedling development in Arabidopsis. *Plant Cell*, 26(8): 3343–3357. doi: 10.1105/tpc.114.128702.

Sewelam N, Kazan K & Schenk PM. 2016. Global plant stress signaling: Reactive oxygen species at the crossroad. *Front Plant Sci*, 7(7): 187.

Shahid M, Dumat C, Khalid S, Schreck E, Xiong T & Niazi NK. 2017. Foliar heavy metal uptake, toxicity and detoxification in plants: A comparison of foliar and root metal uptake. *J Hazard Mat*, 325: 36–58.

Shahid M, Dumat C, Pourrut B, Abbas G, Shahid N & Pinelli E. 2015. Role of metal speciation in lead-induced oxidative stress to Vicia faba roots. *Russ J Plant Physiol*. doi: 10.1134/S1021443715040159.

Shahid M, Khalid S, Abbas G, Shahid N, Nadeem M, Sabir M, Aslam M & Dumat C. 2015. Heavy metal stress and crop productivity. In: Crop Production and Global Environmental Issues. In: Hakeem K (ed.), Springer, Cham. (pp 1–25).

Shahid M, Pourrut B, Dumat C, Nadeem M, Aslam M & Pinelli E. 2014. Heavy-metal-induced reactive oxygen species: Phytotoxicity and physicochemical changes in plants. *Rev Environ Contam Toxicol*, 232: 1–44.

Sharma A, Kapoor D, Wang J, Shahzad B, Kumar V, Bali AS, Jasrotia S, Zheng B, Yuan H & Yan D. 2020. Chromium bioaccumulation and its impacts on plants: An Overview. *Plants*, 9(1): 100.

Sharma A, Sidhu GP, Araniti F, Bali AS, Shahzad B, Tripathi DK, Brestic M, Skalicky M & Landi M. 2020. The role of salicylic acid in plants exposed to heavy metals. *Molecules*, 25(3): 540.

Sharma P, Jha AB, Dubey RS & Pessarakli M. 2012. Reactive oxygen species, oxidative damage, and antioxidative defense mechanism in plants under stressful conditions. *J Bot*, 2012: 26.

Sheng X, Xia J, Jiang C, He L & Qian M. 2008. Characterization of heavy metal-resistant endophytic bacteria from rape (*Brassica napus*) roots and their potential in promoting the growth and lead accumulation of rape. *Environ Pollut*, 156(3): 1164–1170.

Shukla A, Srivastava S & Suprasanna P. 2017. Genomics of metal stress-mediated signalling and plant adaptive responses in reference to phytohormones. *Currgenom*, 18(6): 512–522.

Sinegovskaya VT, Terekhova OA, Lavrent'yeva SI, Ivachenko LE & Golokhvast KS. 2020. Effect of heavy metals on oxidative processes in soybean seedlings. *Russ Agricult Sci*, 46(1): 28–32.

Smeets K, Opdenakker K, Remans T, Forzani C, Hirt H, Vangronsveld J & Cuypers A. 2013. The role of the kinase OXI1 in cadmium- and copper-induced molecular responses in *Arabidopsis thaliana*. *Plant Cell Environ*, 36(6): 1228–1238.

Smiri M, Jelali N & El Ghoul J. 2013. Cadmium affects the NADP-thioredoxin reductase/thioredoxin system in germinating pea seeds. *J Plant Interact*, 8(2): 125–133.

Srivastava S, Srivastava AK, Suprasanna P & D'Souza SF. 2013. Identification and profiling of arsenic stress-induced microRNAs in *Brassica juncea*. *J Exp Bot*, 64(1): 303–315.

Srivastava V, Sarkar A, Singh S, Singh P, de Araujo AS & Singh RP. 2017. Agroecological responses of heavy metal pollution with special emphasis on soil health and plant performances. *Front Environ Sci*, 5: 64 .doi: 10.3389/fenvs.2017.00064.

Štolfa I, Pfeiffer TŽ, Špoljarić D, Teklić T & Lončarić Z. 2015. *Heavy Metal-Induced Oxidative Stress in Plants: Response of the Antioxidative System. Reactive Oxygen Species and Oxidative Damage in Plants under Stress.* (pp 127–163). Springer, Cham.

Sun H, Shamy M & Costa M. 2013. Nickel and epigenetic gene silencing. *Genes*, 4(4): 583–595.

Sytar O, Kumari P, Yadav S, Brestic M & Rastogi A. 2019. Phytohormone priming: Regulator for heavy metal stress in plants. *J Plant Growth Regul*, 38(2): 739–752.

Thapa G, Sadhukhan A, Panda SK & Sahoo L. 2012. Molecular mechanistic model of plant heavy metal tolerance. *Biometals*, 25(3): 489–505.

Thounaojam TC, Panda P, Choudhury S, Patra HK & Panda SK. 2014. Zinc ameliorates copper-induced oxidative stress in developing rice (*Oryza sativa* L.) seedlings. *Protoplasma*, 251(1): 61–69.

Triantaphylidès C & Havaux M. 2009. Singlet oxygen in plants: Production, detoxification and signaling. *Trends Plant Sci*, 14(4): 219–228.

Truta EC, Gherghel DN, Bara ICI & Vochita GV. 2013. Zinc-induced genotoxic effects in root meristems of barley seedlings. *Not Bot Horti Agrobot Cluj Napoca*, 1: 150–156.

Upadhyay A, Singh N, Singh R & Rai U. 2016. Amelioration of arsenic toxicity in rice: Comparative effect of inoculation of *Chlorella vulgaris* and Nannochloropsis sp. on growth, biochemical changes and arsenic uptake. *Ecotoxicol Environ Saf*, 124: 68–73.

Vardhan KH, Kumar PS & Panda RC. 2019. A review on heavy metal pollution, toxicity and remedial measures: Current trends and future perspectives. *J Mol Liq*, 290. doi: 10.1016/j.molliq.2019.111197.

Ventrella A, Catucci L, Piletska E, Piletsky S & Agostiano A. 2009. Interactions between heavy metals and photosynthetic materials studied by optical techniques. *Bioelectrochemistry*, 77(1): 19–25.

Wakeel A, Xu M & Gan Y. 2020. Chromium-induced reactive oxygen species accumulation by altering the enzymatic antioxidant system and associated cytotoxic, genotoxic, ultrastructural, and photosynthetic changes in plants. *Int J Mol Sci*, 21(3): 728.

Wang J & Chen C. 2009. Biosorbents for heavy metals removal and their future. *Biotechnol Adv*, 27(2): 195–226.

Welch RM & Shuman L. 1995. Micronutrient nutrition of plants. *Crit Rev Plant Sci*, 14(1): 49–82.

Wilkins KA, Matthus E, Swarbreck SM & Davies JM. 2016. Calcium-mediated abiotic stress signaling in roots. *Front Plant Sci*, 7: 1296. doi: 10.3389/fpls.2016.01296.

Xia XJ, Zhou YH, Shi K, Zhou J, Foyer CH & Yu JQ. 2015. Interplay between reactive oxygen species and hormones in the control of plant development and stress tolerance. *J Exp Bot*, 66(10): 2839–2856.

Xiong TT, Leveque T, Austruy A, Goix S, Schreck E, Dappe V, Sobanska S, Foucault Y & Dumat C. 2014. Foliar uptake and metal (loid) bioaccessibility in vegetables exposed to particulate matter. *Environ Geochem Health*, 36(5): 897–909.

Yeh CM, Chien PS & Huang HJ. 2007. Distinct signalling pathways for induction of MAP kinase activities by cadmium and copper in rice roots. *J Exp Bot*, 58(3): 659–671.

Yruela I. 2013. Transition metals in plant photosynthesis. *Metallomics*, 5(9): 1090–1109.

Yuan HM, Xu HH, Liu WC & Lu YT. 2013. Copper regulates primary root elongation through PIN1-mediated auxin redistribution. *Plant Cell Physiol*, 54(5): 766–778.

Yusuf M, Fariduddin Q, Varshney P & Ahmad A. 2012. Salicylic acid minimizes nickel and/or salinity-induced toxicity in Indian mustard (*Brassica juncea*) through an improved antioxidant system. *Environ Sci Pollut Res Int*, 19(1): 8–18.

Zelinová V, Alemayehu A, Bocová B, Huttová J & Tamás L. 2015. Cadmium-induced reactive oxygen species generation, changes in morphogenic responses and activity of some enzymes in barley root tip are regulated by auxin. *Biologia*, 70(3): 356–364.

Zeng H, Xu L, Singh A, Wang H, Du L & Poovaiah BW. 2015. Involvement of calmodulin and calmodulin-like proteins in plant responses to abiotic stresses. *Front Plant Sci*, 6: 600.

Zhao L, Wang P, Hou H, Zhang H, Wang Y, Yan S, Huang Y, Li H, Tan J, Hu A, Gao F, Zhang Q, Li Y, Zhou H, Zhang W & Li L. 2014. Transcriptional regulation of cell cycle genes in response to abiotic stresses correlates with dynamic changes in histone modifications in maize. *PLOS ONE*, 9(8): e106070.

Zhou J-M, Jiang Z-C, Qin X-Q, Zhang L-K, Huang Q-B & Xu G-L. 2020. Effects and Mechanisms of calcium ion addition on lead removal from water by *Eichhorniacrassipes*. *Int J Environ Res Public Health*, 17(3): 928.

8 Arsenic and Oxidative Stress in Plants
Induction and Scavenging

Shyna Bhalla and Neera Garg

CONTENTS

8.1 INTRODUCTION

Rapid changes in environmental conditions have been observed in recent years, which are believed to have originated from various natural and anthropogenic factors. These factors fluctuate in their intensity as well as the duration and result in the emission of heavy metals (HMs) (Schutzendubel and Polle 2002). HMs are metals having a density greater than 5 g/mL and/or molecular weight over 20 g/mol (Kim et al. 2015). Exposure to HMs contaminates the soil because of their non-degradable nature, which subsequently affects growth and crop production. The severity of contaminating the plants depends on various factors like solubility of metal, its transport and chemical reactivity as well as pH of the medium. Among different HMs, some act as essential nutrients (Manganese-Mn, Zinc-Zn, Iron-Fe, Nickel-Ni, Molybdenum-Mo and Copper-Cu), while others, like arsenic (As), do not possess such function and prove to be toxic to plants even at low concentrations (Nies 1999).

    The metalloid As is a class I carcinogen having characteristics similar to metals (e.g. Bismuth) as well as non-metals (e.g. Nitrogen-N and Phosphorus-P) (Kashyap and Garg 2018).

The contamination by As might result from the continuous use of incorrect watering practices, phosphate fertilizers and industrial wastes. The World Health Organization (WHO) has permitted a threshold of 10 µg/L of As in drinking water (Hettick et al. 2015) and 20 mg/kg in agricultural soils, however, its increasing concentrations in the rhizosphere alter normal physiological and metabolic processes of plants. As is present in different oxidation states, i.e. -3, 0, +3 and +5, in the environment (Sharma et al. 2014). This helps classify As into inorganic and organic forms. As is predominant and more toxic in the inorganic form, where it occurs as arsenate-AsV and arsenite-AsIII, than the organic forms like arsenocholine, arsenobetaine, monomethylarsonic acid (MMA) and dimethylarsinic acid (DMA) (Garg and Singla 2011). This is because of the high solubility and mobility of AsV and AsIII than organic forms. However, AsIII has been regarded as more toxic, as its uptake and translocation are faster than AsV (Garg and Kashyap 2017). Moreover, these two inorganic forms are interconvertible as well as more accessible to plants than organic species and their availability to plants depend on soil pH, organic matter content, redox conditions of the environment and plant species (Sharma et al. 2014). AsV is prevalent in aerated soils with moderate to high redox potentials, whereas AsIII predominantly occurs in non-aerated soils with low redox potential (Danh et al. 2014). Plants absorb As from soil or take up with the help of transporters. AsV, being chemically like phosphate, competes for the same transporters and gets transported with the help of phosphate transporters (PHT1) across the root plasma membrane. In contrast, AsIII uptake in plants is mediated by NIPs-Nodulin 26-like Intrinsic membrane Proteins, belonging to aquaporins (Garg and Kashyap 2017).

Increasing concentrations of As in soils is the major constraint to normal morphological, physiological as well as biochemical processes of plants (Campos et al. 2014). Plants growing in As-contaminated areas exhibit symptoms like chlorosis, stunted plant growth, wilting, necrosis, reduced growth and yield (Campos et al. 2014). Plant roots are the primary site to come in contact with As, where As causes plasmolysis of roots, inhibition of lateral root growth as well as nutrient acquisition and reduction in stellar growth (Garg and Kashyap 2017). Once inside the cells, As disturbs oxidative phosphorylation and inhibits ATP synthesis (Danh et al. 2014). Furthermore, it binds with sulfhydryl groups (-SH), causing dysfunction of proteins and enzymes (Gunes et al. 2009). Other damaging effects are shown in the form of reduced membrane integrity, photosynthesis as well as respiration (Kashyap and Garg 2018). However, the dominant factor determining the toxicity of As is the generation of oxidative stress, which results from the production of reactive oxygen species (ROS) such as superoxide radical ($O_2^{\bullet-}$), hydroxyl radical ($\bullet OH$), hydrogen peroxide (H_2O_2) and singlet oxygen ($^1O_2$) in different cellular organelles (Garg and Manchanda 2009; Tripathi et al. 2012b). These ROS react with cellular components, thus lead to lipid peroxidation, oxidative destruction of proteins as well as DNA (Yadav and Srivastava 2015), increased malondialdehyde-MDA (Siddiqui et al. 2015) and H_2O_2, (Myrene and Devaraj 2013), chlorophyll damage (Sharma et al. 2012) and depletion of reduced glutathione (GSH) (Schutzendubel and Polle 2002). Therefore, they needed to be scavenged to ameliorate their damaging effects on plants. However, ROS is also beneficial to plants at low levels because it is used in the signal transduction pathway as a second messenger for diverse plant physiological processes (Schutzendubel and Polle 2002).

Plants operate a complex mechanism to regulate the production and accumulation of ROS. The mechanism involved in maintaining the level of ROS includes antioxidant biomarkers network comprising of SOD-superoxide dismutase, CAT-catalase, APX-ascorbate peroxidase, GR-glutathione reductase, POX-peroxidase, glutathione, tocopherol, ascorbate, etc. (Schutzendubel and Polle 2002; Myrene and Devaraj 2013). However, plant exposure to severe As-stressed conditions disturbs the activity of antioxidants and they become unable to cope up with increased levels of ROS. Therefore, there is a demanding need for some remedial strategies which could help in lowering the levels of ROS in plants. The chapter, thus, discusses the mechanism of ROS generation in the presence of As and scavenging of these ROS with the help of possible scavenging systems.

8.2 TYPES OF ROS

In its ground state, molecular O_2 is almost unreactive. However, its reduction during electron transfer reactions allows it to produce reactive species including $O_2^{\bullet-}$, H_2O_2, $^1O_2$ and $\bullet OH$ (Kar 2015). Thus, ROS is a term used for the radical and non-radical species which are produced by partial reduction of oxygen (O_2) (Corpas et al. 2015). ROS are highly unstable due to the presence of unpaired electrons (Kostecka-Gugała and Latowski 2018) and different ROS forms may be linked by a cascade of reactions where one form gets converted to another (Halliwell and Gutteridge 1989).

During the process of photosynthesis, inadequate energy dissipation leads to the formation of triplet chlorophyll state which transmits its excitation energy to ground state molecular O_2 and results in the formation of $^1O_2$ (Holt et al. 2005). $^1O_2$ has its outermost electron pair with antiparallel spins ($\uparrow\downarrow$) and is highly reactive as compared to O_2 (Garg and Manchanda 2009; Tripathy and Oelmüller 2012). The energy needed in the excitation of O_2 to form $^1O_2$ is 94 kJ mol^{-1} (Skovsen et al. 2005). The lifetime of $^1O_2$ in water is 4 μs whereas, in non-polar environments, it lasts for 100 μs (Foyer and Harbinson 1994). However, at room temperature, it can last for up to an hour (Saed-Moucheshi et al. 2014). It can diffuse into the cytosol from chloroplasts and has a diffusion rate of approximately 270 nm (Skovsen et al. 2005).

$O_2^{\bullet-}$ radical anion or simply $O_2^{\bullet-}$ radical or $O_2^{\bullet-}$ is an O_2 with one unpaired electron (Garg and Manchanda 2009). It is formed by the reduction of one electron from dioxygen O_2 (Edreva 2005). $O_2^{\bullet-}$ is short-lived with a half-life of approx. 2–4 μs which allows it to diffuse only up to a few micrometers from its production site (Kavdia 2006; Garg and Manchanda 2009). Due to this reason, it is not able to cross biological membranes (Garg and Manchanda 2009) and participates in several reactions, resulting in more dangerous species, e.g. hydroperoxyl radical $HOO\bullet-$ (Demidchik 2015). This $HOO\bullet-$ crosses the membrane and causes auto-oxidation of lipids (Halliwell and Gutteridge 1989). It also reduces Cu^{2+} and Fe^{3+}, which react with H_2O_2 and produce $\bullet OH$ (Pryor and Squadrito 1995).

The moderately reactive ROS is H_2O_2. It is not a free radical as it contains all paired electrons and results from peroxisomal H_2O_2-producing enzymes as well as enzymatic dismutation of $O_2^{\bullet-}$ (Garg and Manchanda 2009; Mishra and Sharma 2019). It is less damaging, most stable and relatively long-lived with a half-life of 1 ms but can form species like $\bullet OH$ and $HOO\bullet-$ in the presence of metal ions, which cause damage to cellular biomolecules (Garg and Manchanda 2009; Saed-Moucheshi et al. 2014). Its lifetime never exceeds 1 s because of the presence of peroxidases and catalases in its environment (Demidchik 2015). Being a weak oxidant, H_2O_2 is not able to completely modify the biomolecules, but it may interact directly with SH-groups and convert binding sites of transition metals to $\bullet OH$ which causes damaging effects on organic substrates (Saed-Moucheshi et al. 2014).

The most reactive of all the ROS is $\bullet OH$ which reacts with biomolecules because cells are not equipped with any enzymatic mechanism for its elimination (Halliwell and Gutteridge 1989). $\bullet OH$ is produced from H_2O_2 by Fenton reaction(s) which were proposed in the 1930s by Fritz Haber and Joseph Weiss, hence called Haber-Weiss reaction (Haber and Weiss 1932; Halliwell and Gutteridge 1989). $\bullet OH$ is short-lived with a half-life of about 1 ns inside cells, diffuses to <1 nm distance, and this distance limits its interaction with biomolecules (Saed-Moucheshi et al. 2014). During oxidative stress, it is the prime cause of damage to lipids, proteins and DNA (Demidchik 2015). Hence, its production in excess ultimately leads to programmed cell death (Karuppanapandian et al. 2011).

8.3 GENERATION SITES OF ROS

The partially reduced or activated derivatives of O_2 are produced continuously in the form of unwanted by-products as a result of normal metabolic processes occurring in different plant organelles. The major contributors to ROS production are chloroplasts, mitochondria and peroxisomes.

8.3.1 CHLOROPLASTS

In the presence of light, chloroplasts are the most susceptible cell organelles for ROS attack due to their abundance and participation in O_2 production during photosynthesis (Corpas et al. 2015). The photosynthetic compartments of the chloroplast, especially photosystem-I, serve as the main site of generating $O_2^{\bullet-}$ with the help of ferredoxin-NADP$^+$ reductase (Edreva 2005). The electron transport chain (ETC) of the chloroplast thylakoid releases electrons which reduces NADP$^+$, and the resultant NADPH enters Calvin cycle to reduce CO_2 for the production of O_2 (Edreva 2005). In addition, protein Q_A, pheophytin and reaction center of photosystem-II (P680) also contribute to the production of $O_2^{\bullet-}$ and $\bullet$OH (Navari-Izzo et al. 1999). During light conditions, water autolysis in the photosystem-II provides O_2, and electrons from photosystem-I ETC reduces it to form $O_2^{\bullet-}$ (Corpas et al. 2015). In the process, reduced ferredoxin (Fd_{red}) gets converted into oxidized form (Fd_{ox}), but under excessive Fd_{red} conditions it reacts with $O_2^{\bullet-}$ and results in the formation of H_2O_2 (Corpas et al. 2015). The H_2O_2 thus produced accumulates in the thylakoids but not in the intact chloroplasts. Besides this, the excitation of O_2 in photosystem-II from triplet state to singlet state produces singlet oxygen ($^1O_2$) (Corpas et al. 2015). Photosynthetic inhibitors also promote the production of radicals. For example, paraquat inhibits photosystem-I and generates $O_2^{\bullet-}$, whereas DCMU [3-(3,4-dichlorophenyl)-1,1-dimethylurea] excites the chlorophyll molecules to form $^1O_2$ by uncoupling electron fluxes of photosystem-II (Corpas et al. 2015).

8.3.2 MITOCHONDRIA

The ETC of mitochondria is the main source of ROS generation because of unavoidable impairments at the inner mitochondrial membranes. Out of total O_2 consumed in the mitochondria, about 2–5% is used in the formation of $O_2^{\bullet-}$ (Corpas et al. 2015). $O_2^{\bullet-}$ is reported to be formed in the complex I and complex III of mitochondrial ETC (Halliwell and Gutteridge 2015) and in complex II under certain conditions of reverse electron transport (Andreyev et al. 2005). In other studies, the use of inhibitors of ETC complexes I and III (rotenone and antimycin, respectively) confirmed the production of the radicals in these complexes (Corpas et al. 2015). Moreover, in complex III, the reduced state of the ubiquinone pool provides electrons to molecular oxygen for the subsequent formation of $O_2^{\bullet-}$ (Navrot et al. 2007). Manganese-containing superoxide dismutase (MnSOD) proficiently converts almost all the $O_2^{\bullet-}$ to H_2O_2, which then takes part in the Fenton reaction and produces $\bullet$OH by reacting with copper or iron ions (Navrot et al. 2007). Mitochondria of pea stem showed the formation of H_2O_2 by oxidation of complex I and II, however, the generation was found to be more in the case of complex II substrates oxidation (Braidot et al. 1999). Mitochondria also produce $^1O_2$ during the Haber-Weiss mechanism (del Río and López-Huertas 2016).

8.3.3 PEROXISOMES

The metabolic pathways occurring in the peroxisomes also result in the production of ROS. Peroxisomes are single membrane cellular organelles and are the chief sites of H_2O_2 production (del Río and López-Huertas 2016). Various processes including glyoxylate cycle, ureides metabolism, β-oxidation of fatty acids, photorespiration cycle, presence of acyl CoA oxidase, flavin oxidases, xanthine oxidase, dismutation of $O_2^{\bullet-}$ as well as catalases contribute to H_2O_2 generation (Corpas et al. 2015). Depending on the tissue origin and organisms, peroxisomes contain different H_2O_2-producing flavin oxidases including xanthine oxidase (XOD), glycolate oxidase (GOX), polyamine oxidase, sulfite oxidase, diamine oxidase and sarcosine oxidase (del Río and López-Huertas 2016). During β-oxidation process, acyl-CoA gets converted into trans-2-enyl-CoA with the release of H_2O_2 (Corpas et al. 2015). It has been reported that peroxisomal GOX produces about 2- and 50- times more H_2O_2 as compared to chloroplasts and mitochondria, respectively (del Río and López-Huertas 2016). $^1O_2$ is also reported to be formed in peroxisomes. Another enzyme, xanthine oxidoreductase

(XOR), converts the xanthine and hypoxanthine to uric acid and as a result, releases $O_2^{\bullet-}$. In addition, the small ETC of the peroxisomal membrane presents another source of $O_2^{\bullet-}$ formation, which uses NADH as an electron donor (Corpas et al. 2015). Peroxisomes reported from pea leaves confirmed the existence of $O_2^{\bullet-}$ formation sites in the matrix as well as organelle membrane (del Río et al. 2002). Urate oxidase of the mitochondrial matrix also produces $O_2^{\bullet-}$ with intensity lesser than that produced by XOD of peroxisomes (del Río and López-Huertas 2016).

Besides this, some other cellular components also produce ROS (Kostecka-Gugała and Latowski 2018). Cytochrome P_{450} in the endoplasmic reticulum (ER) catalyzes the detoxification reactions, which result in the generation of $O_2^{\bullet-}$ (Sharma et al. 2012). RH-, the organic substrate, reacts with Cytochrome P_{450} and gets reduced to form an intermediate (Cytochrome P_{450}R−), which can react with triplet oxygen and forms Cytochrome P_{450}-ROO−. This complex decomposes and produces $O_2^{\bullet-}$ with the action of Cytochrome b (Sharma et al. 2012). Also, ROS generation occurs at the plasma membrane, cell wall as well as in apoplast (Gill and Tuteja 2010). Plasma membranes of etiolated soybean and maize roots were reported to have $O_2^{\bullet-}$. In soybean, the production of $O_2^{\bullet-}$ was attributed to two enzymes – NADPH oxidase and quinone reductase (menadione) (Heyno et al. 2011). Under stressed conditions, NADPH plays an important role in the generation as well as the accumulation of ROS (Sharma et al. 2012). Cell wall peroxidases and amine oxidases produce H_2O_2 in the apoplast (Gill and Tuteja 2010). Cell wall peroxidases have also been reported to generate •OH in the apoplast (Heyno et al. 2011). Diamine oxidases produce ROS in the cell wall using diamine and different polyamines which reduce a quinone and result in the formation of peroxides (Sharma et al. 2012).

8.4 ROS AND ITS EFFECTS UNDER AS TOXICITY

Although ROS are produced continuously as a result of normal metabolic pathways in plants, their production increases in response to As exposure, which causes irreversible damage to cellular macromolecules and compartments. The increased production of ROS by As may be due to its ability to displace the cation binding sites as well as inactivation of key enzymes that are involved in chloroplast and mitochondrial ETC, along with inhibition of biosynthetic pathway of essential molecules involved in cellular homeostasis (Mishra et al. 2019). However, the major contributor to ROS generation under As stress could be the conversion of AsV to AsIII (Chandrakar et al. 2016). There are many pieces of evidence pointing towards the involvement of As in enhancing ROS production.

The levels of H_2O_2 greatly increased under As exposure in some plant species including grass *Holcus lanatus*, *Pteris ensiformis*, mung bean and pigeonpea (Hartley-Whitaker et al. 2001; Singh et al. 2006; Garg and Kashyap 2019). Besides this, enhanced production of $O_2^{\bullet-}$ has been reported by the action of NAD(P)H oxidase during As mediated upregulation of p22phax subunit (Hunt et al. 2014). Garg and Kashyap (2019) also reported an oxidative burst because of increased $O_2^{\bullet-}$ levels in As-treated *Cajanus cajan*. A similar rise in ROS production has been reported in rice (Choudhury et al. 2011), *Zea mays* (Mallick et al. 2011), chickpea (Gunes et al. 2009), *Vigna radiata* (Upadhyaya et al. 2014), *Spinacia oleracea* (Shahid et al. 2017), *Triticum aestivum* (Hasanuzzaman and Fujita 2013) and *Myracrodruon urundeuva* (Gomes et al. 2013) under As exposure.

Increase in ROS production leads to a decrease in membrane stability, respiration, photosynthesis as well as redox balance (Kashyap and Garg 2018). Increased membrane disintegration as a result of As-induced oxidative stress has been reported in pigeonpea and pea plants (Garg et al. 2015). There are also reports where As disrupts the energy balance in the chloroplasts (Srivastava et al. 2013). Under As-stress, ROS affect photosynthetic metabolism, which ultimately decreases the net photosynthetic rates (Vinit-Dunand et al. 2002). Exposure to As has been reported to impair thylakoid ETC which resulted in abnormal chloroplasts (Farnese et al. 2017). The normal pathway of $NADP^+$ reduction has also been reported to be obstructed by AsV stress in rice seedlings, which resulted in the overproduction of ROS (Mishra et al. 2017). A remarkable decline in chlorophyll pigment synthesis was reported in As-stressed maize (Emamverdian et al. 2015), lettuce (Suneja 2014)

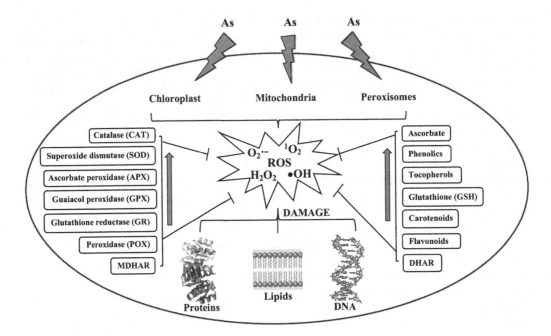

FIGURE 8.1 Induction and scavenging of ROS: As exposure leads to generation of ROS in different plant organelles, which cause damage to cellular components including proteins, lipids and DNA. Consequently, the inherent antioxidative system activates and counterbalances the effects of increased ROS by inhibiting its excessive production.

and red clover (Wang et al. 2011). Moreover, AsV replaces the Pi (inorganic phosphate) from key processes and disrupts the production of ATP, thus lowers the energy levels of the plant cell (as reviewed by Abbas et al. 2018). Besides this, ROS also lead to lipid peroxidation, damaged proteins and changed DNA structure (Møller et al. 2007) (Figure 8.1).

8.4.1 As Stress and Lipid Peroxidation

Peroxidation of lipids has been reported to be induced in As exposed *Pteris vittata* (Singh et al. 2006). Lipid peroxidation is estimated on the scale of MDA content and electrolyte leakage (EL) (Kostecka-Gugała and Latowski 2018). Gupta et al. (2013) reported an increased MDA content in *Arabidopsis* leaves under five day As exposure. In the studies carried out with *Ceratophyllum demersum* (Mishra et al. 2008), *Trigonella foenumgraecum* (Talukdar 2013), *Bacopa monnieri* (Mishra et al. 2013), *Ocimum tenuiflorum* (Siddiqui et al. 2015), soybean (Chandrakar et al. 2016), maize (Anjum et al. 2016), sunflower (Saidi et al. 2017), tomato (Marmiroli et al. 2017), peanut (Bianucci et al. 2017) and *Artemisia annua* (Kumari et al. 2018), increased levels of both MDA and EL were observed when the plants were treated with As. Recently, an increase in lipid peroxidation by-product, TBARS (thiobarbituric acid reactive substances) along with MDA, has been reported in *Vetiveria zizanoides* shoots and roots treated with As (Singh et al. 2017 b). The enhanced EL and MDA content shows the extent of membrane damage (Mishra et al. 2019).

8.4.2 As Induced Protein Damage

The by-products formed as a result of lipid peroxidation damage proteins (Yamauchi et al. 2008). Besides this, proteins can be damaged by modulation in their activity through reactions like carbonylation, glutathionylation, nitrosylation and disulfide bond formation (Yamauchi et al. 2008). All these activities result in amino acid modification, protein adducts formation and increase the susceptibility of protein to peptide bond breakage (Møller and Kristensen 2004). There are reports which show that As stress results in protein carbonylation, and thus forms carbonyls (Parkhey et al. 2014). Under normal conditions, hydrolysis of proteins to amino acids occurs with the help of proteases, but stressed conditions in mung bean and red clover showed the decreased activity of proteases which ultimately suppressed the growth of plants (Ismail 2012). Oxidation of amino acids as a result of As induced ROS generation also causes the release of carbonyl groups which alters or inhibits the activities of proteins (Rajjou et al. 2008). Moreover, thiol groups and sulfur-containing amino acids are more prone to attack due to the ability of AsIII to bind with their sulfhydryl (-SH) groups. As-induced damaging effects on proteins have been reported in *Pteris* sp. (Singh et al. 2006), maize (Stoeva and Bineva 2003), rice (Singh et al. 2009), mung bean (Ismail 2012) and wheat (Liu et al. 2005). (See Figure 8.1.)

8.4.3 As Effects on DNA

Increased accumulation of ROS under As-exposure results in DNA base modifications, deletion of nucleotides, breakage in back-bone, substituted pyrimidinic or purinic sites as well as the exchange of sister chromatids (Mucha et al. 2019). The base modification results in the release of urea, hydroxymethyl urea, thymine glycol and 8-hydroxyguanine (8-OHdG). Among these, 8-OHdG is a highly mutagenic compound that results in G:C to T:A transversion mutations (Okamoto et al. 2008). This modification in G:C sites is indicative of ROS attack on DNA (Sharma et al. 2012). ROS also induce DNA methylation, telomere dysfunction, mitotic arrest, alteration in gene expression and apoptosis (Rao et al. 2017). *Pisum sativum* treated with As showed damaged mitotic microtubules that resulted in impaired growth (Dho et al. 2010). The changes in DNA structure under stressed conditions have been reported in faba bean and maize (Duquesnoy et al. 2010). All these reactions lead to inhibited growth, lowered biomass, yield and ultimately to cell death (Talukdar 2011).

8.5 SCAVENGING OF ROS

Under low levels of stress, ROS produced in plant cells are maintained at steady levels by back-and-forth actions of ROS generating and scavenging systems (Gupta et al. 2013). However, during the commencement of As stress, the level of ROS increases inside the plant body and damages the cellular components. Therefore, in order to ameliorate the damaging effects posed by ROS on the physiology of plants, they are needed to be harmonically dismutated (Tripathi et al. 2012b). ROS effects are alleviated with the help of the plant's inherent antioxidative enzymatic as well as non-enzymatic mechanisms. The occurrence of different mechanisms involved in the scavenging of ROS is usually near the site of ROS production and they work in a coordinated manner (Sharma et al. 2012). Different antioxidants with their ameliorative mechanisms are given in Table 8.1.

8.5.1 Enzymatic Antioxidants

The earliest induced action of plants in response to ROS is the activation of SOD (Gechev et al. 2006). SOD is a ubiquitous metalloenzyme and catalyzes the conversion of $O_2^{\bullet-}$ to H_2O_2 and O_2 (Sharma 2012; Mishra and Sharma 2019). SOD occurs in association with metal cofactors in the form of Cu/Zn-SOD, Fe-SOD and Mn-SOD (Garg and Manchanda 2009; Sharma 2012). They are distributed in the plant cells with localization of Cu/Zn-SOD in chloroplasts and cytosol, Fe-SOD in plastids, Mn-SOD in peroxisomes and mitochondria (Sharma 2012). Ni-SOD has also been reported from *Streptomyces* (Kim et al. 1996). The upregulation of SOD activity in response to stress plays an important role in plant survival (Tiwari and Sarangi 2015). The decline in lipid peroxidation due to stimulation of SOD action has been reported in As-exposed mung bean (Singh et al. 2007), *Z. mays* (Requejo and Tena 2005) and *H. lanatus* (Hartley-Whitaker et al. 2001). The function of SOD in minimizing As concentrations was evidenced from the fronds of *P. vittata*, as its activity was higher

TABLE 8.1

Antioxidants along with the Reactions They Catalyze or Their Functions

Enzymatic Antioxidants	Reactions Catalyzed/Functions
Superoxide dismutase (SOD)	$2O_2^{\cdot-} + 2H^+ \rightarrow H_2O_2 + O_2$
Catalase (CAT)	$H_2O_2 \rightarrow H_2O + {}^1/_2\,O_2$
Peroxidases (POXs)	$H_2O_2 + 2e^- \rightarrow H_2O$
Non-Enzymatic Antioxidants	**Reactions Catalyzed/Functions**
Carotenoids	Quenches ${}^1O_2$ and dissipates excess energy as heat
Tocopherol	Protects against photo-oxidation and quenches ${}^1O_2$
Phenolics	Interferes with lipid peroxidation
Antioxidants of Ascorbate-Glutathione Cycle	**Reactions Catalyzed/Functions**
Ascorbate Peroxidase (APX)	$H_2O_2 + 2AsA \rightarrow H_2O + 2MDHA$
Ascorbate	Detoxifies H_2O_2 via action of APX
Monodehydroascorbate reductase (MDHAR)	$MDHA + NADPH \rightarrow AsA + DHA + NADP^+$
Dehydroascorbate reductase (DHAR)	$DHA + 2GSH \rightarrow AsA + GSSG$
Glutathione	Acts as substrate for carrying out reduction of DHA
Glutathione Reductase (GR)	$GSSG + 2NADPH \rightarrow 2GSH + 2NADP^+$
Guaiacol Peroxidase (GPX)	$2GSH + H_2O_2 \rightarrow GSSG + 2H_2O$

(46–61%) in treated plants than controls (Srivastava et al. 2005; Singh et al. 2010). Shri et al. (2009) reported an up-regulation of Cu/Zn-SOD activity in As stressed rice seedlings. *Panax notoginseng* showed increased SOD activity under As exposure (Zu et al. 2016). The activity of SOD was found to be increased up to 42% in rice seedlings exposed to 25 µM AsV (Mishra et al. 2017). Yadav et al. (2014) noticed an enhanced activity of SOD enzymes in *Helianthus annuus*. However, some have reported differential expression of SOD at varying concentrations of As. In maize plants, Mylona et al. (1998) reported that *Sod* genes expression was increased when exposed to the low concentration of As and decreased at high levels of AsV and AsIII. The reason for decreased SOD activity might be attributed to its inability to fight against high levels of exposed As (Tiwari and Sarangi 2015).

The detoxification reactions catalyzed by SOD result in the enhanced production of H_2O_2, which at higher concentrations can further increase the oxidative damage. To combat this situation, rapid action of H_2O_2 scavenging system is needed (Tiwari and Sarangi 2015). The increase in H_2O_2 levels stimulates the augmented production of detoxifying enzymes such as CAT and POX (Gill and Tuteja 2010). CAT is mainly found in the cytosol, peroxisomes, mitochondria, glyoxisomes and root nodules (Sharma 2012). It is a heme-containing, tetrameric enzyme whose function is to degrade H_2O_2 without involving any electron donor (Tiwari and Sarangi 2015). CAT dismutases H_2O_2 and results in the formation of water and O_2 (Garg and Manchanda 2009; Tiwari and Sarangi 2015). Multiple isozymes of CAT have been reported in plants with their number being two and six in castor bean (Ota et al. 1992) and *Arabidopsis* (Frugoli et al. 1996), respectively. An increase in CAT activity has been reported in several plants in response to stress. Maize showed increased activity of CAT under As exposure (Mylona et al. 1998). Zu et al. (2016) also reported enhancement in CAT activity in As-treated *P. notoginseng* plants. Rice seedlings exposed to even 100 µM As showed 35% increase in CAT activity (Mishra et al. 2017). In contrast, As exposed *Taxithelium nepalense* and mung bean showed a decline in CAT action (Singh et al. 2007). However, Duquesnoy et al. (2010) reported enhanced CAT activity even at high concentrations of both AsV and AsIII in *Z. mays* roots and leaves.

There is a group of heme-containing oxidoreductases in most plants that use H_2O_2 as an electron acceptor for catalyzing several reactions. These are ubiquitously called peroxidases (POXs). The enzyme is stable thermally and is present in soluble as well as covalently/ionically bound forms (Kothari and Gayathri 2018). Under stressed conditions, POXs are the primary ones to alter their activity. These enzymes act on H_2O_2 and give a π-cation radical, porphyrin. In this reaction, H_2O_2 accepts two electrons to get reduced to water and the enzyme itself gets oxidized (Garg and Manchanda 2009). Under As-stress, enhanced activity of POXs was reported in rice, which increased stress tolerance in plants (Shri et al. 2009; Kidwai et al. 2019).

8.5.2 Non-Enzymatic Antioxidants

In the thylakoid membrane of chloroplasts, a fat-soluble, accessory pigment is present that has light-harvesting and antioxidative properties (Sharma 2012). The pigment is carotenoids, a C_{40} isoprenoid (Garg and Manchanda 2009). Carotenoids quench triplet chlorophyll and prevent the formation of $^1O_2$, thus protect the photosynthetic apparatus (Saed-Moucheshi et al. 2014). It also scavenges $^1O_2$ by dissipating excess energy as heat. Carotenoid content has been found to decrease in oats grown in As-treated soils because of AsV mediated inflammation of the thylakoid membrane (Sharma 2012). However, an increase in carotenoid content was reported in As-stressed *P. vittata* (Singh et al. 2006).

Being susceptible to oxidative stress, biological membranes are the primary target of ROS. To minimize the damaging effects, plants contain tocopherols as an essential component of their membranes (Ma et al. 2020). Tocopherols are lipophilic molecules with antioxidative as well as non-antioxidative functions (Caverzan et al. 2016). In plants, four types of tocopherols are known, i.e. α, β, γ and δ (Saed-Moucheshi et al. 2014). The four isomers differ in their antioxidant activity in the order $\alpha > \beta > \gamma > \delta$, which could be attributed to the methylation pattern of the phenolic ring. The presence of three methyl groups is responsible for the highest activity of α-tocopherol (Garg and Manchanda 2009) and their significant existence in chloroplasts helps in protection against photo-oxidation (Saed-Moucheshi et al. 2014). It also quenches $^1O_2$ via charge transfer (Fryer 1992). α-tocopherol scavenges peroxyl radicals of lipid bilayers (Caverzan et al. 2016), thus protects from lipid peroxidation (Saed-Moucheshi et al. 2014). Enhanced content of α-tocopherol was reported in wheat when plants were treated with As (Ghosh and Biswas 2017).

The antioxidant activity is also shown by phenolic compounds. Phenolics are abundant secondary metabolites in plants (Grace and Logan 1996) and act as a non-enzymatic antioxidant for scavenging ROS (Tiwari and Sarangi 2015). These compounds have been shown to possess better antioxidant properties as compared to ascorbate and tocopherols (Balasundram et al. 2006). The ability of phenolics to chelate metal ions via terminating Fenton reaction and delocalization of unpaired electrons via chain-breaking could be related to its effective antioxidative activity (Garg and Manchanda 2009; Singh et al. 2017a). These compounds also interfere with lipid oxidation and can directly scavenge ROS. Their accumulation has been reported to increase under As-stressed conditions (Julkunen-Tiito et al. 2015). Flavonoids also act as scavenger molecules for ROS as they identify and neutralize the radicals before they start their action (Tiwari and Sarangi 2015). Flavonoids have been reported to alter lipid packing and fluidity in membranes which resist diffusion of radicals and peroxidation reactions (Arora et al. 2002). They also alleviate the damages posed on chloroplasts by scavenging $^1O_2$ (Das and Roychoudhury 2014). The level of phenols has been reported to increase when the As-treated rice plants were supplemented with selenium (Se) (Chauhan et al. 2017).

8.5.3 Antioxidants of Ascorbate-Glutathione Cycle

It is essential that the amount of H_2O_2 produced in the chloroplasts is reduced, as it may inhibit the enzymes of the Calvin cycle (Sharma 2012). However, the absence of CAT in chloroplasts increased the need of plant cells to adopt some alternative mechanism of H_2O_2 detoxification. The process has been taken up by peroxidase of the ascorbate-glutathione pathway (Sharma 2012). This peroxidase requires a reductant, and ascorbate (AsA) is a well-known reductant that reduces H_2O_2 in chloroplasts (Garg and Manchanda 2009), hence, is known as ascorbate peroxidase (APX). It is a heme-containing enzyme and plays an important role in maintaining the redox status of cells (Tiwari and Sarangi 2015). APX in the presence of two molecules of AsA reduces H_2O_2 to water and two monodehydroascorbate (MDHA) molecules (Sharma 2012). APX comprises different isozymes in chloroplasts (tAPX- thylakoid membrane-bound APX and sAPX- stromal APX), membranes (mAPX- membrane-bound APX), peroxisomes and glyoxisomes, cytosol (cAPX) as well as mitochondria (mitAPX) (De Leonardis et al. 2000). APX shows a higher affinity towards H_2O_2 as compared to other H_2O_2 scavenging enzymes. Srivastava et al. (2005) observed an increase in the activity of APX in different parts of As-treated *P. vittata* and *P. ensiformis*. Enhancement in APX action has also been reported in a variety of other plants including mung bean (Singh et al. 2007) and rice (Shri et al. 2009) exposed to As.

Another enzyme, GPX (Guaiacol peroxidase), also catalyzes H_2O_2 into the water using GSH and converts it to oxidized glutathione (GSSG) (Sharma 2012). GPX is a member of the peroxidase family and exists in the cytoplasm or bound form with the cell wall. Three isozymes of GPX have been reported: glutathione transferases showing GPX activity (GST-GPX), selenium-dependent GPX and non-selenium dependent phospholipid hydroperoxide GPX (PHGPX) (Garg and Manchanda 2009). It serves as an internal defensive tool in counterattacking oxidative stress in plants by detoxifying lipid peroxidation products (Srivastava et al. 2005). Singh et al. (2007) found an increase in GPX activity by approx. 26% at 50 μM of As-treatment over control of *Phaseolus aureus* plants. In As supplemented *P. vittata*, altered GPX activity was observed by Cao et al. (2004). Enhancement in GPX activity under As exposure has also been reported in *Z. mays* (Shri et al. 2009), rice (Mishra et al. 2017) and soybean (Chandrakar et al. 2016).

The most abundant antioxidant molecule found in different ROS generating sites in plants is AsA (Sharma 2012). It plays an important role in catalyzing the detoxification reactions of ROS, as it donates electrons in both enzymatic as well as non-enzymatic reactions (Caverzan et al. 2016). In chloroplasts, the AsA is present in its reduced form and protects photosynthetic machinery by eliminating H_2O_2, $O_2^{\bullet-}$, •OH and by quenching $^1O_2$ (Singh et al. 2006). It also plays an important role in regulating cell cycle and cell division (Caverzan et al. 2016) as well as regenerating α-tocopherols and carotenoids by ascorbate-glutathione pathway (Sharma 2012). It acts as a co-factor of APX enzyme where it converts H_2O_2 into water and MDHA which subsequently produces dehydroascorbate (DHA) and AsA with the help of NADPH as an electron donor (Sharma 2012; Caverzan et al. 2016). Under As mediated oxidative stress, an increase in AsA concentration, as well as AsA/DHA ratio, was more prominent in *P. vittata* than *P. ensiformis* as reported by Singh et al. (2006). Czech (2008) also observed an increase in AsA content in hypocotyls of As-treated cucumber plants.

AsA, the reductant, is maintained in a reduced form via monodehydroascorbate reductase (MDHAR) and dehydroascorbate reductase (DHAR). MDHAR is a flavin adenine dinucleotide (FAD) enzyme that regenerates AsA along with DHA from the short-lived molecule MDHA (Das and Roychoudhury 2014). In this reaction, NADPH serves as an electron donor (Caverzan et al. 2016). Different isoforms of MDHAR have been confined to cytosol, mitochondria, peroxisomes, chloroplasts and glyoxisomes (Sharma et al. 2012). DHAR is a thiol enzyme and uses GSH (reduced glutathione) as an electron donor for catalyzing the reduction of DHA to AsA (Das and Roychoudhury 2014). Overexpression of the enzyme in tobacco, potato and maize increased the AsA content which suggests the role of DHAR in maintaining the cellular AsA pool (Sharma et al. 2012).

Another important molecule of the ascorbate-glutathione pathway is glutathione. It is a low molecular weight, non-enzymatic antioxidant thiol molecule involved in the detoxification of metalloids (Gill and Tuteja 2010; Yadav et al. 2016). Glutathione exists in two forms: GSSG and GSH (Yadav et al. 2016). Under stressed conditions, GSH acts as a substrate for reducing DHA back to AsA (Garg and Manchanda 2009) and itself gets oxidized. The levels of GSH have been observed to increase in the case of metal exposure (Tiwari and Sarangi 2015). *Pteris* sp. grown under As stress showed substantial enhancement in the synthesis of GSH (Kertulis-Tartar et al. 2009). GSH serves as an electron donor in the reduction reaction of AsV, where Arsenate reductase (AR) accepts electrons from GSH and reduces AsV to AsIII (Yadav et al. 2016). Srivastava et al. (2005) observed that As exposure resulted in increased GSH content in plants like *Hydrilla verticillata*. Earlier, Cao et al. (2004) also reported significantly increased GSH content in As exposed Chinese brake fern. Rice seedlings grown under As application have been reported to eliminate oxidative stress when supplemented with GSH (Shri et al. 2009).

GSH level in cells is maintained by the activity of glutathione reductase (GR) (Sharma 2012). GR occurs in chloroplasts, cytosol, mitochondria as well as plastids and peroxisomes (Edwards et al. 1990). It mediates the NADPH-dependent catalytic reduction of GSSG to GSH. Besides this, it also scavenges H_2O_2 inside plant cells (Tiwari and Sarangi 2015). Stimulated GR activity in As-treated *Pteris* sp. (Srivastava et al. 2005) and rice (Shri et al. 2009) showed its importance in reducing the oxidative burden of plants.

Besides the occurrence of the efficient detoxifying antioxidant system, sometimes the effectiveness of such molecules under increased ROS levels gets lowered down. This causes an imbalance between their generation and scavenging. Some remedial strategies are, thus, needed to lower the enhanced levels of ROS in plants.

8.6 STRATEGIES TO COUNTERACT ROS GENERATION

The encounter with abiotic stresses triggers the formation of endogenous signaling molecules which act as secondary messengers in plants (Begara-Morales et al. 2015). However, the exogenous application of such molecules has gained an interest in lowering the production of ROS. Moreover, supplementing the plants with mineral elements also proves to be an important strategy for the remediation of ROS.

8.6.1 NITRIC OXIDE (NO)

NO is a secondary messenger that has gained interest with regard to the regulation of antioxidants during the production of ROS (Hasanuzzaman and Fujita 2013). Being a lipophilic molecule, NO diffuses the cell membranes and mediates the post-translational modifications (PTMs) (Begara-Morales et al. 2015). The exogenous application of NO has shown effective mechanisms in protecting the plants from damaging effects of HMs by decreasing their accumulation as well as by attenuating the generation of oxidative stress (Hasanuzzaman and Fujita 2013). Under stressed conditions, NO modulates the function of the ascorbate-glutathione cycle (Begara-Morales et al. 2015). The key enzymes of the cycle have been reported to be inactivated or activated by NO as these are potential targets of NO-induced PTMs (Begara-Morales et al. 2015). Proteomic analysis identified APX as PTM target in *Arabidopsis* (Lozano-Juste et al. 2011), *Solanum tuberosum* (Kato et al. 2013) and *Citrus aurantium* (Tanou et al. 2012). Also, NO stimulates the synthesis of GSH which regulates the level of methylglyoxal (MG) in plants by serving as a cofactor for the detoxifying enzyme Glyoxalase I, thus enhances the plant's tolerance to oxidative stresses (Hasanuzzaman and Fujita 2013). In fescue (*Festuca arundinacea*) leaves, As stress alleviating effects of NO have been investigated in the form of increased activities of SOD, CAT and APX as well as reduced MDA content, H_2O_2, $O_2^{•-}$ accumulation and EL (Jing et al. 2010). Application of sodium nitroprusside (SNP) also enhanced the AsA and GSH content in As-treated wheat seedlings and maintained a higher GSH/GSSG ratio compared to control (Hasanuzzaman and Fujita 2013). The activities of DHAR, MDHAR, GR and glyoxalases I and II were also reported to be higher in SNP treated wheat seedlings as compared to only As-stressed ones (Hasanuzzaman and Fujita 2013). These studies support the protective and alleviating role of NO during oxidative stress.

8.6.2 SALICYLIC ACID (SA)

Another signaling molecule that is gaining interest for its involvement in the plant's defense is SA. It is an endogenous signal molecule and a promising growth regulator with metal-chelating properties (Chandrakar et al. 2016). It is produced naturally in plants when they encounter stressed conditions and induces systemic resistance (Saidi et al. 2017). However, its exogenous application has also been reported to confer tolerance to plants. There are reports where SA contributes to cellular homeostasis by regulating antioxidant system response. SA is a direct scavenger of •OH, thus inhibits its generation as well as its influencing effects (Saidi et al. 2017). In *Glycine max*, the activities of SOD, CAT, APX and GPX were reported to be higher under As stress and decreased when the plants were supplied with SA (Chandrakar et al. 2016). The co-application of As and SA has been demonstrated to exert more beneficial effects than pre-treatment of SA (Singh et al. 2015). In As-treated *Halianthus annuus*, the activities of both AsA and GSH increased, however, the increase was reduced with the application of SA (Saidi et al. 2017). This suggests the possible role of SA in diminishing the negative effects posed on plants under stressful conditions.

8.6.3 SELENIUM (SE)

Besides the signaling molecules, Se has also been reported to exert beneficial effects on plants (Malik et al. 2012). It exists mainly in the form of selenite (Kumar et al. 2014) and coexists with As in contaminated soils because of its chemical similarity with As (Pandey et al. 2015). Hu et al. (2014) observed As accumulation to be affected by Se in plants. In mung bean plants, the increased endogenous levels of H_2O_2 and MDA during As-induced oxidative stress were decreased by the application of Se as compared to plants with non-Se treatments (Malik et al. 2012). This decrease may be attributed to an increase in antioxidant activities, suppression of lipid peroxidation and elimination of ROS (Han et al. 2015). Similar results were observed in rice by Singh et al. (2018). Furthermore, Se addition to rice maintained the activities of SOD, CAT, GPX and AsA to provide tolerance against As stress (Singh et al. 2018).

8.6.4 SILICON (SI)

Si is another promising element in providing amelioration to oxidative stress. It is a quasi-essential element (Garg and Bhandari 2016; Garg and Singh 2018) which shares a common uptake pathway with AsIII (Garg and Kashyap 2017). Si reduces the As-induced oxidative stress by inducing the activities of antioxidants in plants (Tripathi et al. 2012a). Enhancement in antioxidant activities including MDHAR, DHAR, GR, APX and SOD was observed in As-treated rice seedlings. Si supplementation also decreased lipid peroxidation (Tripathi et al. 2013). Pandey et al. (2016) reported that Si maintained the level of antioxidants for the alleviation of oxidative stress in *Brassica* seedlings. Moreover, Si decreased the production of $O_2^{•-}$ (Pandey et al. 2016). The activities of GSH regenerative metabolite GR was observed to be enhanced in As-treated *Cajanus cajan* supplemented with Si. This maintained the cellular GSH pool and improved GSH/GSSG ratio (Garg and Kashyap 2019). The combined application of Si and Si nano-particles (SiNp) resulted in reduced production of H_2O_2 and $O_2^{•-}$ radicals under AsV stressed maize plants, moreover, Si and SiNp enhanced the activities of all antioxidant enzymes studied, irrespective of the AsV treatment (Tripathi et al. 2016). In terms of non-enzymatic antioxidant molecules, the joint application of Si and SiNp attenuated the diminishing effects of AsV in maize and enhanced the reduced forms of AsA and GSH as well as maintained reduced/oxidized ratios (Tripathi et al. 2016). Similar results were found in rice when Si was applied in the presence of As (Liu et al. 2014).

8.7 CONCLUSION AND FUTURE PERSPECTIVES

Arsenic, being a carcinogen, results in disturbed plant health which ultimately causes harmful effects on human beings. The major damaging effects of As in plants are increased production of ROS that results in dysfunction of cellular compartments as well as biomolecules. Under normal or limited conditions of stress, plants can combat ROS generation with their inherent tolerance mechanisms (antioxidant system). However, under exposure to extreme conditions of stress, the antioxidant system is unable to stand effectively against the increased levels of ROS. Various strategies, such as the use of NO, SA, Se and Si, have been used in recent years to alleviate the damaging effects of ROS as well as As. Further studies need to be focused on understanding the detailed mechanisms which help in modulating the uptake, transport as well as accumulation of As in plant organs.

ACKNOWLEDGMENT

The authors highly acknowledge the financial assistance provided by Council for Scientific and Industrial Research (CSIR), New Delhi and Department of Biotechnology (DBT), Ministry of Science and Technology, Government of India for carrying out related research.

REFERENCES

Abbas G, Murtaza B, Bibi I, Shahid M, Niazi N, Khan MI, Amjad M, Hussain M, Natasha (2018) Arsenic uptake, toxicity, detoxification, and speciation in plants: Physiological, biochemical, and molecular aspects. *Int J Environ Res Public Health* 15(1):59–104

Andreyev AY, Kushnareva YE, Starkov AA (2005) Mitochondrial metabolism of reactive oxygen species. *Biochemistry (Mosc)* 70(2):200–214

Anjum SA, Tanveer M, Hussain S, Shahzad B, Ashraf U, Fahad S, Hassan W, Jan S, Khan I, Saleem MF, Bajwa AA, Wang L, Mahmood A, Samad RA, Tung SA (2016) Osmoregulation and antioxidant production in maize under combined cadmium and arsenic stress. *Environ Sci Pollut Res* 23(12):11864–11875

Arora A, Sairam RK, Srivastava GC (2002) Oxidative stress and antioxidative system in plants. *Curr Sci* 82(10):1227–1238

Balasundram N, Sundram K, Samman S (2006) Phenolic compounds in plants and Agri-Industrial By-Products: Antioxidant activity, occurrence, and potential uses. *Food Chem* 99(1):191–203

Begara-Morales JC, Sánchez-Calvo B, Chaki M, Valderrama R, Mata-Pérez C, Padilla MN, Corpas FJ, Barroso JB (2015) Modulation of the ascorbate–glutathione cycle antioxidant capacity by posttranslational modifications mediated by nitric oxide in abiotic stress situations. In: Gupta DK, Palma JM, Corpas FJ (eds) *Reactive Oxygen Species and Oxidative Damage in Plants Under Stress*. Springer, Cham, pp 305–320

Bianucci E, Furlan A, Tordable M, Hernández LE, Carpena-Ruiz RO, Castro S (2017) Antioxidant responses of peanut roots exposed to realistic groundwater doses of arsenate: Identification of glutathione S-transferase as a suitable biomarker for metalloid toxicity. *Chemosphere* 181:551–561

Braidot E, Petrussa E, Vianello A, Macri F (1999) Hydrogen peroxide generation by higher plant mitochondria oxidizing complex I or complex II substrates. *FEBS Lett* 451(3):347–350

Campos NV, Loureiro ME, Azevedo AA (2014) Differences in phosphorus translocation contributes to differential arsenic tolerance between plants of *Borreria verticillata* (Rubiaceae) from mine and non-mine sites. *Environ Sci Pollut Res* 21(8):5586–5596

Cao X, Ma LQ, Tu C (2004) Antioxidative responses to arsenic in the arsenic-hyperaccumulator Chinese brake fern (*Pteris vittata* L.). *Environ Pollut* 128(3):317–325

Caverzan A, Casassola A, Brammer SP (2016) Reactive oxygen species and antioxidant enzymes involved in plant tolerance to stress. *Embrapa Trigo-Capítulo em Livro Científico (ALICE)*. http://dx.doi.org/10.5772/61368

Chandrakar V, Dubey A, Keshavkant S (2016) Modulation of antioxidant enzymes by salicylic acid in arsenic exposed *Glycine max* L. *J Soil Sci Plant Nutr* 16(3):662–676

Chauhan R, Awasthi S, Tripathi P, Mishra S, Dwivedi S, Niranjan A, Mallick S, Tripathi P, Pande V, Tripathi RD (2017) Selenite modulates the level of phenolics and nutrient element to alleviate the toxicity of arsenite in rice (*Oryza sativa* L.). *Ecotoxicol Environ Safe* 138:47–55

Choudhury B, Chowdhury S, Biswas AK (2011) Regulation of growth and metabolism in rice (*Oryza sativa* L.) by arsenic and its possible reversal by phosphate. *J Plant Interact* 6(1):15–24

Corpas FJ, Gupta DK, Palma JM (2015) Production sites of reactive oxygen species (ROS) in organelles from plant cells. In: Gupta DK, Palma JM, Corpas FJ (eds) *Reactive Oxygen Species and Oxidative Damage in Plants under Stress*. Springer, Cham, pp 1–22

Czech V (2008) Investigation of arsenate phytotoxicity in cucumber plants. *Acta Biol Szeged* 52(1):79–80

Danh LT, Truong P, Mammucari R, Foster N (2014) A critical review of the arsenic uptake mechanisms and phytoremediation potential of *Pteris vittata*. *Int J Phytoremediation* 16(5):429–453

Das K, Roychoudhury A (2014) Reactive oxygen species (ROS) and response of antioxidants as ROS-scavengers during environmental stress in plants. *Front Environ Sci* 2(53):1–13

De Leonardis S, Dipierro N, Dipierro S (2000) Purification and characterization of an ascorbate peroxidase from potato tuber mitochondria. *Plant Physiol Biochem* 38(10):773–779

del Río LA, Corpas FJ, Sandalio LM, Palma JM, Gómez M, Barroso JB (2002) Reactive oxygen species, antioxidant systems and nitric oxide in peroxisomes. *J Exp Bot* 53(372):1255–1272

del Río LA, López-Huertas E (2016) ROS generation in peroxisomes and its role in cell signaling. *Plant Cell Physiol* 57(7):1364–1376

Demidchik V (2015) Mechanisms of oxidative stress in plants: From classical chemistry to cell biology. *Environ Exper Bot* 109:212–228

Dho S, Camusso W, Mucciarelli M, Fusconi A (2010) Arsenate toxicity on the apices of *Pisum sativum* L. seedling roots: Effects on mitotic activity, chromatin integrity and microtubules. *Environ Exper Bot* 69(1):17–23

Duquesnoy I, Champeau GM, Evray G, Ledoigt G, Piquet-Pissaloux A (2010) Enzymatic adaptations to arsenic-induced oxidative stress in *Zea mays* and genotoxic effect of arsenic in root tips of *Vicia faba* and *Zea mays*. *C R Biol* 333(11–12):814–824

Edreva A (2005) Generation and scavenging of reactive oxygen species in chloroplasts: A submolecular approach. *Agric Ecosyst Environ* 106(2–3):119–133

Edwards EA, Rawsthorne S, Mullineaux PM (1990) Subcellular distribution of multiple forms of glutathione reductase in leaves of pea (*Pisum sativum* L.). *Planta* 180(2):278–284

Emamverdian A, Ding Y, Mokhberdoran F, Xie Y (2015) Heavy metal stress and some mechanisms of plant defense response http://dx.doi.org/10.1155/2015/756120

Farnese FS, Oliveira JA, Paiva EA, Menezes-Silva PE, da Silva AA, Campos FV, Ribeiro C (2017) The involvement of nitric oxide in integration of plant physiological and ultrastructural adjustments in response to arsenic. *Front Plant Sci* 8(516):1–14

Foyer CH, Harbinson JC (1994) Oxygen metabolism and the regulation of photosynthetic electron transport. In: Foyer CH, Mullineaux PM (eds) *Causes of Photooxidative Stress and Amelioration of Defense Systems in Plant*. CRC Press, Boca Raton, FL, pp 1–42

Frugoli JA, Zhong HH, Nuccio ML, McCourt P, McPeek MA, Thomas TL, McClung CR (1996) Catalase is encoded by a multigene family in *Arabidopsis thaliana* (L.) Heynh. *Plant Physiol* 112(1):327–336

Fryer MJ (1992) The antioxidant effects of thylakoid vitamin E (α-tocopherol). *Plant Cell Environ* 15(4):381–392

Garg N, Bhandari P (2016) Interactive effects of silicon and arbuscular mycorrhiza in modulating ascorbate-glutathione cycle and antioxidant scavenging capacity in differentially salt-tolerant *Cicer arietinum* L. genotypes subjected to long-term salinity. *Protoplasma* 253(5):1325–1345

Garg N, Kashyap L (2017) Silicon and *Rhizophagus irregularis*: Potential candidates for ameliorating negative impacts of arsenate and arsenite stress on growth, nutrient acquisition and productivity in *Cajanus cajan* (L.) Millsp. genotypes. *Environ Sci Pollut Res* 24(22):18520–18535

Garg N, Kashyap L (2019) Joint effects of Si and mycorrhiza on the antioxidant metabolism of two pigeonpea genotypes under As (III) and (V) stress. *Environ Sci Pollut Res* 26(8):7821–7839

Garg N, Manchanda G (2009) ROS generation in plants: Boon or bane? *Plant Biosyst* 143(1):81–96

Garg N, Singh S (2018) Arbuscular mycorrhiza *Rhizophagus irregularis* and silicon modulate growth, proline biosynthesis and yield in *Cajanus cajan* L. Millsp (pigeonpea) genotypes under cadmium and zinc stress. *J Plant Growth Regul* 37(1):46–63

Garg N, Singla P (2011) Arsenic toxicity in crop plants: Physiological effects and tolerance mechanisms. *Environ Chem Lett* 9(3):303–321

Garg N, Singla P, Bhandari P (2015) Metal uptake, oxidative metabolism, and mycorrhization in pigeonpea and pea under arsenic and cadmium stress. *Turk J Agric For* 39(2):234–250

Gechev TS, Van Breusegem F, Stone JM, Denev I, Laloi C (2006) Reactive oxygen species as signals that modulate plant stress responses and programmed cell death. *BioEssays* 28(11):1091–1101

Ghosh S, Biswas AK (2017) Selenium modulates growth and thiol metabolism in wheat (Triticum aestivum L.) during arsenic stress. *Am J Plant Sci* 08(3):363–389

Gill SS, Tuteja N (2010) Reactive oxygen species and antioxidant machinery in abiotic stress tolerance in crop plants. *Plant Physiol Biochem* 48(12):909–930

Gomes MP, Carneiro MMLC, Nogueira COG, Soares AM, Garcia QS (2013) The system modulating ROS content in germinating seeds of two Brazilian savanna tree species exposed to As and Zn. *Acta Physiol Plant* 35(4):1011–1022

Grace SC, Logan BA (1996) Acclimation of foliar antioxidant systems to growth irradiance in three broad-leaved evergreen species. *Plant Physiol* 112(4):1631–1640

Gunes A, Pilbeam DJ, Inal A (2009) Effect of arsenic–phosphorus interaction on arsenic-induced oxidative stress in chickpea plants. *Plant Soil* 314(1):211–220

Gupta DK, Inouhe M, Rodríguez-Serrano M, Romero-Puertas MC, Sandalio LM (2013) Oxidative stress and arsenic toxicity: Role of NADPH oxidases. *Chemosphere* 90(6):1987–1996

Haber F, Weiss J (1932) On the catalysis of hydroperoxide. *Naturwissenschaften* 20(51):948–950

Halliwell B, Gutteridge JM (2015) *Free Radicals in Biology and Medicine.* Oxford University Press, Oxford

Halliwell B, Gutteridge JMC (1989) *Free Radicals in Biology and Medicine.* Clarendon Press, Oxford

Han D, Xiong S, Tu S, Liu J, Chen C (2015) Interactive effects of selenium and arsenic on growth, antioxidant system, arsenic and selenium species of *Nicotiana tabacum* L. *Environ Exper Bot* 117:12–19

Hartley-Whitaker J, Ainsworth G, Meharg AA (2001) Copper-and arsenate-induced oxidative stress in *Holcus lanatus* L. clones with differential sensitivity. *Plant Cell Environ* 24(7):713–722

Hasanuzzaman M, Fujita M (2013) Exogenous sodium nitroprusside alleviates arsenic-induced oxidative stress in wheat (*Triticum aestivum* L.) seedlings by enhancing antioxidant defense and glyoxalase system. *Ecotoxicology* 22(3):584–596

Hettick BE, Cañas-Carrell JE, French AD, Klein DM (2015) Arsenic: A review of the element's toxicity, plant interactions, and potential methods of remediation. *J Agric Food Chem* 63(32):7097–7107

Heyno E, Mary V, Schopfer P, Krieger-Liszkay A (2011) Oxygen activation at the plasma membrane: Relation between superoxide and hydroxyl radical production by isolated membranes. *Planta* 234(1):35–45

Holt NE, Zigmantas D, Valkunas L, Li XP, Niyogi KK, Fleming GR (2005) Carotenoid cation formation and the regulation of photosynthetic light harvesting. *Science* 307(5708):433–436

Hu Y, Duan GL, Huang YZ, Liu YX, Sun GX (2014) Interactive effects of different inorganic As and Se species on their uptake and translocation by rice (*Oryza sativa* L.) seedlings. *Environ Sci Pollut Res* 21(5):3955–3962

Hunt KM, Srivastava RK, Elmets CA, Athar M (2014) The mechanistic basis of arsenicosis: Pathogenesis of skin cancer. *Cancer Lett* 354(2):211–219

Ismail GSM (2012) Protective role of nitric oxide against arsenic-induced damages in germinating mung bean seeds. *Acta Physiol Plant* 34(4):1303–1311

Jing Wei J, Yue Fei X, Yuan Fang H (2010) Protective effect of nitric oxide against arsenic-induced oxidative damage in tall fescue leaves. *Afr J Biotechnol* 9(11):1619–1627

Julkunen-Tiitto R, Nenadis N, Neugart S, Robson M, Agati G, Vepsäläinen J, Zipoli G, Nybakken L, Winkler B, Jansen MA (2015) Assessing the response of plant flavonoids to UV radiation: An overview of appropriate techniques. *Phytochem Rev* 14(2):273–297

Kar RK (2015) ROS signaling: Relevance with site of production and metabolism of ROS. In: Gupta DK, Palma JM, Corpas FJ (eds) *Reactive Oxygen Species and Oxidative Damage in Plants Under Stress.* Springer, Cham, pp 115–125

Karuppanapandian T, Moon JC, Kim C, Manoharan K, Kim W (2011) Reactive oxygen species in plants: Their generation, signal transduction, and scavenging mechanisms. *Aust J Crop Sci* 5(6):709–725

Kashyap L, Garg N (2018) Arsenic toxicity in crop plants: Responses and remediation strategies. In: Hasanuzzaman M, Nahar K, Fujita M (eds) *Mechanisms of Arsenic Toxicity and Tolerance in Plants.* Springer, Singapore, pp 129–169

Kato H, Takemoto D, Kawakita K (2013) Proteomic analysis of S-nitrosylated proteins in potato plant. *Physiol Plant* 148(3):371–386

Kavdia M (2006) A computational model for free radicals transport in the microcirculation. *Antioxid Redox Signal* 8(7–8):1103–1111

Kertulis-Tartar GM, Rathinasabapathi B, Ma LQ (2009) Characterization of glutathione reductase and catalase in the fronds of two *Pteris* ferns upon arsenic exposure. *Plant Physiol Biochem* 47(10):960–965

Kidwai M, Dhar YV, Gautam N, Tiwari M, Ahmad IZ, Asif MH, Chakrabarty D (2019) *Oryza sativa* class III peroxidase (OsPRX38) overexpression in *Arabidopsis thaliana* reduces arsenic accumulation due to apoplastic lignification. *J Hazard Mat* 362:383–393

Kim FJ, Kim HP, Hah YC, Roe JH (1996) Differential expression of superoxide dismutases containing Ni and Fe/Zn in *Streptomyces coelicolor*. *Eur J Biochem* 241(1):178–185

Kim RY, Yoon JK, Kim TS, Yang JE, Owens G, Kim KR (2015) Bioavailability of heavy metals in soils: Definitions and practical implementation—A critical review. *Environ Geochem Health* 37(6):1041–1061

Kostecka-Gugała A, Latowski D (2018) Arsenic-induced oxidative stress in plants. In: Hasanuzzaman M, Nahar K, Fujita M (eds) *Mechanisms of Arsenic Toxicity and Tolerance in Plants*. Springer, Singapore, pp 79–104

Kothari S, Gayathri R (2018) Production and characterisation of peroxidase from *Raphanus raphanistrum*. *Int J Res Pharm Sci* 9(3):696–699

Kumar A, Singh RP, Singh PK, Awasthi S, Chakrabarty D, Trivedi PK, Tripathi RD (2014) Selenium ameliorates arsenic induced oxidative stress through modulation of antioxidant enzymes and thiols in rice (*Oryza sativa* L.). *Ecotoxicology* 23(7):1153–1163

Kumari A, Pandey N, Pandey-Rai S (2018) Exogenous salicylic acid-mediated modulation of arsenic stress tolerance with enhanced accumulation of secondary metabolites and improved size of glandular trichomes in *Artemisia annua* L. *Protoplasma* 255(1):139–152

Liu C, Wei L, Zhang S, Xu X, Li F (2014) Effects of nanoscale silica sol foliar application on arsenic uptake, distribution and oxidative damage defense in rice (*Oryza sativa* L.) under arsenic stress. *RSC Adv* 4(100):57227–57234

Liu X, Zhang S, Shan X, Zhu YG (2005) Toxicity of arsenate and arsenite on germination, seedling growth and amylolytic activity of wheat. *Chemosphere* 61(2):293–301

Lozano-Juste J, Colom-Moreno R, León J (2011) In vivo protein tyrosine nitration in *Arabidopsis thaliana*. *J Exper Bot* 62(10):3501–3517

Ma J, Qiu D, Pang Y, Gao H, Wang X, Qin Y (2020) Diverse roles of tocopherols in response to abiotic and biotic stresses and strategies for genetic biofortification in plants. *Mol Breed* 40(2):1–15

Malik JA, Goel S, Kaur N, Sharma S, Singh I, Nayyar H (2012) Selenium antagonises the toxic effects of arsenic on mungbean (*Phaseolus aureus* Roxb.) plants by restricting its uptake and enhancing the antioxidative and detoxification mechanisms. *Environ Exper Bot* 77:242–248

Mallick S, Sinam G, Sinha S (2011) Study on arsenate tolerant and sensitive cultivars of *Zea mays* L.: Differential detoxification mechanism and effect on nutrients status. *Ecotoxicol Environ Saf* 74(5):1316–1324

Marmiroli M, Mussi F, Imperiale D, Lencioni G, Marmiroli N (2017) Abiotic stress response to As and As+ Si, composite reprogramming of fruit metabolites in tomato cultivars. *Front Plant Sci* 8(2201):1–17

Mishra P, Sharma P (2019) Superoxide dismutases (SODs) and their role in regulating abiotic stress induced oxidative stress in plants. In: Hasanuzzaman M, Fotopoulos V, Nahar K, Fujita M (eds) *Reactive Oxygen, Nitrogen and Sulfur Species in Plants*, pp 53–88

Mishra RK, Kumar J, Srivastava PK, Bashri G, Prasad SM (2017) PSII photochemistry, oxidative damage and anti-oxidative enzymes in arsenate-stressed *Oryza sativa* L. seedlings. *Chem Ecol* 33(1):34–50

Mishra S, Dwivedi S, Mallick S, Tripathi RD (2019) Redox homeostasis in plants Under arsenic stress. In: Panda SK, Yamamoto YY (eds) *Redox Homeostasis in Plants*. Springer, Cham, pp 179–198

Mishra S, Srivastava S, Dwivedi S, Tripathi RD (2013) Investigation of biochemical responses of *Bacopa monnieri* L. upon exposure to arsenate. *Environ Toxicol* 28(8):419–430

Mishra S, Srivastava S, Tripathi RD, Trivedi PK (2008) Thiol metabolism and antioxidant systems complement each other during arsenate detoxification in *Ceratophyllum demersum* L. *Aquat Toxicol* 86(2):205–215

Møller IM, Jensen PE, Hansson A (2007) Oxidative modifications to cellular components in plants. *Annu Rev Plant Biol* 58:459–481

Møller IM, Kristensen BK (2004) Protein oxidation in plant mitochondria as a stress indicator. *Photochem Photobiol Sci* 3(8):730–735

Mucha S, Berezowski M, Markowska K (2019) Mechanisms of arsenic toxicity and transport in microorganisms. *Postępy Mikrobiologii—Adv Microbiol* 56(1):88–99

Mylona PV, Polidoros AN, Scandalios JG (1998) Modulation of antioxidant responses by arsenic in maize. *Free Radic Biol Med* 25(4–5):576–585

Myrene RDS, Devaraj VR (2013) Oxidative stress biomarkers and metabolic changes associated with cadmium stress in hyacinth bean (*Lablab purpureus*). *Afr J Biotechnol* 12(29):4670–4682

Navari-Izzo F, Pinzino C, Quartacci MF, Sgherri CL (1999) Superoxide and hydroxyl radicai generation, and superoxide dismutase in PII membrane fragments from wheat. *Free Radic Res* 31 (Supl):3–9

Navrot N, Rouhier N, Gelhaye E, Jacquot JP (2007) Reactive oxygen species generation and antioxidant systems in plant mitochondria. *Physiol Plant* 129(1):185–195

Nies DH (1999) Microbial heavy-metal resistance. *Appl Microbiol Biotechnol* 51(6):730–750

Okamoto Y, Chou PH, Kim SY, Suzuki N, Laxmi YR, Okamoto K, Liu X, Matsuda T, Shibutani S (2008) Oxidative DNA damage in XPC-knockout and its wild mice treated with equine estrogen. *Chem Res Toxicol* 21(5):1120–1124

Ota Y, Ario T, Hayashi K, Nakagawa T, Hattori T, Maeshima M, Asahi T (1992) Tissue-specific isoforms of catalase subunits in castor bean seedlings. *Plant Cell Physiol* 33(3):225–232

Pandey C, Khan E, Panthri M, Tripathi RD, Gupta M (2016) Impact of silicon on Indian mustard (*Brassica juncea* L.) root traits by regulating growth parameters, cellular antioxidants and stress modulators under arsenic stress. *Plant Physiol Biochem* 104:216–225

Pandey C, Raghuram B, Sinha AK, Gupta M (2015) miRNA plays a role in the antagonistic effect of selenium on arsenic stress in rice seedlings. *Metallomics* 7(5):857–866

Parkhey S, Naithani SC, Keshavkant S (2014) Protein metabolism during natural ageing in desiccating recalcitrant seeds of *Shorea robusta*. *Acta Physiol Plant* 36(7):1649–1659

Pryor WA, Squadrito GL (1995) The chemistry of peroxynitrite: A product from the reaction of nitric oxide with superoxide. *Am J Physiol Lung Cell Mol Physiol* 268(5):L699–L722

Rajjou L, Lovigny Y, Groot SP, Belghazi M, Job C, Job D (2008) Proteome-wide characterization of seed aging in *Arabidopsis*: A comparison between artificial and natural aging protocols. *Plant Physiol* 148(1):620–641

Rao CV, Pal S, Mohammed A, Farooqui M, Doescher MP, Asch AS, Yamada HY (2017) Biological effects and epidemiological consequences of arsenic exposure, and reagents that can ameliorate arsenic damage in vivo. *Oncotarget* 8(34):57605–57621

Requejo R, Tena M (2005) Proteome analysis of maize roots reveals that oxidative stress is a main contributing factor to plant arsenic toxicity. *Phytochemistry* 66(13):1519–1528

Saed-Moucheshi A, Shekoofa A, Pessarakli M (2014) Reactive oxygen species (ROS) generation and detoxifying in plants. *J Plant Nutr* 37(10):1573–1585

Saidi I, Yousfi N, Borgi MA (2017) Salicylic acid improves the antioxidant ability against arsenic-induced oxidative stress in sunflower (*Helianthus annuus*) seedling. *J Plant Nutr* 40(16):2326–2335

Schützendübel A, Polle A (2002) Plant responses to abiotic stresses: Heavy metal-induced oxidative stress and protection by mycorrhization. *J Exp Bot* 53(372):1351–1365

Shahid M, Rafiq M, Niazi NK, Dumat C, Shamshad S, Khalid S, Bibi I (2017) Arsenic accumulation and physiological attributes of spinach in the presence of amendments: An implication to reduce health risk. *Environ Sci Pollut Res* 24(19):16097–16106

Sharma I (2012) Arsenic induced oxidative stress in plants. *Biologia* 67(3):447–453

Sharma P, Jha AB, Dubey RS (2014) Arsenic toxicity and tolerance mechanisms in crop plants. In: Pessarakli M (eds) *Handbook of Plant and Crop Physiology*. Boca Raton, FL: CRC Press, pp 762–811

Sharma P, Jha AB, Dubey RS, Pessarakli M (2012) Reactive oxygen species, oxidative damage, and antioxidative defense mechanism in plants under stressful conditions. *J Bot* 2012:1–26

Shri M, Kumar S, Chakrabarty D, Trivedi PK, Mallick S, Misra P, Shukla D, Mishra S, Srivastava S, Tripathi RD, Tuli R (2009) Effect of arsenic on growth, oxidative stress, and antioxidant system in rice seedlings. *Ecotoxicol Environ Saf* 72(4):1102–1110

Siddiqui F, Tandon PK, Srivastava S (2015) Arsenite and arsenate impact the oxidative status and antioxidant responses in *Ocimum tenuiflorum* L. *Physiol Mol Biol Plant* 21(3):453–458

Singh AP, Dixit G, Mishra S, Dwivedi S, Tiwari M, Mallick S, Pandey V, Trivedi PK, Chakrabarty D, Tripathi RD (2015) Salicylic acid modulates arsenic toxicity by reducing its root to shoot translocation in rice (*Oryza sativa* L.). *Front Plant Sci* 6(340):1–12

Singh B, Singh JP, Kaur A, Singh N (2017a) Phenolic composition and antioxidant potential of grain legume seeds: A review. *Food Res Int* 101:1–16

Singh HP, Batish DR, Kohli RK, Arora K (2007) Arsenic-induced root growth inhibition in mung bean (*Phaseolus aureus* Roxb.) is due to oxidative stress resulting from enhanced lipid peroxidation. *Plant Growth Regul* 53(1):65–73

Singh HP, Kaur S, Batish DR, Sharma VP, Sharma N, Kohli RK (2009) Nitric oxide alleviates arsenic toxicity by reducing oxidative damage in the roots of *Oryza sativa* (rice). *Nitric Oxide* 20(4):289–297

Singh N, Ma LQ, Srivastava M, Rathinasabapathi B (2006) Metabolic adaptations to arsenic-induced oxidative stress in *Pteris vittata* L. and *Pteris ensiformis* L. *Plant Sci* 170(2):274–282

Singh N, Raj A, Khare PB, Tripathi RD, Jamil S (2010) Arsenic accumulation pattern in 12 Indian ferns and assessing the potential of *Adiantum capillus-veneris*, in comparison to *Pteris vittata*, as arsenic hyperaccumulator. *Bioresour Technol* 101(23):8960–8968

Singh R, Upadhyay AK, Singh DP (2018) Regulation of oxidative stress and mineral nutrient status by selenium in arsenic treated crop plant *Oryza sativa*. *Ecotoxicol Environ Saf* 148:105–113

Singh S, Sounderajan S, Kumar K, Fulzele DP (2017b) Investigation of arsenic accumulation and biochemical response of in vitro developed *Vetiveria zizanoides* plants. *Ecotoxicol Environ Saf* 145:50–56

Skovsen E, Snyder JW, Lambert JD, Ogilby PR (2005) Lifetime and diffusion of singlet oxygen in a cell. *J Phys Chem B* 109(18):8570–8573

Srivastava M, Ma LQ, Singh N, Singh S (2005) Antioxidant responses of hyper-accumulator and sensitive fern species to arsenic. *J Exper Bot* 56(415):1335–1342

Srivastava S, Srivastava AK, Singh B, Suprasanna P, D'souza SF (2013) The effect of arsenic on pigment composition and photosynthesis in *Hydrilla verticillata*. *Biologia Plant* 57(2):385–389

Stoeva N, Bineva T (2003) Oxidative changes and photosynthesis in oat plants grown in As-contaminated soil. *Bulg J Plant Physiol* 29(1–2):87–95

Suneja Y (2014) Physio-Biochemical Responses and Allelic Diversity for Water Deficit Tolerance Related Traits in *Aegilops tauschii* and *Triticum dicoccoides* (Doctoral dissertation)

Talukdar D (2011) Effect of arsenic-induced toxicity on morphological traits of Trigonella foenum-graecum L. and Lathyrus sativus L. during germination and early seedling growth. *Curr Res J Biol Sci* 3(2):116–123

Talukdar D (2013) Arsenic-induced changes in growth and antioxidant metabolism of fenugreek. *Russ J Plant Physiol* 60(5):652–660

Tanou G, Filippou P, Belghazi M, Job D, Diamantidis G, Fotopoulos V, Molassiotis A (2012) Oxidative and nitrosative-based signaling and associated post-translational modifications orchestrate the acclimation of citrus plants to salinity stress. *Plant J* 72(4):585–599

Tiwari S, Sarangi BK (2015) Arsenic and chromium-induced oxidative stress in metal accumulator and non-accumulator plants and detoxification mechanisms. In: Gupta DK, Palma JM, Corpas FJ (eds) *Reactive Oxygen Species and Oxidative Damage in Plants Under Stress*. Springer, Cham, pp 165–189

Tripathi DK, Kumar R, Pathak AK, Chauhan DK, Rai AK (2012a) Laser-induced breakdown spectroscopy and phytolith analysis: An approach to study the deposition and distribution pattern of silicon in different parts of wheat (*Triticum aestivum* L.) plant. *Agric Res* 1(4):352–361

Tripathi DK, Singh S, Singh VP, Prasad SM, Chauhan DK, Dubey NK (2016) Silicon nanoparticles more efficiently alleviate arsenate toxicity than silicon in maize cultiver and hybrid differing in arsenate tolerance. *Front Environ Sci* 4(46):1–14

Tripathi P, Mishra A, Dwivedi S, Chakrabarty D, Trivedi PK, Singh RP, Tripathi RD (2012b) Differential response of oxidative stress and thiol metabolism in contrasting rice genotypes for arsenic tolerance. *Ecotoxicol Environ Saf* 79:189–198

Tripathi P, Tripathi RD, Singh RP, Dwivedi S, Goutam D, Shri M, Trivedi PK, Chakrabarty D (2013) Silicon mediates arsenic tolerance in rice (*Oryza sativa* L.) through lowering of arsenic uptake and improved antioxidant defence system. *Ecol Eng* 52:96–103

Tripathy BC, Oelmüller R (2012) Reactive oxygen species generation and signaling in plants. *Plant Signal Behav* 7(12):1621–1633

Upadhyaya H, Shome S, Roy D, Bhattacharya MK (2014) Arsenic Induced Changes in Growth and Physiological Responses in *Vigna radiata* Seedling: Effect of Curcumin Interaction. *Am J Plant Sci* 05(24):3609–3618

Vinit-Dunand F, Epron D, Alaoui-Sossé B, Badot PM (2002) Effects of copper on growth and on photosynthesis of mature and expanding leaves in cucumber plants. *Plant Sci* 163(1):53–58

Wang CQ, Xu HJ, Liu T (2011) Effect of selenium on ascorbate–glutathione metabolism During PEG-induced water deficit in *Trifolium repens* L. *J Plant Growth Regul* 30(4):436–444

Yadav G, Srivastava PK, Parihar P, Tiwari S, Prasad SM (2016) Oxygen toxicity and antioxidative responses in arsenic stressed *Helianthus annuus* L. seedlings against UV-B. *J Photochem Photobiol B* 165:58–70

Yadav G, Srivastava PK, Singh VP, Prasad SM (2014) Light intensity alters the extent of arsenic toxicity in *Helianthus annuus* L. seedlings. *Biol Trace Elem Res* 158(3):410–421

Yadav RK, Srivastava SK (2015) Effect of arsenite and arsenate on lipid peroxidation, enzymatic and non-enzymatic antioxidants in *Zea mays* Linn. *Biochem Physiol* 4:186 https://dx.doi.org/10.4172/21689652.1000186

Yamauchi Y, Furutera A, Seki K, Toyoda Y, Tanaka K, Sugimoto Y (2008) Malondialdehyde generated from peroxidized linolenic acid causes protein modification in heat-stressed plants. *Plant Physiol Biochem* 46(8–9):786–793

Yannarelli GG, Fernández-Alvarez AJ, Santa-Cruz DM, Tomaro ML (2007) Glutathione reductase activity and isoforms in leaves and roots of wheat plants subjected to cadmium stress. *Phytochemistry* 68(4):505–512

Zu YQ, Sun JJ, He YM, Wu J, Feng GQ, Li Y (2016) Effects of arsenic on growth, photosynthesis and some antioxidant parameters of Panax notoginseng growing in shaded conditions. *Int J Adv Agric Res* 4:78–88

9 The Key Roles of Proline against Heat, Drought and Salinity-Induced Oxidative Stress in Wheat (*Triticum aestivum* L.)

Akbar Hossain, Sourav Garai, Mousumi Mondal and Arafat Abdel Hamed Abdel Latef

CONTENTS

9.1 INTRODUCTION

Wheat (*Triticum aestivum* L.) is considered a major staple food crop globally, ranked third after maize and rice in terms of production. According to FAO (2014), wheat occupies an area of approximately 218.5 million hectares, which gives 600 million tonnes annual production and 3.26 t ha^{-1} productivity. Over the last decades, a global population increase has led to a higher demand for wheat. However, recent abnormal weather conditions create serious abiotic stresses in plant body *viz.*, heat, drought, salinity, waterlogging and heavy metal (Barlow et al. 2015; Herzog et al. 2016), which cause pollen sterility (Chakrabarti et al. 2011; Dong et al. 2017; Abhinandan et al. 2018), disturb photosynthetic apparatus (Brestic et al. 2016), produce shriveled seeds (Rascio et al. 2015) in wheat, and account for the inordinate production of ROS which ultimately affect the growth and yield of the plant (Rizwan et al. 2016; Bali and Sidhu 2019). Therefore to meet the food and nutritional security of an increasing population, it is important to understand the physiological mechanisms of wheat for sustainable wheat production in the current era of climate change.

Plants follow numerous physiochemical mechanisms to survive against hostile effects of abiotic stresses. For example, during stress conditions, tolerant plants can accumulate compatible osmolytes, particularly amino acids such as proline (Kaur and Asthir 2015; Akter and Islam 2017; Tiwari et al. 2017; Abdel Latef et al. 2019 a, b, 2020 a, b, c; Osman et al. 2020). It is well recognized that proline (Pro) is one of the most extensively dispersed harmonious solutes that accrues in plants during antagonistic abiotic constraints and acts as a significant protagonist in plant stress tolerance (Azooz et al. 2004; Abdel Latef et al. 2009; Caverzan et al. 2016). Besides this, the enhancement of Pro-biosynthesis in the chloroplast substitutes as an admirable osmolyte, metal chelator, self-protective antioxidant and signaling molecule, and also preserves redox-potential for metabolism during abiotic stress (Shao et al. 2008; Szabados and Savoure 2010; Rejeb et al. 2014; Abhinandan et al. 2018). It instructs stress tolerance in plants through controlling mitochondrial utilities, stimulates cell multiplication, activates precise gene appearance, inhibits electrolyte leakage and fetches absorptions of ROS within regular arrays, leading to stress reclamation (Joseph et al. 2015; Bali and Sidhu 2019). Pro catabolism in the mitochondria is linked to oxidative respiration and controls dynamism to continue the growth of plants next to stress (Taiz and Zeiger 2010; Vanlerberghe 2013). Scientists found a positive correlation between an enhanced content of Pro and greater biomass under osmotic stress in plants (Su and Wu 2004; Verbruggen and Hermans 2008; Maghsoudi et al. 2018). On the other hand, Pro might converse a defensive outcome by encouraging stress-defending proteins and expression of salt stress reactive genes owning Pro-reactive elements in their promoters (Chinnusamy et al. 2005; Liang et al. 2013; Gupta et al. 2014).

Several studies have attributed that exogenous application of Pro enhances the activity of antioxidative enzymes, which leads to mitigate the adverse effect of environmental effects on wheat (Nayyar 2003; Kamran et al. 2009; Aly and Latif 2011; Ashraf et al. 2012; Li et al. 2013; Khan et al. 2015). Keeping in view the miscellaneous roles of Pro in plants including wheat, the present chapter focuses on mechanisms, metabolism, transport, regulation and key linking pathways of Pro to abiotic stress tolerance.

9.2   THE ADVERSE EFFECT OF ABIOTIC STRESS-INDUCED OXIDATIVE STRESS IN WHEAT

Abiotic stresses generally have negative impacts on plant growth, development and various physiological mechanisms (Figure 9.1) and thus decrease plant yield. In the field, crops are normally exposed to a combination of one or two or multiple abiotic stresses. Globally, the wheat-growing areas mostly encounter various abiotic stresses, which include drought, heat and salinity, and which reduce the potentiality of production (Pradhan et al. 2012). The reproductive stage of wheat is mostly sensitive to this stress compared to the other stages of development (Prasad et al. 2008; Ji et al. 2010; Lott et al. 2011). This section gives an insight into the effects of different abiotic stresses on the growth and physiology of wheat.

9.2.1   HEAT OR HIGH-TEMPERATURE STRESS

Rapid environmental mutation and changing climate in recent decades have altered temperatures and have caused heat stress in the atmosphere. Gibson and Paulsen (1999) and Asseng et al. (2015) observed that each rise in temperature by degree reduces wheat productivity by about 3–6%. High-temperature stress affects almost 7 million ha of wheat in developing countries, and terminal heat stress is a problem in 40% of temperate environments, which cover 36 million ha (Reynolds et al. 2001; Qaseem et al. 2019).

High-temperature stress affects the stability of many proteins, membranes and RNA species as well as alters the efficiency of enzymatic reactions in the cells of wheat, and affects all major

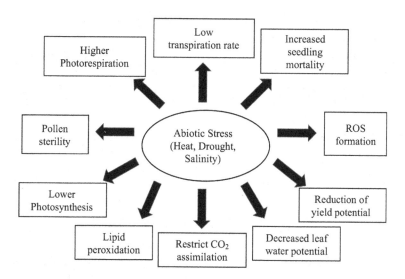

FIGURE 9.1 Abiotic stress-induced ROS generation and its effect on different parameters of wheat.

physiological processes, and these changes create a metabolic imbalance (Hasanuzzaman and Fujita 2013; Brestic et al. 2018; Urban et al. 2018). Some studies found that projected global warming as a consequence of climate change may negatively influence the wheat grain yields and thus increase food insecurity and poverty (Asseng et al. 2015); even though the magnitude and direction of climatic impact on crop yields will vary locally (Tubiello et al. 2000; Modhej et al. 2008; Narayanan et al. 2015). Increment of temperature affects the photosynthesis process by hampering the photosynthesis apparatus in wheat (Mathur et al. 2011; Brestic et al. 2018), such as chlorophyll content, disrupting the chloroplast structure and thylakoid lamellae and disabling chloroplast enzymes (Wang et al. 2010; Farooq et al. 2011; Perdomo et al. 2017). Heat stress also cuts down the activity of RuBisCO, photosynthetic nitrogen use efficiency (NUE) and net photosynthesis in wheat (Khan et al. 2013; Vicente et al. 2018).

The optimum temperature for wheat anthesis and grain filling stages ranges from 12 to 22 °C (Farooq et al. 2011; Hossain and Teixeira da Silva 2012; Nuttall et al. 2018). Heat stress not only hampers the growth and yield by harming the photosynthesis process (Asseng et al. 2011, 2015), promoting the photorespiration (Perdomo et al. 2015), reducing the grain number, size and dry matter production, but also disrupting the flowering processes, mainly pollen viability (Semenov and Shewry 2011; Raja et al. 2019). The germination stage is very sensitive to high temperature (Hossain et al. 2012 a, b). It may result in reduced germination percentage, germination rate and plant emergence, increased abnormal seedlings, decreased seedling vigour, decreased root and shoot growth of germinated seedlings and dry matter production (Wahid et al. 2007; Kumar et al. 2011; Iloh et al. 2014). Also, high-temperature stress during germination to emergence results in heavy seedling mortality and a poor crop stand (Acevedo et al. 2002; Hossain et al. 2013). According to Modhej et al. (2012) and Sattar et al. (2020), leaf water potential and relative leaf water content are reduced by the heat stress in addition to the canopy temperature of stressed plants increasing at vegetative and flowering stages (Siddique et al. 2000). Additionally, reduced grain yield under heat stress might be a consequence of lower assimilates availability, reduced translocation of photosynthates to grain and starch synthesis in developing grain. A temperature of 27–30° C during the anthesis period causes an extensive reduction in the number of grains and loss of yield due to large production of sterile grain (Mitchell et al. 1993; Wheeler et al. 1996).

Some researchers tried to find out which developmental stage is the most affected by high-temperature stress. They showed that photosynthetic capacity decreases rapidly when temperate crops are exposed to high temperatures during reproductive development, and they concluded that high temperatures initially accelerate the breakdown of chloroplast and the plasma membrane coupled with dilation of the thylakoid membrane that follows a similar pattern with a normal senescence plant (Djanaguiraman et al. 2011; Pradhan et al. 2012). Previous literatures on the impact of heat stress at the post-anthesis period showed that high temperatures inhibit biomass production by promoting leaf senescence and reducing radiation use efficiency, which lead to the failure of the assimilate supply to grains and ultimately lowering the grain yield in spring wheat (Acevedo et al. 2002; Kobata et al. 2012; Akter and Islam 2017). Besides this, one of the major consequences of high-temperature stress is the excess generation of ROS such as singlet oxygen, the superoxide radical, hydrogen peroxide and the hydroxyl radical, which leads to plant oxidative stress (Hasanuzzaman et al. 2012; Narayanan et al. 2015).

9.2.2 DROUGHT STRESS

Drought stress is a serious concern in crop production globally and is a major threat to world food security. It not only affects the plant morphology and physiology (Prasad et al. 2011) but also harms at a molecular level. Similar to heat stress, drought stress also limits photosynthesis by altering the chlorophyll content and damaging the photosynthetic apparatus in plants which might be the cause of lower yield potential and storage capacity of wheat (Keyvan 2010). In severe stresses, the wheat plant might suffer at different growth stages, but the reproductive stage, especially flowering and grain development phases, are more critical for drought-induced stress (Farooq et al. 2014). It was well established that the reduction of CO_2 assimilation potential, chlorophyll content, oxidative damage to the chloroplast, lowering transpiration rate and stomatal conductance, spikelet sterility and lower grain filling diminish the accumulation of photosynthates into the plant body and thereby ultimately lower the yield potential of wheat (Prasad et al. 2011; Naveed et al. 2014; Farooq et al. 2014). A similar paradigm also reported by Ihsan et al. (2016) observed that the lower leaf area index, plant growth rate and net assimilation in wheat responded to severe drought stress. Furthermore, drought stress is highly responsible for ROS formation in the plant body due to the interruption of cellular homeostasis (Khan et al. 2015). The higher concentration of ROS poses fatalistic effects on cell organelles and their mechanisms as well as targets the proteins, lipids, RNA and DNA. Ultimately, ROS harm metabolic activities and cause cell mortality (Petrov et al. 2015).

9.2.3 SALINITY STRESS

As with heat and drought stress, salt stress also has been a serious problem for plant growth and development, more particularly in arid and semi-arid zones, as the high temperatures during the summer season cause severe evaporation losses, which leave behind large amounts of salts on the upper soil surface. However, the problem exists even in some of the world's sub-humid and humid regions, especially in coastal areas. Due to vagaries of rainfall, nowadays irrigation demand for most of the crop is increased. Wheat crop is often considered an irrigation-dependent crop with higher water requirements, while heavy rainfall sometimes affected the wheat crop, due to water-logging (Läuchli and Grattan 2007). The FAO expects that over 800 million ha will be affected by salinity in the near future, and considers salinity to be a major limitation to food production for an increasing population (Rengasamy 2006; FAO 2008). The term salinity refers to the presence of the major dissolved inorganic solutes (essentially Na^+, Mg^{2+}, Ca^{2+}, K^+, Cl^-, SO^{4-}, HCO^{3-}, NO^{3-} and CO^{3-}) in water and soil (Bernstein 1975; Tanji 1990; Rhoades et al. 1999). A salinity problem occurs when salt accumulates in the crop root zone to a concentration that causes a severe yield loss (Francois and Maas 1999; Abdel Latef et al. 2017 a, b, c, 2018, 2021).

However, even in regions with sufficient rainfall, salt can accumulate in soil with poor drainage. Another source of soil salinity is the substantial use of salt-bearing fertilizers (es et al. 2013). Atmospheric salt deposition, especially in coastal regions, is also a major source for salinity in soils (Gupta and Abrol 1990). Salinity stress hinders the osmosis process and increases ion toxicity and nutritional disorders (Läuchli and Epstein 1990; Abdel Latef 2011). At the germination stage, wheat seedlings are badly affected by extreme salt stress as a consequence of toxic ion accumulation (Almansouri et al. 2001; Abdel Latef 2010). Plants show a two-phase growth response during salt-induced abiotic stress, namely, growth reduction and inhibition of photosynthetic mechanism due to high salt accumulation in plant tissue (Munns 2002). Chlorophyll content in wheat leaves under salt stress might be attributed to the salt-induced weakening of protein pigment-lipid complex or increased activity of enzyme chlorophyllase (Turan et al. 2007). One of the most detrimental effects of salinity stress is the entry of Na^+ and cl^- ion accumulation in plant cells, and thus severe ion imbalance causes significant physiological disorder(s). Higher uptake of those ions also hinders the uptake of K^+ ions, considered as an essential plant nutrient element, which may lead to lower productivity or even plant mortality (James et al. 2011).

The earlier experiment carried out on two genetically differed wheat cultivars resulted in increased salt concentration, which might be the cause of the production of malondialdehyde (MDA) content in various growth stages of the plant in response to stress (Ashraf et al. 2010). The amino acids production also decreases when the plant is exposed to salinity stress, except for the Pro concentration (El-shintinawy and El-shourbagy 2001). Oxidative stress might be the cause of increasing salt concentration in the plant body, decreasing the chlorophyll, carotenoid and relative water content in leaf tissue (Sairam et al. 2002). This ultimately cuts the growth of seedlings, seedling vigour, shoot and root length, utilization of seed storage and dry weight of wheat crop, all of which may result in lower grain yield (Soltani et al. 2006).

9.3   IMPORTANCE OF PROLINE AND ITS MECHANISMS DURING ABIOTIC STRESS-INDUCED OXIDATIVE STRESS

In the natural habitat, plants experience abiotic stresses in different situations such as in extreme temperature, the occurrence of drought, flood, salinity, heavy metal toxicity, nutrient imbalances, or sometimes by UV-irradiation, which are the prime factors for yield losses of crops globally (Qin et al. 2011; Krasensky and Jonak 2012; Naika et al. 2013). It is also well documented that the frequency of stress has increased in recent times (Lamb 2012). Abiotic stresses cause problems in plant physiological mechanisms by altering the metabolic interface and prohibiting the cellular redox homeostasis (Krasensky and Jonak 2012). As a consequence, the ROS are rapidly accumulated in different cellular organelles and cause oxidative damages in the plant body until the detoxification of ROS has happened at the site of production (Hossain et al. 2012; Hossain and Fujita 2012). The most common phenomenon to counteract abiotic stress in plants is the synthesis of Pro, glycine betaine, trehalose and other compounds, having the characteristics of low molecular weight, and being extremely water-soluble and non-toxic (Verslues and Sharma 2010; Hayat et al. 2012). Previous research studies strongly established that Pro is synthesized in high-stress conditions and has the potential to adjust the osmotic balance and redox potential of cell organelles, remove free radicles, activate the way of detoxification processes, maintain the cell structure and act as a buffering agent and source of energies as well as a signaling molecule (Szabados and Savouré 2010). Electron sink mechanism is another important function of Pro in stress situations (Sharma and Dietz 2006). Some important mechanisms of Pro against specific abiotic stresses are discussed below and summarized in Figure 9.2.

9.3.1   MECHANISM DURING HEAT-INDUCED OXIDATIVE STRESS

The fundamental roles of Pro during heat stress are to maintain the osmotic balance and resist cellular dehydration (Chen et al. 2007). However, the Pro synthesis and its roles are varied with different situations and plant species, for example its bio-synthesis is increased positively in tomato and tobacco leaf (Rivero et al. 2014; Cvikrová et al. 2012) while reported as being decreased in germinating wheat seeds and remaining unchanged in Arabidopsis plants (Rizhsky et al. 2004). This is

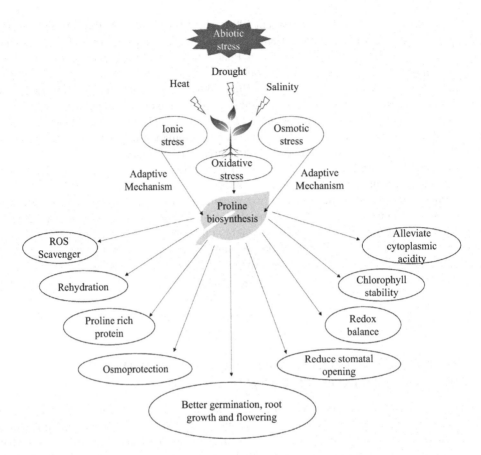

FIGURE 9.2 Importance of proline and its functions in plant growth, development and stress tolerance.

maybe due to the differential expression of genes responsible for Pro biosynthesis. The overexpression of pyrroline-5-carboxylate reductase gene (AtP5CR) in *Arabidopsis* is highly responsible for heat stress tolerance (de Ronde et al. 2004).

Pro favors an increasing protoplasmic concentration to maintain cell turgidity and regular function of cell membrane under heat stress situations. It acts as an osmoprotectant to alleviate the deleterious effect of ROS and stabilize the antioxidant system (Surender Reddy et al. 2015). Wahid et al. (2007) concluded that Pro was responsible for increasing pressure potential in stress-induced sugarcane leaves. Pro also acts as a substitute for water in plant cells through hydrophilic interactions and hydrogen bonding. Additionally, Pro can interact with enzymes to prevent the denaturation of protein under heat stress (Kavi Kishor et al. 2005). In transgenic sugarcane, Pro production was increased in stress situations to protect the chlorophyll and PSII, with higher MDA content being responsible for the better antioxidant system (Molinari et al. 2007). Though the Pro production was increased in the transgenic plant under heat stress conditions, it showed a minor enhancement, as it was used as a source of polyamine synthesis in the early stress period (Cvikrová et al. 2012). The Pro and polyamine have the common precursor, glutamate (Groppa and Benavides 2008).

In contrast, a meaningful enrichment of putrescine content was observed in the leaf tissues of transgenic plants in similar stress conditions, which might be due to the de-novo synthesis. Furthermore, the use of NH_4^+ as a source of N rather than NO_3^- is more effective in the context of heat tolerance, as NH_4^+ is highly related to Pro accumulation (Rivero et al. 2014). It was reported that the enhancement of antioxidative enzymes and glutathione content under high-temperature stress has been stimulated by Pro-induced genetic manipulation in soybean (Kocsy et al. 2005).

9.3.2 MECHANISM DURING DROUGHT-INDUCED OXIDATIVE STRESS

The accumulation of Pro in the host plant facilitates minimizing the osmotic potential for maintaining leaf water potential, organ hydration and turgor pressure to protect the photosynthetic apparatus from stress-induced damage (Wang et al. 2004; Kandowangko et al. 2009). Higher Pro content plays a vital role in sugar modulation, increases photosynthetic efficiency that collectively increases growth and yields under drought stress (Foyer et al. 2017). Earlier studies recognized that Pro minimizes the aerial part of the plant (Doubková et al. 2013). Ali et al. (2008) also revealed that Pro helps to improve the uptake of plant nutrient such as N, P, K and Ca when exposed to drought stress. The NADP$^+$ ions are generated during Pro biosynthesis, while Pro oxidation generates NADPH. This process plays a vital role in buffering redox potential in cytosol and plastids. Thus, in stress tolerance mechanisms, Pro can maintain redox potential by replacement of the NADP$^+$ supply (Hassine et al. 2008; Sharma et al. 2011). The Pro accumulation was reported to be high under drought stress because it acts as an important redox buffering agent (Verslues and Sharma 2010). A similar observation was recorded by Szabados and Savouré (2010).

Exogenous application of Pro as a foliar spray at the critical crop growth stage of wheat enhances the antioxidant activities against the accumulated ROS in plants (Yasmeen et al. 2013). Additionally, it maintains the cell integrity, membrane stability and other macromolecular activities in water deficit situations (Anjum et al. 2011; Nawaz et al. 2016). Seed soaking in Pro just before sowing reduces the stress damage in wheat seedlings (Kamran et al. 2009). A similar stress mitigation effect was also observed in maize plants when Pro was applied externally (Ali et al. 2007). In addition to stress alleviation, storage and transfer of energy are other important mechanisms of Pro in the plant body (Szabados and Savoure 2010; Verslues and Sharma 2010). It supplies reductants, nitrogen in reduced form and C skeletons for post-stress recovery (Vartanian et al. 1992). However, it is also important to decrease Pro concentration after stress release, and it is vital for overall stress alleviation mechanisms (Hayano-Kanashiro et al. 2009).

9.3.3 MECHANISM DURING SALT-INDUCED OXIDATIVE STRESS

In earlier phase of plants, salinity stress lowers the water absorption capacity of the root system and accelerates the leaf water losses due to the higher salt accumulation in soil water and plant body, known as physiological stress/hyperosmotic stress (Munns 2002). It also alters the plant physiological and metabolic processes varying with magnitude and duration of stress (Rahnama et al. 2010). At the initial stage, plant growth is suppressed as a consequence of osmotic stress and then by ion toxicity and dehydration (James et al. 2011). Higher accumulation of Na$^+$ and Cl$^-$ ions in salinity-induced osmotic stress may lead to severe ion imbalance and physiological disorder and even causes plant mortality (James et al. 2011). Under NaCl stress, Pro plays a critical role in protecting photosynthetic activities and the electron transport system in mitochondria (Hamilton and Heckathorn 2001). During the reproductive phase, a higher concentration of Pro is found in seeds and pollen and acts as a compatible solute that guards the cellular structure during dehydration (Lehmann et al. 2010). Additionally, the transpiration factors such as OsNAC5 and ZFP179 regulate the Pro accumulation in the intercellular region under salt stress, which plays a crucial role in stress tolerance (Song et al. 2011). Intercellular Pro accumulation also functions as a source of organic N to plants during stressful situations. The ROS scavenging ability of Pro establishes its antioxidant capacity, preserving the glutathione pool which is considered as a major redox buffer in cellular organelles (Matysik et al. 2002). Pro works as a molecular chaperone, maintains the cell integrity and stabilizes proteins, DNA and enzymes (Matysik et al. 2002; Szabados and Savoure 2010). Krishnan et al. (2008) observed that Pro plays a more pivotal role to stabilize redox homeostasis, exhibiting dual functions as a pro-oxidant via PDH and as a ROS scavenger. In tobacco and mung bean seedlings, Pro improves salt tolerance by boosting the enzymatic activities related to the antioxidant defense system (Hoque et al. 2008; Hossain and Fujita 2012; Hossain et al. 2011). Pro accumulation in chloroplast during salt-induced oxidative stress facilitates the preservation of low NADPH/NADP ratio, maintains redox balance, protects photosynthetic apparatus from oxidative damage and lessens the photo-inhibition (Hare and Cress 1997). However, earlier literature showed that a higher concentration of exogenous Pro application poses toxicity to plants. As an example, Roy et al. (2014) demonstrated that 30 mM Pro application alleviates the salinity stress in terms of higher germination percent and seedling growth while 50 mM resulted in lower growth and nutrient uptake.

9.4 PROLINE BIOSYNTHESIS AND INTER-ORGAN TRANSPORT IN PLANTS UNDER ABIOTIC STRESS-INDUCED OXIDATIVE STRESS

9.4.1 PROLINE BIOSYNTHESIS

Pro biosynthesis occurs via two biosynthetic pathways (Figure 9.3) in higher plants, namely glutamate and ornithine pathways; where glutamate (Glu) and ornithine (Orn) are used as a precursor, respectively (Verslues and Sharma 2010; Hayat et al. 2012). Pro synthesis from glutamate involves two enzymes i.e. Δ1-pyrroline-5-carboxylate synthetase (P5CS) and Δ1-pyrroline-5-carboxylate reductase (P5CR) (Ashraf and Foolad 2007; Burritt 2012). Firstly, glutamate phosphorylates to γ-glutamyl phosphate, which turns into an intermediate product glutamic- 5-semialdehyde (GSA) by reduction process with the help of bifunctional enzyme P5CS, which is continuously converted to Δ1-pyrroline-5-carboxylate (P5C), and thereafter P5C is reduced into Pro by the enzyme P5CR (Burritt 2012). In this process, ATP and NADPH are required for P5CS. In most recent studies, it was established that under stress condition Pro synthesis mainly occurred due to the huge production of glutamate in the plant body (Saadia et al. 2012; Witt et al. 2012) and this glutamate pathway is mostly responsible for Pro accumulation under abiotic stress (Lv et al. 2011). In the second pathway of Pro synthesis, the enzyme ornithine-δ-aminotransferase (δ-OAT) transmits ornithine to GSA in mitochondria and after that converts into Pro via P5C (Szabados and Savoure 2010). In *Arabidopsis thaliana*, P5C exclusively generates for the catabolic branch that produces glutamate ultimately (Funck et al. 2008). Therefore, Pro synthesis only takes place via the glutamate pathway in *Arabidopsis*. In other words, Pro biosynthesis happens in the cytosol under normal conditions

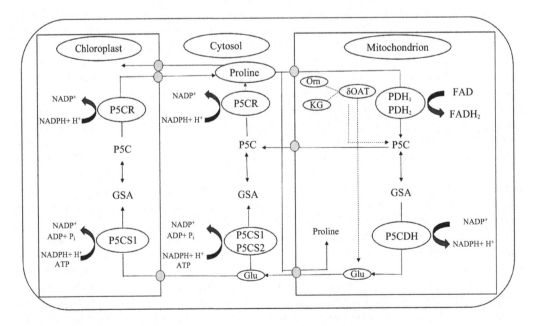

FIGURE 9.3 Proline biosynthesis pathway in the plant cell.

while it takes place in chloroplast also in stress situations and mainly from glutamate (Rejeb et al. 2014). Therefore, it can be concluded that the place of Protein biosynthesis differed with varied environmental conditions.

The biosynthesis enzyme P5CS consists of two genes (P5CS1 and P5CS2) whereas P5CR is comprised of one gene in most plant species (Strizhov et al. 1997), and these enzymes are predicted to be localized in the cytosol of leaf mesophyll cells. The P5CS1 and P5CS2 were detected by using green fluorescent *protein* (GFP) in the cytosol under normal conditions, while in salt or osmotic stress, exclusive accumulation of P5CS1 was detected in chloroplasts (Rayapati et al. 1989). The P5CR activity was found in the cytosol and plastid fraction of *Glycine max*, though it was detected in chloroplasts under osmotic stress (Szoke et al. 1992). The Pro accumulation during stress conditions in intercellular space is mainly determined by its biosynthesis, catabolism and transport between cells and different cellular compartments.

9.4.2 Pro Transport to Uptake and Transport in Inter-Organ of Plants

Pro accumulation and transport take place due to the increased synthesis under stress conditions between intercellular as well as inter-organ, as implied by the compartmentalization of Pro metabolism. It has been identified that amino acid-specific transporters are involved in Pro transport (Rentsch et al. 2007). Amongst these, three Pro specific transporters namely *Prot1*, *Prot2* and AAP_6 belonging to the amino acid/auxin permease (AAP) family have been identified from *Arabidopsis thaliana* based on C-DNA technology and are likely to be responsible for intercellular and long-distance Pro transport (Schwacke et al. 1999; Rentsch et al. 2007). The expression of *Prot1* was found at a higher level during flowering and seed setting in all organs of the plant, more precisely under salt and water stress. Maximum expressions were observed in young flowers, particularly in floral stalk phloem. In a similar situation, *Prot2* is involved in nitrogen distribution, especially under water stress, while AAP_6 transcripts were identified mostly in sink tissues such as roots and cauline leaves (Rentsch et al. 1996). In lucerne, higher Pro accumulation in phloem sap was reported under draught-induced abiotic stress (Girousse et al. 1996). Similarly, Verslues and Sharp (1999) and Raymond and Smirnoff (2002) reported the evidence regarding Pro transport to the root tips of *Zea mays* in the water-scarce situation. In halophytic species, *Limonium latifolium*, Pro sequestration takes place in vacuoles under the stress-free condition, whereas it sequestrates in the cytosol under salt-induced abiotic stress which indicates the importance of *de novo* in both biosynthesis and transportation of Pro (Gagneul et al. 2007). *LeProt1* is involved in the transportation of Pro to the mature as well as germinating pollen (Hossain et al. 2019). Zhang et al. (2014) reported that four *Prot*-homologous genes (*ClProt1–4*) were highly expressed in various organs of *Chrysanthemum lavandulifolium* under stress situations, though the expression of *ClProt2* was restricted in aboveground plant organs. The aforementioned discussion indicates that higher Pro transport occurs under abiotic stress-induced oxidative stress, which also agrees with Kishor et al. (2005).

9.4.3 Pro Transporters and Physiological Mechanisms

Pro transporters belong to the amino acid transporter (ATF)/amino acid/auxin permease (AAAP) family and the amino acid-polyamine-choline (APC) family (Table 9.1) (Rentsch et al. 2007). The ATF/AAAP family has been divided into various sub-families *viz.* amino acid permease (AAP) family, the lysine/histidine transporter (LHT) family and the Pro transporter (*Prot*) family. AAPs facilitate *Prot*on-coupled uptake of glutamate (aspartate) as well as neutral amino acids including Pro (Okumoto et al. 2002; Lee et al. 2007; Schmidt et al. 2007). LHTs are responsible for the transport of neutral and acidic amino acids including Pro with high affinity (Hirner et al. 2006). Apart from the AAP and LHT family, *Prot*s also transport Pro except for other *Prot*einogenic amino acids (Rentsch et al. 1996). Additionally, *Prot*s from *Arabidopsis*, tomato and gray mangrove are responsible for glycine betaine transport (Waditee et al. 2002; Grallath et al. 2005).

TABLE 9.1
Proline Transporters in Different Parts of Plants

Plant Species	Target Proteins	Accession Number	K_m for Proline (μM)	Expression Organs	References
ATF/AAAP (*Amino acid transporter family/amino acid/auxin permease*) gene family					Ueda et al. (2001), Kishor
Prot proline transporter (Substrates: Proline, glycine betaine)					et al. (2005), Verbruggen
Rice	*OsProt*	AB022783	-	All organs	and Hermans (2008),
Arabidopsis	*AtProt1*	At2g39890	427	All organs	Mattioli et al. (2009),
	AtProt2	At3g55740	500	Root, leaf	Lehmann et al. (2010),
	AtProt3	At2g36590	999	Leaf	Hayat et al. (2012),
Tomato	*LeProt1*	AF014808	1900	Pollen	Roychoudhury et al.
Barley	*HvProt1*	AB073084	25	Root	(2015), Kaur and Asthir
	HvProt2	AB545851	246	Root, leaf	(2015), Hossain et al.
Oilpalm	*EgProt1*	AB597035	-	Root, leaf	(2019)
AAP amino acid permease (Substrates: Neutral amino acids and glutamate, aspertate)					
Arabidopsis	*AtAAP1*	At1g58360	601, 1900	Root, flower, endosperm	
	AtAAP2	At5g09220	140	Root, stem, leaf	
	AtAAP3	At1g77380	250	Root	
	AtAAP4	At5g63850	134	Leaf, Stem	
	AtAAP5	At1g44100	500	Root, leaf stem, flower	
	AtAAP6	At5g49630	67	All organs	
LHT (Lysine/Histidine) transporter (Substrates: Neutral and acidic amino acids)					
Arabidopsis	*AtLHT1*	At5g49780	10	Root, leaf, stem	
	AtLHT1	At1g24400	13	Flower	

A *Prot1* from tomato and three *Prots* from *Arabidopsis* transport the γ-aminobutyric acid (GABA), a stress-induced compound, but it was not a compound for gray mangrove Pro transporters (AmTs) (Waditee et al. 2002; Grallath et al. 2005). The affinity of the Pro to At*Prots* is much higher than GABA (Grallath et al. 2005). This discussion on substrate selectivity of AAPs, LHTs and *Prots* reveals that both transporters with low and high affinity for Pro present in plants, demonstrating the functions in nitrogen transfer and Pro-specificity, respectively. Wood et al. (2001) established that the transporters were regulated by higher gene expression as well as their affinity to a particular substrate, and some functioned as osmosensors additionally (Morbach and Kramer 2002; Wood 2006).

The transporter of Pro belonging to AAP and LHT families are not selective for Pro, and therefore they play a great role in the general acquisition and nitrogen allocation in plants. The Arabidopsis knockout mutants, LHT1, AAP8 and AAP1, revealed the multifarious fluctuations in the levels of amino acids (Rolletschek et al. 2005; Hirner et al. 2006; Schmidt et al. 2007; Weigelt et al. 2008; Sanders et al. 2009). In contrast, the specific action of Pro homeostasis in abiotic stress or normal condition in the plant system is indicated by *Prots* selectivity for Pro and other compatible solutes. As an example, the overexpression of *Arabidopsis Prot2* indicates the Pro accumulation during salt-induced stress conditions (Ueda et al. 2001; Waditee et al. 2002). Higher Pro accumulation is also highly correlated with the *AtProt1* and *LeProt1* transcript abundance in pollen and *AtProt3* in the epidermis (Schwacke et al. 1999). However, few pieces of literature are available that establish the direct influence of *Prots* in Pro transportation. Lehman et al. (2010) reveal that the Pro content does not change with the absence or presence of stress, indicating an altered Pro metabolism pathway or reimbursement by alternate transporters. However, it was reported the lower biomass accumulation and Pro concentration in the shoot is influenced by the overexpression of *HvProt* in Arabidopsis, though the higher Pro accumulation has resulted from root tips and thus elongates the root length, confirming the function of Pro in plant organ development (Ueda et al. 2008).

9.5 CONCLUSION

From the overview of earlier studies, it may be concluded that to meet the food and nutritional security of an increasing population across the globe, the productivity of wheat should be increased. However, recent abnormal weather conditions create serious abiotic stresses including heat, drought and salinity, and cause pollen sterility, disturb photosynthetic apparatus and produce shriveled seeds in wheat, ultimately disturbing the growth and yield of wheat. Therefore, to meet the food and nutritional security of an increasing population, it is important to understand the physiological mechanisms of wheat for sustainable wheat production in the current era of climate change. Among the antioxidants responsive to the plant's tolerance to abiotic stress, Pro is considered one of the most advanced harmonious solutes that acts in a defensive role during antagonistic abiotic constraints. Generally, Pro accumulation and intercellular transport occur between cytosol, chloroplasts and mitochondria to act in a leading role in plant tolerance to various abiotic stresses. The evidence revealed that enhancement of Pro-biosynthesis in the chloroplast substitute is an important osmolyte, metal chelator and is a self-protective and signaling molecule and also could preserve the redox-potential for metabolism during abiotic stress. The study also highlighted that Pro also encourages stress-defending proteins through the expression of salt stress reactive genes owning to Pro-reactive elements in their promoters. Besides these roles of Pro, exogenous application of Pro enhances the activity of antioxidative enzymes, which lead to alleviating the adverse effect of environmental stresses. The information of the present study will be helpful for sustainable wheat productivity in the current era of climate change.

REFERENCES

Abdel Latef AA (2010) Changes of antioxidative enzymes in salinity tolerance among different wheat cultivars. *Cereal Research Communications* 38(1):43–55.

Abdel Latef AA (2011) Ameliorative effect of calcium chloride on growth, antioxidant enzymes, protein patterns and some metabolic activities of canola (*Brassica napus* L.) under seawater stress. *Journal of Plant Nutrition* 34(9):1303–1320.

Abdel Latef AA, Kordrostami M, Zakir A, Zaki H, Saleh OM (2019a) Eustress with H_2O_2 facilitates plant growth by improving tolerance to salt stress in two wheat cultivars. *Plants* 8(9):303.

Abdel Latef AA, Abu Alhmad MFA, Ahmad S (2017a) Foliar application of fresh Moringa leaf extract overcomes salt stress in fenugreek (*Trigonella foenum-Graecum*) plants. *Egyptian Journal of Botany* 57:157–179.

Abdel Latef AA, Abu Alhmad MFA, Abdelfattah KE (2017b) The possible roles of priming with ZnO nanoparticles in mitigation of salinity stress in lupine (*Lupinus termis*) plants. *Journal of Plant Growth Regulation* 36(1):60–70.

Abdel Latef AA, Abu Alhmad MFA, Kordrostami M, Abo–Baker AA, Zakir A (2020a). Inoculation with *Azospirillum lipoferum* or *Azotobacter chroococcum* reinforces maize growth by improving physiological activities under saline conditions. *Journal of Plant Growth Regulation* 39(3):1293–1306.

Abdel Latef AA, Dawood MF, Hassanpour H, Rezayian M, Younes NA (2020b) Impact of the static magnetic field on growth, pigments, osmolytes, nitric oxide, hydrogen sulfide phenylalanine ammonia-lyase activity, antioxidant defense system, and yield in lettuce. *Biology Multidisciplinary Digital Publishing Institute* 9(7):172.

Abdel Latef AA, Mostofa MG, Rahman MM, Abdel-Farid IB, Tran L, -SP (2019b) Extracts from yeast and carrot roots enhance maize performance under seawater-induced salt stress by altering physio-biochemical characteristics of stressed plants. *Journal of Plant Growth Regulation* 38(3):966–979.

Abdel Latef AA, Omer AM, Badawy AA, Osman MS, Ragaey MM (2021) Strategy of salt tolerance and inter-active impact of *Azotobacter chroococcum* and/or *Alcaligenes faecalis* inoculation on canola (*Brassica napus* L.) plants grown in saline soil. *Plants* 10(1):110. https://doi.org/10.3390/plants10010110

Abdel Latef AA, Shaddad MA, Ismail AM, Abou Elhamd MF (2009) Benzyladenine can alleviate saline injury of two roselle cultivars via equilibration cytosolutes including anthocyanin. *International Journal of Agriculture and Biolology* 11:151–157.

Abdel Latef AAH, Srivastava AK, El-sadek MSA, Kordrostami M, Tran LP (2018) Titanium dioxide nanopar-ticles improve growth and enhance tolerance of broad bean plants under saline soil conditions. *Land Degradation and Development* 29(4):1065–1073.

Abdel Latef AAH, Srivastava AK, Saber H, Alwaleed EA, Tran LP (2017c) *Sargassum muticum* and *Jania rubens* regulate amino acid metabolism to improve growth and alleviate salinity in chickpea. *Scientific Reports* 7(1):10537.

Abdel Latef AA, Zaid A, Abo-Baker AA, Salem W, Abu Alhmad MF (2020c) Mitigation of copper stress in maize by inoculation with *Paenibacillus polymyxa* and *Bacillus circulans*. *Plants* 9(11):1513.

Abhinandan K, Skori L, Stanic M, Hickerson N, Jamshed M, Samuel MA (2018) Abiotic stress signaling in wheat—an inclusive overview of hormonal interactions during abiotic stress responses in wheat. *Frontiers in Plant Science* 9:734. https://doi.org/10.3389/fpls.2018.00734.

Acevedo E, Silva P, Silva H (2002) Wheat growth and physiology. In: Curtis, B.C., Rajaram, S. and Macpherson, H.G. (Eds.), *Bread Wheat Improvement and Production, FAO Plant Production and Protection Series, No. 30*. Food and Agriculture Organization of the United Nations, Rome, Italy. http://www.fao.org/3/y4011e/y4011e06.htm

Akter N, Rafiqul Islam MR (2017) Heat stress effects and management in wheat: a review. *Agronomy for Sustainable Development* 37(5):37. https://doi.org/10.1007/s13593-017-0443-9.

Ali Q, Ashraf M, Athar HUR (2007) Exogenously applied proline at different growth stages enhances growth of two maize cultivars grown under water deficit conditions. *Pakistan Journal of Botany* 39: 1133–1144.

Ali Q, Ashraf M, Shahbaz M, Humera H (2008) Ameliorating effect of foliar applied proline on nutrient uptake in water stressed maize (Zea mays L.) plants. *Pakistan Journal of Botany* 40(1):211–219.

Almansouri M, Kinet J-M, Lutts S (2001) Effect of salt and osmotic stresses on germination in durum wheat (Triticum durum Desf.). *Plant and Soil* 231(2):243–254.

Aly AA, Latif HH (2011) Differential effects of paclobutrazol on water stress alleviation through electrolyte leakage, phytohormones, reduced glutathione and lipid peroxidation in some wheat genotypes (Triticum aestivum L.) grown in-vitro. *Romanian Biotechnology Letters* 6:6710–6721.

Anjum SA, Wang LC, Farooq M, Hussain M, Xue LL, Zou CM (2011) Brassinolide application improves the drought tolerance in maize through modulation of enzymatic antioxidants and leaf gas exchange. *Journal of Agronomy and Crop Science* 197(3):177–185.

Azooz MM, Shaddad MA, Abdel-Latef AA (2004) The accumulation and compartmentation of proline in relation to salt tolerance of three sorghum cultivars. *Indian Journal of Plant Physiology* 9:1–8.

Ashraf MA, Ashraf M, Ali Q (2010) Response of two genetically diverse wheat cultivars to salt stress at differ-ent growth stages: leaf lipid peroxidation and phenolic contents. *Pakistan Journal of Botany* 42:559–565.

Ashraf MA, Ashraf M, Shahbaz M (2012) Growth stage-based modulation in antioxidant defense system and proline accumulation in two hexaploid wheat (Triticum aestivum L.) cultivars differing in salinity toler-ance. *Flora—Morphology, Distribution, Functional Ecology of Plants* 207(5):388–397.

Ashraf M, Foolad MR (2007) Roles of glycine betaine and proline in improving plant abiotic stress resistance. *Environmental and Experimental Botany* 59(2):206–216.

Asseng S, Foster IAN, Turner NC (2011) The impact of temperature variability on wheat yields. *Global Change Biology* 17(2):997–1012.

Asseng S, Ewert F, Martre P, Rötter RP, Lobell DB, Cammarano D et al. (2015) Rising temperatures reduce global wheat production. *Nature Climate Change* 5(2):143–147.

Bali AS, Sidhu GPS (2019) Abiotic stress-induced oxidative stress in wheat. In: *Wheat Production in Changing Environments*. Springer, Singapore, pp. 225–239. https://doi.org/10.1007/978-981-13-6883-7_10.

Barlow KM, Christy BP, O'leary GJ, Riffkin PA, Nuttall JG (2015) Simulating the impact of extreme heat and frost events on wheat crop production: a review. *Field Crops Research* 171(109): 109–119.

Ben Hassine AB, Ghanem ME, Bouzid S, Lutts S (2008) An inland and a coastal population of the Mediterranean xero-halophyte species Atriplex halimus L. differ in their ability to accumulate proline and glycinebeta-ine in response to salinity and water stress. *Journal of Experimental Botany* 59(6):1315–1326.

Ben Rejeb KB, Abdelly C, Savouré A (2014) How reactive oxygen species and proline face stress together. *Plant Physiology and Biochemistry: PPB* 80:278–284.

Bernstein L (1975) Effects of salinity and sodicity on plant growth. *Annual Review of Phytopathology* 13(1):295–312.

Brestic M, Zivcak M, Hauptvogel P, Misheva S, Kocheva K, Yang X, Li X, Allakhverdiev SI (2018) Wheat plant selection for high yields entailed improvement of leaf anatomical and biochemical traits including tolerance to non-optimal temperature conditions. *Photosynthesis Research* 136(2):245–255.

Brestic M, Zivcak M, Kunderlikova K, Allakhverdiev SI (2016) High temperature specifically affects the photoprotective responses of chlorophyll b-deficient wheat mutant lines. *Photosynthesis Research* 130(1–3):251–266.

Burritt DJ (2012) Proline and the cryopreservation of plant tissues: functions and practical applications. In: Katkov, I. (Ed.), *Current Frontiers in Cryopreservation*. INTECH-Open Access Publisher, Croatia, pp. 415–430.

Caverzan A, Casassola A, Brammer SP (2016) Antioxidant responses of wheat plants under stress. *Genetics and Molecular Biology* 39(1):1–6.

Chakrabarti B, Singh SD, Nagarajan S, Aggarwal PK (2011) Impact of temperature on phenology and pollen sterility of wheat varieties. *Australian Journal of Crop Science* 5:1039–1043.

Chen Z, Cuin TA, Zhou M, Twomey A, Naidu BP, Shabala S (2007) Compatible solute accumulation and stress-mitigating effects in barley genotypes contrasting in their salt tolerance. *Journal of Experimental Botany* 58(15–16):4245–4255.

Chinnusamy V, Jagendorf A, Zhu JK (2005) Understanding and improving salt tolerance in plants. *Crop Science* 45(2):437–448.

Cvikrová M, Gemperlová L, Dobrá J, Martincová O, Prásil IT, Gubis J, Vanková R (2012) Effect of heat stress on polyamine metabolism in proline-over-producing tobacco plants. *Plant Science: An International Journal of Experimental Plant Biology* 182:49–58.

De Ronde JA, Cress WA, Krüger GHJ, Strasser RJ, Van Staden J (2004) Photosynthetic response of transgenic soybean plants, containing an Arabidopsis P5CR gene, during heat and drought stress. *Journal of Plant Physiology* 161(11):1211–1224.

Djanaguiraman M, Prasad PVV, Boyle DL, Schapaugh WT (2011) High-temperature stress and soybean leaves: leaf anatomy and photosynthesis. *Crop Science* 51(5):2125–2131.

Dong B, Zheng X, Liu H, Able JA, Yang H, Zhao H, Zhang M, Qiao Y, Wang Y, Liu M (2017) Effects of drought stress on pollen sterility, grain yield, abscisic acid and protective enzymes in two winter wheat cultivars. *Frontiers in Plant Science* 8:1008. https://doi.org/10.3389/fpls.2017.01008.

Doubková P, Vlasáková E, Sudová R (2013) Arbuscular mycorrhizal symbiosis alleviates drought stress imposed on Knautia arvensis plants in serpentine soil. *Plant and Soil* 370(1–2):149–161. https://doi.org/10.1007/s11104-013-1610-7.

El-Shintinawy F, El-Shourbagy MN (2001) Alleviation of changes in protein metabolism in NaCl-stressed wheat seedlings by thiamine. *Biologia Plantarum* 44(4):541–545.

FAO (2008) *FAO Land and Plant Nutrition Management Services*. FAO Soil Portal, Soil Management, Salt-affected Soils. http://www.fao.org/ag/agl/agll/spush.

FAO (2014) *Crop prospects and food situation*. Food and Agriculture Organization, Global Information and Early Warning System, Trade and Markets Division (EST), Rome.

Farooq M, Bramley H, Palta JA, Siddique KH (2011) Heat stress in wheat during reproductive and grain-filling phases. *Critical Reviews in Plant Sciences* 30(6):491–507.

Farooq M, Hussain M, Siddique KH (2014) Drought stress in wheat during flowering and grain-filling periods. *Critical Reviews in Plant Sciences* 33(4):331–349.

Foyer CH, Ruban AV, Nixon PJ (2017) Photosynthesis solutions to enhance productivity. *Philosophical Transactions of the Royal Society of London. Series B, Biological Sciences* 372(1730). https://doi.org/10.1098/rstb.2016.0374. https://www.ncbi.nlm.nih.gov/pubmed/28808094.

Francois LE, Maas EV (1999) Crop response and management wheat of salt-affected soils. In: Pessarakli, M. (Ed.) *Handbook of Plant and Crop Stress*, 2nd Ed. Marcel Dekker, Inc., New York, pp. 169–201.

Funck D, Stadelhofer B, Koch W (2008) Ornithine-δ-aminotransferase is essential for Arginine catabolism but not for proline biosynthesis. *BMC Plant Biology* 8(1):40.

Gagneul D, Aïnouche A, Duhazé C, Lugan R, Larher FR, Bouchereau A (2007) A reassessment of the function of the so-called compatible solutes in the halophytic Plumbaginaceae *Limonium latifolium*. *Plant Physiology* 144(3):1598–1611.

Gibson LR, Paulsen GM (1999). Yield components of wheat grown under high temperature stress during reproductive growth. *Crop Science* 39(6):1841–6.

Girousse C, Bournoville R, Bonnemain JL (1996) Water deficit-induced changes in concentrations in proline and some other amino acids in the phloem sap of alfalfa. *Plant Physiology* 111(1):109–113.

Grallath S, Weimar T, Meyer A, Gumy C, Suter-Grotemeyer M, Neuhaus JM, Rentsch D (2005) The AtProT family. Compatible solute transporters with similar substrate specificity but differential expression patterns. *Plant Physiology* 137(1):117–126. https://doi.org/10.1104/pp.104.055079.

Groppa MD, Benavides MP (2008) Polyamines and abiotic stress: recent advances. *Amino Acids* 34(1):35–45.

Gupta N, Thind SK, Bains NS (2014) Glycine betaine application modifies biochemical attributes of osmotic adjustment in drought stressed wheat. *Plant Growth Regulation* 72(3):221–228.

Gupta RK, Abrol IP (1990) Salt-affected soils: their reclamation and management for crop production. *Advances in Soil Science* 11:223–288.

Hamilton EW, Heckathorn SA (2001) Mitochondrial adaptations to NaCl. Complex I is protected by antioxidants and small heat shock proteins, whereas complex II is protected by proline and betaine. *Plant Physiology* 126(3):1266–1274.

Hare PD, Cress WA (1997) Metabolic implications of stress induced proline accumulation in plants. *Plant Growth Regulation* 21(2):79–102.

Hasanuzzaman M, Fujita M (2013) Exogenous sodium nitroprusside alleviates arsenic-induced oxidative stress in wheat (Triticum aestivum L.) seedlings by enhancing antioxidant defense and glyoxalase system. *Ecotoxicology* 22(3):584–596.

Hasanuzzaman M, Nahar K, Alam MM, Fujita M (2012) Exogenous nitric oxide alleviates high temperature induced oxidative stress in wheat (*Triticum aestivum* L.) seedlings by modulating the antioxidant defense and glyoxalase system. *Australian Journal of Crop Science* 6:1314–1323.

Hayano-Kanashiro C, Calderón-Vázquez C, Ibarra-Laclette E, Herrera-Estrella L, Simpson J (2009) Analysis of gene expression and physiological responses in three Mexican maize landraces under drought stress and recovery irrigation. *PLOS ONE* 4(10):e7531.

Hayat S, Hayat Q, Alyemeni MN, Wani AS, Pichtel J, Ahmad A (2012) Role of proline under changing environments: a review. *Plant Signaling and Behavior* 7(11):1456–1466.

Herzog M, Striker GG, Colmer TD, Pedersen O (2016) Mechanisms of waterlogging tolerance in wheat—a review of root and shoot physiology. *Plant, Cell and Environment* 39(5):1068–1086.

Hirner A, Ladwig F, Stransky H, Okumoto S, Keinath M, Harms A, Frommer WB, Koch W (2006) Arabidopsis LHT1 is a high-affinity transporter for cellular amino acid uptake in both root epidermis and leaf mesophyll. *Plant Cell* 18(8):1931–1946. https://doi.org/10.1105/tpc.106.041012.

Hossain A, da Silva JAT, Lozovskaya MV, Zvolinsky VP, Mukhortov VI (2012b) High temperature combined with drought affect rainfed spring wheat and barley in south-eastern Russia: yield, relative performance and heat susceptibility index. *Journal of Plant Breeding and Crop Science* 4(11):184–196.

Hossain A, Sarker MAZ, Saifuzzaman M, Teixeira da Silva JA, Lozovskaya MV, Akhter MM (2013) Evaluation of growth, yield, relative performance and heat susceptibility of eight wheat (Triticum aestivum L.) genotypes grown under heat stress. *International Journal of Plant Production* 7(3):615–636.

Hossain A, Teixeira da Silva JA (2012) Phenology, growth and yield of three wheat (Triticum aestivum L.) varieties as affected by high temperature stress. *Notulae Scientia Biologicae* 4(3):97–109. https://doi.org/10.15835/nsb437879.

Hossain A, Teixeira da Silva JAT, Lozovskaya MV, Zvolinsky VP (2012a) High temperature combined with drought affect rainfed spring wheat and barley in South-Eastern Russia: I. Phenology and growth. *Saudi Journal of Biological Sciences* 19(4):473–487.

Hossain MA, Fujita M (2012) Regulatory role of components of ascorbate-glutathione (AsAGSH) pathway in plant tolerance to oxidative stress. In: Anjum, N.A., Umar, S. and Ahmed, A. (Eds.), *Oxidative Stress in Plants: Causes, Consequences and Tolerance*. IK International Publishing House Pvt. Ltd., India, pp. 81–147.

Hossain MA, Kumar V, Burritt DJ, Fujita M, Mäkelä PSA (2019) Osmoprotectant- mediated abiotic stress tolerance in plants. In: Trovato, M., Forlani, G., Signorelli, S. and Funck, D. (Eds.), *Proline Metabolism and Its Functions in Development and Stress Tolerance*. *Springer Nature Switzerland*, pp. 41–72. https://doi.org/10.1007/978-3-030-27423-8.

Hossain MA, Teixeira da Silva JA, Fujita M (2011) Glyoxalase system and reactive oxygen species detoxification system in plant abiotic stress response and tolerance: an intimate relationship. In: Shanker, A.K. and Venkateswarlu, B. (Eds.), *Abiotic Stress/Book 1*. INTECH Open Access Publisher, Rijeka, Croatia, pp. 235–266.

Ihsan MZ, El-Nakhlawy FS, Ismail SM, Fahad S, Daur I (2016) Wheat phenological development and growth studies as affected by drought and late season high temperature stress under arid environment. *Frontiers in Plant Science* 7:795.

Iloh AC, Omatta G, Ogbadu GH, Onyenekwe PC (2014) Effects of elevated temperature on seed germination and seedling growth on three cereal crops in Nigeria. *Scientific Research and Essays* 9(18):806–813.

James RA, Blake C, Byrt CS, Munns R (2011) "Major genes for Na+ exclusion, Nax1 and Nax2 (wheatHKT1;4 and HKT1;5), decrease Na+ accumulation in bread wheat leaves under saline and waterlogged conditions." *Journal of Experimental Botany* 62(8):2939–2947.

Ji X, Shiran B, Wan J, Lewis DC, Jenkins CLD, Condon AG, Richards RA, Dolferus R (2010) Importance of pre-anthesis anther sink strength for maintenance of grain number during reproductive stage water stress in wheat. *Plant, Cell and Environment* 33(6):926–942.

Joseph EA, Radhakrishnan VV, Mohanan KV (2015) A study on the accumulation of proline – An osmoprotectant amino acid under salt stress in some native rice cultivars of North Kerala, India. – Univ. *Journal of Agricultural Research* 3:15–22.

Kamran M, Shahbaz M, Ashraf M, Akram NA (2009) Alleviation of drought-induced adverse effects in spring wheat (Triticum aestivum L.) using proline as a pre-sowing seed treatment. *Pakistan Journal of Botany* 41:621–632.

Kandowangko NY, Suryatmana G, Nurlaeny N, Simanungkalit RDM (2009) Proline and abscisic acid content in droughted corn plant inoculated with Azospirillum sp. and arbuscular mycorrhizae fungi. *HAYATI Journal of Biosciences* 16(1):15–20. https://doi.org/10.4308/hjb.16.1.15.

Kaur G, Asthir BJ (2015) Proline: a key player in plant abiotic stress tolerance. *Biologia Plantarum* 59(4):609–619.

Kavi Kishor PB, Sangam S, Amruth RN, Sri Laxmi P, Naidu KR, Rao KRSS, et al. (2005) Regulation of proline biosynthesis, degradation, uptake and transport in higher plants: its implications in plant growth and abiotic stress tolerance. *Current Science* 88:424–438.

Keyvan S (2010) The effects of drought stress on yield, relative water content, proline, soluble carbohydrates and chlorophyll of bread wheat cultivars. *Journal of Animal and Plant Sciences* 8:1051–1060.

Khan MI, Nazir F, Asgher M, Per TS, Khan NA (2015) Selenium and sulfur influence ethylene formation and alleviate cadmium-induced oxidative stress by improving proline and glutathione production in wheat. *Journal of Plant Physiology* 173:9–18.

Khan MIR, Iqbal N, Masood A, Per TS, Khan NA (2013) Salicylic acid alleviates adverse effects of heat stress on photosynthesis through changes in proline production and ethylene formation. *Plant Signaling and Behavior* 8(11): e26374.

Kishor PK, Sangam S, Amrutha RN, Laxmi PS, Naidu KR, Rao KS, Rao S, Reddy KJ, Theriappan P, Sreenivasulu N (2005) Regulation of proline biosynthesis, degradation, uptake and transport in higher plants: its implications in plant growth and abiotic stress tolerance. *Current Science* 10: 424–438.

Kobata T, Koç M, Barutçular C, Matsumoto T, Nakagawa H, Adachi F, Ünlü M (2012) Assimilate supply as a yield determination factor in spring wheat under high temperature conditions in the Mediterranean zone of South-East Turkey. *Plant Production Science* 15(3):216–227.

Kocsy G, Laurie R, Szalai G, Szilagyi V, Simon-Sarkadi L, Galiba G, de Ronde JA (2005) Genetic manipulation of proline levels affects antioxidants in soybean subjected to simultaneous drought and heat stresses. *Physiologia Plantarum* 124(2):227–235.

Krasensky J, Jonak C (2012) Drought, salt, and temperature stress-induced metabolic rearrangements and regulatory networks. *Journal of Experimental Botany* 63(4):1593–1608.

Krishnan N, Dickman MB, Becker DF (2008) Proline modulates the intracellular redox environment and protects mammalian cells against oxidative stress. *Free Radical Biology and Medicine* 44(4):671–681.

Kumar S, Kaur R, Kaur N, Bhandhari K, Kaushal N, Gupta K, Bains TS, Nayyar H (2011) Heat-stress induced inhibition in growth and chlorosis in mungbean (Phaseolus aureus Roxb.) is partly mitigated by ascorbic acid application and is related to reduction in oxidative stress. *Acta Physiologiae Plantarum* 33(6):2091–2101.

Lamb RS (2012) Abiotic stress responses in plants: a focus on the SRO family. In: Montanaro, G. (Ed.), *Advances in Selected Plant Physiology Aspects*. INTECH-Open Access Publisher, Rijeka, Croatia, pp. 1–21.

Läuchli A, Epstein E (1990) Plant responses to saline and sodic conditions. In: Tanji, K.K. (Ed.), *Agricultural Salinity Assessment and Management, ASCE Manuals and Reports on Engineering Practice*. ASCE, New York, pp. 113–137.

Läuchli A, Grattan SR (2007) Plant growth and development under salinity stress. In: enks MA, Hasegawa PM, Jain SM (eds), *Advances in Molecular Breeding Toward Drought and Salt Tolerant Crops*. Springer, Dordrecht, pp. 1–32.

Lee YH, Foster J, Chen J, Voll LM, Weber APM, Tegeder M (2007) AAP1 transports uncharged amino acids into roots of *Arabidopsis*. *Plant Journal: For Cell and Molecular Biology* 50(2):305–319. https://doi.org/10.1111/j.1365-313X.2007.03045.x.

Lehmann S, Funck D, Szabados L, Rentsch D (2010) Proline metabolism and transport in plant development. *Amino Acids* 39(4):949–962.

Lewandowski I, Härdtlein M, Kaltschmitt M (1999) Sustainable crop production: definition and methodological approach for assessing and implementing sustainability. *Crop Science* 39(1):184–193.

Li X, Yang Y, Jia L, Chen H, Wei X (2013) Zinc-induced oxidative damage, antioxidant enzyme response and proline metabolism in roots and leaves of wheat plants. *Ecotoxicology and Environmental Safety* 89:150–157.

Liang X, Zhang L, Natarajan SK, Becker DF (2013) Proline mechanisms of stress survival. *Antioxidants and Redox Signaling* 19(9):998–1011.

Lott N, Ross T, Smith A, Houston T, Shein K (2011) *Billion Dollar US Weather Disasters, 1980–2010*. National Climatic Data Center, Asheville. http://www.ncdc.noaa.gov/oa/reports/billionz.html.

Lv WT, Lin B, Zhang M, Hua XJ (2011) Proline accumulation is inhibitory to Arabidopsis seedlings during heat stress. *Plant Physiology* 156(4):1921–1933.

Maghsoudi K, Emam Y, Niazi A, Pessarakli M, Arvin MJ (2018) P5CS expression level and proline accumulation in the sensitive and tolerant wheat cultivars under control and drought stress conditions in the presence/absence of silicon and salicylic acid. *Journal of Plant Interactions* 13(1):461–471.

Mathur S, Jajoo A, Mehta P, Bharti S (2011) Analysis of elevated temperature-induced inhibition of photosystem II using chlorophyll a fluorescence induction kinetics in wheat leaves (*Triticum aestivum*). *Plant Biology* 13(1):1–6.

Mattioli R, Costantino P, Trovato M (2009) Proline accumulation in plants: not only stress. *Plant Signaling and Behavior* 4(11):1016–1018.

Mitchell RAC, Mitchell VJ, Driscoll SP, Franklin J, Lawlor DW (1993) Effects of increased CO2 concentration and temperature on growth and yield of winter wheat at two levels of nitrogen application. *Plant, Cell and Environment* 16(5):521–529.

Modhej A, Naderi A, Emam Y, Aynehband A, Normohamadi G (2012) Effects of post-anthesis heat stress and nitrogen levels on grain yield in wheat (*T. durum* and *T. aestivum*) genotypes. *International Journal of Plant Production* 2:257–268.

Molinari HBC, Marur CJ, Daros E, De Campos MKF, De Carvalho JFRP, Filho JCB, Pereira LFP, Vieira LGE (2007) Evaluation of the stress-inducible production of proline in transgenic sugarcane (Saccharum spp.): osmotic adjustment, chlorophyll fluorescence and oxidative stress. *Physiologia Plantarum* 130(2):218–229.

Morbach S, Krämer R (2002) Body shaping under water stress: osmosensing and osmoregulation of solute transport in bacteria. *Chembiochem: A European Journal of Chemical Biology* 3(5):384–397. https://doi.org/10.1002/1439-7633(20020503)3:5<384::AID-CBIC384>3.0.CO;2-H.

Munns R (2002) Comparative physiology of salt and water stress. *Plant, Cell and Environment* 25(2):239–250.

Naika M, Shameer K, Mathew OK, Gowda R, Sowdhamini R (2013) STIFDB2: an updated version of plant Stress-responsive transcription factor database with additional stress signals, stress-responsive transcription factor binding sites and stress-responsive genes in Arabidopsis and rice. *Plant and Cell Physiology* 54(2):e8.

Narayanan S, Prasad PVV, Fritz AK, Boyle DL, Gill BS (2015) Impact of high night-time and high daytime temperature stress on winter wheat. *Journal of Agronomy and Crop Science* 201(3):206–218.

Naveed M, Hussain MB, Zahir ZA, Mitter B, Sessitsch A (2014) Drought stress amelioration in wheat through inoculation with *Burkholderia phytofirmans* strain PsJN. *Plant Growth Regulation* 73(2): 121–131.

Nawaz H, Yasmeen A, Anjum MA, Hussain N (2016) Exogenous application of growth enhancers mitigate water stress in wheat by antioxidant elevation. *Frontiers in Plant Science* 7:597.

Nayyar H (2003) Accumulation of osmolytes and osmotic adjustment in water-stressed wheat (Triticum aestivum) and maize (Zea mays) as affected by calcium and its antagonists. *Environmental and Experimental Botany* 50(3):253–264.

Nuttall JG, Barlow KM, Delahunty AJ, Christy BP, O'Leary GJ (2018) Acute high temperature response in wheat. *Agronomy Journal* 110(4):1296–1308.

Okumoto S, Schmidt R, Tegeder M, Fischer WN, Rentsch D, Frommer WB, Koch W (2002) High affinity amino acid transporters specifically expressed in xylem parenchyma and developing seeds of *Arabidopsis. Journal of Biological Chemistry* 277(47):45338–45346. https://doi.org/10.1074/jbc.M207730200.

Osman MS, Badawy AA, Osman AI, Abdel Latef AAHA (2020) Ameliorative impact of an extract of the halophyte Arthrocnemum macrostachyum on growth and biochemical parameters of soybean under salinity stress. *Journal of Plant Growth Regulation*. https://doi.org/10.1007/s00344-020-10185-2.

Perdomo JA, Capó-Bauçà S, Carmo-Silva E, Galmés J (2017) RuBisCO and RuBisCO Activase play an important role in the biochemical limitations of photosynthesis in rice, wheat, and maize under high temperature and water deficit. *Frontiers in Plant Science* 8:490. https://doi.org/10.3389/fpls.2017.00490.

Perdomo JA, Conesa MÀ, Medrano H, Ribas-Carbó M, Galmés J (2015) Effects of long-term individual and combined water and temperature stress on the growth of rice, wheat and maize: relationship with morphological and physiological acclimation. *Physiologia Plantarum* 155(2):149–165.

Petrov V, Hille J, Mueller-Roeber B, Gechev TS (2015) ROS-mediated abiotic stress-induced programmed cell death in plants. *Frontiers in Plant Science* 6:69.

Plaut Z, Edelstein M, Ben-Hur M (2013). Overcoming salinity barriers to crop production using traditional methods. *Critical Reviews in Plant Sciences* 32(4):250–291.

Pradhan GP, Prasad PV, Fritz AK, Kirkham MB, Gill BS (2012) Effects of drought and high temperature stress on synthetic hexaploid wheat. *Functional Plant Biology: FPB* 39(3):190–198.

Prasad PVV, Pisipati SR, Momčilović I, Ristic Z (2011) Independent and combined effects of high temperature and drought stress during grain filling on plant yield and chloroplast EF-Tu expression in spring wheat. *Journal of Agronomy and Crop Science* 197(6):430–441.

Prasad PVV, Pisipati SR, Ristic Z, Bukovnik U, Fritz AK (2008) Impact of nighttime temperature on physiology and growth of spring wheat. *Crop Science* 48(6):2372–2380.

Qaseem MF, Qureshi R, Shaheen H (2019) Effects of pre-anthesis drought, heat and their combination on the growth, yield and physiology of diverse wheat (Triticum aestivum L.) genotypes varying in sensitivity to heat and drought stress. *Scientific Reports* 9(1):6955.

Qin F, Shinozaki K, Yamaguchi-Shinozaki K (2011) Achievements and challenges in understanding plant abiotic stress responses and tolerance. *Plant and Cell Physiology* 52(9):1569–1582.

Rahnama A, James RA, Poustini K, Munns R (2010) Stomatal conductance as a screen for osmotic stress tolerance in durum wheat growing in saline soil. *Functional Plant Biology* 37(3):255–63.

Raja MM, Vijayalakshmi G, Naik ML, Basha PO, Sergeant K, Hausman JF, Khan PSSV (2019) Pollen development and function under heat stress: from effects to responses. *Acta Physiologiae Plantarum* 41(4):47. https://doi.org/10.1007/s11738-019-2835-8.

Rascio A, Picchi V, Naldi JP, Colecchia S, De Santis G, Gallo A, Carlino E, Lo Scalzo RL, De Gara L (2015) Effects of temperature increase, through spring sowing, on antioxidant power and health-beneficial substances of old and new wheat varieties. *Journal of Cereal Science* 61:111–118.

Rayapati PJ, Stewart CR, Hack E (1989) Pyrroline-5- carboxylate reductase is in pea (*Pisum sativum* L.) leaf chloroplasts. *Plant Physiology* 91(2):581–586.

Raymond MJ, Smirnoff N (2002) Proline metabolism and transport in maize seedlings at low water potential. *Annals of Botany* 89:813–823.

Rengasamy P (2006) World salinization with emphasis on Australia. *Journal of Experimental Botany* 57(5):1017–1023.

Rentsch D, Hirner B, Schmelzer E, Frommer WB (1996) Salt stress-induced proline transporters and salt stress-repressed broad specificity amino acid permeases identified by suppression of a yeast amino acid permease-targeting mutant. *The Plant Cell* 8(8):1437–1446.

Rentsch D, Schmidt S, Tegeder M (2007) Transporters for uptake and allocation of organic nitrogen compounds in plants. *FEBS Letters* 581(12):2281–2289.

Reynolds MP, Nagarajan S, Razzaque MA, Ageeb OAA (2001) Heat tolerance. In: Reynolds, M.P., Ortiz-Monasterio, J.I. and McNab, A. (Eds.), *Application of Physiology in Wheat Breeding. Mexico.* CIMMYT, DF, pp. 124–134.

Rhoades JDF, Chanduwi S, Lesch (1999) *Soil Salinity Assessment.* Food and Agriculture Organization of the Unites Nations. FAO, pp. 5254–5284.

Rivero RM, Mestre TC, Mittler R, Rubio F, Garcia-Sanchez F, Martinez V (2014) The combined effect of salinity and heat reveals a specific physiological, biochemical and molecular response in tomato plants. *Plant, Cell and Environment* 37(5):1059–1073.

Rizhsky L, Liang H, Shuman J, Shulaev V, Davletova S, Mittler R (2004) When defense pathways collide. The response of Arabidopsis to a combination of drought and heat stress. *Plant Physiology* 134(4):1683–96.

Rizwan M, Ali S, Abbas T, Zia-ur-Rehman M, Hannan F, Keller C, Al-Wabel MI, Ok YS (2016) Cadmium minimization in wheat: a critical review. *Ecotoxicology and Environmental Safety* 130:43–53.

Rolletschek H, Hosein F, Miranda M, Heim U, Götz KP, Schlereth A, Borisjuk L, Saalbach I, Wobus U, Weber H (2005) Ectopic expression of an amino acid transporter (VfAAP1) in seeds of Vicia narbonensis and pea increases storage proteins. *Plant Physiology* 137(4):1236–1249. https://doi.org/10.1104/pp.104.056523.

Roy SJ, Negrão S, Tester M (2014) Salt resistant crop plants. *Current Opinion in Biotechnology* 26:115–124.

Roychoudhury A, Banerjee A, Lahiri V (2015) Metabolic and molecular-genetic regulation of proline signaling and itscross-talk with major effectors mediates abiotic stress tolerance in plants. *Turkish Journal of Botany* 39(6):887–910.

Saadia M, Jamil A, Akram NA, Ashraf M (2012) A study of proline metabolism in canola (*Brassica napus* L.) seedlings under salt stress. *Molecules* 17(5):5803–5815.

Sairam RK, Rao KV, Srivastava GC (2002) Differential response of wheat genotypes to long term salinity stress in relation to oxidative stress, antioxidant activity and osmolyte concentration. *Plant Science* 163(5):1037–1046.

Sanders A, Collier R, Trethewy A, Gould G, Sieker R, Tegeder M (2009) AAP1 regulates import of amino acids into developing *Arabidopsis* embryos. *Plant Journal: For Cell and Molecular Biology* 59(4):540–552. https://doi.org/10.1111/j.1365-313X.2009.03890.x.

Sattar A, Sher A, Ijaz M, Ul-Allah S, Rizwan MS, Hussain M, Jabran K, Cheema MA (2020) Terminal drought and heat stress alter physiological and biochemical attributes in flag leaf of bread wheat. *PLOS ONE* 15(5):e0232974.

Schmidt R, Stransky H, Koch W (2007) The amino acid permease AAP8 is important for early seed development in *Arabidopsis thaliana*. *Planta* 226(4):805–813. https://doi.org/10.1007/s00425-007-0527-x.

Schwacke R, Grallath S, Breitkreuz KE, Stransky E, Stransky H, Frommer WB, Rentsch D (1999) LeProT1, a transporter for proline, glycine betaine, and gamma-amino butyric acid in tomato pollen. *Plant Cell* 11(3):377–392. https://doi.org/10.1105/tpc.11.3.377.

Semenov MA, Shewry PR (2011) Modelling predicts that heat stress, not drought, will increase vulnerability of wheat in Europe. *Scientific Reports* 1:1–5.

Shao HB, Chu LY, Shao MA, Jaleel CA, Mi HM (2008) Higher plant antioxidants and redox signaling under environmental stresses. *Comptes Rendus Biologies* 331(6):433–441.

Sharma S, Villamor JG, Verslues PE (2011) Essential role of tissue-specific proline synthesis and catabolism in growth and redox balance at low water potential. *Plant Physiology* 157(1):292–304.

Sharma SS, Dietz KJ (2006) The significance of amino acids and amino acid-derived molecules in plant responses and adaptation to heavy metal stress. *Journal of Experimental Botany* 57(4):711–726.

Siddique MRB, Hamid A, Islam MS (2000) Drought stress effects on water relations of wheat. *Botanical Bulletin of Academia Sinica* 41:35–39.

Soltani A, Gholipoor M, Zeinali E (2006) Seed reserve utilization and seedling growth of wheat as affected by drought and salinity. *Environmental and Experimental Botany* 55(1–2):195–200.

Song SY, Chen Y, Chen J, Dai XY, Zhang WH (2011) Physiological mechanisms underlying OsNAC5-dependent tolerance of rice plants to abiotic stress. *Planta* 234(2):331–45.

Strizhov N, Abrahám E, Okrész L, Blickling S, Zilberstein A, Schell J, Koncz C, Szabados L (1997) Differential expression of two P5CS genes controlling proline accumulation during salt-stress requires ABA and is regulated by ABA1, ABI1 and AXR2 in Arabidopsis. *The Plant Journal: For Cell and Molecular Biology* 12(3):557–569.

Su J, Wu R (2004) Stress-inducible synthesis of proline in transgenic rice confers faster growth under stress conditions than that with constitutive synthesis. *Plant Science* 166(4):941–948.

Surender Reddy P, Jogeswar G, Rasineni GK, Maheswari M, Reddy AR, Varshney RK, Kavi Kishor PB (2015) Proline over-accumulation alleviates salt stress and protects photosynthetic and antioxidant enzyme activities in transgenic sorghum [*Sorghum bicolor* (L.) Moench]. *Plant Physiology and Biochemistry: PPB* 94:104–113.

Szabados L, Savouré A (2010) Proline: a multifunctional amino acid. *Trends in Plant Science* 15(2):89–97.

Szoke A, Miao GH, Hong Z, Verma DPS (1992 Subcellular location of delta-pyrroline-5-carboxylate reductase in root/nodule and leaf of soybean. *Plant Physiology* 99(4):1642–1649.

Taiz L, Zeiger E(ed.) (2010) *Plant Physiology*. Sinauer Associates, Sunderland.

Tanji KK (1990) Nature and extent of agricultural salinity. In: Tanji, K.K. (Ed.), *Agricultural Salinity Assessment and Management. Manuals and Reports on Engineering Practices No. 71*. American Society of Civil Engineers, New York, pp. 1–17.

Tiwari V, Mamrutha HM, Sareen S, Sheoran S, Tiwari R, Sharma P, Singh C, Singh G, Rane J (2017) Managing abiotic stresses in wheat. In: Minhas P, Rane J, Pasala R (eds), *Abiotic Stress Management for Resilient Agriculture*. Springer, Singapore, pp. 313–337.

Tubiello FN, Rosenzweig C, Goldberg RA, Jagtap S, Jones JW (2000) U.S. national assessment technical report effects of climate change on U.S. crop production. Agriculture Sector Assessment Working Paper. http://www.nacc.usgcrp.gov/sectors/agriculture.

Turan MA, Katkat V, Taban S (2007) Variations in proline, chlorophyll and mineral elements contents of wheat plants grown under salinity stress. *Journal of Agronomy* 6:137–141.

Ueda A, Shi W, Sanmiya K, Shono M, Takabe T (2001) Functional analysis of salt-inducible proline transporter of barley roots. *Plant and Cell Physiology* 42(11):1282–1289. https://doi.org/10.1093/pcp/pce166.

Ueda A, Shi W, Shimada T, Miyake H, Takabe T (2008) Altered expression of barley proline transporter causes different growth responses in *Arabidopsis*. *Planta* 227(2):277–286. https://doi.org/10.1007/s00425-007-0615-y.

Urban O, Hlaváčová M, Klem K, Novotná K, Rapantová B, Smutná P, Horáková V, Hlavinka P, Škarpa P, Trnka M (2018) Combined effects of drought and high temperature on photosynthetic characteristics in four winter wheat genotypes. *Field Crops Research* 223:137–149.

Vanlerberghe GC (2013) Alternative oxidase: a mitochondrial respiratory pathway to maintain metabolic and signaling homeostasis during abiotic and biotic stress in plants. *International Journal of Molecular Sciences* 14(4):6805–6847.

Vartanian N, Hervochon P, Marcotte L, Larher F (1992) Proline accumulation during drought rhizogenesis in *Brassica napus* var. oleifera. *Journal of Plant Physiology* 140(5):623–628.

Verbruggen N, Hermans C (2008) Proline accumulation in plants: a review. *Amino Acids* 35(4):753–759.

Verslues P, Sharma S (2010) Proline metabolism and its implications for plant-environment interaction. *Arabidopsis Book* 8:e0140. https://doi.org/10.1199/tab.0140.

Verslues PE, Sharp RE (1999) Proline accumulation in maize (*Zea mays* L.) primary roots at low water potentials. II: metabolic source of increased proline deposition in the elongation zone. *Plant Physiology* 119(4): 1349–1360.

Vicente R, Martínez-Carrasco R, Pérez P, Morcuende R (2018) New insights into the impacts of elevated CO2, nitrogen, and temperature levels on the regulation of C and N metabolism in durum wheat using network analysis. *New Biotechnology* 40(B):192–199.

Waditee R, Hibino T, Tanaka Y, Nakamura T, Incharoensakdi A, Hayakawa S, Suzuki S, Futsuhara Y, Kawamitsu Y, Takabe T, Takabe T (2002) Functional characterization of betaine/proline transporters in betaine-accumulating mangrove. *Journal of Biological Chemistry* 277(21):18373–18382. https://doi.org/10.1074/jbc.M112012200.

Wahid A, Gelani S, Ashraf M, Foolad MR (2007) Heat tolerance in plants: an overview. *Environmental and Experimental Botany* 61(3):199–223.

Wang FY, Liu RJ, Lin XG, Zhou JM (2004) Arbuscular mycorrhizal status of wild plants in saline-alkaline soils of the Yellow River Delta. *Mycorrhiza* 14(2):133–137. https://doi.org/10.1007/s00572-003-0248-3.

Wang GP, Li F, Zhang J, Zhao MR, Hui Z, Wang W (2010) Overaccumulation of glycine betaine enhances tolerance of the photosynthetic apparatus to drought and heat stress in wheat. *Photosynthetica* 48(1):30–41.

Weigelt K, Küster H, Radchuk R, Müller M, Weichert H, Fait A, Fernie AR, Saalbach I, Weber H (2008) Increasing amino acid supply in pea embryos reveals specific interactions of N and C metabolism, and highlights the importance of mitochondrial metabolism. *Plant Journal: For Cell and Molecular Biology* 55(6):909–926. https://doi.org/10.1111/j.1365-313X.2008.03560.x.

Wheeler TR, Hong TD, Ellis RH, Batts GR, Morison JIL, Hadley P (1996) The duration and rate of grain growth, and harvest index, of wheat (Triticum aestivum L.) in response to temperature and CO 2. *Journal of Experimental Botany* 47(5): 623–630.

Witt S, Galicia L, Lisec J, Cairns J, Tiessen A, Araus JL, Palacios-Rojas N, Fernie AR (2012) Metabolic and phenotypic responses of greenhouse-grown maize hybrids to experimentally controlled drought stress. *Molecular Plant* 5(2):401–417.

Wood JM (2006) Osmosensing by bacteria. *Science's STKE: Signal Transduction Knowledge Environment* 2006(357):pe43. https://doi.org/10.1126/stke.3572006pe43.

Wood JM, Bremer E, Csonka LN, Kraemer R, Poolman B, van der Heide T, Smith LT (2001) Osmosensing and osmoregulatory compatible solute accumulation by bacteria. *Comparative Biochemistry and Physiology. Part A, Molecular and Integrative Physiology* 130(3):437–460. https://doi.org/10.1016/S1095-6433(01)00442-1.

Yasmeen A, Basra SMA, Farooq M, Rehman Hu, Hussain N, Athar HuR (2013) Exogenous application of Moringa leaf extract modulates the antioxidant enzyme system to improve wheat performance under saline conditions. *Plant Growth Regulation* 69(3):225–233.

Zhang M, Huang H, Dai S (2014) Isolation and expression analysis of proline metabolism-related genes in Chrysanthemum lavandulifolium. *Gene* 537(2):203–13.

10 Oxidative Stress Alleviation by Modulation of the Antioxidant System under Environmental Stressors

Farhan Ahmad and Aisha Kamal

CONTENTS

10.1 INTRODUCTION

About 2.7–3.8 billion years ago, the introduction of molecular oxygen in the earth's atmosphere by photosynthetic efficient organism triggered the formation of activated oxygen molecules regarded as ROS (reactive oxygen species) during aerobic metabolism (Dowling and Simmons 2009; Demidchik 2015). ROS are produced constantly as by-products but confined in specific cell-compartment such as mitochondria, chloroplast, peroxisomes, plasma membrane and apoplast (Panieri et al. 2013). In normal metabolic conditions, unavoidable production of ROS is unable to cause any damage due to lower toxicity and being readily metabolized by robust antioxidant molecules; thus equilibrium between the generation and detoxification of ROS is maintained (Foyer and Noctor 2005). But under abiotic stress, such as higher accumulation of salts (Ahmad et al. 2019), drought (Hussain et al. 2019), extreme temperature variation (Goraya et al. 2017), metal toxicity (Kohli et al. 2017) or chemical-pesticides (Fatma et al. 2018), an imbalance between generation and scavenging process is caused and eventually leads to a physiological condition known as oxidative stress (Moller et al. 2007). The negative consequences of uncontrolled ROS-accrued oxidative stress encompass the destruction of biomolecules (lipid, protein and nucleic acid), inactivation of metabolic pathway enzymes, the halt of cellular metabolism and finally triggered programmed cell death (Anjum et al. 2012). But before accusing ROS, it is important to mention that ROS play a dual role as cell signaling molecules and assist redox homeostasis in a concentration-dependent manner (Mittler 2017). The lower concentration of ROS acts as a secondary messenger for signaling cascade to regulate germination, gravitropism, senescence and stomatal movement, but any deviation in ROS concentration acts as an alarm signal for stress condition (Gapper and Dolan 2006; Foyer and Noctor 2013). However, ROS signaling is a highly complicated and regulated process in specific-cellular compartments, but complete underlying mechanism is unclear (Mittler 2017).

To ensure survival under harsh environmental condition with optimal metabolic activity, the plant defense system is equipped with a dynamic antioxidant machinery system which comprises of both enzymatic as well as non-enzymatic components to maintain redox homeostasis (Gill and Tuteja 2010; Osman et al. 2020). The major component of the enzymatic section includes superoxide dismutase, catalase, peroxidase, guaiacol peroxidase and monodehydroascorbate reductase, while the non-enzymatic section comprises of low molecular compounds, important secondary metabolites and osmolytes. Expression shifts of these enzymatic genes or any adjustment of antioxidant metabolites, used as alarm signals transmitted in cells, confirms the state of stress (Mittler 2017). To some extent, these antioxidants are sufficient to bounce back against oxidation or ionic stress by modulating their components as per requirements but fall apart in extreme conditions. To face adverse conditions more adeptly, plants have a diverse range of phytohormones mainly responsible for the growth and development process under normal conditions (Davies 2013). Even though they are involved in physiological processes, these plant growth regulators (PGRs) are highly coordinated and interlinked with antioxidant machinery to curtail the overproduction of ROS intermediates in stress conditions (Wani et al. 2016). The overall organization of the defense system is to use antioxidant machinery in a well-organized manner to detoxify diverse ROS intermediates without affecting the normal biosynthetic pathway. In this study, we will discuss the detailed process of photochemical generation of ROS, hotspot cellular compartments and easy targets of oxidative stresses. The study further deals with defense machinery involved in counterattacking the overproduction of ROS and also recapitulates phytohormone mediated-modulation of antioxidant enzymes under unfavorable conditions.

10.2 PHYSIOCHEMICAL MODE OF ROS FORMATION IN BIOLOGICAL SYSTEM

ROS are mostly short-lived free radicals produced as by-products of redox reactions. The term ROS means elements encompassing actuated atoms of oxygen (except H_2O_2 which is not a radical). Free radicals are any chemical species that exist freely and have unpaired electrons (exceptions are carbon-centered radicals or transitional metal) (Cheeseman and Slater 1993). The most well-studied reactive species involved in the biological system includes $^1O_2$ (singlet oxygen), H_2O_2 (hydrogen peroxide), $O^{\bullet-}_2$ (superoxide radical) and $OH^\bullet$ (hydroxyl radical), while less known are alkoxyl and hydroperoxyl radicals, peroxynitrite, ozone and hypochlorous acid (Apel and Hirt 2004; Phaniendra et al. 2015). Due to the presence of an unpaired electron, these molecules are highly reactive and promote oxidation of cellular components to impose oxidative stress (Møller et al. 2007).

Molecular oxygen exists as a diatomic molecule (O_2) and contains a set of unpaired electrons and exists as a free radical. The similar orientation of spin numbers they showed exhibited the phenomenon of spin restriction (Adelhelm et al. 2010). This restriction synthesizes less chemically active and non-toxic oxygen to govern aerobic metabolism (Demidchik 2015). The activation of triplet oxygen, which makes O_2 more "reactive", requires a high input of energy from excited electrons and belongs to the electron transport system to overcome spin restriction (Salin 1988). Usually, oxidative stress initiates with the activation of triplet oxygen (O_2) which makes O_2 more "reactive" and is often regarded as "reactive oxygen intermediates", "oxygen-derived species", and "free oxygen radicals" (Kim et al. 2002; Demidchik 2015). At the ground state having low energy levels with exhibiting ferromagnetic behavior, ROS intermediates are harmless, non-toxic and unable to mutilate any vital biochemical pathway unless they become excited (Fridovich 1998; Suzen et al. 2017).

Fundamentally, energy absorption and sequential reduction of the monovalent process are two underlying mechanisms known for the formation and activation of these low energy ROS intermediate(s) into toxic components from atomic oxygen (Sharma et al. 2012). The energy transfer and sequential reduction of diatomic oxygen favor the formation of singlet oxygen ($^1O_2$), superoxide anion ($O_2^{\bullet-}$), H_2O_2 and hydroxyl radical (OH (Apel and Hirt 2004; Sharma et al. 2012). During the light period, the chlorophyll triplet state formation occurs by utilizing the photosynthetic energy capturing the process of photosystem II that starts the formation of highly toxic ROS known as singlet oxygen, so it is important to mention that singlet oxygen formation is not related to electron transfer to O_2 (Halliwell 2006). Further enhancement of singlet oxygen is related to the limited supply of intracellular CO_2 due to stomatal closure as a severe condition of drought and salinity. Even though $^1O_2$ has only 3 µs of life span (Hatz et al. 2007), it can diffuse 100 nanometers and promotes oxidation of lipids, fatty acids, pigments, proteins and nucleic acid (Davies 2003; Fischer et al. 2013). It is also well documented that singlet oxygen is the only one which is solely responsible for the destruction of photosystem II and cell death (Krieger-Liszkay et al. 2008).

During photosynthesis, generated oxygen accepted four electrons to produce water molecules but accept only one electron at a time due to structural restriction (spin). This single transfer of electron O_2 causes the formation of other intermediate superoxide radicals ($O_2^{\bullet-}$) (Giba et al. 1998). The superoxide radical is moderately active and less toxic with a life span of 4 µs, and not so destructive as singlet oxygen. However, the chemical modification and transformation process (Fenton reaction) changes it into other more toxic and destructive reactive species such as hydroxyl radical and hydrogen peroxide (Halliwell 2006).

Simultaneous reduction and protonation of the nucleophilic reactant of the above-mentioned superoxide radical facilitate the formation of moderately active species to form H_2O_2 (Sharma et al. 2012). Dismutation of superoxide radical can be easily achieved either enzymatically by superoxide dismutase, NADPH oxidase, cell wall peroxidase, amino oxidase or oxalate oxidase, or by non-enzymatic mode (Neil et al. 2002). Photorespiration elevates about 70% of H_2O_2 levels during drought conditions because of higher oxygenation levels than CO_2, due to stomatal closure (Noctor and Mhamdi 2017). In-plant cells, the main source of H_2O_2 production, involve an electron transport system of mitochondria, chloroplast, membrane, β-oxidation of fatty acid and photorespiration as well as photooxidation reaction catalyzed by NADPH oxidase (Sharma et al. 2012). Due to moderately active and high diffusion rates, H_2O_2 can only travel across the membrane and cause oxidative damage far from the site of production and facilitate the formation of other ROS, including hypochlorous acid and $OH^\bullet$ (Bienert et al. 2007; Teotia and Singh 2014).

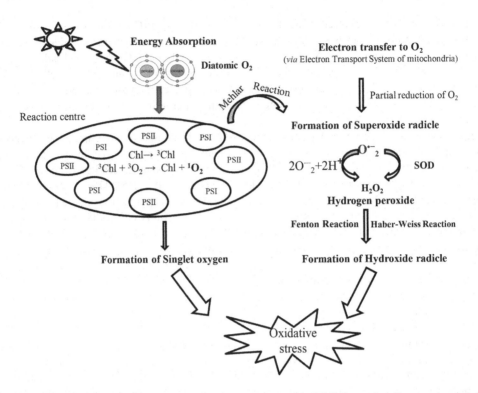

FIGURE 10.1   Schematic representation of reactive oxygen species (ROS) formation. Energy absorption by antenna molecule of the reaction center of Photosystem II (PSII) to produce singlet oxygen radical through chlorophyll triplet state. In the second method, the electron transport chain (ETC) of mitochondria (or non-cyclic photophosphorylation process) provides electrons for partial reduction of O_2 to produce superoxide radical. The superoxide dismutase enzyme (SOD) catalyzes the dismutation process required for the formation of hydrogen peroxide. Oxidation of hydrogen peroxide converts it into hydroxyl radical. Peroxisome (XO), endoplasmic reticulum (CytP450), apoplast (NADPH oxidase) also produce reactive oxygen species.

Highly reactive and toxic ROS are hydroxyl radicals, and they cause complete cellular damage. At neutral pH, hydroxyl radical (OH•) is generated by Fenton reaction between superoxide radical and hydrogen peroxide, catalyzed by transition metal (Das and Roychoudhury 2014). Another method for the formation of OH• is the Haber-Weiss reaction between H_2O_2 and $O_2^{•-}$ by a two-step reaction (Sharma et al. 2012). The first reaction involves the reduction of Fe^{3+} followed by oxidation of dihydrogen peroxide (Fenton reaction) using metal as a catalyst. OH• is considered as a potent intermediate that severely affects cellular components by lipid peroxidation (LPO), protein damage and membrane destruction, as no scavenging mechanism is available in biological systems (Pinto et al. 2003). The schematic representation of chemical modification and transformation of ROS intermediates has been shown in Figure 10.1.

10.3   TARGETS OF OXIDATIVE STRESS

Ideally, a well-coordinated and tight regulatory system has been established between the production and detoxification of ROS to avoid oxidative stress. Unfortunately, these systems are very delicate and unstable, and collapse as soon as biotic and abiotic stressors become dominant or exceed their limit; this leads to overproduction of ROS. An elevated level of ROS imposed oxidative stress which triggered the destruction of biomolecules (Ahmad 2014). The lipid and protein are the main components of the plasma membrane, which allows only selective movement of elements between the cell and the outer changing environment (Lieleg and Ribbeck 2011). Overproduction of ROS hits the minimum threshold point to trigger the process of lipid peroxidation. The polyunsaturated fatty acids, an important component of the plasma membrane, have been considered as a hotspot for attack by free radicals to form lipid hydroperoxides, which causes loss of membrane integrity and makes the cell susceptible to the outer environment (Belhadj et al. 2014). Furthermore, chemical covalent modification of protein is the outcome of excessive formation of toxic intermediates and results in protein oxidation. These modifications have been accomplished through carbonylation (site-specific modification at arginine, lysine and threonine), nitrosylation (addition of NO group of cysteine) and glutathionylation, and interaction with lipid radical evolved as a chain reaction of lipid peroxidation (Møller et al. 2007; Grimm et al. 2012). Overall modification and interaction induce proteolytic cleavage and endure proteasomal degradation Protein inactivation is also linked to oxidative damage of DNA (include mitochondria and chloroplast DNA) instigated by ROS generation. At the molecular level, the multifaceted impact of ROS includes DNA melting, oxidation of deoxyribose sugar, alteration in nucleotide bases and formation of nonfunctional DNA-protein complex and ceases translation/transcription processes (Ahmad et al. 2010).

Without delay, plants deploy every possible mechanism to minimize oxidative stress as soon as the signal is received regarding the loss of equilibrium between production and scavenging of ROS. The first line of defense includes the tight and organized regulation of the antioxidant system deprived of any restriction, having a dual-mode expression of enzymatic as well as non-enzymatic components. These enzymes operate in diverse subcellular compartments that respond and are performed on exposure to oxidative stress. The second line of defense requires complicated phytohormone-mediated modulation of the antioxidant system to achieve the pre-set goal of detoxification in stressed plants.

10.4 ENZYMATIC COMPONENT OF ANTIOXIDANT SYSTEM

10.4.1 SUPEROXIDE DISMUTASE (SOD-E.C.1.15.1.1)

SODs are the representative of metalloproteins that catalyze the dismutation of superoxide radicals into O_2 and H_2O_2 under stress conditions; hence regarded as the first line of defense. Based upon specific location and affinity to bind with metal cofactor, they are typically classified as Fe-SOD (chloroplast), CuZn-SOD (plastid and cytosol) and Mn-SOD (mitochondria) isoform (Corpas et al. 2006; Asensio et al. 2012). The elevated level of SOD slows the rate of conversion of superoxide radical (O_2^-) into caustic hydroxyl radical (Boguszewska and Zagdańska 2012).

10.4.2 CATALASE (CAT-E.C.1.11.1.6)

CAT is first identified and characterized encoded by the nuclear gene. It is a heme-containing tetrameric protein responsible for the cellular level of H_2O_2 into O_2 and H_2O by dismutation reaction especially in peroxisome and glycosomes due to the presence of oxidase enzyme. These cellular compartments are the main hub to carry out energetic-metabolic pathways (photorespiration and β-oxidation of fatty acids) and generate H_2O_2 at a higher rate (Mittler et al. 2004). High turnover rate and it not being essential in reducing elements represented CAT as an effective detoxifying agent of H_2O_2 in an energy-efficient way (Mallick et al. 2002; Sharma et al. 2012).

10.4.3 ASCORBATE PEROXIDASE (APX-E.C.1.1.11.1)

APX is also a heme-containing enzyme and exists in different isoforms based mainly on location *viz*, sAPX (stroma of chloroplasts), tAPX (thylakoid), gmAPX (membrane of glyoxisome) and cAPX (cytosol). APX catalyzes the reduction of hydrogen peroxide into water molecules using ascorbate as a reducing agent – the first step of the Ascorbate-Glutathione cycle (Sharma and Dubey 2005;). CAT and APX act simultaneously in different locations of the cellular compartment, but widely distributed APX are considered as more efficient scavengers than CAT in stress conditions due to high affinity toward H_2O_2.

10.4.4   DEHYDROASCORBATE REDUCTASE (DHAR-EC.1.8.5.1)

DHAR, chemically a monomeric enzyme with thiol in side-chain, reduces dehydroascorbate to regenerate ascorbate. The reduction reaction is initiated by accepting an electron from reduced glutathione as a reducing substrate (Eltayeb et al. 2007). Oxidation-reduction of redox biology facilitates the excessive accumulation of ascorbate in the apoplast and symplast of cells and consequently provides stress tolerance by maintaining redox homeostasis (Chan and Lam 2014).

10.4.5   MONODEHYDROASCORBATE REDUCTASE (MDHAR-E.C.1.6.5.4)

Flavin adenine dinucleotide is a major constituent of enzyme MDHAR usually dispersed in the chloroplast as well as cytosol. These enzymes are highly specific for monodehydroascorbate (MDHA) and catalyze reversible reactions to regenerate ascorbate using NADPH as the electron donor to increase the pool size of ascorbate (Sharma and Dubey 2005). Direct and indirect, MDHAR interconnection with APX is important in scavenging of H_2O_2 along with AsA level and thus maintaining redox state under oxidative stress (Murshed et al. 2013).

10.4.6   GUAIACOL PEROXIDASE (GPX-E.C.1.11.1.7)

GPX is a heme-containing monomeric enzyme located at intracellular or/and extracellular level to restrict H_2O_2 formation. It assists many processes such as cell wall lignification and ethylene biosynthesis, and also helps in wound healing in the plant; hence it is regarded as a "stress enzyme". In such a process, GPX exploits H_2O_2 to oxidize the substrate by using guaiacol/pyrogallol as a reducing substrate and consumes a leftover portion of peroxide in a constructive manner (Asada 1999).

10.4.7   GLUTATHIONE REDUCTASE (GR-E.C.1.6.4.2)

GR belongs to the class oxidoreductase and transfers electrons from NADPH to glutathione disulfide (GSSG) to generate reduced glutathione (GSH). In a series of the reaction catalyzed by MDHAR, DHAR and APX, GSH is said to remove hydrogen peroxide. In a plant cell, a high ratio of GSH/GSSG is crucial for providing tolerance under stress (Foyer et al. 1997). The other important enzymatic antioxidants include glutathione s-transferases (GST), methionine sulfoxide reductase (MSR), glutaredoxin (GRX), thioredoxin (TRX) and peroxiredoxins (PRXs).

10.5   NON-ENZYMATIC COMPONENT OF ANTIOXIDANTS SYSTEM

10.5.1   ASCORBATE (AsA)

The most abundant and extensively studied antioxidants synthesized in mitochondria are then transported to other cellular compartments through facilitated diffusion (Foyer and Noctor 2011). Fundamentally, AsA played a dynamic role in the defense system as they retract H_2O_2 and other toxic entities of oxygen generated through various aerobic metabolic pathways. The defensive mechanism adopted by AsA scavenges these reactive moieties (H_2O_2) to water by using the enzymatic mode of APX, MDHAR and DHAR, already discussed in an earlier section. The benefits of direct scavenging of H_2O_2, $OH^{\bullet}$ and $O^{\bullet-}_2$ via regeneration of tocopherol include oxidation of protection of membrane (Noctor and Foyer 1998). It is also reported that exogenous application of AsA regulates the up-regulation of many antioxidant enzymes to overcome oxidative stress (Dolatabadian et al. 2008).

10.5.2 Tocopherols (TOCs)

Tocopherols are lipophilic antioxidants that predominate in all photosynthetic organisms (Havaux and García-Plazaola 2014). α tocopherol has strong potential to quench singlet oxygen among all isoforms (β, γ and δ) and protects the chloroplast membrane from photo-oxidation (Diplock et al. 1989). Tocopherols are known to protect lipids and other membrane components by physically quenching and chemically reacting with O_2 in chloroplasts, thus protecting the structure and function of PSII (Murata et al. 2012). It also protects membranous polyunsaturated fatty acids (PUFAs) from autoxidation by trapping free radicals involved in the chain reaction of lipid peroxidation (Ledford and Niyogi 2005). Expansion of tocopherols in association with AsA is perceived as the primary response due to exposure to abiotic stress (Munné-Bosch et al., 1999).

10.5.3 Proline (Pro) and Glycine Betaine (GB)

Pro has a multidimensional role: as an antioxidant to protect against oxidative stress, as an osmo-protectant to overcome ionic stress, as well as being a protein stabilizer and chelating agent, a stress signaling molecule, and so on (Ashraf et al. 2007; Kaur and Asthir 2015; Abdel Latef et al. 2017, 2020a,b). Pro acts as a potent scavenger for detoxification of highly reactive and destructive oxygen intermediates (hydroxyl radical and singlet oxygen) species which are actively involved in the oxidation process of lipid peroxidation (Ben Rejeb et al. 2014). The overexpression of the pyrroline-5-carboxylate synthase (P5CS) gene (the coded enzyme required in Pro synthesis) is responsible for the accumulation of Pro in abiotic stress to provide tolerance (Kumar et al. 2010).

GB is a quaternary amine, synthesized by oxidation of choline carried out by choline monooxy-genase and aldehyde dehydrogenase; these enzymes are mainly located in the stromal chloroplast (Roychoudhury and Banerjee 2016). The osmoprotective role of GB includes osmotic adjustments where their accumulation alters the osmotic pressure and absorbs water from surroundings during salinity or drought conditions (Gupta and Huang 2014). A direct involvement of GB in the detoxification of ROS accelerates the repair system involved in the stabilization of protein-protein interaction and inhibits photoinhibition (Li et al. 2013). Some reports present details of the scavenging of ROS, but the detailed mechanism is missing. The insertion of GB gene in transgenic plants enhances the level of GB so the improvement of abiotic stress can be accomplished (Wei et al. 2017).

10.5.4 Polyamines (PAs)

PAs are positively charged organic molecules omnipresent in all life forms. Being structurally simple and unanimously distributed in all cellular compartments, they are involved in diverse processes, including biomolecule stabilization to modulation of the stress response (Ahmad et al. 2020a). PA classification is based on interaction; putrescine and spermidines are found freely in nature, while others are soluble-conjugated and insoluble-bound forms (Minocha et al. 2014). The accumulation of conjugated and free polyamines in plants is very important for their protection against oxidative stress induced by abiotic factors (Merewitz 2016). The steps taken against oxidative stress include detoxification of ROS via modulation of enzymatic enzymes, acting as osmolytes in association with other osmoprotectants (Pro and GB), regulation of nitric oxide production, and interaction with macromolecules (DNA, RNA and protein) at translation and transcription process (Minocha et al. 2014). This holistic approach contributes to the stress tolerance of plants by avoiding oxidative trauma.

10.5.5 Carotenoids

Like tocopherol, carotenoids are the most abundant lipophilic antioxidant found in diverse groups of tissue ranging from photosynthetic and non-photosynthetic tissue to microbes (Havaux 2014). Being a part of the photosynthetic machinery, they have the ability to delocalize unpaired electron form conjugated double-bonded structure and interlinked with β-carotene and zeaxanthin to satiate singlet oxygen without disturbing any structural and functional properties of compound (Ramel et al. 2012). The photoprotective behavior of carotenoids eludes an efficient mechanism to disintegrate excess energy in a productive way (Stahl and Sies 2012). The quenching of triplet chlorophyll curtails the production of harmful reactive intermediates such as singlet oxygen, thus reducing oxidative stress and lipid peroxidation (Gostyukhina et al. 2013). Extensive works of literature are available explaining that the higher accumulation of carotenoid is linked with phytohormone (ABA) signaling, responsible for the up-regulation of antioxidant enzyme in stress conditions (Ruiz-Sola et al. 2014).

10.5.6 FLAVONOIDS

Flavonoids are phenolic secondary metabolites produced in all parts of plants. They have proximity with dihydroxy B-ring-substituted structure that provides the ability to tolerate strong wavelengths of UV-B and UV-A, thus upholds photoprotection (Agati et al. 2013). Flavonoids with antioxidant properties are sited in the nucleus of mesophyll cells to quench H_2O_2 and hydroxyl radicals (Perron and Brumaghim 2009). Interacting with various protein kinases, flavonoids act as signaling molecules and show a strong correlation with auxins to promote cell growth and differentiation (Hou and Kumamoto 2010). Many flavonoid biosynthetic genes are induced under stress conditions (Wang et al. 2016).

10.6 MANAGEMENT OF OXIDATIVE STRESS BY MODULATING ANTIOXIDANT SYSTEM

To cope with abiotic stress and ensure the survival of the plant, diverse as well as contradictory responses of antioxidant enzymes have been reported in a series of studies, including positive or negative correlation with other molecules but beneficial for plant growth and development. In this section, we will highlight the response of antioxidant enzyme(s) or molecules and also phytohormone mediated modulation under various abiotic stresses.

Increased expression of SOD enzyme in specific organs has been reported in halophyte *Cakile maritime* (Houmani et al. 2016), maize seedlings (AbdElgawad et al. 2016) and *Broussonetia papyrifera* (Zhang et al. 2013) under salinity stress. It was also reported that SOD, POD and CAT increased in the stem at high salinity levels, whereas in roots, a significant decline in SOD and POD treated with the same salt concentration occurred without affecting CAT activity (Zhang et al. 2013). However, some findings exhibited that elevated levels of SOD, APX and GPX are responsible for down-regulation of CAT activity in leaves of *Bruguiera parviflora* upon NaCl exposure (Parida et al. 2004). Higher accumulation of proline (as osmoprotectant) was reported in leaves in salt-tolerant genotypes compared to stem, roots or other parts in salt-sensitive genotypes, which maintained turgor pressure of the chlorophyll-containing cell (Gharsallah et al. 2016). In salinity stress, accumulation of Na^+ creates oxidative stress and leads to the elevated level of MDA which results in disruptions of membrane fluidity by altering the degree of saturation of fatty acid and sterols (Chawla et al. 2013). Interestingly, nitrogen supplementations to wheat exposed to salt stress enhanced the antioxidant enzymes SOD, CAT, APX GR and DHAR and up-regulated the level of AsA and GSH levels, simultaneously limiting the generation of H_2O_2 and $O_2^{\cdot-}$ levels and providing salt tolerance to crops (Ahanger et al. 2019).

The drought-resistant genotype of *Cerasus humilis* seedlings showed a noteworthy intensification of DHAR enzyme, but no fluctuations in GR and MDAR activity, while the drought-sensitive genotype showed intensive elevation of only GR enzymes (Ren et al. 2016). The drought-tolerant varieties of many plants have been reported with a low level of MDA (malondialdehyde), indicating the ability to protect oxidative stress by maintaining redox potential (Huseynova et al. 2014). Overexpression of APX and SOD isoform (Cu/ZnSOD in the chloroplast) enhanced the recovery rate and photosynthetic efficiency in drought-tolerant species of *Ipomoea batata* (Lu et al. 2010), but deficient in wild types. In contrast to the drought-prone wheat genotype, the drought-tolerant genotype comprises higher AsA content and APX, and CAT minimizes H_2O_2 and MDA content (Yin et al. 2017). The induction of APX activity to provide tolerance during salinity, drought, metal toxicity and other abiotic stress has been extensively studied and documented (Yannarelli et al. 2006; Sofo et al. 2015; Farooq et al. 2019). Higher H_2O_2 content and enhanced activity level of SOD, CAT, APX, AsA and GSH and elevated tocopherol levels reported in *Solanum lycopersicum* indicate increased antioxidant capacity to tolerate drought-induced oxidative stress (Hasanuzzaman et al. 2020). Application of tri-mixture of nitric oxide and hydrogen sulfide aspirin (NOSH-aspirin) leads to improved performance in *Medicago sativa* plants under severe drought stress through the regulation of CAT and SOD activity, as well as cAPX, Cu/ZnSOD and FeSOD transcripts (Antoniou et al. 2020). Among all antioxidant molecules, GST in collaboration with GSH binds with metal and works as the cytosolic precursor of phytochelatins to transport toxic metalloid from cytosol to less toxic components in the vacuole, thus reducing metal toxicity (Chakravarthi et al. 2006; Kumar and Trivedi 2018). Uniform up-regulation of GSH under Arsenic-stress (roots of *Oryza sativa*), salinity stress (chloroplast of *Oryza sativa*), heat stress (turfgrass) and UV-B exposure in association with AsA maintains lower production of ROS and H_2O_2, and consequently takes out the oxidative stress by increasing GSH/GSSG ratio (Hasanuzzaman et al. 2013). Comparative analysis of root maize exposed to Al-toxicity indicates that sensitive varieties undergo oxidative stress and membrane oxidation processes due to decreased expression of SOD and POD, but no such alteration is reported in the tolerant genotype (Giannakoula et al. 2010). Up-regulation of *BrMDHAR* and *BrDHAR* when they exhibited co-expression protects the degradation of chlorophyll pigments, lowers H_2O_2 levels and avoids oxidative trauma challenged by freezing in *Brassica rapa* (Shin et al. 2013). Membrane lipidomic data with microstructural analysis confirm the role of proline and phospholipase D in membrane shrinkage, destruction of the plasmalemma and cellular injury imposed by low temperature (Kong et al. 2018). Exposure to UV-B radiation causes a recurrent boost of the enzymatic component through SOD, GR, AsA and GPX in cotyledons of cucumbers. The dose-dependent application of tocopherols has repressed the enzymatic response except for GPX (Jain et al. 2004). Interestingly, an unusual response of the antioxidant system was reported on, regarding the individual, sequential and combined exposure of UV-/B or Cd, UV-B+ Cd and/or UV-BCd or CdUV-B. Among all of these, Cd~Cd+UV proved a less lethal combination due to superior cross-adaptation and controlled formation of MDA and H_2O_2 (Xuemei et al. 2011). However, spontaneous exposure of UV-B+Cd and UV-B+Ni significantly increased the MDA content (Li et al. 2012). Activation of the antioxidant defense system is highly regulated to cope with high and low temperature stress in plants but depends on genotype sensitivity. Rice seedlings growing under high temperature exhibited a decreased level of SOD and CAT linked with down expression of OsSOD, OsCAT and $OsAPX_2$, consequently leading to hydrogen peroxide accumulation. Contrary to this, pre-soaking of cotton seeds with hydrogen peroxide solution increased the content of SOD and CAT when exposed to high temperature (Sarwar et al. 2018). A quick alteration in GSH/GSSG and AsA/DHA ratio significantly increased, which relied on the antioxidant system in *Citrullus lanatus* to provide stress tolerance in chilling stress (Cheng et al. 2016). Not only temperature fluctuation but also flood conditions alter the antioxidant enzymes' ability to retain proper functioning of growth and metabolism. Higher SOD, POD and CAT levels were recorded in maize genotypes exposed to two days of waterlogged conditions (Li et al. 2018). In another study, increased GSH, GSSG and H2O2 content were noted in sesame seedlings growing in waterlogged conditions with lower AsA profile (Anee et al. 2019).

10.7 PHYTOHORMONE-MEDIATED MODULATION OF ANTIOXIDANT REGULATION IN STRESS

Plants' growth and development is highly synchronized by diverse groups of the biochemical messenger, secondary messenger, by-products and so on, considered as plant growth regulator/phytohormones (PGRs) to govern normal functioning under optimal/stressed conditions (Wani et al.

2016; Ahmad et al. 2019). The important PGRs include auxins, gibberellins, ethylene, abscisic acid, salicylic acid, brassinosteroids (BRs) and nitric oxide. In this section, efforts have been made to recapitulate findings associated with phytohormone-mediated antioxidant response exhibited in stressful condition (Table 10.1).

10.7.1 Auxins (AUXs)

AUX provides stress tolerance to plants by either modulating an antioxidant system or through the destabilization of DELLA protein (Tognetti et al. 2012). Exogenous application of 2-aminoethanol mitigated drought grown *Hordeum vulgare* and induced overexpression of Cu-Zn SOD, CAT and POD to neutralize the H_2O_2 by mediating AUX homeostasis (Mascher et al. 2005). AUX signaling mutant (*tir1afb2*) showed a reduced level of hydrogen peroxide and superoxide radicals along with synchronous enhancement of APX and CAT, which reduced oxidative stress in salinity stress (Iglesias et al. 2010). Exogenous application of IAA in wild type or endogenous modulation of AUX in transgenic line of *Arabidopsis* showed increased efficiency of SOD to convert superoxide radical to hydrogen peroxide and elevated levels of POD and CAT, further catalyzing H_2O_2 into the water; and enhanced Gactivity required for GSH/GSSG to maintain redox stabilization (Shi et al. 2014). In contradiction to this, mutant *aux1* are more subtle than the wild type, reduce in H_2O_2 and confer AUX involvement on its generation in arsenic toxicity (Krishnamurthy and Rathinasabapathi 2013). As has been recently shown, AUX (IAA) in association with SA promotes the efficiency of APX and POD to degenerate H_2O_2 during heat stress in rice spikelets (Zhang et al. 2017). More recently, endogenous applications of AUX showed pronounced modulation of acyl-CoA oxidase (ACX), xanthine oxidase (XOD) and lipoxygenase (LOX) enzymes, consequently activating the oxidation of membrane protein and lipids exposed to pesticide toxicity (Pinto 2016).

10.7.2 Ethylene (ETH)

ETH reduces stress-induced oxidative stress by modulating and regulating antioxidant enzymes or by regulating important osmolytes (Iqbal et al. 2015). Photosynthetic protection was also attributed to the ethylene-induced antioxidant system to reduce H_2O_2 and prevent oxidative stress, using amplification of SOD, APX, GR and GSH enzyme in Ni/Zn stressed mustard (Khan and Khan 2014). Authors also reported that ethylene regulated proline production; improved photosynthesis; reduced content of H_2O_2 and thiobarbituric acid reactive substances (TBARS); and conversely up-regulated activity of glucokinase (GK), Pyrroline-5-carboxylate synthase (P5CS) and proline oxidase in salt-sensitive mustard upon higher nitrogen assimilation (Iqbal et al. 2015). An evident correlation between ethylene production, polyamines and ROS response-targeting polyamines (PAs) was established during osmotic stress in wheat seedlings (Alcázar et al. 2011). Ethephon, a systemic plant growth regulator of phosphate family initates the productions of cuticular wax to prevent the water loss and retain membrane stability by curtailing ROS production becuause of enhanced activites of SOD aPOD and CAT (Yu et al. 2017). Pre-condition salinity treatment with the synergistic application of ethephon (ethylene stimulator) and silver thiosulphate (ethylene inhibitor) on hydroponically grown tomato seedlings reduced the level of MDA and H_2O_2 (Karni et al. 2010). The mitigating roles of ethylene and GB have been extensively studied in many crops exposed to cold, heat, drought, salinity and light-induced stress as potent osmoprotectants, thus avoiding oxidative strain (Khan et al. 2012; Tso et al. 2016).

10.7.3 Abscisic Acid (ABA)

ABA mainly deals with drought and salinity stress through overexpression of genes linked to antioxidant genes and/or biosynthetic pathways of the osmoprotectant. For example, overexpression of *RrANR* (anthocyanidin reductase) resulted in the accumulation of PAs that further enhanced

TABLE 10.1

Plant Growth Regulator-Mediated Modulation of Antioxidant Enzyme/Molecules in Stress

Plant Growth Regulators	Stress Conditions	Plant Name	Antioxidant Enzymes	Response	References
Salicylic acid	Salinity	*Pisum sativum*	Superoxide dismutases, peroxidase, catalases, proline, carotenoids	+	Ahmad et al. (2017)
	Chilling	*Citrus limon*	Phenylalanine ammonia-lyase activity, phenol content	+	Siboza et al. (2014)
	Cadmium-toxicity	*Glycine max*	Superoxide dismutases, glutathione activity	+	Noriega et al. (2012)
Jasmonic acid	Drought	*Brassica species*	Glutathione reductase, glyoxalase I and glyoxalase II, glutathione peroxidases, monodehydroascorbate reductase, dehydroascorbate reductase	+	Alam et al. (2014)
	Salinity	*Glycine max*	Superoxide dismutases, ascorbate peroxidase, malondialdehyde	+	Farhangi-Abriz and Ghassemi-Golezani. (2018)
	Cd-toxicity	*Capsicum frutescens*	Guaiacol peroxidase, glutathione peroxidase, catalse	+ −	Yan et al. (2013)
	High temperature	*Eriobotrya japonica*	Superoxide dismutase, ascorbate peroxidase, catalase	+	Jin et al. (2014)
			Phenylalanine ammonia-lyase activity, peroxidase	−	
	Osmotic	*Matricaria chamomilla*	Catalase, superoxide dismutases, ascorbate peroxidase, proline	+	Salimi et al. (2016)
Brassinosteroids	Salinity	*Lycopersicon esculentum*	Catalase, superoxide dismutases, guaiacol peroxidases, malondialdehyde	+ −	Shummu et al. (2012)
	Water deficit	*Vigna unguiculata*	Superoxide dismutase, catalases, ascorbate peroxidase	+	Lima and Lobato (2017)
	Drought	*Prunus persicae*	Superoxide dismutase, peroxidase, catalases, ascorbate peroxidase, glutathione reductase	+	Wang et al. (2019)
Nitric oxide / Salicylic acid	Metal (Nickel)	*Brassica napus*	Ascorbate peroxidase, glutathione peroxidase, catalase,	+	Kazemi et al. (2010)
			hydrogen peroxide, proline	−	
	Salinity	*Pisum sativum*	Superoxide dismutase, peroxidase, catalase	+	Yadu et al. (2017)
	Chilling	*Triticum aestivum*	Superoxide dismutase, peroxidase, catalase	+	Esim and Atici (2015)

(Continued)

TABLE 10.1 (CONTINUED)

Plant Growth Regulator-Mediated Modulation of Antioxidant Enzyme/Molecules in Stress

Plant Growth Regulators	Stress Conditions	Plant Name	Antioxidant Enzymes	Response	References
Gibberellic acid	Salinity	*Pisum sativum*	Superoxide dismutase, peroxidase, proline	+	Ahmad et al. (2020b)
	Drought	*Sorghum bicolor*	Ascorbate peroxidase, catalase	+	Roghayyeh et al. (2014)
Cytokinins	Metal (Cadmium)	*Acutodesmus obliquus*	Ascorbic acid, proline, superoxide dismutase, catalae, glutathione reductase	+	Piotrowska-Niczyporuk et al. (2018)
			Hydrogen peroxide, lipid peroxidation	-	

(+) Up-regulation/ increased; (-) down-regulation/decreased

the expression of gene-related *ANR* and *ABF* to ABA signaling and ROS, providing tolerance by scavenging ROS in salinity stressed tobacco plants (Luo et al. 2016). ABA-mediated up-regulation of beta-hydroxybutyrate dehydrogenase (BDH) (regulatory enzyme in GB synthesis) provided drought tolerance by retaining water potential, even though it was in the presence of fluoride, a potent inhibitor of BDH (Zhang et al. 2012; Golestan Hashemi et al. 2018). In a recent study, higher expression of *AnnSp2* (acting as Ca^{2+} transporter of the membrane) arbitrates water retention through stomatal movement and raised the action of SOD, POD, CAT and proline levels to reconcile membrane integrity damaged in salinity and drought stress (Ijaz et al. 2017). Increased proline biosynthesis regulated through ABA signaling pathways to maintain osmotic balance has been reported in many crops experiencing drought, salinity or metal toxicity stress (Burbulis et al. 2010; Hayat et al. 2013).

10.7.4 SALICYLIC ACID (SA)

Dose-dependent applications of SA resulted in the modulation of antioxidant enzymes to improve abiotic tolerance (Fayez and Bazaid 2014; Fatma et al. 2018). Salt-accrued oxidative stress was alleviated by 0.5 mM SA priming of *Pisum sativum* seeds and resulted in enhanced SOD, CAT and POD activities, improved photosynthetic pigments and eventually a lowering of ROS and MDA content (Ahmad et al. 2017). Enhanced expression of GPX (isoform), GS (isoform), DHAR and MDAHR antioxidant gene was mediated by 0.5mM SA salt strength in wheat (Li et al. 2013). Short-term polyethylene glycol (PEG) derived through drought damage in the mustard plant was minimized by supplementing 50µM SA to increase activities of DHAR, GPX, APX and MDAR, and led to a retention of GSH/GSSG ratio and glyoxalase system (Alam et al. 2013). Exogenous application of 10^{-6} M SA induces activation of APX, GR, CAT and SOD, deferred leaf rolling and finally lowered superoxide radicals during long-term drought (Kadioglu et al. 2010). Not only exogenous, but the increased endogenous level was also reported in *Aradiopsis* exposed to Cd-toxicity, which shielded the membrane from ROS intermediates (Guan et al. 2015). Chilling tolerance in banana leaves was reported by spraying 0.5 mM SA mediated-modulation of H_2O_2-metabolizing enzymes (Kang et al. 2003). It has been shown that besides antioxidant enzymes, phenolic and other non-enzymatic antioxidant molecules provide stress tolerance by inactivating lipid free radicals, simultaneously increasing flavonoids and DPPH antioxidant activity (Ahmad et al. 2020c).

Co-application of SA with other phytohormones to strengthen tolerance in stress-sensitive crops was used as an efficient tool, but a detailed underlying mechanism is not yet known (Esim and Atici 2015; Ahmad et al. 2020b)

10.7.5 Brassinosteroids (BRs)

BRs impart survival strength by modulations of antioxidants systems in plants challenged by severe abiotic stress and are critically discussed (Vardhini and Anjum 2015). Increased accumulation of Pro, AsA, soluble sugars and enhanced activities of APX, POD, CAT and SOD as protective mechanisms were mediated by brassinolide (BL-0.2 mg/l), inhibiting electrolyte leakage in water deficit *Xanthoceras sorbifolia* (Li and Feng 2011). Supplementation of 24-epiBL in a combination of Putrescine improves growth traits, phenolic component, stress protein, inorganic elements (N, P, K), up-regulation of SOD, CAT and POD, and simultaneously lowers polyphenol oxidase (PPO) and proline levels under drought-grown cotton (Ahmed et al. 2017). Co-expression of the salt responsive gene (SalT) and BRs (OsBRI1) resulted in improved physiological parameters, increased level of GR, MDHAR and SOD, and proline content was reported by application of 24-epiBL in Indian rice (*Pusa Basmati*) exposed to salinity (Sharma et al. 2013). The membrane stability suppression and retention of water potential were attributed to application of 28-homoBL (homobrassinolide) as well as a decreased level of malondialdehyde, hydrogen peroxide and nitric oxide in temperature-sensitive mustard crop (Sirhindi et al. 2016). A very similar response has been reported, demonstrating the active potential of BRs in improving stress tolerance to wide ranges of the crop (Bajguz and Hayat 2009; Bartwal et al. 2013; Sharma et al. 2017) but this will not be explored here (Table 10.1).

10.8 NITRIC OXIDE (NO)

During the last few decades, NO as a signaling molecule has been extensively studied, including its regulatory action coupled with overall growth and development of plant from the early stages of seed germination to flowering under normal as well as stressful conditions (Freschi 2013; Sami et al. 2018). NO scavenges ROS and modulates the antioxidant system directly by regulating the endogenous level of NO and/or its crosstalk between NO, ROS and different plant growth regulators, and by mitigating nitro and oxidative stress (Hasanuzzaman et al. 2017). NO also reduces the toxicity of ROS through MAPK mediated phytohormones and establishes equilibrium between generations and detoxification of ROS by maintaining membrane stability and providing cell stability (Hasanuzzaman et al. 2017). At the cellular level, NO and H_2O_2 accumulation acts as an indicator of stress and helps in facilitating cellular adaptation to stress conditions. This organelle adaptation system removing over-burden under nitro-oxidative conditions reduces cell harm and cell demise as well (Corpas and Palma 2018).

Exogenous application of SNP declines hydrogen peroxide level, reduces lipid peroxidation level and maintains membrane integrity under arsenic treatment in mung bean (Ismail 2012). Under salinity stress, NO acts as a donor in S-nitroso-N-acetylpenicillamine (SNAP), and enhances the development parameters such as leaf relative water content, photosynthetic pigment levels of osmolytes and antioxidant defense systems in *Cicer aritinum* (Ahmad et al. 2016). Additionally, exogenous NO upgraded salt resistance by moderating the oxidative harm, invigorating proton-pump and Na^+/H^+ antiport action in the tonoplast and K^+/Na^+ proportion (Santisree et al. 2015). Reports indicate that SNP application proves to be very useful in maintaining carbon and nitrogen metabolism by increasing total soluble protein through the enhancement of endopeptidase and carboxypeptidase activity under salinity (Kong et al. 2016). In the roots of *Phaseolus vulgaris*, glucose-6-phosphate dehydrogenase is involved in NR-mediated NO biosynthesis under salt stress (Kong et al. 2016). The increased concentration of NO emerged as a vital element in drought tolerance through improving the cell reinforcement frameworks, rising proline levels, managing ROS detoxification and accumulating osmolytes (Shi et al. 2014). NO accumulation under water scarcity is directly linked with a decrease in the frequency of stomata appearance and closure to avoid water loss in *Vitis vinifera* (Patakas et al. 2010). Interestingly, it also stated that the exogenous application of NO as a donor stimulates the synthesis of ABA (Zhang et al. 2011). The stomata closure is also initiated through ABA stress hormones coupled with NO by activating signaling pathway via mitogen-activated protein kinase (MAPK), cyclic guanosine monophosphate (cGMP) and Ca^{2+} messenger at the molecular level (Gayatri et al. 2013). Exogenously, NO promotes ABA by up-regulating specific gene expression (9-cis-epoxycarotenoid dioxygenase) and also shows feedback inhibition on ABA affectability, thus improving plant growth and showing enhanced resistance to drought pressure (Santisree et al. 2015).

10.9 CONCLUSION AND FUTURE PERSPECTIVES

ROS are an uninterrupted product of aerobic metabolism, and dually play the role of signaling molecules and stress markers under optimal conditions. The metabolic fate includes detoxification of reactive intermediates into water molecules and other non-toxic stable radicals specifically into cellular compartments. Prolonged abiotic stress induces prompt and abnormal ROS generation to disturb metabolic detoxification equilibrium, which leads to oxidative damage. This loss and damage manifests as complete or partial oxidation of nucleic acid, inhibition of genetic information flow, loss of membrane integrity, inactivation of protein and enzymes, reduced cellular activities and apoptosis. However, well-organized deployment of the antioxidant warrior of defense systems can counter oxidative stress and provide stress tolerance. Any loophole in the defense system directly compromizes plant tolerance efficiency and makes plants highly susceptible to oxidative stress. Modulation of the antioxidant system mediated-phytohormones is well known, but the contradictory response of synergistic applications of PGRs in dose-dependent and associated signaling pathways to manage ROS detoxification is proving to be a popular area of research. A collaborative approach of proteomics, metabolomics and molecular studies coupled with advanced imaging techniques will unravel the mystery of ROS interaction and response during environmental conditions. Genetic engineering of transgenic crops co-expressing antioxidant enzymes at higher levels can also be used as an efficient tool/strategy to cope with adverse conditions in the future.

ACKNOWLEDGMENT

The authors are thankful to Integral University Publication Cell (IUPC), Dean Research and Development Committee for allotting manuscript number **(yet to be alloted)**

REFERENCES

AbdElgawad, H., Zinta, G., Pandey, R., Asard, H., & Abuelsoud, W. (2016). High salinity induces different oxidative stress and antioxidant responses in maize seedlings organs. *Frontiers in Plant Science*, 7, 276.

Abdel Latef, A. A. H., Abu Alhmad, M. F., & Abdelfattah, K. E. (2017). The possible roles of priming with ZnO nanoparticles in mitigation of salinity stress in lupine (Lupinus termis) plants. *Journal of Plant Growth Regulation*, 36(1), 60–70.

Abdel Latef, A. A. H., Srivastava, A. K., Saber, H., Alwaleed, E. A., & Tran, L. P. (2017c). *Sargassum muticum* and *Jania rubens* regulate amino acid metabolism to improve growth and alleviate salinity in chickpea. *Scientific Reports*, 7(1), 10537.

Adelhelm, M., Aristov, N., & Habekost, A. (2010). The properties of oxygen investigated with easily accessible instrumentation: The "one-photon-two-molecule" mechanism revisited. *Journal of Chemical Education*, 87(1), 40–44.

Agati, G., Brunetti, C., Di Ferdinando, M., Ferrini, F., Pollastri, S., & Tattini, M. (2013). Functional roles of flavonoids in photoprotection: New evidence, lessons from the past. *Plant Physiology and Biochemistry*, 72, 35–45.

Ahanger, M. A., Aziz, U., Alsahli, A. A., Alyemeni, M. N., & Ahmad, P. (2019). Influence of exogenous salicylic acid and nitric oxide on growth, photosynthesis, and ascorbate-glutathione cycle in salt stressed *Vigna angularis. Biomolecules*, 10(1), 42. doi: 10.3390/biom10010042.

Ahmad, F., Kamal, A., Singh, A., Ashfaque, F., Alamri, S., Siddiqui, M. H., & Khan, M. I. R. (2020a). Seed priming with gibberellic acid induces high salinity tolerance in *Pisum sativum* through antioxidants, secondary metabolites and up-regulation of antiporter genes. *Plant Biology Journal.* doi: 10.1111/plb.13187.

Ahmad, F., Kamal, A., Singh, A., et al. (2020b). Salicylic acid modulates antioxidant system, defense metabolites, and expression of salt transporter genes in Pisum sativum under salinity stress. *Journal of Plant Growth Regulation.* doi: 10.1007/s00344-020-10271-5.

Ahmad, F., Singh, A., & Kamal, A. (2017). Ameliorative effect of salicylic acid in salinity stressed pisum sativum by improving growth parameters, activating photosynthesis and enhancing antioxidant defense system. *Bioscience Biotechnology Research Communications, 10,* 481–489.

Ahmad, F., Singh, A., & Kamal, A. (2019). Salicylic acid–mediated defense mechanisms to abiotic stress tolerance. In M. Iqbal R. Khan, Palakolanu Sudhakar Reddy, Antonio Ferrante, Nafees A. Khan (Eds.), *Plant Signaling Molecules* (pp. 355–369). Woodhead Publishing.

Ahmad, F., Singh, A., & Kamal, A. (2020c). Osmoprotective role of sugar in mitigating abiotic stress in plants (pp. 53–70). In A. Roychoudhury and D.K. Tripathi (Eds.), *Protective Chemical Agents in the Amelioration of Plant Abiotic Stress.* Wiley.

Ahmad, P. (Ed.) (2014). *Oxidative Damage to Plants: Antioxidant Networks and Signaling.* Academic Press.

Ahmad, P., Abdel Latef, A. A., Hashem, A., Abd_Allah, E. F, Gucel, S., Tran, L. S. (2016). Nitric oxide mitigates salt stress by regulating levels of osmolytes and antioxidant enzymes in chickpea. *Frontiers in Plant Science, 31*(7), 347.

Ahmad, P., Jaleel, C. A., Salem, M. A., Nabi, G., & Sharma, S. (2010). Roles of enzymatic and nonenzymatic antioxidants in plants during abiotic stress. *Critical Reviews in Biotechnology, 30*(3), 161–175.

Ahmed, Ahmed H., Hanafy, Essam Darwish, & Alobaidy, Mohammad G. (2017). Impact of putrescine and 24-epibrassinolide on growth, yield and chemical constituents of cotton (*Gossypium barbadense* L.) plant grown under drought stress conditions. *Asian Journal of Plant Sciences, 16,* 9–23.

Alam, M. M., Hasanuzzaman, M., Nahar, K., & Fujita, M. (2013). Exogenous salicylic acid ameliorates short-term drought stress in mustard (*Brassica juncea* L.) seedlings by up-regulating the antioxidant defense and glyoxalase system. *Australian Journal of Crop Science, 7*(7), 1053.

Alam, M. M., Nahar, K., Hasanuzzaman, M., & Fujita, M. (2014). Exogenous jasmonic acid modulates the physiology, antioxidant defense and glyoxalase systems in imparting drought stress tolerance in different Brassica species. *Plant Biotechnology Reports, 8*(3), 279–293.

Alcázar, R., Cuevas, J. C., Planas, J., Zarza, X., Bortolotti, C., Carrasco, P., Salinas, J., Tiburcio, A. F., & Altabella, T. (2011). Integration of polyamines in the cold acclimation response. *Plant Science, 180*(1), 31–38.

Anee, T. I., Nahar, K., Rahman, A., Mahmud, J. A., Bhuiyan, T. F., Alam, M. U., Fujita, M., & Hasanuzzaman, M. (2019). Oxidative damage and antioxidant defense in sesamum indicum after different waterlogging durations. *Plants, 8*(7), 196.

Antoniou, C., Xenofontos, R., Chatzimichail, G., Christou, A., Kashfi, K., & Fotopoulos, V. (2020). Exploring the potential of nitric oxide and hydrogen sulfide (NOSH)-releasing synthetic compounds as novel priming agents against drought stress in Medicago sativa plants. *Biomolecules, 10*(1), 120.

Anjum, N. A., Umar, S., & Ahmad, A. (2012). *Oxidative Stress in Plant: Causes, Consequences and Tolerance.* IK International Publishing House Pvt. Ltd., New Delhi.

Apel, K., & Hirt, H. (2004). Reactive oxygen species: Metabolism, oxidative stress, and signal transduction. *Annual Review of Plant Biology, 55,* 373–399.

Asada, K. (1999). The water-water cycle in chloroplasts: Scavenging of active oxygens and dissipation of excess photons. *Annual Review of Plant Physiology and Plant Molecular Biology, 50*(1), 601–639.

Asensio, A. C., Gil-Monreal, M., Pires, L., Gogorcena, Y., Aparicio-Tejo, P. M., & Moran, J. F. (2012). Two Fe-superoxide dismutase families respond differently to stress and senescence in legumes. *Journal of Plant Physiology, 169*(13), 1253–1260.

Ashraf, M., & Foolad, M. R. (2007). Roles of glycine betaine and proline in improving plant abiotic stress resistance. *Environmental and Experimental Botany, 59*(2), 206–216.

Bajguz, A., & Hayat, S. (2009). Effects of brassinosteroids on the plant responses to environmental stresses. *Plant Physiology and Biochemistry, 47*(1), 1–8.

Bartwal, A., Mall, R., Lohani, P., Guru, S. K., & Arora, S. (2013). Role of secondary metabolites and brassinosteroids in plant defense against environmental stresses. *Journal of Plant Growth Regulation, 32*(1), 216–232.

Ben Rejeb, K. B., Abdelly, C., & Savouré, A. (2014). How reactive oxygen species and proline face stress together. *Plant Physiology and Biochemistry, 80,* 278–284.

Bienert, G. P., Møller, A. L., Kristiansen, K. A., Schulz, A., Møller, I. M., Schjoerring, J. K., & Jahn, T. P. (2007). Specific aquaporins facilitate the diffusion of hydrogen peroxide across membranes. *Journal of Biological Chemistry, 282*(2), 1183–1192.

Boguszewska, D., & Zagdańska, B. (2012). ROS as signaling molecules and enzymes of plant response to unfavorable environmental conditions. In Volodymyr Lushchak and Halyna M. Semchyshyn (Eds.), *Oxidative Stress–Molecular Mechanisms and Biological Effects* (pp. 341–362). Rijeka, Croatia: InTechOpen.

Burbulis, N., Jonytienė, V., Kuprienė, R., Blinstrubienė, A., & Liakas, V. (2010). Effect of abscisic acid on cold tolerance in *Brassica napus* shoots cultured in vitro. *Journal of Food, Agriculture and Environment, 8*(3&4), 698–701.

Chakravarthi, S., Jessop, C. E., & Bulleid, N. J. (2006). The role of glutathione in disulphide bond formation and endoplasmic-reticulum-generated oxidative stress. *EMBO Rep., 7*(3), 271–275.

Chan, C., & Lam, H. M. (2014). A putative lambda class glutathione S-transferase enhances plant survival under salinity stress. *Plant and Cell Physiology, 55*(3), 570–579.

Chawla, S., Jain, S., & Jain, V. (2013). Salinity induced oxidative stress and antioxidant system in salt-tolerant and salt-sensitive cultivars of rice (*Oryza sativa* L.). *Journal of Plant Biochemistry and Biotechnology, 22*, 27–34. doi: 10.1007/s00709-011-0365-3.

Cheeseman, K. H., & Slater, T. F. (1993). An introduction to free radical biochemistry. *British Medical Bulletin, 49*(3), 481–493.

Cheng, F., Lu, J., Gao, M., Shi, K., Kong, Q., Huang, Y., & Bie, Z. (2016). Redox signaling and CBF-responsive pathway are involved in salicylic acid-improved photosynthesis and growth under chilling stress in watermelon. *Frontiers in Plant Science, 7*, 1519.

Corpas, F. J., & Palma, J. M. (2018). Nitric oxide on/off in fruit ripening. *Plant Biology, 20*(5), 805 7.

Corpas, F. J., Fernández-Ocaña, A., Carreras, A., Valderrama, R., Luque, F., Esteban, F. J., Rodríguez-Serrano, M., Chaki, M., Pedrajas, J. R., Sandalio, L. M., del Río, L. A., & Barroso, J. B. (2006). The expression of different superoxide dismutase forms is cell-type dependent in olive (*Olea europaea* L.) leaves. *Plant and Cell Physiology, 47*(7), 984–994.

Das, K., & Roychoudhury, A. (2014). Reactive oxygen species (ROS) and response of antioxidants as ROS-scavengers during environmental stress in plants. *Frontiers in Environmental Science, 2*, 53.

Davies, M. J. (2003). Singlet oxygen-mediated damage to proteins and its consequences. *Biochemical and Biophysical Research Communications, 305*(3), 761–770.

Davies, Peter J., ed. (2013). *Plant Hormones: Physiology, Biochemistry and Molecular Biology.* Springer Science & Business Media.

Demidchik, V. (2015). Mechanisms of oxidative stress in plants: from classical chemistry to cell biology. *Environmental and Experimental Botany, 109*, 212–228.

Diplock, T., Machlin, L. J., Packer, L., & Pryor, P. W. (1989). "Vitamin E: Biochemistry and health implications." *Annals of the New York Academy of Sciences, 570*, 372–378.

Dolatabadian, A., Sanavy, S. A. M. M., & Chashmi, N. A. (2008). The effects of foliar application of ascorbic acid (vitamin C) on antioxidant enzymes activities, lipid peroxidation and proline accumulation of canola (*Brassica napus* L.) under conditions of salt stress. *Journal of Agronomy and Crop Science, 194*(3), 206–213.

Dowling, D. K., & Simmons, L. W. (2009). Reactive oxygen species as universal constraints in life-history evolution. *Proceedings. Biological Sciences, 276*(1663), 1737–1745.

Eltayeb, A. E., Kawano, N., Badawi, G. H., Kaminaka, H., Sanekata, T., Shibahara, T., Inanaga, S., & Tanaka, K. (2007). Overexpression of monodehydroascorbate reductase in transgenic tobacco confers enhanced tolerance to ozone, salt and polyethylene glycol stresses. *Planta, 225*(5), 1255–1264.

Esim, N., & Atici, Ö. (2015). Effects of exogenous nitric oxide and salicylic acid on chilling-induced oxidative stress in wheat (*Triticum aestivum*). *Frontiers in Life Science, 8*(2), 124–130.

Farhangi-Abriz, Salar, & Ghassemi-Golezani, Kazem (2018). How can salicylic acid and jasmonic acid mitigate salt toxicity in soybean plants? *Ecotoxicology and Environmental Safety, 147*, 1010–1016.

Farooq, M. A., Saqib, Z. A., Akhtar, J., Bakhat, H. F., Pasala, R. K., & Dietz, K. J. (2019). Protective role of silicon (Si) against combined stress of salinity and boron (B) toxicity by improving antioxidant enzymes activity in rice. *Silicon, 11*(4), 2193–2197.

Fatma, F., Kamal, A., & Srivastava, A. (2018). Exogenous application of salicylic acid mitigates the toxic effect of pesticides in *Vigna radiata* (L.) Wilczek. *Journal of Plant Growth Regulation, 37*(4), 1185–1194.

Fayez, K. A., & Bazaid, S. A. (2014). Improving drought and salinity tolerance in barley by application of salicylic acid and potassium nitrate. *Journal of the Saudi Society of Agricultural Sciences, 13*(1), 45–55.

Fischer, B. B., Hideg, É., & Krieger-Liszkay, A. (2013). Production, detection, and signaling of singlet oxygen in photosynthetic organisms. *Antioxidants and Redox Signaling, 18*(16), 2145–2162.

Foyer, C. H., Lopez-Delgado, H., Dat, J. F., & Scott, I. M. (1997). Hydrogen peroxide-and glutathione-associated mechanisms of acclimatory stress tolerance and signalling. *Physiologia Plantarum, 100*(2), 241–254.

Foyer, C. H., & Noctor, G. (2005). Redox homeostasis and antioxidant signaling: A metabolic interface between stress perception and physiological responses. *Plant Cell, 17*(7), 1866–1875. doi: 10.1105/tpc.105.033589. PMID: 15987996; PMCID: PMC1167537.

Foyer, C. H., & Noctor, G. (2011). Ascorbate and glutathione: The heart of the redox hub. *Plant Physiology, 155*(1), 2–18.

Foyer, C. H., & Noctor, G. (2013). Redox signaling in plants. *Antioxidants & Redox Signaling, 18*, 2087–2090. doi: 10.1089/ars.2013.5278

Freschi, L. (2013). Nitric oxide and phytohormone interactions: Current status and perspectives. *Frontiers in Plant Science, 4*(4), 398.

Fridovich, I. (1998). Oxygen toxicity: A radical explanation. *Journal of Experimental Biology, 201*(8), 1203–1209.

Gapper, C., & Dolan, L. (2006). Control of plant development by reactive oxygen species. *Plant Physiology, 141*(2), 341–345.

Gayatri, G., Agurla, S., & Raghavendra, A. S. (2013). Nitric oxide in guard cells as an important secondary messenger during stomatal closure. *Frontiers in Plant Science, 4*, 425.

Gharsallah, C., Fakhfakh, H., Grubb, D., & Gorsane, F. (2016). Effect of salt stress on ion concentration, proline content, antioxidant enzyme activities and gene expression in tomato cultivars. *AoB Plants, 8*, plw055. doi: 10.1093/aobpla/plw055

Giannakoula, A., Moustakas, M., Syros, T., & Yupsanis, T. (2010). Aluminum stress induces up-regulation of an efficient antioxidant system in the Al-tolerant maize line but not in the Al-sensitive line. *Environmental and Experimental Botany, 67*(3), 487–494.

Giba, Z., Todorović, S., Grubišić, D., & Konjević, R. (1998). Occurrence and regulatory roles of superoxide anion radical and nitric oxide in plants. *Iugoslavica Physiologica et Pharmacologica Acta, 34*, 447–461.

Gill, S. S., & Tuteja, N. (2010). Reactive oxygen species and antioxidant machinery in abiotic stress tolerance in crop plants. *Plant Physiology and Biochemistry, 48*(12), 909–930.

Golestan Hashemi, F. S., Ismail, M. R., Rafii, M. Y., Aslani, F., Miah, G., & Muharam, F. M. (2018). Critical multifunctional role of the betaine aldehyde dehydrogenase gene in plants. *Biotechnology and Biotechnological Equipment, 32*(4), 815–829.

Goraya, G. K., Kaur, B., Asthir, B., Bala, S., Kaur, G., & Farooq, M. (2017). Rapid injuries of high temperature in plants. *Journal of Plant Biology, 60*(4), 298–305.

Gostyukhina, O. L., Soldatov, A. A., Golovina, I. V., & Borodina, A. V. (2013). Content of carotenoids and the state of tissue antioxidant enzymatic complex in bivalve mollusc Anadara inaequivalis Br. *Journal of Evolutionary Biochemistry and Physiology, 49*(3), 309–315.

Grimm, S., Höhn, A., & Grune, T. (2012). Oxidative protein damage and the proteasome. *Amino Acids, 42*(1), 23–38.

Guan, C., Ji, J., Jia, C., Guan, W., Li, X., Jin, C., & Wang, G. (2015). A GSHS-like gene from Lycium chinense maybe regulated by cadmium-induced endogenous salicylic acid and overexpression of this gene enhances tolerance to cadmium stress in *Arabidopsis. Plant Cell Reports, 34*(5), 871–884.

Gupta, B., & Huang, B. (2014). Mechanism of salinity tolerance in plants: Physiological, biochemical, and molecular characterization. *International Journal of Genomics, 2014*, 18. doi: 10.1155/2014/701596

Halliwell, B. (2006). Reactive species and antioxidants. Redox biology is a fundamental theme of aerobic life. *Plant Physiology, 141*(2), 312–322.

Hasanuzzaman, M., Bhuyan, M. H. M., Zulfiqar, F., Raza, A., Mohsin, S. M., Mahmud, J. A., Fujita, M., & Fotopoulos, V. (2020). Reactive oxygen species and antioxidant defense in plants under abiotic stress: Revisiting the crucial role of a universal defense regulator. *Antioxidants, 9*(8), 681.

Hasanuzzaman, M., Nahar, K., & Fujita, M. (2013). Extreme temperature responses, oxidative stress and antioxidant defense in plants. In K. Vahdati & C. Leslie (Eds.), *Abiotic Stress-Plant Responses and Applications in Agriculture* (pp. 169–205). Rijeka: InTech.

Hasanuzzaman, M., Nahar, K., Hossain, M. S., Anee, T. I., Parvin, K., & Fujita, M. (2017). Nitric oxide pretreatment enhances antioxidant defense and glyoxalase systems to confer PEG-induced oxidative stress in rapeseed. *Journal of Plant Interactions, 12*(1), 323–331.

Hatz, S., Lambert, J. D. C., & Ogilby, P. R. (2007). "Measuring the lifetime of singlet oxygen in a single cell: Addressing the issue of cell viability." *Photochemical and Photobiological Sciences* , *6*(10), 1106–1116.

Havaux, M. (2014). Carotenoid oxidation products as stress signals in plants. *The Plant Journal: For Cell and Molecular Biology, 79*(4), 597–606.

Havaux, M., & García-Plazaola, J. I. (2014). Beyond non-photochemical fluorescence quenching: The overlapping antioxidant functions of zeaxanthin and tocopherols. In Demmig-Adams B., Garab G., Adams III W., Govindjee (eds) *Non-Photochemical Quenching and Energy Dissipation in Plants, Algae and Cyanobacteria* (pp. 583–603). Springer, Dordrecht.

Hayat, S., Hayat, Q., Alyemeni, M. N., & Ahmad, A. (2013). Proline enhances antioxidative enzyme activity, photosynthesis and yield of *Cicer arietinum L.* exposed to cadmium stress. *Acta Botanica Croatica, 72*(2), 323–335.

Hou, D. X., & Kumamoto, T. (2010). Flavonoids as protein kinase inhibitors for cancer chemoprevention: Direct binding and molecular modeling. *Antioxidants and Redox Signaling, 13*(5), 691–719.

Houmani, H., Rodríguez-Ruiz, M., Palma, J. M., Abdelly, C., & Corpas, F. J. (2016). Modulation of superoxide dismutase (SOD) isozymes by organ development and high long-term salinity in the halophyte *Cakile maritima. Protoplasma, 253*(3), 885–894.

Huseynova, I. M., Aliyeva, D. R., & Aliyev, J. A. (2014). Subcellular localization and responses of superoxide dismutase isoforms in local wheat varieties subjected to continuous soil drought. *Plant Physiology and Biochemistry, 81*, 54–60.

Hussain, S., Rao, M. J., Anjum, M. A., Ejaz, S., Zakir, I., Ali, M. A., ... & Ahmad, S. (2019). Oxidative stress and antioxidant defense in plants under drought conditions. In *Plant Abiotic Stress Tolerance* (pp. 207–219). Springer, Cham.

Iglesias, M. J., Terrile, M. C., Bartoli, C. G., D'Ippólito, S., & Casalongué, C. A. (2010). Auxin signaling participates in the adaptive response against oxidative stress and salinity by interacting with redox metabolism in *Arabidopsis. Plant Molecular Biology, 74*(3), 215–222.

Ijaz, R., Ejaz, J., Gao, S., Liu, T., Imtiaz, M., Ye, Z., & Wang, T. (2017). Overexpression of annexin gene AnnSp2, enhances drought and salt tolerance through modulation of ABA synthesis and scavenging ROS in tomato. *Scientific Reports, 7*(1), 1-14.

Iqbal, N., Umar, S., & Khan, N. A. (2015). Nitrogen availability regulates proline and ethylene production and alleviates salinity stress in mustard (*Brassica juncea). Journal of Plant Physiology, 178*, 84–91.

Ismail, G. S. (2012 Jul 1). Protective role of nitric oxide against arsenic-induced damages in germinating mung bean seeds. *Acta Physiologiae Plantarum, 34*(4), 1303–1311.

Jain, K., Kataria, S., & Guruprasad, K. N. (2004). Effect of UV-B radiation on antioxidant enzymes and its modulation by benzoquinone and α-tocopherol in cucumber cotyledons. *Current Science, 87*(1), 87–90.

Jin, P., Duan, Y., Wang, L., Wang, J., & Zheng, Y. (2014). Reducing chilling injury of loquat fruit by combined treatment with hot air and methyl jasmonate. *Food and Bioprocess Technology, 7*(8), 2259–2266.

Kadioglu, A., Saruhan, N., Sağlam, A., Terzi, R., & Acet, T. (2011). Exogenous salicylic acid alleviates effects of long term drought stress and delays leaf rolling by inducing antioxidant system. *Plant Growth Regulation, 64*(1), 27–37.

Kang, G., Wang, C., Sun, G., & Wang, Z. (2003). Salicylic acid changes activities of H2O2-metabolizing enzymes and increases the chilling tolerance of banana seedlings. *Environmental and Experimental Botany, 50*(1), 9–15.

Karni, L., Aktas, H., Deveturero, G., & Aloni, B. (2010). Involvement of root ethylene and oxidative stress-related activities in pre-conditioning of tomato transplants by increased salinity. *The Journal of Horticultural Science and Biotechnology, 85*(1), 23–29.

Kaur, G., & Asthir, B. J. B. P. (2015). Proline: A key player in plant abiotic stress tolerance. *Biologia Plantarum, 59*(4), 609–619.

Kazemi, N., Khavari-Nejad, R. A., Fahimi, H., Saadatmand, S., & Nejad-Sattari, T. (2010). Effects of exogenous salicylic acid and nitric oxide on lipid peroxidation and antioxidant enzyme activities in leaves of *Brassica napus* L. under nickel stress. *Scientia Horticulturae, 126*(3), 402–407.

Khan, M. I. R., & Khan, N. A. (2014). Ethylene reverses photosynthetic inhibition by nickel and zinc in mustard through changes in PS II activity, photosynthetic nitrogen use efficiency, and antioxidant metabolism. *Protoplasma, 251*(5), 1007–1019.

Khan, M. I. R., Iqbal, N., Masood, A., & Khan, N. A. (2012). Variation in salt tolerance of wheat cultivars: Role of glycinebetaine and ethylene. *Pedosphere, 22*(6), 746–754.

Kim, S. O., Merchant, K., Nudelman, R., Beyer, W. F. J., Keng, T., DeAngelo, J., Hausladen, A., & Stamler, J. S. (2002). OxyR: A molecular code for redox-related signaling. *Cell, 109*(3), 383–396.

Kohli, S. K., Handa, N., Gautam, V., Bali, S., Sharma, A., Khanna, K., ... & Kolupaev, Y. E. (2017). ROS signaling in plants under heavy metal stress. In *Reactive Oxygen Species and Antioxidant Systems in Plants: Role and Regulation under Abiotic Stress* (pp. 185–214). Springer, Singapore.

Kong, X., Wang, T., Li, W., Tang, W., Zhang, D., & Dong, H. (2016 Mar 1). Exogenous nitric oxide delays salt-induced leaf senescence in cotton (Gossypium hirsutum L.). *Acta Physiologiae Plantarum, 38*(3), 61.

Kong, X., Wei, B., Gao, Z., Zhou, Y., Shi, F., Zhou, X., Zhou, Q., & Ji, S. (2018). Changes in membrane lipid composition and function accompanying chilling injury in bell peppers. *Plant and Cell Physiology, 59*(1), 167–178.

Krieger-Liszkay, A., Fufezan, C., & Trebst, A. (2008). Singlet oxygen production in photosystem II and related protection mechanism. *Photosynthesis Research, 98*(1–3), 551–564.

Krishnamurthy, A., & Rathinasabapathi, B. (2013). Auxin and its transport play a role in plant tolerance to arsenite-induced oxidative stress in *Arabidopsis thaliana*. *Plant, Cell and Environment, 36*(10), 1838–1849.

Kumar, S., & Trivedi, P. K. (2018). Glutathione S-transferases: Role in combating abiotic stresses including arsenic detoxification in plants. *Frontiers in Plant Science, 9*, 751.

Kumar, V., Shriram, V., Kavi Kishor, P. B., Jawali, N., & Shitole, M. G. (2010). Enhanced proline accumulation and salt stress tolerance of transgenic indica rice by over-expressing P5CSF129A gene. *Plant Biotechnology Reports, 4*(1), 37–48.

Ledford, H. K., & Niyogi, K. K. (2005). Singlet oxygen and photo-oxidative stress management in plants and algae. *Plant, Cell and Environment, 28*(8), 1037–1045.

Li, G., Peng, X., Wei, L., & Kang, G. (2013). Salicylic acid increases the contents of glutathione and ascorbate and temporally regulates the related gene expression in salt-stressed wheat seedlings. *Gene, 529*(2), 321–325.

Li, K. R., & Feng, C. H. (2011). Effects of brassinolide on drought resistance of *Xanthoceras sorbifolia* seedlings under water stress. *Acta Physiologiae Plantarum, 33*(4), 1293–1300.

Li,W., Mo,W., Ashraf, U., Li, G., Wen, T., Abrar, M., Gao, L., Liu, J., & Hu, J. (2018). Evaluation of physiological indices of waterlogging tolerance of different maize varieties in South China. *Applied Ecology and Environmental Research, 16*(2), 2059–2072.

Li, X., Zhang, L., Li, Y., Ma, L., Bu, N., & Ma, C. (2012). Changes in photosynthesis, antioxidant enzymes and lipid peroxidation in soybean seedlings exposed to UV-B radiation and/or Cd. *Plant and Soil, 352*(1–2), 377–387.

Lieleg, O., & Ribbeck, K. (2011). Biological hydrogels as selective diffusion barriers. *Trends in Cell Biology, 21*(9), 543–551.

Lima, J. V., & Lobato, A. K. S. (2017). Brassinosteroids improve photosystem II efficiency, gas exchange, antioxidant enzymes and growth of cowpea plants exposed to water deficit. *Physiology and Molecular Biology of Plants , 23*(1), 59–72.

Lu, Y. Y., Deng, X. P., & Kwak, S. S. (2010). Over expression of CuZn superoxide dismutase (CuZn SOD) and ascorbate peroxidase (APX) in transgenic sweet potato enhances tolerance and recovery from drought stress. *African Journal of Biotechnology, 9*(49), 8378–8391.

Luo, P., Shen, Y., Jin, S., Huang, S., Cheng, X., Wang, Z., Li, P., Zhao, J., Bao, M., & Ning, G. (2016). Overexpression of *Rosa rugosa* anthocyanidin reductase enhances tobacco tolerance to abiotic stress through increased ROS scavenging and modulation of ABA signaling. *Plant Science , 245*, 35–49.

Mallick, N., Mohn, F. H., Soeder, C. J., & Grobbelaar, J. U. (2002). Ameliorative role of nitric oxide on H_2O_2 toxicity to a chlorophycean alga *Scenedesmus obliquus*. *The Journal of General and Applied Microbiology, 48*(1), 1–7.

Mascher, R., Nagy, E., Lippmann, B., Hörnlein, S., Fischer, S., Scheiding, W., Neagoe, A., & Bergmann, H. (2005). Improvement of tolerance to paraquat and drought in barley (*Hordeum vulgare* L.) by exogenous 2-aminoethanol: Effects on superoxide dismutase activity and chloroplast ultrastructure. *Plant Science, 168*(3), 691–698.

Merewitz, E. (2016). Role of polyamines in abiotic stress responses. *CAB Reviews: Perspectives in Agriculture, Veterinary Science, Nutrition and Natural Resources, 11*(3), 1–11.

Minocha, R., Majumdar, R., & Minocha, S. C. (2014). Polyamines and abiotic stress in plants: A complex relationship1. *Frontiers in Plant Science, 5*, 175.

Mittler, R. (2017). ROS are good. *Trends in Plant Science, 22*(1), 11–19.

Mittler, R., Vanderauwera, S., Gollery, M., & Van Breusegem, F. (2004). Reactive oxygen gene network of plants. *Trends in Plant Science, 9*(10), 490–498.

Møller, I. M., Jensen, P. E., & Hansson, A. (2007). Oxidative modifications to cellular components in plants. *Annual Review of Plant Biology, 58*, 459–481.

Munné-Bosch, S., Schwarz, K., & Alegre, L. (1999). Enhanced formation of α-tocopherol and highly oxidized abietane diterpenes in water-stressed rosemary plants. *Plant Physiology, 121*(3), 1047–1052.

Murata, N., Allakhverdiev, S. I., & Nishiyama, Y. (2012). The mechanism of photoinhibition in vivo: Re-evaluation of the roles of catalase, α-tocopherol, non-photochemical quenching, and electron transport. *Biochimica et Biophysica Acta (BBA)-Bioenergetics, 1817*(8), 1127–1133.

Murshed, R., Lopez-Lauri, F., & Sallanon, H. (2013). Effect of water stress on antioxidant systems and oxidative parameters in fruits of tomato (Solanum lycopersicon L, cv. Micro-tom). *Physiology and Molecular Biology of Plants, 19*(3), 363–378.

Neil, S., Desikan, R., & Hancock, J. (2002). Hydrogen peroxide signalling. *Current Opinion in Plant Biology, 5*(5), 388–395.

Noctor, G., & Foyer, C. H. (1998). Ascorbate and glutathione: Keeping active oxygen under control. *Annual Review of Plant Physiology and Plant Molecular Biology, 49*(1), 249–279.

Noctor, G., & Mhamdi, A. (2017). Climate change, CO2, and defense: The metabolic, redox, and signaling perspectives. *Trends in Plant Science, 22*(10), 857–870.

Noriega, G., Caggiano, E., Lecube, M. L., Cruz, D. S., Batlle, A., Tomaro, M., & Balestrasse, K. B. (2012). The role of salicylic acid in the prevention of oxidative stress elicited by cadmium in soybean plants. *Biometals, 25*(6), 1155–1165.

Osman, M. S., Badawy, A. A., Osman, A. I., & Abdel Latef, A. A. (2020). Ameliorative impact of an extract of the halophyte *Arthrocnemum macrostachyum* on growth and biochemical parameters of soybean under salinity stress. *Journal of Plant Growth Regulation.* https://doi.org/10.1007/s00344-020-10185-2.

Panieri, E., Gogvadze, V., Norberg, E., Venkatesh, R., Orrenius, S., & Zhivotovsky, B. (2013). Reactive oxygen species generated in different compartments induce cell death, survival, or senescence. *Free Radical Biology and Medicine, 57*, 176–187.

Parida, A. K., Das, A. B., & Mohanty, P. (2004). Defense potentials to NaCl in a mangrove, *Bruguiera parviflora*: Differential changes of isoforms of some antioxidative enzymes. *Journal of Plant Physiology, 161*(5), 531–542.

Patakas, A. A., Zotos, A., & Beis, A. S. (2010). Production, localisation and possible roles of nitric oxide in drought-stressed grapevines. *Australian Journal of Grape and Wine Research, 16*(1), 203–209.

Perron, N. R., & Brumaghim, J. L. (2009). A review of the antioxidant mechanisms of polyphenol compounds related to iron binding. *Cell Biochemistry and Biophysics, 53*(2), 75–100.

Phaniendra, A., Jestadi, D. B., & Periyasamy, L. (2015). Free radicals: Properties, sources, targets, and their implication in various diseases. *Indian Journal of Clinical Biochemistry , 30*(1), 11–26.

Pinto, E., Sigaud-Kutner, T. C. S., Leitão, M. A. S.,Okamoto, O. K., Morse, D., & Colepicolo, P. (2003). Heavy metal-induced oxidative stress in algae. *Journal of Phycology, 39*(6), 1008–1018.

Piotrowska-Niczyporuk, A., Bajguz, A., Zambrzycka-Szelewa, E., & Bralska, M. (2018). Exogenously applied auxins and cytokinins ameliorate lead toxicity by inducing antioxidant defence system in green alga *Acutodesmus obliquus*. *Plant Physiology and Biochemistry, 132*, 535–546.

Ramel, F., Birtic, S., Cuiné, S., Triantaphylidès, C., Ravanat, J. L., & Havaux, M. (2012). Chemical quenching of singlet oxygen by carotenoids in plants. *Plant Physiology, 158*(3), 1267–1278.

Ren, J., Sun, L. N., Zhang, Q. Y., & Song, X. S. (2016). Drought tolerance is correlated with the activity of antioxidant enzymes in *Cerasus humilis* seedlings. *BioMed Research International, 2016*, 9851095. doi: 10.1155/2016/9851095. PMID: 27047966; PMCID: PMC4800087.

Roghayyeh, S., Saeede, R., Omid, A., & Mohammad, S. (2014). The effect of salicylic acid and gibberellin on seed reserve utilization, germination and enzyme activity of *sorghum (Sorghum bicolor* L.) seeds under drought stress. *Journal of Stress Physiology and Biochemistry, 10*(1), 5–13.

Roychoudhury, A., & Banerjee, A. (2016). Endogenous glycine betaine accumulation mediates abiotic stress tolerance in plants. *Tropical Plant Research, 3*, 105–111.

Ruiz-Sola, M. Á., Arbona, V., Gómez-Cadenas, A., Rodríguez-Concepción, M., & Rodríguez-Villalón, A. (2014). A root specific induction of carotenoid biosynthesis contributes to ABA production upon salt stress in *Arabidopsis. PLOS ONE, 9*(3), e90765. doi: 10.1371/journal.pone.0090765. PMID: 24595399; PMCID: PMC3942475.

Salimi, F., Shekari, F., & Hamzei, J. (2016). Methyl jasmonate improves salinity resistance in German chamomile (*Matricaria chamomilla L.*) by increasing activity of antioxidant enzymes. *Acta Physiologiae Plantarum, 38*(1), 1.

Salin, M. L. (1988). Toxic oxygen species and protective systems of the chloroplast. *Physiologia Plantarum*, 72(3), 681–689.

Sami, F., Faizan, M., Faraz, A., Siddiqui, H., Yusuf, M., & Hayat, S. (2018 Feb 28). Nitric oxide-mediated integrative alterations in plant metabolism to confer abiotic stress tolerance, NO crosstalk with phytohormones and NO-mediated post translational modifications in modulating diverse plant stress. *Nitric Oxide: Biology and Chemistry*, 73, 22–38.

Santisree, P., Bhatnagar-Mathur, P., & Sharma, K. K. (2015 Oct 1). NO to drought-multifunctional role of nitric oxide in plant drought: Do we have all the answers? *Plant Science*, 239, 44–55.

Sarwar, M., Saleem, M. F., Ullah, N., Rizwan, M., Ali, S., Shahid, M. R., Alamri, S. A., Alyemeni, M. N., & Ahmad, P. (2018). Exogenously applied growth regulators protect the cotton crop from heat-induced injury by modulating plant defense mechanism. *Sci. Rep.*, 8(1), 17086.

Sharma, I., Bhardwaj, R., & Pati, P. K. (2013). Stress modulation response of 24-epibrassinolide against Imidacloprid in an elite indica rice variety *Pusa Basmati-1*. *Pesticide Biochemistry and Physiology*, 105(2), 144–153.

Sharma, I., Kaur, N., & Pati, P. K. (2017). Brassinosteroids: A promising option in deciphering remedial strategies for abiotic stress tolerance in rice. *Frontiers in Plant Science*, 8, 2151.

Sharma, P., & Dubey, R. S. (2005). Drought induces oxidative stress and enhances the activities of antioxidant enzymes in growing rice seedlings. *Plant Growth Regulation*, 46(3), 209–221.

Sharma, P., Jha, A. B., Dubey, R. S., & Pessarakli, M. (2012). Reactive oxygen species, oxidative damage, and antioxidative defense mechanism in plants under stressful conditions. *Journal of Botany*, 2012, 217037. doi: 10.1155/2012/217037

Shi, H., Ye, T., Zhu, J. K., & Chan, Z. (2014 May 27). Constitutive production of nitric oxide leads to enhanced drought stress resistance and extensive transcriptional reprogramming in Arabidopsis. *Journal of Experimental Botany*, 65(15), 4119–4131.

Shi, H., Chen, L., Ye, T., Liu, X., Ding, K., & Chan, Z. (2014). Modulation of auxin content in *Arabidopsis* confers improved drought stress resistance. *Plant Physiology and Biochemistry*, 82, 209–217.

Shin, S. Y., Kim, M. H., Kim, Y. H., Park, H. M., & Yoon, H. S. (2013). Co-expression of monodehydroascorbate reductase and dehydroascorbate reductase from *Brassica rapa* effectively confers tolerance to freezing-induced oxidative stress. *Molecules and Cells*, 36(4), 304–315.

Shummu, S., Anil, S., & Sikander Pal, C. (2012). Influence of exogenously applied epibrassinolide and putrescine on protein content, antioxidant enzymes and lipid peroxidation in *Lycopersicon esculentum* under salinity stress. *American Journal of Plant Sciences*, 3(6), 714–720.

Siboza, X. I., Bertling, I., & Odindo, A. O. (2014). Salicylic acid and methyl jasmonate improve chilling tolerance in cold-stored lemon fruit (Citrus limon). *Journal of Plant Physiology*, 171(18), 1722–1731.

Sirhindi, G., Kumar, M., Kumar, S., & Bhardwaj, R. (2016). Brassinosteroids: Physiology and stress management in plants. In N. Tuteja and S.S. Gill *Abiotic Stress Response in Plants* (pp. 279–314). Wiley.

Sofo, A., Scopa, A., Nuzzaci, M., & Vitti, A. (2015). Ascorbate peroxidase and catalase activities and their genetic regulation in plants subjected to drought and salinity stresses. *International Journal of Molecular Sciences*, 16(6), 13561–13578.

Stahl, W., & Sies, H. (2012). Photoprotection by dietary carotenoids: Concept, mechanisms, evidence and future development. *Molecular Nutrition and Food Research*, 56(2), 287–295.

Suzen, S., Gurer-Orhan, H., & Saso, L. (2017). Detection of reactive oxygen and nitrogen species by electron paramagnetic resonance (EPR) technique. *Molecules*, 22(1), 181.

Teotia, S., & Singh, D. (2014). Oxidative stress in plants and its management. In Gaur, R. K., & Sharma, P. (Eds.), *Approaches to Plant Stress and Their Management* (pp. 227–253). Springer, New Delhi.

Tognetti, V. B., Mühlenbock, P. E. R., & Van Breusegem, F. (2012). Stress homeostasis—The redox and auxin perspective. *Plant, Cell and Environment*, 35(2), 321–333.

Tso, K. K. S., Leung, K. K., Liu, H. W., & Lo, K. K. W. (2016). Photoactivatable cytotoxic agents derived from mitochondria-targeting luminescent iridium (iii) poly (ethylene glycol) complexes modified with a nitrobenzyl linkage. *Chemical Communications*, 52(24), 4557–4560.

Vardhini, B. V., & Anjum, N. A. (2015). Brassinosteroids make plant life easier under abiotic stresses mainly by modulating major components of antioxidant defense system. *Frontiers in Environmental Science*, 2, 67.

Wang, F., Kong, W., Wong, G., Fu, L., Peng, R., Li, Z., & Yao, Q. (2016). AtMYB12 regulates flavonoids accumulation and abiotic stress tolerance in transgenic *Arabidopsis thaliana*. *Molecular Genetics and Genomics*, 291(4), 1545–1559.

Wang, X., Gao, Y., Wang, Q., Chen, M., Ye, X., Li, D., Chen, X., Li, L., & Gao, D. (2019). 24-Epibrassinolide-alleviated drought stress damage influences antioxidant enzymes and autophagy changes in peach (*Prunus persicae* L.) leaves. *Plant Physiology and Biochemistry*, 135, 30–40.

Wani, S. H., Kumar, V., Shriram, V., & Sah, S. K. (2016). Phytohormones and their metabolic engineering for abiotic stress tolerance in crop plants. *The Crop Journal*, *4*(3), 162–176.

Wei, D., Zhang, W., Wang, C., Meng, Q., Li, G., Chen, T. H., & Yang, X. (2017). Genetic engineering of the biosynthesis of glycinebetaine leads to alleviate salt-induced potassium efflux and enhances salt tolerance in tomato plants. *Plant Science* , *257*, 74–83.

Yadu, S., Dewangan, T. L., Chandrakar, V., & Keshavkant, S. (2017). Imperative roles of salicylic acid and nitric oxide in improving salinity tolerance in *Pisum sativum* L. *Physiology and Molecular Biology of Plants*, *23*(1), 43–58.

Yan, Z., Chen, J., & Li, X. (2013). Methyl jasmonate as modulator of Cd toxicity in *Capsicum frutescens var. fasciculatum* seedlings. *Ecotoxicology and Environmental Safety*, *98*, 203–209.

Yannarelli, G. G., Gallego, S. M., & Tomaro, M. L. (2006). Effect of UV-B radiation on the activity and isoforms of enzymes with peroxidase activity in sunflower cotyledons. *Environmental and Experimental Botany*, *56*(2), 174–181.

Yin, M., Wang, Y., Zhang, L., Li, J., Quan, W., Yang, L., Wang, Q., & Chan, Z. (2017). The *Arabidopsis* Cys2/His2 zinc finger transcription factor ZAT18 is a positive regulator of plant tolerance to drought stress. *Journal of Experimental Botany*, *68*(11), 2991–3005.

Yu, H., Zhang, Y., Xie, Y., Wang, Y., Duan, L., Zhang, M., & Li, Z. (2017). Ethephon improved drought tolerance in maize seedlings by modulating cuticular wax biosynthesis and membrane stability. *Journal of Plant Physiology*, *214*, 123–133.

Zhang, C. X., Feng, B. H., Chen, T. T., Zhang, X. F., Tao, L. X., & Fu, G. F. (2017). Sugars, antioxidant enzymes and IAA mediate salicylic acid to prevent rice spikelet degeneration caused by heat stress. *Plant Growth Regulation*, *83*(2), 313–323.

Zhang, L., Gao, M., Hu, J., Zhang, X., Wang, K., & Ashraf, M. (2012). Modulation role of abscisic acid (ABA) on growth, water relations and glycinebetaine metabolism in two maize (*Zea mays* L.) cultivars under drought stress. *International Journal of Molecular Sciences*, *13*(3), 3189–3202.

Zhang, M., Dong, J. F., Jin, H. H., Sun, L. N., & Xu, M. J. (2011). Ultraviolet-B-induced flavonoid accumulation in Betula pendula leaves is dependent upon nitrate reductase-mediated nitric oxide signaling. *Tree Physiology*, *31*(8), 798–807.

Zhang, M., Fang, Y., Ji, Y., Jiang, Z., & Wang, L. (2013). Effects of salt stress on ion content, antioxidant enzymes and protein profile in different tissues of *Broussonetia papyrifera*. *South African Journal of Botany*, *85*, 1–9.

11 Antioxidant Enzymes and Salt Stress in Plants

Siamak Shirani Bidabadi, Parisa Sharifi and Arafat Abdel Hamed Abdel Latef

CONTENTS

11.1 INTRODUCTION

Due to the increasing need for irrigation in arid and semi-arid regions, which has intensified over the last 20 years, salinity on farmland, as one of the major abiotic constraints on agriculture around the world, has become a major challenge (Acosta-Motos et al. 2016, 2017; Cirillo et al. 2016). Salinity is one of the biggest environmental threat factors that limits the metabolism of plants and causes limitations on their growth and yield, and the negative effects of this stress are exacerbated by water shortage (Negrao et al. 2017; Isayenkov and Maathuis 2019; Abdel Latef et al. 2021). Salinity causes the accumulation of toxic Na^+ ions in the soil as well as the toxic effect of Na^+ and Cl^- ions on plants and is associated with reduced water availability. In terms of response to salinity, some plants are classified as halophytes due to their specialized mechanisms for dealing with salinity stress in saline environments. However, most crops are sensitive to salt and are classified as glycophytes. In this group of plants, finding beneficial solutions to increase their yield in saline agricultural lands will lead to the production of plants that tolerate salinity (Munns and Tester 2008; Flowers et al. 2015; Isayenkov and Maathuis 2019).

Osmotic stress created in plants grown under salinity stress disrupts cell division and expansion, causes changes in stomatal closure, reduces cellular turgor and causes changes in cell homeostasis (Miller et al. 2010; Abdel Latef and Chaoxing 2011, 2014; Abdel Latef et al. 2021). Besides this, under these stressful conditions, reactive oxygen species (ROS) such as H_2O_2, O_2^- and singlet oxygen are generated in stressed plant cells (Hossain and Dietz 2016; Osman et al. 2020). At the cellular levels, ROS cause serious damage to nucleic acids, and may manifest themselves in the oxidation of deoxyribose, nucleotides consumption, bases alterations, lipid peroxidation and protein oxidation. The process of programmed cell death will occur as the production of ROS continues to increase (Mittler 2017; Abdel Latef et al. 2019 a).

To scavenge ROS generation, plants are equipped with an extensive antioxidant system to eliminate the negative effects of oxidative stress. These biomolecule oxidation inhibitors are classified into two categories: enzymatic or non-enzymatic antioxidant defense systems. Enzymatic antioxidant systems work by increasing the activity of some enzymes such as superoxide dismutase, catalase, ascorbate peroxidase, guaiacol peroxidase, glutathione reductase, monodehydroascorbate reductase and dehydroascorbate reductase, and non-enzymatic antioxidant defense systems work by enhancing antioxidant compounds such as glutathione, ascorbic acid, carotenoids, tocopherols and flavonoids, leading to enhanced plant resistance to salt stress (Ashraf 2009; Abdel Latef 2010, 2011; Abdel Latef et al. 2018).

This chapter provides an overview of recent findings regarding the effects of oxidative stress on plants under saline conditions as well as how antioxidant defenses in plants can overcome the negative effects of these stresses.

11.2   GENERAL ASPECTS OF PLANT SALT STRESS

Salt stress causes major changes in the morphological, physiological and biochemical responses of plants (Sevengor et al. 2011; Abdel Latef et al. 2017 a, b, c, 2019 b). It has been clearly shown that salt stress in plants occurs in both ionic and osmotic stresses (Tester and Davenport 2003). In general, Na^+ intake and sensing, which mainly occur through nonselective cation channels (NSCCs), are the plant's first response to salinity stress (Demidchik and Tester 2002; Demidchik and Maathuis 2007). Because sodium accumulates more in the shoots than in the roots, the leaves are more vulnerable than the roots (Tester and Davenport 2003). Thus, increased Na^+ by roots not only is followed by damage to leaf tissues and necrosis of older leaves but also is associated with deficiency of other nutrients by hampering the uptake of other nutrients (Silberbush and Ben-Asher 2001; Tester and Davenport 2003; Munns 2005). As reviewed by Parvaiz and Satyawati (2008), the efficiency of some transporters in the root plasma membrane, including K^+- selective ion channels, is disrupted by the accumulation of Na^+, which in turn prevents the absorption of other important nutrients such as P, Fe or Zn. Decreased K^+/Na^+ ratio in plant cytosol due to impaired K^+/Na^+ homeostasis occurs in the cytoplasm and is common in plants when exposed to salinity stress. The excessive entry of Na^+ and Cl^- toxic ions into the roots of the plant will upset the ionic balance in the plant, which will result in a lack of K^+ (Zhu 2003). The first stage of Na^+ accumulation on growth retardation occurs from a few minutes to a few days after exposure to salinity, during which the stomata are closed and the elongation of cells, especially in the shoots, is prevented (Isayenkov 2012). In the second stage, which occurs during days or even weeks, due to the continued accumulation of toxic ions (Na^+ and Cl^-) due to salinity, the process of plant metabolism slows down, which leads to premature aging and eventually plant cell death (Munns and Tester 2008; Roy et al. 2014). However, another reference has reported that plants experience osmotic changes in the first stage instead of sodium ions, and sodium responses in the next step occur through the toxic effects of sodium in leaves (Munns and Tester 2008). Osmotic damage can be caused by high concentrations of Na^+ in the leaf apoplast, as Na^+ enters the leaves in the xylem stream and is outstripped as water evaporates (Flowers et al. 1991). Recently, rapid salt reactions including sodium-specific calcium waves have been also reported in the roots of plants under salinity stress (Choi et al. 2014). In both ionic and osmotic stress, the process of tolerance takes place through various physiological and molecular mechanisms, including acquisition through osmotic tolerance, ionic tolerance and even tissue tolerance (Rajendran et al. 2009; Roy et al. 2014). As salt enters the root system, signal cascades are activated that enhance ionic tolerance by restricting the influx of Na^+ into the root and reducing Na^+ translocation. Finally, tissue tolerance is increased by channeling the flow of toxic ions to the vacuoles to prevent their destructive effects on the cell cytoplasm. Many plants use the above-mentioned strategies to increase salinity tolerance so that the difference in tolerance between glycophytes and halophytes is mainly due to the greater strength of the mechanisms used in the latter strategy (Maathuis et al. 2014; Flowers and Colmer 2015). Further understanding of the mechanisms by which ion toxicity occurs in plants may violate current beliefs and lead us to the fact that the role of ion toxicity in the cytosol may be minor and more prevalent in vacuoles. In this regard, the interaction of ions such as K^+ cannot be ignored (Isayenkov and Maathuis 2019).

11.3 OXIDATIVE STRESS IN PLANTS UNDER SALINITY

11.3.1 ROS GENERATION

As mentioned earlier, high concentrations of salt in the soil impose both ionic and osmotic stress on stressed plants, and what happens in this destructive event is the production of ROS, which ultimately causes oxidative stress (Gill and Tuteja 2010; Tuteja et al. 2012; Osman et al. 2020). Oxygen is normally necessary for living organisms for respiration and survival, but when it receives extra energy or electrons, it becomes ROS, which are very dangerous and damage various cell components, including lipids, proteins, carbohydrates and nucleic acids. Different types of ROS include singlet oxygen ($^1O_2$), hydrogen peroxide (H_2O_2), superoxide anions (O^-_2) and hydroxyl radicals ($OH^•$) (Krieger-Liszkay 2005; Apel and Hirt 2004; Miller et al. 2010; Bhattacharjee 2014).

In the process of photosynthesis, light energy is collected by the photosystem II and I, which is used to stimulate electrons in photosynthetic reactions and ends after the transfer of electrons in the NADPH. However, even under normal conditions, about 10% of photosynthetic electrons become active and destructive, such as O^-_2 (Foyer and Noctor 2000). Plants show a response to drought stress by first closing the stomata to prevent excess water loss. As the pores close, the internal concentration of carbon dioxide decreases, thus slowing down the reduction of carbon dioxide by the Calvin cycle. This phenomenon causes the oxidized $NADP^+$ (final electron acceptor in photosystem I) to be depleted, which eventually generates radical oxygen O^-_2 (Hsu and Kao 2003). Lipid peroxidation caused by enhanced ROS generation at the cellular level has three different stages characterized as initiation, progression and termination. In the first stage, the rupture of membranes and unsaturated fatty acids occurs with the accumulation of hydrogen or the addition of oxygen radicals. In the second stage, the ROS react with methyl groups of unsaturated fatty acids. In the last stage, called termination, different lipid dimers are formed due to radicals from different lipids (Bhattacharjee 2014). Protein oxidation is also involved through four steps including metal-catalyzed oxidation, amino acid oxidation, oxidation induced cleavage and conjugation of lipid peroxidation products (Ahmad et al. 2017). The ROS damage to DNA sometimes occurs in the form of oxidation, alkylation and loss of bases, intra-strand cross-links, DNA photoproducts and single-strand DNA breaks (SSBs), and in other cases, where the severity of the injury is much greater, changes are made in both DNA strands like inter-strand cross-links and double-strand DNA breaks (DSBs) (Manova and Gruszka 2015).

11.3.2 SITES OF ROS GENERATION

The major sites of ROS production in cells are chloroplasts, peroxisomes, mitochondria and apoplast (Dietz et al. 2016; Liu and He 2016). Reactions that lead to the production of ROS occur in photosystems I and II, which are located in chloroplasts (Asada 2006). In conditions where the efficiency of photosynthesis decreases (environmental stresses), the rate of electron fluxes from photosynthesis increase, but the electron acceptors in the photosystem II (primary electron-accepting plastoquinone, secondary electron-accepting plastoquinone and the plastoquinone pool) decrease and the possibility of producing ROS increases (Dietz et al. 2016). Other important sites for ROS production are peroxisomes wherein ROS production takes place mainly through photorespiration and fatty acid β-oxidation pathways (Corpas 2015). The generation of $O_2^{\bullet-}$ is performed by various enzymes such as Nicotinamide adenine dinucleotide/Nicotinamide adenine dinucleotide phosphate (NADH/NADPH), which are involved in a short electron chain located in the membrane of peroxisomes, or by xanthine oxidoreductase (XOD/XDH) and uricase in peroxisome matrix. However, in terms of H_2O_2 generation, metabolic processes leading to the production of H_2O_2 are mainly mediated by the photorespiratory glycolate oxidase (GOX) reaction, the enzyme of fatty acid beta-oxidation (acyl-CoA oxidase) and the enzymatic reaction of flavin oxidases (Wang et al. 2015). Mitochondria lack the electron transport chain as the main site of ROS production, so these cell compartments are of lower importance compared to chloroplasts and peroxisomes. In mitochondria, the process of ROS production is regulated by ATP hydrolysis (Das et al. 2015). Another site for ROS production may be plant apoplasts. Production of ROS in this location takes place through NADPH oxidases (Liu and He 2016).

11.4 PLANT RESPONSES TO SALT STRESS: BIOCHEMICAL ADAPTATION MECHANISMS

11.4.1 Antioxidant Machinery

Plants equipped with antioxidant systems contain both enzymatic and non-enzymatic defense systems, which when exposed to salinity, are activated to scavenge ROS and reduce the intensity of oxidative stress caused by salinity (Gill and Tuteja 2010; Miller et al. 2010). The reactions of different plant species and even cultivars in one plant species in response to environmental stresses are very different. Plants that have higher levels of antioxidants have been reported to be more resistant to oxidative stress. Reports indicate that the rate of oxidative cell damage in plants exposed to abiotic stress is controlled by the capacity of their antioxidant defense systems (Silvana et al. 2003; Amirjani 2010; Siringam et al. 2011). According to various reports (Hernandez et al. 2001; Mittova et al. 2002, 2003; Acosta-Motos et al. 2017), salinity stress causes the accumulation of O_2^- and H_2O_2 radicals in various cell chambers, including chloroplasts, mitochondria and peroxisome. These radicals are responsible for increasing some parameters related to oxidative stress such as lipid peroxidation (Figure 11.1). Antioxidant defense responses of salinity-resistant cultivars differ from those that are sensitive to salinity. Antioxidant defenses in salt-sensitive species show an unchanged or even diminished reaction, and these species show lower activities of antioxidant enzymes than salt-tolerant species (Hernández et al. 2001; Mittova et al. 2003; Rubio et al. 2009; López-Gómez et al. 2007). Enhanced antioxidant defenses of different salt-tolerant species have been previously reported (Hernández et al. 1993). Some researchers suggest that increased antioxidant activity is one of the mechanisms involved in salinity tolerance responses (Acosta-Motos et al. 2017; Kamran et al. 2019). Other researchers have linked salt tolerance to high levels of specific antioxidant enzymes (Acosta-Motos et al. 2017; Yang and Guo 2018).

11.4.2 Enzymatic Antioxidants

As shown in Figure 11.2, there are many enzymatic antioxidants in plants, such as superoxide dismutase (SOD), catalase (CAT), ascorbate peroxidase (APX), guaiacol peroxidase (GPX), glutathione reductase (GR), monodehydroascorbate reductase (MDHAR) and dehydroascorbate reductase (DHAR).

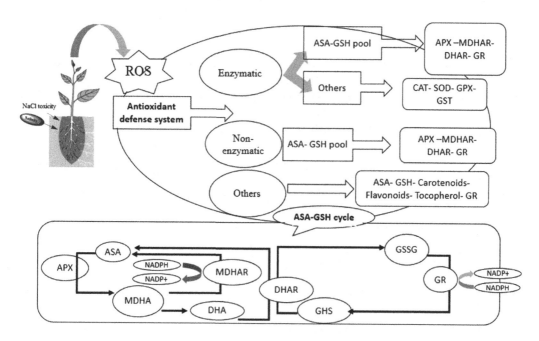

FIGURE 11.1 Main intracellular sites of ROS generation under salt stress.

Table 11.1 indicates current references about the effects of salinity at enzymatic antioxidant levels in different plant species.

11.4.2.1 Superoxide Dismutase

Superoxide dismutases (SODs) are among the most important antioxidant enzymes that form the first line of defense against ROS in the face of a variety of environmental stresses. They detoxify superoxide radicals generated in the biological system by catalyzing their dismutation to H_2O_2 and ultimately to H_2O and O_2 by catalase and peroxidase (Berwal and Ram 2018). The most prevalent

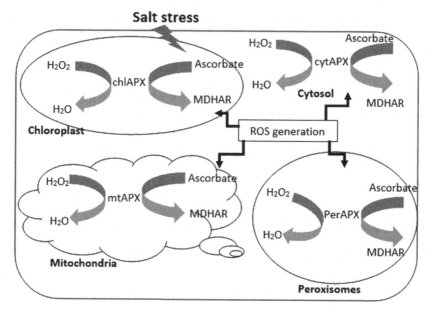

FIGURE 11.2 Enzymatic and non-enzymatic defense system in plant responses to salt stress.

TABLE 11.1

Enzymatic Antioxidant Activity in some Currently Tested Plant Species under Salt Stress

Plant Species Tested	Antioxidant Enzyme Alterations	Reference
Oryza sativa	SOD[+], CAT[+]	Vighi et al. (2017)
Gossypium hirsutum	SOD[+], CAT[+], APX[+], GR[+]	Zhang et al. (2014)
Maize	CAT[+], APX[+], SOD[−]	Abdel Latef et al. (2019)
Gypsophila blanceolata	SOD[+], CAT[+], APX[+]	Sekmen et al. (2012)
Chickpea	POD[+], APX[+−], CAT[+], SOD[+]	Abdel Latef et al. (2017)
Cicer arietinum	SOD[+], APX[+], CAT[+]	Rasool et al. (2013)
Cichorium intybus	SOD[+], APX[+], CAT[+], POD[+]	Sergio et al. (2012)
Nerium oleander	SOD[+], APX[+], CAT[+], GR[+]	Kumar et al. (2017)
Zea mays	SOD[+], APX[+], CAT[+]	Hussain et al. (2013)
Brassica juncea	SOD[+], CAT[+]	Wani et al. (2013)
Sorghum sp.	SOD[+], APX[+], CAT[+], GR[+]	Temizgul et al. (2016)
Paniculum antidotale	SOD[+], APX[+], CAT[+], GR[+]	Hussain et al. (2015)
Desmostachya bipinnata	SOD[+], CAT[+]	Adnan et al. (2016)
Daucus carota	SOD[+], CAT[+]	Bano et al. (2014)
Anacardium occidentale	APX[+], CAT[−]	Ferreira- Silva et al. (2012)
Glycine max	SOD[−], APX[−], CAT[−], GR[−]	Dogan (2011)
Brassica napus	CAT, POD, APX[+]	Abdel Latef et al. (2017)
Eurya emarginata	SOD[+], CAT[+], GPX[+]	Zheng et al. (2016)
Salvia miltiorrhiza	CAT[+], SOD[−]	Gengmao et al. (2014)
Salvia nemorosa	CAT[+], POD[+], GR[+], SOD[+]	Sharifi and Shirani Bidabadi (2020)

[+] and [−] indicate increase and decrease in enzyme activity, respectively

aims of SOD (known as an oxidoreductase) under salt stress is to convert O_2^- to H_2O_2 in the chloroplast, mitochondrion, cytoplasm, apoplast and peroxisome (Choudhury et al. 2017). The activity of SOD determines the concentrations of $O_2^{\bullet-}$ and H_2O_2, thus it has a key role in the antioxidant defense mechanism (Berwal and Ram 2018). SOD has three types of Fe-containing types that are present in the chloroplast, Mn-containing types that are present in peroxisomes and mitochondria, and Cu or Zn-containing types that are present in chloroplast and cytosol (Miller 2012). Houmani et al. (2016) reported the presence of ten types of SOD isoenzymes in their studied plants which include all three types of this antioxidant enzyme. Jing et al. (2015) studied the changes of Cu/Zn-containing SOD against salinity and concluded that this enzyme reduced the damages in the chloroplast under high NaCl levels. Berwal et al. (2016) studied the SOD isozymes pattern of coconut genotypes and they observed a significant variation in SOD enzyme activity as well as in SOD isoforms pattern. Using real-time quantitative reverse transcription-polymerase chain reaction, Soleimani et al. (2017) reported the enhanced activity of Fe-containing SOD of cumin in response to salinity stress. Many studies (Table 11.1) have also shown a positive correlation between SOD activity and the tolerance level of plants. Higher SOD activity along with more SOD isoforms result in better scavenging of the ROS generated during salt stress. Therefore, the SOD isozyme profile can be applied as a suitable biochemical marker for selecting crop germplasm for salt stress tolerance.

11.4.2.2 Ascorbate Peroxidase and Guaiacol Peroxidase

Among peroxidases, ascorbate peroxidases (APX) and guaiacol peroxidase (GPX), using ascorbate and guaiacol electron donors, respectively, are very important for their role in H_2O_2 detoxification in plants stress (Sofo et al. 2015). APX belonging to the family of heme-containing peroxidase enzymes, and scavenges H_2O_2 by converting H_2O_2 into H_2O in various subcellular organelles; plants

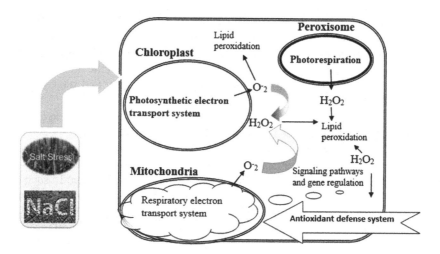

FIGURE 11.3 Localization of APX enzymes and scavenging ROS depending upon its subcellular localization.

apply ascorbic acid as a specific electron donor (Caverzan et al. 2014; Pandey et al. 2017). This antioxidant enzyme is divided into four groups (APX isozymes) based on its location inside the cell (Figure 11.3). The first group is placed in the chloroplast (chlAPX), the second group in mitochondria (mtAPX), the third group in the cytosol (cytAPX) and the last group in the peroxisomes (perAPX) (Caverzan et al. 2014; Pandey et al. 2017; Garcia-Caparros et al. 2019). The differential isozymes of APX based on their location inside the cell and organelles indicate their significant role in developmental processes and plant stress tolerance (Shigeoka et al. 2002). These APX isozymes exhibit various structural and kinetic properties, including the presence of specific conserved domains and signal peptides. The isoforms located in the chloroplast (chlAPX) are reported to be the first ones to interrupt an H_2O_2 molecule because they are adjacent to the acceptor of photosystem I (Huseynova et al. 2014). However, of all four APX (Figure 11.3), the cytosolic type (cytAPX) has been reported to be the most active one (Shigeoka et al. 2002). Overexpression of genes related to the APX through genetic engineering on the salt tolerance of transgenic plums showed that the cytosolic antioxidant APX involved a high degree of salt tolerance in these plants (Diaz-Vivancos et al. 2013).

Various APX isoforms exhibit a differential response to environmental stresses, depending upon their subcellular location and the presence of particular regulatory elements of the related genes (Pandey et al. 2017). APX activity has been reported to enhance in presence of other antioxidant enzymes such as SOD and GR, indicating a positive correlation among different antioxidant enzymes (Pandey et al. 2017). From the above literature, it can be concluded that APX is one of the most important antioxidant defense components in plant cells when exposed to environmental stresses while it plays an essential role in regulating ROS levels in plants when exposed to salt stress (Pandey et al. 2017). However, the puzzle of different isoforms of APX needs to be resolved, and further studies are required to determine the genetic regulation of APX in response to salt stress for sketching superior tactics for plant salinity tolerance. GPX, also as heme-containing enzymes, are of great importance in plant physiology due to their participation in the generation of lignin, defense against biotic and abiotic stresses to reduce the H_2O_2 generation and their role in the breakdown of indole-3-acetic acid (IAA) (Dar et al. 2017).

11.4.2.3 Catalase

Catalase (CAT), a member of the family of heme-containing enzymes, is responsible for the conversion of hydrogen peroxide into water and oxygen. This enzyme is known as a major cellular antioxidant enzyme that defends against oxidative stress (Mhamdi et al. 2010; Saraf

2013). Unlike other enzymes, catalase is activated anywhere in the cell where H_2O_2 is generated (Sofo et al. 2015). Nevertheless, catalase is most abundant in three areas of the plant, including the tissues involved in the photosynthesis process, vascular tissues and tissues involved in the reproduction process (Anjum et al. 2016). CAT has been reported as the most salinity-sensitive enzyme in plants, and its activity varies depending on the plant part and stage of development (Jakovljevic et al. 2017).

However, very contradictory results have been reported regarding the increase or decrease of catalase activity under salinity stress (Table 11.1). Due to the important role of CAT in photorespiration, the CAT catalysis pathway can be considered to be affected under drought and salt stress conditions. Under salt stress, photorespiration acts as an energy sink where it inhibits the over-reduction of the photosynthetic electron transport chain and photoinhibition. In such conditions, mobilization of CAT activity in leaves of stressed plants results in the removal of photorespiratory H_2O_2. Based on the above facts, photorespiration and the CAT pathway seem to be very important processes and are essential components of photosynthesis that play an important part in stress reactions of plant crops to prevent the accumulation of ROS (Bauwe et al. 2012; De Pinto et al. 2013; Voss et al. 2013). Under normal conditions, ROS are efficiently omitted by non-enzymatic and enzymatic antioxidant defense systems, while during saline conditions, the ROS generation overpasses the capacity of the antioxidant systems to eliminate them, causing oxidative stress. In such conditions, CAT isoforms scavenge ROS to eschew the oxidative damage induced by the stressor (Mittler et al. 2011; Vanderauwera et al. 2012; Ajithkumar and Paneerselvam 2014). Overall, H_2O_2 generation increased antioxidant enzyme activities, and CAT is the most important enzyme in charge of H_2O_2 scavenging. Increased CAT activity seems to be correlated with gene expression regulation, and lower oxidative damage has been observed in plants with higher CAT activity, indicating the protective effect of this enzyme (Sofo et al. 2015). Yong et al. (2017) identified a peroxisomal catalase encoding gene, IbCAT2. According to their results, IbCAT2 might play an important role in stress responses. In their study, IbCAT2 expression was also affected by drought or salt treatment in sweet potato, and the expression level and degree of expression change of IbCAT2 varied in different tissues.

11.4.2.4 Glutathione Reductase

Glutathione reductase (GR) is an antioxidant enzyme that is mostly present in areas of the cell such as the cytosol, chloroplasts and mitochondria (Trivedi et al. 2013). The conversion of oxidized glutathione (GSSG) to glutathione (GSH) is presented in Figure 11.2, and has an essential function in the process of scavenging of H_2O_2 through the ascorbate-glutathione cycle, achieved by GR; an important part of the process of scavenging H_2O_2 is done through this cycle (Hasanuzzaman et al. 2017). An increase in the total GSH level is correlated with heightened GR activity. Indeed, one of the most important functions of the antioxidant defense system in plants is related to the glutathione-ascorbate pathway, in which the enzyme GR plays a key role in catalyzing, resulting in increased plant resistance to salinity stress (Rossatto et al. 2017; Gaafar and Seyam 2018). Vighi et al. (2017) asserted that some genes encoding GR isoforms exhibited enhanced expression patterns in the tolerant genotype, in response to salt stress.

11.4.2.5 Monodehydroascorbate Reductase and Dehydroascorbate Reductase

A flavin adenine dinucleotide (FAD) enzyme with a thiol group involved in the reduction of phenoxyl radicals and the production of reduced ascorbate, is monodehydroascorbate reductase (MDHAR). The activity of MDHAR has been observed in different organelles of the cell, including glyoxysomes, mitochondria, chloroplasts and leaf peroxisomes (Leterrier et al. 2005). With increasing plant resistance to salinity stress, the activity of MDHAR has been reported to increase significantly (Eltayeb et al. 2007). Li et al. (2012) reported that the enhanced activity of MDHAR in salt-stressed tomato plants resulted in a higher level of photosynthesis, reduced oxidative damages and increased plant performance. Dehydroscorbate reductase (DHAR) activity maintains high levels of ascorbic acid (AsA) through oxidation. This enzymatic antioxidant is mostly found in seeds and roots as well as in high internodes containing shoots (Chang et al. 2017). Ushimaru et al. (2006) also reported that the higher DHAR activity in *Arabidopsis* plants showed more tolerance against salt stress. The results of a study in rice also showed that increasing the activity of DHAR has a very positive and close correlation with increasing ASA and thus increases plant resistance to salinity (Kim et al. 2014).

11.4.3 NON-ENZYMATIC ANTIOXIDANTS

The non-enzymatic antioxidants viz. ascorbic acid, glutathione, tocopherol, carotenoids and flavonoids were also reported as a protection against oxidative stress caused by salinity (Kojo 2004; Parvaiz and Satyawati 2008; Garcia-Caparros et al. 2019).

11.4.3.1 Ascorbic Acid

Ascorbic acid (ASA) is considered to be one of the most important non-enzymatic antioxidants in plants, possessing the noticeable ability of both scavenging ROS and alleviating oxidative stress damage to plants grown under salt stress (Akram et al. 2017; Abdelgawad et al. 2019). Cell cytoplasm is considered as the most important source of AsA and has a vital role in stress perception, redox homeostasis and subsequent regulation of oxidative stress and plant physio-biochemical responses under normal as well as different abiotic stresses (Zechmann 2011; Venkatesh and Park 2012). Figure 11.4 shows the biosynthesis pathway through which ASA is formed (Akram et al. 2017). During ASA biosynthesis, D-glucose is converted to GDPD-mannose through being catalyzed by the enzymes phosphoglucoisomerase, mannose 6-phosphate isomerase and GDP-D-mannose phosphorylase. Then GDP-D-mannose is converted into L-galactose by the action of enzyme L-galactose dehydrogenase and finally into L-galactono-1, 4-lactone (final precursor of AsA). In another pathway that involves the cell wall's pectin degradation, methyl-galacturonate is formed which is then converted into L-galactonate via two reactions catalyzed by methyl esterase and D-galacturonate reductase. Finally, the enzyme aldono lactonase catalyzes the conversion of L-galactonate into L-galactono-1, 4- lactone and is finally used in ascorbate synthesis. Another biosynthesis pathway involves myo-inositol. Myo-inositol is converted to L-gulono-1, 4-lactone via canalization by myo-inositol oxygenase, glucuronate reductase and aldono lactonase. The gulono-1, 4-lactone is ultimatly used in ascorbate synthesis (Valpuesta and Botella 2004; Akram et al. 2017).

It is believed that AsA can efficiently manage antioxidative metabolism in plants, and by foliar spraying, seed pretreatment or drench application of ASA, its internal content in the plant also improves (Athar et al. 2008; Anjum et al. 2014; Noctor et al. 2014). There are some reports of the improving effect of AsA on antioxidant defense metabolism of canola (Bybordi 2012), *Abelmochus esculentus* (Raza et al. 2013), *Hordeum vulgare* (Agami 2014) and soybean (Kataria et al. 2019), grown under salt stress conditions. One of the most important roles of ASA is participation in AsA-GSH cycle to scavenge ROS through its capacity to donate electrons (Hasanuzzaman et al. 2019).

11.4.3.2 Glutathione

Among non-enzymatic antioxidants, glutathione (GSH), a small intracellular thiol molecule which is considered as a strong non-enzymatic antioxidant, is the most plentiful soluble antioxidant in plant crops and has a fundamental task as an electron donor and scavenger of ROS via AsA-GSH cycle. This antioxidant also acts as a signal transmitter, redox buffer, enzyme cofactor, and cell division and growth synchronizer (Khan and Ashraf 2008; Diaz-Vivancos et al. 2015; Akram et al. 2017; Hasanuzzaman et al. 2019). GHS acts as –SH groups of some antioxidant enzymes via the oxidation of some compounds or through the reparation of –SH groups using the GSH-disulfide exchange reaction (Gill et al. 2013). According to previous reports, the increase in GHS in plants under salinity stress indicates that the plant is more tolerant to stress damage (Hong et al. 2009; Jaleel 2009; Sharifi and Shirani Bidabadi 2020).

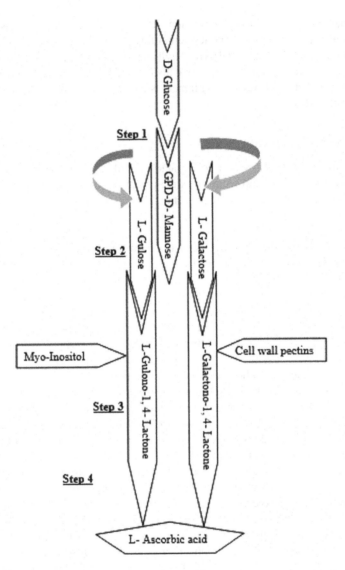

FIGURE 11.4 Several pathways and important precursor molecules are involved in ascorbic acid biosynthesis in plants.

11.4.3.3 Tocopherol

Tocopherols are considered lipidic molecules that have a high solubility and scavenge $^1O_2$ and lipid peroxyl radicals, thereby protecting the chloroplast and maintaining photosynthesis. They have physiological roles in salt tolerance but the mechanisms are still unknown (Falk and Munne-Bosch 2010; Kumar et al. 2013). Mostafa et al. (2015) and Orabi and Abdelhamid (2016) reported that the exogenously applied tocopherol in soybean and *Vicia faba* plants, respectively, caused enhanced photosynthesis and improved antioxidant activity under saline conditions. Higher efficiency of scavenging ROS also has been reported in tocopherol treated *Carex leucochlora* grown under salt-stressed conditions (Ye et al. 2017). The results of a study showed that tocopherols as an antioxidant could activate the antioxidants in plants to enable them to mitigate the oxidative damage, leading to enhanced plant tolerance against salt stress conditions (Semida et al. 2014; Sadiq et al. 2019).

11.4.3.4 Carotenoids

Carotenoids are a huge category of isoprenoid pigments, containing more than 600 different combinations (Zhang et al. 2014). Carotenoid pigments, which are present in all photosynthetic organisms, can be synthesized through two pathways, one through the cytoplasmic mevalonate pathway and the other through the plastid-located pathway. Carotenoid biosynthesis is achieved mostly in the plastids, chloroplasts and chromoplasts (Moise et al. 2014; Nisar et al. 2015). Carotenoids are in charge of the harvesting of light and protect the photosynthetic machinery under high-light environments (Liang et al. 2017; Felemban et al. 2019). Lycopene β-cyclase (LCYB) has been reported to be an essential enzyme that catalyzes the conversion of lycopene into α-carotene and β-carotene in the carotenoid biosynthesis pathway. Various carotenoid genes such as LCYB gene have been reported to be expressed in plants under saline conditions (Abdallah et al. 2016; Kang et al. 2018). Carbonyl carotenoid products are called apocarotenoids. Plant apocarotenoids include important phytohormones such as abscisic acid and strigolactones, most of which play a very important role in reducing the severity of environmental stresses or act as antioxidants (Felemban et al. 2019).

11.4.3.5 Flavonoids

Flavonoids are polyphenolic compounds with a benzo γ pyrone structure and are produced in large quantities in plants (Mierziak et al. 2014). The hydroxyl groups in these compounds scavenge free radicals, and also the capacity of chelation with metals in these compounds prevents the production of ROS and its damage to biological molecules (Schultz et al. 2016). Flavonoids in higher plants can serve as antioxidants in response to environmental stresses including salinity (Agati et al. 2012). Shoeva and Khletskina (2015) reported that imposing salinity to wheat seedlings resulted in an enhanced level of flavonoids concentration. Taibi et al. (2016) also attributed the decrease in salinity damage in *Phaseolus vulgaris* plants to the increase in flavonoid levels in the stressed plant. As shown in Figure 11.2, there is a closed correlation between the non-enzymatic and enzymatic antioxidant systems to scavenge ROS generation. Increasing the activity of antioxidant enzymes and the amounts of total flavonoids result in better protection against oxidative damages under high salinity, which allows maintaining higher yield even in stress conditions (Taibi et al. 2016).

11.5 SALINITY STRESS RESPONSE AND 'OMICS' APPROACHES FOR IMPROVING SALINITY STRESS TOLERANCE IN PLANTS

Due to the dramatic reduction in genetic diversity of plant crops (Massawe et al. 2016), the importance of novel allelic diversity in breeding programs to exploit genetic variations related to various resilience traits such as tolerance to salinity stress becomes more apparent (Brozynska et al. 2016). Plants modulate several genes and proteins involved in salinity tolerance (Arif et al. 2020). Simultaneous use of different omics methods such as transcriptomics, proteomics and metabolomics have been recently used as a widespread technique to deal with salinity and increase plant adaptation to this environmental stress. Indeed, omics' strategies have an important role in providing tolerance to plants against salt stress (Ismail and Horie 2017; Jha et al. 2019). Gene modifications of transcription factors involved in the production of mRNA and protein have also an important role in salinity tolerance (Nongpiur et al. 2016). Acosta-Motos et al. (2017) recently reported signaling genes and calcium pathways in salt acclimation which maintain cell homeostasis. Histone deacetylation, occurring in certain genes such as salt overly sensitive (SOS) signaling genes binding to Ca^{2+}, has been reported to play a vital role in a plant for salinity tolerance (Ismail and Horie 2017). The role of osmosensors has also been considered. Osmosensors regulate genes involved in osmosensing leading to osmoregulation by synthesizing osmolytes and increasing water potential in alleviating the toxic effect of salinity. In this regard, different kinase families are involved in promoting stress-responsive genes (Amirbakhtiar et al. 2019; Guo et al. 2019). New efforts in the field of gene sequencing to decode whole genomes of plants has ignited new interest in novel alleles or genomic information pertaining to salinity tolerance (Brozynska et al. 2016; Munoz et al. 2017). Improving crop growth attributes under salinity stress requires more use of crop wild relatives in plant breeding programs to allow not used genetic variation from crop wild relatives to elite agronomic bases (Wang et al. 2017). Genetic engineering also appears to be a device allowing the development of salinity tolerant plant species. However, further investigations on genomics, transcriptomics, proteomics and metabolomics mechanisms is needed to elucidate the tolerance mechanism under salinity stress.

11.6 CONCLUSION AND FUTURE PERSPECTIVES

Salt stress is one of the biggest factors in inhibiting growth and yield, in which plants suffer from oxidative stress, resulting in substantial damage such as lipid peroxidation, protein oxidation and DNA damage or even cell death. The generation of ROS is caused by superoxide radical, singlet oxygen, hydroxyl radical and hydrogen peroxide, which trigger oxidative stress. Enzymatic and non-enzymatic antioxidant defense system in plants is essential to diminish the damages caused by ROS. The literature suggests that enzymatic and non-enzymatic antioxidant systems have a protective role against salt stress and that antioxidant defense machinery is effective in providing tolerance against salt stress in plants. Great progress has been made in the last decades on the role of the antioxidant defense system in conferring salt-tolerance and sensitivity in plant species, but up to now, many of the basic processes that contribute to tolerance are only partially understood. Therefore, further studies are urgently needed to unravel the details of the antioxidant defense system in plants when facing salt stress. Furthermore, a detailed understanding of the salinity response and tolerance mechanisms in plant crops seems to require further investigation in the field of genomics and proteomics.

REFERENCES

Abdel Latef, A.A. 2010. Changes of antioxidative enzymes in salinity tolerance among different wheat cultivars. *Cereal Research Communications* 38(1): 43–55.

Abdel Latef, A.A. 2011. Ameliorative effect of calcium chloride on growth, antioxidant enzymes, protein patterns and some metabolic activities of canola (*Brassica napus* L.) under seawater stress. *Journal of Plant Nutrition* 34(9): 1303–1320.

Abdel Latef, A.A., A.M. Omer, A.A. Badawy, M.S. Osman, and M.M. Ragaey. 2021. Strategy of salt tolerance and interactive impact of *Azotobacter chroococcum* and/or *Alcaligenes faecalis* inoculation on canola (*Brassica napus* L.) plants grown in saline soil. *Plants* 10(1): 110.

Abdel Latef, A.A., and H. Chaoxing. 2011. Effect of arbuscular mycorrhizal fungi on growth, mineral nutrition, antioxidant enzymes activity and fruit yield of tomato grown under salinity stress. *Scientia Horticulturae* 127(3): 228–233.

Abdel Latef, A.A., and H. Chaoxing. 2014. Does inoculation with *Glomus mosseae* improve salt tolerance in pepper plants? *Journal of Plant Growth Regulation* 33(3): 644–653.

Abdel Latef, A.A., A.K. Srivastava, H. Saber, E.A. Alwaleed, and L.P. Tran. 2017c. *Sargassum muticum* and *Jania rubens* regulate amino acid metabolism to improve growth and alleviate salinity in chickpea. *Scientific Reports* 7(1): 10537.

Abdel Latef, A.A., M. Kordrostami, A. Zakir, H. Zaki, and O.M. Saleh. 2019a. Eustress with H_2O_2 facilitates plant growth by improving tolerance to salt stress in two wheat cultivars. *Plants* 8(9): 303.

Abdel Latef, A.A., M.F.A. Abu Alhmad, and S. Ahmad. 2017a. Foliar application of fresh Moringa leaf extract overcomes salt stress in fenugreek (*Trigonella foenum-Graecum*) plants. *Egyptian Journal of Botany* 57: 157–179.

Abdel Latef, A.A., M.F.A. Abu Alhmad, and K.E. Abdelfattah. 2017b. The possible roles of priming with ZnO nanoparticles in mitigation of salinity stress in lupine (*Lupinus termis*) plants. *Journal of Plant Growth Regulation* 36(1): 60–70.

Abdel Latef, A.A., M.G. Mostofa, M.M. Rahman, I.B. Abdel-Farid, and L.P. Tran. 2019b. Extracts from yeast and carrot roots enhance maize performance under seawater-induced salt stress by altering physio-biochemical characteristics of stressed plants. *Journal of Plant Growth Regulation* 38(3): 966–979.

Abdel Latef, A.A.H., A.K. Srivastava, M.S.A. El-sadek, M. Kordrostami, and L.P. Tran. 2018. Titanium dioxide nanoparticles improve growth and enhance tolerance of broad bean plants under saline soil conditions. *Land Degradation and Development* 29(4): 1065–1073.

Abdel Latef, A.A.H., M.F. Abu Alhmad, M. Kordrostami, A.A. Abo–Baker, and A. Zakir. 2020a. Inoculation with *Azospirillum lipoferum* or *Azotobacter chroococcum* reinforces maize growth by improving physiological activities under saline conditions. *Journal of Plant Growth Regulation* 39(3): 1293–1306.

Abdelgawad, K.F., M.M. El-Mogy, M.I.A. Mohamed, C. Garchery, and R.G. Stevens. 2019. Increasing ascorbic acid content and salinity tolerance of cherry tomato plants by suppressed expression of the ascorbate oxidase gene. *Agronomy* 9(2): 51. doi: 10.3390/agronomy9020051.

Acosta-Motos, J.R., J.A. Hernández, S. Álvarez, G. Barba-Espín, and M.J. Sánchez-Blanco. 2017. The long-term resistance mechanisms, critical irrigation threshold and relief capacity shown by Eugenia myrtifolia plants in response to saline reclaimed water. *Plant Physiology and Biochemistry: PPB* 111: 244–256.

Acosta-Motos, J.R., M.F. Ortuño, S. Álvarez, M.F. López-Climent, A. Gómez-Cadenas, and M.J. Sánchez-Blanco. 2016. Changes in growth, physiological parameters and the hormonal status of *Myrtus communis* L. plants irrigated with water with different chemical compositions. *Journal of Plant Physiology* 191: 12–21.

Adnan, M.Y., T. Hussain, M.Z. Ahmed, B. Gul, M. Ajmal Khan, and B.L. Nielsen. 2021. Growth regulation of *Desmostachya bipinnata* by organ-specific biomass, water relations, and ion allocation responses to improve salt resistance. *Acta Physiologiae Plantarum* 43: 38. https://doi.org/10.1007/s11738-021-03211-7

Agami, R.A. 2014. Applications of ascorbic acid or proline increase resistance to salt stress in barley seedlings. *Biologia Plantarum* 58(2): 341–347. doi: 10.1007/s10535-014-0392-y.

Agati, G., E. Azzarello, S. Pollastri, and M. Tattini. 2012. Flavonoids as antioxidants in plants: Location and functional significance. *Plant Science: An International Journal of Experimental Plant Biology* 196: 67–76. doi: 10.1016/j.plantsci.2012.07.014.

Ahmad, S., H. Khan, U. Shahab, S. Rehman, Z. Rafi, M.Y. Khan, A. Ansari, Z. Siddiqui, J.M. Ashraf, S.M.S. Abdullah, S. Habib, and M. Uddin. 2017. Protein oxidation: An overview of metabolism of sulphur containing amino acid, cysteine. *Frontiers in Bioscience* 9: 71–87.

Ajithkumar, I.P., and R. Panneerselvam. 2014. ROS scavenging system, osmotic maintenance, pigment and growth status of *Panicum sumatrense* Roth. under drought stress. *Cell Biochemistry and Biophysics* 68(3): 587–595.

Akram, N.A., F. Shafiq, and M. Ashraf. 2017. Ascorbic acid-a potential oxidant scavenger and its role in plant development and abiotic stress tolerance. *Frontiers in Plant Science* 8: 613. doi: 10.3389/fpls.2017.00613.

Amirbakhtiar, N., A. Ismaili, M.R. Ghaffari, F. Nazarian Firouzabadi, and Z.S. Shobbar. 2019. Transcriptome response of roots to salt stress in a salinity-tolerant bread wheat cultivar. *PLOS ONE* 14(3): e0213305.

Amirjani, M.R. 2010. Effect of salinity stress on growth, mineral composition, proline content, antioxidant enzymes of soybean. *American Journal of Plant Physiology* 5(6): 350–360.

Anjum, N.A., P. Sharma, S.S. Gill, M. Hasanuzzaman, E.A. Khan, K. Kachhap, A.A. Mohamed, P. Thangavel, G.D. Devi, P. Vasudhevan, A. Sofo, N.A. Khan, A.M. Misra, A.S. Lukatkin, H.P. Singh, E. Pereira, and N. Tuteja. 2016. Catalase and ascorbate peroxidase—representative H2O2-detoxifying heme enzymes in plants. *Environmental Science and Pollution Research* 23: 19002–19029. https://doi.org/10.1007/s11356-016-7309-6.

Anjum, N.A., S.S. Gill, R. Gill, M. Hasanuzzaman, A.C. Duarte, E. Pereira, I. Ahmad, R. Tuteja, and N. Tuteja. 2014. Metal/metalloid stress tolerance in plants: Role of ascorbate, its redox couple, and associated enzymes. *Protoplasma* 251(6): 1265–1283. doi: 10.1007/s00709-014-0636-x.

Apel, K., and H. Hirt. 2004. Reactive oxygen species: Metabolism, oxidative stress, and signal transduction. *Annual Review of Plant Biology* 55: 373–399.

Arif, Y., P. Singh, H. Siddiqui, A. Bajguz, and S. Hayat. 2020. Salinity induced physiological and biochemical changes in plants: An omic approach towards salt stress tolerance 156: 64–77.

Asada, K. 2006. Production and scavenging of reactive oxygen species in chloroplasts and their functions. *Plant Physiology* 141(2): 391–396.

Ashraf, M. 2009. Biotechnological approach of improving plant salt tolerance using antioxidants as markers. *Biotechnology Advances* 27(1): 84–93.

Athar, H., A. Khan, and M. Ashraf. 2008. Exogenously applied ascorbic acid alleviates salt-induced oxidative stress in wheat. *Environmental and Experimental Botany* 63(1–3): 224–231.

Bano, A., M. Ahmad, T.B. Hadda, A. Saboor, S. Sultana, M. Zafar, M.P.Z. Khan, M. Arshad, and M.A. Ashraf. 2014. Quantitative ethnomedicinal study of plantsused in the skardu valley at high altitude ofKarakoram-Himalayan range, Pakistan. *Journal of Ethnobiology and Ethnomedicine* 10: 43.

Bauwe, H., M. Hagemann, R. Kern, and S. Timm. 2012. Photorespiration has a dual origin and manifold links to central metabolism. *Current Opinion in Plant Biology* 15(3): 269–275.

Berwal, M.K., and C. Ram. 2018. Superoxide dismutase: A stable biochemical marker for abiotic stress tolerance in higher plants. In: *Abiotic and Biotic Stress in Plants*, eds. A.B. de Oliveira, IntechOpen: 1–10. doi: 10.5772/intechopen.82079.

Berwal, M.K., P. Sugatha, V. Niral, and K. Hebbar. 2016. Variability in superoxide dismutase isoforms in tall and dwarf cultivars of coconut (*Cocos nucifera* L.) leaves. *Indian Journal of Agricultural Biochemistry* 29(2): 184–188.

Bhattacharjee, S. 2014. Membrane lipid peroxidation and its conflict of interest: The two faces of oxidative stress. *Current Science* 107: 1811–1823.

Brozynska, M., A. Furtado, and R.J. Henry. 2016. Genomics of crop wild relatives: Expanding the gene pool for crop improvement. *Plant Biotechnology Journal* 14(4): 1070–1085.

Bybordi, A. 2012. Effect of ascorbic acid and silicium on photosynthesis, antioxidant enzyme activity, and fatty acid contents in canola exposure to salt stress. *Journal of Integrative Agriculture* 11(10): 1610–1620.

Caverzan, A., A. Bonifacio, F.E.L. Carvalho, C.M. Andrade, G. Passaia, M. Schünemann, Fdos S.. Maraschin, M.O. Martins, F.K. Teixeira, R. Rauber, R. Margis, J.A.G. Silveira, and M. Margis-Pinheiro. 2014. The knockdown of chloroplastic ascorbate peroxidases reveals its regulatory role in the photosynthesis and protection under photo-oxidative stress in rice. *Plant Science: An International Journal of Experimental Plant Biology* 214: 74–87.

Chang, H.Y., S.T. Lin, T.P. Ko, S.M. Wu, T.H. Lin, Y.C. Chang, K.F. Huang, and T.M. Lee. 2017. Enzymatic characterization and crystal structure analysis of *Chlamydomonas reinhardtii* dehydroascorbate reductase and their implications for oxidative stress. *Plant Physiology and Biochemistry: PPB* 120: 144–155.

Choi, W.G., M. Toyota, S.H. Kim, R. Hilleary, and S. Gilroy. 2014. Salt stress–induced Ca^{2+} waves are associated with rapid, long-distance root-to-shoot signaling in plants. *Proceedings of the National Academy of Sciences of the United States of America* 111(17): 6497–6502.

Choudhury, F.K., R.M. Rivero, E. Blumwald, and R. Mittler. 2017. Reactive oxygen species, abiotic stress and stress combination. *The Plant Journal: For Cell and Molecular Biology* 90(5): 856–867. doi: 10.1111/tpj.13299.

Cirillo, C., Y. Rouphael, R. Caputo, G. Raimondi, M.I. Sifola, and S. De Pascale. 2016. Effects of high salinity and the exogenous application of an osmolyte on growth, photosynthesis, and mineral composition in two ornamental shrubs. *Journal of Horticultural Science and Biotechnology* 91(1): 14–22.

Corpas, F.J. 2015. What is the role of hydrogen peroxide in plant peroxisomes? *Plant Biology* 17(6): 1099–1103. doi: 10.1111/plb.12376

Dar, A.A., A.R. Choudhury, P.K. Kancharla, and N. Arumugam. 2017. The FAD2 gene in plants: Occurrence, regulation,and role. *Frontiers in Plant Science* 8: 1–16.

Das, P., K.K. Nutan, S.L. Singla-Pareek, and A. Pareek. 2015. Oxidative environment and redox homeostasis in plants: Dissecting out significant contribution of major cellular organelles. *Frontiers in Environmental Science* 2: 70.

Demidchik, V., and F.J.M. Maathuis. 2007. Physiological roles of nonselective cation channels in plants: From salt stress to signalling and development. *New Phytologist* 175(3): 387–404.

Demidchik, V., and M. Tester. 2002. Sodium fluxes through nonselective cation channels in the plasma membrane of protoplasts from Arabidopsis roots. *Plant Physiology* 128(2): 379–387.

De Pinto, M.C., V. Locato, A. Sgobba, Mdel C Romero-Puertas, C. Gadaleta, M. Delledonne, and L. de Gara. 2013. S-nitrosylation of ascorbate peroxidase is part of programmed cell death signaling in tobacco Bright Yellow-2 cells. *Plant Physiology* 163(4): 1766–1775.

Diaz-Vivancos, P., M. Faize, G. Barba-Espin, L. Faize, C. Petri, J.A. Hernández, and L. Burgos. 2013. Ectopic expression of cytosolic superoxide dismutase and ascorbate peroxidase leads to salt stress tolerance in transgenic plums. *Plant Biotechnology Journal* 11(8): 976–985. doi: 10.1111/pbi.12090.

Dietz, K.J., I. Turkan, and A. Krieger-Liszkay. 2016. Redox- and reactive oxygen species-dependent signaling into and out of the photosynthesizing chloroplast. *Plant Physiology* 171(3): 1541–1550.

Dogan, M. 2011. Antioxidative and proline potentials as a protectivemechanism in soybean plants under salinity stress. *African Journal of Biotechnology* 10: 5972–5978.

Eltayeb, A.E., N. Kawano, G.H. Badawi, H. Kaminaka, T. Sanekata, T. Shibahara, S. Inanaga, and K. Tanaka. 2007. Overexpression of monodehydroascorbate reductase in transgenic tobacco confers enhanced tolerance to ozone, salt and polyethylene glycol stresses. *Planta* 225(5): 1255–1264.

Falk, J., and S. Munné-Bosch. 2010. Tocochromanol functions in plants: Antioxidation and beyond. *Journal of Experimental Botany* 61(6): 1549–1566.

Felemban, A., J. Braguy, M.D. Zurbriggen, and S. Al-Babili. 2019. Apocarotenoids involved in plant development and stress response. *Frontiers in Plant Science* 10: 1168. doi: 10.3389/fpls.2019.01168.

Ferreira-Silva, S.L., E.L. Voigt, E.N. Silva, J.M. Maia, T.C.R. Aragao, and J.A.G. Silveira. 2012. Partial oxidative protection by enzymatic and non-enzymatic components in cashew leaves under high salinity. *Biologia Plantarum* 56: 172–176.

Flowers, T.J., M.A. Hajibagherp, and A.R. Yeo. 1991. Ion accumulation in the cell walls of rice plants growing under saline conditions: Evidence for the Oertli hypothesis. *Plant, Cell and Environment* 14(3): 319–325.

Flowers, T.J., R. Munns, and T.D. Colmer. 2015. Sodium chloride toxicity and the cellular basis of salt tolerance in halophytes. *Annals of Botany* 115(3): 419–431.

Flowers, T.J., and T.D. Colmer. 2015. Plant salt tolerance: Adaptations in halophytes. *Annals of Botany* 115(3): 327–331.

Foyer, C.H., and G. Noctor. 2000. Tansley review no. 112: Oxygen processing in photosynthesis: Regulation and signalling. *New Phytologist* 146(3): 359–388.

Gaafar, R.M., and M.M. Seyam. 2018. Ascorbate–glutathione cycle confers salt tolerance in Egyptian lentil cultivars. *Physiology and Molecular Biology of Plants: An International Journal of Functional Plant Biology* 24(6): 1083–1092.

Garcia-Caparros, P., M. Hasanuzzaman, and M.T. Lao. 2019. Oxidative stress and antioxidant defense in plants under salinity. In: *Reactive Oxygen, Nitrogen and Sulfur Species in Plants: Production, Metabolism, Signaling.* 1st Edition, eds. M. Hasanuzzaman, V. Fotopoulos, K. Nahar, and M. Fujita, John Wiley & Sons Ltd: 291–309.

Gengmao, Z., S. Quanmei, H. Yu, L. Shihui, and W. Changhai. 2014. The physiological and biochemical responses of a medicinal plant (*Salvia miltiorrhiza* L.) to stress caused by various concentrations of NaCl. *PloS one* 9: e89624.

Gill, S.S., and N. Tuteja. 2010. Reactive oxygen species and antioxidant machinery in abiotic stress tolerance in crop plants. *Plant Physiology and Biochemistry: PPB* 48(12): 909–930.

Gill, S.S, N.A. Anjum, M. Hasanuzzaman, R. Gill, D.K. Trivedi, I. Ahmad, E. Pereira, and N. Tuteja. 2013. Glutathione and glutathione reductase: A boon in disguise for plant abiotic stress defense operations. *Plant Physiology and Biochemistry* 70: 204–212. doi: 10.1016/j.plaphy.2013.05.032

Guo, S.M., Y. Tan, H.J. Chu, M.X. Sun, and J.C. Xing. 2019. Transcriptome sequencing revealed molecular mechanisms underlying tolerance of Suaeda salsa to saline stress. *PLOS ONE* 14(7): e0219979.

Hasanuzzaman, M., K. Nahar, T.I. Anee, and M. Fujita. 2017. Glutathione in plants: Biosynthesis and physiological role in environmental stress tolerance. *Physiology and Molecular Biology of Plants: An International Journal of Functional Plant Biology* 23(2): 249–268.

Hasanuzzaman, M., M. Bhuyan, T.I. Anee, K. Parvin, K. Nahar, J.A. Mahmud, and M. Fujita. 2019. Regulation ofascorbate-glutathione pathway in mitigating oxidative damage in plants under abiotic stress. *Antioxidants* 8(9): 384.

Hernández, J.A., F.J. Corpas, M. Gómez, L.A. del Río, and F. Sevilla. 1993. Salt-induced oxidative stress mediated by activated oxygen species in pea leaf mitochondria. *Physiologia Plantarum* 89(1): 103–110.

Hernández, J.A., M.A. Ferrer, A. Jiménez, A.R. Barceló, and F. Sevilla. 2001. Antioxidant systems and O(2) (.-)/H(2)O(2) production in the apoplast of pea leaves. Its relation with salt-induced necrotic lesions in minor veins. *Plant Physiology* 127(3): 817–831.

Houmani, H., M. Rodríguez-Ruiz, J.M. Palma, C. Abdelly, and F.J. Corpas. 2016. Modulation of superoxide dismutase (SOD) isozymes by organ development and high long-term salinity in the halophyte *Cakile maritima*. *Protoplasma* 253(3): 885–894.

Hsu, S.Y., and C.H. Kao. 2003. Differential effect of sorbitol and polyethylene glycol on antioxidant enzymes in rice leaves. *Plant Growth Regulation* 39(1): 83–90.

Huseynova, I.M., D.R. Aliyeva, and J.A. Aliyev. 2014. Subcellular localization and responses of superoxide dismutase isoforms in local wheat varieties subjected to continuous soil drought. *Plant Physiology and Biochemistry: PPB* 81: 54–60. doi: 10.1016/j.plaphy.2014.01.018.

Hussain, I., M.A. Ashraf, F. Anwar, R. Rasheed, M. Niaz, and A. Wahid. 2013. Biochemical characterization of maize (*Zea mays* L.) for salt tolerance. *Plant Biosystems* 148(5): 1016–1026.

Hussain, M.I., D.A. Lyra, M. Farooq, N. Nikoloudakis, and N. Khalid. 2015. Salt and drought stresses in safflower: A review. *Agronomy for Sustainable Development* 36(4) (2016): 4. https://doi.org/10.1007/s13593-015-0344-8

Isayenkov, S.V. 2012. Physiological and molecular aspects of salt stress in plants. *Cytology and Genetics* 46: 302–318. https://doi.org/10.3103/S0095452712050040

Isayenkov, S.V., and F.J.M. Maathuis. 2019. Plant salinity stress: Many unanswered questions remain. *Frontiers in Plant Science* 10: 80. doi: 10.3389/fpls.2019.00080.

Ismail, A.M., and T. Horie. 2017. Genomics, physiology, and molecular breeding approaches for improving salt tolerance. *Annual Review of Plant Biology* 68: 405–434.

Jakovljević, D.Z., M.D. Topuzović, M.S. Stanković, and B.M. Bojović. 2017. Changes in antioxidant enzyme activity in response to salinity-induced oxidative stress during early growth of sweet basil. *Horticulture, Environment, and Biotechnology* 58(3): 240–246.

Jha, U.C., A. Bohra, R. Jha, and S.K. Parida. 2019. Salinity stress response and 'omics' approaches for improving salinity stress tolerance in major grain legumes. *Plant Cell Reports* 38(3): 255–277.

Jing, X., P. Hou, Y. Lu, S. Deng, N. Li, R. Zhao, J. Sun, Y. Wang, Y. Han, T. Lang, M. Ding, X. Shen, and S. Chen. 2015. Overexpression of copper/zinc superoxide dismutase from mangrove *Kandelia candel* in tobacco enhances salinity tolerance by the reduction of reactive oxygen species in chloroplast. *Frontiers in Plant Science* 6: 23. doi: 10.3389/fpls.2015.00023.

Kamran, M., A. Parveen, S. Ahmar, Z. Malik, S. Hussain, M.S. Chattha, M.H. Saleem, M. Adil, P. Heidari, and J.T. Chen. 2019. An overview of hazardous impacts of soil salinity in crops, tolerance mechanisms, and amelioration through selenium supplementation. *International Journal of Molecular Sciences* 21(1): 148. doi: 10.3390/ijms21010148.

Kang, C., H. Zhai, L. Xue, N. Zhao, S. He, and Q. Liu. 2018. A lycopene β-cyclase gene, IbLCYB2, enhances carotenoid contents and abiotic stress tolerance in transgenic sweetpotato. *Plant Science: An International Journal of Experimental Plant Biology* 272: 243–254.

Kataria, S., L. Baghel, M. Jain, and K.N. Guruprasad. 2019. Magnetopriming regulates antioxidant defense system in soybean against salt stress. *Biocatalysis and Agricultural Biotechnology* 18. doi: 10.1016/j.bcab.2019.101090. http://www.ncbi.nlm.nih.gov/pubmed/101090.

Kim, Y.S., I.S. Kim, S.Y. Shin, T.H. Park, H.M. Park, Y.H. Kim, G.S. Lee, H.G. Kang, S.H. Lee, and H.S. Yoon. 2014. Overexpression of dehydroascorbate reductase confers enhanced tolerance to salt stress in rice plants (*Oryza sativa* L. japonica). *Journal of Agronomy and Crop Science* 200(6): 444–456.

Krieger-Liszkay, A. 2005. Singlet oxygen production in photosynthesis. *Journal of Experimental Botany* 56(411): 337–346.

Kumar, D., M.Al. Al Hassan, M.A. Naranjo, V. Agrawal, M. Boscaiu, and O. Vicente. 2017. Effects of salinity and drought on growth, ionic relations, compatible solutes and activation of antioxidant systems in oleander (*Nerium oleander* L.). *PLOS ONE* 12(9): e0185017.

Kumar, S., P. Thakur, N. Kaushal, J.A. Malik, P. Gaur, and H. Nayyar. 2013. Effect of varying high temperatures during reproductive growth on reproductive function, oxidative stress and seed yield in chickpea genotypes differing in heat sensitivity. *Archives of Agronomy and Soil Science* 59(6): 823–843.

Leterrier, M., F.J. Corpas, J.B. Barroso, L.M. Sandalio, and L.A. del Río. 2005. Peroxisomal monodehydroascorbate reductase. Genomic clone characterization and functional analysis under environmental stress conditions. *Plant Physiology* 138(4): 2111–2123.

Li, F., Q.Y. Wu, M. Duan, X.C. Dong, B. Li, and Q.W. Meng. 2012. Transgenic tomato plants overexpressing chloroplastic monodehydroascorbate reductase are resistant to salt- and PEG-induced osmotic stress. *Photosynthetica* 50(1): 120–128.

Liu, Y., and C. He. 2016. Regulation of plant reactive oxygen species (ROS) in stress responses: Learning from AtRBOHD. *Plant Cell Reports* 35(5): 995–1007.

López-Gómez, E., M.A. Sanjuán, P. Diaz-Vivancos, J. Mataix-Beneyto, M.F. García-Legaz, and J.A. Hernández. 2007. Effect of salinity and rootstocks on antioxidant systems of loquat plants (*Eriobotrya japonica* Lindl.): Response to supplementary boron addition. *Environmental and Experimental Botany* 160: 151–158.

Maathuis, F.J.M., I. Ahmad, and J. Patishtan. 2014. Regulation of Na(+) fluxes in plants. *Frontiers in Plant Science* 5: 467. doi: 10.3389/fpls.2014.00467.

Manova, V., and D. Gruszka. 2015. DNA damage and repair in plants—From models to crops. *Frontiers in Plant Science* 6: 885.

Massawe, F., S. Mayes, and A. Cheng. 2016. Crop diversity: An unexploited treasure trove for food security. *Trends in Plant Science* 21(5): 365–368.

Mierziak, J., K. Kostyn, and A. Kulma. 2014. Flavonoids as important molecules of plant interactions with the environment. *Molecules* 19(10): 16240–16265.

Miller, A.F. 2012. Superoxide dismutases: Ancient enzymes and new insights. *FEBS Letters* 586(5): 585–595. doi: 10.1016/j.febslet.2011.10.048

Miller, G., N. Suzuki, S. Ciftci-Yilmaz, and R. Mittler. 2010. Reactive oxygen species homeostasis and signalling during drought and salinity stresses. *Plant, Cell and Environment* 33(4): 453–467.

Mittler, R. 2017. ROS are good. *Trends in Plant Science* 22(1): 11–19.

Mittler, R., S. Vanderauwera, N. Suzuki, G. Miller, V.B. Tognetti, K. Vandepoele, M. Gollery, V. Shulaev, and F. van Breusegem. 2011. ROS signaling: The new wave? *Trends in Plant Science* 16(6): 300–309.

Mittova, V., M. Tal, M. Volokita, and M. Guy. 2002. Salt stress induces up-regulation of an efficient chloroplast antioxidant system in the salt-tolerant wild tomato species Lycopersicon pennellii but not in the cultivated species. *Physiologia Plantarum* 115(3): 393–400.

Mittova, V., M. Tal, M. Volokita, and M. Guy. 2003. Up-regulation of the leaf mitochondrial and peroxisomal antioxidative systems in response to salt-induced oxidative stress in the wild salt-tolerant tomato species Lycopersicon pennellii. *Plant, Cell and Environment* 26(6): 845–856.

Moise, A.R., S. Al-Babili, and E.T. Wurtzel. 2014. Mechanistic aspects of carotenoid biosynthesis. *Chemical Reviews* 114(1): 164–193.

Munns, R. 2005. Genes and salt tolerance: Bringing them together. *New Phytologist* 167(3): 645–663.

Munns, R., and M. Tester. 2008. Mechanisms of salinity tolerance. *Annual Review of Plant Biology* 59: 651–681.

Muñoz, N., A. Ailin Liu, L. Kan, M.W. Li, and H.M. Lam. 2017. Potential uses of wild germplasms of grain legumes for crop improvement. *International Journal of Molecular Sciences* 18(2): 328.

Negrão, S., S.M. Schmöckel, and M. Tester. 2017. Evaluating physiological responses of plants to salinity stress. *Annals of Botany* 119(1): 1–11.

Noctor, G., A. Mhamdi, and C.H. Foyer. 2014. The roles of reactive oxygen metabolism in drought: Not so cut and dried. *Plant Physiology* 164(4): 1636–1648. doi: 10.1104/pp.113.233478.

Nongpiur, R.C., S.L. Singla-Pareek, and A. Pareek. 2016. Genomics approaches for improving salinity stress tolerance in crop plants. *Current Genomics* 17(4): 343–357.

Orabi, S.A., and M.T. Abdelhamid. 2016. Protective role of α-tocopherol on two Vicia faba cultivars against seawater-induced lipid peroxidation by enhancing capacity of anti-oxidative system. *Journal of the Saudi Society of Agricultural Sciences* 15: 145–154.

Osman, M.S., A.A. Badawy, A.I. Osman, and A.A.H. Abdel Latef. 2020. Ameliorative impact of an extract of the halophyte arthrocnemum macrostachyum on growth and biochemical parameters of soybean under salinity stress. *Journal of Plant Growth Regulation* 39: 1–12. doi: 10.1007/s00344-020-10185-2

Pandey, S., D. Fartyal, A. Agarwal, T. Shukla, D. James, T. Kaul, Y.K. Negi, S. Arora, and M.K. Reddy. 2017. Abiotic stress tolerance in plants: Myriad roles of ascorbate peroxidase. *Frontiers in Plant Science* 8: 581. doi: 10.3389/fpls.2017.00581.

Parvaiz, A., and S. Satyawati. 2008. Salt stress and phyto-biochemical responses of plants—A review. *Plant, Soil and Environment* 54(3): 89–99.

Rasool, S., A. Ahmad, T.O. Siddiqi, and P. Ahmad. 2013. Changes in growth, lipid peroxidation and some key antioxidant enzymes in chickpea genotypes under salt stress. *Acta Physiologiae Plantarum* 35(4): 1039–1050.

Raza, S.H., F. Shafiq, M. Chaudhary, and I. Khan. 2013. Seed invigoration with water, ascorbic and salicylic acid stimulates development and biochemical characters of okra (*Ablemoschus esculentus*) under normal and saline conditions. *International Journal of Agriculture and Biology* 15: 486–492.

Rossatto, T., M.N. do Amaral, L.C. Benitez, I.L. Vighi, E.J.B. Braga, A.M. de Magalhães Júnior, M.A.C. Maia, and L. da Silva Pinto. 2017. Gene expression and activity of antioxidant enzymes in rice plants, cv. BRS AG, under saline stress. *Physiology and Molecular Biology of Plants: An International Journal of Functional Plant Biology* 23(4): 865–875. doi: 10.1007/s12298-017-0467-2.

Roy, S.J., S. Negrão, and M. Tester. 2014. Salt resistant crop plants. *Current Opinion in Biotechnology* 26: 115–124. doi: 10.1016/j.copbio.2013.12.004.

Rubio, M.C., P. Bustos-Sanmamed, M.R. Clemente, and M. Becana. 2009. Effects of salt stress on the expression of antioxidant genes and proteins in the model legume Lotus japonicus. *New Phytologist* 181(4): 851–859.

Sadiq, M., N.A. Akram, M. Ashraf, F. Al-Qurainy, and P. Ahmad. 2019. Alpha-tocopherol-induced regulation of growth and metabolism in plants under non-stress and stress conditions. *Journal of Plant Growth Regulation* 38(4): 1325–1340.

Saraf, N. 2013. Enhancement of catalase Activity underSalt Stress in Germinating Seeds of *Vigna radiate*. *Asian Journal of Biomedical and Pharmaceutical Science* 3(17): 6–8.

Schulz, E., T. Tohge, E. Zuther, A.R. Fernie, and D.K. Hincha. 2016. Flavonoids are determinants of freezing tolerance and cold acclimation in *Arabidopsis thaliana*. *Scientific Reports* 6: 34027.

Sekmen, A.H., I. Turkan, Z.O. Tanyolac, C. Ozfidan, and A. Dinc. 2012. Different antioxidant defense responses to salt stress during germination and vegetative stages of endemic halophyte Gypsophila oblanceolata Bark. *Environmental and Experimental Botany* 77: 63–76.

Semida, W.M., R.S. Taha, M.T. Abdelhamid, and M.M. Rady. 2014. Foliar-applied α-tocopherol enhances salt-tolerance in Vicia faba L. plants grown under saline conditions. *South African Journal of Botany* 95: 24–31.

Sergio, L., A. De Paola, V. Cantore, M. Pieralice, N.A. Cascarano, V.V. Bianco, and D. Di Venere. 2012 .Effect of salt stress on growth parameters, enzymatic antioxidant system, and lipid peroxidation in wild chicory (*Cichorium intybus* L.). *Acta Physiologiae Plantarum* 34(6): 2349–2358.

Sevengor, S., F. Yasar, S. Kusvuran, and S. Ellialtioglu. 2011. The effect of salt stress on growth, chlorophyll content, lipid peroxidation and antioxidative enzymes of pumpkin seedling. *African Journal of Agricultural Research* 6(21): 4920–4924.

Sharifi, P., and S. Bidabadi. 2020. Strigolactone could enhances gas-exchange through augmented antioxidant defense system in Salvia nemorosa L. plants subjected to saline conditions stress. *Industrial Crops and Products* 151. doi: 10.1016/j.indcrop.2020.112460. http://www.ncbi.nlm.nih.gov/pubmed/112460.

Shigeoka, S., T. Ishikawa, M. Tamoi, Y. Miyagawa, T. Takeda, Y. Yabuta, and K. Yoshimura. 2002. Regulation and function of ascorbate peroxidase isoenzymes. *Journal of Experimental Botany* 53(372): 1305–1319. doi: 10.1093/jxb/53.372.1305.

Silberbush, M., and J. Ben-Asher. 2001. Simulation study of nutrient uptake by plants from soilless cultures as affected by salinity buildup and transpiration. *Plant and Soil* 233(1): 59–69.

Siringam, K., N. Juntawong, S. Cha-Um, and C. Kirdmanee. 2011. Salt stress induced ion accumulation, ion homeostasis, membrane injury and sugar contents in salt-sensitive rice (*Oryza sativa* L. spp. indica) roots under isoosmotic conditions. *African Journal of Biotechnology* 10: 1340–1346.

Sofo, A., A. Scopa, M. Nuzzaci, and A. Vitti. 2015. Ascorbate peroxidase and catalase activities and their genetic regulation in plants subjected to drought and salinity stresses. *International Journal of Molecular Sciences* 16(6): 13561–13578. doi: 10.3390/ijms160613561.

Soleimani, Z., A.S. Afshar, and F.S. Nematpour. 2017. Responses of antioxidant gene and enzymes to salinity stress in the Cuminum cyminum L *Russian Journal of Plant Physiology* 64(3): 361–367.

Taïbi, K., F. Taïbi, L. Ait Abderrahim, A. Ennajah, M. Belkhodja, and J.M. Mulet. 2016. Effect of salt stress on growth, chlorophyll content, lipid peroxidation and antioxidant defence systems in Phaseolus vulgaris L *South African Journal of Botany* 105: 306–312.

Temizgul, R., M. Kaplan, R. Kara, and S. Yilmaz. 2016. Effects of salt concentrations on antioxidant enzyme activity of grain sorghum. *Current Trends in Natural Sciences* 5: 171–178.

Tester, M., and R. Davenport. 2003. Na$^+$ tolerance and Na+ transport in higher plants. *Annals of Botany* 91(5): 503–527.

Trivedi, D.K., S.S. Gill, S. Yadav, and N. Tuteja. 2013. Genome-wide analysis of glutathione reductase (GR) genes from rice and Arabidopsis. *Plant Signaling and Behavior* 8(2): e23021.

Ushimaru, T., T. Nakagawa, Y. Fujioka, K. Daicho, M. Naito, Y. Yamauchi, H. Nonaka, K. Amako, K. Yamawaki, and N. Murata. 2006. Transgenic Arabidopsis plants expressing the rice dehydroascorbate reductase gene are resistant to salt stress. *Journal of Plant Physiology* 163(11): 1179–1184.

Valpuesta, V., and M.A. Botella. 2004. Biosynthesis of L-ascorbic acid in plants: New pathways for an old antioxidant. *Trends in Plant Science* 9(12): 573–577. doi: 10.1016/j.tplants.2004.10.002.

Vanderauwera, S., K. Vandenbroucke, A. Inzé, B. van de Cotte, P. Mühlenbock, R. de Rycke, N. Naouar, T. van Gaever, M.C., van Montagu, and F. van Breusegem. 2012. AtWRKY15 perturbation abolishes the mitochondrial stress response that steers osmotic stress tolerance in Arabidopsis. *Proceedings of the National Academy of Sciences of the United States of America* 109(49): 20113–20118.

Venkatesh, J., and S.W. Park. 2012. Plastid genetic engineering in *Solanaceae*. *Protoplasma* 249(4): 981–999. doi: 10.1007/s00709-012-0391-9.

Vighi, I.L., L.C. Benitez, M.N. Amaral, G.P. Moraes, P.A. Auler, G.S. Rodrigues, S. Deuner, L.C. Maia, and E.J.B. Braga. 2017. Functional characterization of the antioxidant enzymes in rice plants exposed to salinity stress. *Biologia Plantarum* 61(3): 540–550.

Voss, I., B. Sunil, R. Scheibe, and A.S. Raghavendra. 2013. Emerging concept for the role of photorespiration as an important part of abiotic stress response. *Plant Biology* 15(4): 713–722.

Wang, C., S. Hu, C. Gardner, and T. Lübberstedt. 2017. Emerging avenues for utilization of exotic germplasm. *Trends in Plant Science* 22(7): 624–637.

Wani, A.S., A. Ahmad, S. Hayat, and Q. Fariduddin. 2013. Salt-induced modulation in growth, photosynthesis and antioxidant system in two varieties of *Brassica juncea*. *Saudi Journal of Biological Sciences* 20(2): 183–193. doi: 10.1016/j.sjbs.2013.01.006

Yang, Y., and Y. Guo. 2018. Elucidating the molecular mechanisms mediating plant salt-stress responses. *New Phytologist* 217(2): 523–539.

Ye, H., S. Liu, B. Tang, J. Chen, Z. Xie, T.M. Nolan, H. Jiang, H. Guo, H.Y. Lin, L. Li, Y. Wang, H. Tong, M. Zhang, C. Chu, Z. Li, M. Aluru, S. Aluru, P.S. Schnable, and Y. Yin. 2017. RD26 mediates cross-talk between drought and brassinosteroid signalling pathways. *Nature Communications* 8: 14573. doi: 10.1038/ncomms14573

Yong, B., X. Wang, P. Xu, H. Zheng, X. Fei, Z. Hong, Q. Ma, Y. Miao, X. Yuan, Y. Jiang, and H. Shao. 2017. Isolation and abiotic stress resistance analyses of a catalase gene from Ipomoea batatas (L.) Lam. *BioMed Research International* 2017: 1–10. https://doi.org/10.1155/2017/6847532.

Zechmann, B. 2011. Subcellular distribution of ascorbate in plants. *Plant Signaling and Behavior* 6(3): 360–363. doi: 10.4161/psb.6.3.14342.

Zhang, J., Z. Sun, P. Sun, T. Chen, and F. Chen. 2014. Microalgal carotenoids: Beneficial effects and potential in human health. *Food and Function* 5(3): 413–425. doi: 10.1039/c3fo60607d.

Zheng, J.L., L. Zeng, M.Y. Xu, B. Shen, and C.W. Wu. 2017. Different effects of low- and high-dose water-borne zinc on Zn accumulation, ROS levels, oxidative damage and antioxidant responses in the liver of large yellow croaker *Pseudosciaena crocea. Fish Physiology and Biochemistry* 43: 153–163.

Zhu, J.K. 2003. Regulation of ion homeostasis under salt stress. *Current Opinion in Plant Biology* 6(5): 441–445.

12 Antioxidant Enzymes and Their Genetic Mechanism in Alleviating Drought Stress in Plants

Yasser S. Moursi, Mona F. A. Dawood, Ahmed Sallam, Samar G. Thabet and Ahmad M. Alqudah

CONTENTS

12.1   INTRODUCTION

Due to their immobile nature, plants experience different kinds of threats during their life cycle. Water is the essence of life that plays the most prominent role during plant growth and development. Plants need water for all developmental and biological processes, starting from seed imbibition, seed germination, root and shoot elongation, leaves emergence, flowering until seed set including photosynthesis, nutrient uptake, and transportation, maintaining turgidity and lowering canopy temperature. Drought stress is one of the threats that affect plants at any developmental phase (Thabet and Alqudah 2019). Under drought stress, plants make substantial developmental, morphological, physiological and biochemical changes to reduce the harmful effect of drought stress. Drought is a dynamic, unpredictable, growing and severe stress that threatens the plant growth and development, and diminishes the crop productivity larger than any other stresses (Sallam et al. 2019b). The impact of drought relies on many factors; the genotype, time, and intensity of drought, as well as the plant developmental stage (Shao et al. 2009). Drought occurs either when plants grow in a water-deficit environment, or when the water supply to roots is limited, or when the transpiration rate exceeds the physiological limits.

Drought negatively affects all developmental stages; this had been evidenced by reduction and delay of seed imbibition and seed germination, the plant height, stem diameter (Sallam et al. 2015); days to flowering and leaf area being reduced (Samarah et al. 2009). Many leaf-related parameters – leaf biomass, the number of leaves per plant, leaf size, leaf area – were diminished under drought stress. Plants can cope with drought via three main mechanisms, drought escape, drought avoidance and drought tolerance (Thabet and Alqudah 2019). Sometimes, plants employ a combination of these mechanisms concurrently to resist drought (Sallam et al. 2019 a). Drought escape means the ability of plants to complete their life cycles in a shorter time before the occurrence of severe soil and plant water deficits. Drought avoidance is the ability of plants to keep high tissue water potential even under low soil-moisture and high-water loss conditions. Drought tolerance involves the maintenance of turgor through osmotic adjustment by inducing the accumulation of solute in the cell.

Drought induces severe oxidative stress through the production of reactive oxygen species (ROS). Several studies reported the induction of oxidative stress by water deficit (W Wang et al. 2016). Under drought stress, the concentration of ROS goes up dramatically compared to normal conditions; this sparks the enzymatic antioxidant system (Figure 12.1). ROS include singlet oxygen (O_2^1), superoxide ions (O_2^-) peroxides, and the prominent species hydrogen peroxide (H_2O_2); the elevated concentrations of these species are toxic (Triantaphylides et al. 2008). Therefore, to protect these components, plants turn on the synthesis of several metabolites and proteins with a protective function (Parvaiz Ahmad et al. 2008). The enzymatic antioxidants (superoxide dismutase (SOD), catalase (CAT), ascorbate peroxidase (APX), glutathione reductase (GR), and nonenzymatic ascorbic acid (AsA), tocopherols, glutathione and phenolic compounds antioxidants play a key role to vanquish the deleterious effects of ROS (Ashraf 2009). The induction of antioxidative enzymes, like SOD, peroxidase (POD) and CAT, was found to be associated with tolerance to drought stress (Campos et al. 2019; He et al. 2020). Besides this, drought imposes several alterations in almost all biological processes across the whole plant phenology; photosynthesis rate and net assimilation rate were declined under water deficit (Liang et al. 2019). Similarly, changes were observed in the suppression or the overexpression of genes coding stress proteins and transcription factors (Farooq, Wahid et al. 2009).

A plethora of studies dealt with the induction of both reactive oxygen species and antioxidant systems in plants under different abiotic stresses. In this chapter, we will discuss the alterations of antioxidants under drought stress, but more attention will be paid to the enzymatic antioxidant systems, as well as the genetic attributes of the antioxidant systems under drought stress. Furthermore,

Drought

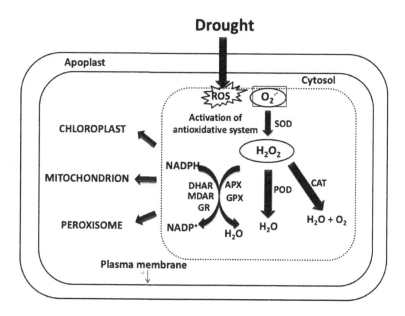

FIGURE 12.1 Induction of reactive oxygen species (ROS) and activation of enzymatic antioxidants under drought stress. **ROS** reactive oxygen species; **DHA** Dehydroascorbate; **DHAR** Dehydroascorbate reductase; **MDHA** Monodehydroascorbate; **MDHAR** Monodehydroascorbate reductase; acid; **GSSG** Oxidized glutathione; **GSH** Reduced glutathione; **GR** Glutathione reductase; **GST** Glutathione-S-transferase; **AsA** Ascorbic; **APX** Ascorbate peroxidase; **GPX** Glutathione peroxidase; **O$_2$·** Superoxide radical; **H$_2$O$_2$** Hydrogen peroxide; **SOD** Superoxide dismutase; **CAT** Catalase; **POD** Peroxidase; **NADPH** Nicotinamide adenine dinucleotide phosphate.

we shed light on the genetic control of the antioxidant enzymes to expand our understanding of the mechanism of these enzymes to alleviate the effect of drought stress on plants.

12.2 REACTIVE OXYGEN SPECIES AND ANTIOXIDANT SYSTEM

Abiotic stress in plants can cause overflow, deregulation or even disruption of electron transport chains in plant organelles such as chloroplasts and mitochondria. Under these conditions molecular oxygen (O$_2$) pairs with a free electron, sparking the accumulation of different types of ROS. Singlet oxygen (^{1}O$_2$), the hydroxyl radical (OH$^-$), the superoxide radical (O$_2^-$) and hydrogen peroxide (H$_2$O$_2$) all strongly oxidize compounds and therefore are very harmful to cell integrity (Sallam et al. 2019 a). Antioxidant systems include antioxidant enzymes and non-enzymatic compounds, which play critical parts in detoxifying ROS induced by drought stress. In this chapter, we will shed more light on the role of the enzymatic antioxidant system in scavenging ROS produced under drought stress. A brief description of the role of non-enzymatic antioxidants under drought conditions will be presented at the end of this chapter.

12.2.1 The Antioxidant Enzymes in Response to Drought Stress in Plants

One of the well-established defense mechanisms in plants under water-deficit is the stimulation of antioxidants enzymes such as CAT, SOD, POD, APX, monodehydroascorbate reductase (MDHAR), dehydroascorbate reductase (DHAR) and GR and glutathione peroxidase (GPX). It is well known that the antioxidant systems are upregulated at different levels. For example, APX is upregulated very significantly in the post-transcriptional stage, while CAT, GPX and APX are upregulated at the transcriptional stage (Laxa et al. 2019 a). In wheat at the seedling stage, a significant turnover in the activities of CAT, POD and APX was found in the drought-tolerant genotypes compared with the sensitive ones (Devi et al. 2011). Similarly, in wild wheat the enzymatic antioxidants (CAT, APX, SOD and guaiacol peroxidase) exhibited greater activities than in the cultivated wheat at the seedling stage (Ahmadi et al. 2018). In a comparative study between nine *Aegilops tauschii* and three *Triticum dicoccoides* grown under rain-fed and well-watered conditions, the enzymatic activities of CAT, APX and SOD showed a dramatic increase in *Aegilops tauschii* grown under rain-fed compared at *Triticum dicoccoides* (Suneja et al. 2017). In other studies, the tolerant wheat genotypes maintained higher antioxidant enzymatic activities of SOD, APX, DHAR, MDHAR, GR and CAT relative to the sensitive ones (Almeselmani et al. 2006). In general, the drought-tolerant genotypes showed higher enzymatic antioxidant potential; more likely, this is due to their genetic architecture that enables better tolerance to oxidative damage under drought stress (Ahmadi et al. 2018; Pour-Aboughadareh et al. 2020). The alterations of these enzymatic antioxidants in various plant species at different growth stages under different drought stress conditions are listed in **Table 12.1**.

12.2.2 Superoxide Dismutase (SOD; EC 1.15.1.1)

In plants, there are three classes of SODs, according to the metal in their active sites; copper and zinc (Cu/ZnSOD), manganese (MnSOD) or iron (FeSOD). The fourth class, Nickel-containing SODs was identified in *Streptomyces* sp (W Wang et al. 2016). SODs are metalloenzyme, which catalyzes the conversion of the superoxide radicals ($O_2^{\bullet-}$) to molecular oxygen (O_2) and finally to H_2O_2. Then, H_2O_2 acts as a signal molecule under drought stress. This mode of action shows that SODs have a dual function, whereby they are reducing firstly the superoxide and then producing a signal molecule. SODs are localized in different plant organelles: chloroplasts, mitochondria, peroxisomes and cytoplasm (Luis et al. 2018).

12.2.2.1 Alteration of SODs under Drought

The elevation of SODs is involved in fighting oxidative stress caused by abiotic stress, indicating their pivotal role in the survival of plants. Yin et al. (2008) reported a significant increase of SODs activity under drought stress in dragon spruce (*Picea asperata Mast.*). In maize, drought resistance was significantly correlated with high levels of Cu/ZnSOD (Malan et al. 1990). Similarly, the activities of chloroplastic and cytosolic Cu/ZnSODs were elevated after exposing pea plants to drought (Mittler and Zilinskas 1994). More likely, the induction of SODs activation is genotype-dependent. Drought-tolerant *Sorghum bicolor* has a higher SOD activity than drought-sensitive varieties (Dat et al. 2000). The induction of SOD activity is not only genotype-dependent but also organ-dependent; in *Lotus japonicas*, under drought, in roots and leaves three types of SODs – MnSOD, a FeSOD and two Cu/ZnSODs (I and II) were identified. In roots, the three types increased under drought, while in leaves they remained unchangeable (Signorelli et al. 2013).

This wide distribution of SODs in various cellular organelles explains their implementation in scavenging ROS during different plant growth stages and improving plant performance under drought. This has been shown as an improvement in physiological (photosynthesis and stomatal conductance) and morphological (yield) parameters. Moreover, they are present in different plant species, which suggests that there is a master mechanism that regulates the increment of SOD activities under drought stress in plant species. It is likely that the SODs in different cellular organelles are cross-talking or they are in a cascade. For example, both the cytosolic and chloroplastic SODs' activities were elevated under drought stress. Besides this, it seems that SODs can independently confer/improve drought tolerance in plants in the absence of other antioxidant enzymes. SOD activation is highly affected by the genetic background, as the tolerant genotypes exhibited higher activities of SODs under drought than the sensitive ones did.

12.2.3 Catalase (CAT; EC 1.11.1.6)

This is an enzyme found in the peroxisomes of plants and other aerobic organisms. It is well-known as a key enzyme involved in the catalytic conversion of H_2O_2 into water and oxygen. The elevation

TABLE 12.1

Response of Antioxidant Enzymes of Different Crop Plants under Drought

Plant Species	Drought Stress Applied	Enzymatic Antioxidant Response	Reference
mung bean (*Vigna radiata*)	12% PEG-8000	CAT ↑, APX ↑ (except line 97006), POD and SOD (↑ or ↓ cultivar dependent),	Ali et al. (2018)
Wheat	Three levels: 75–80% FC, 55–60% FC and 35–40% FC	CAT ↑, SOD ↑, APX ↑	Abid et al. (2018)
	50% of crop evapotranspiration	CAT ↑, SOD —, POD ↓, APX ↓	Elkeilsh et al. (2019)
	10% PEG solution	APX ↑, DHAR ↑, GR ↑, MDHAR ↑ (relative expression and activity)	Shan et al. (2018)
	60% field capacity	CAT ↑, SOD ↑, POD ↑	Habib et al. (2020)
Sugar beet	75% and 50% of the water requirement	CAT ↑, APX ↑, POD ↑	Ghaffari et al. (2019)
Brassica napus L.	10 and 20% PEG	APX ↑, GST ↑, CAT ↓, GR — (10% PEG) or ↓(20%PEG), GPX ↑ (10% PEG) or — (20%PEG),	M Hasanuzzaman et al. (2018)
Vicia faba L.	Soil relative water content (40%)	POD ↑, SOD ↑, CAT ↑, APX ↑, GR↑	Khan et al. (2020)
Durum wheat genotypes)	326 g L⁻¹ PEG-6000	POD ↑, SOD ↑, CAT ↑, APX ↑	Pour-Aboughadareh et al. (2020)
Ocimum basilicum of	50% of soil water holding capacity	SOD ↑, CAT ↑, POD ↑	Taha et al. (2020)
Capsicum annuum	50% of soil water holding capacity	SOD —, CAT ↑, GR ↓, APX↑	Yildizli et al. (2018)
Barely (*Hordeum vulgare*)	75% FC, 50% FC, 25% FC	CAT ↓, APX ↓, POD —, SOD [↑ (75% FC), ↓ (50% FC)],— 25% FC],	Ghorbanpour et al. (2020)
Triticale genotypes	Holding irrigation 40 days from early heading stage to harvest time	SOD ↑, CAT ↑, POD ↑, APX ↑	Saed-Moucheshi et al. (2019)
Maize	15% PEG-6000	SOD ↑, CAT ↑, CAT ↓, APX ↑, GR ↑, DHAR ↑, DHAR ↑	T. Xie et al. (2018)
	30% of water holding capacity	SOD ↑, CAT ↑, POD ↑	Usmani et al. (2020)
Lepidium draba	3, 6, 9 and 12% PEG-6000	CAT [↑ (3-6 % PEG), ↓ (6-9% PEG), POD [↑ (3-6 % PEG),↓ (6-9% PEG), APX [↑ (3-6 % PEG),↓ (6-9% PEG), (activities and expression levels)	Goharrizi et al. (2020)
Cauliflower (*Brassica oleracea* L. var. botrytis)	60% field capacity	SOD ↑, CAT —, POD —	Latif et al. (2016)
Amaranthus tricolor	60% and 30% field capacity	CAT↑, SOD ↑, POD ↑, MDHAR ↑, DHAR ↑, APX ↑, GR ↑	Sarker and Oba (2018)
Salvia miltiorrhiza	Withholding water for 9 days	SOD ↑, GR ↑, GPX ↑, APX ↑	L. Zhang et al. (2018)
	Withholding water for 9 days	SOD ↑, CAT ↑, POD ↑	Wei et al. (2016)

(Continued)

TABLE 12.1 (CONTINUED)

Response of Antioxidant Enzymes of Different Crop Plants under Drought

Plant Species	Drought Stress Applied	Enzymatic Antioxidant Response	Reference
Tomato (*Solanum lycopersicon* L)	50% and 25 % of the evapo-transpired water	CAT ↑, SOD ↑, APX ↑, MDHAR ↑, DHAR ↑, GR ↑	Murshed et al. (2013)
Lycium ruthenicum	Stop watering for 7, 14 and 28 days	CAT ↑, POD ↑, SOD ↑	Guo et al. (2018)
Silybum marianum (L.) Gaertn	75% and 50% of water capacity	SOD ↑, CAT ↑, POD ↑, APX ↑, GR ↑	ElSayed et al. (2019)
Maclura pomifera	75%, 50% and 30% of field capacity	SOD ↑, DHAR ↑, APX ↑, GR ↑	Khaleghi et al. (2019)
Tobacco (*Nicotiana tabacum* L.)	20% PEG-6000	SOD ↑, CAT ↑, APX ↑, GR ↑	Jiayang Xu et al. (2020)
Rice		SOD ↑, POD ↑, CAT ↑, APX ↑	X. Wang et al. (2019)
	5%, 10%, 15% and 20% PEG-6000	SOD ↑ (↓ in sensitive lines at an early stage), POD ↑ (↓ at an early stage), CAT ↑, GR ↑, APX ↑ (↓ in sensitive lines at an early stage)	S. Nahar et al. (2018)
Eggplant	15% PEG-6000	SOD ↑, CAT ↑, APX ↑, GR ↑	Kiran et al. (2019)
Camellia sinensis L		SOD ↑, TRXs ↑, POD ↑,	Rahimi et al. (2020)

*** — unaffected, ↓ decrease, ↑ increase

of catalase antioxidant activity was reported as a response to different abiotic stresses in plants. Without this elevation in catalase activity, that growth would be very severe. Catalase converts the excess hydrogen peroxide into water and oxygen, according to the equation $2H_2O_2$ to $O_2 + 2H_2O$. Several plants have multiple catalase forms, for example, two in castor bean and six in *Arabidopsis* (Frugoli et al. 1996). Plant catalases can be classified into three classes. Class I is abundant in the photosynthetic organs and responsible for the removal of H_2O_2 produced upon photorespiration. Class II is prominent in vascular tissues; their role is not clearly understood and they might play a role in lignification. Class III is highly abundant in seeds and young seedlings and it removes the excessive H_2O_2 produced during fatty acid degradation (Willekens et al. 1997).

Catalases are the main scavenging enzymes that can remove H_2O_2 under stress. This leading role could be because there is a continuous generation of peroxisomes during stress that might help in the removal of H_2O_2 diffusing from the cytosol (Lopez-Huertas et al. 2000). The elevation of catalase activity helps the cells to repair the oxidative damage by reducing toxic levels of H_2O_2 (Vital et al. 2008).

12.2.3.1 Alteration of CAT Antioxidants under Drought

Under drought stress, CAT activity notably increased in dragon spruce (*Picea asperata* Mast.) seedlings exposed to both drought and high light intensity (Yan Yang et al. 2008). Similarly, CAT activity markedly increased under drought stress in alfalfa (Tina et al. 2017), in chickpea (Dalvi et al. 2018) and common bean (Mohammadi et al. 2020). In *Arabidopsis thaliana*, CAT3 is also present in the peroxisome of guard cells and is correlated with drought stress resistance. In this study, the wild plants resisted 22 days of drought, but could not resume growth after three days of re-watering. The mutant plants lacking CAT3 did not survive 22 days of drought tolerance and did not restore growth after three days of re-watering. Conversely, the plants which overexpressed CAT3 showed drought tolerance and recovered after three days of re-watering (Zou et al. 2015).

There are three categories of CATs distributed in different plant organs such as young seedlings, vascular system and photosynthetic organs. This indicates that CATs represent a cornerstone in the antioxidation machine in plants. They are expressed across the whole plant ontogeny, improving seedling development and photosynthesis rates. This is might be attributed to their continuous regeneration concomitantly with peroxisomes. Even in the absence of other detoxification components such as APX, their activation was sufficient to provoke a tolerance against drought as well as drought-stressed plant recovery after re-watering.

12.2.3.2　Guaiacol Peroxidases (POD; EC 1.11.1.7)

Guaiacol peroxidases detoxify H_2O_2. The peroxiredoxins proteins are heme-containing enzymes that belong to class III or the "secreted plant peroxidases". In addition to their main duty as a peroxidative agent, they are also involved in a different cycle, namely the hydroxylic cycle. This widens their role to be involved in several processes, extending from seed germination to senescence, including auxin metabolism, cell wall elongation and protection against pathogens (Passardi et al. 2004).

12.2.4　Ascorbate Peroxidase (APX; EC 1.11.1.1)

APXs reduce H_2O_2 into the water using ascorbate as an electron donor during photosynthesis. APX family includes five isoforms distributed in different sites: thylakoid, microsomes, stroma, cytosol and apoplast (Noctor and Foyer 1998). In addition to APX, the ascorbate-glutathione (ASC-GSH) cycle includes several components such as monodehydroascorbate reductase (MDAR), dehydroascorbate reductase (DHAR), glutathione-S-transferase (GST) and GR. They are working collaboratively; however, they are compartmentalized in different cellular organelles (Laxa et al. 2019 a).

12.2.4.1　Alteration of APX under Drought Stress

Several studies reported the importance of APX for plants to tolerate drought. *Arabidopsis* APX1 mutants showed hypersensitivity to different abiotic stresses including drought (Maruta et al. 2012). Davletova et al. (2005) reported a protective role for cytosolic APX (cAPX) compared with the chloroplastic APX (cpAPX), because the cytosol shielded the cellular organelles. Thus, cAPX prevents the flow of H_2O_2 inside he cell. In rice, drought stress elevated APX two-fold compared to control conditions (N. Singh et al. 2020). In alfalfa nodules, the mRNA of cytosolic ascorbate peroxidase was up-regulated under drought stress (Naya et al. 2007). Tobacco plants overexpressed 9-cisepoxycarotenoid dioxygenase (NCED), and gene SgNCED1 exhibited high APX activity and normal growth under 0.1M mannitol induced drought stress (Y. Zhang et al. 2008). Both PODs and APXs are important for conferring drought tolerance. It seems that the activity of APXs is site-dependent where cAPX conferred better drought tolerance than cpAPX did.

12.2.5　Monodehydroascorbate Reductase (MDHAR)

Monodehydroascorbate reductase is a flavin adenine dinucleotide enzyme that maintains the regeneration of AsA using monodehydroascorbate radical as a precursor and Nicotinamide adenine dinucleotide phosphate NAD(P)H as an electron donor. Thus, MDAR is a key player in the plant antioxidant system for maintaining high AsA levels (Asada 1999). Like other enzymatic antioxidants, several forms of MDAR were found to be present in different cellular compartments such as chloroplasts, the cytosol, peroxisomes and mitochondria (Leterrier et al. 2005). The regeneration capacity of AsA greatly depends on the amendment of MDHAR and DHAR activities (Palma et al. 2015). Both enzymes MDHAR and DHAR are prominently involved in the recycling of AsA harnessing NADPH and GSH as electron donors (Foyer and Noctor 2005).

12.2.6   Dehydroascorbate Reductase (DHAR)

Dehydroascorbate reductase is a thiol enzyme that helps cells to keep the reduced AsA pool. For this purpose, DHAR uses the reduced glutathione form (GSH) as a substrate to reduce dehydroascorbate (DHA) to AsA (Gratão et al. 2005). DHAR catalyzes the reduction of DHA to AsA in the presence of glutathione (GSH) as an electron donor before the lineal hydrolysis of DHA to permanent molecule 2,3-diketogulonic acid (Do et al. 2016). It occurs in different plant organs, and its mode of action has been reported in many plant species (Anjum et al. 2014).

12.2.6.1   Alteration of MDHAR and MDAR under Drought Stress

Drought stress reduced the efficiency of AsA recycling in maize; this is might be attributed to a decline in the activity of MDHAR and DHAR (T. Xie et al. 2018). It has been evidenced that the high levels of MDHAR can mitigate oxidative stress (Li et al. 2012). DHAR exerts tolerance against various types of stresses including drought stress; this is attributed to keeping high levels of AsA (C. Shan et al. 2018). Similarly, DHAR elevated the levels of AsA in rice (Kim et al. 2013). The *Oryza sativa* DHAR markedly improved the growth of *Synechococcus elongatus* PCC 7942 under oxidative stress by turning over GSH/GSSG ration (Kim et al. 2020). DHAR was shown to be associated with increment of grain yield and biomass in transgenic species such as tobacco, and *Arabidopsis* under drought (Ushimaru et al. 2006; Kwon et al. 2003). *Arabidopsis* plants with high levels of DHAR3 revealed an increment in seed germination, root length and elevated contents of SOD, APX and DHAR activity, as well as AsA content relative to the wild plants. On the other hand, they expressed less H_2O_2 and malondialdehyde (MDA) contents (Wenbin ; Wang et al. 2020).

Both MDHAR and DHAR are constituents of the ASC-GSH cycle. They work coordinately to help the plant to maintain high levels of AsA under drought . MDHAR converts monodehydroascorbate into AsA using NADPH as an electron donor, while MDAR catalyzes the same reaction by converting dehydroascorbate into AsA using GSH as an electron donor. The high levels of SODs and APXs following the induction of DHAR3 in *Arabidopsis* indicate that a cross-matching resistance acts in a cascade pathway. Taken together, these findings demonstrated the pivotal role of MDHAR and DHAR in conferring/improving drought tolerance by retaining high levels of AsA.

12.2.7   Glutathione Peroxidases (GPXs; EC 1.11.1.9)

Glutathione peroxidases are a large group of isozymes with a wide substrate spectrum and act as antioxidant enzymes. In the plant cell, they are present in many organelles such as the cytosol, chloroplasts, mitochondria and the endoplasmic reticulum. They protect the cellular biomolecules by catalyzing the detoxification of H_2O_2 and lipid hydroperoxides using GSH (Labudda and Azam 2014).

12.2.8   Alteration of GPXs under Drought Stress

The high levels of GPXs under drought and their ameliorative role was reported in several plant species. Pourtaghi et al. (2011) reported a dramatic increase in GPX activity in relation to the drought intensity in sunflower and soybean. Besides this, they found high positive significant correlations between the seed yield and GPX activity in sunflower. Therefore, the authors suggested that GPXs are a reasonable marker to select for drought tolerance under both mild and severe drought. Likewise, in leaves of sugar beet, a dramatic increase was observed under drought stress, suggesting a protective role of GPXs against oxidative stress (Sayfzadeh and Rashidi 2011). It has been reported that in *Arabidopsis thaliana*, ATGPX3 is involved in two biological processes. ATGPX3 equilibrates the levels of H_2O_2 and indirectly modulates water transpiration via regulating the crosstalk between abscisic acid (ABA) and H_2O_2 signaling during stomatal closure (Miao et al. 2006).

These results showed that GPXs efficiently detoxify hydroperoxides than prevent lipid peroxidation. The positive and significant correlation between GPXs and seed yield indicated that GPXs might be indirectly used to select for high seed yield. Similar to other enzymatic antioxidants, GPXs induced drought tolerance during different growth stages of plant development. Additionally, GPXs indirectly participate in regulating water transpiration via controlling the interaction between two of the biggest signaling pathways, H_2O_2, and ABA pathways under drought stress.

12.2.9 GLUTATHIONE REDUCTASE (GR; EC 1.6.4.2)

Glutathione reductase is a highly conserved enzyme that had been isolated from several plant species. The high GR activity in plants builds up high concentrations of glutathione (GSH), which confers tolerance to plants. GR is essential to maintain the reduced form of glutathione by catalyzing the NADPH-dependent reaction of a disulfide bond of GSSG. Besides this, it has an antioxidative effect by scavenging H_2O_2. GR is one of the ascorbate/glutathione cycle enzymes. It has been evidenced that maintaining high levels of GSH to GSSG is of high importance to protect plants against oxidative damage imposed by drought (Foyer and Noctor 2005). GR is a NAD(P)H-dependent enzyme. GR is the last enzyme in the ASC-GSH cycle. GR is a bifunctional enzyme that protects cells against oxidative damage induced under drought, and GR ensures high levels of GSH by catalyzing the conversion of the oxidized form of glutathione (GSSG) to the reduced form (GSH) (Caverzan et al. 2016).

12.2.9.1 Alteration of GRs under Drought Stress

In three species belong to the family Malvaceace – *Gossypium hirsutum*, *Gossypium barbadense*, and *Anoda cristata* – the activity of GR was found to be higher during recovery of mild drought rather than during drought (Ratnayaka et al. 2003). The authors proposed that mild drought may help in acclimation for further extreme drought or be involved in the tolerance to other stresses. The induction of GR activity is genotype-specific. Furthermore, a significantly high positive correlation was observed in wheat roots and leaves during water deficit (Singh et al. 2012). Such situation triggers a higher demand for antioxidants to reduce oxidative stress (Voss et al. 2013). In line with the high activity of other antioxidants, the high activity of GR marked the drought-tolerant varieties in several plant species. Drought stress enhanced GR activity in many plant species, including maize (Hu et al. 2010), mung bean (*Vigna radiata*) (K. Nahar et al. 2015), wheat (C. Shan et al. 2018) and rice (X. Wang et al. 2019). Interestingly, cold and heat stress induced a cross-tolerance to other abiotic stress including drought by promoting GR activity (Hossain et al. 2018). Surprisingly, the drought-induced GR activity in cotton and cowpea had been completely reversed, and a remarkable decline in GR activity was observed after rehydration (Abid et al. 2018).

The induction of GR activity enhanced drought tolerance in different plant species at different growth stages. GR is a main enzyme of the ASC-GSH cycle to detoxify the ROS under drought. The GR activity ensures high contents of GSH, and a high GSH/GSSG ratio that is essential for plants to withstand drought. Additionally, GSH acts as an antioxidant by itself or as an electron donor in the cellular regeneration of AsA. GRs' high activity conferred tolerance to multiple abiotic stresses suggesting a master mechanism controls the plant tolerance for various abiotic stresses such as cold, heat and drought, that can be improved by inducing GPXs and/or GRs activity. The decline in GR activity after re-watering drought-stressed plants indicates that the increment was temporal and ROS concentration-dependent.

12.2.10 GLUTATHIONE S-TRANSFERASES (GSTs, EC 2.5.1.18)

Glutathione S-transferases (GSTs) belong to phase II, GSH-dependent ROS-detoxification enzymes. They were detected in several cellular parts such as apoplast, cytosol, chloroplast, mitochondria as well as in the nucleus. GSTs increase the affinity of GSH to the electrophilic sites on several phytotoxic substances. Presently, the contribution of GSTs in plants' response to drought is poorly understood (Labudda and Azam 2014). Kojić et al. (2012) reported a significant increase in the activity of GSTs under drought conditions in the roots of maize (*Zea mays*). The GSTs activity increased from 255.5 to 711.6 U/mg protein under drought stress. The researchers assumed that the significant increment in GSTs activity is attributed to the oxidative damage imposed by drought. The expression of *Le*GSTU2 from tomato conferred drought tolerance in *Arabidopsis*, this tolerance has been indicated by high levels of proline and high activation of enzymatic antioxidants (Jing Xu et al. 2015).

Unlike GPXs and GRs, few studies dealt with the contribution of GSTs in drought tolerance in plants, thus, their mode of action is not clearly understood. This intrigued the curiosity for further studies to get more insights about their contribution to drought tolerance.

12.2.11   Peroxiredoxins (PRX; E.C. 1.11.1.15)

PRXs are simple proteins having peroxidase activity on a set of peroxide substrates, including hydrogen peroxide (H_2O_2), alkyl hydroperoxides (ROOH) and peroxynitrite (ONOOH). PRXs are present in all organisms having thiol-based catalytic activity. Cells contain multi forms distributed in organelles such as cytosol, mitochondria, peroxisomes, nucleus and chloroplasts. According to the number of cysteine residues that participate in the catalytic mechanism, PRXs can be categorized into two categories: 1-Cys and 2-Cys (Gupta et al. 2018). Thiol peroxidases are related to the NADPH-thioredoxin reductase (NTR), ferredoxin-dependent TRX reductase (FTR) and GSH/GRX systems (Sevilla et al. 2015). The balanced production and scavenging of ROS in plants by efficient antioxidant systems were found to be essential for plants to withstand various abiotic stresses (Gill and Tuteja 2010). Recently, a new role was uncovered for thiol peroxidases in redox regulation in the chloroplast, known as TRX oxidases (Vaseghi et al. 2018).

12.2.11.1   Alteration of Peroxiredoxins under Drought Stress

In wheat, drought stress increased the GSH content and the GR activity (Zagdańska and Wiśniewski 1996). Surprisingly, cold stress provoked a down-regulation of most Trx-coding genes while drought stress caused an upregulation (Zagorchev et al. 2013). The chloroplastic drought induction stress protein (CDSP32) is extensively studied; firstly, it had been identified in the stromal membrane of *Solanum tuberosum*. Additionally, it has also been isolated several plants, such as *Oryza sativa*, *Arabidopsis thaliana*, *Nicotine tabacum* and *Hordeum vulgare* (Dietz 2003). The *Solanum Tuberosum* CDSP32 mutants conferred hypersensitivity to drought, accumulating high levels of hydrogen peroxide and over-oxidation of 2-Cys Prx (Broin and Rey 2003). In *Brassica juncea*, a drought-tolerant cultivar, TrxCDSP32 was upregulated compared with the sensitive one (Sharma et al. 2017). Unlike other peroxidases that were upregulated under various abiotic stressors, TRXs are mostly drought-specific as cold downregulated the expression of TRX while drought upregulated them. Therefore, they can be used as selection markers for drought.

12.3   GENETIC REGULATION OF ANTIOXIDANT ENZYMES UNDER DROUGHT STRESS

As drought tolerance has a complex genetic basis and diverse mechanisms, hundreds to thousands of genes or proteins are up- or down-regulated under drought stress conditions, mostly on a subcellular scale and for a short time in specific organs. Therefore, in this part of the chapter, we will shed light on the various levels of the genetic control of antioxidant enzymes under drought stress, including the proteomic and the transgenic approaches.

Under drought conditions, ROS occur in various cellular organelles, which transmit signals, leading to reprogramming the cell performance via gene expression (Laxa et al. 2019 b). Plants have developed the underlying mechanisms, namely retrograde and anterograde signaling pathways, to

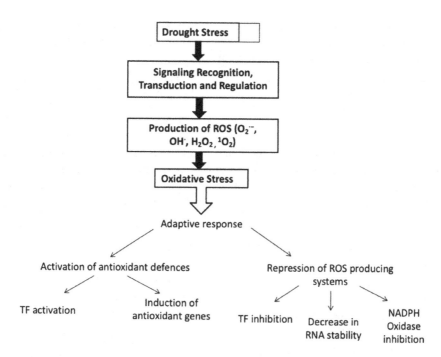

FIGURE 12.2 Differential responses of the drought-responsive genes under oxidative stress provoked by drought.

control the accumulation of ROS in response to drought stress (Suzuki et al. 2014). The antioxidant defense system includes low molecular weight compounds such as GSH, AsA, α-tocopherol, carotenoids and enzymes including APX, CAT, SOD and the thiol peroxidases of the PRX and GPX (Hussain et al. 2016).

12.3.1 THE INDUCTION OF ANTIOXIDANT GENES UNDER DROUGHT STRESS

The up-regulation of antioxidant systems occurs at both the transcriptional and post-transcriptional levels upon drought stress (Figure 12.2). Antioxidative enzymes, namely APX, CAT and GPX, represent the principal ROS scavengers in plants. Drought stress-induced the expression of SOD isoforms (Mn-SOD, Fe-SOD and Cu/ZnSOD) in different plant species. In alfalfa, the expression of the chloroplastic FeSOD (FSD) was down-regulated under drought stress while the chloroplastic and cytosolic Cu/ZnSOD (CSD) were upregulated (Kang and Udvardi 2012). The transgenic plants overexpressing Cu/ZnSOD are more tolerant in response to drought stress (Wu et al. 2016).

APX is induced strongly on the post-transcriptional level and is also regulated on the transcriptional level (Laxa et al. 2019 b). Moreover, cytosolic, chloroplastic and peroxisomal APXs' coding genes' activities are concurrently elevated in all plant species. In *Arabidopsis*, the *alx8* mutant (altered expression of APX2) revealed improved drought stress tolerance (Wilson et al. 2009). Additionally, the transcript level of *APX* showed distinct changes in two genotypes of wheat, one drought-tolerant and the other drought-sensitive. In response to drought stress, cytosolic *APX1* expression levels increased in both genotypes, while cytosolic *APX2* was significantly up-regulated only in the drought-tolerant genotype. Also, the thylakoid *APX* transcript levels increased only in the drought-tolerant genotype (Sečenji et al. 2010). Similarly, *APX* gene expression patterns were studied in rice after 15 days of drought. Peroxisomal *OsAPX3* gene was not affected, while *OsAPX4* was down-regulated (Rosa et al. 2010). Transgenic tobacco overexpressed the peroxisomal or cytosolic APX from poplar showed better plant growth under drought (Cao et al. 2017).

Moreover, Gallé et al. (2009) estimated the change in the two GSTs-related genes' activity in the flag leaf of two drought-tolerant cultivars versus two drought-sensitive cultivars of wheat during grain filling under drought stress. All cultivars had high-expression levels for both genes, but extremely high levels were detected in the drought-tolerant cultivars relative to the sensitive ones. The authors reported an earlier induction of GST-associated genes was observed in the tolerant cultivars, indicating that the tolerant ones acclimate to drought faster than the sensitive ones.

Chen et al. (2012) reported that *Arabidopsis thaliana* glutathione S-transferases U17 (*AtGSTU17*) negatively regulates the signal transduction pathways induced by drought. The plants mutated for *AtGSTU17* revealed more drought tolerance than the wild-type plants of *Arabidopsis* ecotype Columbia and maintained higher concentrations of GSH and ABA than the wild ones.

This action exerted more prominent morphological, biochemical and physiological modifications on the mutated plants. They showed hyposensitivity to ABA upon seed germination, smaller stomatal pores, lower transpiration rates, the vigorous development of primary and secondary root systems, and a longer vegetative growth period compared to the wild ones. Some studies stated that the overexpression of GST-associated genes conferred drought tolerance in plants. A chloroplastic GST from *Prosopis juliflora* and wild soybean enhanced drought tolerance in tobacco (George et al. 2010).

Other components of the ASC-GSH cycle, namely MDHAR, DHAR, GST and GR, work synergistically in different cell compartments. Activation of MDHAR, DHAR, GST and GR transcripts are predominantly induced under drought stress conditions (Laxa et al. 2019 b). Among these four, GR is mainly activated strongest. In wheat, up-regulation of the ASC-GSH metabolism and associated enzymes efficiently counteract enhanced H_2O_2 during drought stress (Lou et al. 2018).

Taken together, these data indicate that APXs showed an ambiguous redox response at their production sites in different plant species under drought stress. Additionally, they respond differently, as some genes were upregulated while others were downregulated, suggesting an independent gene expression control. Another possibility is that their activation is site and genotype-dependent where APXs genes were upregulated in the tolerant genotypes and at specific organelles but the remaining downregulated. Presumably, to save energy under drought stress, there was no need to upregulate all genes at once, since the upregulated ones were adequate to confer/improve drought tolerance. Similarly, MDHAR, DHAR and GR were induced under drought stress with more induction of GR.

Generally, plants respond promptly to drought stress by inducing the expression of the ROS scavenging machinery, particularly those of the ASC-GSH cycle. The activation of the other mechanisms which tolerate drought, such as biosynthesis of compatible solute, is time-consuming and very costly for plants. Overall, all of the ASC-GSH cycle coding genes were induced under drought stress; this induction depends on several factors such as plant species, site of production and genotype.

Compared to APX activation, Laxa et al. (2019b) stated that stimulation of CAT is moderate in several plant species. However, CAT activation seems predominantly to occur on the post-transcriptional level; complex regulation of CAT activity is induced under severe drought including gene expression, translation and protein turnover (Luna et al. 2005). Sofo et al. (2015) observed that *CAT2* gene plays a crucial role in plants exposed to severe drought stress. Previous studies have revealed variable responses to different durations of drought stress in several plant species. For instance, *CAT* gene was upregulated in wheat subjected to mild drought stress for 7 hours (Luna et al. 2005), *Macrotyloma uniflorum* for 78 hours (Reddy et al. 2008) and barley after 2 days of stress (Harb et al. 2015). Furthermore, high levels of *CAT* expression were observed in *Cleome gynandra* and *Cleome spinosa* exposed to prolonged drought conditions (ten days) (Uzilday et al. 2012). In contrast, a lower *CAT* transcript level was observed in wheat after ten days (Sheoran et al. 2015) and in barley after nine and sixteen days of drought stress (Harb et al. 2015).

Briefly, CAT-genes expression increased upon short-term exposure to drought as a prompt plant response, while the prolonged exposure to drought repressed the CAT-genes expression. Several mechanisms figure out this decline in the CAT transcript. CAT is expressed in a concentration-dependent manner i.e. the higher the H_2O_2 concentration the higher the CAT activity, and vice versa. More likely, this decline is a consequence of the decline in SOD that converts the superoxide radicals into H_2O_2, which stimulates the CAT activity. Probably, this decline in CAT activity resulted from the activation of another antioxidant enzyme that detoxifies H_2O_2 such as peroxidases. The last mechanism that can explain the decline in CAT activity; high levels of H_2O_2 are needed to trigger a downstream expression of H_2O_2-dependent genes.

The last constituents of the antioxidants enzymatic system are thioredoxins (TRXs), which are usually highly activated proteins that are induced under drought stress. TRXs are localized in different cellular compartments. In rice, the transcripts encoding TRX were up-regulated in the drought tolerant-cultivar (Shankar et al. 2016). Similarly, a number of TRX genes were up-regulated in tobacco leaves under drought (H. Xie et al. 2016). Also, overexpressing *NTRA* exhibits high drought tolerance with high survival rates, low water loss and reduced ROS accumulation when compared to wild-type and *ntra*-knock out plants (Cha et al. 2014). However, TRX transcripts and activity measurements also indicate a down-regulation of some TRX members in date palm under drought stress (Safronov et al. 2017).

Recently, Laxa et al. (2019 b) reported changes in the activation of the antioxidant defense system between sensitive and tolerant plant species. For instance, the activation of the major scavenger APX and CAT is stronger in drought-tolerant species than the sensitive ones, and the opposite is true in the case of GPX. Tolerant species mainly activate the ascorbate-dependent scavenging system and moderately the glutathione-dependent system. Oppositely, sensitive plant species showed stronger activation of the glutathione-dependent scavenging system compared to the ascorbate-dependent system. Moreover, tolerant species showed inactivation in the TRX-dependent scavenging system. Therefore, understanding the hypothesized role of the ascorbate-dependent scavenging system in tolerant and sensitive genotypes can be a target for improving plant tolerance towards drought stress in the biotechnological application.

12.3.2 Proteomic Alterations under Drought Stress

Proteomic analysis has been successfully used to assess the changes of the drought-responsive proteins in several plant species under drought stress. Several drought-responsive proteins have been identified in the root proteome of drought-tolerant and drought-sensitive rice varieties, such as ABA- and stress-responsive protein, ascorbate peroxidase (Salekdeh et al. 2002). The functions of these proteins include but are not limited to regulation of metabolic processes, biomass production, photorespiration, antioxidant defense system and operation of ion channels in response to drought stress. The application of exogenous salicylic acid can also decrease drought-induced oxidative stress in many plant species (Rozita et al. 2012). In response to drought stress, site-specific alterations in the activity of antioxidant enzymes have been reported by Devi et al. (2012), as CAT stimulates in root and shoots, and APX in the endosperm, GR and POD are up-regulated in shoots of drought-tolerant wheat genotypes.

The expression of the drought-responsive protein-coding gene is always triggered via an increased endogenous ABA level of plants (Hirayama and Shinozaki 2007). Water deficiency also leads to H_2O_2 production in photosynthetic organisms. Interestingly, drought stress-induced ABA may control the increase of H_2O_2 by inducing the catalase isozyme B (CATB) production (Hu et al. 2005). Ye et al. (2011) reported that ABA controls H_2O_2 accumulation through the induction of *OsCATB* in rice leaves under water stress. Furthermore, H_2O_2 acts as a signaling molecule in the plant cell and activates mitogen-activated protein kinase (MAPK) cascade for stress-responsive gene expression (Ye et al. 2011). Other components also involved in the stress signal transduction pathway, namely serine/threonine-protein kinase, histidine kinase and tyrosine phosphatase, have been recognized by nuclear proteome analyses. Maruyama et al. (2004) reported that drought stress led to the upregulation of the transcription factor (TF) proteins, specifically RF2b, bZIP, HB3 and homeobox-leucine zipper protein. Rapid scavenging of ROS, such as superoxide radicals and hydrogen peroxide that are formed by NADPH oxidase activity, is essential to protect oxidative damage of DNA (P Ahmad et al. 2011). Antioxidant defense enzyme SOD is capable of scavenging superoxide radical whereas hydrogen peroxide is scavenged by APX or GPX. The upregulation of SOD, APX, GPX and GST proteins is considered as the main defense system in plants exposed to environmental stress conditions (P Ahmad et al. 2011). Drought-mediated overproduction of the electron transport chain may lead to superoxide radicals and singlet oxygen production in the mitochondria and chloroplast of pea plants, which was accompanied by overproduction of Cu/ZnSOD in the mitochondrial proteome of the pea plant (Mittler and Zilinskas 1994). Drought-induced upregulation of APX, SOD and MDAR was also noted in cell walls of plant cells (Taylor et al. 2005). In general, using the proteomic approach is essential to fulfilling knowledge gaps about the role of stress-responsive proteins during drought stress.

12.3.3 Genetic Engineering and Transgenic Ability of Antioxidant Enzymes

In plants, a lot of studies that dealt with the antioxidant genes' expression mainly focused on the mRNA level, but further functional studies are limited. In recent years, transgenic plants have been developed with a high tendency to express antioxidant enzymes, and these plants showed tolerance against various stresses such as salinity, thermal stress and drought stress (Chakradhar et al. 2017).

The tremendous progress in the identification of the genes involved in drought stress tolerance and transgenic techniques coupled with the development of molecular cloning technology have enabled deep investigation of the transcription of the ROS-scavenging enzymatic genes (Sunkar et al. 2006). After this, Xiong et al. (2018) reported that overexpression (OE) of *Os*LG3 (an ERF family transcription factor) in rice showed high drought tolerance and modulated ROS levels via up-regulation in the OE lines and down-regulation in the RNAi lines of the expression of ten ROS scavenging-related genes. In *Agrostis stolonifera*, overproduction of cytokinin alleviated drought damage and promoted root growth through overexpression of isopentenyl transferase (ipt) (Xu et al. 2020). In tobacco, transgenic plants having *Ah*Cu/ZnSOD showed an enhanced salinity and drought tolerance (Negi et al. 2015).

Remarkably, overexpression of a single gene could mainly increase plant tolerance to different biotic and abiotic stresses. Moreover, overexpression of *Kandelia candel* KcCSD (a Cu/ZnSOD) in tobacco exhibited salinity tolerance in the aspect of lipid peroxidation, root growth and survival rate and enhanced SOD and CAT activity compared to wild type (Jing et al. 2015). SOD and CAT are functionally linked, as SOD converts the superoxide radical into H_2O_2 which induces the CAT activity. Therefore, a concomitant overexpression boosts the performance of the excess cellular H_2O_2 removal. APXs are the main enzymes in the ASC-GSH cycle, with a potential role in the excess ROS removal. The overexpression of these antioxidants' coding genes conferred tolerance against many abiotic stresses, demonstrating the complexity of the antioxidant machinery in the plant. Transgenic tobacco plants showed drought tolerance by the overexpression of monodehydroascorbate reductase under drought stress (Eltayeb et al. 2007). It has been demonstrated that the overexpression of DHAR was associated to increment of grain yield and biomass in transgenic tobacco plants under various stresses including drought (Eltelib et al. 2011). Under mannitol-induced drought, Arabidopsis plants overexpressing sweet potato DHAR3 revealed an increment in seed germination and root length coupled with elevated contents of other antioxidants genes, SOD, APX and DHAR activity, as well as AsA content compared to the wild plants. On the other hand, they retained less H_2O_2 and MDA contents (Wang et al. 2020).

Transgenic lines of wheat, sorghum and rice have been developed by transferring the genes involved in the production of late embryogenesis abundant (LEA) proteins to improve drought stress tolerance (Cheng et al. 2002). Accumulated embryo of the LEA proteins helps plants in maintaining the cell membrane structure and ionic balance under drought stress (Browne et al. 2002). The improved tolerance against drought stress can also be achieved by overexpression of the LEA proteins coding genes through genetic manipulations (Gosal et al. 2009).

The over-expression of GPX-2 of *Synechocystis* PCC 6803 in the chloroplasts (ApGPX2) and cytosol (AcGPX2) in transgenic *Arabidopsis* seedlings reduced the levels of hydroperoxide content relative to the wild-type plants. However, the levels of lipid peroxidation increased in both transgenic and wild-type seedlings. The results reported in this work demonstrated that the overexpression of GPX-2 of S. PCC 6803 in transgenic plants had induced tolerance to oxidative damage evolved by drought (Gaber et al. 2006).

Arabidopsis plants overexpressing ZAT18 (a C2H2 zinc finger protein) exhibited less leaf water loss, lower H_2O_2 content, higher leaf water content and higher POD and CAT activities after drought treatment compared to the wild type (Yin et al. 2017). The double overexpression of APX and Cu/ZnSOD improved the capacity of drought tolerance in sweet potato chloroplasts by stimulating photosynthetic activity compared to wild type (Lu et al. 2010).

Generally, the development of transgenic plants was found to be an efficient approach to develop drought-tolerant plants. The identification, characterization and utilization of more candidate genes to improve drought stress will make this approach more powerful.

12.4 NON-ENZYMATIC ANTIOXIDANTS UNDER DROUGHT STRESS

Non-enzymatic antioxidants are several organic compounds that metabolize ROS. Generally, there are two classes of non-enzymatic antioxidants: lipid-soluble membrane-associated antioxidants (e.g. α-tocopherol and β-carotene) and water-soluble reductants (e.g. ascorbate, phenolics, glutathione, etc.).

12.4.1 Ascorbic Acid (Vitamin C)

This is one of the most intensively studied antioxidants; it exists in all plant cell types, organelles and apoplast. Naturally, ascorbate exists in the reduced form of ascorbic acid (AsA), which represents 90% of the ascorbate pool. The ascorbate concentration ranges from 20 mM in the cytosol to 300 mM in the chloroplast stroma, and it is produced in the mitochondria and translocated into other cellular compartments via proton-electrochemical gradient or diffusion (Seminario et al. 2017).

Ascorbic acid (AsA) acts as the transport of antioxidants and electrons. It is an enzyme cofactor and retains physiological and signaling pathways regulated by phytohormones (Akram et al. 2017), neutralizes ROS directly through the use of secondary antioxidants during the reduction of the oxidized form of α-tocopherol (Mohamed et al. 2018) and is a significant plant metabolite and acts as a cell signaling modulator in many physiological processes such as mitosis (Podgórska et al. 2017). In addition, treatment with AsA increased the growth of quinoa plants and alleviated the harmful effects of drought stress (Aziz et al. 2018). Khazaei et al. (2020) reported that drought stress has a positive effect on increasing AsA content in sweet pepper plants relative to control plants. Increasing the AsA in the plant may occur due to the breakdown of proteins, as well as the response to osmotic changes in cellular content. Farooq et al. (2020) reported that drought stress (60% field capacity) caused enhanced content of AsA in four safflower cultivars. Seminario et al. (2017) displayed that the exacerbation of AsA content was associated with water stress tolerance in tomato. Thus, Farooq et al. (2020) inferred that an increase in AsA level has a significant role in improving plant fresh and dry biomass of all safflower cultivars grown under water-limited conditions. Zhu et al. (2020) found that the ascorbate content was elevated under moderate and severe drought conditions of four cassava cultivars. In a study of El-Beltagi et al. (2020) using water intervals as drought-induced stress, AsA was reduced after 14 days of irrigation and enhanced after 28 days compared to 7 days irrigated plants. Drought stress resulted in a decline in AsA content in leaves and stems, while the content of DHA was only significantly reduced in stems which revealed that the retardation of AsA levels during drought is not explained by conversion to DHA. Although changes in cell wall permeability during drought cannot be discarded, results suggest that the observed reduction in AsA biosynthesis under water deficit is likely not related to a substrate limitation, and there are possibilities of regulation at the transcriptional or post-transcriptional level (Seminario et al. 2017). The effect of drought stress (60% FC) for 20 days caused an increment of the AsA content in *S. lycopersicum* plants (Rady et al. 2020). Conversely, ElSayed et al. (2019) found that ascorbic acid content was reduced under drought stress at 50% water capacity relative to control plants in the leaves of milk thistle (*Silybum marianum* (L.) Gaertn plants. Briefly, AsA minimizes the oxidative damage of various species of the ROS nonenzymatically via reducing $^1O_2$, $O_2^{-\bullet}$ and $^\bullet OH$ coupled with producing another antioxidant (tocopherol). In the ascorbate-glutathione cycle, two AsA molecules act as electron donors to convert H_2O_2 to water. This indicates that AsA is directly involved as an antioxidant by itself and indirectly as a precursor to producing another antioxidant (tocopherol).

12.4.2 Monodehydroascorbate (MDHA) and Dehydroascorbate (DHA)

This is a short lifetime molecule that breaks down into dehydroascorbate and ascorbic acid. MDHA is formed in the ascorbate-glutathione cycle concurrently with the reduction of H_2O_2 to water at the expense of two molecules of ascorbate. In the water cycle in the chloroplast, monodehydroascorbate reductase (MDHAR)/ferredoxin catalyzes the reduction of MDHA into reduced AsA using the Nicotinamide-adenine dinucleotide phosphate (NADPH) as an electron donor (Noctor et al. 2018). Niu et al. (2013) demonstrated the potentiality of the reduced form of GSH to confer drought tolerance, where small amounts of natural ascorbate and nonfunctional ascorbate-glutathione cycle decrease the levels of GSH, which resulted in sensitivity to drought stress in *Arabidopsis* mutants lacking sufficient ascorbate. Having sufficient AsA pool and high GSH/GSSG is essential to quench the toxicity of ROS that evolved under various biotic and abiotic stresses. MDHA is an intermediate component in the biosynthesis of the reduced GSH reserve, which is essential to confer drought tolerance in plants by maintaining a high GSH/GSSG ratio. The recycling of MDHA into GSH indicates that plants efficiently convert secondary products into valuable ones. MDHA is a radical with a short lifetime that, if not rapidly reduced, disproportionates to form ascorbate and DHA (Noctor et al. 2018). Although there is the possibility of enzymic and non-enzymic regeneration of ascorbate directly from MDHA, the rapid disintegration of the MDHA radical revealed that some DHA is always produced when ascorbate is oxidized in leaves and other tissues. DHA is reduced to ascorbate by the action of dehydroascorbate reductase (DHAR), using glutathione (GSH) as the reducing substrate (Asthir et al. 2020). DHA contents increased under moderate and severe drought in two maize hybrids (Hamim et al. 2017). In *Vigna radiata* the ratio of AsA/DHA reduced in response to exposure to 5% PEG for 48 hours (Nahar et al. 2017). Another study of Yildiztugay et al. (2020) revealed that drought stress experienced by PEG application increased the cellular content of DHA, whilst the reduction in AsA/DHA ratio in *Phaseolus vulgaris* was reported. The induced DHA was regenerated to AsA by the catalysis of DHAR. The GSH regeneration required for the activity of DHAR was provided by GR which regenerates GSH and so maintains the GSH pool (Shan et al. 2020).

12.4.3 Reduced Glutathione (GSH)

Glutathione is a thiol derivative that exists in all living organisms and acts as a prominent compound for ROS detoxification. Glutathione has been localized in all cell compartments. Glutathione acts as an antioxidant in several ways. GSH reacts non-enzymatically with some ROS: singlet oxygen, superoxide radicals and hydroxyl radicals. GSH keeps the H_2O_2 levels at the optimal concentrations (Sahoo et al. 2017). In the ascorbate-glutathione cycle, GSH is employed to reduce DHA (dehydroascorbate) and is itself converted to GSSG (oxidized glutathione). A balanced ratio between the reduced status to the oxidized status GSH/GSSG, during H_2O_2 scavenging, is essential for redox signaling pathways (Hasanuzzaman et al. 2019). Hasanuzzaman et al. (2018) reported that application of 10% PEG resulted in increased AsA, GSH and GSSG contents, while the ratio of GSH/GSSG reduced in *Brassica napus* plants. Under drought stress (60% and 40%), field capacity significantly increased the contents of reduced glutathione (GSH), oxidized glutathione (GSSG) and total glutathione (GT) compared to control plants in two *Salvia* species (Shirani et al. 2020). Similarly, reduced glutathione level in chickpea plants was elevated due to moderate and severe drought stress (El-Beltagi et al. 2020). Zhu et al. (2020) found that the reduced glutathione was increased in three tolerant cassava cultivars and reduced for the fourth cultivar under severe drought conditions (20% FC). The effect of drought stress (60% FC) for 20 days caused an increment of the GSH content in *S. lycopersicum* plants (Rady et al. 2020). On the other hand, Pyngrope et al. (2013) found that the drought-sensitive cultivar of indica rice (*Oryza sativa* L.) has a higher reduction of shoot and root contents of GSH due to water deficiency induced by 30% (v/v) polyethylene glycol-6000. Similarly, ElSayed et al. (2019) deduced that GSH content in milk thistle leaves was significantly reduced under drought stress at 50% water capacity compared to well-watered plants. As a component of the glutathione-ascorbate cycle, GSH is of high importance for plants under drought stress, as GSH can act as an antioxidant by itself or reduce DHA into AsA.

12.4.4 Homoglutathione (hGSH)

This is a homologue of GSH in the family *Fabaceae*, with the C-terminal glycine is replaced by β-alanine and with the same functions of GSH (Noctor et al. 2012). Primarily, hGSH contributes to the regulation of nodulation, nitrogen fixation, and symbiotic interactions; additionally, it has a potential antioxidant activity to translocate the reduced sulfur (Zagorchev et al. 2013). In a drought-tolerant cultivar of cowpea *Vigna unguiculata*, high levels of hGGS were reported during water deficit and senescence (de Carvalho et al. 2010). Conversely, the levels of GSH and hGSH in the nodules of alfalfa (*Medicago sativa* L.) did not change significantly under drought stress (Naya et al. 2007). The suppression of GSH and hGSH coding genes inhibited root nodulation in *Medicago truncatula* (Frendo et al. 2005). De Carvalho et al. (2010) showed that the foliar content of hGSH synthetase mRNA was elevated in drought-tolerant cultivar cowpea (*Vigna unguiculata* (L.) Walp.). But Naya et al. (2007) reported that drought stress had no impact on GSH and hGSH contents in the nodules of alfalfa (*Medicago sativa* L.). The study of Marquez-Garcia et al. (2015) revealed that total glutathione (GSH plus GSSG) and homoglutathione (hGSH plus hGSSG) contents reduced in leaves due to drought stress relative to well-watered controls. These studies indicate that GSH and hGSH contribute to the plant antioxidant machinery as well as in various plant developmental processes, especially the translocation of reduced sulfur.

12.4.5 S-Glutathionylation

S-Glutathionylation (thiolation) is generated by the formation of disulfide between a cysteine residue and glutathione in a reversible post-transcriptional modification. It has been shown that this protein S-glutathionylation is induced by both reactive oxygen species and nitrogen species as well. Additionally, it occurs under normal conditions. This modification may act to modulate various biological processes by controlling protein function and inhibiting the irreversible oxidation of protein thiols (Mailloux et al. 2020). It regulates a set of enzymes that are involved in the Calvin cycle such as phosphoribulokinase, glyceraldehyde-3-phosphate dehydrogenase, ribose- 5-phosphate isomerase and phosphoglycerate kinase in the green alga (*Chlamydomonas reinhardtii* P.A. Dang.) cultured under oxidative stress (Zaffagnini et al. 2012). In desiccation tolerant plants, the enormous changes in the water content of tissues during wetting and drying cycles are accompanied by equally extreme fluctuations in their cellular redox state (Labudda and Azam 2014). S-glutathionylation of proteins is a biochemical factor that is likely to contribute towards protection mechanisms that confer desiccation tolerance (Colville and Kranner 2010). Together, thiol-containing antioxidants are diverse and multifunctional molecules. They contribute to various biological processes in plants including plant development and plant adaptation to drought as a key player in the non-enzymatic antioxidant machinery.

12.4.6 α-Tocopherols (Vitamin E)

These are non-enzymatic antioxidants that are produced in the green parts of plants and algae (Srivalli et al. 2003). They exist in four isoforms (α, β, γ and δ), but α-tocopherols are the most active isoform. They act as antioxidants in the chloroplast membrane. A single molecule of α-tocopherol can quench 120 singlet oxygen molecules. α-tocopherol scavenges ROS released upon oxidative stress through interaction with the acyl group of lipids (Munné-Bosch 2005). An increase of one- to three-fold α-tocopherol levels was observed in some grass species (Pourcel et al. 2007). In drought-stressed *Vigna* plants, high levels of α-tocopherols were detected (Manivannan et al. 2007). The oxidative stress triggers the expression of the genes coding the α-tocopherols production (Gong et al. 2012). *Arabidopsis* plants with high levels of α-tocopherol exhibited higher water content, elevated photosynthesis rate and lower oxidative stress as evidenced by less lipid peroxidation and delayed leaf senescence (Espinoza et al. 2013). The effect of drought stress (60% field capacity) for 20 days caused an increment of the tocopherol content in *S. lycopersicum* plants (Rady et al. 2020). Also, Rezayian et al. (2020) reported that the level of tocopherol enhanced in response to three levels of drought stress induced by PEG (5%, 10% and 15%) for three weeks in *Glycine max*. These findings indicate that α-tocopherol is essential for ameliorating the oxidative damage provoked by ROS, via preventing lipid peroxidation. α-tocopherol is a powerful nonenzymatic antioxidant since one molecule can scavenge 120 molecules of ROS.

12.4.7 Carotenoids (Car)

Carotenoids are pigments that exist in plants and microorganisms. In nature, there are more than 600 species of carotenoids. β-carotene possesses powerful reducing characters. It neutralizes singlet oxygen without degradation and reacts with free radicals such as peroxyl ($ROO^{\bullet}$), hydroxyl ($^{\bullet}OH$) and superoxide ($O_2^{\bullet-}$). Carotenoids can quench triplet chlorophylls, the source of $^1O_2$ in leaves (Ramel et al. 2012). The levels of carotenoids are highly affected by the duration and intensity of drought stress as well as on plant species where the concentrations of carotenoids decreased under mild drought and increased under severe drought (Sudrajat et al. 2015). The increase of carotenoids under prolonged drought has been observed in alpine plants (Buchner et al. 2017). The elevation of carotenoid concentrations during drought stress might be attributed to the activation of the xanthophyll cycle (Schweiggert et al. 2017). As drought stress intensified, the quantity of photosynthetic pigments as carotenoids increased. This increase is not significantly different between the control and mild drought stress conditions, but significant under severe drought stress in two varieties of *Cynanchum thesioides* (Zhang et al. 2020). This may be as a result of compensation and over-compensation effects and was similar to the results of Zhang et al. (2017) in blackberry. On the other hand, drought stress significantly reduced carotenoid content, and the decrement was intensified as the exposure time increased from 3 to 12 days in rice plants (Nasrin et al. 2020). Sadak et al. (2020) presented that subjecting *Moringa oleifera* plants to drought stress resulted in decrement of photosynthetic pigments like chlorophyll a, chlorophyll b, carotenoids and total pigments as compared with control plants. Similar reductions of total carotenoids were reported under drought stress at all growth stages in soybean (Basal et al. 2020). As pigments, carotenoids strictly work in chloroplasts, therefore, they are involved in detoxification of the photooxidative stress. There is interdependency between carotenoids and xanthophyll under drought stress; the higher the activities of the xanthophyll cycle the higher the levels of carotenoids.

12.4.8 Phenolic Compounds

One of the most important non-enzymatic antioxidants in plants are phenols, which are a part of the antioxidant system of plants (Sahitya et al. 2018). Phenolic compounds, such as flavonoids, tannins, hydroxylcinnamate esters and lignin, are versatile secondary metabolites. They are involved in ROS detoxification. They are abundant in different plant tissues. Phenolic antioxidants (PhOH) prevent the oxidation of lipids and other molecules by donating a hydrogen atom to radicals $ROO^\bullet + PhOH \rightarrow ROOH + PhO^\bullet$. The high potential of phenolic compounds in ROS metabolism is to donate hydrogen, stabilize the free electrons and chelate transition metal ions conferred their antioxidants properties (Schroeter et al. 2002). Khazaei et al. (2020) stated that phenolic compounds increased considerably in sweet pepper plants under moderate drought stress (60% field capacity) and severe drought stress (30% field capacity). Similar enhancement of phenolic content was reported by Farooq et al. (2020) in response to drought stress of four safflower plants. Lama et al. (2016) reported that flavonoids concentration increased in *Jatropha* seedlings under artificially-induced high drought (200 mm year^{-1}) and artificial damage (50%) conditions. The authors also suggested that flavonoids were responsible for the protection against oxidative stress and photodamage in *Jatropha* leaves. On the other hand, total phenols and total flavanols were decreased in *Agave salmiana* plants under in vitro drought stress imposed via 10%, 20% and 30% PEG (Puente-Garza et al. 2017). The variation of phenolic compounds in plants explains their profound effect on the antioxidative system in plants. Besides this, their wide distribution in different plant organs/tissues widens their effect across the whole plant life cycle, especially under stressful growth conditions such as drought.

12.5 CONCLUSION

The antioxidant systems are strongly involved in plant tolerance mechanisms against oxidative damage provoked by drought stress. Both enzymatic and non-enzymatic antioxidant systems occupy a wide space in plant defense against drought with an upper hand for the enzymatic system. It has become clear that the different enzymatic antioxidants, SOD, APX, DHAR, MDHAR, GR and CAT, operate coordinately even if they are produced in different cellular organelles. Several studies demonstrated that enhancing the activities of these enzymatic systems can confer drought tolerance to various plant species. The turnover of antioxidants under drought stress in many plant species was found to be directly correlated with drought tolerance indicators such as high yield, indicating that these enzymatic antioxidants can be used as reliable markers to select for drought tolerance. More likely, the induction of the antioxidant defense system is genotype-specific, where the drought-tolerant genotypes showed higher activities of antioxidants such as SOD, APX, DHAR, MDHAR, GR and CAT, implying that the genetic architecture plays a key role in drought tolerance via activating these enzymes. The newly developed DNA/protein-manipulation and molecular techniques paved the way to monitor the role of the genes coding antioxidants under various stresses, including drought. The overexpression of genes associated with different components of the enzymatic antioxidant system was found to improve drought tolerance in different plant species. The crosstalk between the different enzymes was evidenced as the overexpression of one type inducing the activity of other ones, suggesting a synergistic operation and lineal activation. Probably, plants favorably activate the oxidoreductase machinery rather than activating other mechanisms such as biosynthesis of compatible solutes because the latter is time-consuming and energetically expensive.

REFERENCES

Abid, Muhammad, Shafaqat Ali, Lei Kang Qi, Rizwan Zahoor, Zhongwei Tian, Dong Jiang, John L Snider, and Tingbo Dai. 2018. "Physiological and biochemical changes during drought and recovery periods at tillering and jointing stages in wheat (Triticum aestivum L.)." *Scientific Reports* 8(1): 1–15.

Ahmad, P, G Nabi, CA Jeleel, and S Umar. 2011. "Free radical production, oxidative damage and antioxidant defense mechanisms in plants under abiotic stress." In: Ahmad P, Umar S (eds), *Oxidative Stress: Role of Antioxidants in Plants*, Studium Press, New Delhi, 19–53.

Ahmad, Parvaiz, Maryam Sarwat, and Satyawati Sharma. 2008. "Reactive oxygen species, antioxidants and signaling in plants." *Journal of Plant Biology* 51(3): 167–173.

Ahmadi, Jafar, Alireza Pour-Aboughadareh, Sedigheh Fabriki Ourang, Ali Ashraf Mehrabi, and Kadambot HM Siddique. 2018. "Wild relatives of wheat: Aegilops–Triticum accessions disclose differential antioxidative and physiological responses to water stress." *Acta Physiologiae Plantarum* 40(5): 90.

Akram, Nudrat A, Fahad Shafiq, and Muhammad Ashraf. 2017. "Ascorbic acid-A potential oxidant scavenger and its role in plant development and abiotic stress tolerance." *Frontiers in Plant Science* 8(613). https://doi.org/10.3389/fpls.2017.00613; https://www.frontiersin.org/article/10.3389/fpls.2017.00613.

Ali, Qasim, Muhammad Tariq Javed, Ali Noman, Muhammad Zulqarnain Haider, Muhammad Waseem, Naeem Iqbal, Muhammad Waseem, Muhammad Shahzad Shah, Faisal Shahzad, and Rashida Perveen. 2018. "Assessment of drought tolerance in mung bean cultivars/lines as depicted by the activities of germination enzymes, seedling's antioxidative potential and nutrient acquisition." *Archives of Agronomy and Soil Science* 64(1): 84–102.

Almeselmani, Moaed, PS Deshmukh, RK Sairam, SR Kushwaha, and TP Singh. 2006. "Protective role of antioxidant enzymes under high temperature stress." *Plant Science: An International Journal of Experimental Plant Biology* 171(3): 382–388.

Anjum, Naser A, Sarvajeet S Gill, Ritu Gill, Mirza Hasanuzzaman, Armando C Duarte, Eduarda Pereira, Iqbal Ahmad, Renu Tuteja, and Narendra Tuteja. 2014. "Metal/metalloid stress tolerance in plants: Role of ascorbate, its redox couple, and associated enzymes." *Protoplasma* 251(6): 1265–1283.

Asada, Kozi. 1999. "The water-water cycle in chloroplasts: Scavenging of active oxygens and dissipation of excess photons." *Annual Review of Plant Physiology and Plant Molecular Biology* 50(1): 601–639.

Ashraf, M. 2009. "Biotechnological approach of improving plant salt tolerance using antioxidants as markers." *Biotechnology Advances* 27(1): 84–93. https://www.sciencedirect.com/science/article/abs/pii/S0734975008001018?via%3Dihub.

Asthir, Bavita, Gurpreet Kaur, and Balraj Kaur. 2020. "Convergence of pathways towards Ascorbateâ—Glutathione for stress mitigation." *Journal of Plant Biology* 63: 243–257.

Aziz, Aniqa, Nudrat Aisha Akram, and Muhammad Ashraf. 2018. "Influence of natural and synthetic vitamin C (ascorbic acid) on primary and secondary metabolites and associated metabolism in quinoa (Chenopodium quinoa Willd.) plants under water deficit regimes." *Plant Physiology and Biochemistry: PPB* 123: 192–203.

Basal, Oqba, A Szabó, and Szilvia Veres. 2020. "Physiology of soybean as affected by PEG-induced drought stress." *Current Plant Biology*. http://www.ncbi.nlm.nih.gov/pubmed/100135.

Broin, Mélanie, and Pascal Rey. 2003. "Potato plants lacking the CDSP32 plastidic thioredoxin exhibit overoxidation of the BAS1 2-cysteine peroxiredoxin and increased lipid peroxidation in thylakoids under photooxidative stress." *Plant Physiology* 132(3): 1335–1343. https://www.ncbi.nlm.nih.gov/pmc/articles/PMC167073/pdf/1321335.pdf.

Browne, John, Alan Tunnacliffe, and Ann Burnell. 2002. "Anhydrobiosis: Plant desiccation gene found in a nematode." *Nature* 416(6876): 38–38.

Buchner, Othmar, Thomas Roach, Joy Gertzen, Stephanie Schenk, Matthias Karadar, Wolfgang Stöggl, Ramona Miller, Clara Bertel, Gilbert Neuner, and Ilse Kranner. 2017. "Drought affects the heat-hardening capacity of alpine plants as indicated by changes in xanthophyll cycle pigments, singlet oxygen scavenging, α-tocopherol and plant hormones." *Environmental and Experimental Botany* 133: 159–175.

Campos, CN, RG Ávila, KRD de Souza, LM Azevedo, and JD Alves. 2019. "Melatonin reduces oxidative stress and promotes drought tolerance in young Coffea arabica L. plants." *Agricultural Water Management* 211: 37–47.

Cao, S, X-H Du, L-H Li, Y-D Liu, L Zhang, X Pan, Y Li, H Li, and H Lu. 2017. "Overexpression of Populus tomentosa cytosolic ascorbate peroxidase enhances abiotic stress tolerance in tobacco plants." *Russian Journal of Plant Physiology* 64(2): 224–234.

Caverzan, Andréia, Alice Casassola, and Sandra Patussi Brammer. 2016. "Reactive oxygen species and antioxidant enzymes involved in plant tolerance to stress." In: *Abiotic and Biotic Stress in Plants-Recent Advances and Future Perspectives*, edited by A.K. Shanker, and C. Shanker, 463–480. Publisher InTech.

Cha, JY, JY Kim, IJ Jung, MR Kim, A Melencion, SS Alam, DJ Yun, SY Lee, MG Kim, and WY Kim. 2014. "NADPH-dependent thioredoxin reductase A (NTRA) confers elevated tolerance to oxidative stress and drought." *Plant Physiology and Biochemistry: PPB* 80: 184–191.

Chakradhar, Thammineni, Srikrishna Mahanty, Ramesha A Reddy, Kummari Divya, Palakolanu Sudhakar Reddy, and Malireddy K Reddy. 2017. "Biotechnological perspective of reactive oxygen species (ROS)-mediated stress tolerance in plants." In: *Reactive Oxygen Species and Antioxidant Systems in Plants: Role and Regulation under Abiotic Stress*, edited by M. Iqbal, R. Khan, and Nafees A. Khan, 53–87. Springer: Singapore.

Chen, JH, HW Jiang, EJ Hsieh, HY Chen, CT Chien, HL Hsieh, and TP Lin. 2012. "Drought and salt stress tolerance of an Arabidopsis glutathione S-transferase U17 knockout mutant are attributed to the combined effect of glutathione and abscisic acid." *Plant Physiology* 158(1): 340–351.

Cheng, Zaiquan, Jayaprakash Targolli, Xingqi Huang, and Ray Wu. 2002. "Wheat LEA genes, PMA80 and PMA1959, enhance dehydration tolerance of transgenic rice (Oryza sativa L.)." *Molecular Breeding* 10(1–2): 71–82.

Colville, Louise, and Ilse Kranner. 2010. "Desiccation tolerant plants as model systems to study redox regulation of protein thiols." *Plant Growth Regulation* 62(3): 241–255.

Dalvi, US, RM Naik, and PK Lokhande. 2018. "Antioxidant defense system in chickpea against drought stress at pre- and post- flowering stages." *Indian Journal of Plant Physiology* 23(1): 16–23.

Dat, J, Steven Vandenabeele, EVMM Vranová, Marc Van Montagu, Dirk Inzé, and Frank Van Breusegem. 2000. "Dual action of the active oxygen species during plant stress responses." *Cellular and Molecular Life Sciences: CMLS* 57(5): 779–795. https://link.springer.com/content/pdf/10.1007/s000180050041 .pdf.

Davletova, Sholpan, Ludmila Rizhsky, Hongjian Liang, Zhong Shengqiang, David J Oliver, Jesse Coutu, Vladimir Shulaev, Karen Schlauch, and Ron Mittler. 2005. "Cytosolic ascorbate peroxidase 1 is a central component of the reactive oxygen gene network of Arabidopsis." *The Plant Cell* 17(1): 268–281. https://www.ncbi.nlm.nih.gov/pmc/articles/PMC544504/pdf/tpc1700268.pdf.

de Carvalho, Maria H Cruz, Judicaëlle Brunet, Jérémie Bazin, Ilse Kranner, Agnès d'Arcy-Lameta, Yasmine Zuily-Fodil, and Dominique Contour-Ansel. 2010. "Homoglutathione synthetase and glutathione synthetase in drought-stressed cowpea leaves: Expression patterns and accumulation of low-molecular-weight thiols." *Journal of Plant Physiology* 167(6): 480–487.

Devi, R, AK Gupta, and N Kaur. 2011. "A comparative study of antioxidant enzymes in wheat seedlings of landrace and cultivated genotypes." *Indian Journal of Agricultural Biochemistry* 24(2): 105–109.

Devi, R, N Kaur, and AK Gupta. 2012. "Potential of antioxidant enzymes in depicting drought tolerance of wheat (Triticum aestivum L.)." *Indian J Biochem Biophys* 49:257–265

Dietz, Karl-Josef. 2003. "Plant peroxiredoxins." *Annual Review of Plant Biology* 54(1): 93–107.

Do, Hackwon, Il-Sup Kim, Byoung Wook Jeon, Chang Woo Lee, Ae Kyung Park, AR Wi, SC Shin, H Park, YS Kim, and HS Yoon. 2016. "Structural understanding of the recycling of oxidized ascorbate by dehydroascorbate reductase (OsDHAR) from Oryza sativa L. japonica." *Scientific Reports* 6: 19498.

El-Beltagi, Hossam S, Heba I Mohamed, and Mahmoud R Sofy. 2020. "Role of ascorbic acid, glutathione and proline applied as singly or in sequence combination in improving chickpea plant through physiological change and antioxidant defense under different levels of irrigation intervals." *Molecules* 25(7): 1702.

Elkeilsh, A, YM Awad, MH Soliman, A Abu-Elsaoud, MT Abdelhamid, and IM El-Metwally. 2019. "Exogenous application of β-sitosterol mediated growth and yield improvement in water-stressed wheat (Triticum aestivum) involves up-regulated antioxidant system." *Journal of Plant Research* 132(6): 881–901.

ElSayed, Abdelaleim I, Mohamed AM El-hamahmy, Mohammed S Rafudeen, Azza H Mohamed, and Ahmad A Omar. 2019. "The impact of drought stress on antioxidant responses and accumulation of flavonolignans in milk thistle (Silybum marianum (l.) gaertn)." *Plants* 8(12): 611.

Eltayeb, Amin Elsadig, Naoyoshi Kawano, Ghazi Hamid Badawi, Hironori Kaminaka, Takeshi Sanekata, Toshiyuki Shibahara, Shinobu Inanaga, and Kiyoshi Tanaka. 2007. "Overexpression of monodehydroascorbate reductase in transgenic tobacco confers enhanced tolerance to ozone, salt and polyethylene glycol stresses." *Planta* 225(5): 1255–1264.

Eltelib, Hani A, Adebanjo A Badejo, Yukichi Fujikawa, and Muneharu Esaka. 2011. "Gene expression of monodehydroascorbate reductase and dehydroascorbate reductase during fruit ripening and in response to environmental stresses in acerola (Malpighia glabra)." *Journal of Plant Physiology* 168(6): 619–627.

Espinoza, A, A San Martín, M López-Climent, S Ruiz-Lara, A Gómez-Cadenas, and JA Casaretto. 2013. "Engineered drought-induced biosynthesis of α-tocopherol alleviates stress-induced leaf damage in tobacco." *Journal of Plant Physiology* 170(14): 1285–1294. https://www.sciencedirect.com/science/ article/pii/S0176161713001624?via%3Dihub.

Farooq, Ayesha, Shazia Anwer Bukhari, Nudrat A Akram, Muhammad Ashraf, Leonard Wijaya, Mohammed Nasser Alyemeni, and Parvaiz Ahmad. 2020. "Exogenously applied ascorbic acid-mediated changes in osmoprotection and oxidative defense system enhanced water stress tolerance in different cultivars of safflower (Carthamus tinctorious L.)." *Plants* 9(1): 104.

Farooq, M, A Wahid, N Kobayashi, D Fujita, and SMA Basra. 2009. "Plant drought stress: Effects, mechanisms and management." *Agronomy for Sustainable Development* 29(1): 185–212.

Foyer, Christine H, and Graham Noctor. 2005. "Redox homeostasis and antioxidant signaling: A metabolic interface between stress perception and physiological responses." *The Plant Cell* 17(7): 1866–1875.

Frendo, Pierre, Judith Harrison, Christel Norman, MJ Hernández Jiménez, G Van de Sype, A Gilabert, and A Puppo. 2005. "Glutathione and homoglutathione play a critical role in the nodulation process of Medicago truncatula." *Molecular Plant-Microbe Interactions: MPMI* 18(3): 254–259.

Frugoli, Julia A, Hai Hong Zhong, Michael L Nuccio, Peter McCourt, Mark A McPeek, Terry L Thomas, and C McClung. 1996. "Catalase is encoded by a multigene family in Arabidopsis thaliana (L.) Heynh.. " *Plant Physiology* 112(1): 327–336. http://www.plantphysiol.org/content/plantphysiol/112/1/327.full.pdf.

Gaber, Ahmed, Kazuya Yoshimura, Takashi Yamamoto, Yukinori Yabuta, Toru Takeda, Hitoshi Miyasaka, Yoshihisa Nakano, and Shigeru Shigeoka. 2006. "Glutathione peroxidase-like protein of synechocystis PCC 6803 confers tolerance to oxidative and environmental stresses in transgenic Arabidopsis." *Physiologia Plantarum* 128(2): 251–262.

Gallé, Á, Jolán Csiszár, Maria Secenji, Adrienn Guóth, László Cseuz, Irma Tari, János Györgyey, and László Erdei. 2009. "Glutathione transferase activity and expression patterns during grain filling in flag leaves of wheat genotypes differing in drought tolerance: Response to water deficit." *Journal of Plant Physiology* 166(17): 1878–1891.

George, Suja, Gayatri Venkataraman, and Ajay Parida. 2010. "A chloroplast-localized and auxin-induced glutathione S-transferase from phreatophyte Prosopis juliflora confer drought tolerance on tobacco." *Journal of Plant Physiology* 167(4): 311–318.

Ghaffari, H, MR Tadayon, M Nadeem, M Cheema, and J Razmjoo. 2019. "Proline-mediated changes in antioxidant enzymatic activities and the physiology of sugar beet under drought stress." *Acta Physiologiae Plantarum* 41(2): 23.

Ghorbanpour, M, H Mohammadi, and K Kariman. 2020. "Nanosilicon-based recovery of barley (Hordeum vulgare) plants subjected to drought stress." *Environmental Science: Nano* 7.2 (2020): 443-461.

Gill, Sarvajeet Singh, and Narendra Tuteja. 2010. "Reactive oxygen species and antioxidant machinery in abiotic stress tolerance in crop plants." *Plant Physiology and Biochemistry: PPB* 48(12): 909–930.

Gong, Chunmei, Pengbo Ning, and Juan Bai. 2012. "Responses of antioxidative protection to varying drought stresses induced by micro-ecological fields on desert C4 and C3 plants in Northwest China." *Life Science Journal* 9(4): 2006–2016.

Gosal, Satbir S, Shabir H Wani, and Manjit S Kang. 2009. "Biotechnology and drought tolerance." *Journal of Crop Improvement* 23(1): 19–54.

Gratão, Priscila L, Andrea Polle, Peter J Lea, and Ricardo A Azevedo. 2005. "Making the life of heavy metal-stressed plants a little easier." *Functional Plant Biology: FPB* 32(6): 481–494.

Guo, YY, HY Yu, MM Yang, DS Kong, and YJ Zhang. 2018. "Effect of drought stress on lipid peroxidation, osmotic adjustment and antioxidant enzyme activity of leaves and roots of Lycium ruthenicum Murr. seedling." *Russian Journal of Plant Physiology* 65(2): 244–250.

Gupta, Dharmendra K, José M Palma, and Francisco J Corpas. 2018. *Antioxidants and Antioxidant Enzymes in Higher Plants*. Berlin: Springer International Publishing.

Habib, N, Q Ali, S Ali, MT Javed, M Zulqurnain Haider, R Perveen, MR Shahid, M Rizwan, MM Abdel-Daim, A Elkelish, and M Bin-Jumah. 2020. "Use of nitric oxide and hydrogen peroxide for better yield of wheat (Triticum aestivum L.) under water deficit conditions: Growth, osmoregulation, and antioxidative defense mechanism." *Plants* 9(2): 285.

Hamim, H, V Violita, T Triadiati, and M Miftahudin. 2017. "Oxidative stress and photosynthesis reduction of cultivated (Glycine max L.) and wild soybean (G. Tomentella L.) exposed to drought and paraquat." *Asian Journal of Plant Sciences* 16(2): 65–77.

Harb, Amal, Dalal Awad, and Nezar Samarah. 2015. "Gene expression and activity of antioxidant enzymes in barley (Hordeum vulgare L.) under controlled severe drought." *Journal of Plant Interactions* 10(1): 109–116.

Hasanuzzaman, M, K Nahar, TI Anee, MIR Khan, and M Fujita. 2018. "Silicon-mediated regulation of antioxidant defense and glyoxalase systems confers drought stress tolerance in Brassica napus L.." *South African Journal of Botany* 115: 50–57.

Hasanuzzaman, M, Bhuyan MHM, Anee TI, Parvin K, Nahar K, Mahmud JA, and Fujita M. 2019. "Regulation of ascorbate-glutathione pathway in mitigating oxidative damage in plants under abiotic stress." *Antioxidants* 8(9): 384.

He, J, Y Zou, Q Wu, and K Kuča. 2020. "Mycorrhizas enhance drought tolerance of trifoliate orange by enhancing activities and gene expression of antioxidant enzymes." *Scientia Horticulturae* 262. http://www.ncbi.nlm.nih.gov/pubmed/108745.

Hirayama, Takashi, and Kazuo Shinozaki. 2007. "Perception and transduction of abscisic acid signals: Keys to the function of the versatile plant hormone ABA." *Trends in Plant Science* 12(8): 343–351.

Hossain, Mohammad Anwar, Zhong-Guang Li, Tahsina Sharmin Hoque, David J Burritt, Masayuki Fujita, and Sergi Munné-Bosch. 2018. "Heat or cold priming-induced cross-tolerance to abiotic stresses in plants: Key regulators and possible mechanisms." *Protoplasma* 255(1): 399–412. https://link.springer .com/content/pdf/10.1007/s00709-017-1150-8.pdf.

Hu, Xiuli, Mingyi Jiang, Aying Zhang, and Jun Lu. 2005. "Abscisic acid-induced apoplastic H 2 O 2 accumulation up-regulates the activities of chloroplastic and cytosolic antioxidant enzymes in maize leaves." *Planta* 223(1): 57.

Hu, Xiuli, Ruixia Liu, Yanhui Li, Wei Wang, Fuju Tai, Ruili Xue, and Chaohai Li. 2010. "Heat shock protein 70 regulates the abscisic acid-induced antioxidant response of maize to combined drought and heat stress." *Plant Growth Regulation* 60(3): 225–235.

Hussain, T, B Tan, Y Yin, F Blachier, MCB Tossou, and N Rahu. 2016. "Oxidative stress and inflammation: What polyphenols can do for us?." *Oxidative Medicine and Cellular Longevity* 5, 97–104.

Jamshidi Goharrizi, K, SS Moosavi, F Amirmahani, F Salehi, and M Nazari. 2020. "Assessment of changes in growth traits, oxidative stress parameters, and enzymatic and non-enzymatic antioxidant defense mechanisms in Lepidium draba plant under osmotic stress induced by polyethylene glycol." *Protoplasma* 257(2): 459–473.

Jing, Xiaoshu, Peichen Hou, Yanjun Lu, Shurong Deng, Niya Li, Rui Zhao, Jian Sun, Yang Wang, Yansha Han, Tao Lang, Mingquan Ding, Xin Shen, and Shaoliang Chen. 2015. "Overexpression of copper/ zinc superoxide dismutase from mangrove Kandelia candel in tobacco enhances salinity tolerance by the reduction of reactive oxygen species in chloroplast." *Frontiers in Plant Science* 6: 23–23. https:// doi.org/10.3389/fpls.2015.00023; https://pubmed.ncbi.nlm.nih.gov/25657655; https://www.ncbi.nlm.nih .gov/pmc/articles/PMC4302849/.

Kang, Yun, and Michael Udvardi. 2012. "Global regulation of reactive oxygen species scavenging genes in alfalfa root and shoot under gradual drought stress and recovery." *Plant Signaling and Behavior* 7(5): 539–543.

Khaleghi, Alireza, Rohangiz Naderi, Cecilia Brunetti, Bianca Elena Maserti, Seyed Alireza Salami, and Mesbah Babalar. 2019. "Morphological, physiochemical and antioxidant responses of Maclura pomifera to drought stress." *Scientific Reports* 9(1): 1–12.

Khan, M Nasir, Mazen A AlSolami, Riyadh A Basahi, Manzer H Siddiqui, Asma A Al-Huqail, Zahid Khorshid Abbas, Zahid H Siddiqui, Hayssam M Ali, and Faheema Khan. 2020. "Nitric oxide is involved in nano-titanium dioxide-induced activation of antioxidant defense system and accumulation of osmolytes under water-deficit stress in Vicia faba L." *Ecotoxicology and Environmental Safety* 190: 110152.

Khazaei, Zahra, Behrooz Esmaielpour, and Asghar Estaji. 2020. "Ameliorative effects of ascorbic acid on tolerance to drought stress on pepper (Capsicum annuum L) plants." *Physiology and Molecular Biology of Plants: An International Journal of Functional Plant Biology* 26(8): 1649–1662.

Kim, YS, IS Kim, MJ Bae, YH Choe, YH Kim, HM Park, HG Kang, and HS Yoon. 2013. "Homologous expression of cytosolic dehydroascorbate reductase increases grain yield and biomass under paddy field conditions in transgenic rice (Oryza sativa L. japonica)." *Planta* 237(6): 1613–1625.

Kim, YS, SI Park, JJ Kim, JS Boyd, J Beld, A Taton, KI Lee, IS Kim, JW Golden, and HS Yoon. 2020. "Expression of heterologous OsDHAR gene improves glutathione (GSH)-dependent antioxidant system and maintenance of cellular redox status in Synechococcus elongatus PCC 7942." *Frontiers in Plant Science* 11: 231.

Kiran, S, C Ates, S Kusvuran, M Talhouni, and SS Ellialtioglu. 2019. "Antioxidative response of grafted and non-grafted eggplant seed-lings under drought and salt stresses." *Agrochimica* 63(2): 123–137.

Kojić, Danijela, Slobodanka Pajević, A Jovanović-Galović, Jelena Purać, E Pamer, S Škondrić, Snežana Milovac, Z Popović, and G Grubor-Lajšić. 2012. "Efficacy of natural aluminosilicates in moderating drought effects on the morphological and physiological parameters of maize plants (Zea mays L.)." *Journal of Soil Science and Plant Nutrition* 12(1): 113–123.

Kwon, Suk-Yoon, Sun-Mee Choi, Young-Ock Ahn, Haeng-Soon Lee, Hae-Bok Lee, Yong-Mok Park, and Sang-Soo Kwak. 2003. "Enhanced stress-tolerance of transgenic tobacco plants expressing a human dehydroascorbate reductase gene." *Journal of Plant Physiology* 160(4): 347–353.

Labudda, Mateusz, and Fardous Mohammad Safiul Azam. 2014. "Glutathione-dependent responses of plants to drought: A review." *Acta Societatis Botanicorum Poloniae* 83(1).

Lama, Ang Dawa, Jorma Kim, Olli Martiskainen, Tero Klemola, Juha-Pekka Salminen, Esa Tyystjärvi, Pekka Niemelä, and Timo Vuorisalo. 2016. "Impacts of simulated drought stress and artificial damage on concentrations of flavonoids in Jatropha curcas (L.), a biofuel shrub." *Journal of Plant Research* 129(6): 1141–1150. https://link.springer.com/content/pdf/10.1007/s10265-016-0850-z.pdf.

Latif, M, NA Akram, and M Ashraf. 2016. "Regulation of some biochemical attributes in drought-stressed cauliflower (Brassica oleracea L.) by seed pre-treatment with ascorbic acid." *The Journal of Horticultural Science and Biotechnology* 91(2): 129–137.

Laxa, Miriam, Michael Liebthal, Wilena Telman, Kamel Chibani, and Karl-Josef Dietz. 2019a. "The role of the plant antioxidant system in drought tolerance." *Antioxidants* 8(4): 94. https://www.ncbi.nlm.nih.gov/pmc/articles/PMC6523806/.

Leterrier, Marina, Francisco J Corpas, Juan B Barroso, Luisa M Sandalio, and LA del Río. 2005. "Peroxisomal monodehydroascorbate reductase. Genomic clone characterization and functional analysis under environmental stress conditions." *Plant Physiology* 138(4): 2111–2123.

Li, F, QY Wu, M Duan, XC Dong, B Li, and QW Meng. 2012. "Transgenic tomato plants overexpressing chloroplastic monodehydroascorbate reductase are resistant to salt- and PEG-induced osmotic stress." *Photosynthetica* 50(1): 120–128.

Liang, Dong, Zhiyou Ni, Hui Xia, Yue Xie, Xiulan Lv, Jin Wang, Lijin Lin, Qunxian Deng, and Xian Luo. 2019. "Exogenous melatonin promotes biomass accumulation and photosynthesis of kiwifruit seedlings under drought stress." *Scientia Horticulturae* 246: 34–43.

Lopez-Huertas, Eduardo, Wayne L Charlton, Barbara Johnson, Ian A Graham, and Alison Baker. 2000. "Stress induces peroxisome biogenesis genes." *The EMBO Journal* 19(24): 6770–6777. https://www.ncbi.nlm.nih.gov/pmc/articles/PMC305880/pdf/cdd600.pdf.

Lou, Lili, Xiaorui Li, Junxiu Chen, Yue Li, Yan Tang, and Jinyin Lv. 2018. "Photosynthetic and ascorbate-glutathione metabolism in the flag leaves as compared to spikes under drought stress of winter wheat (Triticum aestivum L.)." *PLoS One*, 13(3), e0194625.

Lu, Yan-Yuan, Xi-Ping Deng, and Sang-Soo Kwak. 2010. "Over expression of CuZn superoxide dismutase (CuZn SOD) and ascorbate peroxidase (APX) in transgenic sweet potato enhances tolerance and recovery from drought stress." *African Journal of Biotechnology* 9(49): 8378–8391.

Luis, A, Francisco J Corpas, Eduardo López-Huertas, and José M Palma. 2018. "Plant superoxide dismutases: Function under abiotic stress conditions." In: *Antioxidants and Antioxidant Enzymes in Higher Plants*, 1–26. Springer: Berlin, Germany.

Luna, Celina M, Gabriela M Pastori, Simon Driscoll, Karin Groten, Stephanie Bernard, and Christine H Foyer. 2005. "Drought controls on H2O2 accumulation, catalase (CAT) activity and CAT gene expression in wheat." *Journal of Experimental Botany* 56(411): 417–423.

Mailloux, Ryan J, Robert Gill, and Adrian Young. 2020. "Protein S-glutathionylation and the regulation of cellular functions." In: *Oxidative Stress*, 217–247. Elsevier.

Malan, C, MM Greyling, and J Gressel. 1990. "Correlation between CuZn superoxide dismutase and glutathione reductase, and environmental and xenobiotic stress tolerance in maize inbreds." *Plant Science* 69(2): 157–166.

Manivannan, P, C Abdul Jaleel, A Kishorekumar, B Sankar, R Somasundaram, R Sridharan, and R Panneerselvam. 2007. "Changes in antioxidant metabolism of Vigna unguiculata (L.) Walp. by propiconazole under water deficit stress." *Colloids and Surfaces. B, Biointerfaces* 57(1): 69–74. https://www.sciencedirect.com/science/article/abs/pii/S0927776507000215?via%3Dihub.

Marquez-Garcia, B, D Shaw, JW Cooper, B Karpinska, MD Quain, EM Makgopa, K Kunert, and CH Foyer. 2015. "Redox markers for drought-induced nodule senescence, a process occurring after drought-induced senescence of the lowest leaves in soybean (Glycine max)." *Annals of Botany* 116(4): 497–510.

Maruta, T, T Inoue, M Noshi, M Tamoi, Y Yabuta, K Yoshimura, T Ishikawa, and S Shigeoka. 2012. "Cytosolic ascorbate peroxidase 1 protects organelles against oxidative stress by wounding- and jasmonate-induced H2O2 in Arabidopsis plants." *Biochimica et Biophysica Acta* 1820(12): 1901–1907.

Maruyama, Kyonoshin, Yoh Sakuma, Mie Kasuga, Yusuke Ito, Motoaki Seki, Hideki Goda, Yukihisa Shimada, Shigeo Yoshida, Kazuo Shinozaki, and Kazuko Yamaguchi-Shinozaki. 2004. "Identification of cold-inducible downstream genes of the Arabidopsis DREB1A/CBF3 transcriptional factor using two microarray systems." *The Plant Journal: For Cell and Molecular Biology* 38(6): 982–993.

Miao, Yuchen, Dong Lv, Pengcheng Wang, Xue-Chen Wang, Jia Chen, Chen Miao, and Chun-Peng Song. 2006. "An Arabidopsis glutathione peroxidase functions as both a redox transducer and a scavenger in abscisic acid and drought stress responses." *The Plant Cell* 18(10): 2749–2766.

Mittler, Ron, and Barbara A Zilinskas. 1994. "Regulation of pea cytosolic ascorbate peroxidase and other antioxidant enzymes during the progression of drought stress and following recovery from drought." *The Plant Journal: For Cell and Molecular Biology* 5(3): 397–405. https://onlinelibrary.wiley.com/doi/abs/10.1111/j.1365-313X.1994.00397.x?sid=nlm%3Apubmed.

Mohamed, Heba I, Samia A Akladious, and Hossam S El-Beltagi. 2018. "Mitigation the harmful effect of salt stress on physiological, biochemical and anatomical traits by foliar spray with trehalose on wheat cultivars." *Fresenius Environment Bulletin* 27: 7054–7065.

Mohammadi, M, A Tavakoli, M Pouryousef, and E Mohseni Fard. 2020. "Study the effect of 24-epibrassinolide application on the Cu/Zn-SOD expression and tolerance to drought stress in common bean." *Physiology and Molecular Biology of Plants* 26(3): 459–474.

Munné-Bosch, Sergi. 2005. "The role of α-tocopherol in plant stress tolerance." *Journal of Plant Physiology* 162(7): 743–748.

Murshed, R, F Lopez-Lauri, and H Sallanon. 2013. "Effect of water stress on antioxidant systems and oxidative parameters in fruits of tomato (Solanum lycopersicon L, cv. Micro-tom)." *Physiology and Molecular Biology of Plants: An International Journal of Functional Plant Biology* 19(3): 363–378.

Nahar, K, M Hasanuzzaman, MM Alam, A Rahman, JA Mahmud, T Suzuki, and M Fujita. 2017. "Insights into spermine-induced combined high temperature and drought tolerance in mung bean: Osmoregulation and roles of antioxidant and glyoxalase system." *Protoplasma* 254(1): 445–460.

Nahar, Kamrun, Mirza Hasanuzzaman, Md Alam, and Masayuki Fujita. 2015. "Glutathione-induced drought stress tolerance in mung bean: Coordinated roles of the antioxidant defence and methylglyoxal detoxification systems." *AoB Plants* 7, plv069.

Nahar, Shamsun, Lakshminarayana R Vemireddy, Lingaraj Sahoo, and Bhaben Tanti. 2018. "Antioxidant protection mechanisms reveal significant response in drought-induced oxidative stress in some traditional rice of Assam, India." *Rice Science* 25(4): 185–196.

Nasrin, Shamima, Shukanta Saha, Hasna Hena Begum, and Rifat Samad. 2020. "Impacts of drought stress on growth, protein, proline, pigment content and antioxidant enzyme activities in rice (Oryza sativa L. var. BRRI dhan-24)." *Dhaka University Journal of Biological Sciences* 29(1): 117–123.

Naya, L, R Ladrera, J Ramos, EM González, C Arrese-Igor, FR Minchin, and M Becana. 2007. "The response of carbon metabolism and antioxidant defenses of alfalfa nodules to drought stress and to the subsequent recovery of plants." *Plant Physiology* 144(2): 1104–1114. http://www.plantphysiol.org/content/plantphysiol/144/2/1104.full.pdf.

Negi, NP, DC Shrivastava, V Sharma, and NB Sarin. 2015. "Overexpression of CuZnSOD from Arachis hypogaea alleviates salinity and drought stress in tobacco." *Plant Cell Reports* 34(7): 1109–1126. https://doi.org/10.1007/s00299-015-1770-4.

Niu, Yue, Yuping Wang, Ping Li, Feng Zhang, Heng Liu, and Guochang Zheng. 2013. "Drought stress induces oxidative stress and the antioxidant defense system in ascorbate-deficient vtc1 mutants of Arabidopsis thaliana." *Acta Physiologiae Plantarum* 35(4): 1189–1200.

Noctor, Graham, Amna Mhamdi, Sejir Chaouch, YI Han, Jenny Neukermans, B Marquez-Garcia, Guillaume Queval, and Christine H Foyer. 2012. "Glutathione in plants: An integrated overview." *Plant, Cell and Environment* 35(2): 454–484.

Noctor, Graham, and Christine H Foyer. 1998. "Ascorbate and glutathione: Keeping active oxygen under control." *Annual Review of Plant Physiology and Plant Molecular Biology* 49(1): 249–279.

Noctor, G, JP Reichheld, and CH Foyer. 2018. "ROS-related redox regulation and signaling in plants." *Seminars in Cell and Developmental Biology* 80: 3–12. https://doi.org/10.1016/j.semcdb.2017.07.013.

Palma, José M, Francisca Sevilla, Ana Jiménez, Luis A del Río, Francisco J Corpas, Paz Álvarez de Morales, and Daymi M Camejo. 2015. "Physiology of pepper fruit and the metabolism of antioxidants: Chloroplasts, mitochondria and peroxisomes." *Annals of Botany* 116(4): 627–636.

Pandey, Prachi, Jitender Singh, V Achary, and Mallireddy K Reddy. 2015. "Redox homeostasis via gene families of ascorbate-glutathione pathway." *Frontiers in Environmental Science* 3: 25.

Passardi, Filippo, David Longet, Claude Penel, and Christophe Dunand. 2004. "The class III peroxidase multigenic family in rice and its evolution in land plants." *Phytochemistry* 65(13): 1879–1893.

Podgórska, Anna, Maria Burian, and Bożena Szal. 2017. "Extra-cellular but extra-ordinarily important for cells: Apoplastic reactive oxygen species metabolism." *Frontiers in Plant Science* 8: 1353.

Pour-Aboughadareh, Alireza, Alireza Etminan, Mostafa Abdelrahman, Kadambot HM Siddique, and LP Tran. 2020. "Assessment of biochemical and physiological parameters of durum wheat genotypes at the seedling stage during polyethylene glycol-induced water stress." *Plant Growth Regulation*, 92(1), 81–93.

Pourcel, Lucille, Jean-Marc Routaboul, Véronique Cheynier, Loïc Lepiniec, and Isabelle Debeaujon. 2007. "Flavonoid oxidation in plants: From biochemical properties to physiological functions." *Trends in Plant Science* 12(1): 29–36. https://www.cell.com/trends/plant-science/fulltext/S1360-1385(06)00314-1?_returnURL=https%3A%2F%2Flinkinghub.elsevier.com%2Fretrieve%2Fpii%2FS1360138506003141%3Fshowall%3Dtrue.

Pourtaghi, Alireza, Farokh Darvish, Davod Habibi, Gorbane Nourmohammadi, and Jahanfar Daneshian. 2011. "Effect of irrigation water deficit on antioxidant activity and yield of some sunflower hybrids." *Australian Journal of Crop Science* 5(2): 197.

Puente-Garza, César A, Cristina Meza-Miranda, Desiree Ochoa-Martínez, and Silverio García-Lara. 2017. "Effect of in vitro drought stress on phenolic acids, flavonols, saponins, and antioxidant activity in Agave salmiana." *Plant Physiology and Biochemistry: PPB* 115: 400–407.

Pyngrope, Samantha, Kumari Bhoomika, and RS Dubey. 2013. "Reactive oxygen species, ascorbate-glutathione pool, and enzymes of their metabolism in drought-sensitive and tolerant indica rice (Oryza sativa L.) seedlings subjected to progressing levels of water deficit." *Protoplasma* 250(2): 585–600.

Rady, Mostafa M, Hussein EE Belal, Farouk M Gadallah, and Wael M Semida. 2020. "Selenium application in two methods promotes drought tolerance in Solanum Lycopersicum plant by inducing the antioxidant defense system." *Scientia Horticulturae* 266. http://www.ncbi.nlm.nih.gov/pubmed/109290.

Rahimi, Mehdi, Mojtaba Kordrostami, Mojtaba Mortezavi, and Sanam SafaeiChaeikar. 2020. "Identification of drought-responsive proteins of sensitive and tolerant tea (Camellia sinensis L) clones under normal and drought stress conditions." *Current Proteomics* 17(3): 227–240.

Ramel, Fanny, Simona Birtic, Stéphan Cuiné, Christian Triantaphylidès, Jean-Luc Ravanat, and Michel Havaux. 2012. "Chemical quenching of singlet oxygen by carotenoids in plants." *Plant Physiology* 158(3): 1267–1278. https://www.ncbi.nlm.nih.gov/pmc/articles/PMC3291260/pdf/1267.pdf.

Ratnayaka, HH, WT Molin, and TM Sterling. 2003. "Physiological and antioxidant responses of cotton and spurred anoda under interference and mild drought." *Journal of Experimental Botany* 54(391): 2293–2305.

Reddy, P Chandra Obul, G Sairanganayakulu, M Thippeswamy, P Sudhakar Reddy, MK Reddy, and Chinta Sudhakar. 2008. "Identification of stress-induced genes from the drought tolerant semi-arid legume crop horsegram (Macrotyloma uniflorum (Lam.) Verdc.) through analysis of subtracted expressed sequence tags." *Plant Science* 175(3): 372–384.

Rezayian, Maryam, Hassan Ebrahimzadeh, and Vahid Niknam. 2020. "Nitric oxide stimulates antioxidant system and osmotic adjustment in soybean Under drought stress." *Journal of Soil Science and Plant Nutrition*: 1–11.

Rosa, Sílvia B, Andréia Caverzan, Felipe K Teixeira, Fernanda Lazzarotto, Joaquim AG Silveira, Sérgio Luiz Ferreira-Silva, João Abreu-Neto, Rogério Margis, and Márcia Margis-Pinheiro. 2010. "Cytosolic APx knockdown indicates an ambiguous redox responses in rice." *Phytochemistry* 71(5–6): 548–558.

Rozita, Kabiri, Farahbakhsh Hassan, and Nasibi Fatemeh. 2012. "Salicylic acid ameliorates the effects of oxidative stress induced by water deficit in hydroponic culture of Nigella sativa." *Journal of Stress Physiology and Biochemistry* 8(3).

Sadak, MS, AR Abd El-Hameid, FSA Zaki, MG Dawood, and ME El-Awadi. 2020. "Physiological and biochemical responses of soybean (Glycine max L.) to cysteine application under sea salt stress." *Bulletin of the National Research Centre* 44(1): 1–13.

Saed-Moucheshi, Armin, Hooman Razi, Ali Dadkhodaie, Masoud Ghodsi, and Manoochehr Dastfal. 2019. "Association of biochemical traits with grain yield in triticale genotypes under normal irrigation and drought stress conditions." *Australian Journal of Crop Science* 13(2): 272.

Safronov, Omid, Jürgen Kreuzwieser, Georg Haberer, Mohamed S Alyousif, Waltraud Schulze, Naif Al-Harbi, Leila Arab, Peter Ache, Thomas Stempfl, and Joerg Kruse. 2017. "Detecting early signs of heat and drought stress in Phoenix dactylifera (date palm)." *PLoS One*, 12(6), e0177883.

Sahitya, U Lakshmi, MSR Krishna, R Deepthi, G Shiva Prasad, and D Kasim. 2018. "Seed antioxidants interplay with drought stress tolerance indices in Chilli (Capsicum annuum L) seedlings." *BioMed Research International* 2018.

Sahoo, Smita, Jay Prakash Awasthi, Ramanjulu Sunkar, and Sanjib Kumar Panda. 2017. "Determining glutathione levels in plants." In: *Plant Stress Tolerance*, 273–277. Springer.

Salekdeh, Gh Hosseini, Joel Siopongco, Leonard J Wade, Behzad Ghareyazie, and John Bennett. 2002. "Proteomic analysis of rice leaves during drought stress and recovery." *PROTEOMICS: International Edition* 2(9): 1131–1145.

Sallam, Ahmed, Ahmad M Alqudah, Mona FA Dawood, P Stephen Baenziger, and Andreas Börner. 2019a. "Drought stress tolerance in wheat and barley: Advances in physiology, breeding and genetics research." *International Journal of Molecular Sciences* 20(13): 3137.

Sallam, Ahmed, Ahmed Amro, Ammar Elakhdar, Mona FA Dawood, Yasser S Moursi, and P Stephen Baenziger. 2019b. "Marker–trait association for grain weight of spring barley in well-watered and drought environments." *Molecular Biology Reports* 46(3): 2907–2918. https://doi.org/10.1007/s11033-019-04750-6.

Sallam, Ahmed, Mervat Hashad, El-Sayed Hamed, and Mohamed Omara. 2015. "Genetic variation of stem characters in wheat and their relation to kernel weight under drought and heat stresses." *Journal of Crop Science and Biotechnology* 18(3): 137–146. https://doi.org/10.1007/s12892-015-0014-z.

Samarah, NH, AM Alqudah, JA Amayreh, and GM McAndrews. 2009. "The effect of late-terminal drought stress on yield components of four barley cultivars." *Journal of Agronomy and Crop Science* 195(6): 427–441.

Sarker, Umakanta, and Shinya Oba. 2018. "Catalase, superoxide dismutase and ascorbate-glutathione cycle enzymes confer drought tolerance of Amaranthus tricolor." *Scientific Reports* 8(1): 1–12.

Sayfzadeh, Saeed, and Majid Rashidi. 2011. "Response of antioxidant enzymes activities of sugar beet to drought stress." *Journal of Agricultural and Biological Science* 6(4): 27–33.

Schroeter, Hagen, Clinton Boyd, Jeremy PE Spencer, Robert J Williams, Enrique Cadenas, and Catherine Rice-Evans. 2002. "MAPK signaling in neurodegeneration: Influences of flavonoids and of nitric oxide." *Neurobiology of Aging* 23(5): 861–880. https://www.sciencedirect.com/science/article/abs/pii/S0197458002000751?via%3Dihub.

Schweiggert, Ralf M, Jochen U Ziegler, Ehab MR Metwali, Fouad H Mohamed, Omar A Almaghrabi, Naif M Kadasa, and Reinhold Carle. 2017. "Carotenoids in mature green and ripe red fruits of tomato (Solanum Lycopersicum L.) grown under different levels of irrigation." *Archives of Biological Sciences* 69(2): 305–314.

Sečenji, Maria, Eva Hideg, Attila Bebes, and János Györgyey. 2010. "Transcriptional differences in gene families of the ascorbate-glutathione cycle in wheat during mild water deficit." *Plant Cell Reports* 29(1): 37–50.

Seminario, A, L Song, A Zulet, HT Nguyen, EM González, and E Larrainzar. 2017. "Drought stress causes a reduction in the biosynthesis of ascorbic acid in soybean plants." *Frontiers in Plant Science* 8: 1042.

Sevilla, F, D Camejo, A Ortiz-Espín, A Calderón, JJ Lázaro, and A Jiménez. 2015. "The thioredoxin/peroxiredoxin/sulfiredoxin system: Current overview on its redox function in plants and regulation by reactive oxygen and nitrogen species." *Journal of Experimental Botany* 66(10): 2945–2955.

Shan, CJ, YY Jin, Y Zhou, and H Li. 2020. "Nitric oxide participates in the regulation of ascorbate-glutathione cycle and water physiological characteristics of Arabidopsis thaliana by NaHS." *Photosynthetica* 58(1): 80–86.

Shan, Changjuan, Shengli Zhang, and Xingqi Ou. 2018. "The roles of H 2 S and H 2 O 2 in regulating AsA-GSH cycle in the leaves of wheat seedlings under drought stress." *Protoplasma* 255(4): 1257–1262.

Shankar, Rama, Annapurna Bhattacharjee, and Mukesh Jain. 2016. "Transcriptome analysis in different rice cultivars provides novel insights into desiccation and salinity stress responses." *Scientific Reports* 6: 23719.

Shao, Hong-Bo, Li-Ye Chu, C Abdul Jaleel, P Manivannan, R Panneerselvam, and Ming-An Shao. 2009. "Understanding water deficit stress-induced changes in the basic metabolism of higher plants—Biotechnologically and sustainably improving agriculture and the ecoenvironment in arid regions of the globe." *Critical Reviews in Biotechnology* 29(2): 131–151. https://doi.org/10.1080/07388550902869792; https://www.tandfonline.com/doi/full/10.1080/07388550902869792.

Sharma, Rahul, Parivartan Vishal, Sanjana Kaul, and Manoj K Dhar. 2017. "Epiallelic changes in known stress-responsive genes under extreme drought conditions in Brassica juncea (L.) Czern.. " *Plant Cell Reports* 36(1): 203–217.

Sheoran, Sonia, Vidisha Thakur, Sneh Narwal, Rajita Turan, HM Mamrutha, Virender Singh, Vinod Tiwari, and Indu Sharma. 2015. "Differential activity and expression profile of antioxidant enzymes and physiological changes in wheat (Triticum aestivum L.) under drought." *Applied Biochemistry and Biotechnology* 177(6): 1282–1298.

Shirani, M, KN Afzali, S Jahan, V Strezov, and M Soleimani-Sardo. 2020. "Pollution and contamination assessment of heavy metals in the sediments of Jazmurian playa in southeast Iran." *Scientific Reports* 10(1), 1–11.

Signorelli, Santiago, Francisco J Corpas, Omar Borsani, Juan B Barroso, and Jorge Monza. 2013. "Water stress induces a differential and spatially distributed nitro-oxidative stress response in roots and leaves of Lotus japonicus." *Plant Science: An International Journal of Experimental Plant Biology* 201–202: 137–146. https://www.sciencedirect.com/science/article/abs/pii/S016894521200249X?via%3Dihub.

Singh, Namrata, Sanjay Singh, and CAK Singh. 2020. "Stress physiology and metabolism in hybrid rice. III. Puddling and soil compaction regulate water status and internal metabolism during drought." *Journal of Crop Improvement* 34(6), 741–766.

Singh, S, AK Gupta, and N Kaur. 2012. "Differential responses of antioxidative defence system to long-term field drought in wheat (Triticum aestivum L.) genotypes differing in drought tolerance." *Journal of Agronomy and Crop Science* 198(3): 185–195.

Sofo, Adriano, Antonio Scopa, Maria Nuzzaci, and Antonella Vitti. 2015. "Ascorbate peroxidase and catalase activities and their genetic regulation in plants subjected to drought and salinity stresses." *International Journal of Molecular Sciences* 16(6): 13561–13578.

Srivalli, B, Viswanathan Chinnusamy, and Renu Khanna-Chopra. 2003. "Antioxidant defense in response to abiotic stresses in plants." *Journal of Plant Biology-New Delhi* 30(2): 121–140.

Sudrajat, Dede J, Iskandar Z Siregar, Nurul Khumaida, Ulfah J Siregar, and Irdika Mansur. 2015. "Adaptability of white jabon (Anthocephalus cadamba MIQ.) seedling from 12 populations to drought and waterlogging." *AGRIVITA, Journal of Agricultural Science* 37(2): 130–143.

Suneja, Yadhu, Anil Kumar Gupta, and Navtej Singh Bains. 2017. "Bread wheat progenitors: Aegilops tauschii (DD genome) and Triticum dicoccoides (AABB genome) reveal differential antioxidative response under water stress." *Physiology and Molecular Biology of Plants: An International Journal of Functional Plant Biology* 23(1): 99–114.

Sunkar, Ramanjulu, Avnish Kapoor, and Jian-Kang Zhu. 2006. "Posttranscriptional induction of two Cu/Zn superoxide dismutase genes in Arabidopsis is mediated by downregulation of miR398 and important for oxidative stress tolerance." *The Plant Cell* 18(8): 2051–2065. https://doi.org/10.1105/tpc.106 .041673. https://www.ncbi.nlm.nih.gov/pubmed/16861386; https://www.ncbi.nlm.nih.gov/pmc/articles/ PMC1533975/; https://www.ncbi.nlm.nih.gov/pmc/articles/PMC1533975/pdf/tpc1802051.pdf.

Suzuki, Nobuhiro, Rosa M Rivero, Vladimir Shulaev, Eduardo Blumwald, and Ron Mittler. 2014. "Abiotic and biotic stress combinations." *New Phytologist* 203(1): 32–43.

Taha, RS, HF Alharby, AA Bamagoos, RA Medani, and MM Rady. 2020. "Elevating tolerance of drought stress in Ocimum basilicum using pollen grains extract; a natural biostimulant by regulation of plant performance and antioxidant defense system." *South African Journal of Botany* 128: 42–53.

Taylor, Nicolas L, Joshua L Heazlewood, David A Day, and A Harvey Millar. 2005. "Differential impact of environmental stresses on the pea mitochondrial proteome." *Molecular and Cellular Proteomics: MCP* 4(8): 1122–1133.

Thabet, Samar G, and Ahmad M Alqudah. 2019. "Crops and drought." In: *eLS*, 1–8. John Wiley & Sons, Ltd.

Tina, RR, XR Shan, Y Wang, SY Guo, B Mao, W Wang, HY Wu, and TH Zhao. 2017. "Response of antioxidant system to drought stress and re-watering in alfalfa during branching." IOP Conference Series. Earth and Environmental Science.

Triantaphylidès, Christian, Markus Krischke, Frank Alfons Hoeberichts, Brigitte Ksas, Gabriele Gresser, Michel Havaux, Frank Van Breusegem, and Martin Johannes Mueller. 2008. "Singlet oxygen is the major reactive oxygen species involved in photooxidative damage to plants." *Plant Physiology* 148(2): 960–968. https://www.ncbi.nlm.nih.gov/pmc/articles/PMC2556806/pdf/pp1480960.pdf.

Ushimaru, Takashi, Tomofumi Nakagawa, Yuko Fujioka, Katsue Daicho, Makiko Naito, Yuzo Yamauchi, Hideko Nonaka, Katsumi Amako, Kazuki Yamawaki, and Norio Murata. 2006. "Transgenic Arabidopsis plants expressing the rice dehydroascorbate reductase gene are resistant to salt stress." *Journal of Plant Physiology* 163(11): 1179–1184.

Usmani, Muhammad Munir, Fahim Nawaz, Sadia Majeed, Muhammad Asif Shehzad, Khawaja Shafique Ahmad, Gulzar Akhtar, Muhammad Aqib, and Rana Nauman Shabbir. 2020. "Sulfate-mediated drought tolerance in Maize involves regulation at physiological and biochemical levels." *Scientific Reports* 10(1): 1–13.

Uzilday, B, I Turkan, AH Sekmen, R Ozgur, and HC Karakaya. 2012. "Comparison of ROS formation and antioxidant enzymes in Cleome gynandra (C4) and Cleome spinosa (C3) under drought stress." *Plant Science* 182: 59–70.

Vaseghi, Mohamad-Javad, Kamel Chibani, Wilena Telman, Michael Florian Liebthal, Melanie Gerken, Helena Schnitzer, Sara Mareike Mueller, and Karl-Josef Dietz. 2018. "The chloroplast 2-cysteine peroxiredoxin functions as thioredoxin oxidase in redox regulation of chloroplast metabolism." *eLife* 7: e38194.

Vital, Shantel A, Rocky W Fowler, Alvarro Virgen, Dalton R Gossett, Stephen W Banks, and Juan Rodriguez. 2008. "Opposing roles for superoxide and nitric oxide in the NaCl stress-induced upregulation of antioxidant enzyme activity in cotton callus tissue." *Environmental and Experimental Botany* 62(1): 60–68.

Voss, I, B Sunil, R Scheibe, and AS Raghavendra. 2013. "Emerging concept for the role of photorespiration as an important part of abiotic stress response." *Plant Biology* 15(4): 713–722. https://onlinelibrary.wiley .com/doi/full/10.1111/j.1438-8677.2012.00710.x.

Wang, W, MX Xia, J Chen, R Yuan, FN Deng, and FF Shen. 2016. "Gene expression characteristics and regulation mechanisms of superoxide dismutase and its physiological roles in plants under stress." *Biochemistry. Biokhimiia* 81(5): 465–480. https://link.springer.com/content/pdf/10.1134/S00062979160 50047.pdf.

Wang, Wenbin, Xiangpo Qiu, Ho Soo Kim, Yanxin Yang, Dianyun Hou, Xuan Liang, and Sang-Soo Kwak. 2020. "Molecular cloning and functional characterization of a sweetpotato chloroplast IbDHAR3 gene in response to abiotic stress." *Plant Biotechnology Reports* 14(1): 9–19.

Wang, Xinpeng, Hualong Liu, F Yu, Bowen Hu, Yan Jia, Hanjing Sha, and Hongwei Zhao. 2019. "Differential activity of the antioxidant defence system and alterations in the accumulation of osmolyte and reactive oxygen species under drought stress and recovery in rice (Oryza sativa L.) tillering." *Scientific Reports* 9(1): 8543.

Wei, Tao, Kejun Deng, Dongqing Liu, Yonghong Gao, Yu Liu, Meiling Yang, Lipeng Zhang, Xuelian Zheng, Chunguo Wang, Wenqin Song, C Chen, and Y Zhang. 2016. "Ectopic expression of DREB transcription factor, AtDREB1A, confers tolerance to drought in transgenic Salvia miltiorrhiza." *Plant and Cell Physiology* 57(8): 1593–1609.

Willekens, Hilde, Sangpen Chamnongpol, Mark Davey, Martina Schraudner, Christian Langebartels, Marc Van Montagu, Dirk Inzé, and Wim Van Camp. 1997. "Catalase is a sink for H2O2 and is indispensable for stress defence in C3 plants." *The EMBO Journal* 16(16): 4806–4816. https://www.ncbi.nlm.nih.gov/pmc/articles/PMC1170116/pdf/004806.pdf.

Wilson, Pip B, Gonzalo M Estavillo, Katie J Field, Wannarat Pornsiriwong, Adam J Carroll, Katharine A Howell, Nick S Woo, Janice A Lake, Steven M Smith, A Harvey Millar, S von Caemmerer, and BJ Pogson. 2009. "The nucleotidase/phosphatase SAL1 is a negative regulator of drought tolerance in Arabidopsis." *The Plant Journal: For Cell and Molecular Biology* 58(2): 299–317.

Wu, Yi-Ru, Yi-Chen Lin, and Huey-wen Chuang. (2016). "Laminarin modulates the chloroplast antioxidant system to enhance abiotic stress tolerance partially through the regulation of the defensin-like gene expression." *Plant Science* 247: 83–92.

Xie, H, DH Yang, H Yao, G Bai, YH Zhang, and BG Xiao. 2016. "iTRAQ-based quantitative proteomic analysis reveals proteomic changes in leaves of cultivated tobacco (Nicotiana tabacum) in response to drought stress." *Biochemical and Biophysical Research Communications* 469(3): 768–775.

Xie, Tenglong, Wanrong Gu, Liguo Zhang, Lijie Li, Danyang Qu, Caifeng Li, Yao Meng, Jing Li, Shi Wei, and Wenhua Li. 2018. "Modulating the antioxidant system by exogenous 2-(3, 4-dichlorophenoxy) triethylamine in maize seedlings exposed to polyethylene glycol-simulated drought stress." *PLOS ONE* 13(9), e0203626.

Xiong, Haiyan, J Yu, Jinli Miao, Jinjie Li, Hongliang Zhang, Xin Wang, Pengli Liu, Yan Zhao, Chonghui Jiang, Zhigang Yin, Y Li, Y Guo, B Fu, W Wang, Z Li, J Ali, and Z Li. 2018. "Natural variation in OsLG3 increases drought tolerance in rice by inducing ROS scavenging." *Plant Physiology* 178(1): 451–467.

Xu, Jiayang, Yuyi Zhou, Zicheng Xu, Zheng Chen, and Liusheng Duan. 2020. "Combining physiological and metabolomic analysis to unravel the regulations of coronatine alleviating water stress in tobacco (Nicotiana tabacum L.)." *Biomolecules* 10(1): 99.

Xu, Jing, Xiao-Juan Xing, Yong-Sheng Tian, Ri-He Peng, Yong Xue, Wei Zhao, and Quan-Hong Yao. 2015. "Transgenic Arabidopsis plants expressing tomato glutathione S-transferase showed enhanced resistance to salt and drought stress." *PLOS ONE* 10(9), e0136960.

Yang, Yan, Chao Han, Qing Liu, Bo Lin, and Jianwen Wang. 2008. "Effect of drought and low light on growth and enzymatic antioxidant system of Picea asperata seedlings." *Acta Physiologiae Plantarum* 30(4): 433–440.

Yang, Yang, Chuntao Yin, Weizhi Li, and Xudong Xu. 2008. "α-Tocopherol is essential for acquired chill-light tolerance in the cyanobacterium Synechocystis sp. strain PCC 6803." *Journal of Bacteriology* 190(5): 1554–1560. https://www.ncbi.nlm.nih.gov/pmc/articles/PMC2258665/pdf/1577-07.pdf.

Ye, Nenghui, Guohui Zhu, Yinggao Liu, Yingxuan Li, and Jianhua Zhang. 2011. "ABA controls H$_2$O$_2$ accumulation through the induction of OsCATB in rice leaves under water stress." *Plant and Cell Physiology* 52(4): 689–698.

Yildizli, Aytunç, Sertan Çevik, and Serpil Ünyayar. 2018. "Effects of exogenous myo-inositol on leaf water status and oxidative stress of Capsicum annuum under drought stress." *Acta Physiologiae Plantarum* 40(6): 122.

Yildiztugay, Evren, Ceyda Ozfidan-Konakci, Mustafa Kucukoduk, and Ismail Turkan. 2020. "Flavonoid naringenin alleviates short-term osmotic and salinity stresses through regulating photosynthetic machinery and chloroplastic antioxidant metabolism in Phaseolus vulgaris." *Frontiers in Plant Science* 11: 682.

Yin, Mingzhu, Yanping Wang, Lihua Zhang, Jinzhu Li, Wenli Quan, Li Yang, Qingfeng Wang, and Zhulong Chan. 2017. "The Arabidopsis Cys2/His2 zinc finger transcription factor ZAT18 is a positive regulator of plant tolerance to drought stress." *Journal of Experimental Botany* 68(11): 2991–3005.

Zaffagnini, Mirko, Mariette Bedhomme, Hayam Groni, Christophe H Marchand, Carine Puppo, Brigitte Gontero, Corinne Cassier-Chauvat, Paulette Decottignies, and SD Lemaire. 2012. "Glutathionylation in the photosynthetic model organism Chlamydomonas reinhardtii: A proteomic survey." *Molecular and Cellular Proteomics: MCP* 11(2), M111–014142.

Zagdańska, Barbara, and Krzysztof Wiśniewski. 1996. "Changes in the thiol/disulfide redox potential in wheat leaves upon water deficit." *Journal of Plant Physiology* 149(3–4): 462–465.

Zagorchev, Lyuben, Charlotte E Seal, Ilse Kranner, and Mariela Odjakova. 2013. "A central role for thiols in plant tolerance to abiotic stress." *International Journal of Molecular Sciences* 14(4): 7405–7432. https://res.mdpi.com/d_attachment/ijms/ijms-14-07405/article_deploy/ijms-14-07405.pdf.

Zhang, C, H Yang, W Wu, and W Li. 2017. "Effect of drought stress on physiological changes and leaf surface morphology in the blackberry." *Brazilian Journal of Botany* 40(3): 625–634.

Zhang, X, Z Yang, Z Li, F Zhang, and L Hao. 2020. "Effects of drought stress on physiology and antioxidative activity in two varieties of Cynanchum thesioides." *Brazilian Journal of Botany* 43(1): 1–10. https://doi.org/10.1007/s40415-019-00573-8.

Zhang, L, M Wu, Y Teng, S Jia, D Yu, T Wei, C Chen, and W Song. 2018. "Overexpression of the Glutathione Peroxidase 5 (RcGPX5) Gene From Rhodiola crenulata Increases Drought Tolerance in Salvia miltiorrhiza." *Frontiers in Plant Science* 9: 1950.

Zhang, Yiming, Jinfen Yang, Shaoyun Lu, Jiongliang Cai, and Zhenfei Guo. 2008. "Overexpressing SgNCED1 in tobacco increases ABA level, antioxidant enzyme activities, and stress tolerance." *Journal of Plant Growth Regulation* 27(2): 151–158.

Zhu, Yanmei, Xinglu Luo, Gul Nawaz, Jingjing Yin, and Jingni Yang. 2020. "Physiological and Biochemical Responses of four cassava cultivars to drought stress." *Scientific* Reports 10(1): 1–12.

Zou, Jun-Jie, Xi-Dong Li, Disna Ratnasekera, Cun Wang, Wen-Xin Liu, Lian-Fen Song, Wen-Zheng Zhang, and Wei-Hua Wu. 2015. "Arabidopsis calcium-dependent protein Kinase8 and Catalase3 function in abscisic acid-mediated signaling and H2O2 homeostasis in stomatal guard cells under drought stress." *The Plant Cell* 27(5): 1445–1460. http://www.plantcell.org/content/plantcell/27/5/1445.full.pdf.

13 Extreme Temperature-Induced Osmotic and Oxidative Stress
Defensive Role of Organic Solutes and Antioxidants

Amandeep Cheema and Neera Garg

CONTENTS

13.1 INTRODUCTION

Among the various environmental factors, temperature is an important aspect which significantly affects all the metabolic processes in plants. Plant species have their temperature range (also called the optimum range) within which they show maximum growth and development. Temperatures above or below this optimum range substantially affect the productivity of various agricultural and horticultural crops. These sub-optimum ranges differ from plant to plant and are usually categorized into two types: (i) cold stress, comprising of a) chilling stress (lower than freezing) and b) freezing stress (more than freezing), and (ii) heat stress (Źróbek-Sokolnik 2012). Plants growing under high temperature display less biomass as compared to the ones at low temperature (Kim et al. 2007).

Plants get affected at physiological, biochemical and molecular levels under temperature stress. During cold stress, the soil water freezes and water which is present in the plant cells expands and turns to ice, which can cause the cell membrane to rupture which directs towards cell death. During high temperatures, there is a loss of water from the soil by evaporation, decreasing the availability of water to plant roots. Moreover, heat in the surroundings also enhances the process of transpiration in plants, resulting in water loss. In such conditions, cell walls get disrupted, the permeability of the membrane changes and protoplasmic streaming gets decreased, resulting in more electrolytic leakage causing an imbalance in osmosis and water-deficit (Farooq et al. 2009; Wang et al. 2016; Hussain et al. 2018). These disruptions consequently cause degradation of chlorophyll, reduction in net photosynthesis and decrease in carbon dioxide exchange. Moreover, under stressed conditions, the generation of reactive oxygen species (ROS) cause unspecific oxidations of proteins and damage of membrane lipids, or lead to injury in DNA (Kamiński et al. 2012). These reactive radicals include molecules that can be either free, such as hydroxyl radical ($\bullet$OH) and superoxide anion ($O_2^{\bullet-}$), or non-free, such as hydrogen peroxide (H_2O_2) and singlet oxygen ($^1O_2$).

To sustain the proper cell volume and turgor pressure as well as protect intracellular macromolecules from oxygen radicals, plants accumulate low molecular weight organic solutes (or osmolytes) (Lee et al. 2004; Suprasanna et al. 2016). These molecules can be classified into four main classes, as (a) amino acids, (b) betaines and related compounds, (c) polyols and non-reducing sugars and (d) polyamines. In addition to these solutes, plants have an antioxidant defensive system, which either breaks down and removes free radicals from the cell (enzymatic) or interrupts free radical chain reactions (non-enzymatic) thus protecting cells from damage. Both of these arms work in harmony to safeguard the normal functioning of cell organelles. Enzymatic antioxidants involve superoxide dismutase (SOD), catalase (CAT), ascorbate peroxidase (APX), mono-dehydroascorbate reductase (MDHAR), dehydroascorbate reductase (DHAR) and glutathione reductase (GR), whereas non-enzymatic antioxidants comprise ascorbic acid (AA), glutathione (GSH), alpha-tocopherol, carotenoids and phenolics. Enzymatic antioxidants play a crucial part in overcoming stress-induced responses by converting more toxic forms to lesser ones, whereas non-enzymatic antioxidants protect the photosynthetic apparatus by scavenging free oxygen radicals (Sharma et al. 2012). The plants which have acclimatized themselves in stressed conditions have been found to maintain significantly high levels of antioxidants as well as responsible gene expression. It is therefore important to understand the relationship between antioxidant enzyme activity and the gene encoding these enzymes to know about molecular mechanisms that contribute towards temperature tolerance (Ahanger et al. 2017).

Therefore, this chapter emphasizes the defensive roles of organic solutes and antioxidants in enabling plants to endure these extreme temperature regimes.

13.2 TEMPERATURE AND ITS EXTREMES

The metabolism of plants gets affected by various environmental factors, of which temperature plays a major role. The global temperature rise, as reported by the United States Environmental Protection Agency (EPA), was revealed as being an increase in temperatures (0.85 °C between 1880 and 2012 in the ocean and land surface) during the last 30 years (IPCC 2007). Every plant species has a specific temperature limit, in which they can grow as well as develop profusely and is represented by optimum, minimum and maximum (Hatfield et al. 2011). Temperature fluctuations occur naturally during growth and reproduction in plants. However, extreme variations (either too low or too high), result in the alteration of physiological, biochemical and molecular processes of plants as reported by various researchers (Barlow et al. 2015; Yu et al. 2020). Clarifications of mechanisms by which temperature stress causes damage are important to know, as well as responses by which plants cope with adverse temperature conditions. Therefore, the stresses and their physiological effects on plants are explained below.

13.2.1 COLD STRESS

Cold stress involves the effects of freezing, i.e. below 0°C, and chilling i.e. between 0–15°C, which are represented by various alterations in plants at physiological, biochemical and cytological levels. (Abdel Latef and Chaoxing 2011; Liu et al. 2018). Ding et al. (2019) reported physiological effects of chilling in *Arabidopsis thaliana* as a reduction in plant growth, stems splitting, abnormal curling, surface lesions on leaves and fruits, internal discoloration, etc. Due to cold stress-induced dehydration, membrane structure and composition changes, which lower the protoplasmic streaming and enhance electrolyte leakage as well as plasmolysis (Yadav 2010). These stress-induced changes can be reversible or irreversible depending upon the length of time of exposure to low temperature. Crops of tropical, sub-tropical and temperate regions, such as *Oryza sativa*, *Zea mays* and *Cicer* sp, are more sensitive to low temperature, which is due to their adaptation in above 0° temperatures as compared to below it (Sanghera et al. 2011). Moreover, other crops like *Cucurbita* sp, *Glycine max* Linn. and various cereal crops suffered high losses in yield under cold stress which, according to Thakur et al. (2010), is due to delay in the heading, thus resulting in pollen sterility and decline in grain yield of crops. Low temperatures have been found to induce changes in the expression of many coded and non-coded genes. For example, in *Ipomea batatas*, 51,989 unigenes were found to be differentially expressed during chilling stress and recovery, 3,216 differentially expressed genes (DEGs) were identified in tuberous roots stored at 13 °C and 4 °C, and 27,636 unigenes were found to differentially expressed during cold storage (Ji et al. in 2017, 2019, 2020). Genes involved in lipid metabolism were also found to get differentially expressed under low temperature (Xie et al. 2017).

13.2.2 HEAT STRESS

Heat stress is referred to as a rise in temperature more than the threshold value for a period of time, which is enough to make an unalterable decline in plant growth (Wahid et al. 2007). High temperatures negatively affect physiological processes, while extreme damage occurs in the form of the death of cells, tissues, organs and finally the whole plant (Argosubekti 2020). The rising concentration of greenhouse gases (30% in carbon dioxide and 150% in methane over the past 250 years) is the main cause behind a rise of 0.2 °C in global temperatures every ten years (Friedlingstein et al. 2014). These conditions result in a decline in growth as well as productivity of plants greater than any of the other biotic or abiotic factors (Wahid et al. 2007; Porter 2005). For example, with a rise in each degree Celsius of global temperature, wheat production was found to fall by 6% (Asseng et al. 2015). High temperature stress may cause a disturbance in the photosynthetic process by increasing chlorophyllase activity, damaging photosynthetic pigments, reducing the activity of PSII and impairing the regeneration capacity of RuBP (Wang et al. 2018). It also destroys chlorophyll content and reduces the activities of the enzymes involved in sugar metabolism. These damages minimize relative growth as well as the rate of net assimilation in sugarcane, millet and maize and under high temperature as reported by Wahid et al. (2007) and Ashraf and Hafeez (2007). Moreover, Maestri et al. (2002) found a substantial reduction in protein, oil and starch contents in cereals under high temperatures. Furthermore, high temperatures had been observed to negatively affect stomatal density, pore spaces and stomatal conductance, resulting in a major loss of water from plants (Sharma et al. 2015). Besides this, a significant decrease in the stages of meiosis, fertilization and growth of fertilized embryo was observed by Camejo et al. (2005) in *Lycopersicum esculentum*, resulting in yield reduction. The reproductive phase of development is a sensitive stage to high temperature stress in several crops such as wheat, chickpea, maize and Sorghum (Cairns et al. 2012; Kaushal et al. 2013; Djanaguiraman et al. 2014; Singh et al. 2016; Dwivedi et al. 2017). The reproductive processes involving the viability of pollen and stigma, growth of pollen tube and early development of embryo were found to be sensitive to heat stress (Giorno et al. 2013).

13.3 OSMOTIC AND OXIDATIVE STRESS INDUCED BY TEMPERATURE

13.3.1 OSMOTIC STRESS

Water is commonly considered the most vital molecule for the sustenance of life. The soil water, which is available to plants, can be found in the porous regions between soil particles. Under high temperatures, this water evaporates or freezes at -2°C, under freezing stress, making the plant deprived of its water source. Moreover, under temperature extremes, either high or low, internal water also gets lost during the excretion of wastes or from osmosis into concentrated aqueous surroundings (Yancey 2005). Plants lose water under high temperatures due to transpiration or freezing causing disruption of the structure as well as the function of cells and finally dehydration (Yadav 2010). Prolonged exposure to heat stress causes a decrease in chlorophyll content and electron transport. Moreover, it damages proteins, sets free magnesium and calcium ions, increases amylolytic activity and causes disruption in the transport of assimilates (Kozłowska et al. 2007; Zandalinas et al. 2016). On the opposite side, low temperature causes abnormal organization of phospholipids thus interfering with normal membrane structure and change in the uptake as well as absorption of water by the plant, thus declining the hydraulic conductance and water potential of the cells (Masouleh et al. 2019).

13.3.2 Oxidative Stress

Oxidative stress is an outcome of most of the abiotic calamities, which is initiated by the formation of ROS. These reactive species include singlet oxygen ($^1O_2$), superoxide radical ($O_2\bullet-$), hydrogen peroxide (H_2O_2) and hydroxyl radical (OH$\bullet$). These ROS are produced in the chloroplast or mitochondria, either enzymatically or non-enzymatically, and damage lipids, proteins as well as nucleic acids (Apel and Hirt 2004). Temperature stress either increases the synthesis of ROS or decreases the ability of the cell to scavenge these reactive species, resulting in disturbance in homeostatic balance inside the cell and promotion of peroxidation of lipid (Jewell et al. 2010; Tuteja et al. 2012). Lipid peroxidation results in the generation of lipid-derived radicals which exaggerates the stressed condition and damaged protein as well as nucleic acids. ROS mainly attack at these two sites of phospholipid membrane: first at the double bond between two carbon atoms and second ester linkage amid glycerol and fatty acids. Polyunsaturated fatty acids (PUFA) are the main attacking spots for ROS damage in the plasma membrane of plants, and their peroxidation leads to breakage of the chain which makes the membrane more fluid and permeable, affecting the turgidity (Das and Roychoudhary 2014). Reaction of oxygen with unsaturated lipids generates a variety of oxidation products of which malondialdehyde (MDA) has been examined as the most mutagenic. MDA content was found to progressively accumulate in chickpea cultivar (Canıtez 87) under chilling stress (Turan and Ekmekci 2011), which was also in accordance with reports on chickpea (Nayyar et al. 2005) and rice (Morsy et al. 2007) seedlings. Moreover, the increase in lipid membrane damage through increased MDA contents was observed under high temperature in cotton leaves (Sarwar et al. 2019). Various researchers had reported alteration of permeability in the membrane, chloroplast structure disintegration and membrane injury in chickpea, banana and soybean due to oxidative stress in low temperature conditions (Croser et al. 2003; Nayyar et al. 2005; Tambussi et al. 2004; Chinnusamy et al. 2007). Exposure of high temperature to six varieties of *Lens culnaris* resulted in peroxidation of lipid and loss in membrane integrity (Chakraborty and Pradhan 2012; Kipp and Boyle 2013). ROS cause protein oxidation by causing modification in proteins either directly or indirectly under stressful conditions. ROS modulate proteins' activity directly through the formation of the disulfide bond, nitrosylation, glutathionylation, carbonylation or indirectly by conjugating with fatty acid degraded products (Yamauchi et al. 2008). As a result of ROS generation, manipulation of a specific site in amino acid occurs, causing alteration of electric charge, aggregation of cross-linked reaction products and fragmentation of the peptide chain, thereby increasing the vulnerability of proteins for oxidation. In general, greater concentrations of carbonylated proteins are produced in oxidative stress as compared to normal conditions (Romero-Puertas et al. 2002; Møller and Kristensen 2004).

Nuclear DNA of plants is well protected by histone proteins, however, mitochondrial and chloroplast DNA lack these proteins and are closely located to the ROS generating system, therefore more prone to oxidative injury as compared to nuclear DNA (Richter 1992). Under stressed conditions, both sugar and nitrogenous base parts of DNA get exposed to damage by ROS. ROS attack

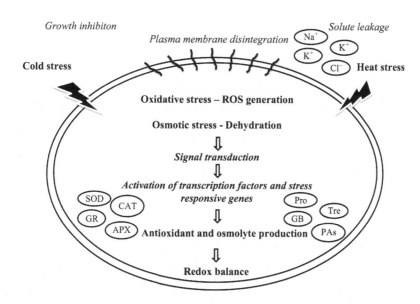

FIGURE 13.1 Schematic representation of stressed conditions created by temperature extremes in plant cells – inhibition of growth, disintegration of plasma membrane, leakage of solutes, etc. Oxidative stress is caused by reactive oxygen species (ROS), and osmotic stress is caused by dehydration due to loss of water and freezing of water during heat and cold stress respectively. These stressed conditions transduce signals for the expression of antioxidants and osmolyte production, which thereby maintain the redox balance by reversing the process of osmosis, scavenging oxygen radicals or converting highly toxic forms of ROS into less toxic ones. SOD, superoxide dismutase; CAT, catalase; GR, glutathione reductase; APX, ascorbate peroxidase; Pro, Proline; GB, Glycine Betaine; Tre, Trehalose; PA, Polyamines; Na^+, Sodium ion; K^+, Potassium ion; Cl^-, Chloride ion.

selectively modify G:C sites, thus breaking the strand and altering the sequence (Figure 13.1). Besides this, ROS also cause strand breakage due to the production of a radical called deoxyribose by extracting H atom from the fourth carbon atom of deoxyribose (Evans et al. 2004).

13.4 DEFENSE RESPONSES

To overcome stressed conditions, plants possess a defense system consisting of large quantities of different types of organic solutes (or osmolytes) to regulate osmotic balance and antioxidants (enzymatic and non-enzymatic) to regulate redox homeostasis. The role of these systems has been briefly described below.

13.4.1 ROLE OF ORGANIC SOLUTES IN DEFENSE RESPONSE

Organic solutes have low molecular weight and are highly soluble as well as non-toxic at high cellular concentrations. These solutes protect plants from oxidative damage by preventing electrolyte leakage, regulating cellular turgor and stabilizing membranes, thereby bringing concentrations of ROS within normal ranges (Hayat et al. 2012). Chemically, these molecules are categorized into four main classes: amino acids and their derivatives, betaines and related compounds, sugars, and polyamines (Kumar et al. 2018). Concentrations of these osmolytes were found to be higher in cold-tolerant seedlings of *Pinus halepensis* than the sensitive ones (Taïbi et al. 2018).

13.4.1.1 Amino Acids and Their Derivatives

These include proline (Pro), alanine, glycine, ornithine, γ-aminobutyric acid (GABA), aspargine, pipecolic acid and glutamate citrulline, which act as precursors for most of the osmolytes by preventing damage to membrane and ionic-toxicity. They are widely found in many of the angiosperm plant families including some important crop families like Poaceae, Brassicaceae and Fabaceae (Suprasanna et al. 2016). Amongst all amino acids, Pro has been credited as the most vital amino acid molecule, playing a highly beneficial role in unfavorable conditions. Pro is a type of amino acid that has been recorded to serve as an osmolyte, usually biosynthesized in high amounts in response to stresses. Stressed conditions consequently over-synthesize Pro in plants, which is found to be closely connected with improved tolerance to stress via maintaining turgor of cells, membrane stability and effective controlling over the production and scavenging of ROS, which eventually protects the plant cells from oxidative damage (Kumar et al. 2015; Harsh et al. 2016). Due to its zwitterion character, under stressful conditions Pro accumulates to a high concentration in cell cytoplasm as a result of either or both de novo synthesis and decline in degradation (dos Reis et al. 2012). Posmyk and Janas (2007) reported overcoming the chilling induced oxidative stress in *Vigna radiate* by supplementing Pro exogenously, which acts as a source of nitrogen (N) and carbon (C), scavenging oxygen radicals and stabilizing membranes, thus improving seedling growth. Besides this, Pro acts as a primer for the activity of C3 cycle and assists plants in recovering from osmotic stress. Yuan et al. (2011) reported thermotolerance in pretreated seedlings of Freesia due to a high level of proline which could maintain osmotic balance, stabilize different cellular compartments and scavenge free oxygen radicals. Moreover, Pro keeps the ratio of $NADP^+$ to NADPH high and increases oxidative pentose phosphate activity which is needed during stressed conditions (Shetty 2004). Likewise, a significant accumulation of Pro content was observed in *Vigna aconitifolia* under heat stress (Harsh et al. 2016). Furthermore, Pro content of *Prunus persica* shoots got enhanced when two heat treatments were given to cold-acclimated plants but remained unchanged during the prevailing cold treatment (Shin et al. 2016). Overexpression of P5CSF129A, which is a mutated form of *Vigna* P5CS in transgenic chickpea, resulted in proline accumulation with elongation of roots and increase in biomass by providing the required energy in the form of NAD(P)H and $FADH_2$ under water-stressed conditions (Ghanti et al. 2011). In 2012, Zhan et al. identified a protein rich in proline SICKLE (SIC), vital for growth and tolerance to stress in *Arabidopsis*. In addition, the plants without this protein showed sickle-like leaf margin, delayed flowering and reduced height (Zhan et al. 2012).

13.4.1.2   Betaines and Related Compounds

Betaines comprise the carboxylic acid group and are classified as quaternary ammonium compounds. They are considered as fully methylated amino or imino acids. Diverse forms of betaines occur, such as proline betaine, β-alanine betaine and glycine betaine, based on the plant species and environmental stress. (Nahar et al. 2016). Amongst the mentioned types, glycine betaine (GB) (N,N,N-trimethylglycine), which was discovered for the first time in *Beta vulgaris*, is the most common entity, is diversely distributed, occurs during dehydrating conditions and plays a role in maintaining photosystem II (PSII) efficacy by accumulating mainly in the thylakoid membrane of chloroplast (Ashraf and Foolad 2007). In GB-accumulator plant species, tolerant genotypes usually show higher GB accumulation than sensitive genotypes in response to stress. Also, the application of GB is found to prevent structural injuries in organelles that produce ROS, i.e. chloroplasts and mitochondria (Upadhyaya et al. 2013). Shan et al., in 2016, found a reduction in chilling damage in peaches under cold storage with the exogenous treatment of GB. In genetically engineered rice, protection of the photosynthetic process is more under cold stress, when the chloroplast-targeted *codA* gene is transferred into plants non-targeted *codA* gene, which yields its products in the cytosol. This genotype can biosynthesize 500% more GB. Spraying of low levels (>0.09 µmol g^{-1} f.wt.) of GB in plant tissues plays a protecting role from chilling temperatures. The amount of GB was high in the meristematic organs like buds and apices, which is thought to be vital for preserving plants and stimulating recovery in growth after cold stresses (Park et al. 2006). Quan et al. (2004) observed improvement in chilling stress tolerance in maize with the accumulation of GB as a result of *bet A* gene transfer. Moreover, the introduction of *BADH* gene (responsible for synthesizing betaine aldehyde dehydrogenase) from *Spinacia oleracea* in *Nicotiana tabaccum*, helped in preserving the activation of Rubisco and inhibited the accumulation of Rubisco activase, thereby imparting heat stress tolerance (Yang et al. 2005). Expression of the stress regulator gene *Osmyb4* in genetically modified apple resulted in the accumulation of GB and was connected to improved cold tolerance by Pasquali et al. (2008). Wang et al. (2010) reported over-accumulation of GB to be the reason behind increased photosynthesis under heat stress. Park et al. (2016) observed higher H_2O_2 levels, catalase gene CAT1 expression and catalase activity in GB treated plant than control.

13.4.1.3 Sugars

Sugars are known to be the energy source for plants, as they can supply carbon. They are considered to be the best osmolytes, because of their ability to protect the plants in stressed conditions by safeguarding various metabolic processes (photosynthesis, respiration and phosphogluconate pathways) from generating as well as scavenging systems of ROS (Lippert and Galinski 1992). For instance, mannitol is found to protect the chloroplastic apparatus from the attack of •OH radicals in the conditions of stress (Upadhyaya et al. 2013). Besides this, raffinose, an oligosaccharide is associated with mitigating the effects of dehydration, incorporating endurance in the barren state and assisting with substrates for the generation of energy at the germination state (ElSayed et al. 2014). Galactinol synthase (GolS) is an enzyme responsible for the synthesis of oligosaccharides related to the raffinose family, which act as osmoprotectants in plants. The transcription of GolS 1, 2, 4 and raffinose synthase 2 (RS2) was highly stimulated in *Arabidopsis*, over-expressing heat shock transcription factor A2 (HsfA2) and increasing the contents of galactinol and raffinose, thus imparting chilling tolerance (Nishizawa et al. 2008). Trehalose (Tre) is a sugar that is non-reducing and composed of two glucose units joined with glycosidic bonds (Richards et al. 2002). Tre also acts as a stabilizer of membrane or molecules as it can substitute water by making H bonds in membrane or macromolecules under freezing stress (Crowe 2007). Moreover, Tre can act as a molecule in the signaling process and is reported to control the carbon as well as abscisic acid (ABA) metabolism under osmotic stress (Upadhyaya et al. 2013). In addition, Tre is the only molecule under saccharides, which can exist as a glass-like state when completely dehydrated during temperature stress (Elbein et al. 2003). Thus, under stressful conditions, sugars get accumulated in plants and protect the membrane from disintegration (Valluru and Van den Ende 2008).

13.4.1.4 Polyamines

Polyamine –PA accumulation is generally taken as response of the plant to abiotic stresses, but the relationship remains unclear between the cause of PA accumulation and protection of biochemical processes. Levels of PA were found to vary from species to species, and the three most ample PAs – Putrescine (Put), spermidine (Spd) and spermine (Spm) – have been observed to show a significant increase during stressed conditions (Liu et al. 2007; Yang et al. 2007). Exogenous supply of Spd in rice seedlings ameliorated heat-stimulated damages, enhanced antioxidant activities and reduced H_2O_2 accumulation (Mostofa et al. 2014). Likewise, soybean, tobacco, pistachio and cucumber showed heat tolerance after the accumulation of PA (Xu et al. 2011; Radhakrishnan and Lee 2013). Champa et al. (2015) reported a reduction in damage induced by chilling stress after the supplementation of 1 mM Spm exogenously which led to the maintenance of grape berries' quality and life span at the period of storage in low temperature. PAs can improve cold resistance by binding at the phospholipid site in the cell membrane thus preventing cytolysis (Li and He 2012). Exogenous treatment of Spm sustained high levels of Spm and Spd endogenously, which as a result prevented the accumulation of Put and decreased the damage caused by chilling (Roy and Wu 2002). Sun et al. (2018) observed the result of different Put and D-Arg treatments on various parameters of *Anthurium andraeanum* in chilling stress. They found an increase in activities of antioxidants and roots; metabolism of nitrogen and content of chlorophyll and proline; and decrease in chilling damage with Put addition. Similarly, PA supplementation enhanced tolerance in Stevia plants towards cold conditions (Moradi Peynevandi et al. 2018). Kasukabe et al. (2007) found a substantial rise in

Spd synthase, SPDs' activity and content in leaves of transgenic *Arabidopsis*, when a SPDs cDNA was transferred from *Cucurbita ficifolia* resulting in improved tolerance to chilling and freezing. Genes responsible for the biosynthesis of PAs, like arginine decarboxylase (ADC), ornithine decarboxylase (ODC), S-adenosylmethionine decarboxylase (SAMDC) or Spd synthase (SPDS), were found to impart tolerance of stress in plants (Liu et al. 2007), such that the transgenic plants over-expressing these genes have been found to tolerate multiple stresses including temperature stress (Kasukabe et al. 2007; Kasukabe et al. 2006; Wi et al. 2006; Prabhavathi and Ranjam 2007; Wen et al. 2008).

13.4.1.5 Proteins

Proteins have a vital role in response to plant stress and are directly involved in forming novel phenotypes by adjusting physiological traits. A whole of proteins in a particular organism at a particular time is called the proteome. The role of proteomics in extreme temperatures has been reported by various researchers (Ahmad et al. 2016; Johnová et al. 2016). Plants respond to water-deficit conditions by an enhanced accumulation of several osmolytes and hydrophilic proteins (e.g. LEA proteins), induction of several heat-shock transcription factors (HSFs) and downstream heat-shock proteins (HSPs) (Kosova et al. 2018). Liu et al. (2016) observed thermotolerance in transgenic tobacco, yeast (GS115) and *E. coli* (BL21) through the overexpression of Group 3 Lea protein ZmLEA3 which was able to protect lactate dehydrogenase (LDH) activity at low temperatures.

Heat shock proteins (HSPs) are molecular chaperones, which bind and stabilize proteins at intermediate stages of folding, assembly, degradation and translocation across membranes. Five evolutionarily conserved groups of HSPs are: HSP100, HSP90, HSP70, HSP60 and small HSPs (sHSPs) (Krishna 2003). Although all major HSPs have related functions, however, each major HSP family has a unique mode of action. Some promote the degradation of misfolded proteins, whereas others prevent them from aggregating by binding to different types of folding intermediates (HSP70 and HSP60) and promote the reactivation (HSP100) of proteins that have already aggregated (Yadav et al. 2020). Wang et al. (2014) observed up-regulation in expressions of HSPs (HSP101, HSP90, HSP70 and sHSPs) in bentgrass, indicating cooperation between different HSPs under heat stress.

13.4.2 Role of Antioxidants in Modulating Defense Responses

To prevent ROS injury, plants have an antioxidant defense system consisting of enzymatic and non-enzymatic components, which are synthesized in different organelles of plants: peroxisomes, mitochondria and chloroplasts. Various studies indicate an increase in antioxidants to be accountable for plant tolerance against temperature stress (Gill and Tuteja 2010; Kumar et al. 2011; Hasanuzzaman et al. 2013; Soengas et al. 2018). Details of these two components are below.

13.4.2.1 Enzymatic Antioxidants

Enzymatic antioxidants prevent ROS damage by breaking down and eliminating free radicals from the cells. These comprise superoxide dismutase (SOD), catalase (CAT), ascorbate peroxidase (APX), mono-dehydroascorbate reductase (MDHAR), dehydroascorbate reductase (DHAR), glutathione reductase (GR) and guaiacol peroxidase (GPX). Antioxidants are found to be expressed superiorly in the stress-tolerant genotypes (Kumar et al. 2013) due to the better efficiency of these genotypes in scavenging oxygen radicals (Huang and Guo 2005; Almeselmani et al. 2009).

13.4.2.1.1 Superoxide Dismutase

Superoxide dismutase – SOD, E.C.15.1.1 – is a metalloenzymes which forms the front line of defense against ROS. It can catalyze the disproportion of $O_2^{\bullet-}$ to O_2 and H_2O_2. In plants, three isozymes of SOD are reported: first, copper/zinc SOD, which exists in three isoforms, located in the cytosol, peroxisome, chloroplast and mitochondria; second, manganese SOD, which can be found in mitochondria; and third, iron SOD in chloroplasts (Jackson et al. 1978). Copper/zinc SOD in eukaryotes is cyanide sensitive and occurs in form of a dimer, whereas the other two are tolerant to cyanide and exist as dimers or tetramers SODs and reduce the risk of OH• generation, by eliminating $O_2^{•-}$ using metal-catalyzed Haber-Weiss-type reaction which is 10,000 times faster than the spontaneous dismutation (Del Río et al. 1998; Gill and Tuteja 2010). This enzyme plays a central role in the antioxidant defense mechanism, as SOD activity regulates the concentrations of the two Haber-Weiss reaction substrates $O_2^{•-}$ and H_2O_2 (Kehrer 2000Nocter). Moieni-Korbekandi et al. (2014) reported an increase in SOD activity under cold conditions in two cultivars of *Brassica napus*. Likewise, Soengas et al. (2018) found a noteworthy increase of SOD in conditions of heat and cold stress in *B. oleoraca*, as compared to control temperature, implying oxidative stress in plants under stressful conditions. Moreover, out of ten *CsSODs* genes, most of the SODs were found to get induced in *Camellia sinensis* under cold stress by Zhou et al. (2019). Furthermore, 29 SOD genes in *B. juncea* and 18 genes in *B. rapa* were found to get upregulated under abiotic stress environments (Verma et al. 2019) (Table 13.1).

TABLE 13.1

The Type of Stress and the Mechanism Employed by Them for Thermotolerance in Different Plant Species

Plant Species	Stress	Mechanism of Tolerance	References
Arabis paniculata	Heat	Increased lipid saturation and content of soluble sugar	Tang et al. (2016)
Cabbage (*Brassica oleracea* L. var. *capitata* L.)	Cold	Increased activity of SS and SPS and content of soluble sugars	Yang et al. (2017)
Tomato (*Lycopersicon esculentum* Mill.)	Cold	Increase in polyamine content	Diao et al. (2017)
Hevea brasiliensis	Cold	Increase in Antioxidant content	Mai et al. (2009)
Camellia sinensis	Cold	Upregulation most of *CsSOD* genes	Zhou et al. (2019)
Brassica juncea and *B. rapa*	Cold	Upregulation of SOD genes	Verma et al. (2019)
Oryza sativa	Cold	Higher expression of *OsCATA* and *OsCATB*	Vighi et al. (2016)
Brassica rapa	Cold (freezing)	Coexpression of *BrMDHAR* and *BrDHAR*	Shin et al. (2013)
Zea mays	Cold	Modifying *ZmDHAR* expression, which encodes enzyme DHAR, essential in the recycling pathway of AA.	Xiang et al. (2019)
Citrullus lanatus	Cold	Noteworthy expression of six *ClGPX* genes	Zhou et al. (2018)
E. coli	Heat and Cold	Expression of *PgGR* genes	Achary et al. (2015)
Solanum lycopersicon	Heat and light	Upregulation of α-toc producing genes –*vte5*	Spicher et al. (2017)
Dacus carota	Cold (chilling)	Early response of plastid terminal oxidase *DcPTOX* genes	Campos et al. (2016)
Brassica oleraca	Heat	Higher antioxidants and osmolytes production	Soengas et al. (2018)

SS and SPS – Sucrose synthase	ZmDHAR – *Zea mays* DHAR
SPS – Sucrose phosphate synthase	AA – Ascorbic acid
SOD – Superoxide dismutase	ClGPX – *Citrullus lanatus* Guaiacol peroxidase
CsSOD – *Camellia sinensis* SOD	DcPTOX – *Dacus carota* plastid terminal oxidase
CAT – Catalase	BrMDHAR – *Brassica rapa* MDHAR
BrDHAR – *Brassica rapa* DHAR	MDHAR – Monodehydroascorbate reductase
OsCAT – *Oryza sativa* CAT	DHAR – Dehydroascorbate reductase

13.4.2.1.2 Catalase

Among enzymatic antioxidants, catalase – CAT, E.C.1.11.1.6 – was the first enzyme to be discovered and characterized. It is a universal enzyme and has a tetrameric structure. It catalyzes the reaction, in which two molecules of H_2O_2 get converted to water and oxygen. Willekens et al. (1995) gave a classification of CAT, based upon the gene expression profile of tobacco plant. According to them, CATs of Class I are regulated by light and are expressed in green tissues, whereas CATs of Class II are highly active in vascular tissues, while Class III CATs are found abundantly in seeds as well as young seedlings. H_2O_2 is effectively degraded by CAT when the cells are lacking in energy and are producing H_2O_2 at a rapid rate through catabolic processes (Mallick and Mohn 2000). Almeselmani et al. (2006) found a significant increase in SOD and CAT activity in the five genotypes of wheat under high temperature and positively correlated these with the content of chlorophyll and negatively with the index of membrane injury. Similarly, Sairam et al. (2000) reported an increase in SOD and CAT activities under elevated temperatures in *Triticum aestivum* and correlated this with high relative water and AsA content, as well as less H_2O_2 and malondialdehyde content. Vighi et al. (2016) stated higher expression of OsCATA and OsCATB in the sensitive genotype of *Oryza sativa* under cold stress, making the sensitive genotype able to maintain a better balance between lipid peroxidation and CAT activity than the resistant genotype.

13.4.2.1.3 Ascorbate-Glutathione Cycle

In addition to SOD and CAT, there occurs an ascorbate-glutathione (AA-GSH) cycle (also referred to as Asada-Halliwell pathway) to detoxify H_2O_2, which involves continuous reactions of AA, GSH and NADPH (Nicotinamide adenine dinucleotide phosphate hydrogen) catalyzed by the enzymes APX, MDHAR, DHAR and GR. Out of these, APX forms the central component of AA-GSH cycle and uses two molecules of AA for reducing H_2O_2 into the water with a simultaneous generation of two MDHA molecules. Depending upon a sequence, five chemically and enzymatically different APX isoenzymes are found in cytosol, stroma, thylakoids, mitochondria and peroxisomes (Nakano and Asada 1987). Several researchers have reported an increase in APX activity under high temperature in wheat, sorghum, lentil and soybean (Almeselmani et al. 2009; Djanaguiraman et al. 2010; Chakraborty and Pradhan 2012; D'Souza and Devaraj 2013). Badawi et al. (2007) reported a positive correlation between greater SOD, APX and CAT activities with heat tolerance. Similarly, Rani et al. (2013) noted more adaptation in relatively tolerant genotypes of *B. juncea* due to higher antioxidant activities under high temperature. Turan and Ekmekci (2011) reported enhanced activity of APX to be the reason behind the acclimation of chickpea to 2 and 4 °C. Pod walls of chickpea showed high APX activity, which could be due to an increase in GSH translocation from walls to seeds, thus helping in scavenging ROS and providing tolerance to pod wall under low temperature stress (Kaur et al. 2009). Co-expression of BrMDHAR and BrDHAR in *Brassica rapa* via hybridization helped in providing resistance to freezing (Shin et al. 2013). Besides this, other crops such as coffee, tomato, cucumber, grapes and alfalfa showed tolerance to cold stress which was correlated to the increase in antioxidant content (Wang and Li 2006; Ibrahim and Bafeel 2008; Zhao et al. 2009; Yang et al. 2011; Hasanuzzaman et al. 2013).

The second major enzyme of AA-GSH cycle is mono-dehydroascorbate reductase – MDHAR, EC 1.6.5.4. It is a FAD (flavin adenine dinucleotide) enzyme and acts as a catalyst for regenerating AA from MDHA radicals by means of NAD(P)H which donates electrons (e⁻s) (Hossain and Asada 1985). MDHAR isoenzymes are reported to be present in cytosol as well as organelles including mitochondria, peroxisomes and chloroplasts (Hossain et al. 1984; Jim´enez et al. 1998). The third enzyme is dehydroascorbate reductase – DHAR, EC 1.8.5.1 – which is found profusely in dry seeds, roots and green and etiolated shoots. Ushimaru et al. (1997) reported DHAR to be monomeric in structure, catalyzing DHA reduction to AA, in which GSH acts as a reducing agent and also helps in sustaining a reduced form of AA. DHAR has been isolated by researchers from chloroplast and non-chloroplast regions in spinach leaves and potato tuber (Hossain et. al. 1984; Dipierro and Borraccino 1991). Contents of MDHAR, DHAR, GR and other antioxidants were reported to be enhanced under cold stress in *Zea mays*, indicating their role in imparting tolerance (Hasanuzzaman et al. 2013). In sweet corn seedlings, Xiang et al. (2019) found that cold stressed conditions altered AA content by modifying *ZmDHAR* expression, which encodes enzyme DHAR and is essential in the recycling pathway of AA.

13.4.2.1.4 Guaiacol Peroxidase

Guaiacol peroxidase – GPX, E.C. 1.11.1.7 – is a protein which contains heme and oxidizes aromatic e⁻ donors, like guaiacol and pyrogallol, by preference, at the cost of H_2O_2. It is extensively found in microbes, plants and animals. GPX consists of four conserved disulfide bridges and two Ca^{2+} ions (Schuller et al. 1996). Various GPX isoenzymes are present in plant tissues, specifically located in the cell wall, cytosol and vacuoles (Asada 1994). GPX functions as an effective quencher of excess O_2 forms, which are reactive, and peroxy radicals during abiotic stresses (Vangronsveld and Clijsters 1994). Zhou et al. (2018) reported a noteworthy expression of six *ClGPX* genes in watermelon under cold stress. A family member of GPX gene family, RcGPX5, was introduced into *Salvia miltiorrhiza* after extracting it from *Rhodiola crenulata*. The genetically engineered plant demonstrated higher tolerance to oxidative and osmotic stress than the wild type by reducing the synthesis of malondialdehyde (Zhang et al. 2018).

13.4.2.1.5 Glutathione reductase

Glutathione reductase (GR) is a NAD(P)H-dependent enzyme that reduces glutathione disulfide (GSSG) to glutathione reduced (GSH), thus sustaining a high cellular ratio of GSH to GSSG (Ghisla and Massey 1989). Even though GR is present in the cytosol, mitochondria, chloroplasts and peroxisomes, chloroplastic isoforms account for around 80% of its activity in photosynthetic organs (Edwards et al. 1990). Many researchers have reported a rise in the activity of GR under temperature extremes. For example, Almeselmani et al. (2006) positively correlated an increase in GR activity in five wheat genotypes with chlorophyll (Chl) content and negatively with the index of membrane injury under heat stress. Yadav et al. (2013) found a correlation between an increase in enzymatic antioxidants and the extent of tolerance in *B. juncea* under heat stress. Moreover, Liu et al. (2016) observed an increase in GR activity in cucumber under low temperature, which helped in improving Chl content in the plant. Furthermore, Achary et al. (2015) reported enhanced stress tolerance in transgenic *E. coli*, in which *PgGR* cDNA isolated from *Pennisetum glaucum* (L.) was introduced. In addition to enzymatic antioxidants, there exist a second arm of defense – non-enzymatic antioxidants – which are described below.

13.4.2.2 Non-Enzymatic Antioxidants

Non-enzymatic antioxidants work in coordination with the enzymatic arm to ameliorate oxidative stress most efficiently. Non-enzymatic antioxidants protect the cell damage by interrupting the chain reactions of free radicals and help in cellular growth by altering processes of cell division and elongation, senescence as well as death (de Pinto and De Gara 2004). Non-enzymatic antioxidants include ascorbic acid (AA), reduced glutathione (GSH), alpha tocopherols, carotenoids and phenolics.

13.4.2.2.1 Ascorbic Acid

Ascorbic acid (AA) has a low molecular weight and is crucial for protecting plants against ROS generated oxidative stress. It is the most profused antioxidant and has a main role in plants growth, differentiation as well as metabolism. Although, the major portion of AA is present in the cytosol (>90%), however, it gets localized into the apoplast in millimolar concentration and protects the cell from oxidative attack (Barnes et al. 2002). In the apoplastic region, AA protects the membrane by directly reacting with free radicals and regenerating alpha-tocopherols from tocopheroxyl radical, via AA-GSH cycle (Noctor and Foyer 1998). Under stressful conditions, the levels of AA alter depending upon the balance between demand, AA biosynthesis and turnover (Chaves et al. 2002).

For example, Soengas et al. (2018) observed a rise in AA content under cold stress in *B. oleracea* L. var. capitata. L. Xiang et al. (2019) reported alteration in content of AA in *Zea mays* due to modulation in the expression of *Zm DHAR* (essential for AA recycle pathway) under temperature stress.

13.4.2.2.2 Reduced Glutathione

Reduced glutathione (GSH) is a nonprotein thiol, with low molecular weight and plays a crucial role in defense against loss caused by ROS intracellularly. It works as an antioxidant by reacting with the free radicals and can be found in almost every cellular compartment, cytosol, vacuoles, mitochondria, chloroplasts and endoplasmic reticulum (Foyer and Nocter 2003). It protects against cell damage by i) regenerating AA via AA-GSH cycle, ii) reacting directly with electrophiles, or iii) donating a proton to organic free radicals and producing GSSG (Asada 1994). Xu et al. (2012) reported low ROS generation which was found to be due to enhanced production of AA in heat-adapted turf grass. Levels of GSH and the ratio of GSH to GSSG were found to increase under high temperature (40 °C) in chilling-sensitive maize genotypes, whereas the activity of GR, as well as the content of total hydroxyl methylglutathione, were enhanced in both chilling tolerant and sensitive genotypes (Kocsy et al. 2002). The same authors also reported the accumulation of GSH levels in wheat, due to an increase in enzymatic activity, responsible for the synthesis of GSH and high GSH/GSSG ratio under heat stress. In 2012, Kumar et al. found noteworthy higher levels of AA and GSH at high temperature in *Zea mays* than in *Oryza sativa* genotypes. This could be due to the better ability of *Z. mays* genotypes to overcome oxidative damage than *O. sativa* genotypes. Recently, Soengas et al. (2018) recorded a substantial rise in glutathione level in broccoli seedlings under cold stress.

13.4.2.2.3 Tocopherols

Tocopherols (Toc) are antioxidants that can scavenge O_2^- and lipid peroxy radicals. It has four isomers – alpha (α), beta (β), gamma (γ) and delta (δ) – based on methylation pattern and number of methyl groups linked to the phenolic ring. Out of these, α isomer shows a maximum proportion of antioxidant activity followed by β, γ and δ *in vivo* (Fukuzawa et al. 1982; Diplock 1989). Toc are located in the green parts of plants only and shield the structure as well as functions of PSII in chloroplasts by quenching and reacting with oxygen (Ivanov and S. Khorobrykh 2003). In this manner, Toc forms an effective free radical trap by interrupting chain reactions in the lipid autooxidation process. Chennupati et al. (2011) reported 752% increase in α-Toc levels under high temperature in soybean, the reverse being the case for γ and δ. Spicher et al. (2017) found the upregulation of α-Toc in tomato, which contributes towards protecting PSII from the damage of light under temperature stress. Export of photoassimilate in vte2 simultaneously lowered down with the deposition of callose in transfer cell walls of phloem parenchyma. These results together illustrate the restricted role of Toc in photoprotection and phloem loading under low-temperature (Maeda et al. 2006). Spicher et al. (2017) illustrated the role of VTE5 in supporting α-Toc production in tomato and protecting PSII from damage by finding strong light and heat inhibition in plants with reduced VTE5 expression as well as α-Toc levels. These findings illustrate the role of Toc in scavenging free radicals and safeguarding the process of photosynthesis in plants (Table 13.1).

13.4.2.2.4 Carotenoids

Carotenoids are grouped under lipophilic antioxidants which are capable of neutralizing different forms of ROS (Young and Lowe 2001). In the case of plants, carotenoids capture the light in the region 400–550 nm and allow the absorbed energy to pass to the Chl (Sieferman-Harms 1987). Carotenoids scavenge $^1O_2$ and prevent oxidative damage by quenching triplet form (3Chl*) as well as the exciting form (Chl*) of the chlorophyll molecule, thus protecting the photosynthetic apparatus. Carotenoids have a chain of isoprene residues that bear many alternated double bonds that easily allow uptake and release of extra energy as heat (Mittler 2002). Also, carotenoids act as precursors in the synthesis of abscisic acid and strigolactone which are involved in high temperature tolerance (Nambara and Marion-Poll 2005; Xie et al. 2010). Moreover, Havaux (2014), reported that stress-induced responses in plants lead to oxidation of carotenoids, generating electrophile species. These species are known to be very reactive, which induce such alteration in expressions of the gene that lead to acclimation for stress conditions. Quantitative real-time polymerase chain reaction expression analysis of genes involved in carotenoid biosynthesis showed that gene expression and metabolite accumulation are specific for tissues and respond differently to the increasing temperature stress in *Fragaria vesca* (Jackson et al. 1978)). Campos et al. (2016) reported immediate but temporary early response of plastid terminal oxidase (DcPTOX) genes in *Dacus carota*, which was involved in carotenoid biosynthesis under chilling stress. These studies indicate the potential role of carotenoids as an antioxidant for stress adaptation.

13.4.2.2.5 *Phenolics*

Phenolics are one of the secondary metabolites – flavonoids, lignin, hydroxycinnamate esters, and tannins – which are profusely found in plants and possess antioxidant properties (Grace and Logan 2000). Structurally, polyphenols are composed of an aromatic ring with -OH (hydroxyl) or OCH_3 (methoxy) group attached to it, which in combination are responsible for its antioxidant activity. They have a high capability to donate e^- or H atom, due to which, they show better antioxidant properties *in vitro* antioxidant assays as compared to AA and α-Toc. They can: a) directly scavenge O_2; b) trap lipid alkoxy radicals and prevent lipid damage; and c) modify the order of lipid packing and lessen membrane fluidity (Arora et al. 2000). Soengas et al. (2018) observed higher phenolic content at 32 °C in *Brassica oleraca* as compared to the control conditions.

13.5 CONCLUSION AND PROSPECTS

Temperature extremes have been affecting food security to a major extent, leading to an excess generation of ROS in the plants. ROS extensively cause damage to lipids, proteins and DNA, destroying cellular integrity and causing the death of the plant. For protecting themselves, plants have evolved osmolytes and antioxidants as defense systems which bring the disturbed metabolic state to normal either by improving water content in the tissues or upregulating the expression of genes and transcribed proteins which are involved in scavenging oxygen radicals. However, when the stress level rises above the threshold limits the defense system of plants needs more forces to work potentially. Since plants which have more native or induced antioxidant activity show more tolerance towards stresses, genetic engineering of responsible genes or the use of certain soil microbes can play a beneficial role in this context.

ACKNOWLEDGMENTS

The authors are thankful for providing financial assistance to the Department of Biotechnology (DBT), Government of India, helping in undertaking the related research.

REFERENCES

Abdel Latef AA, Chaoxing H (2011) Arbuscular mycorrhizal influence on growth, photosynthetic pigments, osmotic adjustment and oxidative stress in tomato plants subjected to low temperature stress. *Acta Physiol Plant* 33(4):1217–1225

Achary VM, Reddy CS, Pandey P, Islam T, Kaul T, Reddy MK (2015) Glutathione reductase a unique enzyme: Molecular cloning, expression and biochemical characterization from the stress adapted C4 plant, *Pennisetum glaucum* (L.) R. Br. *Mol Biol Rep* 42(5):947–962

Ahanger MA, Akram NA, Ashraf M, Alyemeni MN, Wijaya L, Ahmad P (2017) Plant responses to environmental stresses-from gene to biotechnology. *AoB Plants* 9(4). doi: 10.1093/aobpla/plx025

Ahmad P, Abdel Latef AAH, Rasool S, Akram NA, Ashraf M, Gucel S (2016) Role of proteomics in crop stress tolerance. *Front Plant Sci* 7:1336. doi: 10.3389/fpls.2016.01336

Almeselmani M, Deshmukh P, Sairam R (2009) High temperature stress tolerance in wheat genotypes: Role of antioxidant defence enzymes. *Acta Agronomica Hungarica* 57(1):1–14

Almeselmani M, Deshmukh PS, Sairam RK, Kushwaha SR, Singh TP (2006) Protective role of antioxidant enzymes under high temperature stress. *Plant Sci* 171(3):382–388

Apel K, Hirt H (2004) Reactive oxygen species: Metabolism, oxidative stress, and signal transduction. *Annu Rev Plant Biol* 55:373–399. doi: 10.1146/annurev.arplant.55.031903.141701

Argosubekti N (2020) A review of heat stress signaling in plants. *IOP Conf Ser.: Earth Environ Sci* 484(1):012041

Arora A, Byrem TM, Nair MG, Strasburg GM (2000) Modulation of liposomal membrane fluidity by flavonoids and isoflavonoids. *Arch Biochem Biophys* 373(1):102–109

Asada K (1994) Production and action of active oxygen species in photosynthetic tissues. In: *Causes of Photooxidatve Stress and Amelioration of Defense System in Plants* CH Foyer, PM Mullineaux (eds). CRC Press, Boca Raton, FL. pp. 77–103

Ashraf MF, Foolad MR (2007) Roles of glycine betaine and proline in improving plant abiotic stress resistance. *Environ Exp Bot* 59(2):206–216

Asseng S, Ewert F, Martre P, Rötter RP, Lobell DB, Cammarano D et al. (2015) Rising temperatures reduce global wheat production. *Nat Clim Change* 5(2):143

Badawi M, Danyluk J, Boucho B, Houde M, Sarhan F (2007) The CBF gene family in hexaploid wheat and its relationship to the phylogenetic complexity of cereal CBFs. *Mol Genet Genomics* 277(5):533–554. doi: 10.1007/s00438-006-0206-9

Barlow KM, Christy BP, O'leary GJ, Riffkin PA, Nuttall JG (2015) Simulating the impact of extreme heat and frost events on wheat crop production: A review. *Field Crops Res* 171:109–119

Barnes JD, Zheng Y, Lyons TM (2002) Plant resistance to ozone: The role of ascorbate in air pollution and plant biotechnology. In: *Journal of Microbiology* K Omasa, H Saji, S Youssefian, N Kondo (eds). Springer, Tokyo, Japan. pp. 235–254

Cairns JE, Sonder K, Zaidi PH, Verhulst N, Mahuku G, Babu R et al. (2012) Maize production in a changing climate: Impacts, adaptation, and mitigation strategies. *Adv Agron* 114:1–58

Camejo D, Rodríguez P, Morales MA, Dell'Amico JM, Torrecillas A, Alarcón JJ (2005) High temperature effects on photosynthetic activity of two tomato cultivars with different heat susceptibility. *J Plant Physiol* 162(3):281–289

Campos MD, Nogales A, Cardoso HG, Campos C, Grzebelus D, Velada I, Arnholdt-Schmitt B (2016) Carrot plastid terminal oxidase gene (*dcptox*) responds early to chilling and harbors intronic pre-mirnas related to plant disease defense. *Plant Gene* 7:21–25

Chakraborty U, Pradhan B (2012) Oxidative stress in five wheat varieties (*Triticum aestivum* L.) exposed to water stress and study of their antioxidant enzyme defense system, water stress responsive metabolites and H_2O_2 accumulation. *Braz J Plant Physiol* 24(2):117–130

Chaves MM, Pereira JS, Maroco J, Rodrigues ML, Ricardo CP, Osório ML, Carvalho I, Faria T, Pinheiro C (2002) How plants cope with water stress in the field? Photosynthesis and growth. *Ann Bot* 89(7):907–916

Chennupati P, Seguin P, Liu W (2011) Effects of high temperature stress at different development stages on soybean isoflavone and tocopherol concentrations. *J Agric Food Chem* 59(24):13081–13088

Chinnusamy V, Zhu J, Zhu JK (2007) Cold stress regulation of gene expression in plants. *Trends Plant Sci* 12(10):444–451

Croser JS, Clarke HJ, Siddique KHM, Khan TN (2003) Low-temperature stress: Implications for chickpea (*Cicer arietinum* L.) improvement. *Crit Rev Plant Sci* 22(2):185–219. doi: 10.1080/713610855

Crowe JH (2007) Trehalose as a "chemical chaperone". In: *Molecular Aspects of the Stress Response: Chaperones, Membranes and Networks* P Csermely, L Vígh (eds). Springer, New York, NY. pp. 143–158

Das K, Roychoudhury A (2014) Reactive oxygen species (ROS) and response of antioxidants as ROS-scavengers during environmental stress in plants. *Front Environ Sci* 2:53. doi: 10.3389/fenvs.2014.00053

De Pinto MC, De Gara L (2004) Changes in the ascorbate metabolism of apoplastic and symplastic spaces are associated with cell differentiation. *J Exp Bot* 55(408):2559–2569

del Río LA, Sandalio LM, Corpas FJ, López-Huertas E, Palma JM, Pastori GM (1998) Activated oxygen-mediated metabolic functions of leaf peroxisomes. *Physiol Plant* 104(4):673–680

Diao Q, Song Y, Shi D, Qi H (2017) Interaction of polyamines, abscisic acid, nitric oxide, and hydrogen peroxide under chilling stress in tomato (*Lycopersicon esculentum* Mill.) Seedlings. *Front Plant Sci*:203. doi: 10.3389/fpls.2017.00203

Ding Y, Lv J, Shi Y, Gao J, Hua J, Song CP, Gong Z, Yang S (2019) EGR2 phosphatase regulates OST1 kinase activity and freezing tolerance in *Arabidopsis*. *EMBO J* 38(1):e99819. doi: 10.15252/embj.201899819

Dipierro S, Borraccino G (1991) Dehydroascorbate reductase from potato tubers. *Phytochemistry* 30(2):427–429

Diplock AT (1989) *Vitamin E: Biochemistry and Health Implications*. New York Academy of Sciences, New York

Djanaguiraman M, Prasad PVV, Seppanen M (2010) Selenium protects sorghum leaves from oxidative damage under high temperature stress by enhancing antioxidant defense system. *Plant Physiol Biochem* 48(12):999–1007

Djanaguiraman M, Prasad PVV, Murugan M, Perumal M, Reddy UK (2014) Physiological differences among sorghum (Sorghum bicolor L. Moench) genotypes under high temperature stress. *Environ Exp Bot* 100:43–54. doi: 10.1016/j.envexpbot.2013.11.013

Dos Reis SP, Lima AM, De Souza CRB (2012) Recent molecular advances on downstream plant responses to abiotic stress. *Int J Mol Sci* 13(7):8628–8647

D'Souza M, Devaraj VR (2013) Induction of thermotolerance through heat acclimation in lablab bean (*Dolichos lablab*). *Afr J Biotechnol* 12:5695–5704. doi: 10.5897/AJB2013.13074

Dwivedi R, Prasad S, Jaiswal B, Kumar A, Tiwari A, Patel S, Pandey G, Pandey G (2017) Evaluation of wheat genotypes (*Triticum aestivum* L.) at grain filling stage for heat tolerance. *Int J Pure App Biosci* 5(2):971–975

Edwards EA, Rawsthorne S, Mullineaux PM (1990) Subcellular distribution of multiple forms of glutathione reductase in leaves of pea (*Pisum sativum* L.). *Planta* 180(2):278–284

Elbein AD, Pan YT, Pastuszak I, Carroll D (2003) New insights on trehalose: A multifunctional molecule. *Glycobiology* 13(4):17R–27R

ElSayed AI, Rafudeen MS, Golldack D (2014) Physiological aspects of raffinose family oligosaccharides in plants: Protection against abiotic stress. *Plant Biol* 16(1):1–8

Evans MD, Dizdaroglu M, Cooke MS (2004) Oxidative DNA damage and disease: Induction, repair and significance. *Mutat Res* 567(1):1–61

Farooq M, Wahid A, Basra SMA, Islam-ud-Din IU (2009) Improving water relations and gas exchange with brassinosteroids in rice under drought stress. *J Agron Crop Sci* 195(4):262–269. doi: 10.1111/j.1439-037X.2009.00368.x

Foyer CH, Noctor G (2003) Redox sensing and signalling associated with reactive oxygen in chloroplasts, peroxisomes and mitochondria. *Physiol Plant* 119(3):355–364

Friedlingstein P, Andrew RM, Rogelj J, Peters GP, Canadell JG, Knutti R, Luderer G, Raupach MR, Schaeffer M, van Vuuren DP, Le Quéré C (2014) Persistent growth of CO_2 emissions and implications for reaching climate targets. *Nat Geosci* 7(10):709–715

Fukuzawa K, Tokumura A, Ouchi S, Tsukatani H (1982) Antioxidant activities of tocopherols on Fe2+-ascorbate-induced lipid peroxidation in lecithin liposomes. *Lipids* 17(7):511–513

Ghanti SK, Sujata KG, Kumar BV, Karba NN, Janardhan Reddy K, Rao MS, Kishor PK (2011) Heterologous expression of P5CS gene in chickpea enhances salt tolerance without affecting yield. *Biol Plant* 55(4):634–640

Ghisla S, Massey V (1989) Mechanisms of flavoprotein-catalyzed reactions. *Eur J Biochem* 181(1):1–17

Gill SS, Tuteja N (2010) Reactive oxygen species and antioxidant machinery in abiotic stress tolerance in crop plants. *Plant Physiol Biochem* 48(12):909–930. doi: 10.1016/j.plaphy.2010.08.016

Giorno F, Wolters-Arts M, Mariani C, Rieu I (2013) Ensuring reproduction at high temperatures: The heat stress response during anther and pollen development. *Plants* 2(3):489–506

Grace SC, Logan BA (2000) Energy dissipation and radical scavenging by the plant phenylpropanoid pathway. *Philos Trans R Soc Lond B Biol Sci* 355(1402):1499–1510

Harindra Champa WA, Gill MIS, Mahajan BVC, Bedi S (2015) Exogenous treatment of spermine to maintain quality and extend postharvest life of table grapes (*Vitis vinifera* L.) cv. Flame seedless under low temperature storage. *LWT Food Sci Technol* 60(1):412–419

Harsh A, Sharma YK, Joshi U, Rampuria S, Singh G, Kumar S, Sharma R (2016) Effect of short-term heat stress on total sugars, proline and some antioxidant enzymes in moth bean (*Vigna aconitifolia*). *Ann Agric Sci* 61(1):57–64

Hasanuzzaman M, Nahar K, Fujita M (2013) Extreme temperatures, oxidative stress and antioxidant defense in plants. In *Abiotic Stress-Plant Responses and Applications in Agriculture* K Vahdati and C Leslie (eds). InTech, Rijeka. pp. 169–205

Hatfield JL, Boote KJ, Kimball BA, Ziska LH, Izaurralde RC, Ort D, Thomson AM, Wolfe D (2011) Climate impacts on agriculture: Implications for crop production. *Agron J* 103(2):351–370

Havaux M (2014) Carotenoid oxidation products as stress signals in plants. *Plant J* 79(4):597–606

Hayat S, Hayat Q, Alyemeni MN, Wani AS, Pichtel J, Ahmad A (2012) Role of proline under changing environments: A review. *Plant Signal Behav* 7(11):1456–1466

Hossain MA, Asada K (1985) Monodehydroascorbate reductase from cucumber is a flavin adenine dinucleotide enzyme. *J Biol Chem* 260(24):12920–12926

Hossain MA, Nakano Y, Asada K (1984) Monodehydroascorbate reductase in spinach chloroplasts and its participation in regeneration of ascorbate for scavenging hydrogen peroxide. *Plant Cell Physiol* 25(3):385–395

Huang M, Guo Z (2005) Responses of antioxidative system to chilling stress in two rice cultivars differing in sensitivity. *Biologia Plant* 49(1):81–84

Hussain HA, Hussain S, Khaliq A, Ashraf U, Anjum SA, Men S, Wang L (2018) Chilling and drought stresses in crop plants: Implications, cross talk, and potential management opportunities. *Front Plant Sci* 9:393. doi: 10.3389/fpls.2018.00393

Ibrahim MM, Bafeel SO (2008) Photosynthetic efficiency and pigment contents in alfalfa (*Medicago sativa*) seedlings subjected to dark and chilling conditions. *Int J Agric Biol* 10(3):306–310

Intergovernmental Panel Climate Change (IPCC) (2007) 2007: Impacts, adaptation and vulnerability: Contribution of Working Group II to the Fourth Assessment Report of the Intergovernmental panel on climate change. Clim Change. Cambridge University Press, Cambridge, UK and New York, NY

Ivanov BN, Khorobrykh S (2003) Participation of photosynthetic electron transport in production and scavenging of reactive oxygen species. *Antioxid Redox Signal* 5(1):43–53

Jackson C, Dench J, Moore AL, Halliwell B, Foyer CH, Hall DO (1978) Subcellular localisation and identification of superoxide dismutase in the leaves of higher plants. *Eur J Biochem* 91(2):339–344

Jewell MC, Campbell BC, Godwin ID (2010) Transgenic plants for abiotic stress resistance. In: *Transgenic Crop Plants* C Kole, CH Michler, AG Abbott, TC Hall (eds). Springer, Berlin, Heidelberg. pp. 67–132

Ji C, Bian X, Lee CJ, Kim HS, Kim SE, Park SC, Xie Y, Guo X, Kwak SS (2019) De novo transcriptome sequencing and gene expression profiling of sweet potato leaves during low temperature stress and recovery. *Gene* 700:23–30. doi: 10.1016/j.gene.2019.02.097

Ji CY, Chung WH, Kim HS, Jung WY, Kang L, Jeong JC, Kwak SS (2017) Transcriptome profiling of sweetpotato tuberous roots during low temperature storage. *Plant Physiol Biochem* 112:97–108. doi:10.1016/j.plaphy.2016.12.021

Ji CY, Kim HS, Lee CJ, Kim SE, Lee HU, Nam SS, Li Q, Ma DF, Kwak SS (2020) Comparative transcriptome profiling of tuberous roots of two sweetpotato lines with contrasting low temperature tolerance during storage. *Gene* 727:144244. doi: 10.1016/j.gene.2019.144244

Jiménez A, Hernández JA, Pastori G, del Río LA, Sevilla F (1998) Role of the ascorbate-glutathione cycle of mitochondria and peroxisomes in the senescence of pea leaves. *Plant Physiol* 118(4):1327–1335

Johnová P, Skalák J, Saiz-Fernández I, Brzobohatý B (2016) Plant responses to ambient temperature fluctuations and water-limiting conditions: A proteome-wide perspective. *Biochim Biophys Acta* 1864(8):916–931. doi: 10.1016/j.bbapap.2016.02.007

Kamiński P, Koim-Puchowska B, Puchowski P, Jerzak L, Wieloch M, Bombolewska K (2012) Enzymatic antioxidant responses of plants in saline anthropogenic environments. In: *Plant Science* NK Dhal, SC Sahu (eds). InTech, Rijeka, pp 35–64. doi: 10.5772/51149

Kasukabe Y, He L, Nada K, Misawa S, Ihara I, Tachibana S (2007) Overexpression of spermidine synthase enhances tolerance to multiple environmental stresses and upregulates the expression of various stress regulated genes in transgenic *Arabidopsis thaliana*. *Plant Cell Physiol* 45(6):712–722

Kasukabe Y, He L, Watakabe Y, Otani M, Shimada T, Tachibana S (2006) Improvement of environmental stress tolerance of sweet potato by introduction of genes for spermidine synthase. *Plant Biotechnol* 23(1):75–83

Kaur S, Gupta AK, Kaur N, Sandhu JS, Gupta SK (2009) Antioxidative enzymes and sucrose synthase contribute to cold stress tolerance in chickpea. *J Agron Crop Sci* 195(5):393–397. doi: 10.1111/j.1439-037X.2009.00383.x

Kaushal N, Awasthi R, Gupta K, Gaur P, Siddique KH, Nayyar H (2013) Heat-stress-induced reproductive failures in chickpea (*Cicer arietinum*) are associated with impaired sucrose metabolism in leaves and anthers. *Func Plant Biol* 40(12):1334–1349

Kehrer JP (2000) The Haber–Weiss reaction and mechanisms of toxicity. *Toxicol* 149(1):43–50

Kim SH, Gitz DC, Sicher RC, Baker JT, Timlin DJ, Reddy VR (2007) Temperature dependence of growth, development, and photosynthesis in maize under elevated CO_2. *Environ Exp Bot* 61(3):224–236. doi: 10.1016/j.envexpbot.2007.06.005

Kipp E, Boyle M (2013) The effects of heat stress on reactive oxygen species production and chlorophyll concentration in *Arabidopsis thaliana*. *Res Plant Sci* 1(2):20–23

Kocsy G, Szalai G, Galiba G (2002) Induction of glutathione synthesis and glutathione reductase activity by abiotic stresses in maize and wheat. *Sci World J* 2:1699–1705. doi: 10.1100/tsw.2002.812

Kosová K, Vítámvás P, Urban MO, Prášil IT, Renaut J (2018) Plant abiotic stress proteomics: The major factors determining alterations in cellular proteome. *Front Plant Sci* 9:122

Kozłowska M, Rybus-Zając M, Stachowiak J, Janowska B (2007) Changes in carbohydrate contents of Zantedeschia leaves under gibberellin-stimulated flowering. *Acta Physiol Plant* 29(1):27–32. doi: 10.1007/s11738-006-0004-3

Krishna P (2003) Plant responses to heat stress. In: *Plant Responses to Abiotic Stress* H Hirt, K Shinozaki (eds). Springer, Berlin, Heidelberg. pp. 73–101

Kumar S, Kaur R, Kaur N, Bhandhari K, Kaushal N, Gupta K, Bains TS, Nayyar H (2011) Heat-stress induced inhibition in growth and chlorosis in mungbean (*Phaseolus aureus* Roxb.) is partly mitigated by ascorbic acid application and is related to reduction in oxidative stress. *Acta Physiol Plant* 33(6):2091–2101. doi: 10.1007/s11738-011-0748-2

Kumar S, Singh R, Nayyar H (2013) α-Tocopherol application modulates the response of wheat (*Triticum aestivum* l.) seedlings to elevated temperatures by mitigation of stress injury and enhancement of antioxidants. *J Plant Growth Regul* 32(2):307–314. doi: 10.1007/s00344-012-9299-z

Kumar V, Khare T, Shaikh S, Wani SH (2018) *Compatible Solutes and Abiotic Stress Tolerance in Plants: Metabolic Adaptations in Plants during Abiotic Stress*. Taylor & Francis (CRC Press), USA. pp. 213–220

Kumar V, Shriram V, Hossain MA, Kishor PK (2015) Engineering proline metabolism for enhanced plant salt stress tolerance. In: *Managing Salt Tolerance in Plants: Molecular and Genomic Perspectives* Wani SH, MA Hussain (eds.) Taylor, Francis, UK. pp. 353. doi:10.1201/b19246-20

Lee SB, Kwon HB, Kwon SJ, Park SC, Jeong MJ, Han SE, Byun MO, Daniell H (2004) Accumulation of trehalose within transgenic chloroplasts confers drought tolerance. *Mol Breed* 11:1–13

Li Y, He J (2012) Advance in metabolism and response to stress of polyamines in plant. *Acta Agric Boreali Sin* 27:240–245

Lippert K, Galinski EA (1992) Enzyme stabilization be ectoine-type compatible solutes: Protection against heating, freezing and drying. *Appl Microbiol Biotechnol* 37(1):61–65. doi: 10.1007/BF00174204

Liu J, Shi Y, Yang S (2018) Insights into the regulation of C-repeat binding factors in plant cold signaling. *J Integr Plant Biol* 60(9):780–795

Liu JH, Kitashiba H, Wang J, Ban Y, Moriguchi T (2007) Polyamines and their ability to provide environmental stress tolerance to plants. *Plant Biotechnol* 24(1):117–126

Liu Y, Liang J, Sun L, Yang X, Li D (2016 Jul 14) Group 3 LEA protein, ZmLEA3, is involved in protection from low temperature stress. *Front Plant Sci* 7:1011

Maeda H, Song W, Sage TL, DellaPenna D (2006) Tocopherols play a crucial role in low-temperature adaptation and phloem loading in Arabidopsis. *Plant Cell* 18(10):2710–2732

Maestri E, Klueva N, Perrotta C, Gulli M, Nguyen HT, Marmiroli N (2002) Molecular Genetics of heat tolerance and heat shock proteins in cereals. *Plant Mol Biol* 48(5–6):667–681

Mai J, Herbette S, Vandame M, Kositsup B, Kasemsap P, Cavaloc E, Julien JL, Améglio T, Roeckel-Drevet P (2009) Effect of chilling on photosynthesis and antioxidant enzymes in *Hevea brasiliensis* Muell. Arg. *Trees Struct Funct* 23(4):863–874. doi: 10.1007/s00468-009-0328-x

Mallick N, Mohn FH (2000) Reactive oxygen species: Response of algal cells. *J Plant Physiol* 157(2):183–193

Masouleh SSS, Aldine NJ, Sassine YN (2019) The role of organic solutes in the osmotic adjustment of chilling-stressed plants (vegetable, ornamental and crop plants). *Ornam Hortic* 25(4):434–442

Meehl GA, Stocker TF, Collins WD, Friedlingstein P, Gaye T, Gregory JM, Kitoh A, Knutti R, Murphy JM, Noda A, Raper SC, Watterson AJ, Weaver Z, Zhao JC (2007) The physical science basis: Contribution of working group I to the fourth assessment report of the intergovernmental panel on climate change global climate projections. In: *IPCC, 2007: Climate Change* Solomon S, D Qin, M Manning, Z Chen, M Marquis, KB Averyt, M Tignor, HL Miller (eds). Cambridge University Press, Cambridge, UK and New York, NY. pp. 747–846

Mittler R (2002) Oxidative stress, antioxidants and stress tolerance. *Trends Plant Sci* 7(9):405–410

Moieni-Korbekandi Z, Karimzadeh G, Sharifi M (2014) Cold-induced changes of proline, malondialdehyde and chlorophyll in spring canola cultivars. *J Plant Physiol Breed* 4(1):1–11

Møller IM, Kristensen BK (2004) Protein oxidation in plant mitochondria as a stress indicator. *Photochem Photobiol Sci* 3(8):730–735

Moradi Peynevandi K, Razavi SM, Zahri S (2018) The ameliorating effects of polyamine supplement on physiological and biochemical parameters of *Stevia rebaudiana* Bertoni under cold stress. *Plant Prod Sci* 21(2):123–131

Morsy MR, Jouve L, Hausman JF, Hoffmann L, Stewart JM (2007) Alteration of oxidative and carbohydrate metabolism under abiotic stress in two rice (*Oryza sativa* L.) genotypes contrasting in chilling tolerance. *J Plant Physiol* 164(2):157–167

Mostofa MG, Yoshida N, Fujita M (2014) Spermidine pretreatment enhances heat tolerance in rice seedlings through modulating antioxidative and glyoxalase systems. *Plant Growth Regul* 73(1):31–44

Nahar K, Hasanuzzaman M, Rahman A, Alam M, Mahmud JA, Suzuki T, Fujita M (2016) Polyamines confer salt tolerance in mung bean (*Vigna radiata* L.) by reducing sodium uptake, improving nutrient homeostasis, antioxidant defense, and methylglyoxal detoxification systems. *Front Plant Sci* 7:1104

Nakano Y, Asada K (1987) Purification of ascorbate peroxidase in spinach chloroplasts; its inactivation in ascorbate-depleted medium and reactivation by monodehydroascorbate radical. *Plant Cell Physiol* 28(1):131–140

Nambara E, Marion-Poll A (2005) Abscisic acid biosynthesis and catabolism. *Annu Rev Plant Biol* 56:165–185

Nayyar H, Bains TS, Kumar S (2005) Chilling stressed chickpea seedlings: Effect of cold acclimation, calcium and abscisic acid on cryoprotective solutes and oxidative damage. *Environ Exp Bot* 54(3):275–285. doi: 10.1016/j.envexpbot.2004.09.007

Nishizawa A, Yabuta Y, Shigeoka S (2008) Galactinol and raffinose constitute a novel function to protect plants from oxidative damage. *Plant Physiol* 147(3):1251–1263

Noctor G, Foyer C (1998) Ascorbate and glutathione: Keeping active oxygen under control. *Annu Rev Plant Physiol Plant Mol Biol* 49(1):249–279. doi: 10.1146/annurev.arplant.49.1.249

Park EJ, Jeknic Z, Chen TH (2006) Exogenous application of glycinebetaine increases chilling tolerance in tomato plants. *Plant Cell Physiol* 47(6):706–714

Park YS, You SY, Cho S, Jeon HJ, Lee S, Cho DH, Kim JS, Oh JS (2016) Eccentric localization of catalase to protect chromosomes from oxidative damages during meiotic maturation in mouse oocytes. *Histochem Cell Biol* 146(3):281–288

Pasquali G, Biricolti S, Locatelli F, Baldoni E, Mattana M (2008) Osmyb4 expression improves adaptive responses to drought and cold stress in transgenic apples. *Plant Cell Rep* 27(10):1677–1686. doi: 10.1007/s00299-008-0587-9

Porter JR (2005) Rising temperatures are likely to reduce crop yields. *Nature* 436(7048):174–174

Posmyk MM, Janas KM (2007) Effects of seed hydropriming in presence of exogenous proline on chilling injury limitation in *Vigna radiata* L. seedlings. *Acta Physiol Plant* 29(6):509–517

Prabhavathi VR, Rajam MV (2007) Polyamine accumulation in transgenic eggplant enhances tolerance to multiple abiotic stresses and fungal resistance. *Plant Biotechnol* 24(3):273–282

Quan R, Shang M, Zhang H, Zhao Y, Zhang J (2004) Engineering of enhanced glycine betaine synthesis improves drought tolerance in maize. *Plant Biotechnol J* 2(6):477–486

Radhakrishnan R, Lee IJ (2013) Spermine promotes acclimation to osmotic stress by modifying antioxidant, abscisic acid, and jasmonic acid signals in soybean. *J Plant Growth Regul* 32(1):22–30. doi: 10.1007/s00344-012-9274-8

Rani T, Yadav RC, Yadav NR, Rani A, Singh D (2013) A genetic transformation in oilseed brassicas- a review. *Indian J Agri Sci* 83(4):367–373

Richards AB, Krakowka S, Dexter LB, Schmid H, Wolterbeek APM, Waalkens-Berendsen DH, Kurimoto M (2002) Trehalose: A review of properties, history of use and human tolerance, and results of multiple safety studies. *Food Chem Toxicol* 40(7):871–898

Richter C (1992) Reactive oxygen and DNA damage in mitochondria. *Mutat Res* 275(3–6):249–255

Romero-Puertas MC, Palma JM, Gómez M, Del Río LA, Sandalio LM (2002) Cadmium causes the oxidative modification of proteins in pea plants. *Plant Cell Environ* 25(5):677–686

Roy M, Wu R (2002) Overexpression of S-adenosylmethionine decarboxylase gene in rice increases polyamine level and enhances sodium chloride-stress tolerance. *Plant Sci* 163(5):987–992. doi: 10.1016/S0168-9452(02)00272-8

Sairam RK, Srivastava GC, Saxena DC (2000) Increased antioxidant activity under elevated temperatures: A mechanism of heat stress tolerance in wheat genotypes. *Biologia Plant* 43(2):245–251

Sanghera GS, Wani SH, Hussain W, Singh NB (2011) Engineering cold stress tolerance in crop plants. *Curr Genomics* 12(1):30

Sarwar M, Saleem MF, Ullah N, Ali S, Rizwan M, Shahid MR, Alyemeni MN, Alamri SA, Ahmad P (2019) Role of mineral nutrition in alleviation of heat stress in cotton plants grown in glasshouse and field conditions. *Sci Rep* 9(1):1–7

Schuller DJ, Ban N, van Huystee RB, McPHERSON A, Poulos TL (1996) The crystal structure of peanut peroxidase. *Structure* 4(3):311–321

Shan T, Jin P, Zhang Y, Huang Y, Wang X, Zheng Y (2016) Exogenous glycine betaine treatment enhances chilling tolerance of peach fruit during cold storage. *Postharvest Biol Tec* 114:104–110. doi: 10.1016/j.postharvbio.2015.12.005

Sharma DK, Andersen SB, Ottosen CO, Rosenqvist E (2015) Wheat cultivars selected for high Fv/Fm under heat stress maintain high photosynthesis, total chlorophyll, stomatal conductance, transpiration and dry matter. *Physiol Plant* 153(2):284–298

Sharma P, Jha AB, Dubey RS, Pessarakli M (2012) Reactive oxygen species, oxidative damage, and antioxidative defense mechanism in plants under stressful conditions. *J Bot*. doi: 10.1155/2012/217037

Shetty K (2004) Role of proline-linked pentose phosphate pathway in biosynthesis of plant phenolics for functional food and environmental applications: A review. *Process Biochem* 39(7):789–804

Shin H, Oh S, Arora R, Kim D (2016) Proline accumulation in response to high temperature in winter-acclimated shoots of Prunus persica: A response associated with growth resumption or heat stress? *Can J Plant Sci* 96(4):630–638

Shin SY, Kim MH, Kim YH, Park HM, Yoon HS (2013) Co-expression of monodehydroascorbate reductase and dehydroascorbate reductase from *Brassica rapa* effectively confers tolerance to freezing-induced oxidative stress. *Mol Cells* 36(4):304–315

Siefermann-Harms D (1987) The light-harvesting and protective functions of carotenoids in photosynthetic membranes. *Physiol Plant* 69(3):561–568

Singh V, Nguyen CT, Yang Z, Chapman SC, van Oosterom EJ, Hammer GL (2016 Jul) Genotypic differences in effects of short episodes of high-temperature stress during reproductive development in sorghum. *Crop Sci* 56(4):1561–1572

Soengas P, Rodríguez VM, Velasco P, Cartea ME (2018) Effect of temperature stress on antioxidant defenses in *Brassica oleracea*. *ACS Omega* 3(5):5237–5243

Spicher L, Almeida J, Gutbrod K, Pipitone R, Dörmann P, Glauser G, Rossi M, Kessler F (2017) Essential role for phytol kinase and tocopherol in tolerance to combined light and temperature stress in tomato. *J Exp Bot* 68(21–22):5845–5856

Sun X, Wang Y, Tan J, Al E (2018) Effects of exogenous putrescine and D-Arg on physiological and biochemical indices of anthurium under chilling stress. *Jiangsu J Agric Sci* 34:152–157

Suprasanna P, Nikalje GC, Rai AN (2016) Osmolyte accumulation and implications in plant abiotic stress tolerance. In: *Osmolytes and Plants Acclimation to Changing Environment: Emerging Omics Technologies* Iqbal N, R Nazar, A Khan, N (eds). Springer, New Delhi. pp. 1–12

Taïbi K, Del Campo AD, Vilagrosa A et al. (2018) Distinctive physiological and molecular responses to cold stress among cold-tolerant and cold-sensitive Pinus halepensis seed sources. *BMC Plant Biol* 18(1):236. doi: 10.1186/s12870-018-1464-5

Tambussi EA, Bartoli CG, Guiamet JJ, Beltrano J, Araus JL (2004) Oxidative stress and photodamage at low temperatures in soybean (*Glycine max* L. Merr.) leaves. *Plant Sci* 167(1):19–26

Tang T, Liu P, Zheng G, Li W (2016) Two phases of response to long-term moderate heat: Variation in thermotolerance between *Arabidopsis thaliana* and its relative *Arabis paniculata*. *Phytochemistry* 122:81–90. doi: 10.1016/j.phytochem.2016.01.003

Thakur P, Kumar S, Malik JA, Berger JD, Nayyar H (2010) Cold stress effects on reproductive development in grain crops: An overview. *Environ Exp Bot* 67(3):429–443. doi: 10.1016/j.envexpbot.2009.09.004

Turan O, Ekmekçi Y (2011) Activities of photosystem II and antioxidant enzymes in chickpea (*Cicer arietinum* L.) cultivars exposed to chilling temperatures. *Acta Physiol Plant* 33(1):67–78. doi: 10.1007/s11738-010-0517-7

Tuteja N, Tiburcio AF, Gill SS, Tuteja R (2012) *Improving Crop Resistance to Abiotic Stress*, 1st edn. Wiley-VCH Verlag GmbH & Co. KgaA, New Hamsphire

Upadhyaya H, Sahoo L, Panda SK (2013) Molecular physiology of osmotic stress in plants. In: *Molecular Stress Physiology of Plants* GR Rout, AB Das (ed). Springer, India. pp. 179–192

Ushimaru T, Maki Y, Sano S, Koshiba K, Asada K, Tsuji H (1997) Induction of enzymes involved in the ascorbate-dependent antioxidative system, namely, ascorbate peroxidase, monodehydroascorbate reductase and dehydroascorbate reductase, after exposure to air of rice (*Oryza sativa*) seedlings germinated under water. *Plant Cell Physiol* 38(5):541–549

Valluru R, Van den Ende W (2008) Plant fructans in stress environments: Emerging concepts and future prospects. *J Exp Bot* 59(11):2905–2916

Vangronsveld J, Clijsters H (1994) Toxic effects of metals. In: *Plants and the Chemical Elements. Biochemistry, Uptake, Tolerance and Toxicity, Verlagsgesellschaft, Weinheim* Farago ME (ed). VCH, Publishers, Weinheim, Germany. pp. 150–177

Verma D, Lakhanpal N, Singh K (2019) Genome-wide identification and characterization of abiotic-stress responsive SOD (superoxide dismutase) gene family in *Brassica juncea* and *B. rapa*. *BMC Genomics* 20(1):227. doi: 10.1186/s12864-019-5593-5

Vighi IL, Benitez LC, do Amaral MN, Auler PA, Moraes GP, Rodrigues GS, Da Maia LC, Pinto LS, Braga EJ (2016) Changes in gene expression and catalase activity in Oryza sativa L. under abiotic stress. *Genet Mol Res* 15(4):gmr15O4897

Wahid A, Gelani S, Ashraf M, Foolad MR (2007) Heat tolerance in plants: An overview. *Environ Exp Bot* 61(3):199–223 https://doi.org/10.1016/j.envexpbot.2007.05.011

Wang GP, Hui Z, Li F, Zhao MR, Zhang J, Wang W (2010) Improvement of heat and drought photosynthetic tolerance in wheat by overaccumulation of glycinebetaine. *Plant Biotechnol Rep* 4(3):213–222

Wang K, Zhang X, Goatley M, Ervin E (2014) Heat shock proteins in relation to heat stress tolerance of creeping bentgrass at different N levels. *PLOS ONE* 9(7):e102914

Wang LJ, Li SH (2006) Salicylic acid-induced heat or cold tolerance in relation to Ca²⁺ homeostasis and antioxidant systems in young grape plants. *Plant Sci* 170(4):685–694

Wang QL, Chen JH, He NY, Guo FQ (2018) Metabolic reprogramming in chloroplasts under heat stress in plants. *Int J Mol Sci* 19(3):849

Wang W, Chen Q, Hussain S, Mei J, Dong H, Peng S, Huang J, Cui K, Nie L (2016) Pre-sowing seed treatments in direct-seeded early rice: Consequences for emergence, seedling growth and associated metabolic events under chilling stress. *Sci Rep* 6:19637. doi: 10.1038/srep19637

Wen X-P, Pang X-M, Matsuda N, Kita M, Inoue H, Hao Y-J, Honda C, Moriguchi T (2008) Over-expression of the apple spermidine synthase gene in pear confers multiple abiotic stress tolerance by altering polyamine titers. *Transgen Res* 17(2):251–263

Wi SJ, Kim WT, Park KY (2006) Overexpression of carnation S-adenosylmethionine decarboxylase gene generates a broad-spectrum tolerance to abiotic stresses in transgenic tobacco plants. *Plant Cell Rep* 25(10):1111–1121

Xiang N, Li C, Li G, Yu Y, Hu J, Guo X (2019) Comparative evaluation on vitamin E and carotenoid accumulation in sweet corn (*Zea mays* L.) seedlings under temperature stress. *J Agric Food Chem* 67(35):9772–9781

Xie X, Yoneyama K, Yoneyama K (2010) The strigolactone story. *Annu Rev Phytopathol* 48:93–117. doi: 10.1146/annurev-phyto-073009-114453

Xie Z, Wang A, Li H, Yu J, Jiang J, Tang Z, Ma D, Zhang B, Han Y, Li Z (2017) High throughput deep sequencing reveals the important roles of microRNAs during sweet potato storage at chilling temperature. *Sci Rep* 7(1):16578. doi: 10.1038/s41598-017-16871-8

Xu M, Dong J, Zhang M, Xu X, Sun L (2012) Cold-induced endogenous nitric oxide generation plays a role in chilling tolerance of loquat fruit during postharvest storage. *Postharvest Biol Tec* 65:5–12. doi: 10.1016/j.postharvbio.2011.10.008

Xu S, Hu J, Li Y, Ma W, Zheng Y, Zhu SJ (2011) Chilling tolerance in Nicotiana tabacum induced by seed priming with putrescine. *Plant Growth Regul* 63(3):279–290. doi: 10.1007/s10725-010-9528-z

Xu Y, Zhan C, Huang B (2011) Heat shock proteins in association with heat tolerance in grasses. *Int J Proteomics:*529648. doi: 10.1155/2011/529648

Yadav A, Singh J, Ranjan K, Kumar P, Khanna S, Gupta M, Kumar V, Wani SH, Sirohi A (2020) Heat shock proteins: Master players for heat-stress tolerance in plants during climate change. *HEAT Stress Tolerance Plants Physiol Mol Genet Perspect* 23:189–211

Yadav AK, Singh D, Arya RK (2013) Morphological characterization of Indian mustard (*Brassica juncea*) genotypes and their application for DUS testing. *Indian J Agric Sci* 83(12):29–41

Yadav SK (2010) Cold stress tolerance mechanisms in plants: A review. *Agron Sustain Dev* 30(3):515–527

Yamauchi Y, Furutera A, Seki K, Toyoda Y, Tanaka K, Sugimoto Y (2008) Malondialdehyde generated from peroxidized linolenic acid causes protein modification in heat-stressed plants. *Plant Physiol Biochem* 46(8–9):786–793

Yancey PH (2005) Organic osmolytes as compatible, metabolic and counteracting cytoprotectants in high osmolarity and other stresses. *J Exp Biol* 208(15). doi: 10.1242/jeb.01730

Yang JC, Zhang JH, Liu K, Wang ZQ, Liu LJ (2007) Involvement of polyamines in the drought resistance of rice. *Jexp Bot* 58(6):1545–1555. doi: 10.1093/jxb/erm032

Yang R, Guo L, Wang J, Wang Z, Gu Z (2017) Heat shock enhances isothiocyanate formation and antioxidant capacity of cabbage sprouts. *J Food Process Preserv* 41(4):1–10

Yang X, Liang Z, Lu C (2005) Genetic engineering of the biosynthesis of glycinebetaine enhances photosynthesis against high temperature stress in transgenic tobacco plants. *Plant Physiol* 138(4):2299–2309

Yang X, Yang YN, Xue LJ, Zou MJ, Liu JY, Chen F, Xue HW (2011) Rice ABI5-Like1 regulates abscisic acid and auxin responses by affecting the expression of ABRE-containing genes. *Plant Physiol* 156(3):1397–1409. doi: 10.1104/pp.111.173427

Young AJ, Lowe GM (2001) Antioxidant and prooxidant properties of carotenoids. *Arch Biochem Biophys* 385(1):20–27. doi: 10.1006/abbi.2000.2149

Yu J, Su D, Yang D, Dong T, Tang Z, Li H, Han Y, Li Z, Zhang B (2020) Chilling and heat stress-induced physiological changes and microRNA-related mechanism in Sweetpotato (*Ipomoea batatas* L.). *Front Plant Sci* 11:687

Yuan Y, Qian H, Yu Y, Lian F, Tang D (2011) Thermotolerance and antioxidant response induced by heat acclimation in Freesia seedlings. *Acta Physiol Plant* 33(3):1001–1009. doi: 10.1007/s11738-010-0633-4

Zandalinas SI, Rivero RM, Martínez V, Gómez-Cadenas A, Arbona V (2016) Tolerance of citrus plants to the combination of high temperatures and drought is associated to the increase in transpiration modulated by a reduction in abscisic acid levels. *BMC Plant Biol* 16(1):105

Zhan X, Wang B, Li H, Liu R, Kalia RK, Zhu JK, Chinnusamy V (2012) Arabidopsis proline-rich protein important for development and abiotic stress tolerance is involved in microRNA biogenesis. *Proc Natl Acad Sci* 109(44):18198–18203

Zhang L, Wu M, Teng Y, Jia S, Yu D, Wei T, Chen C, Song W (2018) Overexpression of the Glutathione Peroxidase 5 (RcGPX5) Gene From Rhodiola crenulata Increases Drought Tolerance in Salvia miltiorrhiza. *Front Plant Sci* 9:1950

Zhao J, Sun Z, Zheng J, Guo X, Dong Z, Huai J, Gou M, He J, Jin Y, Wang J, Wang G (2009) Cloning and characterization of a novel CBL-interacting protein kinase from maize. *Plant Mol Biol* 69(6):661–674. doi: 10.1007/s11103-008-9445-y

Zhou Y, Li J, Wang J, Yang W, Yang Y (2018) Identification and characterization of the glutathione peroxidase (GPX) gene family in watermelon and its expression under various abiotic stresses. *Agronomy* 8(10):206

Zhou C, Zhu C, Fu H, Li X, Chen L, Lin Y, Lai Z, Guo Y (2019) Genome-wide investigation of superoxide dismutase (SOD) gene family and their regulatory miRNAs reveal the involvement in abiotic stress and hormone response in tea plant (*Camellia sinensis*). *PLOS ONE* 14(10):e0223609

Źróbek-Sokolnik A (2012) Temperature stress and responses of plants. In: *Environmental Adaptations and Stress Tolerance of Plants in the Era of Climate Change* P Ahmad, M Prasad (eds). Springer, New York. pp. 113–134

14 Insights into the Enzymatic Antioxidants and Their Genetic Expressions Responses of Plants to Heavy Metals

Mona F.A. Dawood, Amira M.I. Mourad, Dalia Z. Alomari and Arafat Abdel Hamed Abdel Latef

CONTENTS

14.1 INTRODUCTION

Modern life is characterized by excessive anthropogenic activities, rapid industrialization, and modern agricultural practices, which are the main causes of heavy metal burst spread in the environment. Various practices such as the use of pesticides, fertilizers, municipal and compost wastes, as well as the emission of polluted metals from smelting industries and metalliferous mines, release harmful metals into the environment (Zaid et al. 2019, 2020; Shirani 2020). Metals/metalloids include lead (Pb), cadmium (Cd), mercury (Hg), arsenic (As), chromium (Cr), copper (Cu), selenium (Se), nickel (Ni), silver (Ag) and zinc (Zn). Other less common metallic contaminants include aluminum (Al), cesium (Cs), cobalt (Co), manganese (Mn), molybdenum (Mo), strontium (Sr) and uranium (U) (Singh et al. 2011; Dhalaria et al. 2020), and tungsten (W) (Dawood and Azooz 2019), antimony (Sb) (Ortega et al. 2017) are of the soil and water concomitants. As reviewed by Nagajyoti et al. (2010), Fe, Cu and Zn are considered to be essential for plants and animals. The availability of elementary metals varies, and metals such as Cu, Zn, Fe, Mn, Mo, Ni and Co are essentially considered to be micronutrients, whose uptake in excess to the plant requirements results in toxic effects. Some heavy metals, such as Cd, Cr, Pb, Al, Hg, etc., although being non-essential and without physiological function, are very toxic even at very low concentrations (Dubey et al. 2018). Other heavy metals are also called trace elements due to their presence in trace (10 mg kg^{-1} or mg L^{-1}) or ultra-trace (1 μg kg^{-1} or μg L^{-1}) quantities in the environmental matrices (Singh et al. 2015; Sami et al. 2020).

Biological organisms are unable to degrade metals, so metals remain in their body parts and environment, thus having a major consequence for health. Furthermore, irrespective of the origin of the metals in the soil, high concentrations of many metals cause disintegration of soil quality, attenuate crop yield, lead to poor quality of agricultural products and pose significant hazards to human, animal and ecosystem health (Singh et al. 2011). In plants, the exposure to heavy metals mediates multi-physiological and biochemical alterations, and plants respond by inducing and/or adopting a series of mechanisms to be able to withstand the negative consequences of heavy metal toxicity. As heavy metals act as external stimuli, various mechanisms have been triggered by plants, including (i) sensing of external stress stimuli, (ii) signal transduction and transmission of a signal into the cell and (iii) inducing appropriate measures to counterbalance the adverse impacts of stress stimuli by mediating the physiological, biochemical and molecular status of the cell (Singh et al. 2015). Ara and Sinha (2014) displayed that the heavy metal stress responses in plants include mitogen-activated protein kinase cascades which might alter Ca level in the cytosol and finally transcriptional mediation of different stress-responsive genes. Generally, heavy metals adversely affect plants through (i) resemblance with the nutrient cations causing competition for absorption at root surface; for instance, As, W and Cd compete with P, Mo and Zn, respectively, for their absorption; (ii) explicit interplay of heavy metals with sulfhydryl group (-SH) of functional proteins, which deactivates their structure and function, and thus causes their inactivation; (iii) displacement of essential cations from specific binding sites, thereby losing their function and (iv) generation of reactive oxygen species (ROS), which consequently damage the macromolecules (DalCorso et al. 2013; Singh et al. 2015). Consequently, it is crucial to elucidate plants' responses to heavy metal stress in order to search for new techniques of resilience and quantitatively and qualitatively find mechanisms to lessen the impact of their presence on crops.

Once inside the cell, heavy metals change metabolic pathways, accounting for the reduction of growth and lower biomass accumulation on different plants species (El-Amier et al. 2019; Sami et al. 2020; Ahmad et al. 2020; Dawood and Azooz 2019, 2020; Choudhary et al. 2020; da Silva Rodrigues et al. 2020; Espinosa-Vellarino et al. 2020). Heavy metal toxicity also induces stunted stem and root length, and leads to chlorosis in younger leaves that can extend to the older leaves after prolonged exposure (Gangwar and Singh 2011; Gangwar et al. 2011; Srivastava et al. 2012; Dawood and Azooz 2019, 2020). These toxic effects fluctuate depending on the heavy metal applied, the concentration, plant tested, growth stage, etc. At the cellular and molecular levels, heavy metal toxicity affects plants in many ways. For instance, it alters the key physiological and biochemical processes such as seed germination, pigment synthesis, photosynthesis, gas exchanges, respiration, inactivation and denaturation of enzymes, blocks functional groups of metabolically important molecules, affects hormonal balance, nutrient assimilation and protein synthesis, and can lead to cell death and DNA replication (Nagajyoti et al. 2010; Keunen et al. 2011; He et al. 2012; Hossain et al. 2012; Singh et al. 2013, 2015; Imtiaz et al. 2018; Małecka et al. 2019). Trace metal element toxicity includes changes in the chlorophyll concentration in leaves and damage of the photosynthetic apparatus, inhibition of transpiration and destruction of carbohydrate metabolism, as well as nutrition and oxidative stress, which collectively affect plant development and growth (Bankaji et al. 2014; Malecka et al. 2015, 2019; Imtiaz et al. 2018; Dawood and Azooz 2019, 2020; Jan et al. 2020).

Heavy metals influence the machinery of PSI and PSII and alter stomatal function, which accounts for the reduction of photosynthesis (Appenroth 2010). In this respect, heavy metals mainly affect two key enzymes of CO_2 fixation, ribulose 1, 5-bisphosphate carboxylase (RuBisCO), and phosphoenolpyruvate carboxylase (PEPC). Heavy metals as Cd^{2+} ions lower the activity of RuBPC and damage its structure by substituting for Mg^{2+} ions, which are important cofactors of carboxylation reactions, and may also shift RuBisCO activity towards oxygenation reactions (Krantev et al. 2008). Reduction in chlorophyll content, net photosynthetic rate, stomatal conductance and water use efficiency was reported (Singh et al. 2008). At toxic concentrations of heavy metals, many studies recorded retardation in the respiration of several plants (Sandalio et al. 2001; Lösch 2004; Mahmood et al. 2013). Phytohormone homeostasis is another response of plants which experienced heavy metal stress (Kanwar et al. 2015). Increased levels of abscisic acid or ethylene were reported by Zengin (2006) and Maksymiec (2011) under heavy metal stress. Also, Maksymiec (2011) reported the contribution of jasmonic acid in *Phaseolus coccineus* as an early response to cadmium metal stress. Hossain et al. (2012) reported that Cu toxicity induced ethylene synthesis, thereby caused senescence via reducing cell growth and accelerated cell wall rigidity through lignification. Kanwar et al. (2015) reported that As(V) induced the synthesis of 4BRs, castasterone, teasterone, 24-epibrassinolide and typhasterol in mustard plants. Heavy metal resistance includes various modes of action such as immobilization, exclusion, chelation and metal sequestration in the vacuole, and the expression of stress-inducible proteins related to (Dubey et al. 2018). It is well-known fact that chelation is one of the prevalent pathways to deactivate heavy metals where low molecular weight chelators such as reduced glutathione, phytochelatins and metallothionein bind to heavy metals and assist their translocation into the vacuoles (Shri et al. 2014; Viehweger 2014; Dubey et al. 2018; Dawood and Azooz 2020). Generally, multiple pieces of the research proposed mechanisms of transition metal accumulation by plants such as phytoaccumulation, phytoextraction, phytovolatilization, phytodegradation and phytostabilization (Poschenrieder et al. 2006; Singh et al. 2011).

Alongside the aforementioned detrimental effects, heavy metals are induced through the exacerbation of ROS, causing oxidative stress resulting in clear imbalances of redox homeostasis. Besides this, stress conditions also trigger the synthesis of reactive nitrogen species (RNS) including nitric oxide (NO), peroxynitrite ($ONOO^-$) and S-nitroso-glutathione (GSNO) (Wang et al. 2013; Espinosa-Vellarino et al. 2020), causing nitrosative stress in the cells (Corpas and Barroso 2013; Sallam et al. 2019; Bashandy et al. 2020), and the interaction between ROS and RNS propagate what is known as nitro-oxidative stress (Sahay and Gupta 2017; Kolbert et al. 2019). Heavy metal and metalloid toxicity promote high levels of ROS and RNS production (Feng et al. 2009; Pan et al. 2011; Corpas and Barroso 2017; Feigl et al. 2015; Chai et al. 2016; Ortega et al. 2017). Consequently, growth and biomass production, the content of photosynthetic pigments and levels of photosynthesis, and the absorption and distribution of other nutrients are all altered (Natasha et al. 2019). As a protection against the damage caused by this ROS and RNS imbalance, plants have developed both enzymatic and non-enzymatic antioxidant systems capable of eliminating these reactive species more efficiently (Sallam et al. 2019; Espinosa-Vellarino et al. 2020). The increased production of ROS and RNS and their interactions may act on the antioxidant systems involved in the response to heavy metals, both inducing the activity of these systems and modifying the expression of the genes involved in them (Sahay and Gupta 2017; Espinosa-Vellarino et al. 2020).

This chapter will discuss the reactive oxygen production and inactivation mechanisms by ROS metabolizing enzymatic and non-enzymatic mechanisms, and we will focus predominately on antioxidants variations under different heavy metals.

14.2 OXIDATIVE STRESS UNDER HEAVY METAL STRESS

Under normal conditions, ROS plays several essential roles in regulating the expression of different genes. Reactive oxygen species control numerous processes like the cell cycle, plant growth, abiotic stress responses, systemic signaling, programmed cell death, pathogen defense and development (Shahid et al. 2014). As has been mentioned, heavy metals disturb redox homeostasis by stimulating the formation of ROS including free radicals (superoxide anion, O_2^-; hydroperoxyl radical, HO_2^-; alkoxy radical, RO^-; carbonate radical, $CO_3^{\bullet-}$; semiquinone, $SQ^{\bullet-}$; and hydroxyl radical, ^-OH; and non-radical molecules (hydrogen peroxide, H_2O_2; ozone, O_3; hypochlorous acid, HOCl; hypobromous acid, HOBr; hypoiodous acid, HOI; and HOC and singlet oxygen, $^1O_2$ in different cell compartments (Hasanuzzaman et al. 2020). Heavy metals tend to generate ROS if they exceed permissible limits (Shahzad et al. 2018; Dawood and Azooz 2020). Based on the physical and biochemical characteristics of bioactive-metals, these metals can be classified into two groups: redox metals like Cr, Cu and Fe, and non-redox metals like Cd, Hg, Ni and Zn. Redox-active metals can produce oxidative injuries in plants via Haber-Weiss and Fenton reactions, which consequently generate ROS and lead to disturbing the balance between prooxidant and antioxidant levels (Jozefczak et al. 2012). However, redox-inactive metals (e.g. Pb, Cd, Ni, Al, Mn and Zn) cannot generate ROS directly by participating in biological redox reactions such as Haber-Weiss/Fenton reactions. However, these metals form covalent bonds with the protein sulfhydryl groups, as these metals can share electrons (Sharma et al. 2020). Also, these metals induce ROS generation via different indirect mechanisms, such as stimulating the activity of NADPH oxidases, displacing essential cations from specific binding sites of enzymes and inhibiting enzymatic activities from their affinity for -SH groups on the enzyme (Shahid et al. 2014). It has been shown in several studies that ROS generated by NADPH oxidases during stress are channeled by the plant to serve as a stress signal to activate acclimation and defense mechanisms, which in turn counteract oxidative stress (Miller et al. 2008; Singh et al. 2015). Therefore, the fate of ROS (i.e. whether it will act as a signaling molecule or damaging one) in the cellular system depends upon the output of many complex processes that involve the antioxidative system, signaling cascades, redox alterations, etc. When the generation of ROS exceeds that of the scavenging potential of antioxidants, oxidative stress occurs (Singh et al. 2015).

The high content of these species as a response to heavy metal toxicity deteriorates the intrinsic antioxidant defense system of cells and causes oxidative stress. This process is one of the primary reasons for the change in plant biology at the physiological level under heavy metal toxicity. Plants may suffer through various drastic physiological changes, which are mainly due to the imbalance in the generation and scavenging of ROS, termed as an oxidative burst (Shahzad et al. 2018; Sharma et al. 2019, 2020). Cells undergo toxic symptoms as a result of the interaction between ROS and biomolecules. Heavy metal-induced ROS causes lipid peroxidation membrane dismantling, and damage to DNA, proteins and carbohydrates. However, the toxic effects of heavy metal-induced ROS on plant macromolecules vary and depend on the duration of exposure, stage of plant development, the concentration of heavy metals tested, the intensity of plant stress and the particular organs studied (Abdel Latef 2013; Abdel Latef and Abu Alhamad 2013; Abdel Latef et al. 2020 a, b; Shahid et al. 2014;). Oxidative stress causes lipid peroxidation, damages nucleic acids and proteins and alters carbohydrate metabolism, resulting in cell dysfunction and death (Hasanuzzaman et al. 2019, 2020).

The primary defense mechanism for heavy metal tolerance is the reduced absorption of these metals into plants or their sequestration in root cells. Secondary heavy metal tolerance mechanisms include activation of antioxidant enzymes and the binding of heavy metals by phytochelatins, glutathione and amino acids. These defense systems work in combination to manage the cascades of oxidative stress and to defend plant cells from the toxic effects of ROS (Shahid et al. 2014).

14.3 ANTIOXIDATIVE RESPONSE TO OXIDATIVE DAMAGE BY HEAVY METALS

To prevent heavy metal-induced ROS toxicity, plants have developed different defense mechanisms by which they can convert ROS into less toxic by-products (Tang et al. 2010; Alvarez et al. 2012). These mechanisms involve blocking metal entrance into plants, elevated root excretion of metals, limiting toxic metal accumulation in sensitive tissue, chelation by organic molecules, metal binding to the cell wall and sequestration in vacuoles. These mechanisms help plants to sustain their cellular redox state and mitigate the noxious impacts caused by oxidative stress (Tang et al. 2010). The majority of these defense mechanisms depend on metabolic mediation of natural compounds such as phytochelatins (PCs), reduced glutathione (GSH), carotenoids and tocopherols, and enzymatic antioxidant systems including catalase (CAT and EC 1.11.1.6), superoxide dismutases (SOD and EC 1.15.1.1), ascorbate peroxidase (APX, EC 1.11.1.11), peroxidase (POD, EC 1.11.1.7), guaiacol peroxidase (POD, EC 1.11.1.7), glutathione reductase (GR, EC 1.6.4.2), monodehydroascorbate reductase (MDHAR, EC 1.6.5.4) and dehydroascorbate reductase (DHAR, EC 1.8.5.1). The increased levels of these metabolic intermediary compounds and antioxidant enzymes lead to increased stress tolerance against heavy-metal-induced ROS (He et al. 2011; Shahid et al. 2014; Espinosa-Vellarino et al. 2020). The effect of essential and non-essential heavy metals on the induction of radical and non-radical reactive oxygen species (ROS) is represented in Figure 14.1.

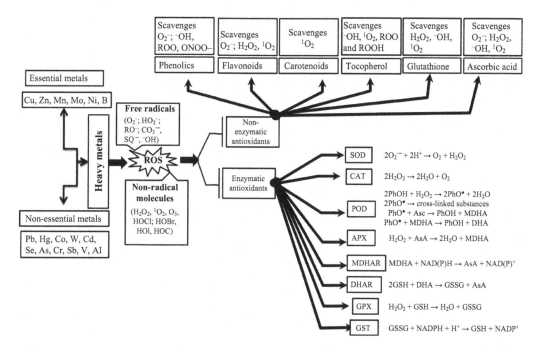

FIGURE 14.1 The effect of essential and non-essential heavy metals on the induction of radical and non-radical reactive oxygen species (ROS). The ROS have been scavenged by different enzymatic and non-enzymatic antioxidants. Essential heavy metals include copper (Cu), Zinc (Zn), manganese (Mn), molybdenum (Mo), nickel (Ni) and boron (B) as well as non-essential heavy metals such as lead (Pb), mercury (Hg), cobalt (Co), tungsten (W), cadmium (Cd), selenium (Se), arsenic (As), chromium (Cr), antimony (Sb), vanadium (V) and aluminum (Al). Reactive oxygen species including free radicals (superoxide anion, O2—; hydroperoxyl radical, HO2—; alkoxy radical, RO—; carbonate radical, CO3•—; semiquinone, SQ•—; and hydroxyl radical, —OH; and non-radical molecules (hydrogen peroxide, H2O2, ozone (O3), hypochlorous acid (HOCl), hypobromous acid (HOBr), hypoiodous acid (HOI), and HOC and singlet oxygen, 1O2. The antioxidant enzymes include catalase (CAT and EC 1.11.1.6), superoxide dismutase (SOD and EC 1.15.1.1), ascorbate peroxidase (APX, EC 1.11.1.11), guaiacol peroxidase (POD, EC 1.11.1.7), glutathione peroxidase (GPX, EC 1.11.1.9), glutathione reductase (GR, EC 1.6.4.2), monodehydroascorbate reductase (MDHAR, EC 1.6.5.4) and dehydroascorbate reductase (DHAR, EC 1.8.5.1).

Different antioxidants responded differently to the applied heavy metals (represented in Table 14.1) based on dose, duration, species and the stage of growth. The following discussion will include detailed antioxidant responses to various heavy metals.

14.3.1 ANTIOXIDANT RESPONSES OF PLANTS TO TOXIC EFFECTS OF ESSENTIAL METALS

14.3.1.1 Copper (Cu)

Cu is steadily accumulated in the soil and water environments due to deregulated industrial discharge from tanneries, electroplating units and rampant agricultural practices such as excessive use of Cu enriched fungicides (Sharma et al. 2019). The study of antioxidant responses of different plants to Cu toxicity showed variable responses. In this regard, Xiong and Wang (2005) indicated that Cu phytotoxicity in *Brassica pekinensis*, where Cu treatments enhanced electrolyte leakage, a membrane damage marker and POD activity, showed a significant correlation between Cu concentration in shoots with electrolyte leakage and POD activity. A study by Mei et al. (2015) reported triggering the activities of POD and SOD in leaf and root of cotton genotypes under Cu stress (100 ppm for four days). On the other hand, Fidalgo et al. (2011) observed that SOD and CAT did not change significantly while APX activity decreased in *Solanum nigrum* exposed to 200 μmol L^{-1} Cu for four weeks. Yadav and co-authors (2018) reported increases in the activities of antioxidant enzymes superoxide dismutase (SOD), dehydroascorbate reductase (DHAR), glutathione reductase (GR), glutathione-S-transferase (GST), glutathione peroxidase (GPX) and ascorbate peroxidase (APX) in *B. juncea* plants treated with 0.0005 M Cu.

Georgiadou et al. (2018) reported that increasing Cu in the form of $CuCl_2$ at concentrations of 200, 500 and 1000 ppm on basil plants increased cellular damage and nitro-oxidative stress (NO and H_2O_2), correlating with an induction in the activity of reactive oxygen and nitrogen species metabolism enzymes (SOD, CAT, APX and NR). The same study also recorded that the treatment with Cu led to increased concentration of the allergenic protein profilin, but a decrease in the concentration of total proteins, which was likely due to proteolysis and antioxidant capacity in terms of phosphomolybdenum and ferric reducing antioxidant power assay (FRAP). A further study by Małecka et al. (2019) noted that the shoots and roots of *Brassica junicea* germinated in 50 μM of $CuSO_4$ increased the activities of APX, SOD and POD. The same study also revealed that CuZnSOD increased for shoot and roots of *B. juncea* but the expression of MnSOD transcript decreased for both organs, and Cu-presence did not affect MnSOD (25 kDa) and CuZnSOD (15 kDa and 20 kDa) proteins subunits. Hossain et al. (2020) stated that the low level of Cu stress (0.3 mM) induced oxidative stress and membrane damage in lentil seedlings without affecting CAT, APX, MDHAR, GR and GST activities, whilst DHAR activity reduced and APX activity stimulated. On the other hand, high Cu-stress (3 mM) retarded CAT and DHAR activities, and MDHAR was not detected at all at this concentration, whilst APX and GR increased their activities and their relative expression markedly compared to non-stressed plants, and GST showed no response to the same level of Cu stress. A recent study by Saleem et al. (2020) revealed that as the concentration of Cu increased in different mixtures of Cu-contaminated soil and natural soil, the activities of POD, SOD, APX and CAT elevated.

Using copper in nanoparticulate (NP) form has been also studied extensively in recent years, which can affect plants positively or negatively. The study of Kim et al. (2012) stated that the application of Cu-NPs at the concentration of 100 mg L^{-1} increased SOD, CAT and POD activities in cucumber plants. Copper NPs were more toxic than copper sulfate in lettuce and decreased CAT and APX activities in both shoots and roots (Trujillo-Reyes et al. 2014). On the other hand, Hong et al. (2015) found that Cu-NPs reduced the activities of antioxidant enzymes in lettuce and alfalfa plants in a dose-dependent manner. Yang et al. (2020) reported that CuO NP at concentrations of 0, 62.5, 125 and 250 mg/L showed a variance of response in rice (*Oryza Sativa* L.) seedlings in terms of concentration applied and organ of the rice that was tested. The CAT activity of the rice leaves

TABLE 14.1
Representation of the Effect of some Heavy Metals on Plant Antioxidative Responses

Heavy Metal	Stress Imposed/Plant	Antioxidative Response	Reference
Cadmium (Cd)	Spinach	CAT ↑, POD ↑, GR↑, APX—	Pinto et al. (2017)
	Triticum aestivum	SOD ↑, POD ↑, CAT ↑	Kaya et al. (2020 b)
	Pisum sativium	SOD ↑, APX ↑, CAT ↑↓, GR ↑	Sager et al. (2020)
Lead (Pb)	*Eichhornia crassipes*	APX ↑, POD ↑, SOD ↑, POD ↑ and then decreased at 1000 mg/L	Malar et al. (2015)
	Coronopus didymus	SOD ↑, CAT ↑, APX ↑, GPX ↑, GR ↑, at the highest Pb concentration, SOD↓ and CAT ↓	Sidhu et al. (2016)
	Phanerochaete chrysosporium	SOD ↑, POD ↓, CAT ↓	Huang et al. (2017)
	Satureja hortensis L.	APX ↑, CAT ↑	Ghotbi-Ravandi et al. (2019)
	Fagopyrum kashmirianum	SOD ↑, APX ↑, POD ↑, GR ↑, GST ↑, CAT ↓	Hakeem et al. (2019)
	Ricinus communis	MDHAR ↓, APX ↓, POD ↑, SOD ↑, CAT ↑	Kiran et al. (2019)
Antimony (Sb)	Sunflower	CAT ↑, POD ↑, GR↑, APX↑, DHAR↑, GST expression↑	Ortega et al. (2017)
Cd	*Pisum saitvum*	SOD ↑, POD ↑, CAT ↑, APX ↑, NADH-oxidase ↑, GR ↑	El-Amier et al. (2019)
	Althaea rosea	SOD ↑, POD ↑, CAT ↑, APX ↑	Huang et al. (2020)
	Hydrilla verticillata	SOD ↑, POD ↑, CAT ↑	Zhang et al. (2020)
Arsenic (As)	Wheat	APX ↑, GST ↑, GR ↑, MDHAR —, CAT —, DHAR↓ ,GPX ↓, GlyI↓, GlyII ↓	Hasanuzzaman and Fujita (2013)
	Hydrilla verticillata	SOD ↑, POD ↑, CAT ↓	Srivastava and Shrivastava (2017)
	Vicia faba	SOD ↑, APX ↑, CAT ↓, GR ↑, GlyI ↑, DHAR↓, MDHAR↓ and GlyII ↓,	Ahmad et al. (2020)
	Barley	SOD ↑, CAT ↑	Zvobgo et al. (2019)
Aluminium (Al)	*Fagopyrum* species	SOD ↑, APX ↑, POD ↓, GR ↑, GST ↑	Pirzadah et al. (2019)
Tungesten (W)	*Brocolli*	SOD ↑, APX ↓, GPX ↓, GST ↓, IPO ↑, SPO ↑, PPO ↑	Dawood and Azooz (2019)
Copper (Cu)	*Allium cepa*	SOD ↑, CAT ↑	Kalefetoğlu Macar et al. (2020)
	Lentil	CAT ↑, APX ↓, MDHAR ↓, DHAR ↓, GR ↑, GST —, GR and APX expression ↑	Hossain et al. (2020)
	Spirodela polyrhiza (L.)	SOD ↑, CAT ↑, POD ↑ at 1 & 10 μM SOD ↓, CAT ↓, POD ↓ at 20 & 40 μM	Singh et al. (2020)
Chromium (Cr)	*Glycine max L.*	CAT ↑, POD ↑, SOD ↑	Bilal et al. (2018)
Molybdnum (Mo)	Soybean	POD ↑, CAT ↑, APX ↑	Xu et al. (2018)
Zinc (Zn)	*Ocimum basilicum*	CAT —, APX —, SOD —	Georgiadou et al., (2018)
	Chenopodium murale	CAT ↑, POD ↑, SOD ↑	Zoufan et al. (2018)
	Solanum nigrum	CAT ↑, APX ↑, SOD ↑	Sousa et al. (2020)
Selenium (Se)	Rice	SOD↑, CAT↓, GPX↓, APX↑, DHAR↑, MDHAR↑ and GR↑, GST ↑	Mostofa et al. (2020)

(Continued)

TABLE 14.1 (CONTINUED)

Representation of the Effect of some Heavy Metals on Plant Antioxidative Responses

Heavy Metal	Stress Imposed/Plant	Antioxidative Response	Reference
Vanadium (V)	*Cicer arietinum*	CAT ↑, POD ↑, SOD ↑	Imtiaz et al. (2018)
	Dog tail grass	SOD ↓, CAT ↓, POD ↓	Aihemaiti et al. (2019)
Mercury (Hg)	*Spirodela polyrhiza* (L.)	SOD ↑, CAT ↑, POD ↑ at 0.1 & 0.2 μM	Singh et al. (2020)
		SOD ↓, CAT ↓, POD ↓ at 0.3 & 0. 4 μM	
	Garlic	SOD ↓, CAT ↓, POD —	Hu et al. (2020)
Nickel (Ni)	*Pisum saitvum*	SOD ↑, POD ↑, CAT ↑, APX ↑, NADH-oxidase ↑, GR ↑	El-Amier et al. (2019)
	Hydrilla verticillata	SOD ↑, POD ↑, CAT ↑	Zhang et al. (2020)
Cobalt (Co)	Barley	SOD ↑, CAT ↑	Lwalaba et al. (2017)

significantly decreased with 125 and 250 mg/L CuO NPs, whilst CAT activity at 62.5 CuO NPs showed no significant change compared to the control. Additionally, unremarkable changes were found in CAT and POD activities between the roots exposed to CuO NPs and the control roots. Also, the SOD activity was significantly upregulated by 166% in the rice roots exposed to 125 mg/L CuO NPs, but no alterations were observed in the rice leaves. The increasing SOD activity in the rice roots exposed to CuO NPs suggested that rice seedlings could generate significant amounts of ROS. However, there was no change in the activity of CAT and POD in rice roots exposed to CuO NPs. This result can probably be explained by the excess H_2O_2 existing in rice roots, which exceeded the maximum quenching activity of CAT and POD. These previous studies declared that many factors have been suggested as potential mechanisms of CuO NPs, causing phytotoxicity, DNA damage, metal ions released from NPs, ROS generation and oxidative stress (Liu et al. 2018) as well as deregulation of antioxidant system and chelation mechanisms.

14.3.1.2　Zinc (Zn)

Zn is an essential micronutrient required for many enzymes included in many physiological and metabolic pathways of plants. However, excess zinc or long-duration exposure causes Zn toxicity to plants. Besides natural resources, industrial activities such as mining, smelting and sewage sludge, as well as persistent use of Zn fertilizers, provide additional sources of Zn contamination (Mateos-Naranjo et al. 2014). Zn pollution can be also generated by anthropogenic activities, such as municipal wastewater discharge, coal-burning power plants, manufacturing processes involving zinc and atmospheric fallout (Wuana and Okieimen 2011; Dong et al. 2012). Zn accumulation negatively affected the growth as well as increased ROS, causing alterations in antioxidants. Previous studies showed that the accumulation of MDHAR was reported under Zn stress in *Phaseolus vulgaris* (Chaoui et al. 1997) and *Brassica juncea* (Prasad et al. 1999). Zhao et al. (2012) detected that *Phytolacca americana*, a metal hyperaccumulator, exposed to higher levels of Zn and even under severe oxidative damage (high lipid peroxidation degree) presented unaltered CAT activity, while at least partially inducing other enzymatic defenses. Other studies on *Raphanus sativus* (Ramakrishna and Rao 2015) and *Pinus sylvestris* (Ivanov et al. 2012) exhibited triggering of APX activity due to Zn stressor. Li et al. (2013) reported that Zn stress differentially affected the antioxidative responses of the tested organs of wheat exposed to various Zn levels. In wheat leaves, Zn excess did not change hydrogen peroxide or membrane lipid peroxidation, but enhanced SOD, POD, CAT, GR and APX activities. On the other hand, roots exhibited higher oxidative burst via increasing H_2O_2 and malondialdehyde (MDA) contents accompanied by the stimulation of SOD and the inhibition of POD and GR. The study of Jayasri and Suthindhiran (2017) showed that the application of zinc at different concentrations (0.5, 5, 10, 15 and 20 mg/L) stimulated the activity of catalase significantly on the fronds of duckweed (*Lemna minor*). Georgiadou et al. (2018) observed that $ZnCl_2$ at (Zn 720 ppm) exerted no significant effect on SOD, APX and CAT activities. Zoufan et al. (2018) reported that various Zn stress (150, 300 and 600 μM) enhanced the activities of CAT, POD and SOD in *Chenopodium murale*. Sousa et al. (2020) reported that when plants were exposed to 500 μM Zn, the activity of SOD and APX was enhanced in roots and shoots of *Solanum nigrum*, whilst Zn stress had no significant effects on the activity of CAT on both organs. Małecka et al. (2019) found that the activities of APX, SOD and POD were increased in response to 50μM of $ZnSO_4$ in the shoots and roots of *Brassica junicea*. Furthermore, a reduction in the expression of the gene encoding CuZnSOD was observed in the roots and leaves of plants treated with Zn, whilst the genes encoding MnSOD exhibited an increment only after 24 hours and 8 hours for shoots and roots, respectively. Also, western blotting indicates that the presence of trace metals does not increase the synthesis of the proteins CuZnSOD and MnSOD in the organs of *Brassica juncea* plants, but induces an increase in their activity.

The POD activity was remarkably inhibited by ZnO nanoparticles (50 mg/L). Zafar et al. (2016) reported that ZnO nanoparticles (500 to 1500 mg/L) negatively affect the *Brassica nigra* seed germination and seedling growth, and also increased antioxidative activities and non-enzymatic antioxidants contents. Zoufan et al. (2020) displayed that using Zn as ZnO nanoparticles induced phytotoxicity on *Chenopodium murale* plants under hydroponic culture via the induction of oxidative stress where lipid peroxidation, leaf H_2O_2 and leaf electrolyte leakage increased significantly compared with the control along with the increment of CAT, POD and SOD activities in the treated plants. Another study by Pavani et al. (2020) revealed that ZnONP's exposure significantly reduced growth parameters in the plants treated with 25 mg/100ml after 60 days compared to the control. This was correlated with elevated hydrogen peroxide level as well as promotion in the activities of POD, GR, CAT as well as ascorbic acid content (ASA) content. Thus, it seems that the performance of the antioxidant network against Zn or in its nano-size is particularly determined by plant species, plant organs, metal concentrations, growth conditions and exposure times.

14.3.1.3 Manganese (Mn)

Mn is an essential structural and catalytic microelement (needed in small amounts) of various proteins important for growth, development and plant tolerance to stress, but it is toxic when present in excess (Sheng et al. 2016). In recent years, with rapid industrialization, the Mn level in soil has been released into the ecosystem owing to the mining process and widespread use of Mn-containing fertilizers and sewage sludge (Luo et al. 2015). At toxic doses of Mn, the ROS have elevated, affecting the antioxidant system positively or negatively based on several factors. CAT showed lower activity in excess Mn treatment, which revealed a possible delay in quenching of H_2O_2 and toxic peroxides mediated by CAT and in turn an enhancement in ROS-mediated lipid peroxidation under Mn toxicity in tobacco (Edreva and Apostolova 1989), while CAT activity was increased by excess Mn in barley (Demirevska-Kepova et al. 2004). Shi et al. (2006) reported that the activities of SOD, APX, POD, GR and DHAR were increased by Mn stress on *Cucumis sativus* plants. In a study by Shi and Zhu (2008), the activities of SOD, POD, APX, DHAR and GR were shown to significantly increase, whilst CAT was reduced in cucumber. Saidi et al. (2014) recorded that Mn stress (100 μM) stimulated the activity of SOD and reduced CAT, APX and GPX activities, hence the levels of H_2O_2 and MDA were enhanced in sunflower leaves. Sheng et al. (2015) reported that Mn stress increased SOD, DHAR and GR activities in roots and leaves of polish wheat, whilst CAT and APX reduced eventually for both organs. Further excess Mn decreased DHAR activity in roots, whereas increased that of leaves. Sheng et al. (2016) reported that Mn stress increased activities of SOD, GR and DHAR and decreased APX in the leaves and roots of Polish wheat. Nazari et al. (2018) evaluated changes in primary and secondary metabolites in *Mentha aquatica* plants exposed to four levels of Mn supplementation (2, 40, 80 and 160 µM) and found significant increases in SOD, CAT and POD levels in plants treated with 80 and 160 µM Mn compared to control treatment (2 µM). A recent work of da Silva Rodrigues and co-workers (2020) found that excess Mn supplied at 2500 µM in hydroponic solution induced significant increases in SOD, CAT, APX and POD levels in soybean plants. In light of the above, the change of antioxidative enzyme activity under Mn toxicity may be related to plant species.

14.3.1.4 Nickel (Ni)

Ni is an essential micronutrient for plants, and in minute amounts enhances plant yield and quality. However, at high concentrations in the soil environment it becomes toxic. Ni is emitted into the environment from both natural and anthropogenic sources. Of particular concern is the elevated concentration of Ni received the agricultural soils by airborne-Ni particles. The primary sources of Ni emissions into the ambient air are the combustion of coal and oil for heat or power generation, Ni mining, steel manufacture and other miscellaneous sources, such as cement manufacture (Bhalerao et al. 2015). The phytotoxic effect of Ni is linked to oxidative stress in plants. Ni cannot produce ROS directly as it is not a redox-active metal. However, it indirectly affects many antioxidant enzymes. Exposure of plants to Ni at low concentrations (0.05 mM) and/or for short times has been shown to increase the activities of SOD, POD, GR and POD to enhance the activation of other antioxidant defenses and hence quench ROS (Freeman et al. 2004; Gomes-Juniora 2006). Application of Ni at 25 and 50 µg/L on barley plant stimulated SOD activity (Parlak 2016). The stimulation of SOD activity has also been reported in several plants exposed to Pb, Cu, Cd, Zn and Ni (Israr et al. 2011; Malecka et al. 2012; Kanwar et al. 2015; Yadav et al. 2018). Enhanced CAT and APX activity has been observed in various plant species after the application of trace metals Pb, Cu, Cd, Zn, Ni and As (Israr et al. 2011; Wang et al. 2009; Malecka et al. 2012; Kanwar et al. 2015; Yadav et al. 2018). In a study by Kumar et al. (2012), the impact of 200 and 400 µM nickel treatments on barley plants did not affect CAT activity in roots while a significant increase in CAT activity was observed only under 200 µM Ni in leaves. Furthermore, the activities of GPX, APX, SOD and GR are enhanced in leaves and roots under both nickel concentrations. On the other hand, Gajewska et al. (2006), examining the effect of Ni in wheat shoots, showed a decline in CAT activity following increased Ni concentration. Based on the fact that CAT is an iron-porphyrin and that high concentrations of Ni have been shown to decrease Fe, it was postulated that reduction in CAT activity in plant tissues subjected to excess Ni may result from the deficiency of metals essential for the biosynthesis of this enzyme molecule (Gajewska et al. 2006). Georgiadou et al. (2018) observed that the application of $NiCl_2.6H_2O$ at different concentrations (100, 210 and 500 ppm) induced the antioxidant enzymes SOD and APX, but not CAT which was lower than control at 500 ppm. El-Amier et al. (2019) stated that the treatment of *Pisum sativum* seedlings with Ni (100 mM $NiCl_2$ for three days) enhanced the activities of NADH-oxidase, APX, GR, CAT and SOD. Hasanuzzaman et al. (2019) observed an enhancement in the GSH and oxidized glutathione (GSSG) contents in *Oryza sativa* seedlings under Ni stress. Moreover, Ni stress induced the enhancement of SOD, GPX, APX, MDHAR, DHAR and GR activities.

Using NiO nanoparticles also affects radish growth and generated oxidative damage when used at concentrations of 0.25, 0.5, 1.0, 1.5 and 2 mg/mL. CAT activity increased markedly up to 1.5 mg/mL compared to control plants; however, it was reduced at the highest concentration of 2.0 mg mL^{-1} NiO nanoparticles. The same study reported that the activity of SOD was higher in NiO nanoparticles up to 1 mg/mL treated plants and then reduced at 1.5 and 2.0 mg/mL (Abdel-Salam et al. 2018). Chahardoli et al. (2020) investigated using different concentrations (0, 50, 100, 1000 and 2500 mg/L) of engineered nickel nanoparticles (NiO-NPs) in the hydroponic culture of *Nigella arvensis*. The antioxidants enzymes APX, CAT, SOD and POD exhibited increment in roots and shoots. By contrast, the antioxidant activities, formation of secondary metabolites and other related physiological parameters, such as the total antioxidant capacity, DPPH (2,2-diphenyl-1-picryl-hydrazyl-hydrate) scavenging activity and total saponin content, were inhibited after the concentration of

NiO-NPs was increased to 100 mg/L. In light of the above studies, when the oxidative burst exceeds the threshold of plant tolerance, plant antioxidant mechanisms and activities are diminished and weakened.

14.3.1.5 Boron (B)

B is an essential micronutrient in higher plants, although it is toxic in excess. Excessive boron may occur naturally in the soil or groundwater or be added to the soil from mine tailings, fertilizers or irrigation water, resulting in B toxicity to plants (Nable et al. 1997). Liu et al. (2017) observed that SOD and POD activities increased in wheat plants at 150 and 300 mg B Kg^{-1}, whilst the highest B level retarded CAT activity. The SOD also displayed higher activity in two cultivars of lentil (*Lens culinaris*) plants exposed to B stress (Tepe and Aydemir 2017). However, boron toxicity reduced the SOD activity of wheat (Genisel et al. 2017; Ozfidan-Konakci et al. 2020). Boron toxicity increased the activity of CAT in wheat (Genisel et al. 2017) and *Puccinellia tenuiflora* (Zhao et al. 2019), but the reduction in CAT activity was reported in wheat (Liu et al. 2017) and lentil (Tepe and Aydemir 2017). The boron-stressed plants stimulated higher POD activity in tomato (Kaya et al. 2011), *Puccinellia tenuiflora* (Zhao et al. 2019) and wheat (Liu et al. 2017). But the reduction of its activity was recorded for wheat (Ozfidan-Konakci et al. 2020). Another H_2O_2-metabolizing enzyme, APX, was found to increase under B stress in different plants, such as lettuce (Sahin et al. 2017), wheat (Yildiztugay et al. 2019) and linseed (Pandey and Verma 2017), and opposite findings revealed that a reduction in APX activities was reported for *Lemna minor* L. and *Lemna gibba* (Gür et al. 2016), wheat (Genisel et al. 2017) and rice (Farooq et al. 2019). One of the AsA-GSH cycle enzymes, GR, was found to be differentially affected by B toxicity. One study revealed increased GR activity (Genisel et al. 2017; Yildiztugay et al. 2019) and another reported diminished activity (Pandey and Verma 2017; Ozfidan-Konakci et al. 2020). The activities of DHAR and MDHAR, apart from the AsA-GSH cycle, were reduced in citrus under boron toxicity (Han et al. 2009), but DHAR activity showed no response to B toxicity in other studies (Yildiztugay et al. 2019; Ozfidan-Konakci et al. 2020). A recent study of Kaya et al. (2020 a) studied the toxicity of different concentrations of boron on pepper where retarded plant growth is associated with oxidative stress and increment of APX, CAT, POD, GR and glyoxalase II, whilst levels of DHAR, MDHAR and glyoxalase I were decreased. Also, Choudhary et al. (2020) stated that different concentrations of B (2.5, 5, 10, 20 and 30 mg/kg) increased the activity of CAT, POD and SOD as compared with the control in *Mentha arvensis* and *Cymbopogon fexuosus* plants. The activities of CAT and POD were maximally increased when 30 mg/kg B was applied to both plants. Thus, the different responses of plants to B toxicity can be seen.

14.3.1.6 Molybdenum (Mo)

Mo is an important micronutrient required by plants. In fields, Mo content in plants is closely related to Mo levels in the soil, and can be elevated by human activities. Industrialization has led to Mo accumulation in soil and groundwater, posing a potential health risk. Mining and processing of Mo also emit a large amount of Mo into the ecosystem. In an agricultural field surrounding a Mo mining site, the Mo concentrations increased, indicating possible pollution and a potential health risk for the ecosystem (Shi et al. 2018). Earlier studies of Singh and Chaudhari (1992) presented that excess Mo decreased POD activity in groundnut leaves. However, Rout and Das (2011) displayed that molybdenum toxicity was associated with a high level of POD and CAT activity in different cultivars of rice. Mo application increased most AsA-GSH cycle pathways, in terms of AsA content and the activities of APX, MDHAR, DHAR and GR in the leaves of Chinese cabbage (*Brassica campestris* L. ssp. *pekinensis*), whilst ascorbate oxidase reduced (Nie et al. 2007). Mo pollution triggered the metal chelators (phytochelatins, cysteine and glutathione) and antioxidant enzyme activities (GR, APX, CAT and SOD) in leaves of *Achilla tenuifolia* (Boojar and Tavakkoli 2011). Under excess Mo stress, the activities of POD, CAT and APX enzymes in roots and shoots were increased which may contribute to Mo tolerance in soybean seedlings (Xu et al. 2018).

14.3.2 Antioxidant Responses of Plants to Toxic Effects of Non-Essential Metals

14.3.2.1 Cadmium (Cd)

Cd is considered to be in the top ten list of hazardous compounds by the Agency for Toxic Substances and Disease Registry (https://www.atsdr.cdc.gov). Cadmium is dispersed in the environment through mining, smelting, phosphate fertilizers, sewage sludge, Ni-Cd batteries, plating, pigments and plastics items. The environmental Cd goes into the soil with rainwater and is taken up by the plant and then enters the food chain. The high uptake of the divalent cations to the aerial parts of the plant shifts the cellular phosphorylation state and causes a range of physiological disturbances and oxidative stresses in the cell (Al-Qurainy et al. 2017). Cadmium stress causes oxidative damage in plants through the production of reactive oxygen species (ROS). However, the antioxidant system plays an important role in the removal of the elevated ROS and provides tolerance to plants under Cd stress. Variance in the responses of plants' antioxidant enzymes will be discussed. DHAR was also reported to be stimulated in response to Cd stress (Gupta et al. 1999; Prasad et al. 1999; Gallego et al. 1999). Dong et al. (2006) reported on the effect of Cd concentration in tomato seedlings and observed that POD and SOD activities were enhanced in plants that were given concentrations from 1–10 µM parallel to MDA accumulation, indicating that oxidative stress was the resultant of Cd stress in tomato plants. Sandalio et al. (2001) studied the effects of cadmium on pea plants where the enhancement of ROS content in cells could be related to a decrease of the detoxification mechanism of enzymatic antioxidants such as SOD. Schützendübel and Polle (2002) reported that the activities of antioxidative enzymes (CAT, POD and SOD) lessened and GSH was retarded for a short period by cadmium. Romero-Puertas et al. (2007) found a substantial decrease in the expression of genes coding for CuZnSOD and no changes in MnSOD in *Pisum sativum* under Cd stress. Other studies record no alterations or low expression of genes coding for SOD. Fidlago et al. (2011) showed no differences in MnSOD-related mRNA accumulation in plant organs (leaves and roots), but CuZnSOD-related transcripts reduced in leaves but remained stable in Cd-treated Solanum nigrum L. roots. Luo et al. (2011) reported that Cd stress induced an upregulated expression of FeSOD, MnSOD, Chl CuZnSOD, Cyt CuZnSOD, APX, GPX, GR and POD between four and twenty-four hours after initial treatment of Lolium perenne, and they stated that the gene transcript profile was related to the enzyme activity under Cd stress. Romero-Puertas et al. (2007) indicated two groups of genes in pea plants treated with Cd. Group 1 includes some elements of the signal transduction cascade that accentuated or attenuated the Cd effect on CAT, MDHAR and CuZnSOD mRNA expression, and the other group was formed by the genes MnSOD, APX and GR, which were not affected by these modulators during the Cd treatment because their expression was not modified compared to control plants. Hasanuzzaman et al. (2017a) studied the effect of Cd toxicity levels (0.5 and 1 mM) for two days where MDA and H_2O_2 contents decreased, and AsA, GSH and GSH/GSSG ratio increased. Also, the increased activity of APX, DHAR, GR, GST, CAT and glyoxalase II was recorded under both Cd stresses, whilst GPX activity increased under severe stress and glyoxalase I activity enhanced under mild stress only. Al Mahmud et al. (2018) reported that the application of 0.5 and 1.0 mM $CdCl_2$ for three days on *Brassica juncea* exhibited a reduction of AsA and an increment of DHA and GSSG with triggering of SOD and GPX activity at severe stress. In Cd-resistant plants, CAT, POD, APX and SOD activity and gene expression are higher than the susceptible ones (Ni et al. 2018). El-Amieret et al. (2019) illustrated that Cd stress induced the antioxidant system of *Pisum sativium*, where the activities of NADH-oxidase, superoxide dismutase, ascorbate peroxidase, catalase and glutathione reductase were enhanced. Małecka et al. (2019) noted that the shoots and roots of *Brassica junicea* which experienced 50 µM of $CdCl_2$ increased the activities of APX, SOD and POD. Furthermore, the presence of cadmium ions had no significant effect on CuZnSOD expression of shoot and roots, whilst the expression of MnSOD transcript showed a general reduction in shoots and roots. On the other hand, the western blot analysis was performed for roots and shoots for detection of MnSOD (25 kDa) and CuZnSOD (15 kDa and 20 kDa), and subunits showed similar signals for control and Cd-treated seedlings. Thus, the metal presence likely did not change the levels of the CuZnSOD and MnSOD proteins. Recently, Sami et al. (2020) found that SOD, APX and POD were increased in response to Cd stress, and CAT was significantly reduced. Ahanger et al. (2020) reported an increment in GSH and tocopherol content along with SOD, GST and DHAR activities with higher H_2O_2 and $O_2^{\bullet-}$ content in Cd-stressed (100 µM CdCl2, 20 d) *Vigna angularis* seedlings, whereas AsA levels and CAT activity declined. In another study, SOD, CAT, POX and GR activities were upregulated with a higher content of H_2O_2 under Cd stress (50 µM $CdCl_2$, 100 d) in two *Mentha arvensis* (cv. Kosi and Kusha) genotypes, pointing out the activation of an antioxidant defense system for conferring Cd toxicity tolerance (Zaid et al. 2020).

Cadmium also may reach the environment sooner or later in the form of nanoparticles. The effect of L-Cysteine-capped CdS nanoparticles in *Spirodela polyrrhiza* plants after 1, 2 and 4 days application showed the SOD activity had significantly decreased and the POD activity had notably increased compared to the control (Khataee et al. 2014). Also, using CdSe-NPs in *Lemna minor* showed an increase in SOD and CAT activities in contrast to POD, which exhibited a decline in its activity, thus ROS generation dominated over being scavenged; therefore the NPs led to oxidative stress in the plant (Tarrahi et al. 2019).

14.3.2.2 Selenium (Se)

The metalloid Se is an essential micronutrient/trace element for humans and certain animals. However, it is beneficial for plants, and its significance in plants remains controversial (Shahid et al. 2018). The essentiality and phytotoxicity of Se may depend on the dose, speciation and target species (Drahonovský et al. 2016). Over the past few decades, Se levels have been rising in agricultural soils and could be toxic to plants, humans and animals (Chen et al. 2014). Its large-scale distribution in the environment could be due to fossil fuel combustion, mining, irrigation and industrial discharge (Winkel et al. 2015). Oxidative burst is a noxious impact of Se, which deregulates or upregulates the antioxidant system. Ulhassan et al. (2019) studied the effect of different concentrations of Se at 50 µM, 100 µM and 200 µM on *Brassica napus* plants and found that SOD and APX activity increased and CAT and GR activity decreased with increasing Se dose. Furthermore, GSH, GSSG, non-protein thiols (NPTs), phytochelatins (PCs) and cysteine in the leaves and roots of *B. napus* plants were increased by Se stress. Also, increased gamma-glutamylcysteine synthase (γ-ECS), glutathione-S-transferase (GST) and phytochelatin synthase (PCS) were indicated. The in vivo supply of sodium selenite (Na_2SeO_3) in sand soils at concentrations of 0, 0.001, 0.01, 0.1 and 1 mM increased CAT activity and reduced that of POD activity (Shrivastava et al. 2016).

14.3.2.3 Cobalt (Co)

Co, a relatively rare heavy metal, is naturally present in the crust of the Earth primarily in the form of sulfides, oxides and arsenides (Suh et al. 2016). Cobalt is important for many industries such as paint, magnetic products and hard metal production. Nowadays, the demand for Co has increased due to its widespread use in lithium-ion batteries (Campbell 2020). Co ions in excessive doses are mortal risks for the antioxidant balance of the cells due to elevated ROS, as reported by several studies. Macar et al. (2020) reported that Co-treatment triggered a noticeable rise in the activities of SOD and CAT enzymes, MDA content and the abnormalities in the meristematic cells of *Allium cepa*. Gopal (2014) studied the use of cobalt (10 to 400 µM) on pigeon pea (Cajanus cajan Mill) where the reduction of chlorophyll and carotenoid content on exposure to excess Co was associated with decreased activity of CAT and SOD, revealing the antiperoxidative nature of high Co. However, a marked enhancement in the activities of APX and POD and high cysteine and non-protein thiols indicated the induction of antioxidants at an excessive dose of Co. Samet (2020) showed that excessive cobalt concentration triggered an elevated level of lipid peroxidation and CAT activity along with a reduced fresh weight in roots of *Lactuca sativa* L. Walaba et al. (2020) showed that the joint application of Co and Cu increased the expression of HvFeSOD, down-regulated HvCuZnSOD and did not affect HvMnSOD. HvCuZnSOD expression should be higher under the Cu and Co co-treatment, as it was up-regulated in the treatment of Cu alone. Thus its down-regulation could be interpreted by the presence of Co in the combined treatment. Indeed, as a divalent cation, Co has been confirmed to reduce the activity of CuZnSOD by inhibiting its biosynthesis (Romero-Puertas et al. 2002). Also, Co competitively reduces the uptake of Zn at the root surface, as clearly demonstrated by lower accumulation of Zn upon Co stress, resulting in reduced CuZnSOD activity (Aravind and Prasad 2003). The same hypothesis would also be true for HvMnSOD. Conversely, the increased expression of HvFeSOD would be the result of the enzymatic scavenging regulation to compensate for the inactivation of these two HvSOD isoforms (Jozefczak et al. 2014).

14.3.2.4 Lead (Pb)

Pb is classified as a non-essential heavy metal, and it causes crucial environmental problems through soil and water pollution owing to its high toxicity and persistence (Bertoli et al. 2011; Soares et al. 2020). Apart from natural weathering processes, Pb environmental pollution origi-nates from various sources such as mining and smelting processes, Pb-containing paints, gaso-line and explosives, and from the disposal of Pb enriched municipal sewage sludge (Capelo et al. 2012). Lead stress induces ROS production and imbalance of antioxidants. In this regard, nor-mally, the increase or decrease in antioxidant enzymatic activities depends on plant genotype, the extent of stress to which plants are exposed, lead concentration, plant stage and duration of metal exposure. A specific Pb level might be inhibitory for one enzyme but promotive for the others; however, these dynamics in their activities vary from plant to plant. Some researchers declared that Pb induced an increase in the antioxidative defense system (Reddy et al. 2005; Ashraf et al. 2017) while others found significant reductions in their activities at higher Pb levels (Chen et al. 2007; Ashraf et al. 2017). In this context, Kıran et al. (2014) evaluated the effects of Pb stress on lettuce (*Lactuca sativa*) where SOD and GR were increased with oxidative stress. The study of Jayasri and Suthindhiran (2017) showed that the application of lead (Pb^{+2}) (1, 2, 4, 6 and 8 mg/l) increased the activity of catalase significantly on the fronds of duckweed (*Lemna minor*). Also, Alamri et al. (2018) stated that the application of 2 mM Pb on growth medium enhanced the activities of SOD, CAT and GR on wheat. On the other hand, Pb stress using $Pb(NO_3)_2$ at dif-ferent concentrations (i.e. 400, 800 and 1,200 ppm) affected the rice cultivars differentially. In this respect, Pb up to 800 ppm triggered the activities of SOD, POD, CAT and APX and reduced at the concentration of 1, 200 ppm in all rice cultivars at both heading and maturity stage in scented rice. Overall, the enzymatic activities remained comparatively higher at heading than the maturity stage in all rice cultivars (Ashraf et al. 2017). Hasanuzzaman et al. (2018) reported that the application of 0.5 and 1.0 mM $Pb(NO_3)_2$ for two days induced triggering of APX, SOD and GST activities, but CAT, GPX MDHAR and DHAR decreased; GR increased initially and then declined on wheat. Małecka et al. (2019) noted that exposure of *Brassica junicea* to 50 μM of $Pb(NO_3)_2$ increased the activities of APX, SOD and POD in the shoots and roots. Furthermore, there was a reduction in the expression of the gene encoding CuZnSOD and MnSOD in roots and shoots of Pb-treated plants except for an increment in the activity of CuZnSOD roots after eight hours, and an approximately two-fold increase in the level of MnSOD transcript was found in plant roots after 24 hours of Pb in comparison to the control. Recently, Soares et al. (2020) stated that the defense mechanism of *Brassica juncea* L. seeds and seedlings under Pb stress was evaluated by antioxidant enzyme activity which was different between the two stages. In the seeds, APX and CAT activity showed no differences from the control (lead-free treatment) whilst the activities of POD and SOD were reduced. In seedlings, the activities of APX, CAT, POD and SOD were dramatically reduced in response to lead stress. Although ROS are toxic to plants, there is tolerance to ROS production via antioxidants enzymes, which limit the degree of plant susceptibility to oxidative stress. The reduction in antioxidant system activity is more frequently observed at higher concentrations or longer periods of exposure to heavy metals. However, in some cases, the activity of these defense enzymes may increase, induced by heavy metal toxicity.

Therefore, the toxic effects of heavy metals probably overcome the antioxidant capacity to defend plant growth against abiotic stress.

14.3.2.5 Tungsten (W)

W is a scarce transition heavy metal and presents naturally in soil, but emerging anthropogenic activities facilitate its omittance in the environment (Dawood and Azooz 2019, 2020). The main sources of tungsten in the environment are traffic, smelting or mining (Chibuike and Obiora 2014), other industrial applications such as manufacture of light bulbs, golf clubs, electronics and specialized components of modern technology (Koutsospyros et al. 2006) as well as its addition to phosphate fertilizers or other additions to soil such as sewage water (Dawood and Azooz 2019). Tungstate, as well as other heavy metals, affects the production of reactive oxygen species and the antioxidant system. Few studies have described the antioxidant system responses under W stress. In this respect, Kumar and Aery (2011) noted that the activity of peroxidase enzyme decreased with the application of W at concentrations of 3 and 9 mg/kg. Higher administration of tungsten (27–243 mg/kg) resulted in increased total phenols, free proline and activity of enzyme peroxidase. Dawood and Azooz (2019) stated that application of a toxic level of tungstate (100 mg/kg) as a soluble form of W on clay soil of potted broccoli plants increased lipoxygenase (LOX), phenylalanine ammonia-lyase (PAL), polyphenol oxidase (PPO), SOD, ionic and soluble peroxidases activities, and CAT, APX, GST and GPX activities reduced dramatically. A recent study by Dawood and Azooz (2020) on broccoli plants during the germination stage revealed that low levels of tungstate at 1, 5, 10 and 50 mg L^{-1}) enhanced the activities of SOD, CAT, APX, GST and GPX and reduced that of soluble and ionic peroxidases, whilst the higher concentration 100 mg L^{-1} increased CAT activity, but reduced SOD, GST and GPX activities for 15 days on broccoli seedlings. Thus, the dose of tungstate and plant stage are limiting factors to determine the response of antioxidants to tungsten stress.

14.3.2.6 Arsenic (As)

Contamination of the environment with the metalloid As occurs in the environment through geological activities, smelting operations, fossil fuel combustion and the use of pesticides and herbicides (Kanwar et al. 2015; Ghosh et al. 2016; Naeem et al. 2020). The activity of antioxidant enzymes showed a variance in response in terms of species, variety, concentration and the duration of exposure. In this regard, Gupta et al. (2009) showed higher activities of SOD, CAT and GPX at the lower concentration of As (50 mM and 150 mM) followed by a decrease at higher concentration (300 mM) compared to controls in the two varieties of *Brassica juncea*. Similarly, Khan et al. (2009) reported that exposure of 20-day-old plants of *B. juncea* to 5 and 25 mM As for 96 hours in hydroponic cultures led to reduced plant growth, an accumulated MDA content and increased activities of SOD, APX and GR. Similar findings indicating higher antioxidative potential were noticed under As stress in other plants (Dave et al. 2013; Mallick et al. 2014). Kanwar et al. (2015) stated that As (V) toxicity significantly increased the activities of antioxidant enzymes like SOD, CAT, POD, APX, GR and MDHAR in 30-day-old *B. juncea* plants, whilst DHAR activity exhibited no particular trend. In 60-day-old *B. juncea* plants, the activities of SOD and GR were increased in response to As(V) relative to control plants. A decreasing trend in the activity of CAT, DHAR and MDHAR was noticed with the increasing concentration of As. But POD and APX activity increased under 0.1 mM of As-treated plants and decreased for 0.3 mM of As-treated plants as compared to untreated ones. Another study by Rodríguez-Ruíz et al. (2019) studied the impact of As toxicity on hydroponically grown pea plants. The As toxicity caused an increase in APX and GR activities in both roots and leaves, the DHAR and MDAR activities decreased in roots but were unaffected in leaves, and, in both organs, the GSH and GSSG content decreased, the AsA content was unaltered and there was a marked increase in phytochelatins. Naeem et al. (2020) found that As-affected plants enhanced the antioxidant system in terms of POD, SOD, APX and CAT, and the content of artemisinin on two varieties of *Artemisia annua* and the tolerant variety showed a better activity of these enzymes over the sensitive one. Thus, hyper activities of the antioxidative enzymes may be a strategy adopted by the plant to tolerate As toxicity, especially for tolerant varieties.

14.3.2.7 Chromium (Cr)

Cr is among the most widespread toxic trace elements found in agricultural soils and results from various anthropogenic activities. Combustion of oil, coal and waste from chemical and metallurgy industries and industrial effluent from tanneries result in high input of Cr to the ecosystem (Antonkiewicz et al. 2019). Excessive Cr levels in plants affect plants drastically and induce oxidative stress in plants, while alterations in antioxidant enzyme activity also occurred. Pandey et al. (2005) found that high doses of Cr(VI) decreased CAT, SOD and APX activities in the root or leaf tissue of *B. juncea* seedlings, whereas low doses increased their activities. Dey et al. (2009) reported that higher Cr concentrations (50 and 100 μM) decreased SOD and CAT activities in both the shoots and roots of wheat. Following a short-term treatment for one hour at concentrations of 50–200 μM, Cr(VI) was reported to induce the dose-dependent increase of CAT, SOD, APX and GR activities in week-old seedlings of *Brasicca juncea* and *Vigna radiata* (Diwan et al. 2010). Other studies have shown that Cr (0, 0.25 and 0.5 mM) treatments in the form of $K_2Cr_2O_7$ negatively affected antioxidant enzyme activities (CAT and APX) in leaves and roots of wheat (Ali et al. 2015 a, b).

Treatment of *Allium cepa* L. bulbs with Cr(VI) at different concentrations for five days (120 h) showed that a low dose (12.5 μM) had a stimulatory effect on root growth and high doses (25–200 μM) inhibited root growth, indicating hormesis phenomenon. Inhibition of root growth was correlated with the dose-dependent increase in the generation of ROS, cell death, lipid peroxidation, repression of antioxidative enzymes (catalase, superoxide dismutase, ascorbate peroxidase), induction of DNA damage, chromosome aberrations or micronuclei in root cells. The above effects were, however, reversed when the duration of Cr(VI) treatment was limited to 3–24 hours followed by recovery in tap water for four days, which resulted in the dose-dependent stimulation of root growth, mitosis and increased activity of the antioxidative enzymes that eliminated oxidative stress and genotoxicity (Patnaik et al. 2013). Gill et al. (2015) reported that Cr exposure reduced APX, whilst CAT activity increased in *Brassica napus* plants. Mahmud et al. (2017) confirmed that due to Cr stress, the few components of the AsA-GSH pathway were enhanced in *Brassica juncea*. They found that five days duration of exposure to two concentrations of Cr (0.15 mM and 0.3 mM) in form of K_2CrO_4 induced decrement of AsA content, MDHAR and DHAR activities, but did not affect GSH content and upregulated the activities of APX and GR. They also revealed that the higher APX and GR activity might play a function in scavenging excess ROS. Jan et al. (2020) recently reported that Cr-stressed tomato plants showed a rise in the activities of SOD, CAT, GST, APX, Gly I and GR activities, whilst that of MDHAR, Gly II and DHAR activities were reduced. Ahmad et al. (2020) observed that the activities of antioxidant enzymes, such as APX, SOD, POD and CAT, enhanced in response to different concentrations of Cr on cauliflower plants. Zaheer et al. (2020) declared that using different levels of tannery water as a source of Cr negatively affected the growth of rapeseed (*Brassica napus* L.) plants and retarded the antioxidant system in terms of SOD, POD, CAT and APX of the roots. Thus, different responses of antioxidants were recorded based on duration and plant stage, but generally before the threshold and tolerant cultivars exhibited a high antioxidant system.

14.3.2.8 Antimony (Sb)

Antimony stress is dominantly due to anthropogenic activities and the extensive applications of Sb compounds (Brun et al. 2008). Antimony is vastly used in multiple commercial products such as flame retardants, semiconductors, alloys and tracer bullets (Filella et al. 2002). Metal mining, smelting and burning of municipal waste have also caused severe Sb contamination (He 2007). The studies on antimony have been seldom (Feng et al. 2013; Vaculiková et al. 2014; Peško et al. 2016; Ortega et al. 2017). Feng et al. (2009) studied the effect of Sb-stress in the hydroponic culture at two concentration of 5 and 20 mg L^{-1} on four species of fern plants, including *Pteris cretica* (PCA), *Microlepia hancei* (MH), *Cyrtomium fortune* (CYF) and *Cyclosorus dentatus* (CYD). The greater activities of POD, CAT and APX were found in the fronds of the PCA than the other ferns, suggesting that PCA was more capable of eliminating excess H_2O_2 and stabilizing ROS homeostasis. The suppression of GR activity for the fern plants suggested that the ability of GR to regenerate GSH was impaired which was observed for the tolerant fern PCA indicating that PCA was more capable of Sb tolerance. SOD did not change for tolerant fern PCA and reduced for sensitive ones. Benhamdi et al. (2014) collected soils and two wild steppic plants (*Hedysarum pallidum* Desf. and *Lygeum spartum* L) of different zones from the abandoned antimony mining area of Djebel Hamimat (Algeria) where the activity of SOD was higher in *H. pallidum* than in *L. spartum*, whilst that of GST, CAT, POD and APX were generally higher in *L. spartum* than in *H. pallidum*. For both species, APX and GST are overall more active in the upper parts than in the roots, while it is the reverse for SOD and CAT. Vaculíkova et al. (2014) stated that the activity of APX and CAT increased in maize roots with increasing concentration of Sb in the media; however, the activity of POD decreased. Ortega et al. (2017) described that the application of 0.5 and 1 mM Sb induced an increment of APX, POD and non-enzymatic antioxidants (phenols, flavonoids, phenylpropanoid glycosides, AsA and only in leaves for the GSH), but GR activity increased only in leaves at the 1.0 mM Sb and did not change in roots. DHAR activity increased in roots only at 1.0 mM Sb treatment, whereas in leaves DHAR activity was very comparable to or lower than the controls. Furthermore, the GSNOR (S-nitrosoglutathione) activity increased for roots and leaves, but much more so for roots under Sb stress. Moreover, Sb stress increases SOD activity, but lipid peroxidation is only increased significantly at 1 mM Sb. This may be due to the observed activation in the expression of GSTs and CuZn-SOD I and II which may attenuate hydroperoxides. This would reduce Sb-induced oxidative damage, and maintain protein functionality and redox homeostasis. Sb stress also significantly increased the activity of DHAR (belonging to the group of GSTs), especially in roots, which would contribute to increased antioxidant activity through the formation of AsA. In the plants subjected to Sb stress, the authors also reported that increased GST expression and GSH in leaves may be involved in Sb detoxification via the formation of GSH-metal complexes and by activating the binding of GSH to toxic compounds which are indicative of their participation in mechanisms of Sb tolerance.

Espinosa-Vellarino et al. (2020) stated that the Sb toxicity induced changes in the SOD, POX, APX and GR antioxidant activities, which show a clear activation in the roots. In leaves, only the SOD and APX increased. The DHAR, POD and GR activities were inhibited in roots but showed no changes in the leaves. Ascorbate increased while GSH decreased in the roots. The total AsA + DHA content increased in the roots, but the total GSH + GSSG content decreased, while neither is altered in the leaves. Sb toxicity also upregulated the expression of the SOD, APX and GR genes, while the expression of GST decreased dramatically in roots but increased in leaves. The authors also recorded changes in the pattern of the elongation zone growth with smaller and disorganized cells. This could be associated with the ability of the Sb to form complexes with thiol groups (GSH), altering both redox homeostasis and auxin content of the roots and the quiescent center. The accumulation of O_2^- and triggering of SOD gene expression and SOD activity revealed the induction of the antioxidant system. Besides this, in roots, there is a sharp decrease in the expression of GST, while in leaves it increased. In roots with high amounts of accumulated Sb, the inhibition of GST expression, the increase in GR activity and the decrease in DHAR activity favored the availability of GSH for the formation of chelates with Sb. Also, Feng et al. (2019) reported an increment of SOD activity in rice plants under different forms of Sb stress which correlated to a low incidence of Sb toxicity on membrane lipid peroxidation. The increases in SOD, POX and APX activity reflect the involvement of the antioxidant machinery in the response to toxicity from Sb (Feng et al. 2009, 2019; Benhamdi et al. 2014; Vaculiková et al. 2014; Ortega et al. 2017). Thus, higher antioxidative enzyme activities and higher scavenging capacities than normal plants are the main strategies for defending against Sb stress.

14.3.2.9   Aluminum (Al)

Al ranks as the third most abundant element in the Earth's crust, after oxygen and silicon. Al bioavailability, and, in consequence, toxicity is predominately associated with acid soils. Some agricultural practices, such as removal of agricultural products from the farm, leaching of nitrogen below the plant root zone, inappropriate use of nitrogenous fertilizers and build-up in organic matter, account for further acidification of cultivated soils (Silva 2012). One of the toxic effects of Al toxicity in plants is the overproduction of reactive oxygen species which affects the antioxidative system of the plants. Some Al-sensitive cultivars register lower levels of antioxidant enzymes, such as CAT, SOD, GST and POD, as compared to Al-resistant cultivars, resulting in high accumulation of ROS in Al-sensitive cultivars (Darkó et al. 2004). In Al-resistant plants, the enzyme activity of CAT, POD, APX and SOD, and gene expression is higher (Wu et al. 2013). Exposure of ryegrass plants to Al stress enhanced SOD, CAT, POD and APX activities, whilst reduced the gene expression of Fe-SOD and Cu/Zn-SOD in shoots, whereas no changes in the expression pattern of these genes were recorded in roots. Moreover, the expression of the Mn-SOD gene was up-regulated in shoots and roots exposed to Al (Pontigo et al. 2017). Nahar et al. (2017) found that the exposure of *Vigna radiata* to 0.5 mM AlCl$_3$ for two and three days reduced AsA content, but increased GSH and GSSG. Besides this, the activities of SOD, GST, GPX, APX and GR were increased, but MDHAR, DHAR and CAT decreased. Qian et al. (2018) found that the activities of root SOD, POD and CAT initially increased and subsequently decreased in rice plants. However, the activities of these enzymes significantly decreased under 1000 µmol/L Al stress. Awasthi et al. (2019) studied the effect of aluminum stress (100 µM AlCl$_3$, 48 h) on two cultivars of Indian rice, where the tolerant cultivar increased AsA content in both roots and shoots, enhanced the GSH content in shoots, elevated the levels of APX, MDHAR, DHAR and GR activities compared to sensitive ones which exhibited higher accumulation of AsA in both roots and shoots, reduced the GSH content in roots while shoots content was unaltered, increased APX, MDHAR and DHAR activities and slightly stimulated GR activities in response to high accumulation of H$_2$O$_2$ compared to the tolerant one. Sami et al. (2020) found that SOD, CAT, APX and POD were increased in response to Al stress (25 µM) in *Brassica napus*.

Al in nanoscale form has been also tested by several studies. (Yanık and Vardar 2015) have reported that Al$_2$O$_3$ elevated the peroxidase activity in *Triticium aestivum*. Chahardoli et al. (2020) tested the impact of different concentrations (0, 50, 100, 1000 and 2500 mg/L) of engineered Al$_2$O$_3$ NPs in the hydroponically grown tissues of *Nigella arvensis* L. and found that the antioxidant enzymes APX, CAT, SOD and POD increased at beneficial and toxic doses in roots and shoots compared to the control plants.

14.3.2.10   Mercury (Hg)

Hg is a toxic heavy metal element. Natural sources include geological parent material, rock outcroppings, wind-blown dust, volcanic eruptions, marine aerosols and forest fires, whilst anthropogenic sources include mining, coal burning and unsafe disposal of industrial solid/liquid wastes. Modern industrialization and urbanization has led to the release of Hg into ecosystems throughout the world (Larssen 2010; Yang et al. 2018). Mercury at low concentrations dramatically induces oxidative burst and impacts the defense mechanism of the antioxidant system. Zhou et al. (2007) observed an increment of POD activity in alfalfa (*Medicago sativa* L) up to 20 µM Hg and the same finding was recorded in *Brassica juncea* (Shiyab et al. 2009). The phytotoxicity and oxidative damage of Hg at different concentrations (0.0, 2.5, 5.0, 10 and 25 µM) were evaluated in wheat plants where CAT, APX, POD and SOD activities were increased in the roots and leaves of plants with increasing Hg concentration up to 10 µM, and low activities of these enzymes were observed at 25 µM Hg (Sahu et al. 2012). Application of HgCl$_2$ (10, 20, 30, 40, 50 and 60 mg L^{-1}) for ten days induced the antioxidative enzyme activities (SOD, CAT, APX and POD) as the concentration of Hg increased up to 40 mg L^{-1} while a slight decline at higher doses was reported in *Sesbania grandiflora* (Malar et al. 2015). Malar et al. (2015) stated that the decreased level of SOD activity at higher concentrations of

Hg stress might be attributed to enzyme damage from high H_2O_2 levels or its poisonous ROS derivatives. In vivo supply of Hg ($HgCl_2$) in sand soils at different concentrations (0, 0.001, 0.01, 0.1 and 1 mM) reduced CAT and POD activities (Shrivastava et al. 2016). Recently, Singh et al. (2020) found that the activities of ROS-scavenging enzymes (SOD, CAT and POD) were evaluated under various concentrations of Hg^{+2} (0.0, 0.1, 0.2, 0.3 and 0.4 μM) on *Spirodela polyrhiza*, where an enhancing trend of these antioxidants was detected in plants submitted to 0.1 and 0.2 μM Hg^{+2} which then decreased consistently for 0.3 and 0.4 μM Hg^{+2}. Another study by Hu et al. (2020) revealed that under Hg stress (3 mg Hg Kg^{-1}), the activities of SOD and CAT were significantly decreased, but the POD activity was not altered in garlic plants.

14.3.2.11 Vanadium (V)

V is the fifth most available transition metal and is widely present (~0.01%) in the earth's crust (Amorim et al. 2007). The land adjacent to industrial sites has high V content owing to anthropogenic activities such as petrochemical and steel industry emissions, and phosphorite treating factories. The majority of distributed V into the atmosphere is adsorbed by the soil surface particles, thus can easily be taken up by plants, however, its effect on plants depends on the concentration, because V at lower levels enhances the plant growth (Imtiaz et al. 2018), while higher levels induce noxious impacts on growth, ROS accumulation and different responses of antioxidative enzymes. In this respect, Imtiaz et al. (2016) displayed that the antioxidant enzyme activities such as SOD, CAT and POD were increased in both genotypes of chickpea against V stress (15, 30, 60 and 120 mg V L^{-1}) by using ammonium metavanadate (NH_4VO_3). Imtiaz et al. (2018) reported that the enzyme activities SOD, CAT and POD and ion leakage were increased linearly with increasing V concentrations in chickpea (*Cicer arietinum* L.) grown in red soil. However, the enzymatic activities were much elevated at higher levels of V (130, 170 and 200 V mg kg^{-1}) than lower levels (50 and 90 V mg kg^{-1}) in chickpea plants. The increase in enzymes' (SOD, CAT and POD) activities could be due to the result of the direct effect of metal ions, and the indirect effect of induced free oxide radicals. Levels of phytochelatins, glutathione, thiols and some antioxidants, which are capable of protecting important cellular organs and components from oxidative damage and are closely related to plant tolerance and detoxification against metal toxicity, were depressed in corn seedlings in the presence of vanadium (Hou et al. 2019).

In dog tail grass, activities of SOD, POD and CAT were inhibited due to vanadium toxicity, resulting in disruption of the dynamic balance between ROS and antioxidant enzymes. This disruption led to elevated levels of H_2O_2, O_2^- and other ROS, and caused increased content of plant MDA, a product of lipid peroxidation in the cell membrane (Aihemaiti et al. 2019).

14.3.3 Some Trace Elements in the Form of Nanoparticles: Antioxidant Enzyme Responses

Metal and metal oxide toxicity in the form of nanoparticles can potentially include at least three mechanisms: (1) particles may emit toxic substances into exposure culture, (2) surface interactions with the media may give toxic substances and (3) particles or their surfaces may interact directly with biological targets and disrupt them (Ma et al. 2013). The interaction of nanoparticles with organisms or factors present in the environment may induce ROS (Navarro et al. 2008). ROS generation depends on many actions of nanoparticles such as physicochemical properties, biotransformation, size, shape and metal ions emitted from metal and metal oxide nanoparticles (Yang et al. 2017). ROS production and antioxidant imbalance are the main toxic effects of metal oxide nanoparticles (Wang et al. 2017). The antioxidant system response to metal oxide nanoparticles displayed different magnitudes based on type, size, shape, the dose applied and source of the nanoparticles, the plant species and the exposure period of nanoparticles to crops, and plant growth stage. In a study on the physiological effects of anatase TiO_2 nanoparticles on duckweed, a decrement in SOD activity in concentrations up to 200 mg/L was observed (Song et al. 2012). The study of Okupnik and

Pflugmacher (2016) on the effect of different concentrations of titanium oxide nanoparticles (0.01, 0.1, 1 and 10mg/L) on aquatic macrophytes *Hydrilla verticillata* indicated the concentration-dependent increase of CAT and GR. Also, Wen-wen et al. (2010) reported a reduction in SOD activity in HeLa cells exposed to Fe_3O_4 nanoparticles. Wang et al. (2011) reported that Fe_3O_4 nanoparticles induced oxidative stress as compared to Fe_3O_4 bulk particles in the ryegrass and pumpkin roots and shoots, as indicated by increased SOD and CAT activities and lipid peroxidation. Lannone et al. (2016) studied the effect of citric acid-coated iron oxide nanoparticles (5, 10, 15 and 20 mg/L) on wheat plants grown in a hydroponic system where nanoparticle treatment was reported to be responsible for the increment in enzymatic activities, but the degree of increase was dependent on the concentration of the applied nanoparticles. Increased enzymatic activities, namely CAT, APX, POD and SOD, were observed in roots. Zhao et al. (2012) reported that soil contaminated with CeO_2 at 400 and 800 mg/kg levels increased CAT activity in shoots of maize plants grown for 20 days. It seems that the enzyme is activated by CeO_2, but at high concentrations the activity gradually decreased. Similarly, APX activity also declined with 800 mg/kg dose with a concomitant decrease in the H_2O_2 level. However, both the enzymes eliminate the excess of H_2O_2 to prevent lipid damage by CeO_2 nanoparticles. CAT and APX activities in kidney bean roots reduced sharply when treated with 500 mg L^{-1} CeO_2 nanoparticles for 15 days compared with control, and at the same time, the root soluble protein increased by 204% (Majumdar et al. 2014). In another study, SOD and POD activities and MDA levels were retarded by 1000 mg kg^{-1} treatment of CeO_2-NPs. The toxicity could be ascribed to the biotransformation of CeO_2 NPs and the high sensitivity of *Lactuca* plants to the released Ce^{3+} ions (Gui et al. 2015).

Dimkpa et al. (2012) examined the extent of phytotoxicity level generated in sand-grown wheat plants in response to the treatment of nano-oxides of copper and zinc (500 mg/kg). The study showed that copper nanoparticle-treated plants exhibited higher CAT and POD compared to the control plants while such an increase was absent in the case of zinc nanoparticle treatment. Faisal et al. (2013) studied the effects of nickel oxide nanoparticles in tomato plants and observed the higher formation of oxidative stress in protoplasts of roots which was correlated with elevated levels of CAT and SOD as well as in aerial parts. Nair and Chung (2014) showed a gradual increase in lipid peroxidation levels along with an increment in peroxidase levels in the plants. Besides this, mRNA expression levels of Cu/ZnSOD, CAT and POD showed an increment in copper oxide nanoparticle-treated plants compared to the control. Rajeshwari et al. (2015) studied the cytotoxic effects of aluminum oxide nanoparticles in *Allium cepa* when root tips of plants were exposed to 0.01, 1 and 100μg/mL, where a dose-dependent increment in SOD activities was observed.

14.4 GENES AND GENE EXPRESSION IN RESPONSE TO HEAVY METAL STRESS IN PLANTS

In the previous part of this chapter, the role of antioxidant enzymes in response to the different heavy metal stress has been extensively discussed. In this part, we will review the different genes that control antioxidant enzymes. Understanding the genetic control of antioxidant enzymes will help in producing plants with high antioxidant content and hence highly tolerant of heavy metal stress.

Genetic approaches such as genetic engineering and genome-wide association study (GWAS) have played an important role in understanding the genes controlling the heavy metals (Alomari et al. 2018 a, b) and the mechanism of heavy metal resistance in the different plant species. For example, some genes controlling heavy metal tolerance were identified by gene manipulation, one of the genetic engineering techniques. Some of the identified genes are ATP sulfurylase (APS), GR, glyoxalases (glyoxalase I and II), serine acetyltransferase (SAT), phytochelatin synthase (PCS), γ-glutamylcysteine synthetase (GSH1), glutathione synthetase (GSH2) and cystathionine synthase (CTS) (Yadav 2010). Overexpression of these genes in some plants resulted in high tolerance to heavy metals. These enzymatic genes regulate GSH and phytochelatins (PC) levels in the plants

TABLE 14.2

Antioxidant Enzymes Regulated in Plants under Heavy Metal Stress

Plant Species	Gene Controlling the Enzyme	Antioxidative Enzyme	Effect of the Gene in the Plant Parts	Reference
Nicotiana tabacum	Serine acetyltransferase	cysteine precursor O-acetylserine	Increasing Cadmium (Cd) concentration in roots, hence playing an important role in heavy metals tolerance.	Wawrzyński et al. (2006)
	γ-glutamylcysteine synthetase	GSH precursor γ-EC		
	phytochelatin synthase	phytochelatin synthase		
Populus canescens	modified bacterial GSH1 gene (S1ptTECS)	GSH1 protein expression in the shoots.	Heavy metals tolerance.	Bittsánszky et al. (2005)
Arabidopsis thaliana			Full Hg tolerance in the shoots. Partial cadmium (Cd) tolerance in the shoots. Metalloids (As) tolerance in the root.	Li et al. (2006)
Brassica juncea L.	*Escherichia coli* GSH2 gene	glutathione synthetase	Cadmium tolerance at seedling and mature plants.	Liang Zhu et al. (1999)
	Escherichia coli GSH1 gene	γ-ECS		
	AtPCS1 gene	phytochelatin synthase (PCS)	Tolerance to cadmium, As and zinc stresses.	Gasic and Korban (2007a,b)
Avicennia germinans	AvPCS gene	phytochelatin synthase (PCS)	Cd and Cu detoxification	Gonzalez-Mendoza et al. (2007)

and hence provide a high level of heavy metal tolerance. The strategy of how some of these genes control the tolerance in the different plants is summarized in Table 14.2. Based on the results of these studies, the overexpression of GSH, PCS, GSH2, APS and CTS genes seems to be a promising strategy to produce plants with a high level of heavy metal tolerance.

GWAS is a powerful approach that identifies alleles associated with different complex traits by using linkage disequilibrium (LD) and the examination of marker-trait associations (Yang et al. 2015; Alomari et al. 2017; Zhang et al. 2018; Alqudah et al. 2020). Identifying quantitative trait loci (QTLs) and tests for the gene models help in the preliminary identification of genes controlling the different traits (Potkin et al. 2009; Hussain et al. 2017; Mourad et al. 2018 a, 2019). However, more studies are needed to confirm these identified genes. GWAS was used widely to identify genes controlling heavy metal tolerance in different crops. In rapeseed (*Brassica napus*), four candidate genes were identified to be controlling cadmium tolerance at the seedling stage (Chen et al. 2018). Five gene models (GSTUs, BCATs, UBP13, TBR and HIPP01) were identified to have higher expression under lead contamination in rapeseed seedlings (Zhang et al. 2020). In *Aegilops tauschii*, five genes (pdil5-1, Acc-1, DME-5A, TaAP2-D, TaAP2-B) associated with cadmium tolerance were identified (Qin et al. 2015). In the case of maize (*Zea mays*), three candidate genes were reported to control heavy metal tolerance. Out of these three genes, two genes encoded zinc/cadmium-transporting ATPase. The third gene was responsible for a vacuolar ATPase encoding (Zhao et al. 2018).

One of the common approaches to alleviating heavy metal stress factors in the plant is the response of antioxidant enzymes, which help in reducing cellular damage by limiting ROS. On the other hand, plants have a complex antioxidant system which consists of major ROS-scavenging enzymes such as SOD, CAT, APX, GR, MDHAR and DHAR (Ghori et al. 2019).

The metalloenzyme SOD is detected primarily in bovine erythrocytes and then in bacteria, higher plants and vertebrates (Rabinowitch and Sklan 1981). SOD enzyme is the first line of enzymatic defenses against oxidative damage. It catalyzes the conversion of toxic O_2^- radicals to H_2O_2 and molecular O_2. SOD is found in plant parts such as roots, leaves, fruits and seeds (Giannopolitis and Ries 1977). SOD may be categorized based on the presence of metal cofactors, protein folds and or subcellular distribution. However, the main categorization of SOD is based on the presence of metal cofactors. So, SOD is categorized as copper/zinc (Cu/Zn) SODs, iron (Fe) SODs or manganese (Mn) SODs (Filiz and Tombuloğlu 2014). The first plant SOD gene cloned was from *Zea mays* L. In general, the SODs are encoded by a small multigene family (Cannon et al. 1987). While these SOD proteins are encoded by nuclear genes, they are located in different cellular parts. Cu/ZnSODs are mainly found in the cytosol, chloroplasts, peroxisomes and/or the extracellular space, whereas FeSODs are mainly located in chloroplasts and probably the cytosol, and MnSODs are in the mitochondria (Pilon et al. 2011). Different research results reported SOD genes that exhibit different expression patterns based on the actual plant species and the specific environmental stress to which the plant is exposed. Descriptions of the SOD gene family have been reported for various plant species, such as *Arabidopsis thaliana*, *Sorghum bicolor*, *Gossypium raimondii*, *Gossypium arboreum*, *Gossypium hirsutum*, *Musa acuminata* and *Populus trichocarpa* (Kolahi et al. 2020). In *Z. Mays*, nine different isoforms of SOD genes were classified. Six isoforms are Cu/Zn-SODs (Sod1, Sod2, Sod4, Sod4A, Sod5 and Sod9), one is a Fe-SOD (SodB) and four are Mn-SODs (Sod-3.1, Sod-3.2, Sod-3.3 and Sod-3.4) (Scandalios 1997). Two SOD-genes identified as Cu/ZnSOD genes were isolated from tomato leaves. Another nine SOD genes were observed in tomato and classified into two major groups (Cu/ZnSODs and Fe-MnSODs). The Cu/ZnSOD group had four members containing a copper-zinc domain (SlSOD1, 2, 3 and 4), whereas the Fe-MnSOD group involved five members with an iron/manganese SOD alpha-hairpin domain and an iron/manganese SOD, C-terminal domain (SlSOD5, 6, 7, 8 and 9). Two Cu/ZnSODs (SlSOD1 and 2) and one Fe-MnSOD (SlSOD8) are thought to be localized within the cytoplasm. Another Fe-MnSOD (SlSOD9) is present within the mitochondrion and the remaining members in the chloroplast (Alscher et al. 2002; Kolahi et al. 2020). In the grapevine genome, ten SOD genes (six copper/zinc, two iron, and two manganese SODs) were detected on 12 different chromosomes (Hu et al. 2019). Nickel (Ni)SOD as a new type of SOD was first discovered, cloned and characterized in *Streptomyces*, nevertheless, up to now, NiSOD has not been found in plants (Wolfe-Simon et al. 2005; Youn et al. 1996). In *Arabidopsis*, seven genes for SOD were identified and were classified into three CuZnSODs (CSD1, CSD2 and CSD3), three FeSODs (FSD1, FSD2 and FSD3), and one MnSOD (MSD1) (Kliebenstein et al. 1998). CSD2 and FeSOD proteins were found in the *Arabidopsis* chloroplast.

In plant chloroplasts, a key enzyme called APX regulates the hydrogen peroxide detoxification, and it is classified as I-heme-peroxidase. APX is present in higher plants, chlorophytes and red algae, as well as members of the protist kingdom. The APX enzyme isoforms are categorized based on their subcellular localization with the soluble isoforms being primarily located in the cytosol (cAPX), chloroplast stroma (sAPX) and mitochondria (mitAPX) (Shigeoka et al. 2002).

Chloroplastic APX (chlAPX) isoenzymes genes in the plant are classified into two groups: the first group contains single genes that encode two isoenzymes, and it is inclusive of genes from spinach (*S. oleracea*), tobacco (*N. tabacum*), pumpkin (*Cucurbita* sp) and ice plant (*M. crystallium*). The second group consists of individual genes that codify different isoenzymes, and it is inclusive of genes from *Arabidopsis*, rice (*Oryza sativa*) and tomato (*Solanum lycopersicum*) (Kolahi et al. 2020).

Also, studying the expression of the different antioxidant enzyme genes helped in understanding the genetic control of heavy metal tolerance. In wheat, some studies were conducted to measure the concentration of the antioxidant enzymes in the different wheat parts in response to the heavy metals. Contamination with lead in the soil resulted in higher expression of genes controlling some of the antioxidant enzymes such as CAT, SOD, GPX and APX. This higher expression was found in both roots and leaves of the wheat with higher expression in the roots than the leaves tissue

(Navabpour et al. 2020). At the seedling growth stage, higher expression of genes controlling SOD, CAT and POD antioxidant enzymes was found under the contamination of heavy metals like Zn, Cd, Cu, Cr, Ni, Co, Pb and Mn (Murzaeva 2004). In the fenugreek plant (*Trigonella foenum-graecum* L.), high expression of antioxidant enzymes encoding genes (CAT, POD and APX) was noted (Alaraidh et al. 2018).

14.5 CONCLUSION

In the current era, a tremendous increase in environmental toxicity because of extensive application of fertilizer and anthropogenic activities has increased the presence of toxic metals and/or metalloids and their nanoforms in agricultural soil, which especially impacts crops. Therefore, studying the behavior of heavy metal-induced ROS production and the ability of plants to detoxify these ROS are critical outlines for improving the ability to assess risks or improve phytoremediation performance. The reviewed studies provide insights that could assist in enhancing stress tolerance via encoding of antioxidant systems and are employed in breeding and engineering programs aimed at developing plants with new and desired agronomical traits. The provided information are a useful outcomes that can help to decipher and analyze active regulatory networks controlling heavy metal stress responses and tolerance.

REFERENCES

Abdel Latef, A.A. 2013. Growth and some physiological activities of pepper (*Capsicum annuum* L.) in response to cadmium stress and mycorrhizal symbiosis. *Journal of Agricutural Science and Technology* 15: 1437–1448.

Abdel Latef, A.A., Abu Alhmad, M.F. 2013. Strategies of copper tolerance in root and shoot of broad bean (*Vicia faba* L.). *Pakistan Journal of Agricultural Sciences* 50: 323–328.

Abdel Latef, A.A., Zaid, A., Abo-Baker, A.A. et al. 2020b. Mitigation of copper stress in maize by inoculation with *Paenibacillus polymyxa* and *Bacillus circulans*. *Plants* 9(11): 1513.

Abdel Latef, A.A., Zaid, A., Abu Alhmad, M.F., Abdelfattah, K.E. 2020a. The impact of priming with Al$_2$O$_3$ nanoparticles on growth, pigments, osmolytes, and antioxidant enzymes of Egyptian roselle (*Hibiscus sabdariffa* L.) cultivar. *Agronomy* 10(5): 681.

Abdel-Salam, E.M., Qahtan, A.A., Faisal, M. et al. 2018. Phytotoxic assessment of nickel oxide (NiO) nanoparticles in radish. In: *Phytotoxicity of Nanoparticles* Faisal Mohammad, Saquib Quaiser, Alatar Abdulrahman A, Al-Khedhairy Abdulaziz A (eds), 269–284. Springer, Cham.

Ahanger, M.A., Aziz, U., Sahli, A.A. et al. 2020. Combined kinetin and spermidine treatments ameliorate growth and photosynthetic inhibition in *Vigna angularis* by up-regulating antioxidant and nitrogen metabolism under cadmium stress. *Biomolecules* 10(1): 147.

Ahmad, P., Alam, P., Balawi, T.H. et al. 2020. Sodium nitroprusside (SNP) improves tolerance to arsenic (As) toxicity in Vicia faba through the modifications of biochemical attributes, antioxidants, ascorbate-glutathione cycle and glyoxalase cycle. *Chemosphere* 244: 125480.

Ahmad, R., Ali, S., Rizwan, M., et al. 2020. Hydrogen sulfide alleviates chromium stress on cauliflower by restricting its uptake and enhancing antioxidative system. *Physiologia Plantarum* 168(2): 289–300.

Aihemaiti, A., Jiang, J., Blaney, L. et al. 2019. The detoxification effect of liquid digestate on vanadium toxicity to seed germination and seedling growth of dog's tail grass. *Journal of Hazardous Materials* 369: 456–464.

Al Mahmud, J., Hasanuzzaman, M., Nahar, K., et al. 2018. Insights into citric acid-induced cadmium tolerance and phytoremediation in Brassica juncea L.: Coordinated functions of metal chelation, antioxidant defense and glyoxalase systems. *Ecotoxicology and Environmental Safety* 147: 990–1001.

Alamri, S.A., Siddiqui, M.H., Al-Khaishany, M.Y., et al. 2018. Ascorbic acid improves the tolerance of wheat plants to lead toxicity. *Journal of Plant Interactions* 13(1): 409–419.

Alaraidh, I.A., Alsahli, A.A., Abdel Razik, E.S.A. 2018. Alteration of antioxidant gene expression in response to heavy metal stress in *Trigonella foenum-Graecum* L. *South African Journal of Botany* 115: 90–93.

Ali, S., Bharwana, S.A., Rizwan, M. et al. 2015a. Fulvic acid mediates chromium (Cr) tolerance in wheat (Triticum aestivum L.) through lowering of Cr uptake and improved antioxidant defense system. *Environmental Science and Pollution Research International* 22(14): 10601–10609.

Ali, S., Chaudhary, A., Rizwan, M. et al. 2015b. Alleviation of chromium toxicity by glycinebetaine is related to elevated antioxidant enzymes and suppressed chromium uptake and oxidative stress in wheat (*Triticum aestivum* L.). *Environmental Science and Pollution Research International* 22(14): 10669–10678.

Alomari, D.Z., Eggert, K., Von Wirén, N. et al. 2017. Genome-wide association study of calcium accumulation in grains of European wheat cultivars. *Frontiers in Plant Science* 8: 1797.

Alomari, D.Z., Eggert, K., Von Wirén, N. et al. 2018a. Identifying candidate genes for enhancing grain Zn concentration in wheat. *Frontiers in Plant Science* 9: 1313.

Alomari, D.Z., Eggert, K., Von Wirén, N. et al. 2018b. Whole-genome association mapping and genomic prediction for iron concentration in wheat grains. *International Journal of Molecular Sciences* 20(1): 76.

Alqudah, A.M., Sallam, A., Stephen Baenziger, P.S., Börner, A. 2020. GWAS : Fast-forwarding gene identification and characterization in temperate Cereals: Lessons from Barley—A review. *Journal of Advanced Research* 22: 119–135.

Al-Qurainy, F., Khan, S., Tarroum, M. et al. 2017. Biochemical and genetical responses of *Phoenix dactylifera* L. to cadmium stress. *BioMed Research International* 2017: 1–10.

Alscher, R.G., Erturk, N., Heath, L.S. 2002. Role of superoxide dismutases (SODs) in controlling oxidative stress in plants. *Journal of Experimental Botany* 53(372): 1331–1341.

Álvarez, R., Del Hoyo, A., García-Breijo, F. et al. 2012. Different strategies to achieve Pb-tolerance by the two Trebouxia algae coexisting in the lichen *Ramalina farinacea*. *Journal of Plant Physiology* 169(18): 1797–1806.

Amorim, F.A.C., Welz, B., Costa, A.C.S. et al. 2007. Determination of vanadium in petroleum and petroleum products using atomic spectrometric techniques. *Talanta* 72(2): 349–359.

Antonkiewicz, J., Baran, A., Pełka, R. et al. 2019. A mixture of cellulose production waste with municipal sewage as new material for an ecological management of wastes. *Ecotoxicology and Environmental Safety* 169: 607–614.

Appenroth, K.J. 2010. Definition of "heavy metals" and their role in biological systems. In: *Soil Heavy Metals, Soil Biol*, Vol. 9 Varma A, Sherameti I (eds), 19–29. Springer, Berlin.

Ara, H., Sinha, A.K. 2014. Conscientiousness of mitogen activated protein kinases in acquiring tolerance for abiotic stresses in plants. *Proceedings of the Indian National Science Academy* 80(2): 211–219.

Aravind, P., Prasad, M.N. 2003. Zinc alleviates cadmium-induced oxidative stress in Ceratophyllum demersum L.: A free floating freshwater macrophyte. *Plant Physiology and Biochemistry* 41(4): 391–397.

Archana, P.V., Pandey, N. 2017. Changes in ascorbate, non-protein thiols-cysteine in linseed seedlings subjected to boron stress. *Indian Journal of Agricultural Biochemistry* 30(1): 50.

Ashraf, U., Hussain, S., Anjum, S.A. et al. 2017. Alterations in growth, oxidative damage, and metal uptake of five aromatic rice cultivars under lead toxicity. *Plant Physiology and Biochemistry* 115: 461–471.

Awasthi, J.P., Saha, B., Panigrahi, J. et al. 2019. Redox balance, metabolic fingerprint and physiological characterization in contrasting North East Indian rice for aluminum stress tolerance. *Scientific Reports* 9(1): 8681.

Bankaji, I., Sleimi, N., López-Climent, M.F. et al. 2014. Effects of combined abiotic stresses on growth, trace element accumulation, and phytohormone regulation in two halophytic species. *Journal of Plant Growth Regulation* 33(3): 632–643.

Bashandy, S.R., Abd-Alla, M.H., Dawood, M.F.A. 2020. Alleviation of the toxicity of oily wastewater to canola plants by the N_2-fixing, aromatic hydrocarbon biodegrading bacterium *Stenotrophomonas maltophilia*-SR1. *Applied Soil Ecology* 154: http://www.ncbi.nlm.nih.gov/pubmed/103654.

Benhamdi, A., Bentellis, A., Rached, O. et al. 2014. Effects of antimony and arsenic on antioxidant enzyme activities of two steppic plant species in an old antimony mining area. *Biological Trace Element Research* 158(1): 96–104.

Bertoli, A.C., Carvalho, R., Cannata, M.G. et al. 2011. Toxidez do chumbo no teor e translocação de nutrientes em tomateiro. *Biotemas* 24: 7–15.

Bhalerao, S.A., Sharma, A.S., Poojari, A.C. 2015. Toxicity of nickel in plants. *International Journal of Pure and Applied Bioscience* 3: 345–355.

Bilal, S., Khan, A.L., Shahzad, R. et al. 2018. Mechanisms of Cr(VI) resistance by endophytic Sphingomonas sp. LK11 and its Cr(VI) phytotoxic mitigating effects in soybean (*Glycine max* L.). *Ecotoxicology and Environmental Safety* 164: 648–658.

Bittsánszky, A., Kömives, T., Gullner, G. et al. 2005. Ability of transgenic poplars with elevated glutathione content to tolerate zinc(2+) stress. *Environment International* 31(2): 251–254.

Boojar, M.M.A., Tavakkoli, Z. 2011. New molybdenum-hyperaccumulator among plant species growing on molybdenum mine-a biochemical study on tolerance mechanism against metal toxicity. *Journal of Plant Nutrition* 34(10): 1532–1557.

Brun, C.B., Åström, M.E., Peltola, P., Johansson, M. 2008. Trends in major and trace elements in decomposing needle litters during a long-term experiment in Swedish forests. *Plant and Soil* 306(1–2): 199–210.

Campbell, G.A. 2020. The cobalt market revisited. *Mineral Economics* 33(1–2): 1–8.

Cannon, R.E., White, J.A., Scandalios, J.G. 1987. Cloning of cDNA for maize superoxide dismutase 2 (SOD2). *Proceedings of the National Academy of Sciences of the United States of America* 84(1): 179.

Capelo, A., Santos, C., Loureiro, S., et al. 2012. Phytotoxicity of lead on *Lactuca sativa*: Effects on growth, mineral nutrition, photosynthetic activity and oxidant metabolism. *Fresenius Environmental Bulletin* 21: 450–459.

Chahardoli, A., Karimi, N., Ma, X., Qalekhani, F. 2020. Effects of engineered aluminum and nickel oxide nanoparticles on the growth and antioxidant defense systems of *Nigella arvensis* L. *Scientific Reports* 10(1): 1–11.

Chai, L.Y., Mubarak, H., Yang, Z.H. et al. 2016. Growth, photosynthesis, and defense mechanism of antimony (Sb)-contaminated Boehmeria nivea L. *Environmental Science and Pollution Research International* 23(8): 7470–7481.

Chaney, R.L., Ryan, J.A. 1994. Risk based standards for arsenic lead and cadmium in urban soils. Dechema, Frankfurt. Germany.

Chaoui, A., Habib Ghorbal, M.H., El Ferjani, E. 1997. Effects of cadmium-zinc interactions on hydroponically grown bean (*Phaseolus vulgaris* L.). *Plant Science* 126(1): 21–28.

Chen, J., Zhu, C., Li, L. et al. 2007. Effects of exogenous salicylic acid on growth and H2O2-metabolizing enzymes in rice seedlings under lead stress. *Journal of Environmental Sciences* 19(1): 44–49.

Chen, L., Wan, H., Qian, J. et al. 2018. Genome-Wide association study of cadmium accumulation at the seedling stage in rapeseed (*Brassica napus* L .). *Frontiers in Plant Science* 9: 375.

Chen, Y., Mo, H.Z., Hu, L.B. et al. 2014. The endogenous nitric oxide mediates selenium-induced phytotoxicity by promoting ROS generation in *Brassica rapa*. *PLOS ONE* 9(10): e110901.

Choudhary, S., Zehra, A., Naeem, M.. et al. 2020. Effects of boron toxicity on growth, oxidative damage, antioxidant enzymes and essential oil fingerprinting in *Mentha arvensis*and *Cymbopogon fexuosus*. *Chemical and Biological Technologies in Agriculture* 7: 8.

Corpas, F.J., Barroso, J.B. 2013. Nitro-oxidative stress vs oxidative or nitrosative stress in higher plants. *New Phytologist* 199(3): 633–635.

Corpas, F.J., Barroso, J.B. 2017. Lead-induced stress, which triggers the production of nitric oxide (NO) and superoxide anion (O2·-) in Arabidopsis peroxisomes, affects catalase activity. *Nitric Oxide: Biology and Chemistry* 68: 103–110.

Da Silva Rodrigues, W., Pereira, Y.C., De Souza, A.L.M. et al. 2020. Alleviation of oxidative stress induced by 24-epibrassinolide in soybean plants exposed to different manganese supplies: Upregulation of antioxidant enzymes and maintenance of photosynthetic pigments. *Journal of Plant Growth Regulation* 2020: 1–16.

Dalcorso, G., Manara, A., Furini, A. 2013. An overview of heavy metal challenge in plants: From roots to shoots. *Metallomics* 5(9): 1117–1132.

Darkó, É., Ambrus, H., Stefanovits-Bányai, É. et al. 2004. Aluminium toxicity, Al tolerance and oxidative stress in an Al-sensitive wheat genotype and in Al-tolerant lines developed by in vitro microspore selection. *Plant Science* 166(3): 583–591.

Dave, R., Singh, P.K., Tripathi, P. et al. 2013. Arsenite tolerance is related to proportional thiolic metabolite synthesis in rice (*Oryza sativa* L.). *Archives of Environmental Contamination and Toxicology* 64(2): 235–242.

Dawood, M.F.A., Azooz, M.M. 2019. Concentration-dependent effects of tungstate on germination, growth, lignification-related enzymes, antioxidants, and reactive oxygen species in broccoli (Brassica oleracea var. italica L.). *Environmental Science and Pollution Research International* 26(36): 36441–36457.

Dawood, M.F.A., Azooz, M.M. 2020. Insights into the oxidative status and antioxidative responses of germinating broccoli (*Brassica oleracea var. italica L.*) seeds in tungstate contaminated water. *Chemosphere.* https://www.ncbi.nlm.nih.gov/pubmed/32739687.

Demirevska-Kepova, K., Simova-Stoilova, L., Stoyanova, Z. et al. 2004. Biochemical changes in barley plants after excessive supply of copper and manganese. *Environmental and Experimental Botany* 52(3): 253–266.

Dey, S.K., Jena, P.P., Kundu, S. 2009. Antioxidative efficiency of *Triticum aestivum* L. exposed to chromium stress. *Journal of Environmental Biology* 30(4): 539–544.

Dhal, B., Thatoi, H.N., Das, N.N., Pandey, B.D. 2013. Chemical and microbial remediation of hexavalent chromium from contaminated soil and mining/metallurgical solid waste: A review. *Journal of Hazardous Materials* 250–251: 272–291.

Dhalaria, R., Kumar, D., Kumar, H. et al. 2020. Arbuscular mycorrhizal fungi as potential agents in ameliorating heavy metal stress in plants. *Agronomy* 10(6): 815.

Dilek Tepe, H., Aydemir, T. 2017. Effect of boron on antioxidant response of two lentil (*Lens culinaris*) cultivars. *Communications in Soil Science and Plant Analysis* 48(16): 1881–1894.

Dimkpa, C.O., Mclean, J.E., Latta, D.E. et al. 2012. CuO and ZnO nanoparticles: Phytotoxicity, metal speciation, and induction of oxidative stress in sand-grown wheat. *Journal of Nanoparticle Research* 14(9): 1125.

Diwan, H., Khan, I., Ahmad, A., Iqbal, M. 2010. Induction of phytochelatins and antioxidant defence system in *Brassica juncea* and *Vigna radiata* in response to chromium treatments. *Plant Growth Regulation* 61(1): 97–107.

Dong, C.-D., Chen, C.-F., Chen, C.-W. 2012. Contamination of zinc in sediments at river mouths and channel in northern Kaohsiung Harbor, Taiwan. *International Journal of Environmental Science and Development* 3: 517.

Dong, J., Wu, F., Zhang, G. 2006. Influence of cadmium on antioxidant capacity and four microelement concentrations in tomato seedlings (*Lycopersicon esculentum*). *Chemosphere* 64(10): 1659–1666.

Drahonovský, J., Száková, J., Mestek, O. et al. 2016. Selenium uptake, transformation and inter-element interactions by selected wildlife plant species after foliar selenate application. *Environmental and Experimental Botany* 125: 12–19.

Dubey, S., Shri, M., Gupta, A. et al. 2018. Toxicity and detoxification of heavy metals during plant growth and metabolism. *Environmental Chemistry Letters* 16(4): 1169–1192.

Edreva, A., Apostolova, E. 1989. Manganese toxicity in tobacco: A biochemical investigation. *Agrochemica* 33: 441–451.

El-Amier, Y., Elhindi, K., El-Hendawy, S. et al. 2019. Antioxidant system and biomolecules alteration in *Pisum sativum* under heavy metal stress and possible alleviation by 5-aminolevulinic acid. *Molecules* 24(22): 4194.

Eng, R., Lei, L., Su, J. et al. 2019. Toxicity of different forms of antimony to rice plant: Effects on root exudates, cell wall components, endogenous hormones and antioxidant system. *Science of the Total Environment* 711. http://www.ncbi.nlm.nih.gov/pubmed/134589.

Espinosa-Vellarino, F.L., Garrido, I., Ortega, A. et al. 2020. Effects of antimony on reactive oxygen and nitrogen species (ros and rns) and antioxidant mechanisms in tomato plants. *Frontiers in Plant Science* 11: 674.

Faisal, M., Saquib, Q., Alatar, A.A. et al. 2013. Phytotoxic hazards of NiO-nanoparticles in tomato: A study on mechanism of cell death. *Journal of Hazardous Materials* 250–251: 318–332.

Feigl, G., Lehotai, N., Molnár, Á. et al. 2015. Zinc induces distinct changes in the metabolism of reactive oxygen and nitrogen species (ROS and RNS) in the roots of two *Brassica* species with different sensitivity to zinc stress. *Annals of Botany* 116(4): 613–625.

Feng, R., Wei, C., Tu, S. et al. 2013. The uptake and detoxification of antimony by plants: A review. *Environmental and Experimental Botany* 96: 28–34.

Feng, R., Wei, C., Tu, S. et al. 2009. Antimony accumulation and antioxidative responses in four fern plants. *Plant and Soil* 317(1–2): 93.

Fidalgo, F., Freitas, R., Ferreira, R. et al. 2011. *Solanum nigrum* L. antioxidant defence system isozymes are regulated transcriptionally and posttranslationally in Cd-induced stress. *Environmental and Experimental Botany* 72(2): 312–319.

Filella, M., Belzile, N., Chen, Y.W. 2002. Antimony in the environment: A review focused on natural waters: I. Occurrence. *Earth-Science Reviews* 57(1–2): 125–176.

Filiz, E., Tombuloğlu, H. 2014. In silico analysis of DREB transcription factor genes and proteins in grasses. *Applied Biochemistry and Biotechnology* 174(4): 1272–1285.

Freeman, J.L., Persans, M.W., Nieman, K. et al. 2004. Increased glutathione biosynthesis plays a role in nickel tolerance in Thlaspi nickel hyperaccumulators. *Plant Cell* 16(8): 2176–2191.

Gajewska, E., Sklodowska, M., Slaba, M., Mazur, J. 2006. Effect of nickel on antioxidative enzyme activities, proline and chlorophyll contents in wheat shoots. *Biologia Plantarum* 50(4): 653–659.

Gallego, S.M., Benavides, M.P., Tomaro, M.L. 1999. Effect of cadmium ions on antioxidant defense system in sunflower cotyledons. *Biologia Plantarum* 42(1): 49–55.

Gangwar, S., Singh, V.P. 2011. Indole acetic acid differently changes growth and nitrogen metabolism in *Pisum sativum* L. seedlings under chromium (VI) phytotoxicity: Implication of oxidative stress. *Scientia Horticulturae* 129(2): 321–328.

Gangwar, S., Singh, V.P., Srivastava, P.K., Maurya, J.N. 2011. Modification of chromium(VI) phytotoxicity by exogenous gibberellic acid application in *Pisum sativum* (L.) seedlings. *Acta Physiologiae Plantarum* 33(4): 1385–1397.

Gasic, K., Korban, S.S. 2007a. Expression of Arabidopsis phytochelatin synthase in Indian mustard (*Brassica juncea*) plants enhances tolerance for Cd and Zn. *Planta* 225(5): 1277–1285.

Gasic, K., Korban, S.S. 2007b. Transgenic Indian mustard (*Brassica juncea*) plants expressing an Arabidopsis phytochelatin synthase (AtPCS1) exhibit enhanced As and Cd tolerance. *Plant Molecular Biology* 64(4): 361–369.

Genisel, M., Turk, H., Dumlupinar, R. 2017. Exogenous aminolevulinic acid protects wheat seedlings against boron-induced oxidative stress. *Romanian Biotechnological Letters* 22: 12741–12750.

Georgiadou, E.C., Kowalska, E., Patla, K. et al. 2018. Influence of heavy metals (Ni, Cu, and Zn) on nitro-oxidative stress responses, proteome regulation and allergen production in basil (Ocimum basilicum L.) plants. *Frontiers in Plant Science* 9: 862.

Ghori, N.-H., Ghori, T., Hayat, M.Q. et al. 2019. Heavy metal stress and responses in plants. *International Journal of Environmental Science and Technology* 16(3): 1807–1828.

Ghosh, S., Shaw, A.K., Azahar, I. et al. 2016. Arsenate (AsV) stress response in maize (*Zea mays* L.). *Environmental and Experimental Botany* 130: 53–67.

Ghotbi-Ravandi, A.A., Ghaderian, S.M., Azizollahi, Z. 2019. Lead uptake, bioaccumulation and tolerance in summer savory (*Satureja hortensis* L.). *Journal of Plant Process and Function* 8: 63–75.

Giannopolitis, C.N., Ries, S.K. 1977. Superoxide dismutases: I. Occurrence in higher plants. *Plant Physiology* 59(2): 309.

Gill, R.A., Zang, L., Ali, B. et al. 2015. Chromium-induced physio-chemical and ultrastructural changes in four cultivars of Brassica napus L. *Chemosphere* 120: 154– 164.

Gomes-Junior, R.A., Moldes, C.A., Delite, F.S. et al. 2006. Nickel elicits a fast antioxidant response in Coffeaarabica cells. *Plant Physiology and Biochemistry* 44(5–6): 420–429.

Gonzalez-Mendoza, D., Moreno, A.Q., Zapata-Perez, O. 2007. Coordinated responses of phytochelatin synthase and metallothionein genes in black mangrove, *Avicennia germinans*, exposed to cadmium and copper. *Aquatic Toxicology* 83(4): 306–314.

Gopal, R. 2014. Antioxidant defense mechanism in pigeon pea under cobalt stress. *Journal of Plant Nutrition* 37(1): 136 –145.

Gui, X., Zhang, Z., Liu, S. et al. 2015. Fate and phytotoxicity of CeO_2 nanoparticles on lettuce cultured in the potting soil environment. *PLOS ONE* 10(8): e0134261.

Gupta, M., Cuypers, A., Vangronsveld, J., Clijsters, H. 1999. Copper affects the enzymes of the ascorbate-glutathione cycle and its related metabolites in the roots of Phaseolus vulgaris. *Physiologia Plantarum* 106(3): 262–267.

Gupta, M., Sharma, P., Sarin, N.B., Sinha, A.K. 2009. Differential response of arsenic stress in two varieties of *Brassica juncea* L. *Chemosphere* 74(9): 1201–1208.

Gür, N., Türker, O.C., Böcük, H. 2016. Toxicity assessment of boron (B) by *Lemna minor* L. and *Lemna gibba* L. and their possible use as model plants for ecological risk assessment of aquatic ecosystems with boron pollution. *Chemosphere* 157: 1–9.

Hakeem, K.R., Alharby, H.F., Rehman, R. 2019. Antioxidative defense mechanism against lead-induced phytotoxicity in *Fagopyrum kashmirianum*. *Chemosphere* 216: 595–604.

Han, S., Tang, N., Jiang, H.-X. et al. 2009. CO_2 assimilation, photosystem II photochemistry, carbohydrate metabolism and antioxidant system of citrus leaves in response to boron stress. *Plant Science* 176(1): 143–153.

Hasanuzzaman, M., Alam, M.M., Nahar, K. et al. 2019. Silicon-induced antioxidant defense and methylglyoxal detoxification works coordinately in alleviating nickel toxicity in *Oryza sativa* L. *Ecotoxicology* 28(3): 261–276.

Hasanuzzaman, M., Bhuyan, M., Anee, T.I. et al. 2019. Regulation of ascorbate-glutathione pathway in mitigating oxidative damage in plants under abiotic stress. *Antioxidants* 8(9): 384.

Hasanuzzaman, M., Bhuyan, M.H.M., Zulfiqar, F. et al. 2020. Reactive oxygen species and antioxidant defense in plants under abiotic stress: Revisiting the crucial role of a universal defense regulator. *Antioxidants* 9(8): 681.

Hasanuzzaman, M., Fujita, M. 2013. Exogenous sodium nitroprusside alleviates arsenic-induced oxidative stress in wheat (*Triticum aestivum* L.) seedlings by enhancing antioxidant defense and glyoxalase system. *Ecotoxicology* 22(3): 584–596.

Hasanuzzaman, M., Nahar, K., Anee, T.I., Fujita, M. 2017a. Exogenous silicon attenuates cadmium-induced oxidative stress in Brassica napus L. by modulating AsA-GSH pathway and glyoxalase system. *Frontiers in Plant Science* 8: 1061.

Hasanuzzaman, M., Nahar, K., Gill, S.S. et al. 2017b. Hydrogen peroxide pretreatment mitigates cadmium-induced oxidative stress in *Brassica napus* L.: An intrinsic study on antioxidant defense and glyoxalase systems. *Frontiers in Plant Science* 8: 115.

Hasanuzzaman, M., Nahar, K., Rahman, A. et al. 2018. Exogenous glutathione attenuates lead-induced oxidative stress in wheat by improving antioxidant defense and physiological mechanisms. *Journal of Plant Interactions* 13(1): 203–212.

Havaux, M. 2014. Carotenoid oxidation products as stress signals in plants. *Plant Journal : For Cell and Molecular Biology* 79(4): 597–606.

He, H., Zhan, J., He, L., Gu, M. 2012. Nitric oxide signaling in aluminum stress in plants. *Protoplasma* 249(3): 483–492.

He, J., Qin, J., Long, L. et al. 2011. Net cadmium flux and accumulation reveal tissue-specific oxidative stress and detoxification in Populus × canescens. *Physiologia Plantarum* 143(1): 50–63.

He, M.C. 2007. Distribution and phytoavailability of antimony at an antimony mining and smelting area, Hunan, China. *Environmental Geochemistry and Health* 29(3): 209–219.

Hong, J., Rico, C.M., Zhao, L. et al. 2015. Toxic effects of copper-based nanoparticles or compounds to lettuce (*Lactuca sativa*) and alfalfa (*Medicago sativa*). *Environmental Science. Processes and Impacts* 17(1): 177–185.

Hossain, M.A., Piyatida, P., Da Silva, J.A.T., Fujita, M. 2012. Molecular mechanism of heavy metal toxicity and tolerance in plants: Central role of glutathione in detoxification of reactive oxygen species and methylglyoxal and in heavy metal chelation. *Journal of Botany* 2012: 1–37.

Hossain, M.S., Abdelrahman, M., Tran, C.D. et al. 2020. Insights into acetate-mediated copper homeostasis and antioxidant defense in lentil under excessive copper stress. *Environmental Pollution* 258: 113544.

Hou, M., Li, M., Yang, X., Pan, R. 2019. Responses of nonprotein thiols to stress of vanadium and mercury in maize (*Zea mays* L.) seedlings. *Bulletin of Environmental Contamination and Toxicology* 102(3): 425–431.

Hu, X., Hao, C., Cheng, Z.-M., Zhong, Y. 2019. Genome-wide identification, characterization, and expression analysis of the grapevine superoxide dismutase (SOD) family. *International Journal of Genomics* 2019: 7350414.

Hu, Y., Wang, Y., Liang, Y. et al. 2020. Silicon alleviates mercury toxicity in garlic plants. *Journal of Plant Nutrition* 43(16): 2508–2517.

Huang, C., Lai, C., Xu, P. et al. 2017. Lead-induced oxidative stress and antioxidant response provide insight into the tolerance of *Phanerochaete chrysosporium* to lead exposure. *Chemosphere* 187: 70–77.

Huang, Y., Zu, L., Zhang, M. et al. 2020. Tolerance and distribution of cadmium in an ornamental species *Althaea rosea* Cavan. *International Journal of Phytoremediation* 22(7): 713–724.

Hussain, W., Baenziger, P.S., Belamkar, V. et al. 2017. Genotyping-by-sequencing derived high-density linkage map and its application to QTL mapping of Flag leaf traits in bread wheat. *Scientific Reports* 7(1): 1–15.

Iannone, M.F., Groppa, M.D., De Sousa, M.E. et al. 2016. Impact of magnetite iron oxide nanoparticles on wheat (*Triticum aestivum* L.) development: Evaluation of oxidative damage. *Environmental and Experimental Botany* 131: 77–88.

Imtiaz, M., Ashraf, M.U., Rizwan, M. et al. 2018. Vanadium toxicity in chickpea (*Cicer arietinum* L.) grown in red soil: Effects on cell death, ROS and antioxidative systems. *Ecotoxicology and Environmental Safety* 158: 139–144.

Israr, M., Jewell, A., Kumar, D., Sahi, S.V. 2011. Interactive effects of lead, copper, nickel and zinc on growth, metal uptake and antioxidative metabolism of *Sesbania drummondii*. *Journal of Hazardous Materials* 186(2–3): 1520–1526.

Ivanov, Y.V., Savochkin, Y.V., Kuznetsov, V.V. 2012. Scots pine as a model plant for studying the mechanisms of conifers adaptation to heavy metal action: 2. Functioning of antioxidant enzymes in pine seedlings under chronic zinc action. *Russian Journal of Plant Physiology* 59(1): 50–58.

Jan, S., Noman, A., Kaya, C. et al. 2020. 24-Epibrassinolide alleviates the injurious effects of Cr (VI) toxicity in tomato plants: Insights into growth, physio-biochemical attributes, antioxidant activity and regulation of ascorbate–glutathione and glyoxalase cycles. *Journal of Plant Growth Regulation* 39(4): 1587–1604.

Jayasri, M.A., Suthindhiran, K. 2017. Effect of zinc and lead on the physiological and biochemical properties of aquatic plant *Lemna minor*: Its potential role in phytoremediation. *Applied Water Science* 7(3): 1247–1253.

Jiang, M., Ye, Z., Zhang, H., Miao, L. 2019. Broccoli Plants Over-expressing an ERF transcription factor gene BoERF1 facilitates both salt stress and sclerotinia stem rot resistance. *Journal of Plant Growth Regulation* 38(1): 1–13.

Jozefczak, M., Keunen, E., Schat, H. et al. 2014. Differential response of *Arabidopsis* leaves and roots to cadmium: Glutathione-related chelating capacity vs antioxidant capacity. *Plant Physiology and Biochemistry* 83: 1–9.

Jozefczak, M., Remans, T., Vangronsveld, J., Cuypers, A. 2012. Glutathione is a key player in metal-induced oxidative stress defenses. *International Journal of Molecular Sciences* 13(3): 3145–3175.

Kanwar, M.K., Poonam, M., Bhardwaj, R. 2015. Arsenic induced modulation of antioxidative defense system and brassinosteroids in *Brassica juncea* L. *Ecotoxicology and Environmental Safety* 115: 119–125.

Kaya, C., Ashraf, M., Al-Huqail, A.A. et al. 2020a. Silicon is dependent on hydrogen sulphide to improve boron toxicity tolerance in pepper plants by regulating the AsA-GSH cycle and glyoxalase system. *Chemosphere* 127241.

Kaya, C., Ashraf, M., Alyemeni, M.N., Ahmad, P. 2020b. Responses of nitric oxide and hydrogen sulfide in regulating oxidative defence system in wheat plants grown under cadmium stress. *Physiologia Plantarum* 168(2): 345–360.

Kaya, C., Tuna, A.L., Guneri, M., Ashraf, M. 2011. Mitigation effects of silicon on tomato plants bearing fruit grown at high boron levels. *Journal of Plant Nutrition* 34(13): 1985–1994.

Keunen, E., Remans, T., Bohler, S. et al. 2011. Metal-induced oxidative stress and plant mitochondria. *International Journal of Molecular Sciences* 12(10): 6894–6918.

Khan, I., Ahmad, A., Iqbal, M. 2009. Modulation of antioxidant defence system for arsenic detoxification in Indian mustard. *Ecotoxicology and Environmental Safety* 72(2): 626–634.

Khan, M.U., Malik, R.N., Muhammad, S. et al. 2015. Health risk assessment of consumption of heavy metals in market food crops from Sialkot and Gujranwala Districts, Pakistan. *Human and Ecological Risk Assessment* 21(2): 327–337.

Khataee, A., Movafeghi, A., Nazari, F. et al. 2014. The toxic effects of L-cysteine-capped cadmium sulfide nanoparticles on the aquatic plant *Spirodela polyrrhiza*. *Journal of Nanoparticle Research* 16(12): 2774.

Kim, S., Lee, S., Lee, I. 2012. Alteration of phytotoxicity and oxidant stress potential by metal oxide nanoparticles in *Cucumis sativus*.*Water, Air, and Soil Pollution* 223(5): 2799–2806.

Kiran, B.R., Prasad, M.N.V. 2019. Defense manifestations of enzymatic and non-enzymatic antioxidants in *Ricinus communis* L. exposed to lead in hydroponics. *The Eurobiotech Journal* 3(3): 117–127.

Kiran, S., Özkay, F., Kuşvuran, Ş. et al. 2014. The effect of humic acid applied to the plants of lettuce (*Lactuca sativa* var. *crispa*) irrigated with water with high content of lead on some characteristics. *Research Journal of Biological Sciences* 7: 14–19.

Kliebenstein, D.J., Monde, R.-A., Last, R.L. 1998. Superoxide dismutase in Arabidopsis: An eclectic enzyme family with disparate regulation and protein localization. *Plant Physiology* 118(2): 637.

Kolahi, M., Mohajel Kazemi, E., Yazdi, M., Goldson-Barnaby, A. 2020. Oxidative stress induced by cadmium in lettuce (*Lactuca sativa* Linn.): Oxidative stress indicators and prediction of their genes. *Plant Physiology and Biochemistry* 146: 71–89.

Kolbert, Z., Barroso, J.B., Brouquisse, R. et al. 2019. A forty year journey: The generation and roles of NO in plants. *Nitric Oxide : Biology and Chemistry* 93: 53–70.

Krantev, A., Yordanova, R., Janda, T. et al. 2008. Treatment with salicylic acid decreases the effect of cadmium on photosynthesis in maize plants. *Journal of Plant Physiology* 165(9): 920–931.

Kumar, A., Aery, N.C. 2011. Effect of tungsten on growth, biochemical constituents, molybdenum and tungsten contents in wheat. *Plant, Soil and Environment* 57(11): 519–525.

Kumar, H., Sharma, D., Kumar, V. 2012. Nickel-induced oxidative stress and role of antioxidant defence in Barley roots and leaves. *International Journal of Environmental Biology* 2: 121–128.

Larssen, T. 2010. Mercury in Chinese reservoirs. *Environmental Pollution* 158(1): 24–25.

Li, X., Yang, Y., Jia, L. et al. 2013. Zinc-induced oxidative damage, antioxidant enzyme response and proline metabolism in roots and leaves of wheat plants. *Ecotoxicology and Environmental Safety* 89: 150–157.

Li, Y., Dankher, O.P., Carreira, L. et al. 2006. The shoot-specific expression of gamma-glutamylcysteine synthetase directs the long-distance transport of thiol-peptides to roots conferring tolerance to mercury and arsenic. *Plant Physiology* 141(1): 288–298.

Liang Zhu, Y.L., Pilon-Smits, E.A.H., Jouanin, L., Terry, N. 1999. Overexpression of glutathione synthetase in Indian mustard enhances cadmium accumulation and tolerance. *Plant Physiology* 119(1): 73–80.

Liu, C., Lu, W., Ma, Q., Ma, C. 2017. Effect of silicon on the alleviation of boron toxicity in wheat growth, boron accumulation, photosynthesis activities, and oxidative responses. *Journal of Plant Nutrition* 40(17): 2458–2467.

Liu, J., Dhungana, B., Cobb, G.P. 2018. Environmental behavior, potential phytotoxicity, and accumulation of copper oxide nanoparticles and arsenic in rice plants. *Environmental Toxicology and Chemistry* 37(1): 11–20.

Lösch, R. 2004. Plant mitochondrial respiration under the influence of heavy metals. In: *Heavy Metal Stress in Plants: Biomolecules to Ecosystem* 3rd ed Prasad MNV (ed), 182–200. Springer, Berlin.

Luo, H., Li, H., Zhang, X., Fu, J. 2011. Antioxidant responses and gene expression in perennial ryegrass (Lolium perenne L.) under cadmium stress. *Ecotoxicology* 20(4): 770–778.

Luo, Z.,Tian, D., Ning, C. et al. 2015. Roles of *Koelreuteria bipinnata* as a suitable accumulator tree species in remediating Mn, Zn, Pb, and Cd pollution on Mn mining wasteland sin southern China. *Environmental Earth Sciences* 74(5): 4549–4559.

Lwalaba, J.L.W., Zvobgo, G., Fu, L. et al. 2017. Alleviating effects of calcium on cobalt toxicity in two barley genotypes differing in cobalt tolerance. *Ecotoxicology and Environmental Safety* 139: 488–495.

Lwalaba, J.L.W., Louis, L.T., Zvobgo, G. et al. 2020. Physiological and molecular mechanisms of cobalt and copper interaction in causing phyto-toxicity to two barley genotypes differing in Co tolerance. *Ecotoxicology and Environmental Safety* 187: 109866.

Ma, H., Williams, P.L., Diamond, S.A. 2013. Ecotoxicity of manufactured ZnO nanoparticles—A review. *Environmental Pollution* 172: 76–85.

Macar, O., Kalefetoğlu Macar, T., Çavuşoğlu, K., Yalçın, E. 2020. Determination of protective effect of carob (Ceratonia siliqua L.) extract against cobalt(II) nitrate-induced toxicity. *Environmental Science and Pollution Research International* 27(32): 40253–40261.

Macar, T.K., Macar, O., Yalçın, E., Çavuşoğlu, K. 2020. Resveratrol ameliorates the physiological, biochemical, cytogenetic, and anatomical toxicities induced by copper(II) chloride exposure in Allium cepa L *Environmental Science and Pollution Research International* 27(1): 657–667.

Mahmood, S., Ishtiaq, S., Malik, M.I. et al. 2013. Differential growth and photosynthetic responses and pattern of metal accumulation in sunflower (*Heliathus annuus* L.) cultivars at elevated levels of lead and mercury. *Pakistan Journal of Botany* 45: 367–374.

Mahmud, J.A., Hasanuzzaman, M., Nahar, K. et al. 2017. Maleic acid assisted improvement of metal chelation and antioxidant metabolism confers chromium tolerance in Brassica juncea L *Ecotoxicology and Environmental Safety* 144: 216–226.

Maksymiec, W. 2011. Effects of jasmonate and some other signalling factors on bean and onion growth during the initial phase of cadmium action. *Biologia Plantarum* 55(1): 112–118.

Malar, S., Sahi, S.V., Favas, P.J.C., Venkatachalam, P. 2015. Assessment of mercury heavy metal toxicity-induced physiochemical and molecular changes in *Sesbania grandiflora* L. *International Journal of Environmental Science and Technology* 12(10): 3273–3282.

Malar, S., Shivendra Vikram, S., Jc Favas, P., Perumal, V. 2016. Lead heavy metal toxicity induced changes on growth and antioxidative enzymes level in water hyacinths [Eichhornia crassipes (Mart.)]. [Eichhornia crassipes (Mart.)]. *Botanical Studies* 55(1): 1–11.

Małecka, A., Konkolewska, A., Hanć, A. et al. 2019. Insight into the Phytoremediation Capability of Brassica juncea (v. Malopolska): Metal accumulation and antioxidant enzyme activity. *International Journal of Molecular Sciences* 20(18): 4355.

Malecka, A., Kutrowska, A., Piechalak, A., Tomaszewska, B. 2015. High peroxide level may be a characteristic trait of a hyperaccumulator. *Water, Air, and Soil Pollution* 226(4): 84.

Malecka, A., Piechalak, A., Mensinger, A. et al. 2012. Antioxidative defense system in *Pisum sativum* roots exposed to heavy metals (Pb, Cu, Cd, Zn). *Polish Journal of Environmental Studies* 21: 1721–1730.

Mallick, S., Kumar, N., Sinha, S. et al. 2014. H_2O_2 pretreated rice seedlings specifically reduces arsenate not arsenite: Difference in nutrient uptake and antioxidant defense response in a contrasting pair of rice cultivars. *Physiology and Molecular Biology of Plants* 20(4): 435–447.

Mateos-Naranjo, E., Castellanos, E.M., Perez-Martin, A. 2014. Zinc tolerance and accumulation in the halophytic species *Juncus acutus. Environmental and Experimental Botany* 100: 114–121.

Miller, G., Shulaev, V., Mittler, R. 2008. Reactive oxygen signaling and abiotic stress. *Physiologia Plantarum* 133(3): 481–489.

Mostofa, M.G., Rahman, M.M., Siddiqui, M.N. et al. 2020. Salicylic acid antagonizes selenium phytotoxicity in rice: Selenium homeostasis, oxidative stress metabolism and methylglyoxal detoxification. *Journal of Hazardous Materials* 122572.

Mourad, A.M.I., Sallam, A., Belamkar, V. et al. 2018a. Genetic architecture of common bunt resistance in winter wheat using genome-wide association study. *BMC Plant Biology* 18(1): 1–14.

Mourad, A.M.I., Sallam, A., Belamkar, V. et al. 2018b. Genome-wide association study for identification and validation of novel SNP markers for Sr6 stem rust resistance gene in bread wheat. *Frontiers in Plant Science* 9: 380.

Mourad, A.M.I., Sallam, A., Belamkar, V. et al. 2019. Molecular marker dissection of stem rust resistance in Nebraska bread wheat germplasm. *Scientific Reports* 9(1): 1–10.

Munjal, R. 2019. Oxidative stress and antioxidant defense in plants under high temperature. In: *Reactive Oxygen, Nitrogen and Sulfur Species in Plants: Production, Metabolism, Signaling and Defense Mechanisms* Hasanuzzaman Mirza, Fotopoulos Vasileios, Nahar Kamrun, Fujita Masayuki (eds), 337–352.

Murzaeva, S.V. 2004. Effect of heavy metals on wheat seedlings; activation of antioxidant enzymes. *Applied Biochemistry and Microbiology* 40(1): 114–119.

Mustafa, G., Komatsu, S. 2016. Insights into the response of soybean mitochondrial proteins to various sizes of aluminum oxide nanoparticles under flooding stress. *Journal of Proteome Research* 15(12): 4464–4475.

Nable, R.O., Bañuelos, G.S., Paull, J.G. 1997. Boron toxicity. *Plant and Soil* 193(2): 181–198.

Naeem, M., Sadiq, Y., Jahan, A. et al. 2020. Salicylic acid restrains arsenic induced oxidative burst in two varieties of *Artemisia annua* L. by modulating antioxidant defence system and artemisinin production. *Ecotoxicology and Environmental Safety* 202: 110851.

Nagajyoti, P.C., Lee, K.D., Sreekanth, T.V.M. 2010. Heavy metals, occurrence and toxicity for plants: A review. *Environmental Chemistry Letters* 8(3): 199–216.

Nahar, K., Hasanuzzaman, M., Suzuki, T., Fujita, M. 2017. Polyamines-induced aluminum tolerance in mung bean: A study on antioxidant defense and methylglyoxal detoxification systems. *Ecotoxicology* 26(1): 58–73.

Nahar, K., Rahman, M., Hasanuzzaman, M. et al. 2016. Physiological and biochemical mechanisms of spermine-induced cadmium stress tolerance in mung bean (*Vigna radiata* L.) seedlings. *Environmental Science and Pollution Research International* 23(21): 21206–21218.

Nair, P.M.G., Chung, I.M. 2014. Impact of copper oxide nanoparticles exposure on *Arabidopsis thaliana* growth, root system development, root lignification, and molecular level changes. *Environmental Science and Pollution Research International* 21(22): 12709–12722.

Natasha Shahid, M., Niazi, N.K. et al. 2018. A critical review of selenium biogeochemical behavior in soil-plant system with an inference to human health. *Environmental Pollution* 234: 915–934.

Natasha Shahid, M., Khalid, S. et al. 2019. Biogeochemistry of antimony in soil-plant system: Ecotoxicology and human health. *Applied Geochemistry* 106: 45–59.

Navabpour, S., Yamchi, A., Bagherikia, S., Kafi, H. 2020. Lead-induced oxidative stress and role of antioxidant defense in wheat (*Triticum aestivum* L.). *Physiology and Molecular Biology of Plants* 26(4): 793–802.

Navarro, E., Baun, A., Behra, R. et al. 2008. Environmental behavior and ecotoxicity of engineered nanoparticles to algae, plants, and fungi. *Ecotoxicology* 17(5): 372–386.

Nazari, M., Zarinkamar, F., Niknam, V. 2018. Changes in primary and secondary metabolites of *Mentha aquatica* L. exposed to different concentrations of manganese. *Environmental Science and Pollution Research International* 25(8): 7575– 7588.

Ni, J., Wang, Q., Shah, F.A. et al. 2018. Exogenous melatonin confers cadmium tolerance by counterbalancing the hydrogen peroxide homeostasis in wheat seedlings. *Molecules* 23(4): 799.

Nie, Z.J., Hu, C.X., Sun, X.C. et al. 2007. Effects of molybdenum on ascorbate-glutathione cycle metabolism in Chinese cabbage (*Brassica campestris* L. ssp. *pekinensis*). *Plant and Soil* 295(1–2): 13–21.

Okupnik, A., Pflugmacher, S. 2016. Oxidative stress response of the aquatic macrophyte *Hydrilla verticillata* exposed to TiO$_2$ nanoparticles. *Environmental Toxicology and Chemistry* 35(11): 2859–2866.

Ortega, A., Garrido, I., Casimiro, I., Espinosa, F. 2017. Effects of antimony on redox activities and antioxidant defence systems in sunflower (*Helianthus annuus* L.) plants. *PLOS ONE* 12(9): e0183991.

Özfİdan-konakçi, C., Yildiztugay, E., Elbasan, F. et al. 2020. Assessment of antioxidant system and enzyme/nonenzyme regulation related to ascorbate-glutathione cycle in ferulic acid-treated *Triticum aestivum* L. roots under boron toxicity. *Turkish Journal of Botany* 44(1): 47–61.

Pan, X., Zhang, D., Chen, X. et al. 2011. Antimony accumulation, growth performance, antioxidant defense system and photosynthesis of *Zea mays* in response to antimony pollution in soil. *Water, Air, and Soil Pollution* 215(1–4): 517–523.

Pandey, V., Dixit, V., Shyam, R. 2005. Antioxidative responses in relation to growth of mustard (Brassica juncea cv. Pusa Jaikisan) plants exposed to hexavalent chromium) plants exposed to hexavalent chromium. *Chemosphere* 61(1): 40–47.

Patnaik, A.R., Achary, V.M.M., Panda, B.B. 2013. Chromium (VI)-induced hormesis and genotoxicity are mediated through oxidative stress in root cells of *Allium cepa* L. *Plant Growth Regulation* 71(2): 157–170.

Pavani, K.V., Beulah, M., Sai Poojitha, G.U. 2020. The effect of zinc oxide nanoparticles (ZnO NPs) on *Vigna mungo* L. seedling growth and antioxidant activity. *Nanoscience and Nanotechnology-Asia* 10(2): 117–122.

Peško, M., Molnárová, M., Fargašová, A. 2016. Response of tomato plants (*Solanum lycopersicum*) to stress induced by Sb (III). *Acta Environmentalica Universitatis Comenianae* 24(1): 42–47.

Pilon, M., Ravet, K., Tapken, W. 2011. The biogenesis and physiological function of chloroplast superoxide dismutases. *Biochimica et Biophysica Acta (BBA)—Bioenergetics* 1807(8): 989–998.

Pinto, F.R., Mourato, M.P., Sales, J.R. et al. 2017. Oxidative stress response in spinach plants induced by cadmium. *Journal of Plant Nutrition* 40(2): 268–276.

Pirzadah, T.B., Malik, B., Tahir, I. et al. 2019. Aluminium stress modulates the osmolytes and enzyme defense system in *Fagopyrum* species. *Plant Physiology and Biochemistry* 144: 178–186.

Pontigo, S., Godoy, K., Jiménez, H. et al. 2017. Silicon-mediated alleviation of aluminum toxicity by modulation of Al/Si uptake and antioxidant performance in ryegrass plants. *Frontiers in Plant Science* 8: 642.

Poschenrieder, C.H., Tolrà, R., Barceló, J. 2006. Can metals defend plants against biotic stress? *Trends in Plant Science* 11(6): 288–295.

Potkin, S.G., Turner, J.A., Guffanti, G. et al. 2009. Genome-wide strategies for discovering genetic influences on cognition and cognitive disorders: Methodological considerations. *Cognitive Neuropsychiatry* 14(4–5): 391–418.

Prasad, K.V.S.K., Paradha Saradhi, P., Sharmila, P. 1999. Concerted action of antioxidant enzymes and curtailed growth under zinc toxicity in *Brassica juncea*. *Environmental and Experimental Botany* 42(1): 1–10.

Qian, L., Huang, P., Hu, Q. et al. 2018. Morpho-physiological responses of an aluminum-stressed rice variety 'liangyoupei 9'. *Pakistan Journal of Botany* 50: 893–899.

Qin, P., Wang, L., Liu, K. et al. 2015. Genomewide association study of Aegilops tauschii traits under seedling-stage cadmium stress. *Crop Journal* 3(5): 405–415.

Rabinowitch, H.D., Sklan, D. 1981. Superoxide dismutase activity in ripening cucumber and pepper fruit. *Physiologia Plantarum* 52(3): 380–385.

Rajeshwari, A., Kavitha, S., Alex, S. et al. 2015. Cytotoxicity of aluminum oxide nanoparticles on Allium cepa root tip-effects of oxidative stress generation and biouptake. *Environmental Science and Pollution Research International* 22(14): 11057–11066.

Ramakrishna, B., Rao, S.S.R. 2015. Foliar application of brassinosteroids alleviates adverse effects of zinc toxicity in radish (*Raphanus sativus* L.) plants. *Protoplasma* 252(2): 665–677.

Reddy, A.M., Kumar, S.G., Jyothsnakumari, G. et al. 2005. Lead induced changes in antioxidant metabolism of horsegram (*Macrotyloma uniflorum* (Lam.) Verdc.) and bengalgram (*Cicer arietinum* L.). *Chemosphere* 60(1): 97–104.

Rodríguez-Ruíz, M., Aparicio-Chacón, M.V., Palma, J.M., Corpas, F.J. 2019. Arsenate disrupts ion balance, sulfur and nitric oxide metabolisms in roots and leaves of pea (*Pisum sativum* L.) plants. *Environmental and Experimental Botany* 161: 143–156.

Romero-Puertas, M.C., Corpas, F.J., Rodríguez-Serrano, M. et al. 2007. Differential expression and regulation of antioxidative enzymes by cadmium in pea plants. *Journal of Plant Physiology* 164(10): 1346–1357.

Romero-Puertas, M.C., Palma, J.M., Gómez, M. et al. 2002. Cadmium causes the oxidative modification of proteins in pea plants. *Plant, Cell and Environment* 25(5): 677–686.

Rout, G.R., Das, P. 2011. Rapid hydroponic screening for molybdenum tolerance in rice through morphological and biochemical analysis. *Plant, Soil and Environment* 48(11): 505–512.

Sager, S.M.A., Wijaya, L., Alyemeni, M.N. et al. 2020. Impact of different cadmium concentrations on two *Pisum sativum* L. genotypes. *Pakistan Journal of Botany* 52(3): 821–829.

Sahay, S., Gupta, M. 2017. An update on nitric oxide and its benign role in plant responses under metal stress. *Nitric Oxide : Biology and Chemistry* 67: 39–52.

Sahin, S., Kısa, D., Göksu, F., Geboloğlu, N. 2017. Effects of boron applications on the physiology and yield of lettuce. *Annual Research and Review in Biology* 21(6): 1–7.

Sahu, G.K., Upadhyay, S., Sahoo, B.B. 2012. Mercury induced phytotoxicity and oxidative stress in wheat (*Triticum aestivum* L.) plants. *Physiology and Molecular Biology of Plants* 18(1): 21–31.

Saidi, I., Nawel, N., Djebali, W. 2014. Role of selenium in preventing manganese toxicity in sunflower (*Helianthus annuus*) seedling. *South African Journal of Botany* 94: 88–94.

Saleem, M.H., Fahad, S., Adnan, M. et al. 2020. Foliar application of gibberellic acid endorsed phytoextraction of copper and alleviates oxidative stress in jute (*Corchorus capsularis* L.) plant grown in highly copper-contaminated soil of China. *Environmental Science and Pollution Research International* 27(29): 37121–37133.

Sallam, A., Alqudah, A., Dawood, M.F.A. et al. 2019. Drought stress tolerance in wheat and barley: Advances in physiology, breeding and genetics research. *International Journal of Molecular Sciences* 20(13): 3137.

Samet, H. 2020. Alleviation of cobalt stress by exogenous sodium nitroprusside in iceberg lettuce. *Chilean Journal of Agricultural Research* 80(2): 161–170.

Sami, A., Shah, F.A., Abdullah, M. et al. 2020. Melatonin mitigates cadmium and aluminium toxicity through modulation of antioxidant potential in *Brassica napus* L. *Plant Biology* 22(4): 679–690.

Sandalio, L.M., Dalurzo, H.C., Gómez, M. et al. 2001. Cadmium-induced changes in the growth and oxidative metabolism of pea plants. *Journal of Experimental Botany* 52(364): 2115–2126.

Scandalios, J.G. 1997. Molecular Genetics of superoxide dismutases in plants. *Oxidative Stress and the Molecular Biology of Antioxidative Defenses* 134: 527–568.

Schützendübel, A., Polle, A. 2002. Plant responses to abiotic stresses: Heavy metal-induced oxidative stress and protection by mycorrhization. *Journal of Experimental Botany* 53(372): 1351–1365.

Shahid, M., Pourrut, B., Dumat, C. et al. 2014. Heavy-metal-induced reactive oxygen species: Phytotoxicity and physicochemical changes in plants. *Reviews of Environmental Contamination and Toxicology* 232: 1–44.

Shahzad, B., Tanveer, M., Rehman, A. et al. 2018. Nickel; whether toxic or essential for plants and environment—A review. *Plant Physiology and Biochemistry* 132: 641–651.

Sharma, A., Kapoor, D., Wang, J. et al. 2020. Chromium bioaccumulation and its impacts on plants: An overview. *Plants* 9(1): 100.

Sharma, R., Bhardwaj, R., Thukral, A.K. et al. 2019. Oxidative stress mitigation and initiation of antioxidant and osmoprotectant responses mediated by ascorbic acid in Brassica juncea L. subjected to copper (II) stress. *Ecotoxicology and Environmental Safety* 182: 109436.

Sheng, H., Zeng, J., Liu, Y. et al. 2016. Sulfur mediated alleviation of Mn toxicity in polish wheat relates to regulating Mn allocation and improving antioxidant system. *Frontiers in Plant Science* 7: 1382.

Sheng, H., Zeng, J., Yan, F. et al. 2015. Effect of exogenous salicylic acid on manganese toxicity, mineral nutrients translocation and antioxidative system in polish wheat (*Triticum polonicum* L.). *Acta Physiologiae Plantarum* 37(2): 32.

Shi, Q., Zhu, Z. 2008. Effects of exogenous salicylic acid on manganese toxicity, element contents and antioxidative system in cucumber. *Environmental and Experimental Botany* 63(1–3): 317–326.

Shi, Q., Zhu, Z., Xu, M. et al. 2006. Effect of excess manganese on the antioxidant system in Cucumis sativus L. under two light intensities. *Environmental and Experimental Botany* 58(1–3): 197–205.

Shi, Z., Zhang, J., Wang, F. et al. 2018. Arbuscular mycorrhizal inoculation increases molybdenum accumulation but decreases molybdenum toxicity in maize plants grown in polluted soil. *RSC Advances* 8(65): 37069–37076.

Shigeoka, S., Ishikawa, T., Tamoi, M. et al. 2002. Regulation and function of ascorbate peroxidase isoenzymes. *Journal of Experimental Botany* 53(372): 1305–1319.

Shirani, B.S. 2020. The role of Fe-nano particles in scarlet sage responses to heavy metals stress. *International Journal of Phytoremediation* 22(12): 1259–1268.

Shiyab, S., Chen, J., Han, F.X. et al. 2009. Mercury-induced oxidative stress in Indian mustard (*Brassica juncea* L.). *Environmental Toxicology* 24(5): 462–471.

Shri, M., Dave, R., Diwedi, S. et al. 2014. Heterologous expression of *Ceratophyllum demersum* phytochelatin synthase, CdPCS1, in rice leads to lower arsenic accumulation in grain. *Scientific Reports* 4: 5784.

Shrivastava, S., Shrivastav, A., Sharma, J. 2016. Co-exposure effects of selenium and mercury on *Phaseolus vulgaris* excised leaves segment by enhancing the nr, anti-oxidative enzyme activity and detoxification mechanisms. *Advanced Techniques in Biology and Medicine* 4(2): 178.

Sidhu, G.P.S., Singh, H.P., Batish, D.R., Kohli, R.K. 2016. Effect of lead on oxidative status, antioxidative response and metal accumulation in *Coronopus didymus*. *Plant Physiology and Biochemistry* 105: 290–296.

Silva, S. 2012. Aluminium toxicity targets in plants. *Journal of Botany* 2012: 1–7.

Singh, A.L., Chaudhari, V. 1992. Enzymatic studies in relation to micronutrient deficiencies and toxicities in groundnut. *Plant Physiology and Biochemistry* 19: 107–109.

Singh, H., Kumar, D., Soni, V. 2020. Copper and mercury induced oxidative stresses and antioxidant responses of *Spirodela polyrhiza* (L.) Schleid. *Biochemistry and Biophysics Reports* 23: 100781.

Singh, R., Gautam, N., Mishra, A., Gupta, R. 2011. Heavy metals and living systems: An overview. *Indian Journal of Pharmacology* 43(3): 246.

Singh, S., Khan, N.A., Nazar, R. 2008. Photosynthetic traits and activities of antioxidant enzyme in black gram (*Vigna mungo* L. Hepper) under cadmium stress. *American Journal of Plant Physiology* 3: 25–32.

Singh, S., Parihar, P., Singh, R. et al. 2015. Heavy metal tolerance in plants: Role of transcriptomics, proteomics, metabolomics, and ionomics. *Frontiers in Plant Science* 6: 1143.

Singh, V.P., Srivastava, P.K., Prasad, S.M. 2013. Nitric oxide alleviates arsenic-induced toxic effects in ridged Luffa seedlings. *Plant Physiology and Biochemistry* 71: 155–163.

Soares, T.F.S.N., Dias, D.C.F., Oliveira, A.M.S. et al. 2020. Exogenous brassinosteroids increase lead stress tolerance in seed germination and seedling growth of *Brassica juncea* L. *Ecotoxicology and Environmental Safety* 193: 110296.

Song, G., Gao, Y., Wu, H. et al. 2012. Physiological effect of anatase TiO_2 nanoparticles on *Lemna minor. Environmental Toxicology and Chemistry* 31(9): 2147–2152.

Sousa, B., Soares, C., Oliveira, F. et al. 2020. Foliar application of 24-epibrassinolide improves *Solanum nigrum* L. tolerance to high levels of Zn without affecting its remediation potential. *Chemosphere* 244: 125579.

Srivastava, S., Shrivastava, M. 2017. Zinc supplementation imparts tolerance to arsenite stress in *Hydrilla verticillata* (L.f.) Royle. *International Journal of Phytoremediation* 19(4): 353–359.

Srivastava, G., Kumar, S., Dubey, G. et al. 2012. Nickel and ultraviolet-B stresses induce differential growth and photosynthetic responses in *Pisum sativum* L. seedlings. *Biological Trace Element Research* 149(1): 86–96.

Suh, M., Thompson, C.M., Brorby, G.P. et al. 2016. Inhalation cancer risk assessment of cobalt metal. *Regulatory Toxicology and Pharmacology* 79: 74–82.

Tang, K., Zhan, J.-C., Yang, H.-R., Huang, W.D. 2010. Changes of resveratrol and antioxidant enzymes during UV-induced plant defense response in peanut seedlings. *Journal of Plant Physiology* 167(2): 95–102.

Tarrahi, R., Movafeghi, A., Khataee, A. et al. 2019. Evaluating the toxic impacts of cadmium selenide nanoparticles on the aquatic plant *Lemna minor. Molecules* 24(3): 410.

Trujillo-Reyes, J., Majumdar, S., Botez, C.E. et al. 2014. Exposure studies of core-shell Fe/Fe(3)O(4) and Cu/CuO NPs to lettuce (Lactuca sativa) plants: Are they a potential physiological and nutritional hazard? *Journal of Hazardous Materials* 267: 255–263.

Ulhassan, Z., Huang, Q., Gill, R.A. et al. 2019. Protective mechanisms of melatonin against selenium toxicity in Brassica napus: Insights into physiological traits, thiol biosynthesis and antioxidant machinery. *BMC Plant Biology* 19(1): 1–16.

Uruç Parlak, K. 2016. Effect of nickel on growth and biochemical characteristics of wheat (*Triticum aestivum* L.) seedlings. *NJAS—Wageningen Journal of Life Sciences* 76: 1–5.

Vaculiková, M., Vaculík, M., Šimková, L. et al. 2014. Influence of silicon on maize roots exposed to antimony—Growth and antioxidative response. *Plant Physiology and Biochemistry* 83: 279–284.

Viehweger, K. 2014. How plants cope with heavy metals. *Botanical Studies* 55(1): 35.

Wang, D., Zhao, L., Ma, H. et al. 2017. Quantitative analysis of reactive oxygen species photogenerated on metal oxide nanoparticles and their bacteria toxicity: The role of superoxide radicals. *Environmental Science and Technology* 51(17): 10137–10145.

Wang, H., Kou, X., Pei, Z. et al. 2011. Physiological effects of magnetite (Fe_3O_4) nanoparticles on perennial ryegrass (*Lolium perenne* L.) and pumpkin (*Cucurbita mixta*) plants. *Nanotoxicology* 5(1): 30–42.

Wang, S.L., Liao, W.B., Yu, F.Q. et al. 2009. Hyperaccumulation of lead, zinc, and cadmium in plants growing on a lead/zinc outcrop in Yunnan Province, China. *Environmental Geology* 58(3): 471–476.

Wang, Y., Loake, G.J., Chu, C. 2013. Cross-talk of nitric oxide and reactive oxygen species in plant programmed cell death. *Frontiers of Plant Science* 4: 314.

Wawrzyński, A., Kopera, E., Wawrzyńska, A. et al. 2006. Effects of simultaneous expression of heterologous genes involved in phytochelatin biosynthesis on thiol content and cadmium accumulation in tobacco plants. *Journal of Experimental Botany* 57(10): 2173–2182.

Wen-Wen, C., Si-Jia, H., Chen-Xi, W. et al. 2010. Cytotoxicity effects of nano-Fe3O4 on HeLa cells. In: Bioinformatics and Biomedical Engineering (iCBBE) 4th International Conference on, 2010, 1–4. IEEE.

Winkel, L.H.E., Vriens, B., Jones, G.D. et al. 2015. Selenium cycling across soil-plant-atmosphere interfaces: A critical review. *Nutrients* 7(6): 4199–4239.

Wolfe-Simon, F., Grzebyk, D., Schofield, O., Falkowski, P.G. 2005. The role and evolution of superoxide dismutases in algae. *Journal of Phycology* 41(3): 453–465.

Wu, K., Xiao, S., Chen, Q. et al. 2013. Changes in the activity and transcription of antioxidant enzymes in response to Al stress in black soybeans. *Plant Molecular Biology Reporters* 31(1): 141–150.

Wuana, R.A., Okieimen, F.E. 2011. Heavy metals in contaminated soils: A review of sources, chemistry, risks and best available strategies for remediation. *ISRN Ecology* 2011: 1–20.

Xiong, Z.T., Wang, H. 2005. Copper toxicity and bioaccumulation in Chinese cabbage (*Brassica pekinensis* Rupr.). *Environmental Toxicology* 20(2): 188–194.

Xu, S., Hu, C., Tan, Q. et al. 2018. Subcellular distribution of molybdenum, ultrastructural and antioxidative responses in soybean seedlings under excess molybdenum stress. *Plant Physiology and Biochemistry* 123: 75–80.

Yadav, P., Kaur, R., Kanwar, M.K. et al. 2018. Ameliorative Role of Castasterone on Copper Metal Toxicity by Improving Redox Homeostasis in Brassica juncea L *Journal of Plant Growth Regulation* 37(2): 575–590.

Yadav, S.K. 2010. Heavy metals toxicity in plants: An overview on the role of glutathione and phytochelatins in heavy metal stress tolerance of plants. *South African Journal of Botany* 76(2): 167–179.

Yang, H., Li, C., Lam, H.M. et al. 2015. Sequencing consolidates molecular markers with plant breeding practice. *TAG. Theoretical and Applied Genetics. Theoretische und Angewandte Genetik* 128(5): 779–795.

Yang, J., Cao, W., Rui, Y. 2017. Interactions between nanoparticles and plants: Phytotoxicity and defense mechanisms. *Journal of Plant Interactions* 12(1): 158–169.

Yang, J., Li, G., Bishopp, A. et al. 2018. A comparison of growth on mercuric chloride for three Lemnaceae species reveals differences in growth dynamics that effect their suitability for use in either monitoring or remediating ecosystems contaminated with mercury. *Frontiers in Chemistry* 6: 112.

Yang, Z., Xiao, Y., Jiao, T. et al. 2020. Effects of copper oxide nanoparticles on the growth of rice (*Oryza sativa* L.) Seedlings and the Relevant Physiological Responses. *International Journal of Environmental Research and Public Health* 17(4): 1260.

Yanık, F., Vardar, F. 2015. Toxic effects of aluminum oxide (Al_2O_3) nanoparticles on root growth and development in Triticum aestivum. *Water, Air, and Soil Pollution* 226(9): 296.

Yildiztugay, E., Ozfidan-Konakci, C., Karahan, H. et al. 2019. Ferulic acid confers tolerance against excess boron by regulating ROS levels and inducing antioxidant system in wheat leaves (*Triticum aestivum*). *Environmental and Experimental Botany* 161: 193–202.

Youn, H.-D., Kim, E.-J., Roe, J.-H. et al. 1996. A novel nickel-containing superoxide dismutase from *Streptomyces* spp. *Biochemical Journal* 318(3): 889–896.

Zafar, H., Ali, A., Ali, J.S. et al. 2016. Effect of ZnO nanoparticles on Brassica nigra seedlings and stem explants: Growth dynamics and antioxidative response. *Frontiers in Plant Science* 7: 535.

Zaheer, I.E., Ali, S., Saleem, M.H. et al. 2020. Role of iron–lysine on morpho-physiological traits and combating chromium toxicity in rapeseed (*Brassica napus* L.) plants irrigated with different levels of tannery wastewater. *Plant Physiology and Biochemistry* 155: 70–84.

Zaid, A., Mohammad, F., Fariduddin, Q. 2020. Plant growth regulators improve growth, photosynthesis, mineral nutrient and antioxidant system under cadmium stress in menthol mint (*Mentha arvensis* L.). *Physiology and Molecular Biology of Plants* 26(1): 25–39.

Zaid, A., Wani, S.H., Masoodi, K.Z. 2019. Role of nitrogen and sulfur in Mitigating cadmium induced Metabolism. *Journal of Pharmaceutical Sciences and Research* 35(1): 121–141.

Zengin, F.K. 2006. The effects of Co2+ and Zn2+ on the contents of protein, abscisic acid, proline and chlorophyll in bean (Phaseolus vulgaris cv. Strike) seedlings) seedlings. *Journal of Environmental Biology* 27(2): 441.

Zhang, F., Xiao, X., Xu, K. et al. 2020. Genome-wide association study (GWAS) reveals genetic loci of lead (Pb) tolerance during seedling establishment in rapeseed (*Brassica napus* L .). *BMC Genomics* 21(1): 1–12.

Zhang, H., Zhang, L.L., Li, J. et al. 2020. Comparative study on the bioaccumulation of lead, cadmium and nickel and their toxic effects on the growth and enzyme defence strategies of a heavy metal accumulator, Hydrilla verticillata (Lf) Royle. *Environmental Science and Pollution Research* 2020: 1–13.

Zhang, Y., Liu, P., Zhang, X. et al. 2018. Multi-locus genome-wide association study reveals the genetic architecture of stalk lodging resistance-related traits in maize. *Frontiers in Plant Science* 9: 611.

Zhao, L., Peng, B., Hernandez-Viezcas, J.A. et al. 2012. Stress response and tolerance of Zea mays to CeO_2 nanoparticles: Cross talk among H_2O_2, heat shock protein, and lipid peroxidation. *ACS Nano* 6(11): 9615–9622.

Zhao, Q., Sun, Q., Dong, P. et al. 2019. Jasmonic acid alleviates boron toxicity in *Puccinellia tenuiflora*, a promising species for boron phytoremediation. *Plant and Soil* 445(1–2): 397–407.

Zhao, X., Luo, L., Cao, Y. et al. 2018. Genome-wide association analysis and QTL mapping reveal the genetic control of cadmium accumulation in maize leaf. *BMC Genomics* 19(1): 1–13.

Zhou, Z.S., Huang, S.Q., Guo, K. et al. 2007. Metabolic adaptations to mercury-induced oxidative stress in roots of *Medicago sativa* L. *Journal of Inorganic Biochemistry* 101(1): 1–9.

Zoufan, P., Baroonian, M., Zargar, B. 2020. ZnO nanoparticles-induced oxidative stress in *Chenopodium murale* L, Zn uptake, and accumulation under hydroponic culture. *Environmental Science and Pollution Research International* 27(4): 11066–11078.

Zoufan, P., Karimiafshar, A., Shokati, S. et al. 2018. Oxidative damage and antioxidant response in *Chenopodium murale* L. exposed to elevated levels of Zn. *Brazilian Archives of Biology and Technology* 61: e18160758.

Zvobgo, G., Lwalaba, J.L.W., Sagonda, T. et al. 2019. Alleviation of arsenic toxicity by phosphate is associated with its regulation of detoxification, defense, and transport gene expression in barley. *Journal of Integrative Agriculture* 18(2): 381–394.

15 Role of Non-Enzymatic Antioxidant Defense Mechanisms in Imparting Heavy Metal, Salt, Water and Temperature Stress Tolerance in Plants

Kiran Saroy and Neera Garg

CONTENTS

15.1 INTRODUCTION

Various abiotic constraints (heavy metal, salinity, drought and temperature) are the primary cause for the loss of plant yield and productivity worldwide (Awasthi et al. 2014; Zandalinas et al. 2018). These stresses affect about 96.5% of arable land worldwide (salinity – 19.5%, flood –13%, anoxia and drought – 64%) and are challenging researchers to investigate plant responses under the changed growing conditions (Andjelkovic 2018; Yadav et al. 2020). The negative impacts are elicited in the form of overproduction of reactive oxygen species (ROS), particularly hydroxyl radical (OH·), hydrogen peroxide (H_2O_2), superoxide ($O_2^{·-}$) and also singlet oxygen species ($^1O_2$), which harm proteins, lipids, carbohydrates and DNA and finally lead to oxidative stress (Goel and Madan 2014). In order to scavenge ROS, plants upregulate the enzymatic (catalase, CAT; ascorbate peroxidase, APX; and superoxide dismutase, SOD) antioxidant profiling which allows them to survive in adverse conditions (Zandalinas et al. 2018). However, these enzymatic antioxidants are not sufficient to completely reduce ROS generation in plants (Kasote et al. 2015). Therefore, plants deploy an array of non-enzymatic antioxidant defense mechanisms comprising of glutathione, GSH; ascorbic acid, ASA (Vitamin C); α-Tocopherols (Vitamin E); carotenoids, C_t; and phenolic compounds

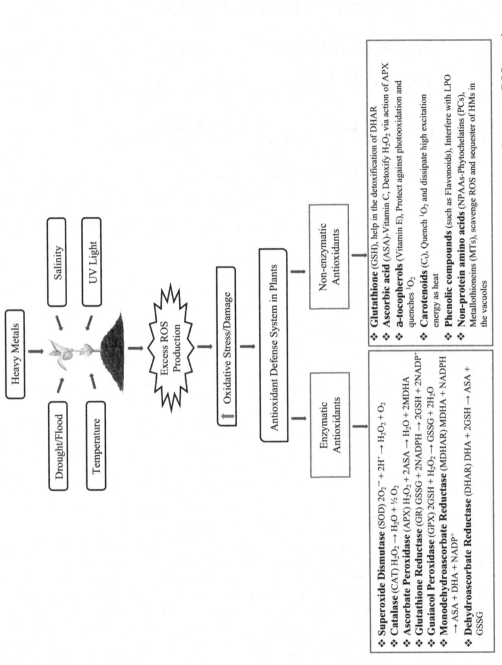

FIGURE 15.1 Diagrammatic representations of the dynamic roles played by different antioxidants in plants against abiotic stresses. ROS, reactive oxygen species; O_2, superoxide radical; H_2O_2, hydrogen peroxide; O_2, oxygen molecule; GSSG, oxidized glutathione; $^1O_2$, singlet oxygen; LPO, lipid peroxidation; HMs, heavy metals.

(namely flavonoids) under various abiotic stresses (Figure 15.1) (Kasote et al. 2015). Besides this, under heavy metal (HM) toxicity, plants activate non-protein amino acids (NPAAs) such as phyto-chelatins, PCs; and metallothioneins, MTs (Verbruggen et al. 2009; Song et al. 2014) that perform important roles in metal sequestration into vacuoles.

GSH is a main water-soluble compound and plays a central part in ROS scavenging through the GSH-ASA cycle (Zagorchev et al. 2013). It is the component of glutathione-S-transferase (GST), that helps in dehydroascorbate reductase (DHAR) detoxification and xenobiotics (Ahmad et al. 2009). Additionally, most eukaryotic organisms generate ASA, a water-soluble antioxidant that controls numerous physiological processes monitoring growth, development and stress tolerance (Akram et al. 2017). Moreover, tocopherols and carotenoids play important photoprotective roles via scattering high excitation energy as heat or by removing ROS and reducing lipid peroxidation (LPO) (Ahmad et al. 2009; Uarrota et al. 2018). Phenolic compounds such as flavonoids are significant constituents of ROS scavenging systems (Garg and Singla 2015) and are synthesized to effectively counter the stress-induced oxidative damage (Di Ferdinando et al. 2012). They also display the potential for heavy metal stress protection by chelating transition metals (viz. Fe, Cu, Ni and Zn) to generate hydroxyl radical (Agati and Tattini 2010) and are effective inhibitors of LPO (Ahmad et al. 2009). Besides this, NPAAs have an essential role in biosynthesis or catabolism of primary metabolites by serving as intermediates, acting as nitrogen/carbon (N/C) storage compounds. PCs and MTs are low molecular weight (LMW) compounds, which scavenge ROS, sequestrate HMs in the vacuoles and activate transcription genes, such as PC synthase (PCS) gene, under HM stress surroundings (Brunetti et al. 2015; Navarrete et al. 2019). PCs are minor cysteine (Cys)-rich, metal-binding peptides found in plants (Fulekar et al. 2009), which accumulate and stabilize HMs in the vacuole of cells via the formation of high molecular weight (HMW) complexes of metal ions (Cd, Zn, Pb, Cu, etc.) with thiol, amino sulfhydryl, carboxyl and hydroxyl groups and molecules carrying these functional groups, being defined as metal chelators (Emamverdian et al. 2015; Hasan et al. 2017). Similarly, Plant MTs are Cys-rich LMW (4-8kDa) metal-binding proteins, which support maintaining metal-ion concentrations in plant cells in a consistent range (Guo et al. 2013). Recently, a lot of importance has been given to non-enzymatic antioxidant defense mechanisms for conferring abiotic strain tolerance in plants. Therefore, this chapter focuses on updated information about the mechanisms adopted by different non-enzymatic antioxidant defense systems for imparting stress tolerance to plants.

15.2 ROLE OF NON-ENZYMATIC ANTIOXIDANTS

15.2.1 Glutathione

Glutathione (GSH), amplifies plant tolerance to diverse abiotic constraints, including salinity, water, temperature, drought and HM (Hasanuzzaman and Fujita 2013). It is a water-soluble, tripeptide γ-glutamylcysteinyl-Gly, non-protein thiol which is reviewed as the most essential intracellular defense molecule opposing ROS-induced oxidative injury in plants (Hasanuzzaman et al. 2017). GSH is responsible for scavenging of OH·, $^1O_2$ and safeguarding the enzymes of thiol groups. It takes place in reduced form (GSH) in plant tissue in almost all cellular partitions such as mitochondria, cytosol, vacuole, endoplasmic reticulum (ER), chloroplast, peroxisomes and also apoplast (Ahmad et al. 2009; Das and Roychoudhury 2014). It regulates various metabolic activities like growth, cell cycle, immune function, gene expression and protein biogenesis as well as amino acids (AAs) transport across membranes under non-stress and stress situations (Shao et al. 2007; Nahar et al. 2015a). It also defends membranes by balancing the reduced form of both α-Toc and zeaxanthin, hampers oxidative denaturation of proteins and helps as a component for both glutathione peroxidase (GPOX) and GST. Besides this, GSH performs a vital function in the ASA-GSH cycle whereby it regenerates ASA and oxidized form GSSG (glutathione disulfide) then GSSG reprocessed back to GSH through the activity of NADPH-dependent glutathione reductase (GR). The equilibrium between two forms of GSH is necessary to retain the cellular redox condition (Ahmad et al. 2009). An additional principal role of GSH is the production of phytochelatins (PCs), which are attached to HMs for secure movement and sequestration in the vacuole (Srivalli and Khanna-Chopra 2008). Increased levels of GSH along with improved GSH/GSSG ratios were observed in the tolerant genotypes of chickpea (Garg and Singla 2015; Garg and Bhandari 2016; Bharti and Garg 2019) and pigeonpea (Pandey and Garg 2017) under salt stress which destroyed the harmful effects of ROS generation. Similarly, increased GSH levels have been associated with improved salt tolerance in the wild type (WT) canola saplings with a three-fold rise in GSH and Cys synthesis which could be related to the presence of AtNHX1 gene when crosschecking with transgenic plants that do not have this gene (Ruiz and Blumwald 2002). Moreover, GSH also helps in the detoxification of methylglyoxal (MG), because it works as a cofactor with two enzymes viz. glyoxalase I (Gly I) and glyoxalase II (Gly II) of the glyoxalase mechanism (Singla-Pareek et al. 2003). Pokkali (salt tolerant) rice cultivars displayed significantly higher GSH levels along with enhanced Gly I and II activities than IR64 (salt sensitive). Besides this, the tolerant cultivar displayed greater GSH/GSSG and ASA/DHA (ascorbate to dehydroascorbate) ratios, concurrently with greater activities of SOD, CAT, APX as well as GPOX which were linked to the alleviation of salt stress, as implied by declined Na/K ratio, ROS and the extent of oxidative DNA damage (El-Shabrawi et al. 2010). Likewise, in tobacco, improved GSH grade, GSH redox form, GPOX, GST as well as Gly I activity was observed under salt stress which could be related to the reduction in protein carbonyl-ation, strengthened antioxidant defense and MG detoxification systems (Islam et al. 2009). Moreover, exogenously used GSH improved abiotic constraint tolerances in the onion bulb by reducing epider-mis harm, improving membrane permeation, preventing protoplasmic protrusion and maintaining cell resilience (Aly-Salama and Al-Mutawa 2009). Significantly increased levels of diverse antioxidant constituents such as GSH, proline, ASA, C_t, phenols and GB have been reported in two types of canola (*Brassica napus* L. cv. Serw and cv. Pactol) under salt stress (Khattab 2007). Moreover, the GSH redox signaling procedure also enhanced salt tolerance stamina in an *Arabidopsis* mutant, where GSH worked as a downline element of signal transduction passages, along with the increased accumula-tion of GSH. This mutant lowered the stomatal aperture along with transpiration rate and carried out improvement in root and vegetative development which was imputed to increase GSH levels (Shao et al. 2007). Exogenic supply of GSH could result in aligned initiation of the antioxidant and Gly systems which efficiently diminished oxidative harm and MG toxicity and enhanced physiological alteration of *Vigna radiata* saplings subjected to temporary salt stress (Nahar et al. 2015b).

Besides this, under HM stressed terms, GSH performs a vital function in safeguarding the plants from toxicity by chelation of HMs for transport and sequestration inside the vacuole or by acting as a precursor for PCs biosynthesis (Hasanuzzaman et al. 2017). Furthermore, it has been observed that under Cd stress, improved antioxidant activity was observed in leaves of *Phragmites australis* plants, with an increase in GSH, pyridine pools and reduced GSH which protected the activity of photosynthetic pigments from thiophilic burst (Pietrini et al. 2003). Under Cd stress, total GSH, non-protein thiols, GSH and PCs increased significantly, while GSSG decreased in *Vigna mungo* plants (Molina et al. 2008).

Excessive temperatures, either high or low, are also the main ecological calamities that restrict plant production, where GSH is found to perform a significant role in reducing oxidative stress (Yogendra and Nirmaljit 2017). The GSH pools, along with temporal and spatial variations, its regulation and importance in redox signaling and defense systems, are important substances for thermal resistance (Szalai et al. 2009). A higher whole GSH and GSH/GSSG proportion conferred heat tolerance (HT) in *Triticum aestivum*, *Zea mays* and *Vigna radiata* (Nieto-Sotelo and Ho 1986; Dash and Mohanty 2002; Nahar et al. 2015a). Similarly, ASA-GSH sequence was improved to counter the elevated temperatures in *Malus domestica* plants exposed to heat stress with the high-est total GSH along with increased activity and expression of DHAR, APOX and GR (Ma et al. 2008). The supplementation of *Vigna radiata* saplings with GSH resulted in improved physiological activities together with enhanced antioxidant and increased Gly I and Gly II activities, endogenic GSH content and GSH/GSSG proportion, with a reduction of GSSG content and MG under short term heat stress (Nahar et al. 2015a). Besides this, low temperatures also triggered a larger upsurge in complete GSH content, GR activity and reduced GSH/GSSG proportion in wheat variety (Kocsy et al. 2002), and increased total GSH pool and SOD activity in *Dunaliella salina* (Haghjou et al. 2009). Greater GSH levels along with the function of enzymes of the ASA-GSH sequence have also been reported in strawberry varieties (Luo et al. 2011).

Besides this, GSH is also engaged in stomatal closure triggered by abscisic acid (ABA), as its appearance in guard cells controls ABA signaling in the *Arabidopsis* cad2-1 mutant (Akter et al. 2012). Exogenously supplied GSH decreased chemicals, namely p-nitrobenzyl chloride (PNBC) and iodomethane (IDM), resulting in the diminution of GSH in stomata. Elevated GSH content has also been associated with the regulation of leaf water content and specifies its regulatory role in leaf rolling (Saruhan et al. 2012). Chen et al. (2012) indicated that GST was overexpressed in *Arabidopsis thaliana* plants and owned a signaling function and controlled plant development by keeping the balance in GSH pools. The mutant plant atgstu17 produced further GSH and ABA levels than wild type (WT). Likewise, the exogenous GSH supply of WT plants resulted in increased ABA content and also better drought tolerance (Table 15.1).

TABLE 15.1
Role of Diverse Antioxidants in Abiotic Stress Management in Various Plant Species

Antioxidant	Plant Species	Abiotic Stress and References
Glutathione	*Vigna radiata*	Salt stress (Nahar et al. 2015b)
	Cicer areitnum	Salt stress (Garg and Singla 2015; Garg and Bhandari 2015; Bharti and Garg 2019)
	Cajanus cajan	Salt stress (Pandey and Garg 2017)
		As stress (Garg and Kashyap 2019)
	Arabidopsis	Cd stress (Xiang et al. 2001; Pietrini et al. 2003; Molina et al. 2008)
	Phargmites australis	
	Vigna mungo	
Ascorbic acid	Flux cultivars	Salinity (El-Hariri et al. 2010)
	Suaeda salsa	High light stress (Pang et al. 2011)
	Lens culnaris Medik.	Salt stress (Alami-Milani and Aghaei-Gharachorlou 2015)
	Vicia faba L.	Pb stress (Fu-Di Xie et al. 2014; Alamri et al. 2018)
	Wheat plants	
	Zea mays L.	Water deficit conditions, low temperature (Darvishan et al. 2013; Ahmad et al. 2014)
	Brassica napus	Drought (Shafiq et al. 2014)
	Cajanus cajan	As stress (Garg and Kashyap 2019)
Tocopherols	*Arabidopsis thaliana*	Cu and Cd stress (Collin et al. 2008)
	Brassica juncea	Salt and $CdCl_2$ stress (Yusuf et al 2010)
		Salt, Cd and osmotic stress (Kumar et al. 2013)
	Wheat plants	As stress (Ghosh and Biwas 2017)
Carotenoids	*Sorghum bicolor*	High light and Salt stress (Sharma and Hall 1992)
	Transgenic sweet potato	Salt stress (Kang et al. 2017)
	Gonyaulax polyedra	Hg^{2+}, Cd^{2+}, Pb^{2+} and Cu^{2+} toxicity (Okamoto et al 2001)
	Helianthus annuus L.	Cu stress (El-Tayeb et al. 2006)
	Phaseolus vulgaris L	HM stress (Ni, Cu, Cr and Zn) (Zengin 2013)
Flavonoids	*Ipomoea purpurea*	Heat stress (Coberly and Raushe 2003)
	Oryza sativa L.	Salinity (Chutipaijit et al. 2009)
	Citrus sinensis	Cold stress (Crifò et al. 2011)
	Glycine max	Salt stress (Qu et al. 2016)
	Solanum nigrum	NaCl toxicity (Abdallah et al. 2016)
	Oryza sativa	As stress (Chauhan et al. 2017)

15.2.2 ASCORBIC ACID

L-ascorbic acid (ASA, Vitamin C) is the main antioxidant in plants which performs a significant function in scavenging of increased cellular ROS entities to protect the photosynthetic parts and other cellular organs from oxidative damage (Mazid et al. 2011; Venkatesh and Park 2014; Akram et al. 2017). It works as a cell-signaling molecule in cell wall growth, cell division, expansion, hormone biosynthesis and regeneration of antioxidants (Hossain et al. 2017). It is a cofactor of various enzymes, namely 1-amino-cyclopropane-1-carboxylic acid (ACC) oxidase (in ethylene synthesis), violaxanthin de-epoxidase (VDE) (in xanthophyll cycle) and 2-oxoacid-dependent dioxygenases (in ABA and GA synthesis) (Smirnoff 2000). Currently, it has been declared that ASA performs a major role in safeguarding against various environmental constraints particularly drought, salinity and temperature stress with altered endogenous ASA content (Larkindale et al. 2005; Hemavathi et al. 2010; Venkatesh and Park 2014). ASA is an unstable compound, which changes the balance of ASA to DHA in plants to maintain the redox balance as well as maintain a higher total ASA pool (Hasanuzzaman et al. 2017). Numerous studies reported ASA regulation of antioxidant defense in various plants under stress, like under salinity stress in canola (Bybordi 2012), *Hordeum vulgare* under salt stress (Agami 2014), *Brassica napus* under drought (Shafiq et al. 2014), pigeonpea under As stress (Garg and Kashyap 2019). *Arabidopsis vtc-1* mutants had low ascorbate content, and low activities of MDAR and DHAR enzymes of ASA-GSH sequence than the WT under salt stress (Huang et al. 2005). Tolerant chickpea genotypes displayed more ASA accumulation than sensitive types and caused redox imbalance by lowering ASA/DHA and GSH/GSSG ratio under salt stress, indicating that redox homeostasis was important for salt tolerance (Bharti and Garg 2019). Besides this, transgenic potato plants expressed *GalUR* gene and strawberry plants expressed *GLOase* gene, along with some fold increase in ASA synthesis and an improved survival under NaCl and drought stresses (Hemavathi et al. 2010). Genetically modified *Arabidopsis* and alfalfa displayed overexpression of the ascorbate peroxidase (*APX*) gene, and *OsAPX2* gene improved tolerance under salt and drought constraints (Guan et al. 2012). Moreover, ASA-deficient mutant vtc1 has a vital function in the form of co-factor for the biosynthesis of gibberellic acid-GA, abscisic acid-ABA and salicylic acid-SA. ASA appears to influence endogenous levels as well as the signaling of these plant hormones and consequently affects+ comebacks against the abiotic environmental constraints (Mazid et al. 2011).

Higher biosynthesis of ascorbate in plants receiving full light and increased photoinhibition along with oxidative injury in ascorbate-lacking plants indicate its function in excess light energy dissemination (Smirnoff 2000; Yabuta et al. 2007). High light constraint results in the initiation of the cytosolic APX and guards the cytosol and additional cellular segments from oxidative stress (Mullineaux and Karpinski 2002). In *Arabidopsis* plants, cytosolic APX1 protects the thylakoid/stromal and mitochondrial APXs by scavenging O_2^- and H_2O_2 content during light stress (Davletova et al. 2005). Besides this, thylakoid membrane-bound APX (tAPX) is a determining constituent of antioxidative systems in photo-oxidative stress in chloroplasts, and increased activity of tAPX under stress maintains the redox balance of ascorbate (Yabuta et al. 2007). Additionally, genetically modified *Arabidopsis* plants overexpressing *Suaeda salsa* chloroplastic stromal APX (sAPX) and tAPX and suggest that transgenic plants showed increased tolerance to oxidative stress caused by high light (Pang et al. 2011). In the majority of plant species, endogenous ASA content is not enough to alleviate the negative impacts of these stresses (Mazid et al. 2011; Shafiq et al. 2014). Thus, exogenous supplementation of ASA has been advised to lessen the negative impact of stress on plants (Shao et al. 2007; Alayafi 2020) by upgrading plant growth, photosynthesis efficiency, oxidative defense potential, transpiration along with photosynthetic pigments. Moreover, exogenous supplementation of ASA alleviated Pb-induced oxidative injury in wheat plants by improving the activities of antioxidant enzymes (SOD, CAT and GR), suppressing the formation of MDA (Malondialdehyde) as well as H_2O_2 content. Moreover, ASA also improved the essential nutrient (K, N, P, Mg and Ca) content, Rubisco activity as well as reduced chlorophyll (Chl) degradation (Alamri et al. 2018). Foliar application of ASA opposed the detrimental effect of salinity and was accompanied by a substantial upsurge in the growth of flux cultivars (El-Hariri et al. 2010). Besides this, foliar treatment of ASA counteracted the negative effect of NaCl by significantly increasing plant growth, IAA contents as well as yield components in wheat with a decrease in total phenol contents (Elhamid et al. 2014).

15.2.3 TOCOPHEROLS

Tocopherols (Tocs), a major vitamin E compound, are lipophilic antioxidants that are produced particularly in photosynthetic organisms, higher plants and algae (Hussain et al. 2013). In plants, normally four types of natural Tocs are known i.e. α, β, γ and δ (Fritsche et al. 2017). In most higher plants, α-Toc accumulates chiefly in photosynthetic tissue and γ-Toc in seeds (Abbasi et al. 2007). The four isomers differ in their antioxidant activity in the order α > β > γ > δ, which could be attributed to the methylation pattern of the phenolic ring. The presence of three methyl groups is responsible for the highest activity of α-Tocs (Fritsche et al. 2017), and their significant existence in chloroplasts helps in protection against photo-oxidation (Munné-Bosch 2005). α-Toc also participates in the detoxification of ROS with GSH and ASA as well as scavenges lipid peroxyl radicals in the thylakoid membrane (Munné-Bosch 2005). It works together with the other antioxidants such as GSH, ASA, C_t, etc. and helps in the maintenance of redox balance inside the plant cell under various adverse environmental conditions (Munné-Bosch 2005; Kapoor et al. 2015). The increase in the amount of Toc is necessary for the biological efficacy of redox recycling and its interdependency while the electron transfer inside a cell (Ahmad et al. 2010). Toc acts as $^1O_2$ scavenger in PSII of *Chlamydomonas reinhardtii* (Grewe et al. 2014), due to its vital role in signal transduction, gene expression regulation as well as the export of photoassimilates (Falk and Munné-Bosch 2010). Under Cu and Cd stress, *A. thaliana* plants had an upsurge in α-Toc and ASA content with a decrease in GSH and other antioxidant enzymes (thiol-based reductases and peroxidases) (Collin et al. 2008). Transgenic *Brassica juncea* plants displayed overexpression of the γ-Toc methyltransferase (γ-TMT) gene which resulted in a six-fold rise in the α-Toc for the improved potential to NaCl, CdCl₂ toxicity (Yusuf et al. 2010). Similarly, in wild type (WT) and α-Toc-enriched transgenic (TR) *Brassica juncea* plants raised under salinity, CdCl₂ and osmotic constraint (mannitol) displayed an increment in Toc content and antioxidative enzymes (SOD, CAT and APX) with a decrease in ASA and GSH content (Kumar et al. 2013). Similarly, enhanced content of α-toc was reported in wheat when plants were treated with As stress (Ghosh and Biswas 2017). Tocs work in seeds by defending the embryo from ROS attack by reducing metabolic activity (glucose level and ion leakage) in the *Oryza sativa* seeds under both accelerated aging as well as stress conditions (Chen et al. 2016).

15.2.4 CAROTENOIDS

In the photosynthetic plants, thylakoid membranes of chloroplasts have a fat-soluble, accessory and ubiquitous pigment, carotenoid (C_t), that has light harvesting as well as non-enzymatic antioxidative properties (Cannella et al. 2016). The pigment carotenoid is C_{40} isoprenoid, absorbs light at a wavelength between 400 and 550 nm, passes the captured energy to Chl and quenches triplet Chl. C_t as a non-enzymatic antioxidant scavenges $^1O_2$ by dissipating excess energy as heat and inhibits the formation of $^1O_2$, thus protecting the photosynthetic apparatus (Kalaji et al. 2017). In a study, *Arabidopsis* showed overexpression of β-carotene hydroxylase (in zeaxanthin biosynthetic cycle) which provided stress tolerance to increased sunlight via reducing leaf chlorosis, LPO and oxidative damage of membranes (Davison et al. 2002). Similarly, transgenic *Arabidopsis* showed high potential to salinity in opposition to WT plants by overexpression of phytoene synthase gene (*SePSY*) of C_t synthesis, and these transgenic lines displayed increment in the photosynthetic efficiency (photosynthetic rate and PS II activity) and antioxidative capacity (SOD and peroxidase) (Han et al. 2008). Under salt stress, transgenic *Ipomoea batatas* (RC plants) sustained high photosystem II efficiency (total C_t and levels) than non-transgenic plants (NT) because RC plants were generated by RNA interference (RNAi) silencing of the IbCHY-β gene and resulted in increased levels of β-carotene, C_t along with salinity tolerance (Kang et al. 2017). Similarly, under salinity, C_t pathway genes were studied in tomato plants where lycopene β cyclase (catalyzes the lycopene to β-carotene) was highly affected than other genes (phytoene synthase, zeta carotene desaturase, phytoene desaturase). The increased gene expression upregulates antioxidant enzymes, which help in scavenging ROS and protect the plants from salt stress (Ann et al. 2011). Besides this, unicellular alga *Gonyaulax polyedra* subjected to toxic metal Pb^{2+}, Hg^{2+}, Cu^{2+} and Cd^{2+} exhibited high β-carotene levels than SOD and APX activities due to higher oxygen uptake inside the cells (Okamoto et al. 2001). In contrast, excess Cu in *Helianthus annuus* L. resulted in reduced fresh as well as dry weights and photosynthetic pigments Chl a, b and C_t concentrations with a substantial upsurge in Chl a/b proportion and LPO (El-Tayeb et al. 2006). Similarly, the C_t and Chl content reduced gradually under high copper sulfate and cadmium dichloride in *Lemna minor* due to inhibition in the intake and transport of other important metal elements such as Mn, Zn and Fe (Hou et al. 2007). C_t content decreased in red clover and bean plants grown in As treated soils since As (V) facilitated inflammation of the thylakoid membrane (Stoeva et al. 2005). In *Phaseolus vulgaris* L., HM stress (Ni, Cu, Cr and Zn) caused oxidative stress and resulted in decreased C_t content (Zengin 2013).

15.2.5 FLAVONOIDS

Flavonoids are LMW, ubiquitous and polyphenolic compounds in plants, and occur naturally either as free aglycones or glycosidic conjugates (Eghbaliferiz and Iranshahi 2016). They are divided into subgroups – flavones, flavonols, flavanonols, isoflavones, flavanones and chalcones (Panche et al. 2016). Flavonoids are most responsive to UV-light, and are the most effective UV-B absorbers among the polyphenols (Di Ferdinando et al. 2012). Flavonoids act as scavenger molecules for ROS as they identify and neutralize the radicals before they start their action (Løvdal et al. 2010). Flavonoids have been reported to alter lipid packing and fluidity in membranes which resist diffusion of radicals and peroxidation reactions (Oteiza et al. 2005). They also alleviate the damages posed on chloroplasts by scavenging $^1O_2$ (Agati et al. 2010). The synthesis of flavonoids is upregulated by a broad range of abiotic constraints, ranging from N/P (nitrogen to phosphorus) to cold and salt as well as drought stress (Agati et al. 2010). Enhanced expression of flavonoid genes (chalcone synthase, phenylalanine ammonia-lyase and flavonol synthase) was induced by NaCl toxicity in *Solanum nigrum* L., and these genes were involved in limiting NaCl induced oxidative injury (Abdallah et al. 2016). In soybean, flavonoid metabolic enzymes (chalcone synthase, cytochrome P450 monooxygenase and chalcone isomerase) have a regulatory role in ROS scavenging and provide potential to salt toxicity (Qu et al. 2016). Moreover, cold stress-induced transcriptomic amendments focused on the upsurge of flavonoid biosynthesis in blood oranges (Crifò et al. 2011). Most flavonols aggregate upon cold exposure, with 26 genes encoding transcription factors and enzymes of the flavonoid synthetic pathway in *A. thaliana* and confirmed their role in the defense against ROS as well as tolerance against low temperature (Schulz et al. 2016). During HMs (Co, Fe, Mn, Mo, Ni, Zn, Cu, etc.) toxicity, phenolic compounds can work as metal chelators and instantly protect molecular species of active oxygen (Michalak 2006). The levels of phenols have been reported to increase when the As-treated rice plants were treated with selenium (Se), which increased the absorption of nutrient elements (Mo, Fe, Zn, Mn and Co) and allowed better plant growth (Chauhan et al. 2017). *A. thaliana* supplied with flavanone naringenin and flavonol quercetin induced root growth and increased sapling weight under HM (Zn and Cd) stress (Keilig and Ludwig-Mueller 2009). Similarly, pretreatment of *Lupinus luteus* L. roots with flavonoids reduced symptoms of Pb toxicity, with a remarkable upsurge in antioxidative activity, and displayed their ability to scavenge ROS (Izbiańska et al. 2014).

15.2.6 PHYTOCHELATINS

Phytochelatins (PCs) are LMW, Cys-rich small polypeptides along with the common structure of $(\gamma$-Glu-Cys$)_n$Gly ($n = 2$–11) (Hasan et al. 2017). Synthesis of PC plays a vital role in providing tolerance to HMs in plants (Emamverdian et al. 2015). PCs form complexes with toxic HM (Cd, Zn, Pb, Cu, etc.) by attaching with sulfhydryl and carboxyl groups in the cytosol and then transporting them into the vacuole for sequestration through transporters (Dago et al. 2014; Yadav et al. 2015). Vacuolar sequestration is necessary for HM equilibrium in plants, which is mediated through ATP-dependent vacuolar pumps (V-ATPase) and via tonoplast transporters (Sharma et al. 2016). The production of PCs has been indicated to increase in *Triticum aestivum* with increasing concentration of Zn and Cd (Sun et al. 2005). Carrier et al. (2003) reported storage of approximately 80% of leaf Cd in cytosol and vacuoles of *Brassica napus* under Cd toxicity. Various PCS (phytochelatin synthase) genes responsible

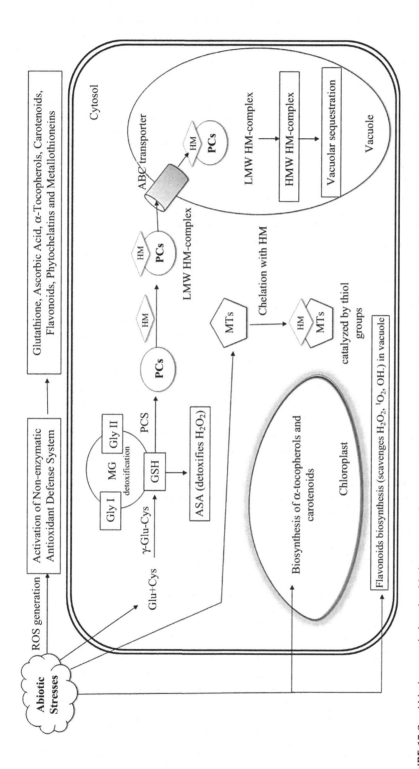

FIGURE 15.2 Abiotic stress-induced oxidative stress, tolerance and detoxification mechanisms in plant cells. Glu, glutamate; Cys, cysteine; MG, methylglyoxal; Gly I-glyoxalase I; Gly H-glyoxalase II; GSH, glutathione; PCS, phytochelatin synthase; PCs, phytochelatins; HM, heavy metal; LMW, low molecular weight; HMW, heavy molecular weights; ASA, ascorbic acid; H_2O_2, hydrogen peroxide; ROS, reactive oxygen species; MTs, metallothionins; $^1O_2$, singlet oxygen; OH, hydroxyl radical.

for PC production have been reported in *Arabidopsis thaliana* (AtPCS1), *Triticum aestivum* (TaPCS1), *Brassica juncea* (BjPCS1) and *Oryza sativa* (OsPCS1) (Lee et al. 2003; Jan and Parray 2016). Moreover, enhanced cadmium and arsenite tolerance was noticed in genetically modified tobacco plants, because of the overexpression of NtPCS1 gene which played important roles in metal(loid) tolerance, growth and development in tobacco plants (Lee and Hwang 2015). Additionally, PCs have an important role in metal ion homeostases such as maintenance of the GSH/GSSG ratio and sulfur (S) metabolism in plant cells (Hasan et al. 2015). In higher plants, groups of sulfate transporters (SULTR) are involved in uptake where low S supply up-regulates the expression of a high-affinity sulfate transporter (Ernst et al. 2008). HM transporter 1 (HMT1) was first discovered in the *Schizosaccharomyces pombe* (in 1995) for Cd tolerance and was termed as cadystins (Mendoza-Cózatl et al. 2011). Compared to free metal ions, the PCs metal compounds are highly consistent and less toxic (Franchi et al. 2014). PC complexes with HM (Cu) have been reported in Cu-hyperaccumulating plant *Nicotiana caerulescenses* (Leitenmaier and Kupper 2013). PC has been stated to display HM tolerance in *Ceratophyllum demersum* under Pb toxicity, where stress leads to an increase in GSH and Cys content, with an upsurge in PC levels which consequently reduced the ROS production (Mishra et al. 2006). Similarly, *Avicennia germinans* also displayed tolerance under Cu and Cd toxicity by increasing the transcriptional level and expression of *AvMt2* due to sequestration in the vacuoles (Yadav et al. 2015) (Figure 15.2).

15.2.7 METALLOTHIONEINS

Metallothioneins (MTs) are Cys-rich HM-binding proteins with LMWs, heat-stable and the best metal detoxifying ligands (Du et al. 2012). Based on the position of Cys remnants, MTs could be categorized into three groups: Group-I (monomers along with two Cys-rich groups segregated by a spacing part), Group-II (translational monomers in that Cys remnants are dispersed in the whole sequence) and Group-III MTs (peptide chains of variable length) (Panda et al. 2016). MTs are rich in thiols, causing them to bind a number of trace metals. MTs can bind metal ions (Cd^{2+}, Zn^{2+}, Cu^{2+}, Co^{2+}, Ag^+ and Hg^{2+}) by the formation of mercaptide bonds with these Cys groups present in proteins (Kisa et al. 2017). The biosynthesis of MTs increased several-fold during oxidative stress, which shielded the cells against cytotoxicity and DNA damage in rice. The most notable purpose of MT is metal detoxification by chelation (Mekawy et al. 2018). MTs are synthesized through mRNA translation to alleviate HM toxicity through HM transport and cellular sequestration (Emamverdian et al. 2015). They have diverse functions in plants e.g. scavenging of ROS, repair of the plasma membrane, maintenance of the redox balance, cell growth and DNA repairing (Wong et al. 2004; Macovei et al. 2010). MTs are metal chelators and also have good ROS scavenging activity along with modulating the oxidative signal. MT genes are shown in different organs, such as leaves, roots, stems, fruits as well as seeds, with MT1 in leaves and roots, MT2 in leaves, stems and developing seeds, MT3 in leaves and maturing fruits and MT4 in growing seeds, leaves and roots of cotton (Kisa et al. 2017). The overexpression of these MT genes improves plant tolerance to HM ions (Cd, Zn and Cu) (Nezhad et al. 2013). Demonstration of MTs genes can be activated by both essential and non-essential metals (Cu, Zn, Ni, Cd, Pb, etc.) (Kumari and Das 2019). In the case of soybean, MT1, MT2 and MT3 were reported to be responsible for Cd detoxification while MT4 was involved in Zn sequestration (Pagani et al. 2012). MTs connected with the down-stream of oxidative signal forms have also been explained in poplar (Smith et al. 2004) and rice (Wong et al. 2004). In addition, MTs are also involved in various processes such as growth, embryonic and microspore expansion, ageing, fruit development, apoptosis and stress reflexes (Hassinen et al. 2011; Duan et al. 2019). In rice, MTs genes (type 1 and type 2) have been noted from the rice MT class (Zhou et al. 2006). Expression and characterization assessment of type 2 MT from grey mangrove variety (*Avicennia marina*), in response to metal (Zn, Cu and Pb) stress, sustain the speculation that *AmMT2* may be engaged in the method of metal equilibrium or tolerance in *Avicennia marina* (Huang and Wang 2010). On the other hand, MT1a performs

TABLE 15.2

Phytochelatins (PCs) and Metallothionins (MTs) Synthesis in Diverse Plant Species under Heavy Metal Stress

Non-Protein Amino Acids	Plant Species	Heavy Metal and References
Phytochelatin	*Arabidopsis thaliana*	Cd and Zn (Lee et al. 2003)
		Cd (Pomponi et al. 2006; Chen et al. 2016)
		As stress (Garg and Kashyap 2019)
	Oryza sativa	Mn (Clemens and Ma 2016)
		Zn and Cd (Takahashi et al. 2012)
	Ceratophyllum demersum	Pb toxicity (Mishra et al. 2006)
	Avicennia germinas	Cu and Cd stress (Yadav et al. 2015)
	Schizosaccharomyces pombe	Cd toxicity (Shine et al. 2015)
	Actodesmus obliquus	Pb stress (Talarek-Karwel et al. 2020)
	Amaranthus hypochondriacus	Cd stress (Xie et al. 2019)
	Cajanus cajan	As toxicity (Garg and Kashyap 2019)
Metallothionein	*Porteresia coarctata*	Zn, Cu and Cd (Usha et al. 2011; Yang et al. 2011)
	Tamarix hispida	
	Pisum sativum	Cu strees (Turchi et al. 2012)
	Tamarix androssowii	Cd stress (Zhou et al. 2014)
	Lycopersicon esculentum	Cu and Pb stress (Kısa et al. 2017)
	Oryza sativa	Cr stress (Yu et al. 2019)
	Cajanus cajan	As toxicity (Garg and Kashyap 2019)

a role in Cu balance in the roots of *Arabidopsis thaliana* because MT1a has expressed abundantly in roots (Guo et al. 2008). In a study, *GhMT3a* were isolated from *Gossypium hirsutum* through cDNA library screening, where genetically modified tobacco plants highly expressing *GhMT3a* displayed augmented tolerance against Cu^{2+} and Zn^{2+} stresses. Recombinant *GhMT3a* protein displayed the capability to bind with these metal ions and function as an efficient ROS scavenger, and its expression could be controlled by stress by ROS signaling (Xue et al. 2009). HM treatments (Pb and Cu) enhanced *MT1* and *MT2* gene expressions comparison to the control plants in *Lycopersicon esculentum* and showed that *MT* transcripts are controlled by both metals and expression pattern variations to *MT* isoforms and tissue kinds (Kısa et al. 2017). In a study, high Cr exposures resulted in enhancement of MTs in rice tissues and unstable transcriptional modifications in *OsMT* genes, which further worked in regulation and as resilience pathways for metal ion chelation and ROS safeguarding (Yu et al. 2019). Similarly, in *A. thaliana* plants, the quartet of the MT genes (*MT1a, MT2a, MT2b* and *MT1c*) were expressed, which provided tolerance to plants against Cu and Cd stress (Zhou and Goldsbrough 1995). Similarly, the type 2 MT gene (*VFMT2*) was up-regulated in response to HM treatments (Cd, Cu and Zn) in *Populus alba* (Macovei et al. 2010). Likewise, type 1 MT protein (CcMT1) of *Cajanus cajan* exhibited greater tolerance to Cu and Cd in *E. coli* and *A. thaliana* (Sekhar et al. 2011). Similarly, expression of MT type 2 gene (*PsMT_{A1}*) from *Pisum sativum* in white poplar plays a role in ROS safeguarding, leading to increased metal tolerance, activation of antioxidative enzyme defenses (SOD, CAT and APX) and sequestration of the enhanced zinc and copper in root and leaf (Turchi et al. 2012). High expression of MT gene *OsMT2c* from *Oryza sativa* in *Arabidopsis* leads to augmented ROS scavenging and shows enhanced tolerance to Cu stress versus WT plants (Liu et al. 2015) (Table 15.2).

15.3 CONCLUSION AND FUTURE PROSPECTS

This chapter summarizes the recent advancement in heavy metal, salinity, water and temperature stress tolerance research, particularly in the alleviation of these stress-induced oxidative damages through non-enzymatic antioxidative pathways. However, all plants survive in a constantly fluctuating atmosphere, as they develop a sequence of processes to stand effectively against environmental stress, which produces additional ROS. Increased ROS cause damage to plants by affecting the physiological, metabolic and biochemical processes. However, plant cells are resourced with antioxidant defense machinery which can detoxify the injurious consequences of ROS to a certain extent. However, in the last few decades, non-enzymatic antioxidants, e.g. GSH, ASA, a-Tocs, C_t, flavonoids, PC and MT, have given commendable results in improving plant tolerance to diverse abiotic constraints in combination with non-enzymatic antioxidants. PC and MT production play key roles in modulating the HM induced stress reflexes in plants through the improved expression of antioxidant genes and balancing the intracellular damage. However, detailed studies regarding various molecular mechanisms underlying stress management in plants need to be undertaken to understand the specific roles and functions of each non-enzymatic antioxidant under various abiotic constraints.

ACKNOWLEDGMENT

We highly acknowledge the financial assistance provided by the University Grants Commission (UGC) and Department of Biotechnology (DBT), Ministry of Science and Technology, Government of India for carrying out related research.

CONFLICT OF INTEREST

There is no conflict of interest between the authors.

REFERENCES

Abbasi AR, Hajirezaei M, Hofius D, Sonnewald U, Voll LM (2007) Specific roles of alpha- and gamma-tocopherol in abiotic stress responses of transgenic tobacco. *Plant Physiol* 143(4):1720–1738

Agami RA (2014) Applications of ascorbic acid or proline increase resistance to salt stress in barley seedlings. *Biologia Plant* 58(2):341–347

Agati G, Tattini M (2010) Multiple functional roles of flavonoids in photoprotection. *New Phytol* 186(4):786–793

Agati G, Brunetti C, Di Ferdinando M, Ferrini F, Pollastri S, Tattini M (2013) Functional roles of flavonoids in photoprotection: New evidence, lessons from the past. *Plant Physiol Biochem* 72:35–45

Ahmad I, Basra SMA, Wahid A (2014) Exogenous application of ascorbic acid, salicylic acid and hydrogen peroxide improves the productivity of hybrid maize at low temperature stress. *Int J Agric Biol* 16(4):825–830

Ahmad P, Jaleel CA, Azooz MM, Nabi G (2009) Generation of ROS and non-enzymatic antioxidants during abiotic stress in plants. *Bot Res Intern* 2(1):11–20

Ahmad P, Jaleel CA, Salem MA, Nabi G, Sharma S (2010) Roles of enzymatic and nonenzymatic antioxidants in plants during abiotic stress. *Crit Rev Biotechnol* 30(3):161–175

Akram NA, Shafiq F, Ashraf M (2017) Ascorbic acid-a potential oxidant scavenger and its role in plant development and abiotic stress tolerance. *Front Plant Sci* 8:613

Akter N, Sobahan MA, Uraji M, Ye W, Hossain MA, Mori IC, Nakamura Y, Murata Y (2012) Effects of depletion of glutathione on abscisic acid- and methyl jasmonate-induced stomatal closure in Arabidopsis thaliana. *Biosci Biotechnol Biochem* 76(11):2032–2037. http://www.ncbi.nlm.nih.gov/pubmed/120384

Alami-Milani M, Aghaei-Gharachorlou P (2015) Effect of ascorbic acid application on yield and yield components of lentil (*Lens culinaris* Medik.) under salinity stress. *Int J Biosci* 6(1):43–49

Alamri SA, Siddiqui MH, Al-Khaishany MY, Nasir Khan M, Ali HM, Alaraidh IA, Alsahli AA, Al-Rabiah H, Mateen M (2018) Ascorbic acid improves the tolerance of wheat plants to lead toxicity. *J Plant Interact* 13(1):409–419

Alayafi AAM (2020) Exogenous ascorbic acid induces systemic heat stress tolerance in tomato seedlings: Transcriptional regulation mechanism. *Environ Sci Pollut Res Int* 27(16):19186–19199

Aly-Salama KH, Al-Mutawa MM (2009) Glutathione-triggered mitigation in salt-induced alterations in plasmalemma of onion epidermal cells. *Int J Agric Biol (Pak)* 11(5):639–642

Andjelkovic V (2018) *Plant, Abiotic Stress and Responses to Climate Change.* In Tech, London, UK

Ann BM, Devesh S, Gothandam KM (2011) Effect of salt stress on expression of carotenoid pathway genes in tomato. *J Stress Physiol Biochem* 7(3): 87–94

Awasthi R, Kaushal N, Vadez V, Turner NC, Berger J, Siddique KH, Nayyar H (2014) Individual and combined effects of transient drought and heat stress on carbon assimilation and seed filling in chickpea. *Funct Plant Biol* 41(11):1148–1167

Ben Abdallah SB, Aung B, Amyot L, Lalin I, Lachâal M, Karray-Bouraoui N, Hannoufa A (2016) Salt stress (NaCl) affects plant growth and branch pathways of carotenoid and flavonoid biosyntheses in Solanum nigrum. *Acta Physiol Plant* 38(3):72

Bharti A, Garg N (2019) SA and AM symbiosis modulate antioxidant defense mechanisms and asada pathway in chickpea genotypes under salt stress. *Ecotoxicol Environ Saf* 178:66–78

Brunetti P, Zanella L, De Paolis A, Di Litta D, Cecchetti V, Falasca G, Barbieri M, Altamura MM, Costantino P, Cardarelli M (2015) Cadmium-inducible expression of the ABC-type transporter AtABCC3 increases phytochelatin-mediated cadmium tolerance in *Arabidopsis. J Exp Bot* 66(13):3815–3829

Bybordi A (2012) Effect of ascorbic acid and silicium on photosynthesis, antioxidant enzyme activity, and fatty acid contents in canola exposure to salt stress. *J Integr Agric* 11(10):1610–1620

Cannella D, Möllers KB, Frigaard NU, Jensen PE, Bjerrum MJ, Johansen KS, Felby C (2016) Light-driven oxidation of polysaccharides by photosynthetic pigments and a metalloenzyme. *Nat Commun* 7(1):11134

Carrier P, Baryla A, Havaux M (2003) Cadmium distribution and micro localization in oilseed rape (*Brassica napus*) after long-term growth on cadmium-contaminated soil. *Planta* 216(6):939–950

Chauhan R, Awasthi S, Tripathi P, Mishra S, Dwivedi S, Niranjan A, Mallick S, Tripathi P, Pande V, Tripathi RD (2017) Selenite modulates the level of phenolics and nutrient element to alleviate the toxicity of arsenite in rice (*Oryza sativa* L.). *Ecotoxicol Environ Saf* 138:47–55

Chen D, Li Y, Fang T, Shi X, Chen X (2016) Specific roles of tocopherols and tocotrienols in seed longevity and germination tolerance to abiotic stress in transgenic rice. *Plant Sci* 244:31–39

Chen J, Yang L, Yan X, Liu Y, Wang R, Fan T, Ren Y, Tang X, Xiao F, Liu Y, Cao S (2016) Zinc-finger transcription factor ZAT6 positively regulates cadmium tolerance through the glutathione-dependent pathway in *Arabidopsis. Plant Physiol* 171(1):707–719

Chen L, Chen Y, Jiang J, Chen S, Chen F, Guan Z, Fang W (2012) The constitutive expression of *Chrysanthemum dichrum* ICE1 in *Chrysanthemum grandiflorum* improves the level of low temperature, salinity and drought tolerance. *Plant Cell Rep* 31(9):1747–1758

Chutipaijit S, Cha-Um S, Sompornpailin K (2009) Differential accumulations of proline and flavonoids in indica rice varieties against salinity. *Pak J Bot* 41(5):2497–2506

Clemens S, Ma JF (2016) Toxic heavy metal and metalloid accumulation in crop plants and foods. *Annu Rev Plant Biol* 67:489–512

Coberly LC, Rausher MD (2003) Analysis of a chalcone synthase mutant in *Ipomoea purpurea* reveals a novel function for flavonoids: Amelioration of heat stress. *Mol Ecol* 12(5):1113–1124

Collin VC, Eymery F, Genty B, Rey P, Havaux M (2008) Vitamin E is essential for the tolerance of *Arabidopsis thaliana* to metal-induced oxidative stress. *Plant Cell Environ* 31(2):244–257

Crifò T, Puglisi I, Petrone G, Recupero GR, Lo Piero ARL (2011) Expression analysis in response to low temperature stress in blood oranges: Implication of the flavonoid biosynthetic pathway. *Gene* 476(1–2):1–9

Dago À, González I, Ariño C, Martínez-Coronado A, Higueras P, Díaz-Cruz JM, Esteban M (2014) Evaluation of mercury stress in plants from the Almadén mining district by analysis of phytochelatins and their Hg complexes. *Environ Sci Technol* 48(11):6256–6263

Darvishan M, Tohidi-Moghadam HR, Zahedi H (2013) The effects of foliar application of ascorbic acid (vitamin C) on physiological and biochemical changes of corn (*Zea mays* L) under irrigation withholding in different growth stages. *Maydica* 58(2):195–200

Das K, Roychoudhury A (2014) Reactive oxygen species (ROS) and response of antioxidants as ROS-scavengers during environmental stress in plants. *Front Environ Sci* 2:53

Dash S, Mohanty N (2002) Response of seedlings to heat-stress in cultivars of wheat: Growth temperature-dependent differential modulation of photosystem 1 and 2 activity, and foliar antioxidant defense capacity. *J Plant Physiol* 159(1):49–59

Davison PA, Hunter CN, Horton P (2002) Overexpression of β-carotene hydroxylase enhances stress tolerance in *Arabidopsis. Nature* 418(6894):203–206

Davletova S, Rizhsky L, Liang H, Shengqiang Z, Oliver DJ, Coutu J, Shulaev V, Schlauch K, Mittler R (2005) Cytosolic ascorbate peroxidase 1 is a central component of the reactive oxygen gene network of *Arabidopsis. Plant Cell* 17(1):268–281

Di Ferdinando M, Brunetti C, Fini A, Tattini M (2012) Flavonoids as antioxidants in plants under abiotic stresses. In: Ahmad P, Prasad M (eds.) *Abiotic Stress Responses in Plants*. Springer, New York. pp. 159–179

Du J, Yang JL, Li CH (2012) Advances in metallotionein studies in forest trees. *Plant OMICS* 5(1):46

Duan L, Yu J, Xu L, Tian P, Hu X, Song X, Pan Y (2019) Functional characterization of a type 4 metallothionein gene (CsMT4) in cucumber. *Hortic Plant J* 5(3):120–128

Eghbaliferiz S, Iranshahi M (2016) Prooxidant activity of polyphenols, flavonoids, anthocyanins and carotenoids: Updated review of mechanisms and catalyzing metals. *Phytother Res* 30(9):1379–1391

El Hariri DM, Sadak MS, El-Bassiouny HMS (2010) Response of flax cultivars to ascorbic acid and α-tocopherol under salinity stress conditions. *Int J Acad Res* 2(6):101–109

Elhamid EMA, Sadak MS, Tawfik MM (2014) Alleviation of adverse effects of salt stress in wheat cultivars by foliar treatment with antioxidant 2—Changes in some biochemical aspects, lipid peroxidation, antioxidant enzymes and amino acid contents. *Agric Sci* 05(13):1269

El-Shabrawi H, Kumar B, Kaul T, Reddy MK, Singla-Pareek SL, Sopory SK (2010) Redox homeostasis, antioxidant defense, and methylglyoxal detoxification as markers for salt tolerance in Pokkali rice. *Protoplasma* 245(1–4):85–96

El-Tayeb MA, El-Enany AE, Ahmed NL (2006) Salicylic acid-induced adaptive response to copper stress in sunflower (*Helianthus annuus* L.). *Plant Growth Regul* 50(2–3):191–199

Emamverdian A, Ding Y, Mokhberdoran F, Xie Y (2015) Heavy metal stress and some mechanisms of plant defense response. *Sci World J* 2015:756120. doi: 10.1155/2015/756120

Ernst WH, Krauss GJ, Verkleij JA, Wesenberg D (2008) Interaction of heavy metals with the sulphur metabolism in angiosperms from an ecological point of view. *Plant Cell Environ* 31(1):123–143

Falk J, Munné-Bosch S (2010) Tocochromanol functions in plants: Antioxidation and beyond. *J Exp Bot* 61(6):1549–1566

Franchi N, Piccinni E, Ferro D, Basso G, Spolaore B, Santovito G, Ballarin L (2014) Characterization and transcription studies of a phytochelatin synthase gene from the solitary tunicate *Ciona intestinalis* exposed to cadmium. *Aquat Toxicol* 152:47–56

Fritsche S, Wang X, Jung C (2017) Recent advances in our understanding of tocopherol biosynthesis in plants: An overview of key genes, functions, and breeding of vitamin E improved crops. *Antioxidants* 6(4):99

Fu-Di Xie CMY, Lian-Ju M (2014) Effects of exogenous application of ascorbic acid on genotoxicity of Pb in *Vicia faba* roots. *Int J Agric Biol* 16(4):831–835

Fulekar MH, Singh A, Bhaduri AM (2009) Genetic engineering strategies for enhancing phytoremediation of heavy metals. *Afr J Biotechnol* 8(4):529–535

Garg N, Bhandari P (2016) Interactive effects of silicon and arbuscular mycorrhiza in modulating ascorbate-glutathione cycle and antioxidant scavenging capacity in differentially salt-tolerant *Cicer arietinum* L. genotypes subjected to long-term salinity. *Protoplasma* 253(5):1325–1345

Garg N, Kashyap L (2019) Joint effects of Si and mycorrhiza on the antioxidant metabolism of two pigeonpea genotypes under As (III) and (V) stress. *Environ Sci Pollut Res* 26(8):7821–7839

Garg N, Singla P (2015) Naringenin- and Funneliformis mosseae-Mediated Alterations in Redox State Synchronize Antioxidant Network to Alleviate Oxidative Stress in Cicer arietinum L. Genotypes Under Salt Stress. *J Plant Growth Regul* 34(3):595–610

Ghosh S, Biswas AK (2017) Selenium Modulates Growth and Thiol Metabolism in Wheat (*Triticum aestivum* L.) during Arsenic Stress. *Am J Plant Sci* 08(3):363–389

Goel S, Madan B (2014) Genetic engineering of crop plants for abiotic stress tolerance. In: Ahmed P, Rasool S (eds.) *Emerging Technologies and Management of Crop Stress Tolerance*. Academic Press. vol 1. Elsevier, Amsterdam. pp. 99–123

Grewe S, Ballottari M, Alcocer M, D'Andrea C, Blifernez-Klassen O, Hankamer B, Mussgnug JH, Bassi R, Kruse O (2014) Light-harvesting complex protein LHCBM9 is critical for photosystem II activity and hydrogen production in *Chlamydomonas reinhardtii*. *Plant Cell* 26(4):1598–1611

Guan Q, Takano T, Liu S (2012) Genetic transformation and analysis of rice OsAPx2 gene in *Medicago sativa*. *PLOS ONE* 7(7):e41233

Guo J, Xu L, Su Y, Wang H, Gao S, Xu J, Que Y (2013) ScMT2-1-3, a metallothionein gene of sugarcane, plays an important role in the regulation of heavy metal tolerance/accumulation. *BioMed Res Int* 2013(904769):1–12

Guo WJ, Meetam M, Goldsbrough PB (2008) Examining the specific contributions of individual *Arabidopsis* metallothioneins to copper distribution and metal tolerance. *Plant Physiol* 146(4):1697–1706

Haghjou MM, Shariati M, Smirnoff N (2009) The effect of acute high light and low temperature stresses on the ascorbate–glutathione cycle and superoxide dismutase activity in two *Dunaliella salina* strains. *Physiol Plant* 135(3):272–280

Han H, Li Y, Zhou S (2008) Overexpression of phytoene synthase gene from *Salicornia europaea* alters response to reactive oxygen species under salt stress in transgenic *Arabidopsis*. *Biotechnol Let* 30(8):1501–1507

Hasan M, Ahammed GJ, Yin L, Shi K, Xia X, Zhou Y, Yu J, Zhou J (2015) Melatonin mitigates cadmium phytotoxicity through modulation of phytochelatins biosynthesis, vacuolar sequestration, and antioxidant potential in *Solanum lycopersicum* L. *Front Plant Sci* 6:601

Hasan M, Cheng Y, Kanwar MK, Chu XY, Ahammed GJ, Qi ZY (2017) Responses of plant proteins to heavy metal stress-a review. *Front Plant Sci* 8:1492

Hasanuzzaman M, Al Mahmud J, Nahar K, Anee TI, Inafuku M, Oku H, Fujita M (2017) Responses, adaptation, and ROS metabolism in plants exposed to waterlogging stress. In: Khan M, Khan N (eds.) *Reactive Oxygen Species and Antioxidant Systems in Plants: Role and Regulation under Abiotic Stress*. Springer, Singapore. pp. 257–281

Hasanuzzaman M, Fujita M (2013) Exogenous sodium nitroprusside alleviates arsenic-induced oxidative stress in wheat (*Triticum aestivum* L.) seedlings by enhancing antioxidant defense and glyoxalase system. *Ecotoxicology* 22(3):584–596

Hasanuzzaman M, Nahar K, Anee TI, Fujita M (2017) Glutathione in plants: Biosynthesis and physiological role in environmental stress tolerance. *Physiol Mol Biol Plants* 23(2):249–268

Hassinen VH, Tervahauta AI, Schat H, Kärenlampi SO (2011) Plant metallothioneins—Metal chelators with ROS scavenging activity? *Plant Bio* 13(2):225–232

Hemavathi , Upadhyaya CP, Akula N, Young KE, Chun SC, Kim DH, Park SW (2010) Enhanced ascorbic acid accumulation in transgenic potato confers tolerance to various abiotic stresses. *Biotechnol Lett* 32(2):321–330

Hossain MA, Munné-Bosch S, Burritt DJ, Diaz-Vivancos P, Fujita M, Lorence A (2017) *Ascorbic Acid in Plant Growth, Development and Stress Tolerance*. Springer

Hou W, Chen X, Song G, Wang Q, Chi Chang CC (2007) Effects of copper and cadmium on heavy metal polluted waterbody restoration by duckweed (*Lemna minor*). *Plant Physiol Biochem* 45(1):62–69

Huang C, He W, Guo J, Chang X, Su P, Zhang L (2005) Increased sensitivity to salt stress in an ascorbate-deficient *Arabidopsis* mutant. *J Exp Bot* 56(422):3041–3049

Huang GY, Wang YS (2010) Expression and characterization analysis of type 2 metallothionein from grey mangrove species (*Avicennia marina*) in response to metal stress. *Aquat Toxicol* 99(1):86–92

Hussain N, Irshad F, Jabeen Z, Shamsi IH, Li Z, Jiang L (2013) Biosynthesis, structural, and functional attributes of tocopherols in planta; past, present, and future perspectives. *J Agric Food Chem* 61(26):6137–6149

Islam MM, Hoque MA, Okuma E, Banu MNA, Shimoishi Y, Nakamura Y, Murata Y (2009) Exogenous proline and glycine betaine increase antioxidant enzyme activities and confer tolerance to cadmium stress in cultured tobacco cells. *J Plant Physiol* 166(15):1587–1597

Izbiańska K, Arasimowicz-Jelonek M, Deckert J (2014) Phenylpropanoid pathway metabolites promote tolerance response of lupine roots to lead stress. *Ecotoxicol Environ Saf* 110:61–67

Jan S, Parray JA (2016) *Approaches to Heavy Metal Tolerance in Plants*. Springer, Singapore. pp. 1–18

Kalaji HM, Schansker G, Brestic M, Bussotti F, Calatayud A, Ferroni L, Goltsev V, Guidi L, Jajoo A, Li P, Losciale P, Mishra VK, Misra AN, Nebauer SG, Pancaldi S, Penella C, Pollastrini M, Suresh K, Tambussi E, Yanniccari M, Zivcak M, Cetner MD, Samborska IA, Stirbet A, Olsovska K, Kunderlikova K, Shelonzek H, Rusinowski S, Bąba W (2017) Frequently asked questions about chlorophyll fluorescence, the sequel. *Photosynth Res* 132(1):13–66

Kang L, Ji CY, Kim SH, Ke Q, Park SC, Kim HS, Lee HU, Lee JS, Park WS, Ahn MJ, Lee HS, Deng X, Kwak SS (2017) Suppression of the β-carotene hydroxylase gene increases β-carotene content and tolerance to abiotic stress in transgenic sweet potato plants. *Plant Physiol Biochem* 117:24–33

Kapoor D, Sharma R, Handa N, Kaur H, Rattan A, Yadav P, Gautam V, Kaur R, Bhardwaj R (2015) Redox homeostasis in plants under abiotic stress: Role of electron carriers, energy metabolism mediators and proteinaceous thiols. *Front Environ Sci* 3:13

Kasote DM, Katyare SS, Hegde MV, Bae H (2015) Significance of antioxidant potential of plants and its relevance to therapeutic applications. *Int J Biol Sci* 11(8):982

Keilig K, Ludwig-Mueller J (2009) Effect of flavonoids on heavy metal tolerance in *Arabidopsis thaliana* seedlings. *Bot Stud* 50(3):311–318

Khattab H (2007) Role of glutathione and polyadenylic acid on the oxidative defense systems of two different cultivars of canola seedlings grown under saline conditions. *Aust J Basic Appl Sci* 1(3):323–334

Kısa D, Öztürk L, Doker S, Gökçe İ (2017) Expression analysis of metallothioneins and mineral contents in tomato (*Lycopersicon esculentum*) under heavy metal stress. *J Sci Food Agric* 97(6):1916–1923

Kocsy G, Szalai G, Galiba G (2002) Induction of glutathione synthesis and glutathione reductase activity by abiotic stresses in maize and wheat. *Sci World J* 2:1699–1705

Kumar D, Yusuf MA, Singh P, Sardar M, Sarin NB (2013) Modulation of antioxidant machinery in α-tocopherol-enriched transgenic *Brassica juncea* plants tolerant to abiotic stress conditions. *Protoplasma* 250(5):1079–1089

Kumari S, Das S (2019) Expression of metallothionein encoding gene bmtA in biofilm-forming marine bacterium *Pseudomonas aeruginosa* N6P6 and understanding its involvement in Pb (II) resistance and bioremediation. *Environ Sci Pollut Res Int* 26(28):28763–28774

Larkindale J, Hall JD, Knight MR, Vierling E (2005) Heat stress phenotypes of *Arabidopsis* mutants implicate multiple signaling pathways in the acquisition of thermotolerance. *Plant Physiol* 138(2):882–897

Lee BD, Hwang S (2015) Tobacco phytochelatin synthase (NtPCS1) plays important roles in cadmium and arsenic tolerance and in early plant development in tobacco. *Plant Biotechnol Rep* 9(3):107–114

Lee S, Moon JS, Ko TS, Petros D, Goldsbrough PB, Korban SS (2003) Overexpression of *Arabidopsis* phytochelatin synthase paradoxically leads to hypersensitivity to cadmium stress. *Plant Physiol* 131(2):656–663

Leitenmaier B, Küpper H (2013) Compartmentation and complexation of metals in hyperaccumulator plants. *Front Plant Sci* 4:374

Liu J, Shi X, Qian M, Zheng L, Lian C, Xia Y, Shen Z (2015) Copper-induced hydrogen peroxide upregulation of a metallothionein gene, OsMT2c, from *Oryza sativa* L. confers copper tolerance in *Arabidopsis thaliana*. *J Hazard Mater* 294:99–108

Løvdal T, Olsen KM, Slimestad R, Verheul M, Lillo C (2010) Synergetic effects of nitrogen depletion, temperature, and light on the content of phenolic compounds and gene expression in leaves of tomato. *Phytochemistry* 71(5–6):605–613

Luo Y, Tang H, Zhang Y (2011) Production of reactive oxygen species and antioxidant metabolism about strawberry leaves to low temperatures. *J Agric Sci* 3(2):89

Ma YH, Ma FW, Zhang JK, Li MJ, Wang YH, Liang D (2008) Effects of high temperature on activities and gene expression of enzymes involved in ascorbate–glutathione cycle in apple leaves. *Plant Sci* 175(6):761–766

Macovei A, Ventura L, Donà M, Faè M, Balestrazzi A, Carbonera D (2010) Effects of heavy metal treatments on metallothionein expression profiles in white poplar (*Populus alba* L.) cell suspension cultures. *An Univ Oradea Fasc Biol* 1:194–198

Mazid M, Khan TA, Khan ZH, Quddusi S, Mohammad F (2011) Occurrence, biosynthesis and potentialities of ascorbic acid in plants. *IJPAES* 1(2):167–184

Mekawy AMM, Assaha DV, Munehiro R, Kohnishi E, Nagaoka T, Ueda A, Saneoka H (2018) Characterization of type 3 metallothionein-like gene (OsMT-3a) from rice, revealed its ability to confer tolerance to salinity and heavy metal stresses. *Environ Exp Bot* 147:157–166

Mendoza-Cózatl DG, Jobe TO, Hauser F, Schroeder JI (2011) Long-distance transport, vacuolar sequestration, tolerance, and transcriptional responses induced by cadmium and arsenic. *Curr Opin Plant Biol* 14(5):554–562

Michalak A (2006) Phenolic compounds and their antioxidant activity in plants growing under heavy metal stress. *Pol J Environ Stud* 15(4):523–530

Mishra S, Srivastava S, Tripathi RD, Kumar R, Seth CS, Gupta DK (2006) Lead detoxification by coontail (*Ceratophyllum demersum* L.) involves induction of phytochelatins and antioxidant system in response to its accumulation. *Chemosphere* 65(6):1027–1039

Molina AS, Nievas C, Chaca MVP, Garibotto F, González U, Marsá SM, Luna C, Giménez MS, Zirulnik F (2008) Cadmium-induced oxidative damage and antioxidative defense mechanisms in *Vigna mungo* L. *Plant Growth Regul* 56(3):285

Mullineaux P, Karpinski S (2002) Signal transduction in response to excess light: Getting out of the chloroplast. *Curr Opin Plant Biol* 5(1):43–48

Munné-Bosch S (2005) The role of α-tocopherol in plant stress tolerance. *J Plant Physiol* 162(7):743–748

Nahar K, Hasanuzzaman M, Alam MM, Fujita M (2015a) Exogenous glutathione confers high temperature stress tolerance in mung bean (*Vigna radiata* L.) by modulating antioxidant defense and methylglyoxal detoxification system. *Environ Exp Bot* 112:44–54

Nahar K, Hasanuzzaman M, Alam MM, Fujita M (2015b) Roles of exogenous glutathione in antioxidant defense system and methylglyoxal detoxification during salt stress in mung bean. *Biologia Plant* 59(4):745–756

Navarrete A, González A, Gómez M, Contreras RA, Díaz P, Lobos G, Brown MT, Sáez CA, Moenne A (2019) Copper excess detoxification is mediated by a coordinated and complementary induction of glutathione, phytochelatins and metallothioneins in the green seaweed *Ulva compressa*. *Plant Physiol Biochem* 135:423–431

Nezhad RM, Shahpiri A, Mirlohi A (2013) Heterologous expression and metal-binding characterization of a type 1 metallothionein isoform (OsMTI-1b) from rice (*Oryza sativa*). *Protein J* 32(2):131–137

Nieto-Sotelo J, Ho THD (1986) Effect of heat shock on the metabolism of glutathione in maize roots. *Plant Physiol* 82(4):1031–1035

Okamoto OK, Pinto E, Latorre LR, Bechara EJH, Colepicolo P (2001) Antioxidant modulation in response to metal-induced oxidative stress in algal chloroplasts. *Arch Environ Contam Toxicol* 40(1):18–24

Oteiza PI, Erlejman AG, Verstraeten SV, Keen CL, Fraga CG (2005) Flavonoid-membrane interactions: A protective role of flavonoids at the membrane surface? *Clin Dev Immunol* 12(1):19–25

Pagani MA, Tomas M, Carrillo J, Bofill R, Capdevila M, Atrian S, Andreo CS (2012) The response of the different soybean metallothionein isoforms to cadmium intoxication. *J Inorg Biochem* 117:306–315

Panche AN, Diwan AD, Chandra SR (2016) Flavonoids: An overview. *J Nutr Sci* 5: E47. doi:10.1017/jns.2016.41

Panda SK, Choudhury S, Patra HK (2016) Heavy-metal-induced oxidative stress in plants: Physiological and molecular perspectives. In: Tuteja N, Gill SS (eds.) *Abiotic Stress Response in Plants* 221–236. Wiley, Hoboken, NJ

Pandey R, Garg N (2017) High effectiveness of *Rhizophagus irregularis* is linked to superior modulation of antioxidant defence mechanisms in *Cajanus cajan* (L.) Millsp. genotypes grown under salinity stress. *Mycorrhiza* 27(7):669–682

Pang CH, Li K, Wang B (2011) Overexpression of SsCHLAPXs confers protection against oxidative stress induced by high light in transgenic *Arabidopsis thaliana*. *Physiol Plant* 143(4):355–366

Pietrini F, Iannelli MA, Pasqualini S, Massacci A (2003) Interaction of cadmium with glutathione and photosynthesis in developing leaves and chloroplasts of *Phragmites australis* (Cav.) Trin. ex Steudel. *Plant Physiol* 133(2):829–837

Pomponi M, Censi V, Di Girolamo V, De Paolis A, Di Toppi LS, Aromolo R, Costantino P, Cardarelli M (2006) Overexpression of Arabidopsis phytochelatin synthase in tobacco plants enhances Cd(2+) tolerance and accumulation but not translocation to the shoot. *Planta* 223(2):180–190

Qu L, Huang Y, Zhu C, Zeng H, Shen C, Liu C, Zhao Y, Pi E (2016) Rhizobia-inoculation enhances the soybean's tolerance to salt stress. *Plant Soil* 400(1–2):209–222

Ruiz J, Blumwald E (2002) Salinity-induced glutathione synthesis in *Brassica napus*. *Planta* 214(6):965–969

Saruhan N, Saglam A, Kadioglu A (2012) Salicylic acid pretreatment induces drought tolerance and delays leaf rolling by inducing antioxidant systems in maize genotypes. *Acta Physiol Plant* 34(1):97–106

Schulz E, Tohge T, Zuther E, Fernie AR, Hincha DK (2016) Flavonoids are determinants of freezing tolerance and cold acclimation in *Arabidopsis thaliana*. *Sci Rep* 6:34027

Sekhar K, Priyanka B, Reddy VD, Rao KV (2011) Metallothionein 1 (CcMT1) of pigeonpea (Cajanus cajan, L.) confers enhanced tolerance to copper and cadmium in Escherichia coli and Arabidopsis thaliana. *Environ Exp Bot* 72(2):131–139

Shafiq S, Akram NA, Ashraf M, Arshad A (2014) Synergistic effects of drought and ascorbic acid on growth, mineral nutrients and oxidative defense system in canola (*Brassica napus* L.) plants. *Acta Physiol Plant* 36(6):1539–1553

Shao HB, Chu LY, Lu ZH, Kang CM (2007) Primary antioxidant free radical scavenging and redox signaling pathways in higher plant cells. *Int J Biol Sci* 4(1):8

Sharma PK, Hall DO (1992) Changes in carotenoid composition and photosynthesis in *Sorghum* under high light and salt stresses. *J Plant Physiol* 140(6):661–666

Sharma SS, Dietz KJ, Mimura T (2016) Vacuolar compartmentalization as indispensable component of heavy metal detoxification in plants. *Plant Cell Environ* 39(5):1112–1126

Shine AM, Shakya VP, Idnurm A (2015) Phytochelatin synthase is required for tolerating metal toxicity in a basidiomycete yeast and is a conserved factor involved in metal homeostasis in fungi. *Fungal Biol Biotech* 2(1):3

Singla-Pareek SL, Reddy MK, Sopory SK (2003) Genetic engineering of the glyoxalase pathway in tobacco leads to enhanced salinity tolerance. *Proc Natl Acad Sci* 100(25):14672–14677

Smirnoff N (2000) Ascorbic acid: Metabolism and functions of a multi-facetted molecule. *Curr Opin Plant Biol* 3(3):229–235

Smith CM, Rodriguez-Buey M, Karlsson J, Campbell MM (2004) The response of the poplar transcriptome to wounding and subsequent infection by a viral pathogen. *New Phytol* 164(1):123–136

Song WY, Yamaki T, Yamaji N, Ko D, Jung KH, Fujii-Kashino M, An G, Martinoia E, Lee Y, Ma JF (2014) A rice ABC transporter, OsABCC1, reduces arsenic accumulation in the grain. *Proc Natl Acad Sci* 111(44):15699–15704

Srivalli S, Khanna-Chopra R (2008) Role of glutathione in abiotic stress tolerance. In: Khan NA, Singh S, Umar S (eds.) *Sulfur Assimilation and Abiotic Stress in Plants*. Springer, Berlin, Heidelberg. pp. 207–225

Stoeva N, Berova M, Zlatev Z (2005) Effect of arsenic on some physiological parameters in bean plants. *Biologia Plant* 49(2):293–296

Sun Q, Wang XR, Ding SM, Yuan XF (2005) Effects of interactions between cadmium and zinc on phytochelatin and glutathione production in wheat (*Triticum aestivum* L.). *Environ Toxicol* 20(2):195–201

Szalai G, Kellős T, Galiba G, Kocsy G (2009) Glutathione as an antioxidant and regulatory molecule in plants under abiotic stress conditions. *J Plant Growth Regul* 28(1):66–80

Takahashi R, Ishimaru Y, Shimo H, Ogo Y, Senoura T, Nishizawa NK, Nakanishi H (2012) The OsHMA2 transporter is involved in root-to-shoot translocation of Zn and Cd in rice. *Plant Cell Environ* 35(11):1948–1957

Talarek-Karwel M, Bajguz A, Piotrowska-Niczyporuk A (2020) 24-Epibrassinolide modulates primary metabolites, antioxidants, and phytochelatins in *Acutodesmus obliquus* exposed to lead stress. *J Appl Phycol* 2020(32):263–276

Turchi A, Tamantini I, Camussi AM, Racchi ML (2012) Expression of a metallothionein A1 gene of *Pisum sativum* in white poplar enhances tolerance and accumulation of zinc and copper. *Plant Sci* 183:50–56

Uarrota VG, Stefen DLV, Leolato LS, Gindri DM, Nerling D (2018) Revisiting carotenoids and their role in plant stress responses: From biosynthesis to plant signaling mechanisms during stress. In: Gupta D, Palma J, Corpas F (eds.) *Antioxidants and Antioxidant Enzymes in Higher Plants*. Springer, Cham. pp. 207–232

Usha B, Keeran NS, Harikrishnan M, Kavitha K, Parida A (2011) Characterization of a type 3 metallothionein isolated from *Porteresia coarctata*. *Biologia Plant* 55(1):119–124

Venkatesh J, Park SW (2014) Role of L-ascorbate in alleviating abiotic stresses in crop plants. *Bot Stud* 55(1):38

Verbruggen N, Hermans C, Schat H (2009) Mechanisms to cope with arsenic or cadmium excess in plants. *Curr Opin Plant Biol* 12(3):364–372

Wong HL, Sakamoto T, Kawasaki T, Umemura K, Shimamoto K (2004) Down-regulation of metallothionein, a reactive oxygen scavenger, by the small GTPase OsRac1 in rice. *Plant Physiol* 135(3):1447–1456

Xiang C, Werner BL, Christensen EM, Oliver DJ (2001) The biological functions of glutathione revisited in *Arabidopsis* transgenic plants with altered glutathione levels. *Plant Physiol* 126(2):564–574

Xie M, Chen W, Lai X, Dai H, Sun H, Zhou X, Chen T (2019) Metabolic responses and their correlations with phytochelatins in *Amaranthus hypochondriacus* under cadmium stress. *Environ Pollut* 252(B):1791–1800

Xue T, Li X, Zhu W, Wu C, Yang G, Zheng C (2009) Cotton metallothionein GhMT3a, a reactive oxygen species scavenger, increased tolerance against abiotic stress in transgenic tobacco and yeast. *J Exp Bot* 60(1):339–349

Yabuta Y, Mieda T, Rapolu M, Nakamura A, Motoki T, Maruta T, Yoshimura K, Ishikawa T, Shigeoka S (2007) Light regulation of ascorbate biosynthesis is dependent on the photosynthetic electron transport chain but independent of sugars in *Arabidopsis*. *J Exp Bot* 58(10):2661–2671

Yadav A, Ram A, Majithiya D, Salvi S, Sonavane S, Kamble A, Ghadigaonkar S, Jaiswar JRM, Gajbhiye SN (2015) Effect of heavy metals on the carbon and nitrogen ratio in *Avicennia marina* from polluted and unpolluted regions. *Mar Pollut Bull* 101(1):359–365

Yadav S, Modi P, Dave A, Vijapura A, Patel D, Patel M (2020) Effect of abiotic stress on crops. *Sustainable Crop Production*. doi: 10.5772/intechopen.88434

Yang J, Wang Y, Liu G, Yang C, Li C (2011) Tamarix hispida metallothionein-like ThMT3, a reactive oxygen species scavenger, increases tolerance against Cd(2+), Zn(2+), Cu(2+), and NaCl in transgenic yeast. *Mol Biol Rep* 38(3):1567–1574

Yogendra KM, Nirmaljit K (2017) Antioxidants: Environmental stress mitigating metabolites. *Curr Trends Biomed Eng Biosci* 5. http://www.ncbi.nlm.nih.gov/pubmed/555660

Yu XZ, Lin YJ, Zhang Q (2019) Metallothioneins enhance chromium detoxification through scavenging ROS and stimulating metal chelation in *Oryza sativa*. *Chemosphere* 220:300–313

Yusuf MA, Kumar D, Rajwanshi R, Strasser RJ, Tsimilli-Michael M, Govindjee , Sarin NB (2010) Overexpression of gamma-tocopherol methyl transferase gene in transgenic Brassica juncea plants alleviates abiotic stress: Physiological and chlorophyll a fluorescence measurements. *Biochim Biophys Acta* 1797(8):1428–1438

Zagorchev L, Seal C, Kranner I, Odjakova M (2013) A central role for thiols in plant tolerance to abiotic stress. *Int J Mol Sci* 14(4):7405–7432

Zandalinas SI, Mittler R, Balfagón D, Arbona V, Gómez-Cadenas A (2018) Plant adaptations to the combination of drought and high temperatures. *Physiol Plant* 162(1): 2–12

Zengin F (2013) Physiological behavior of bean (*Phaseolus vulgaris* L.) Seedlings under metal stress. *Biol Res* 46(1):79–85

Zhou B, Yao W, Wang S, Wang X, Jiang T (2014) The metallothionein gene, TaMT3, from *Tamarix androssowii* confers Cd2+ tolerance in tobacco. *Int J Mol Sci* 15(6):10398–10409

Zhou G, Xu Y, Li J, Yang L, Liu JY (2006) Molecular analyses of the metallothionein gene family in rice (*Oryza sativa* L.). *J Biochem Mol Biol* 39(5):595–606

Zhou J, Goldsbrough PB (1995) Structure, organization and expression of the metallothionein gene family in Arabidopsis. *Mol Gen Genet* 248(3):318–328

16 Regulatory Role of UV-B in Modulating Antioxidant Defence System of Plants

Abreeq Fatima, Madhulika Singh and Sheo Mohan Prasad

CONTENTS

16.1 INTRODUCTION

16.1.1 THE SOURCE OF UV RADIATION IN THE ECOSYSTEM

Boon of life or not? The ultraviolet (UV) spectrum is broadly divided into three regions: UV-C (220- 280 nm), UV-B (280 -320 nm) and UV-A (320 - 400 nm). UV-B is the most energetic component of sunlight reaching the earth's surface; levels of UV-B have increased due to ozone depletion (Strid et al. 1994). The majority of UV-B radiation is filtered out by atmospheric ozone before it reaches the earth's surface. However, the impact of UV on the earth's surface has increased dramatically in the recent past as levels of atmospheric ozone have decreased drastically. This worsening has led to renewed attention and efforts to understand the effects of UV radiation on plants and other organisms. Due to ozone depletion, the enhanced level of UV-B radiation has heightened awareness of the cytotoxic, mutagenic and carcinogenic consequences of UV-irradiation (Rai and Agrawal 2017). The increase in UV-B levels (Bais et al. 2015) is due to an accelerating depletion of the stratospheric ozone layer and is mainly anthropogenic (Singh et al. 2019). Despite its low concentration, ozone plays a critical role in chemical and biological processes by filtering ultraviolet radiation in the UV-B range. Biological systems are vulnerable to wavelengths in the transitional range of 280–320 nm and are thus greatly affected by ozone losses (Rozema et al. 1997). Any perturbation that leads to an increase in UV-B radiation needs careful attention to the possible consequences. As a result, the UV-B region of the ultraviolet spectrum has gained an importance over other environmental factors. The primary concern over ozone depletion is the potential impact on human health and ecosystems due to increased UV exposure. This enhanced exposure of UV-B is potentially detrimental to all living beings but is particularly harmful to plants due to their obligatory requirement of sunlight for survival and their inability to move (Strid et al. 1994).

Lower yields of cash crops may result due to increased UV-B stress (Wang and Frei 2011), and higher UV-B levels in the upper ocean layer may inhibit phytoplankton activities, which can have an impact on the entire marine ecosystem (Hader et al. 2007). An increase in skin cancer, skin aging and cataracts in the human population is expected in a higher UV environment (Duan et al. 2019). In addition to direct biological consequences, indirect effects may arise through changes in atmospheric chemistry. A well-studied response to UV-B radiation is damage to the photosynthetic apparatus (Bashri et al. 2018). Increased UV-B alters photochemical reaction rates in the lower atmosphere that are important in the production of surface layer ozone which has detrimental effects on plants (Tang and Madronich 1995; Mathur and Jajoo 2015). Therefore, one of the earliest responses to UV-B radiation is the activation of the flavonoid biosynthetic pathway. These pigments absorb UV radiation and are generally accumulated in the epidermis, where they help to keep UV radiation away from reaching photosynthetic tissues (Hahlbrock and Scheel 1989) since the epidermis blocks transmittance of 95 to 99% of incoming UV radiation. A number of studies have been carried out to investigate whether an increase in UV-B radiation resulting from ozone depletion would have a significant impact on plants or not.

The increased level of UV-B radiation directly or indirectly affects the growth and productivity of the plant (Thomas and Purthur 2017). It affects the general physiological and biochemical pathway of the cell by damaging the membranous structure, enzyme activity and nucleic acids via inducing the level of ROS (Lidon et al. 2012; Zlatev et al. 2012; Bashri et al. 2018). The production of ROS in the plant cell is a common consequence, but under the adverse condition, it is over-produced and acts as a highly toxic agent, which is primarily generated inside chloroplasts, mitochondria, peroxisomes and cell membranes (Gill and Tuteja 2010; Singh et al. 2019) during the transport of electrons. Plants can develop a self-detoxification mechanism by invoking enzymatic antioxidant system including superoxide dismutase (SOD), peroxidase (POD), catalase (CAT), glutathione-S-transferase (GST), ascorbate peroxidase (APX), etc. and non-enzymatic antioxidant system which includes proline, cysteine, non-protein thiols (NP-SH), etc. (Gill and Tuteja 2010; Singh et al. 2019). In this chapter, the effect of UV-B stress on plants and its role in antioxidant metabolism in the light of recent findings will be discussed.

16.1.2 How Radiation Affects Life on Earth

The perception of light by humans indeed reveals a very colorful world but it may not be the case for other organisms. For instance, the vision of bees, butterflies and other insects extends into the ultraviolet range of the spectrum, and these insects can see the accumulation of UV-absorbing pigments by plants. However, it appears that plants are taking undue advantage of animal vision in the UV range for displaying their flowers but at the same time insects gained further adaptations to see flower patterns that are visible in the UV range only. In many flowers, ultraviolet light uncovers secret paths and "landing strips" that lead to delicious food for the insects (Menzel and Backhaus 1991). These markings are visible only to selected insects while they are hidden from the majority of other animals and humans. Thus, on the evolutionary scale, the color vision of some insects and the spectral properties of flowers have developed into mutual plant-pollinator relationships (Chittka et al. 1994). The benefits of plant-pollinator co-evolution are efficient in reproduction for plants and for the availability of high energy food for pollinators. However, some carnivores use ultraviolet markings for attracting pollinators to the insect traps, by imitating the UV-visible patterns of flowers (Menzel and Backhaus 1991, Menzel and Shmida, 1993). Green plants produce many essential substances and vitamins that benefit us since our bodies are not able to synthesize them, for example vitamin C and antioxidants, which are important components of healthy food. The most important group of chemicals is the phenolics which exhibit a wide variety of beneficial biological roles, including antiviral, antibacterial, immune-stimulating, anti-allergic, anti-inflammatory and anti-carcinogenic (Treutter 2006). They are also powerful antioxidants, scavenging ROS and free radicals that can chelate with metal ions such as iron and copper, enabling our bodies to use these important micro-nutrients. Important sources of phenolics are different herbs (i.e. medicinal plants), fruits, vegetables, grains (i.e. buckwheat and wild rice), tea, coffee, beans and red wine.

Extensive studies on UVR explain that UV-B radiation increases the amount of active substances in many plant species. The synthesis of different phenolic compounds (i.e. flavonoids and stilbenes) and vitamin D production is stimulated by UV-B. Vitamin D is also synthesized in plankton, which is then ingested by fish and can eventually become human food, rich in vitamin D and beneficial to health. UVR stimulation has also been shown to increase plant production of different alkaloids, essential oils and terpenoids with medicinal properties. It is also reported that the low fluence rate of UV-B radiation gives a positive effect on biomass production in cyanobacteria. It enhances the activity of photosynthetic pigments including phycobiliprotein, the efficiency of photosystem II, non-enzymatic antioxidants such as proline, ascorbate, cysteine, non-protein thiol (NPSH) phenolics, and total antioxidant potential of *Nostoc muscorum*, *Phormidium foveolarum* and *Arthrospira platensis*. The results of this study indicate that the application of low fluence rate UV-B radiation can efficiently modify secondary metabolites of the cyanobacterial system which is beneficial for the human being at the economic level. Similarly, in higher plants at a low fluence rate, UV-B radiation enhances photomorphogenic response, secondary metabolites and phenolic compounds (Jenkins 2009). Indeed, there is a lot more to the list of negative impacts of UV-B on living organisms, but it is not absolute. UV-B also plays several beneficial roles under different conditions depending on its amount and the duration of exposure.

16.2 IMPACT OF UV-B ON PLANTS

16.2.1 Positive Role of UV-B in Mediating Plants against Various Stresses

UV-B exposure enhanced hardiness and increased tree growth in eastern New Zealand and lead to agricultural expansion in south eastern South America (Andrady et al. 2015). The modulating effects of UV-B radiation are found to strengthen crop production, increase the level of photoprotection, antioxidative response, production of secondary metabolites and alter the resistance capacity against pests and disease attack (Wargent and Jordan 2013). The enhancement in photosynthetic pigments is an important physiological response under UV-B. On exposure of seeds to UV-B (760 µW/90 min), pigment content was enhanced by 35% in kidney bean varieties, and it was 45% in both cabbage and beet (Dhanya and Thomas 2017). Similarly, when *Vigna mungo* seeds were exposed to UV-B radiation, the anthocyanin pigment content increased about three-fold at 40 min treatment (Shaukat et al. 2013). The anthocyanin content was also found to have increased due to the irradiation of UV-B (350 µW/m^{-2}) in the maize plant (Rudnoy et al. 2015).

Under field conditions, when wheat seeds were exposed to ambient UV-B (5 kJ) radiation, the abscisic acid content, stomatal conductance, transpiration rate and intercellular CO_2 concentration were significantly increased (Li et al. 2010). The combined action of ambient UV-B (0.4Wm^{-2}-0.6Wm^{-2}) and photosynthetically active radiation (PAR) causes an increased level of essential oil production and total monoterpene in peppermint (Behn et al. 2010). It is also documented that the seedlings of cowpea (*Vigna unguiculata*) when treated with UV-B irradiation for about 15, 30, 45 and 60 min along with dimethoate, the level of carotenoids was found to be enhanced (19%) by the treatment with UV-B alone and in combination with 50 ppm dimethoate (25%) (Mishra et al. 2008), and this enhanced level of carotenoid content protects the plant from photodamage. When the rice seedlings in the field were irradiated with ambient UV-B radiation of 9.72 kJ m^{-2}, it upregulated the photosynthetic activity and increased the chlorophyll concentration by 35% compared to the control.

The UV-B irradiated rice seedlings also showed enhanced tolerance to photoinhibition (Xu and Qiu 2007). In another report, when ten different wheat cultivars were irradiated with UV-B radiation in the field with the biologically effective conditions, the cultivars showed significant increases in proteins, total sugar, total amino acid content and grain quality index (Zu et al. 2004). Another study reported that when the rice plants were irradiated with ambient UV-B, total nitrogen content and storage protein known as glutelin was significantly increased in the grains (Hidema et al. 2005). Also, the amylose concentration significantly increased due to the irradiation of rice seedlings with UV-B radiation of 9.7 kJ m^{-2} (Xu and Qiu 2007). Thus, by UV-B irradiation, the protein, amino acids and sugar content are increased in different crops, which ultimately aids in improving the food quality.

16.2.2 Negative Role of UV-B on Plants

The chronic effects of UV-B radiation in disturbing metabolic processes are vast and are well documented in various studies (Santos et al. 2012; Oliveira et al. 2016). The overall growth of the plant and the amount of biomass production is photosynthesis dependent. The detrimental effects of UV-B radiation in most organisms are chiefly caused by free radicals primarily known as oxidative stress (Rastogi et al. 2011; Das and Roychoudhury 2014). For example, over-production and accumulation of reduced oxygen intermediates including singlet oxygen can negatively affect macromolecules such as proteins, DNA and lipids.

16.3 DOES UV-B AFFECT THE MORPHOLOGY AND PHYSIOLOGY OF THE PLANT?

On the morphological front, sensitivity of the plants to UV-B includes seed germination. Seedlings are highly susceptible to UV radiation undergoing the transition from vegetative to reproductive phase (Teramura and Sullivan 1987). Primary effects include reduced or stunted growth and partitioning of growth to different organs such as reduced leaf area and internode, glazing and burning of the leaf of the plants, and ultimately there is a drastic decrease in overall biomass production (Yang et al. 2004). These changes appear differently in different plants by affecting different organs (Teramura 1983). Various recent studies in cyanobacteria show that UV-B radiation results in a wide range of responses at the cellular level, includingmotility of cyanobacteria, protein biosynthesis, photosynthesis, nitrogen fixation, ascorbate glutathione cycle and survival (Zlatev et al. 2012; Singh et al. 2012). The molecular targets include DNA and the photosynthetic apparatus. The phycobiliproteins, which serve as solar energy harvesting antennae, are specifically bleached by UV radiation. However, several studies have demonstrated an adaptation to UV stress, as an increased resistance could also be observed. Long-term exclusion of solar UV radiation decreases photosynthetic competence. Adaptive mutagenesis has also been found in cyanobacteria, which increases their resistance to UV-B.

16.3.1 Effect on Photosynthetic Activity of the Green Plants

Apart from being an energy source, light also acts as a major environmental indication for plants to adjust to its surrounding conditions, since, for photosynthesis, PAR is required which shows large seasonal and diurnal changes, and whose intensity is also influenced by meteorological factors. When the photon intensity is within the range of the utilizing capacity of chloroplasts for CO_2 fixation, photosynthesis runs efficiently without any hindrance caused by biochemical factors (Hideg et al. 2002). But it is well reported in several studies that net photosynthesis is severely inhibited by UV-B radiation (Lidon et al. 2012). Similarly, deterioration in the photosynthesis rate was also observed in green alga by White and Jahnke (2002) and in *Microcystis aeruginosa* by Tang et al. (2018) under the exposure of UV rays. Fukuchi et al. (2004) have demonstrated that UV-A can break

UV-B absorbing pigment, thereby reducing the capacity of the plants to fight off UV-B stress. This disturbance could be due to an increase in ROS generation than its scavenging mechanism. Another well-studied response to UV-B radiation is damage to the photosynthetic apparatus. Because photosystem II is the UV-B sensitive component (Renger et al. 1989), its damage is assessed by measuring the increase in variable chlorophyll fluorescence, and hence an increase in fluorescence can be observed after doses of UV-B radiation in the physiologically relevant range of radiation (Tevini et al. 1991; Bashri et al. 2018).

16.3.2 UV-B Induced Disturbance in the Cellular Homeostasis

Basic life processes including photosynthesis, chlorophyll fluorescence, photomorphogenesis, biosynthesis of cellular components and development are governed by light (Hideg et al. 2002; Bashri et al. 2018). Light requirements of the plant are largely flexible, in terms of intensities, spectral distribution and developmental stage. The scientific community has reported that UV-B might cause ROS generation by disturbing the normal reduction pathway of $NADP^+$ to NADPH during the photosynthetic electron transport chain. According to Bashri et al. (2018), generation of ROS might be due to the oxidation of lipids and proteins. It is reported that UV-B causes increased production of H_2O_2 which results in lipid peroxidation (Figure 16.1). One of its consequence is the excessive generation or accumulation of malondialdehyde (MDA) content (Dwivedi et al. 2015), resulting due to severe disturbances in the membrane integrity which leads to senescence and ultimately death of the plant (Gill and Tuteja 2010; Singh et al. 2019).

Lipid peroxidation and free radical generation are majorly considered as an agent for leaf senescence. When the level of ROS increases, it damages the chlorophyll, leading to chlorosis. Halliwell and Gutteridge (1995) and Singh et al. (2019) justified that the oxidative radicals like hydroxyl ion ($\bullet OH$) and superoxide anion ($O_2\bullet-$) could modify the specific molecular properties of the cell.

16.3.3 Damage to the Macromolecules: DNA and Lipid Membranes

During the early pre-Cambrian era, fluxes of solar UV-B at the surface of the earth were several-fold higher than today due to the lack of oxygen in the atmosphere. Recent studies show that UV-B radiation results in a wide range of responses at the cellular level. DNA absorbs UVR in three spectral regions: I (peak of 260–264 nm), II (peak of 192 nm) and III (below 125 nm). Spectral regions II and III would only have relevance in outer space. Since DNA is one of the most notable targets of UV radiations (Hollosy 2002), it is more susceptible to degradation, resulting in phototransformations, production of cyclobutane pyrimidine dimers (CPDs) and pyrimidine- pyrimidinone dimers (Hader et al. 2007; Rastogi et al. 2010). DNA damage may result in oxidative damage, caused by the direct interaction of ionizing radiation with DNA molecules. DNA protein cross-links, DNA strand breaks and deletion or insertion of base pairs can also be induced by UV radiation (Slieman and Nicholson 2000; Britt et al. 2004; Batista et al. 2009).

It is well reported that an elevated level of UV-B can damage DNA and proteins (Bashri et al. 2018) resulting in the inactivation of photosystem II (PS II) by downregulating the genes of RUBISCO enzymes namely rbcS and rbcL and also the genes of light-harvesting complex (LHC) namely Lhcb and psbA (Zlatev et al. 2012). On the molecular front, UV exposure causes a wide range of responses, one of which is chromosomal aberrations: change in the structure of the chromosome. This causes a massive reduction in the growth rate of organisms. It is reported that reproduction is negatively correlated with UV stress, encompassing sterility, the occurrence of developmental abnormalities or reduction in viability of offspring. In a report on *Nostoc commune*, it was found that UV-B treatment increased the concentration of 493 proteins out of 1350 which was three-fold higher than the terrestrial species of *Nostoc commune* (Ehling-Schulz et al. 2002).

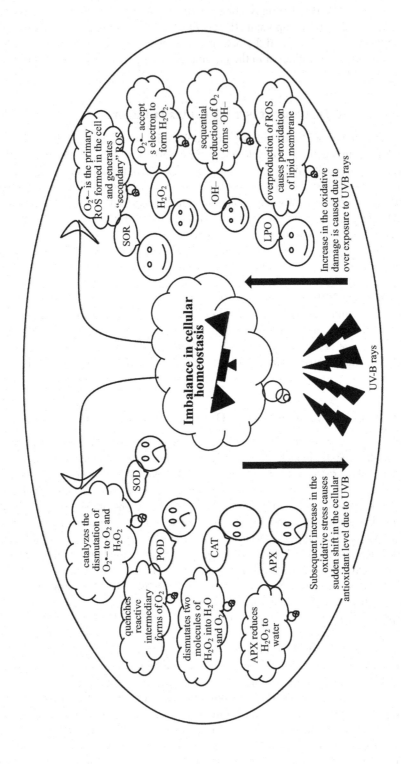

FIGURE 16.1   A pictorial representation shows antioxidant mediated cellular homeostasis against UV-B stress induced oxidative damage.

16.4   ANTIOXIDANTS SYSTEM AS AN INHERENT MITIGATION TOOL OF UV-B RADIATION IN PLANTS

16.4.1   ENZYMATIC ANTIOXIDANT SYSTEM

Any biotic or abiotic stress triggers the overproduction of cellular ROS and damages macromolecules such as DNA, proteins, etc., which has been previously discussed (Lidon et al. 2012; Hideg et al. 2013). The production of ROS and antioxidants (enzymatic and non-enzymatic) goes hand in hand and it is the adaptive measure of plants against abiotic stresses (Table 16.1).

The role of UV-B on the antioxidant activity of bitter gourd (0.4Wm^{-2}/30 min) resulted in higher activity of SOD, POD, CAT and GST. Similarly, when the seedlings were exposed to UV-B they exhibited an increase in the activity of SOD, CAT, GPX and POD by a range of 25–48% over the control (Mishra et al. 2009). Different plants when exposed to UV irradiation exhibited differential responses in terms of antioxidant enzyme activities. In a report of the experiment conducted on different cultivars of wheat, the SOD activity was significantly increased (Zu et al. 2004). Apart from that, phenolic content was also increased under UV-B irradiated maize plants, and UV-B also upregulated the activity of CAT, APX and GST. Of the three antioxidant enzymes, the highest enhancement in isoenzymes or the gene expression level was recorded in the case of GST-1>62% (Rudnoy et al. 2015). At field condition, when two rice varieties (Baijiaolaojing and Yuelianggu) were exposed to UV-B radiation of 2.5, 5.0 and 7.5 kJ, the SOD, CAT and POD activities were found to get upregulated considerably (He et al. 2015). In another study, when the blueberry plants were irradiated with UV-B radiation (0.07, 0.12 and 0.19 W/m^2 for 0–72 h), the SOD activity was upregulated and a six-fold increment was reported in APX (0.12 W/m^2) gene expression (Inostroza-Blancheteau et al. 2016).

In another study, when *Rosularia elymaitica* was subjected to UVA (30 W) stress, SOD activity was found to be slightly affected whereas APX (16:00h) activity was increased. However, the APX activity in the samples taken at 24:00h was slightly or significantly lower than the samples taken at

TABLE 16.1
Summary of the Adaptive Responses Expressed by Plants under UV-B Exposure

Plant Species	Intensity	Radiation	Responses of the Plant	References
Cucumis sativus (Cucumber)	3.4, 5.5 kJ m^{-2} or 10.6 kJ m^{-2}	UV-B	Survival rate of the seedlings improved by 112 and 82% and growth in height increased by 35 and 40%, respectively	Teklemariam and Blake (2003)
V.aconitifolia (Moth bean)	1.2 to 7.2 kJ m^{-2}	UV-B	Increase in ascorbic acid, proline and total phenol content	Dwivedi et al. (2015)
V.mungo (Black gram)	1.2 to 7.2 kJ m^{-2}	UV-B	GPX activity enhanced several folds	Dwivedi et al. (2015)
Momordica charantia (Bitter gourd)	0.4 Wm^{-2}	UV-B	Higher SOD, CAT and POD content	Mishra et al. (2009)
Oryza sativa (Rice)	9.7 kJ m^{-2}	UV-B	Amylase concentration increased	Xu and Qiu (2007)
Oryza sativa (Rice)	9.72 kJ m^{-2}	UV-B	Upregulation of photosynthetic activity, increased chlorophyll content	Xu and Qiu (2007)
Mentha piperata (Peppermint)	0.6 Wm^{-2}	UV-B	Increases essential oil production and total monoterpene content	Behn et al. (2010)
Triticum aestivum (Wheat)	5 kJm^{-2}	UV-B	Enhanced SOD activity due to the overproduction of ROS	Zu et al. (2004)

9:00 and 16:00h (Habibi and Hajiboland 2011). On the other hand, the activity of CAT(12–16 h) was increased by combined treatments of UVA + B and no significant enzymatic changes were observed in control samples.

16.4.2 Non-Enzymatic Antioxidant Defense System

Low levels of UV-B radiation stimulate the accumulation of non-enzymatic antioxidants including different biosynthetic pathways such as phenylpropanoids, cinnamates, flavonoids, anthocyanin and pyridoxine (Hideg et al. 2013; Singh et al. 2019) in plants and cyanobacteria (Hideg et al. 2013). Agati et al. (2011) also reported that under UV-B stress, the plant develops a first line of a defense system in the mesophyll cells of the leaf epidermis and synthesizes flavonoids (UV-absorbing pigments) by phenylpropanoid biosynthetic pathway, and UV-B mediated enhancement in flavonoids content enhances leaf thickness which acts as a selective filter for UV-B radiation. The accumulation of non-enzymatic antioxidants like ascorbic acid, α-tocopherol, glutathione and proline in epidermal layers is a sign of enhanced UV accumulation in plants and is also a defense response that these compounds are actively engaged in the scavenging of free radicals (Gill and Tuteja 2010; Miller et al. 2010; Das and Roychoudhury 2014).

UV-B mediated production of flavonoid content is an effective inbuilt free radical scavenging tool acting as a "sunscreen" for plants (Rice-Evans et al. 1997). In another study when two varieties of kidney bean *Phaseolus vulgaris* and *Phaseolus ellipticus* were subjected to the low level of UV-B, they were found to have an enhanced rate (32–35%) of ascorbic acid and vitamin E (79%) respectively. Similarly, in sugar beet, 90 min exposure of UV-B could enhance the vitamin C content up to 74% which helps the plant to cope up with the deleterious effects of environmental stress (Shaukat et al. 2013). Examination of three varieties of dika wheat and two varieties of rapeseed exposed to UV-B were found to have enhanced proline and flavonoid content (Badridze et al. 2016). In another study on *Vigna mungo*, exposure of UV-B for 40 min resulted in four-fold enhancement in flavonoid and soluble phenol contents (Shaukat et al. 2013). The non-enzymatic antioxidants including proline, ascorbic acid, total proline (TP's) and total flavonoids (TF's) were significantly increased in the leaves of *Vigna acontifolia* when exposed to UV-B, and the accumulation of these secondary compounds was found to be higher in the epidermal layer (Dwivedi et al. 2015). The increasing intensities of UV-B (1.2 to 7.2 kJ) causes a progressive increase in ascorbic acid and proline content in *V. mungo* and *V. acontifolia* respectively and these metabolites have immense ability to scavenge the ROS. The GPX activity was quite enhanced in both species and was subsequently enhanced in *V. mungo* whereas the maximum activity of this enzyme was recorded under the exposure of 4.8 kJ of UV-B in leaves of *V. acontifolia* (Dwivedi et al. 2015). It is also reported that *V. acontifolia* has a higher potential for a faster and greater increase in the accumulation of various antioxidants and assists in enhanced tolerance to UV-B induced damages than *V.mungo* (Dwivedi et al. 2015). It is reported that 21-day old maize seedlings when exposed to UV-B radiation (10 min) a day up to eight days with a continuous increase of ten min per day increased antioxidant enzymes activity on the final day, and it was seen that the activities of APX and GPX increased in roots and leaves up to two- to three-fold in roots (Rudnoy et al. 2015). It is also documented, when ten-day-old seedlings of cucumber (*Cucumis sativus*) were exposed for six days with 3.4, 5.5or 10.6 kJ m^{-2} of ambient UV-B under heat stress, the survival rate of the seedlings was improved by 82% and 112% and the height was increased by 35 and 40% respectively (Teklemariam and Blake 2003). UV-B irradiation upregulated the antioxidant mechanisms in different plants, which indicates that the plants can scavenge ROS, are protected from photoinhibition, DNA damage and lipid peroxidation, and these features ultimately lead to enhancement in the stress tolerance of the plant.

Phenols act as the first barrier in ROS scavenging and complement the antioxidant machinery to control excessive ROS (Agati et al. 2012). They do assist in the antiradical activity of the plant system during ROS production. Phenolic compounds have an immense capacity to scavenge ROS such as $O_2\bullet-, H_2O_2, \bullet OH-$ etc. in the cytosol and vacuoles (Agati et al. 2013;

Das and Roychoudhury 2014). They are the result of the secondary metabolism of the plant system and play a major role in antioxidant activities. The overall increase of antiradical activity and total phenol content (TPC) under UV-B conditions demonstrate that phenols play a pivotal role in UV irradiated leaves of plants.

16.4.3 Signaling Role of UV-B in Mediating Antioxidant Defense Responses in Plants

There are two important aspects of UV-B signaling in plants, (i) non-specific signaling and (ii) UV-B-specific signaling which depends on light called photo morphogenesis-related signaling (Jenkins 2009; Jansen and Bornman 2012). UV-B specific signaling chiefly involves UV resistance locus 8 (UVR8) photoreceptor (Kliebenstein et al. 2002; Rizzini et al. 2011), which regulates the function ofthe photomorphogenesis response regulators, constitutively photomorphogenic 1 (COP1) (Oravecz et al. 2006; Rizzini et al. 2011) and elongated hypocotyl 5 (HY5) (Kliebenstein et al. 2002; Brown et al. 2005) (Figure 16.2).

Various researchers reported that the intrinsic tryptophan residues Trp285 and Trp233 serve as UV-B chromophores (Rizzini et al. 2011; Christie et al. 2012;). These tryptophans and UVR-8 mediate photoinduced proton-coupled electron transfer reactions which causes the monomerization of UVR8 (Mathes et al. 2015). It is variously documented that the main mechanism of UVR8 signaling involves interaction with other proteins, such as COP1, and the activated UVR8 monomer interacts with COP1 (a key component in light signaling) through the C-terminal 27 amino acid region of UVR8 (Oravecz et al. 2006; Rizzini et al. 2011; But at this phase, this interaction with COP1 does not lead to UVR8 degradation but abolishes the E3 ligase function of the COP1/SPA (suppressor of phytochrome A) complex which leads to the accumulation of the E3 targets. This effect of the mechanism is witnessed in the transcription factor HY5 (Brown et al. 2005; Huang et al. 2013) on HY5 mutant seedlings which are UV hypersensitive (Kliebenstein et al. 2002; Brown et al. 2005). Furthermore, a very strong and interchangeable relation persists between HY5 and

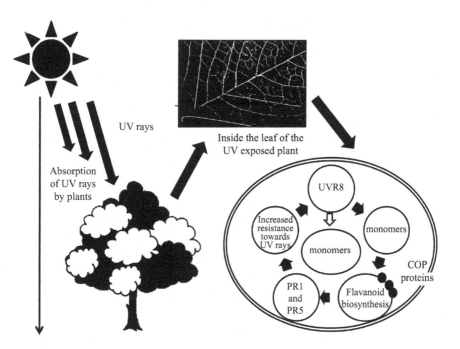

FIGURE 16.2 A hypothetical illustration of UV-B induced signaling mechanism in plants.

COP1 proteins. In the darkness, COP1 ubiquitylates HY5, leading to proteasomal degradation (Osterlund et al. 2000; Lau and Deng 2012) whereas the same, when exposed to UV-B, inhibits HY5 degradation causing HY5 accumulation (Oravecz et al. 2006).

UV-B signaling in higher plants is mediated by UV-B photoreceptor, UVR8It iskown as UV resistance locus 8, present both in cytosol and nucleus (majorly), this promotes acclimation as well as tolerance against UV-B. This UV-B photoreception, in turn, stimulates nuclear accumulation of UVR8 which differs from other known photoreceptors since it is a specific tryptophan amino acid instead of a prosthetic chromophore for light absorption during UV-B photoreception. This UVR8 was found to be upregulating the action of the UV target gene, chalcone synthase (CHS) which is a major regulatory gene in the flavonoid biosynthetic pathway and responds immediately after treating the plants with the UV radiation (Kliebenstein et al. 2002; Rizzini et al. 2011; Jiang et al. 2012; Tilbrook et al. 2013;). The absorption of UV-B dissociates the UVR8 dimer into monomers and initiates signal transduction through interaction, with the COP proteins constitutively photo-morphogenic 1 having a dual role in the regulation of UV-B induced UVR 8 nuclear accumulation. This signaling has been reported earlier by Yin et al. (2016). These two are simultaneously required for UV-B induced expression of the elongated hypocotyl 5 (HY5) transcription factor, which plays a central role in the regulation of genes involved in photomorphogenic UV-B responses (Jenkins et al. 1997). Besides this, UVR8 is also involved in the increased expression of pathogenesis-related (PR1 and PR5) proteins, and these UVR-8 regulated genes are found to encode chloroplast proteins having UVR-8 in photosynthesis under UV-B radiation.

16.5   CONCLUSION AND PROSPECTS

The above mentioned roles of UV-B are interesting as they provide incongruous insights to the adaption and development of the plant to the environmental cues. There are so many reports which point towards the priming role of UV-B radiation. The priming of the seeds is a healthy process that increases the vigor of the plants. It also increases the tolerance and resistance against various abiotic stress such as drought, salinity, metal stress, etc., and this method has also been proven as ecologically safe to improve plant productivity and yield as well as to protect them from various diseases. Hence, this strategy is employed nowadays to get better crop yield which may prove as an alternative source to plant breeding programs, as it is cost-effective. It has several beneficial effects on the growth of plants. In this process, plants are exposed to the low intensity of UV radiation for a definite period of time. Seeds and seedlings when exposed to mild UV-B radiation exhibit physi-ological, biochemical and morphological changes which in turn activate enzymatic and non-enzy-matic protective systems to screen out different stresses. Therefore, it is suggested that the priming of seeds with UV-B will help to bring about enhanced tolerance to environmental stress and will increase the yield of major cash crops. But the mechanisms involved therein must be elucidated with further instrumentation and lab work.

ACKNOWLEDGMENTS

Professor Sheo Mohan Prasad and Abreeq Fatima are very grateful to the UGC, New Delhi and Head Department of Botany, University of Allahabad for providing financial assistance to carry out this work. We also acknowledge the UGC for providing financial support to Madhulika Singh in the form of DSK Post Doc Fellowship (DSKPDF-BL/18-19/0173) in Banaras Hindu University. The authors also declare no conflict of interest.

REFERENCES

Agati, G., Azzarello, E., Pollastri, S., Tattini, M., 2012. Flavonoids as antioxidants in plants: Location and functional significance. *Plant. Sci.* 196, 67–76.

Agati, G., Biricolti, S., Guidi, L., Ferrini, F., Fini, A., Tattini, M., 2011. The biosynthesis of flavonoids is enhanced similarly by UV radiation and root zone salinity in L. *Vulgare* leaves. *J. Plant. Physiol.* 168(3), 204–212.

Agati, G., Brunetti, C., Di Ferdinando, M., Ferrini, F., Pollastri, S., Tattini, M., 2013. Functional roles of flavonoids in photoprotection: New evidence, lessons from the past. *Plant Physiol. Biochem.* 72, 35–45.

Andrady, A.L., Aucamp, P.J., Austin, A., Bais, A.F., Ballaré, C.L., Barnes, P.W., Bernhard, G.H., et al., 2015. Environmental effects of ozone depletion and its interactions with climate change: 2014 assessment executive summary. *Photochem. Photobiol. Sci.* 14(1), 14–18.

Badridze, G., Kacharava, N., Chkhubianishvili, E., Rapava, L., Kikvidze, M., Chanishvili, S., Shakarishvili, N., Mazanishvili, L., Chigladze, L., 2016. Effect of UV radiation and artificial acid rain on productivity of wheat. *Russ. J. Ecol.* 47(2), 158–166.

Bais, A.F., McKenzie, R.L., Bernhard, G., Aucamp, P.J., Ilyas, M., Madronich, S., Tourpali, K., 2015. Ozone depletion and climate change: Impacts on UV radiation. *Photochem. Photobiol. Sci.* 14(1), 19–52.

Bashri, G., Singh, M., Mishra, R.K., Kumar, J., Singh, V.P., Prasad, S.M., 2018. Kinetin regulates UV-B-induced damage to growth, photosystemII photochemistry, and nitrogen metabolism in tomato seedlings. *J. Plant Growth Regul.* 37(1), 233–245.

Batista, L.F.Z., Kaina, B., Meneghini, R., Menck, C.F.M., 2009. How DNA lesions are turned into powerful killing structures: Insights from UV-induced apoptosis. *Mutat. Res.* 681(2–3), 197–208.

Behn, H., Albert, A., Marx, F., Noga, G., Ulbrich, A., 2010. Ultraviolet-B and photosynthetically active radiation interactively affect yield and pattern of monoterpenes in leaves of peppermint (Mentha x piperita L.). *J. Agric. Food Chem.* 58(12), 7361–7367.

Britt, A.B., 2004. Repair of DNA damage induced by solar UV. *Photosynth. Res.* 81(2), 105–112.

Brown, B.A., Cloix, C., Jiang, G.H., Kaiserli, E., Herzyk, P., Kliebenstein, D.J., Jenkins, G.I., 2005. A UV-B-specific signaling component orchestrates plant UV protection. *Proc Natl Acad Sci U S A* 102(50), 18225–18230.

Chittka, L., Shmida, A., Troje, N., Menzel, R., 1994. Ultraviolet as a component of flower reflections, and the colour perception of Hymenoptera. *Vision J* 34(11), 1489–1508.

Christie, J.M., Arvai, A.S., Baxter, K.J., Heilmann, M., Pratt, A.J., O'Hara, A., Kelly, S.M., Hothorn, M., Smith, B.O., Hitomi, K., Jenkins, G.I., Getzoff, E.D., 2012. Plant UVR8 photoreceptor senses UV-B by tryptophan-mediated disruption of cross-dimer salt bridges. *Science* 335(6075), 1492–1496.

Das, K., Roychoudhury, A., 2014. Reactive oxygen species (ROS) and response of antioxidants as ROS-scavengers during environmental stress in plants. *Front. Environ. Sci.* 2, 53.

Dhanya, T., Puthur, J., 2017. UV radiation priming: A means of amplifying the inherent potential for abiotic stress tolerance in crop plants. *Env. Exp. Bot.* 138. doi:10.1016/j.envexpbot.2017.03.003.

Duan, X., Wu, T., Liu, T., Yang, H., Ding, X., Chen, Y., Mu, Y., 2019. Vicenin-2 ameliorates oxidative damage and photoaging via modulation of MAPKs and MMPs signaling in UVB radiation exposed human skin cells. *J Photochem Photobiol B* 190, 76–85.

Dwivedi, R., Singh, V.P., Kumar, J., Prasad, S.M., 2015. Differential physiological and biochemical responses of two Vigna species under enhanced UV-B radiation. *J. Radiat.Res. Appl. Sci.* 8(2), 173–181.

Ehling-Schulz, M., Schulz, S., Wait, R., Görg, A., Scherer, S., 2002. The UV-B stimulon of the terrestrial cyanobacterium *Nostoc commune* comprises early shock proteins and late acclimation proteins. *Mol. Microbiol.* 4, 827–843.

Fukucfi, K., Takahashi, K., Tatsumoto, H., 2004. Impacts of UV- A irradiation on laser-induced fluorescence spectra in UV-B damaged peanut (*Arachis hypogaea* L.) leaves. *Environ. Technol.* 25(11), 1233–1240.

Gill, S.S., Tuteja, N., 2010. Reactive oxygen species and antioxidant machinery in abiotic stress tolerance in crop plants. *Plant Physiol. Biochem.* 48(12), 909–930.

Habibi, G., Hajiboland, R., 2011.Comparison of water stress and uv radiation effects on induction of cam and antioxidative defense in the succulent *Rosularia elymaitica* (Crassulaceae). *Acta Biologica Cracoviensia Series Botanica* 53(2), 15–24.

Häder, D.P., Kumar, H.D., Smith, R.C., Worrest, R.C., 2007. Effects of solar UV radiation on aquatic ecosystems and interactions with climate change. *Photochem. Photobiol. Sci.* 6(3), 267–285.

Hahlbrock, K., Scheel, D., 1989. Physiology and molecular biology of phenylpropanoid metabolism. *Annu. Rev. Plant Physiol. Plant Mol. Biol.* 40(1), 347–369.

Halliwell, B., Gutteridge, J.M., 1995. The definition and measurement of antioxidants in biological systems. *Free Radic Biol Med* 18(1), 125–126.

He, Y., Zhan, F., Zu, Y., Xu, W., Li, Y., 2015. Effects of enhanced UV-B radiation on culm characteristics and lodging index of two local rice varieties in Yuanvang terraces under field condition. *Ying Yong Sheng Tai Xue Bao* 26(1), 39–45.

Hideg, E., Barta, C., Kálai, T., Vass, I., Hideg, K., Asada, K., 2002. Detection of singlet oxygen and superoxide with fluorescent sensors in leaves under stress by photoinhibition or UV radiation. *Plant Cell Physiol.* 43(10), 1154–1164.

Hideg, E., Jansen, M.A., Strid, A., 2013. UV-B exposure, ROS, and stress: Inseparable companions or loosely linked associates? *Trends. Plant. Sci.* 18(2), 107–115.

Hidema, J., Zhang, W.H., Yamamoto, M., Sato, T., Kumagai, T., 2005. Changes in grain size and grain storage protein of rice (Oryza sativa L.) in response to elevated UV-B radiation under outdoor conditions. *J. Radiat. Res.* 46(2), 143–149.

Hollósy, F., 2002. Effects of ultraviolet radiation on plant cells. *Micron* 33(2), 179– 197.

Huang, X., Ouyang, X., Yang, P., Lau, O.S., Chen, L., Wei, N., Deng, X.W., 2013. Conversion from CUL4-based COP1-SPA E3 apparatus to UVR8-COP1-SPA complexes underlies a distinct biochemical function of COP1 under UV-B. *Proc. Natl Acad. Sci. U. S. A.* 110(41), 16669–16674.

Inostroza-Blancheteau, C., Acevedo, P., Loyola, R., Arce-Johnson, P., Alberdi, M., Reyes-Díaz, M., 2016. Short-term UV-B radiation affects photosynthetic performance and antioxidant gene expression in highbush blueberry leaves. *Plant Physiol. Biochem.* 107, 301–309.

Jansen, M.A., Bornman, J.F., 2012. UV-B radiation: From generic stressor to specific regulator. *Physiol Plant* 145(4), 501–504.

Jenkins, G.I., 2009. Signal transduction in responses to UV-B radiation. *Annu. Rev. Plant Biol.* 60, 407–431.

Jenkins, M.E., Suzuki, T.C., Mount, D.W., 1997. Evidence that heat and ultraviolet radiation activate a common stress-response program in plants that is altered in the uvh6 mutant of *Arabidopsis thaliana*. *Plant Physiol.* 115(4), 1351–1358.

Kliebenstein, D.J., Lim, J.E., Landry, L.G., Last, R.L., 2002. *Arabidopsis* UVR8 regulates ultraviolet-B signal transduction and tolerance and contains sequence similarity to human regulator of chromatin condensation 1. *Plant Physiol.*. doi: 10.1104/pp.005041.

Lau, O.S., Deng, X.W., 2012. The photomorphogenic repressors COP1 and DET1: 20 years later. *Trends Plant Sci.* 17(10), 584–593.

Li, Y.A., He, L.L., Zu, Y.Q., 2010. Intraspecific variation in sensitivity to ultraviolet-B radiation in endogenous hormones and photosynthetic characteristics of 10 wheat cultivars grown under field conditions. *S. Afr. J. Bot.* 76(3), 493–498.

Lidon, F.J.C., Teixeira, M., Ramalho, J.C., 2012. Decay of the chloroplast pool of ascorbate switches on the oxidative burst in UV-B-irradiated rice. *J. Agron. Crop Sci.* 198(2), 130–144.

Mathes, T., Heilman, M., Pandit, A., Zhu, J., Ravensbergen, J., Kloz, M., Fu, Y., Smith, B.O., Christie, J.M., Jenkins, G.I., Kennis, J.T.M., 2015. Proton-coupled electron transfer constitutes the proactivation mechanism of the plant photoreceptor UVR8. *J. Am. Chem. Soc.* 137(25), 8113–8120.

Mathur, S., Jajoo, A., 2015. Investigating deleterious effects of ultraviolet (UV) radiations on wheat by a quick method. *Acta. Physiol. Plant.* 37, 1–7.

Menzel, R., Backhaus, W., 1991. Colour vision in insects. In: P. Gouras (Ed.), *Vision and Visual Dysfunction, 6.The Perception of Color*, London, Macmillan, pp. 262–293.

Menzel, R., Shmida, A., 1993. The ecology of flower colours and the natural colour vision of insect pollinators: The Israeli flora as a study case. *Biol. Rev.* 68(1), 81–120.

Miller, G., Suzuki, N., Ciftci-Yilmaz, S., Mittler, R., 2010. Reactive oxygen species homeostasis and signaling during drought and salinity stresses. *Plant Cell Environ.* 33(4), 453–467.

Mishra, V., Srivastava, G., Prasad, S.M., 2009. Antioxidant response of bitter gourd (*Momordica charantia* L.) seedlings to interactive effect of dimethoate and UV-B irradiation. *Sci Hortic* 120(3), 373–378.

Mishra, V., Srivastava, G., Prasad, S.M., Abraham, G., 2008. Growth, photosynthetic pigments and photosynthetic activity during seedling stage of cowpea (*Vigna unguiculata*) in response to UV-B and dimethoate. *Pest. Biochem. Physiol.* 92(1), 30–37.

Oliveira, J.M., Almeida, A.R., Pimentel, T., Andrade, T.S., Henriques, J.F., Soares, A.M., Loureiro, S., Gomes, N.C., Domingues, I., 2016. Effect of chemical stress and ultraviolet radiation in the bacterial communities of zebrafish embryos. *Environ. Poll* 208(B), 626–636. (*Barking, Essex*, 1987).

Oravecz, A., Baumann, A., Máté, Z., Brzezinska, A., Molinier, J., Oakeley, E.J., Adám, E., Schäfer, E., Nagy, F., Ulm, R., 2006.Constitutively photomorphogenic1 is required for the UV-B response in *Arabidopsis*. *Plant Cell* 18(8), 1975–1990.

Osterlund, M.T., Hardtke, C.S., Wei, N., Deng, X.W., 2000. Targeted destabilization of HY5 during light-regulated development of *Arabidopsis*. *Nature* 405(6785), 462–466.

Rai, K., Agrawal, S.B., 2017. Effects of UV-B radiation on morphological, physiological and biochemical aspects of plants: An overview. *J.Sci. Res.* 61, 87–113.

Rastogi, R.P., Richa Kumar, A., Tyagi, M.B., Sinha, R.P., 2010. Molecular mechanisms of ultraviolet radiation induced DNA damage and repair. *J. Nucleic Acids*:592980. doi: 10.4061/2010/592980

Renger, G., Völker, M., Eckert, H.J., Fromme, R., Hohm-Veit, S., Gräber, P., 1989. On the mechanism of photosystem II deterioration by UV-B irradiation. *Photochem. Photobiol.* 49(1), 97–105.

Rice-Evans, C.A., Miller, N.J., Paganga, G., 1997. Antioxidant properties of phenolic compounds. *Trends. Plant. Sci.* 2(4), 152–159.

Rizzini, L., Favory, J.J., Cloix, C., Faggionato, D., O'Hara, A., Kaiserli, E., Baumeister, R., Schäfer, E., Nagy, F., Jenkins, G.I., Ulm, R., 2011. Perception of UV-B by the *Arabidopsis* UVR8 protein. *Science* 332(6025), 103–106.

Rozema, J., van de Staaij, J., Björn, L.O., Caldwell, M., 1997. UV-B as an environmental factor in plant life: Stress and regulation. *Trend. Eco. Evo.* 12(1), 22–28.

Rudnóy, S., Majláth, I., Pál, M., Páldi, K., Rácz, I., Janda, T., 2015. Interactions of S methylmethionine and UV-B can modify the defence mechanisms induced in maize. *Acta. Physiol. Plant.* 37, 1–11.

Santos, J., Sousa, M.J., Leão, C., 2012. Ammonium is toxic for aging yeast cells, inducing death and shortening of the chronological lifespan. *PLOS ONE* 7(5), e37090.

Shaukat, S.S., Farooq, M.A., Siddiqui, M.F., Zaidi, S.A.H.A.R., 2013. Effect of enhanced UVB radiation on germination, seedling growth and biochemical responses of *Vigna mungo* (L.) Hepper. *Pak. J. Bot.* 45, 779–785.

Singh, M., Bashri, G., Prasad, S.M., Singh, V.P., 2019. Kinetin alleviates UV-B-induced damage in *Solanum lycopersicum*: Implications of phenolics and antioxidants. *J. Plant. Growth. Regul.* 38(3), 831–841.

Singh, V.P., Srivastava, P.K., Prasad, S.M., 2012. Differential effect of UV-B radiation on growth, oxidative stress and ascorbate-glutathione cycle in two cyanobacteria under copper toxicity. *Plant. Physiol. Biochem.* 61, 61–70.

Slieman, T.A., Nicholson, W.L., 2000. Artificial and solar UV radiation induces strand breaks and cyclobutane pyrimidine dimers in *Bacillus subtilis* spore DNA. *Appl Environ Microbiol* 66(1), 199–205.

Strid, A., Chow, W.S., Anderson, J.M., 1994. UV-B damage and protection at the molecular level in plants. *Photosynth. Res.* 39(3), 475–489.

Tang, X., Madronich, S., 1995. Effects of increased solar ultraviolet radiation on trophospheric composition and air quality. *Ambio* 24, 188–190.

Tang, X., Krausfeldt, L.E., Shao, K., LeCleir, G.R., Stough, J.M.A., Gao, G., Boyer, G.L., Zhang, Y., Paerl, H.W., Qin, B., Wilhelm, S.W., 2018. Seasonal gene expression and the ecophysiological implications of toxic *Microcystis aeruginosa* blooms in lake Taihu. *Environ. Sci. Technol.* 52(19), 11049–11059.

Teklemariam, T., Blake, T.J., 2003. Effects of UV B preconditioning on heat tolerance of cucumber (*Cucumis sativus* L.). *Environ. Exp. Bot.* 50(2), 169–182.

Teramura, A.H., 1983. Effects of ultraviolet-B radiation on the growth and yield of crop plants. *Physiol. Plant.* 58(3), 415–427.

Teramura, A.H., Sullivan, J.H., 1987. Soybean growth to enhanced levels of ultraviolet radiation under greenhouse conditions. *Am. J. Bot.* 74(7), 975–979.

Tevini, M., Braun, J., Fieser, G., 1991. The protective function of the epidermal layer of rye seedlings against ultraviolet-B radiation. *Photochem. Photobiol.* 53(3), 329–333.

Thomas, T.T.D., Puthur, J.T., 2017. UV radiation priming: A means of amplifying the inherent potential for abiotic stress tolerance in crop plants. *Environmental and Experimental Botany* 138, 57–66.

Tilbrook, K., Arongaus, A.B., Binkert, M., Heijde, M., Yin, R., Ulm, R., 2013. The UVR8 UV-B photoreceptor: Perception, signaling and response in *Arabidopsis*. doi: 10.1199/tab.0164.

Treutter, D., 2006. Significance of flavonoids in plant resistance: A review. *Environ. Chem. Lett.* 4(3), 147–157.

Wang, Y., Frei, M., 2011. Stressed food–the impact of abiotic environmental stresses on crop quality. *Agric. Ecosyst. Environ.* 141(3–4), 271–286.

Wargent, J.J., Jordan, B.R., 2013. From ozone depletion to agriculture: Understanding the role of UV radiation in sustainable crop production. *New Phytol.* 197(4), 1058–1076.

White, A.L., Jahnke, L.S., 2002. Contrasting effects of UV-A and UV-B on photosynthesis and photoprotection of beta-carotene in two *Dunaliella* spp. *Plant. Cell. Physiol.* 43(8), 877–884.

Xu, K., Qiu, B.S., 2007. Responses of superhigh-yield hybrid rice *Liangyoupeijiu* to enhancement of ultraviolet-B radiation. *Plant Sci.* 172(1), 139–149.

Yang, H., Zhao, Z., Qiang, W., An, L., Xu, S., Wang, X., 2004. Effects of enhanced UV-B radiation on the hormonal content of vegetative and reproductive tissues of two tomato cultivars and their relationships with reproductive characteristics. *Plant. Growth. Regul.* 43(3), 251–258.

Yin, R., Skvortsova, M.Y., Loubery, S., Ulm, R., 2016. COP1 is required for UV-B induced nuclear accumulation of the UVR8 photoreceptor. *Proc. Nat. Acad. Sci. USA*. 113, E4415–E4422.

Zlatev, Z.S., Lidon, F.J.C., Kaimakanova, M., 2012. Plant physiological responses to UV-B radiation. *Emir. J. Food. Agric.* 24, 481–501.

Zu, Y.Q., Li, Y., Chen, J.J., Chen, H.Y., 2004. Intraspecific responses in grain quality of 10 wheat cultivars to enhanced UV-B radiation under field conditions. *J Photochem Photobiol B* 74(2–3), 95–100.

17 Ozone and Enzymatic and Non-Enzymatic Antioxidant Enzymes in Plants

Muhammad Shahedul Alam, Arafat Abdel Hamed Abdel Latef, Md. Ashrafuzzaman

CONTENTS

17.1 INTRODUCTION

Atmospheric trace gas ozone (O_3) is the triatomic form of oxygen evolved in two phases in the atmosphere; the stratosphere and troposphere. Unlike the protective ozone naturally formed in the stratosphere, tropospheric ozone is a harmful secondary air pollutant associated with human health problems, as well as crop damage (Camp et al. 1994; Kollner and Krause 2003; Mills et al. 2018; Sharma and Davis 1997). Tropospheric ozone is formed mainly by photochemical reactions between nitrogen oxides, carbon monoxide and non-methane volatile organic compounds, commonly termed precursor's gas. These reactions are highly influenced by the sunlight, precursors' concentration and the ratio between them, and favored by high temperatures and stagnant weather conditions (Ainsworth et al. 2012; The Royal Society 2008). Even with increasing awareness of global change and efforts to mitigate these ozone precursor gases' emissions, the global background level of tropospheric ozone concentrations is rapidly increasing or stable in many parts of the world (Begum et al. 2020; Tarasick et al. 2019).

Ozone precursor gases occur in all areas of the globe due to human activities such as industrialization, urbanization, land use changes, policy changes, deforestation, energy generation, transportation, fossil fuel combustion and rapid population and economic growth (Banerjee and Roychoudhury 2019; Ueda et al. 2015b). Also, downward transport of stratospheric ozone sources through the free troposphere contributes approximately 10% to ground-level ozone (Ainsworth et al. 2012; Banerjee and Roychoudhury 2019). Ozone formation requires photolysis, and intense sunlight and high temperatures can accelerate the production of ozone, and therefore, tropical countries, especially East and South Asia, are the hotspots of tropospheric ozone (Begum et al. 2020; Wang et al. 2017). Plants, being sessile, have to deal with the phytotoxic effect and cannot escape ozone pollution exposure. The strong oxidant ozone is taken up into plants during photosynthetic gas exchange and decomposes into reactive oxygen species (ROS) in the apoplast and leads to losses in photosynthetic capacity due to damage of photosynthetic enzymes and stomatal closure, leading to reduced crop yield and quality (Kangasjärvi et al. 2005; Feng et al. 2008). The plant evolves two major tolerance mechanisms to control against ozone stress, such as control of ozone entry into the leaf by stomatal closure and detoxification of ozone or ozone derived ROS in the apoplast by a recharged antioxidant system, mainly apoplastic ascorbate (AsA) (Frei 2015; Krasensky et al. 2017). However, in crops, tropospheric ozone leads to substantial yield losses, estimated to exceed 200 million tons yearly for the major staple crops maize, wheat, rice and soybean (Mills et al. 2018). Therefore, information on interactions between ozone and different enzymatic and non-enzymatic antioxidant enzymes is essential to figure out all possible ways for the mitigation of ozone-induced crop yield losses.

17.2   GLOBAL DISTRIBUTION SCENARIO OF TROPOSPHERIC OZONE

Globally, the air pollutant ozone is not uniformly distributed over the land surface; concentrations tend to be high during summer and low during winter (Ainsworth 2017). Existing background ozone concentrations in terms of monthly mean in the northern hemisphere range from 35 to 50 ppb, which is considerably higher than the southern hemisphere (Ainsworth et al. 2012). However, ozone and its precursors are distributed in urban to suburban and rural areas, where humans, animals and plants experience higher ozone concentrations (Sicard et al. 2016). The lifetime of ozone molecules is 22 ±2 days in the troposphere, allowing regional and hemispheric-scale transmission (Stevenson et al. 2006). For instance, ozone concentrations differ from approximately 20 ppb over parts of Australia and South America to 55–60 ppb in Asia, the Middle East, Europe and North America. Simultaneously, croplands in China, India and the United States are exposed to higher ozone concentrations than farmlands in Australia or Brazil (Ainsworth 2017; Brauer et al. 2016; Ramankutty et al. 2008).

In the summer, peak daily concentrations occur in the late afternoon; even during heatwaves ozone level can reach above 100 ppb in polluted areas (Felzer et al. 2007; The Royal Society 2008). Mean maximum daily concentrations well above 60 ppb and acute ozone stress episodes exceeding 100 ppb coincide with high plant activity during the day and during the growing season for most major crops in spring and summer (Brauer et al. 2016). Ozone concentration in the United States decreased by 4.7% from 1990 to 2013, though the ozone concentration was 67 ppb. In Bangladesh, ozone concentration increased 21.3% from 1990 to 2013, and levels in India, Brazil, Pakistan and China increased by 20.2%, 17.3%, 16.5% and 13.2%, respectively, over that same period. However, globally, ozone increased by 8.9% over these 23 years (Brauer et al. 2016).

Therefore, although globally, ozone is not distributed uniformly, many of the world's agricultural regions are threatened by the increasing level of concentration, and most sufferers will be in Asia. Even in Europe, by 2050, there is a possibility of increasing ozone by two ppb than the present average ozone concentration (35 ppb) due to the recent urbanisation in parts of Europe, and global warming (Hendriks et al. 2016).

17.3   IMPACT OF OZONE ON PLANTS

The phytotoxic coincidence of high tropospheric ozone concentrations during cultivation seasons leads to significant reductions in crop productivity; consequently, global food security is profoundly affected by ozone (Ashrafuzzaman et al. 2017; Cho et al. 2011). Different approaches were taken while measuring the effect of ozone on crops. Along with field and greenhouse experiments, several modeling studies were conducted to quantify the ozone-related yield gap. All the approaches projected the ozone-induced significant yield loss of crops. For instance, ozone reduces global yield annually by 12.4%, 7.1%, 6.1% and 4.4% for soybean, wheat, maize and rice, respectively (Mills et al. 2018). However, the response to ozone by crops and varieties differs significantly due to the different genetic structure, duration, concentration of ozone exposure, weather conditions and growth stages of crops (Singh 2020; Ainsworth et al. 2012; Cho et al. 2011). Chlorosis and pale color of leaves, necrotic dark brown spots or dead regions on leaves, reduced growth rate and a stunted phenotype are some typical ozone stress symptoms in plants leading to reduced crop yield and quality (Ashrafuzzaman et al. 2018; Ueda et al. 2015 a).

Ozone can negatively affect crop productivity directly through oxidative stress and indirectly by contributing to global warming through its role as a greenhouse gas (Ainsworth 2017). Ozone induced crop damage can be categorized by chronic effects and acute effects. Exposure of plants to an ozone concentration of more than 100 ppb for several hours to a few days leads to the acute effect, characterized by high production of ROS, and subsequent cell death, appearing as brown spots on the leaves (Frei 2015). During the cultivation period, plants can experience a background ozone concentration of ~60–100 ppb, which is expressed by chronic stress and resulting in ROS formation, reduce photosynthesis and stomatal conductance, increased rates of respiration, disturbed metabolism and eventually leads to a significant decrease in yield and grain quality (Banerjee and Roychoudhury 2019; Emberson et al. 2018; Krasensky et al. 2017).

The highly reactive molecule ozone primarily enters into the leaf apoplast through the stomata and degrades rapidly into various ROS that trigger a cellular response and ultimately damage crop productivity (Ainsworth 2017; Krasensky et al. 2017). The ROS in the apoplast include singlet oxygen ($^1O_2$), hydrogen peroxide (H_2O_2), superoxide radicals ($O_2^{\bullet-}$) and hydroxyl radicals ($^{\bullet}OH$), which enhance the oxidative stress and lead to cell death (Ainsworth 2017). These ROS can interfere with various enzymatic processes to damage cell membrane lipids. Different plant hormones determine the duration and extent of ozone-induced cell damage. Salicylic acid causes programmed cell death; jasmonic acid is required for limiting the lesion spreading, ethylene promotes endogenous ROS formation and abscisic acid is expected to affect lesion initiation (Kangasjärvi et al. 2005).

The chlorophyll content is an indicator of plants' stress response under ozone stress (Begum et al. 2020). Chlorophyllase is a primary catabolic enzyme present in photosynthetic organisms to synthesize chlorophyll (Harpaz-Saad et al. 2007). The ROS generated by ozone can attack the chloroplast leading to the destruction of chlorophyll. However, the mode of interaction between ozone and chlorophyllase is still elusive. In non-destructive methods, rice showed decreased chlorophyll content under ozone stress (Ashrafuzzaman et al. 2017; Ueda et al. 2015 a, b), where tolerant genotypes showed less degradation of chlorophyll than susceptible ones. A similar trend was observed in wheat (Begum et al. 2020). A meta-analysis by Feng et al. (2008) showed 40% reduced chlorophyll content in wheat under the elevated concentration of ozone. The total chlorophyll (Chl a + Chl b) content of common bean was affected by ozone exposure. Maize also showed decreased chlorophyll a and b in elevated ozone (Choquette et al. 2020). A significant reduction of approximately 50% of total Chl was recorded in the genotype Fepagro 26, whereas chlorophyll content of other common bean varieties, e.g. Irai, was not affected (Caregnato et al. 2013). Leaf chlorophyll concentration decrease of 14% induced by elevated ozone was reported in soybean (Morgan et al. 2003). Similarly, ozone-induced decrease in chlorophyll was reported in tobacco (Saitanis et al. 2001), soybean (Sun et al. 2014), rapeseed and sugar beet (Kollner and Krause 2003).

Rubisco (Ribulose-1,5-bisphosphate carboxylase/oxygenase) is a copper-containing enzyme involved in the first major step of carbon fixation. It is the central enzyme of photosynthesis and has implications for crop yield, nitrogen and water usage, and the global carbon cycle (Andersson and Backlund 2008). Ozone-derived ROS decreased Rubisco activity and played a vital role in lowering the photosynthetic rate (Fiscus et al. 2005). Ozone causes a decline in the concentration, rather than the activation state of Rubisco (Galmés et al. 2013). Photosynthetic rates decline in five modern cultivars of winter wheat exposed to elevated ozone concentration at 1.5 times the ambient ozone concentration due to non-stomatal factors, e.g. lower maximum carboxylation capacity and electron transport rates (Feng et al. 2016). Rubisco's activity and expression level showed a 10% decrease between the high and ambient ozone exposed soybean leaf samples (Galant et al. 2012). Decreased concentrations and damage of foliar Rubisco were reported in several studies on rice leaves exposed to high ozone levels (Ashrafuzzaman et al. 2018; Sarkar et al. 2015; Cho et al. 2008; Feng et al. 2008). Sensitive maize lines showed more significant ozone-induced senescence and loss of photosynthetic capacity compared to the tolerant line (Choquette et al. 2020). Ozone affected soybean yield mainly via reducing the maximum rate of Rubisco carboxylation and maximum electron transport rates (Sun et al. 2014).

Decreased photosynthetic protein, pigmen and nitrogen content resulted in reduced photosynthetic rates, and increased respiration rates accelerate altered leaf antioxidant balance (Dutilleul et al. 2003). Plant growth rates, total leaf area and biomass are also reduced due to changes in primary metabolism in elevated ozone (Ainsworth 2008; Felzer et al. 2007; Feng et al. 2008). However, the effects of elevated ozone on crop physiological processes also increase over a growing season. For instance, due to elevated ozone, reduction in the photosynthesis percentage was higher in reproductive phases in contrast to vegetative phases of development, partly because the leaves developed during late reproductive phases were exposed to elevated ozone at the top of the canopy for an extended period (Feng et al. 2008; Morgan et al. 2003). At the reproductive phase of crops, ozone directly affects flowering, pollen sterility, grain injury and abortion (Wilkinson et al. 2012).

Malondialdehyde (MDA) is a widely used marker of oxidative lipid injury caused by ozone stress. Peroxidation of membrane lipids is a significant damaging effect of ROS, which is detected by measuring MDA contents (Kong et al. 2016). Studies on rice showed a significant increase in MDA concentration in ozone-sensitive genotypes compared to tolerant genotypes (Ashrafuzzaman et al. 2017; Ueda et al. 2015b). MDA content increases up to 1.4 fold in rice leaves after 24h of fumigation with ozone, implying that an increase in MDA is not caused by ozone directly; instead, it results from other metabolic processes affected by ozone (Ueda et al. 2013).

Lignin is a complex aromatic polymer that confers rigidity to cell walls and is therefore associated with tolerance to abiotic stresses and plants' mechanical stability (Frei 2013). High tropospheric ozone led to increased lignin levels in the above ground parts of several crops. Genes and enzymes involved in lignin biosyntheses, such as phenylalanine ammonia-lyase (PAL) or peroxidases (POX), form part of the defense mechanism to contain cell death (Frei et al. 2010). Analysis of foliar lignin content as an agronomically important parameter may represent apoplastic stress, since the coupling of monolignol molecules requires the oxidation of the hydroxyl group and, therefore, depends on the apoplastic redox status (Frei 2013). In an ozone fumigation experiment, Ueda et al. (2015 a) observed an overall 3.4% lignin content increase in rice. They employed an ozone concentration of 60 ppb for seven hours throughout the growing season plus three additional 150 ppb episodes for one day.

17.4   OZONE TOLERANCE MECHANISMS

Ideal tolerance against ozone stress includes control of ozone entry into the leaf, referred to as stress avoidance, and stress defense can refer to a recharged antioxidant system capable of competing with ROS accumulation by protecting membranes from lipid peroxidation (Banerjee and Roychoudhury 2019; Frei 2015; Krasensky et al. 2017; Ueda et al. 2015 b). Stomatal closure can be the first line of defense against ozone stress, as the stomatal aperture is the main factor controlling the influx of ozone into plant leaves (Ainsworth 2017; Frei 2015). Vahisalu et al. (2008) described a plasma membrane slow anion channel 1 (SLAC1), preferentially expressed in guard cells, which is involved in ozone-induced stomatal closure. The initiation of SLAC1 enhances anion efflux and depolarization of the cell membranes, and subsequently activates potassium-efflux channels. Nagatoshi et al. (2016) reported transcriptional factors golden 2-like 1 and 2 (GLK1, GLK2) as positive regulators of potassium influx channels and stomatal movement, and suppression of GLK1 and GLK2 resulted in decreased stomatal aperture and water loss under ozone stress, resulting in better ozone tolerance. For example, Brosché et al. (2010) stated that stomatal conductance is the most critical factor determining ozone sensitivity for the model plant *Arabidopsis thaliana*, under an acute ozone treatment (350 ppb for seven hours). However, different species prioritize different mechanisms for ozone

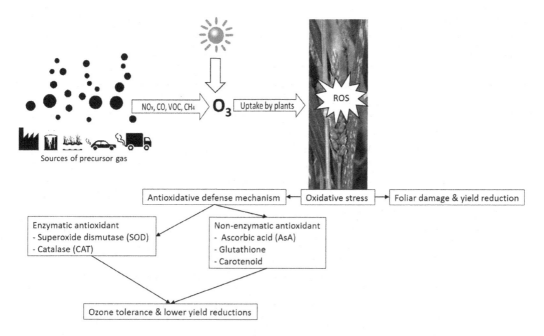

FIGURE 17.1 Generalized illustration of the primary sources, the formation of tropospheric ozone, mode action to plants and major antioxidant defense. NOx, Nitrogen oxides; VOCs, Volatile organic compounds; CO, Carbon monoxide; CH_4, methane; O_3, Ozone; ROS, reactive oxygen species.

defense, for example stomatal conductance showed a high correlation with ozone tolerance in soybean (Chutteang et al. 2016) and wheat (Biswas et al. 2008) but seemed to be of minor importance in poplar (Ryan et al. 2009) and bean (Guidi et al. 2010).

The second line of defense, detoxification of ozone or ozone derived ROS in the apoplast, can be initiated after the entry of ozone into the leaf through the stomata (Frei 2015; Kangasjärvi et al. 2005). Numerous enzymatic and non-enzymatic antioxidants are particularly crucial for ozone defense (Figure 17.1).

17.5 RESPONSES OF NON-ENZYMATIC ANTIOXIDANTS UNDER OZONE STRESS

Ascorbic acid (AsA) content is particularly crucial for ozone defense because of its high concentration and presence in different cellular compartments, including the apoplast (Yendrek et al. 2015). Apoplastic ascorbate detoxifies 30–50% of ozone entering leaves in different species (Moldau et al. 1997; Turcsanyi et al. 2000). Conklin et al. (1996) used ozone fumigation as a screening tool to discover AsA biosynthetic genes in an *A. thaliana* mutant collection and found a hypersensitive mutant containing only 30% of the wild-type AsA, which highlights that the total ascorbate pool is an important factor determining ozone tolerance in plants. Based on this finding, GDP-mannose/L-galactose AsA biosynthetic pathway was discovered in plants (Smirnoff et al. 2001; Wheeler et al. 1998). In a continuation, Chen and Gallie (2005) examined dehydroascorbate reductase (DHAR) overexpressing *A. thaliana* and found improved ozone tolerance under both acute and chronic exposure. Similarly, in a season-long ozone fumigation experiment, Frei et al. (2012) tested a rice TOS17 mutant, lacking expression of an AsA biosynthetic gene (GDP-D-mannose-3′,5′-epimerase) and found about 20 to 30% lower foliar AsA concentration and stronger biomass and yield reductions at ozone concentrations exceeding the ambient level. Greater apoplastic ascorbate contents also reflected in better ozone tolerance in legume crops, especially garden pea, which had a greater capacity for detoxification in an elevated ozone environment (Yendrek et al. 2015). However, a

few studies have also demonstrated that apoplastic ascorbate levels are not related to ozone tolerance (D'Haese et al. 2005), and maybe too low to play a significant role in ozone detoxification
(Sanmartin et al. 2003).

Furthermore, a candidate gene (Os09g0365900) underlying a quantitative trait locus (OzT9),
localized in the apoplast, for ozone stress tolerance in rice, was first identified and characterized
as a potential ozone responsive gene which has putative ascorbate oxidase (AO) activity (Frei et al.
2010). Further studies revealed that the expression patterns of this gene interfere with differential
ozone tolerance in rice but do not possess AO activity and were renamed as ozone-responsive apoplastic protein1 (*OsORAP1*) (Ashrafuzzaman et al. 2020; Ueda et al. 2015 b).

Glutathione, the primary source of nonprotein thiols in most plant cells, is a tripeptide (γ-glutamyl-
cysteinyl-glycine). The chemical reactivity of glutathione makes it particularly suitable to serve a
broad range of biochemical functions in all organisms. It has been detected in all cell compartments
such as cytosol, chloroplasts, endoplasmic reticulum, vacuoles and mitochondria (Millar et al.
2003). The abundance and redox state of the antioxidant glutathione is essential for detoxification of
ozone-induced ROS and to prevent cellular damage and maintain regular metabolic activity (Foyer
and Noctor 2005). In plants, it occurs predominantly in the reduced form (GSH), and the physiological significance of AsA is strictly dependent on the total thiol group's redox state, especially
the GSH molecule (Krasensky et al. 2017; Caregnato et al. 2013). Unlike AsA, in the apoplast, GSH
is found at low concentrations, whereas in the chloroplast, it ranges from 1 to 4.5 mM (Foyer and
Halliwell 1976; Law et al. 1983). The ascorbate-glutathione system, also known as Halliwell-Asada
cycle, is responsible for keeping the cellular redox balance (Foyer and Noctor 2005). Glutathione
redox state was shifted towards a highly reduced form in the ozone treatment and was more reduced
in the ozone tolerant rice line SL41 than in the other genotypes (Frei et al. 2010). The enzymes,
namely ascorbate peroxidase (APX), monodehydroascorbate reductase (MDHAR), dehydroascorbate reductase (DHAR) and glutathione reductase (GR) involved in the ascorbate-glutathione cycle,
responded to ozone treatment with up-regulation and down-regulation. A bi-functional MDHAR
(Os04g0412500) unaffected by ozone treatment was more strongly expressed in SL41 and could
explain the higher level of reduced AsA observed in this genotype. In the garden pea, foliar glutathione content was found higher than soybean, and the common bean thus showed a better ozone
tolerance. Total foliar glutathione content was 2.9- to 4.2-fold higher in ambient ozone and 1.7- to
6.1-fold higher in elevated ozone (Yendrek et al. 2015).

Carotenoid is considered an essential non-enzymatic antioxidant in plants and plays a vital role
in scavenging ROS, especially $^1O_2$. The conjugated double-bonded structure's ability to delocalize unpaired electrons is the primary consequence of carotenoids' antioxidant activity (Mortensen
et al. 2001). It can quench singlet oxygen without degradation and for the chemical reactivity of
β-carotene with free radicals such as the peroxyl (ROO•), •OH and O_2•⁻ (Ahmad et al. 2010). For
the non-invasive diagnosing of ozone stress in rice and wheat, vegetation indices representing the
carotenoid to chlorophyll pigment ratio were particularly ozone-responsive (Ashrafuzzaman et al.
2017; Begum et al. 2020). In many stress situations or during plant senescence, chlorophyll degrades
faster than carotenoids (Penuelas et al. 1995).

17.6   INTERACTIONS OF OZONE AND ENZYMATIC ANTIOXIDANTS

Superoxide dismutase (SOD) can counteract the toxic effects of ROS by converting superoxide
anion O^{2-} into O_2 and H_2O_2 (Krasensky et al. 2017). The metalloenzyme SOD consisting of Cu and
Zn or Mn and are of three different types in plants: Mn-SOD (localized in mitochondria), Fe-SOD
(localized in chloroplasts) and Cu/Zn-SOD (localized in the cytosol, peroxisomes and chloroplasts)
(Mittler 2002). Under environmental stresses, SOD forms the first line of defense against ROS-
induced damages and is upregulated by abiotic stress conditions (Boguszewska et al. 2010). *A.
thaliana* plants treated with 300 ppb ozone daily for six hours resulted in a higher level of cytosolic
Cu/Zn SOD in plants than in ambient air-treated controls. Also, approximately 12 h after ozone

treatment, SOD mRNA levels increased two- to three-fold higher than the controls (Sharma and Davis 1994). The transgenic approach of overexpressing Mn-SOD in the chloroplasts resulted in a three- to four-fold reduction of visible ozone injury in tobacco plants. However, the overproduction of SOD in the mitochondria activity had only a minor effect on ozone tolerance (Camp et al. 1994). Rice, under acute ozone stress, maintained expression level in cytosolic SODs, whereas chloroplastic, peroxisomal and mitochondrial SODs showed down-regulated expression levels under ozone stress (Ueda et al. 2013). The up-regulation of SOD was observed in the wheat genotype Y16, which is identified as tolerant to ozone stress at the booting stage. SOD might facilitate the delay in the effect of ozone on plant senescence. Increasing the apoplastic H_2O_2 content at flowering might also result from the induction of apoplastic SOD activity in Y16 at booting (Wang et al. 2014).

Catalase (CAT) is an antioxidant enzyme presents in almost all major sites of H_2O_2 production in the cellular environment (such as peroxisomes, mitochondria, cytosol and chloroplast) of higher plants and is responsible for reducing hydrogen peroxide to water and molecular oxygen using either iron or manganese as a cofactor (Yang and Poovaiah 2002). For the antioxidant defense mechanism to operate, catalases are closely associated with SOD activity to remove the H_2O_2 produced by SOD (Camp et al. 1994). Increasing CAT activity is observed when two high-yielding rice cultivars Shivani and Malviyadhan 36 are exposed to ozone (Sarkar et al. 2015). CAT induction activity was more in Shivani, which followed SOD induction simultaneously to detoxify H_2O_2. Cho et al. (2008) also reported increasing catalase in rice leaves exposed to ozone for 12 hours and 24 hours. In response to ozone fumigation, CAT also increased two times in Irai, a kidney bean variety of Brazil (Caregnato et al. 2013). However, Frei et al. (2010) reported significant up-regulation of a catalase gene in the sensitive rice genotype SL15 in the ozone treatment and showed a significantly more robust response.

17.7 OZONE AND OSMOLYTES

Plants accumulate small organic molecules called osmolytes (amino acids, sugars and polyols, methylsulfonium compounds, methylamines and urea) under different environmental stress, which have an important role in stress tolerance mechanisms (Iqbal and Nazar 2015). To date, limited information is available in relation to osmolytes and ozone stress management. However, proline accumulation under ozone stress was reported in several previous works. Alfalfa (*Medicago sativa*) showed a sharper increase (80.7%) in proline content when the plants were exposed to 100 ppb compared to 50 ppb ozone. However, Cheeseweed (*Malva parviflora*), which had a better ozone resistance, showed less accumulation of proline in ozone stress (El-Khatib 2003). Similarly, a higher amount of serine and proline was reported in the needles of Scots pine seedlings exposed to the low level of ozone (Manninen et al. 2000).

17.8 TRANSGENIC APPROACHES IN THE BREEDING OF OZONE TOLERANT CROPS

Only a few transgenic approaches have been conducted so far in ozone tolerance studies with a number of limitations. The results in controlled environments (greenhouse condition) can differ with the actual field conditions (Ainsworth 2017). For instance, visual leaf damage showed a correlation with ethylene production in greenhouse trials in transgenic potato with altered ethylene biosynthetic gene, but all transgenics exhibited stunted growth under both clean air and elevated ozone conditions in the field (Sinn et al. 2004). However, some transgenic approaches with altered ascorbate turnover or enhanced ascorbate biosynthesis have been reported with different abiotic stresses, as it is a major antioxidant and can play a vital role as a redox player in the plant, and also as a determining factor of the nutritional value of fruit crops and vegetables (summarized by Macknight et al. 2017). In line with this, a recent transgenic approach was carried out with increased foliar ascorbate levels by incorporating ascorbate biosynthesis gene GDP-L-galactose phosphorylase (*AcGGP*) from kiwifruit (*Actinidia chinensis* Planch.) in a high yielding rice variety, and tested against multiple abiotic stresses including ozone. A remarkably better performance in ozone stress was reported only when ozone coincided with salinity stress (Ali et al. 2019). Unfortunately, the lack of consistent performance both in the greenhouse and actual field conditions can be a major obstacle to introducing transgenic approaches in ozone tolerance breeding, as the mechanisms of plant response to ozone stress may vary.

17.9   CONCLUSION

Rising tropospheric ozone concentrations in many developing parts of the world are becoming a major threat to crop plants, as farmers have almost no management options. Nevertheless, it is getting less attention due to a lack of a simple diagnosing process, and awareness. A better understanding of enzymatic and metabolic pathways involving different antioxidant enzymes can be helpful to cope with ozone stress. Spraying of ascorbic acid and other antiozonant chemicals, transgenic approaches with enhanced ascorbate and other antioxidants have been utilized so far against ozone stress management on a small scale experimental level. However, little information is available in large scale field applications or suitable ozone resistant or tolerant transgenic crop varieties. Therefore, there is an enormous scope to explore and functionally characterize different ozone-responsive enzymatic and non-enzymatic antioxidants to minimize the ozone-induced crop yield loss, thereby contributing to secure the food supply of many highly ozone affected areas around the world.

REFERENCES

Ahmad, P., Jaleel, C.A., Salem, M.A., Nabi, G., Sharma, S., 2010. Roles of enzymatic and nonenzymatic antioxidants in plants during abiotic stress. *Crit. Rev. Biotechnol.* 30(3), 161–175. https://doi.org/10.3109/07388550903524243

Ainsworth, E.A., 2008. Rice production in a changing climate: A meta-analysis of responses to elevated carbon dioxide and elevated ozone concentration. *Glob. Chang. Biol.* 14(7), 1642–1650. https://doi.org/10.1111/j.1365-2486.2008.01594.x

Ainsworth, E.A., 2017. Understanding and improving global crop response to ozone pollution. *Plant J.* 90(5), 886–897. https://doi.org/10.1111/tpj.13298

Ainsworth, E.A., Yendrek, C.R., Sitch, S., Collins, W.J., Emberson, L.D., 2012. The effects of tropospheric ozone on net primary productivity and implications for climate change. *Annu. Rev. Plant Biol.* 63, 637–661. https://doi.org/10.1146/annurev-arplant-042110-103829

Ali, B., Pantha, S., Acharya, R., Ueda, Y., Wu, L., Ashrafuzzaman, M., Ishizaki, T., Wissuwa, M., Bulley, S., Frei, M., 2019. Enhanced ascorbate level improves multi-stress tolerance in a widely grown indica rice variety without compromising its agronomic characteristics. *J. Plant Physiol.* 240, 152998. https://doi.org/10.1016/j.jplph.2019.152998

Andersson, I., Backlund, A., 2008. Structure and function of RuBisCO. *Plant Physiol. Biochem.* 46(3), 275–291. https://doi.org/10.1016/j.plaphy.2008.01.001

Ashrafuzzaman, M., Haque, Z., Ali, B., Mathew, B., Yu, P., Hochholdinger, F., de Abreu Neto, J.B., McGillen, M.R., Ensikat, H.J., Manning, W.J., Frei, M., 2018. Ethylenediurea (EDU) mitigates the negative effects of ozone in rice: Insights into its mode of action. *Plant Cell Environ.* 41(12), 2882–2898

Ashrafuzzaman, M., Lubna, F.A., Holtkamp, F., Manning, W.J., Kraska, T., Frei, M., 2017. Diagnosing ozone stress and differential tolerance in rice (*Oryza sativa* L.) with ethylenediurea (EDU). *Environ. Pollut.* 230, 339–350. https://doi.org/10.1016/j.envpol.2017.06.055

Ashrafuzzaman, M., Ueda, Y., Frei, M., 2020. Natural sequence variation at the OsORAP1 locus is a marker for ozone tolerance in Asian rice. *Environ. Exp. Bot.* 178. http://www.ncbi.nlm.nih.gov/pubmed/104153

Banerjee, A., Roychoudhury, A., 2019. Rice responses and tolerance to elevated ozone, advances in rice research for abiotic stress tolerance. Elsevier Inc. https://doi.org/10.1016/b978-0-12-814332-2.00019-8

Begum, H., Alam, M.S., Feng, Y., Koua, P., Ashrafuzzaman, M., Shrestha, A., Kamruzzaman, M., Dadshani, S., Ballvora, A., Naz, A.A., Frei, M., 2020. Genetic dissection of bread wheat diversity and identification of adaptive loci in response to elevated tropospheric ozone. *Plant Cell Environ.* 2650–2665. https://doi.org/10.1111/pce.13864

Biswas, D.K., Xu, H., Li, Y.G., Liu, M.Z., Chen, Y.H., Sun, J.Z., Jiang, G.M., 2008. Assessing the genetic relatedness of higher ozone sensitivity of modern wheat to its wild and cultivated progenitors/relatives. *J. Exp. Bot.* 59(4), 951–963. https://doi.org/10.1093/jxb/ern022

Boguszewska, D., Grudkowska, M., Zagdańska, B., 2010. Drought-responsive antioxidant enzymes in potato (*Solanum tuberosum* L.). *Potato Res.* 53(4), 373–382. https://doi.org/10.1007/s11540-010-9178-6

Brauer, M., Freedman, G., Frostad, J., Van Donkelaar, A., Martin, R.V., Dentener, F., Dingenen, R. Van, Estep, K., Amini, H., Apte, J.S., Balakrishnan, K., Barregard, L., Broday, D., Feigin, V., Ghosh, S., Hopke, P.K., Knibbs, L.D., Kokubo, Y., Liu, Y., Ma, S., Morawska, L., Sangrador, J.L.T., Shaddick, G., Anderson, H.R., Vos, T., Forouzanfar, M.H., Burnett, R.T., Cohen, A., 2016. Ambient air pollution exposure estimation for the global burden of Disease 2013. *Environ. Sci. Technol.* 50(1), 79–88. https://doi.org/10.1021/acs.est.5b03709

Brosché, M., Merilo, E., Mayer, F., Pechter, P., Puzõrjova, I., Brader, G., Kangasjärvi, J., Kollist, H., 2010. Natural variation in ozone sensitivity among Arabidopsis thaliana accessions and its relation to stomatal conductance. *Plant Cell Environ.* 33(6), 914–925. https://doi.org/10.1111/j.1365-3040.2010.02116.x

Camp, W., Willekens, H., Bowler, C., et al., 1994. Elevated levels of superoxide dismutase protect transgenic plants against ozone damage. *Nat. Biotechnol.* 12, 165–168. https://doi.org/10.1038/nbt0294-165

Caregnato, F.F., Bortolin, R.C., Divan Junior, A.M., Moreira, J.C.F., 2013. Exposure to elevated ozone levels differentially affects the antioxidant capacity and the redox homeostasis of two subtropical *Phaseolus vulgaris* L. varieties. *Chemosphere* 93(2), 320–330. https://doi.org/10.1016/j.chemosphere.2013.04.084

Chen, Z., Gallie, D.R., 2005. Increasing tolerance to ozone by elevating foliar ascorbic acid confers greater protection against ozone than increasing avoidance. *Plant Physiol.* 138(3), 1673–1689. https://doi.org/10.1104/pp.105.062000

Cho, K., Shibato, J., Agrawal, G.K., Jung, Y.H., Kubo, A., Jwa, N.S., Tamogami, S., Satoh, K., Kikuchi, S., Higashi, T., Kimura, S., Saji, H., Tanaka, Y., Iwahashi, H., Masuo, Y., Rakwal, R., 2008. Integrated transcriptomics, proteomics, and metabolomics analyses to survey ozone responses in the leaves of rice seedling. *J. Proteome Res.* 7(7), 2980–2998. https://doi.org/10.1021/pr800128q

Cho, K., Tiwari, S., Agrawal, S.B., Torres, N.L., Agrawal, M., Sarkar, A., Shibato, J., Agrawal, G.K., Kubo, A., Rakwal, R., 2011. Tropospheric ozone and plants: Absorption, responses, and consequences. *Rev. Environ. Contam. Toxicol.* https://doi.org/10.1007/978-1-4419-8453-1_3

Choquette, N.E., Ainsworth, E.A., Bezodis, W., Cavanagh, A.P., 2020. Ozone tolerant maize hybrids maintain RuBisCO content and activity during long-term exposure in the field. *Plant Cell Environ.* https://doi.org/10.1111/pce.13876

Chutteang, C., Booker, F.L., Na-Ngern, P., Burton, A., Aoki, M., Burkey, K.O., 2016. Biochemical and physiological processes associated with the differential ozone response in ozone-tolerant and sensitive soybean genotypes. *Plant Biol.* 18(Suppl 1), 28–36. https://doi.org/10.1111/plb.12347

Conklin, P.L., Williams, E.H., Last, R.L., 1996. Environmental stress sensitivity of an ascorbic acid-deficient Arabidopsis mutant. *Proc. Natl. Acad. Sci. U. S. A.* 93(18), 9970–9974. https://doi.org/10.1073/pnas.93.18.9970

D'Haese, D., Vandermeiren, K., Asard, H., Horemans, N., 2005. Other factors than apoplastic ascorbate contribute to the differential ozone tolerance of two clones of *Trifolium repens* L. *Plant Cell Environ.* 28(5), 623–632. https://doi.org/10.1111/j.1365-3040.2005.01308.x

Dutilleul, C., Garmier, M., Noctor, G., Mathieu, C., Chétrit, P., Foyer, C.H., de Paepe, R., 2003. Leaf mitochondria modulate whole cell redox homeostasis, set antioxidant capacity, and determine stress resistance through altered signaling and diurnal regulation. *Plant Cell* 15(5), 1212–1226. https://doi.org/10.1105/tpc.009464

El-Khatib, A.A., 2003. The response of some common Egyptian plants to ozone and their use as biomonitors. *Environ. Pollut.* 124(3), 419–428. https://doi.org/10.1016/S0269-7491(03)00045-9

Emberson, L.D., Pleijel, H., Ainsworth, E.A., van den Berg, M., Ren, W., Osborne, S., Mills, G., Pandey, D., Dentener, F., Büker, P., Ewert, F., Koeble, R., Van Dingenen, R., 2018. Ozone effects on crops and consideration in crop models. *Eur. J. Agron.* 100, 19–34. https://doi.org/10.1016/j.eja.2018.06.002

Felzer, B.S., Cronin, T., Reilly, J.M., Melillo, J.M., Wang, X., 2007. Impacts of ozone on trees and crops. *C. R. Geosci.* 339(11–12), 784–798. https://doi.org/10.1016/j.crte.2007.08.008

Feng, Z., Kobayashi, K., Ainsworth, E.A., 2008. Impact of elevated ozone concentration on growth, physiology, and yield of wheat (*Triticum aestivum* L.): A meta-analysis. *Glob. Chang. Biol.* 14(11), 2696–2708. https://doi.org/10.1111/j.1365-2486.2008.01673.x

Feng, Z., Wang, L., Pleijel, H., Zhu, J., Kobayashi, K., 2016. Differential effects of ozone on photosynthesis of winter wheat among cultivars depend on antioxidative enzymes rather than stomatal conductance. *Sci. Total Environ.* 572, 404–411. https://doi.org/10.1016/j.scitotenv.2016.08.083

Fiscus, E.L., Booker, F.L., Burkey, K.O., 2005. Crop responses to ozone: Uptake, modes of action, carbon assimilation and partitioning. *Plant Cell Environ.* 28(8), 997–1011. https://doi.org/10.1111/j.1365-3040 .2005.01349.x

Foyer, C.H., Halliwell, B., 1976. The presence of glutathione and glutathione reductase in chloroplasts: A proposed role in ascorbic acid metabolism. *Planta* 133(1), 21–25. https://doi.org/10.1007/BF00386001

Foyer, C.H., Noctor, G., 2005. Oxidant and antioxidant signalling in plants: A re-evaluation of the concept of oxidative stress in a physiological context. *Plant Cell Environ.* 28(8), 1056–1071. https://doi.org/10.1111 /j.1365-3040.2005.01327.x

Frei, M., 2013. Lignin: Characterization of a multifaceted crop component. *Sci. World J.* 2013. https://doi.org /10.1155/2013/436517

Frei, M., 2015. Breeding of ozone resistant rice: Relevance, approaches and challenges. *Environ. Pollut.* 197, 144–155. https://doi.org/10.1016/j.envpol.2014.12.011

Frei, M., Tanaka, J.P., Chen, C.P., Wissuwa, M., 2010. Mechanisms of ozone tolerance in rice: Characterization of two QTLs affecting leaf bronzing by gene expression profiling and biochemical analyses. *J. Exp. Bot.* 61(5), 1405–1417. https://doi.org/10.1093/jxb/erq007

Frei, M., Wissuwa, M., Pariasca-Tanaka, J., Chen, C.P., Südekum, K.H., Kohno, Y., 2012. Leaf ascorbic acid level—Is it really important for ozone tolerance in rice? *Plant Physiol. Biochem.* 59, 63–70. https://doi .org/10.1016/j.plaphy.2012.02.015

Galant, A., Koester, R.P., Ainsworth, E.A., Hicks, L.M., Jez, J.M., 2012. From climate change to molecu-lar response: Redox proteomics of ozone-induced responses in soybean. *New Phytol.* 194(1), 220–229. https://doi.org/10.1111/j.1469-8137.2011.04037.x

Galmés, J., Aranjuelo, I., Medrano, H., Flexas, J., 2013. Variation in RuBisCO content and activity under variable climatic factors. *Photosynth. Res.* 117(1–3), 73–90. https://doi.org/10.1007/s11120-013-9861-y

Guidi, L., Degl'Innocenti, E., Giordano, C., Biricolti, S., Tattini, M., 2010. Ozone tolerance in *Phaseolus vulgaris* depends on more than one mechanism. *Environ. Pollut.* 158(10), 3164–3171. https://doi.org/10 .1016/j.envpol.2010.06.037

Harpaz-Saad, S., Azoulay, T., Arazi, T., Ben-Yaakov, E., Mett, A., Shiboleth, Y.M., Hörtensteiner, S., Gidoni, D., Gal-On, A., Goldschmidt, E.E., Eyal, Y., 2007. Chlorophyllase is a rate-limiting enzyme in chlo-rophyll catabolism and is posttranslationally regulated. *Plant Cell* 19(3), 1007–1022. https://doi.org/10 .1105/tpc.107.050633

Hendriks, C., Forsell, N., Kiesewetter, G., Schaap, M., Schöpp, W., 2016. Ozone concentrations and damage for realistic future European climate and air quality scenarios. *Atmos. Environ.* 144, 208–219. https:// doi.org/10.1016/j.atmosenv.2016.08.026

Iqbal, N., Nazar, R., 2015. Osmolytes and plants acclimation to changing environment: Emerging omics tech-nologies. *Osmolytes Plants Acclim. Chang Environ. Emerg. Omi Technol.*, 1–170. https://doi.org/10 .1007/978-81-322-2616-1

Kangasjärvi, J., Jaspers, P., Kollist, H., 2005. Signalling and cell death in ozone-exposed plants. *Plant Cell Environ.* 28(8), 1021–1036. https://doi.org/10.1111/j.1365-3040.2005.01325.x

Kollner, B., Krause, G.H.M., 2003. Effects of two different ozone exposure regimes on chlorophyll and sucrose contents of leaves and yield parameters of sugar beet (*Beta Vulgaris* L.) and rape (*Brassica Napus* L.). *Water Air Soil Pollut.* 144, 317–332.

Kong, W., Liu, F., Zhang, C., Zhang, J., Feng, H., 2016. Non-destructive determination of malondialdehyde (MDA) distribution in oilseed rape leaves by laboratory scale NIR hyperspectral imaging. *Sci. Rep.* 6, 35393. https://doi.org/10.1038/srep35393

Krasensky, J., Carmody, M., Sierla, M., Kangasjärvi, J., 2017. Ozone and reactive oxygen species. *eLS*, 1–9. https://doi.org/10.1002/9780470015902.a0001299.pub3

Law, M.Y., Charles, S.A., Halliwell, B., 1983. Glutathione and ascorbic acid in spinach (*Spinacia oleracea*) chloroplasts: The effect of hydrogen peroxide and of paraquat. *Biochem. J.* 210(3), 899–903. https://doi .org/10.1042/bj2100899

Macknight, R.C., Laing, W.A., Bulley, S.M., Broad, R.C., Johnson, A.A., Hellens, R.P., 2017. Increasing ascorbate levels in crops to enhance human nutrition and plant abiotic stress tolerance. *Curr. Opin. Biotechnol.* 44, 153–160. https://doi.org/10.1016/j.copbio.2017.01.011

Manninen, A.M., Holopainen, T., Lyytikàinen-Saarenmaa, P., Holopainen, J.K., 2000. The role of low-level ozone exposure and mycorrhizas in chemical quality and insect herbivore performance on Scots pine seedlings. *Glob. Chang. Biol.* 6(1), 111–121. https://doi.org/10.1046/j.1365-2486.2000.00290.x

Millar, A.H., Mittova, V., Kiddle, G., Heazlewood, J.L., Bartoli, C.G., Theodoulou, F.L., Foyer, C.H., 2003. 5Cce80a56309D8E3D44E357413B0889703E98535.Pdf 133, 443–447. https://doi.org/10.1104/pp.103 .028399.kinetics

Mills, G., Sharps, K., Simpson, D., Pleijel, H., Frei, M., Burkey, K., Emberson, L., Uddling, J., Broberg, M., Feng, Z., Kobayashi, K., Agrawal, M., 2018. Closing the global ozone yield gap: Quantification and cobenefits for multistress tolerance. *Glob. Chang. Biol.* 24(10), 4869–4893. https://doi.org/10.1111/gcb .14381

Mittler, R., 2002. Oxidative stress, antioxidants and stress tolerance. *Trends Plant Sci.* 7(9), 405–410. https:// doi.org/10.1016/S1360-1385(02)02312-9

Moldau, H., Padu, E., Bichele, I., 1997. Quantification of ozone decay and requirement for ascorbate in *Phaseolus vulgaris* L. mesophyll cell walls. *Phyt. Ann. REI Bot.* 37, 175–180

Morgan, P.B., Ainsworth, E.A., Long, S.P., 2003. How does elevated ozone impact soybean? A meta-analysis of photosynthesis, growth and yield. *Plant Cell Environ.* 26(8), 1317–1328. https://doi.org/10.1046/j .0016-8025.2003.01056.x

Mortensen, A., Skibsted, L.H., Truscott, T.G., 2001. The interaction of dietary carotenoids with radical species. *Arch. Biochem. Biophys.* 385(1), 13–19. https://doi.org/10.1006/abbi.2000.2172

Nagatoshi, Y., Mitsuda, N., Hayashi, M., Inoue, S.I., Okuma, E., Kubo, A., Murata, Y., Seo, M., Saji, H., Kinoshita, T., Ohme-Takagi, M., 2016. Golden 2-LIKE transcription factors for chloroplast development affect ozone tolerance through the regulation of stomatal movement. *Proc. Natl. Acad. Sci. U. S. A.* 113(15), 4218–4223. https://doi.org/10.1073/pnas.1513093113

Penuelas, J., Baret, F., Filella, I., 1995. Semi-empirical indices to assess carotenoids/chlorophyll a ratio from leaf spectral reflectance. *Photosynthetica* 31, 221–230

Ramankutty, N., Evan, A.T., Monfreda, C., Foley, J.A., 2008. Farming the planet: 1. Geographic distribution of global agricultural lands in the year 2000. *Global Biogeochem. Cycles* 22(1), 1–19. https://doi.org/10 .1029/2007GB002952

Royal Society 2008. Ground-level ozone in the 21st century: Future trends, impacts and policy implications. October.

Ryan, A., Cojocariu, C., Possell, M., Davies, W.J., Hewitt, C.N., 2009. Defining hybrid poplar (*Populus deltoides* x *Populus trichocarpa*) tolerance to ozone: Identifying key parameters. *Plant Cell Environ.* 32(1), 31–45. https://doi.org/10.1111/j.1365-3040.2008.01897.x

Saitanis, C.J., Riga-Karandinos, A.N., Karandinos, M.G., 2001. Effects of ozone on chlorophyll and quantum yield of tobacco (*Nicotiana tabacum* L.) varieties. *Chemosphere* 42(8), 945–953. https://doi.org/10.1016 /S0045-6535(00)00158-2

Sanmartin, M., Drooudi, P.A., Lyons, T., Pateraki, I., Barnes, J., Kanellis, A.K., 2003. Over-expression of ascorbate oxidase in the apoplast of transgenic tobacco results in altered ascorbate and glutathione redox states and increased sensitivity to ozone. *Planta* 216(6), 918–928. https://doi.org/10.1007/s00425 -002-0944-9

Sarkar, A., Singh, A.A., Agrawal, S.B., Ahmad, A., Rai, S.P., 2015. Cultivar specific variations in antioxidative defense system, genome and proteome of two tropical rice cultivars against ambient and elevated ozone. *Ecotoxicol. Environ. Saf.* 115, 101–111. https://doi.org/10.1016/j.ecoenv.2015.02.010

Sharma, Y.K., Davis, K.R., 1997. The effects of ozone on antioxidant responses in plants. *Free Radical. Biol. Med.* 23, 480–488.

Sicard, P., Serra, R., Rossello, P., 2016. Spatiotemporal trends in ground-level ozone concentrations and metrics in France over the time period 1999–2012. *Environ. Res.* 149, 122–144. https://doi.org/10.1016/j .envres.2016.05.014

Sinn, J.P., Schlagnhaufer, C.D., Arteca, R.N., Pell, E.J., 2004. Ozone-induced ethylene and foliar injury responses are altered in 1-aminocyclopropane-1-carboxylate synthase antisense potato plants. *New Phytol.* 164(2), 267–277. https://doi.org/10.1111/j.1469-8137.2004.01172.x

Singh, A., 2020. Impact of air pollutants on plant metabolism and antioxidant machinery, 57–86. https://doi .org/10.1007/978-981-15-3481-2_4

Smirnoff, N., Conklin, P.L., Loewus, F.A., 2001. Biosynthesis of a scorbic a CID in plants: A renaissance. *Annu. Rev. Plant. Physiol. Plant. Mol. Biol.* 52(1), 437–467

Stevenson, D.S., Dentener, F.J., Schultz, M.G., Ellingsen, K., van Noije, T.P.C., Wild, O., Zeng, G., Amann, M., Atherton, C.S., Bell, N., Bergmann, D.J., Bey, I., Butler, T., Cofala, J., Collins, W.J., Derwent, R.G., Doherty, R.M., Drevet, J., Eskes, H.J., Fiore, A.M., Gauss, M., Hauglustaine, D.A., Horowitz, L.W., Isaksen, I.S.A., Krol, M.C., Lamarque, J.-F, Lawrence, M.G., Montanaro, V., Müller, J.-F., Pitari, G., Prather, M.J., Pyle, J.A., Rast, S., Rodriguez, J.M., Sanderson, M.G., Savage, N.H., Shindell, D.T., Strahan, S.E., Sudo, K., Szopa, S., 2006. Multimodel ensemble simulations of present-day and near-future tropospheric ozone. *J. Geophys. Res. Atmos.* 111(D8). https://doi.org/10.1029/2005JD006338

Sun, J., Feng, Z., Ort, D.R., 2014. Impacts of rising tropospheric ozone on photosynthesis and metabolite levels on field grown soybean. *Plant Sci.* 226, 147–161. https://doi.org/10.1016/j.plantsci.2014.06.012

Tarasick, D., Galbally, I.E., Cooper, O.R., Schultz, M.G., Ancellet, G., Leblanc, T., Wallington, T.J., Ziemke, J., Liu, X., Steinbacher, M., Staehelin, J., Vigouroux, C., Hannigan, J.W., García, O., Foret, G., Zanis, P., Weatherhead, E., Petropavlovskikh, I., Worden, H., Osman, M., Liu, J., Chang, K.-L., Gaudel, A., Lin, M., Granados-Muñoz, M., Thompson, A.M., Oltmans, S.J., Cuesta, J., Dufour, G., Thouret, V., Hassler, B., Trickl, T., Neu, J.L., 2019. Tropospheric ozone assessment report: Tropospheric ozone from 1877 to 2016, observed levels, trends and uncertainties. *Elementa: Science of the Anthropocene* 7, 39. https://doi .org/10.1525/elementa.376

Turcsányi, E., Lyons, T., Plöchl, M., Barnes, J., 2000. Does ascorbate in the mesophyll cell walls form the first line of defence against ozone? Testing the concept using broad bean (*Vicia faba* L.). *J. Exp. Bot.* 51(346), 901–910. https://doi.org/10.1093/jexbot/51.346.901

Ueda, Y., Frimpong, F., Qi, Y., Matthus, E., Wu, L., Höller, S., Kraska, T., Frei, M., 2015a. Genetic dissection of ozone tolerance in rice (*Oryza sativa* L.) by a genome-wide association study. *J. Exp. Bot.* 66(1), 293–306. https://doi.org/10.1093/jxb/eru419

Ueda, Y., Siddique, S., Frei, M., 2015b. A novel gene, ozone-responsive apoplastic PROTEIN1, enhances cell death in ozone stress in rice. *Plant Physiol.* 169(1), 873–889. https://doi.org/10.1104/pp.15.00956

Ueda, Y., Uehara, N., Sasaki, H., Kobayashi, K., Yamakawa, T., 2013. Impacts of acute ozone stress on superoxide dismutase (SOD) expression and reactive oxygen species (ROS) formation in rice leaves. *Plant Physiol. Biochem.* 70, 396–402. https://doi.org/10.1016/j.plaphy.2013.06.009

Vahisalu, T., Kollist, H., Wang, Y., Nishimura, N., Valerio, G., Lamminmäki, A., Brosché, M., Moldau, H., Schroeder, J.I., Kangasjärvi, J., 2008. NIH public access 452, 487–491. https://doi.org/10.1038/ nature06608.SLAC1

Wang, J., Zeng, Q., Zhu, J., Chen, C., Liu, G., Tang, H., 2014. Apoplastic antioxidant enzyme responses to chronic free-air ozone exposure in two different ozone-sensitive wheat cultivars. *Plant Physiol. Biochem.* 82, 183–193. https://doi.org/10.1016/j.plaphy.2014.06.004

Wang, T., Xue, L., Brimblecombe, P., Lam, Y.F., Li, L., Zhang, L., 2017. Ozone pollution in China: A review of concentrations, meteorological influences, chemical precursors, and effects. *Sci. Total Environ.* 575, 1582–1596. https://doi.org/10.1016/j.scitotenv.2016.10.081

Wheeler, G., Jones, M., Smirnoff, N., 1998. The biosynthetic pathway of vitamin C in higher plants. *Nature* 393(6683), 365–369. https://doi.org/10.1038/30728

Wilkinson, S., Mills, G., Illidge, R., Davies, W.J., 2012. How is ozone pollution reducing our food supply? *J. Exp. Bot.* 63(2), 527–536. https://doi.org/10.1093/jxb/err317

Yang, T., Poovaiah, B.W., 2002. Hydrogen peroxide homeostasis: Activation of plant catalase by calcium/ calmodulin. *Proc. Natl. Acad. Sci. U. S. A.* 99(6), 4097–4102. https://doi.org/10.1073/pnas.052564899

Yendrek, C.R., Koester, R.P., Ainsworth, E.A., 2015. A comparative analysis of transcriptomic, biochemical, and physiological responses to elevated ozone identifies species-specific mechanisms of resilience in legume crops. *J. Exp. Bot.* 66(22), 7101–7112. https://doi.org/10.1093/jxb/erv404

18 Non-Enzymatic Antioxidants' Significant Role in Abiotic Stress Tolerance in Crop Plants

Bharathi Raja Ramadoss, Usharani Subramanian,
Manivannan Alagarsamy and Manu Pratap Gangola

CONTENTS

18.1 INTRODUCTION

Reactive oxygen species (ROS) are inevitable by-products of aerobic metabolism in plants and are produced in different cellular components such as chloroplasts, mitochondria, peroxisomes and other sites of a cell during photosynthesis, photorespiration and respiration (Gangola and Ramadoss 2018). ROS are highly reactive forms of oxygen and include singlet oxygen, superoxide ion radical, hydrogen peroxide and hydroxyl radicals. ROS are in balance with the antioxidant system's activity in plants during normal growth and development, thus play vital roles in cell cycle/division/expansion, meristem development, seed germination, organ growth, flower development and various other processes. However, plants being sessile organisms must cope with changing environmental conditions called "abiotic stresses" throughout their life, which can trigger the production or accumulation of ROS in plant cells (Gangola and Ramadoss 2018). Abiotic stresses include high temperature, high light intensity, drought, flood, high salinity and mineral deficiencies. The outburst of ROS in plants, termed oxidative stress, refers to an imbalance between ROS production/accumulation and the antioxidant system's activity, affecting plant performance and productivity. In addition, the world population is increasing at an alarming rate, and to feed the growing population, the productivity of crop plants must be increased which can be achieved by improving the status of the antioxidant system in plants.

18.2 WHAT ARE ANTIOXIDANTS?

Antioxidants can be categorized into enzymatic and non-enzymatic antioxidants. The enzymatic antioxidant breaks down to remove the free radicals. The antioxidant enzymes, such as superoxide dismutase (SOD), glutathione peroxidase (GPX), catalase (CAT), guaiacol peroxidase (POX) and peroxiredoxins (Prxs), catalyze ROS degradation in a plant cell. Non-enzymatic antioxidants act in two different ways: (1) directly detoxifying ROS and radicals, and (2) reducing the substrates for antioxidant enzymes. The non-enzymatic antioxidative defense system consists of ascorbate/ascorbic acid, glutathione, tocopherol, carotenoids, flavonoids, proline, raffinose family oligosaccharides and sugar alcohols. Non-enzymatic antioxidants are found to be present in all the cellular compartments and act directly towards the detoxification of radicals and ROS. Some of the non-enzymatic antioxidants can reduce the substrates for antioxidant enzymes. Important non-enzymatic antioxidants, localization and important functions in the plant cell are mentioned in Table 18.1. The different transgenic studies have been carried out to target different genes involved in the antioxidant system. Some of the transgenic crops developed by targeting components of the antioxidant system are presented in Table 18.2. In this chapter, the role of various non-enzymatic antioxidants in abiotic stress tolerance in plants and their pathways to modulate their concentration in different crop plants have been discussed.

18.3   ASADA-HALLIWELL PATHWAY OR ASCORBATE-GLUTATHIONE PATHWAY – AN OVERVIEW

The ROS scavenging in plants is controlled by ascorbate-glutathione cycle (AsA-GSH) in cellular components including chloroplasts, mitochondria, cytosol, apoplast and peroxisomes. This pathway, also called Asada-Halliwell pathway, is mainly involved in the antioxidant defense mechanism that detoxifies hydrogen peroxide (H_2O_2) that is produced as a metabolic waste by-product (Anjum et al. 2010). This cycle involves ascorbate, glutathione and NADPH and enzymes associated with these metabolites. In the first step of the reaction, H_2O_2 is reduced by reacting with ascorbate (electron donor) to water in a reaction catalyzed by ascorbate peroxidase (APX). The monodehydroascorbate (oxidized ascorbate) is regenerated by enzyme

TABLE 18.1

Important Non-Enzymatic Antioxidants, Localization and Key Functions in a Plant Cell

Non-Enzymatic Antioxidants	Functions	Localization
Ascorbate	• Scavenges $^1O_2$, $O_2^{\bullet-}$ and $OH^\bullet$ • Regeneration of tocopherol • Dissipation of excess excitation energy	cyt, chl, vac, mito, cw
Glutathione	• Regulates cellular heavy metal concentration • Induces the expression of stress-responsive genes • Scavenges H_2O_2, $^1O_2$, $O_2^{\bullet-}$ and $OH^\bullet$ • Regenerates ascorbate *via* enzymatic or non-enzymatic pathway	cyt, chl, vac, mito, ER
Tocopherol	• Maintains the integrity of photosystem II • Prevent the chain propagation step of lipid peroxidation • Scavenger of singlet oxygen	chl and lipid rich environment
Carotenoid	• Scavenger of singlet oxygen and prevents its production • Stabilizes light-harvesting complex and thylakoid membrane • Dissipation of excess excitation energy as heat	plastid
RFOs and sugar alcohols	• Detoxification of reactive oxygen species • Act as osmoprotectants	cyt, vac

* chl, mito, cyt, vac, cw and ER stand for chlorophyll, mitochondrion, cytosol, vacuole, cell wall and endoplasmic reticulum, respectively. RFOs stand for raffinose family oligosaccharides.

TABLE 18.2

Transgenic Crops Developed by Targeting Components of Antioxidant System

Crop	Genes Transformed	Source	Result – Improved Tolerance to	References
Potato (*Solanum tuberosum* L.)	*StSOD1*	Potato	Increases low temperature tolerance	Che et al. (2020)
	codA	*Arthrobacter globiformis*	Enhanced drought tolerance	You et al. (2019)
	Glutathione reductase	*Arabidopsis thaliana*	Methyl viologen, drought and cadmium	Eltayeb et al. (2010)
	D-galacturonic acid reductase	Strawberry	Salinity stress	Upadhyaya et al. (2011)Venkatesh et al. (2012)
Sweet potato (*Ipomoea batatas* L.)	Cu/Zn-SOD and APX	*Manihot esculenta* and *Pisum sativum*	Methyl viologen-mediated oxidative stress, chilling stress, dehydration and air pollution stress	Lim et al. (2007)
	swpa4 peroxidase	*Sweet potato*	Enhanced tolerance to high salinity conditions	Kim et al. (2020)
Tobacco (*Nicotiana tabacum*)	*PcAPX*	Populus Tomentosa	Increased AsA content. Decreased lipid peroxidation and H2O2 contents. Increased salt and drought stress tolerance.	Cao et al. (2017)
	CaAPX	*Camellia azalea*	Enhanced MDHAR and DHAR activity. Enhanced cold and HT tolerances.	Wang et al. (2017)
	Glutathione reductase	*Pisum sativum*	Oxidative stress	Broadbent et al. (1995)
	Fe-SOD	*Arabidopsis thaliana*	Methyl viologen causing oxidative stress	Van Camp et al. (1996)
	Glutathione S-transferase	Cotton	Methyl viologen-mediated stress	Yu et al. (2003)
	dehydroascorbate reductase	Human	Methyl viologen, H_2O_2, low temperature and NaCl stress	Kwon et al. (2003)
	Ascorbate peroxidase like-1	*Capsicum annuum*	*Phytophthora nicotianae* and MV-mediated stress	Sarowar et al. (2005)
	DHAR	*Arabidopsis thaliana*	Ozone, drought, salt and polyethylene glycol stresses	Eltayeb et al. (2006)
	MDHAR	*Arabidopsis thaliana*	Ozone, salt and polyethylene glycol stresses	Eltayeb et al. (2007)

(Continued)

TABLE 18.2 (CONTINUED)

Transgenic Crops Developed by Targeting Components of Antioxidant System

Crop	Genes Transformed	Source	Result – Improved Tolerance to	References
	Glutathione transferase	*Trichoderma virens*	Cadmium	Dixit et al. (2011)
Rice (*Oryza sativa* L.)	*Δ¹-pyrroline-5-carboxylate synthetase*	*Vigna aconitifolia* L.	Salt and water stress	Zhu et al. (1998)
	Mn-SOD	*Pisum sativum* L./ yeast	Drought/salt stress	Tanaka et al. (1999)Wang et al. (2005)
	Cu/Zn-SOD	*Avicennia marina*	Methyl viologen-mediated oxidative stress, salinity stress and drought stress	Prashanth et al. (2008)
	γ-glutamylcysteine synthetase	Overexpression under inducible promoter	Methyl viologen- and salt-mediated oxidative stress	Choe et al. (2013)
Canola (*Brassica napus*)	*MnSOD*	*Triticum aestivum*	Aluminum toxicity	Basu et al. (2001)
Sugarcane (*Saccharum* spp.)	*Δ¹-pyrroline-5-carboxylate synthetase*	*Vigna aconitifolia* L.	Water stress	Molinari et al. (2007)
Maize (*Zea mays* L.)	*Mn-SOD*	*Nicotiana lumbaginifolia* L.	Chilling and oxidative stress	Van Breusegem et al. (1999)
Maize (*Zea mays* L.)	*BADH*	*Atriplex micrantha*	Enhanced salinity tolerance	Di et al. (2015)
Wheat (*Tritivumestivum*)	*Glutathione S-transferase*	*Zea mays*	Herbicide stress	Milligan et al. (2001)
	BADH	*Atriplex hortensis* L	Enhanced GB accumulation resulted in improved salt tolerance	Tian et al. (2017)
Tomato (*Lycopersicon esculentum* L.)	*Ascorbate peroxidase*	*Pisum sativum*	Chilling and salt stress	Wang et al. (2005)
Chinese cabbage (*Brassica campestris* L. ssp. Pekinensis)	*Cu/Zn-SOD and CAT*	*Zea mays*	Sulphur dioxide and salt stress	Tseng et al. (2007)
Groundnut *Arachis hypogea*	*AtHDG11*	*Arabidopsis thaliana*	Enhanced drought and salinity tolerance	Banavath et al. (2018)
Cotton *Gossypium hirsutum* L.	*ApGSMT2 g and ApDMT2 g*	*Aphanothecehalophytica*	Enhanced salt tolerance and yield improvement	Song et al. (2018)

monodehydroascorbate reductase (MDHAR). Monodehydroascorbate is radial and it can be reduced to ascorbate and dehydroascorbate. Dehydroascorbate is reduced to ascorbate with the presence of glutathione (GSH) yielding oxidized glutathione. Finally, GSSG regenerates GSH by the activity of GR by consuming an electron from NADPH (Gill and Tuteja 2010). The final products, AsA and GSH, are both strong antioxidants. However, the balance and maintenance

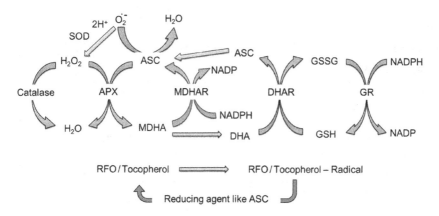

FIGURE 18.1 Asada-Halliwell/ascorbate-glutathione pathway of hydrogen peroxide scavenging and ascorbic acid regeneration involving various antioxidant enzymes. (H_2O_2 – hydrogen peroxide, APX – ascorbate peroxidase, ASC – ascorbate, MDHAR – monodehydroascorbate reductase, GSH – reduced glutathione, GSSG – oxidized glutathione, MDHA – monodehydroascorbate, DHA – dehydroascorbate), GR – glutathione reductase.

of their redox state are essential to confer resistance in crop plants (Anjum et al. 2010, 2014; Gill and Tuteja 2010) (Figure 18.1).

18.4 NON-ENZYMATIC COMPONENTS OF THE ANTIOXIDANT SYSTEM

18.4.1 Ascorbic Acid (AsA)

Ascorbate (also known as vitamin C or L-*threo*-hexenon-1,4-lactone) is a well-known non-enzymatic and soluble antioxidant abundantly found in the cytosol, chloroplast, vacuole, mitochondria and cell wall of a plant cell (Smirnoff 1996; Potters et al. 2002; Shao et al. 2007). It is also ascribed as one of the most powerful antioxidants due to its electron-donating capacity in various enzymatic and non-enzymatic reactions (Winkler et al. 1994; Foyer and Noctor 2011). In leaves, ascorbate shares >10% of soluble carbohydrates which is even more than chlorophyll (Noctor and Foyer 1998). The protective effect of ascorbic acid under various stress conditions with a different method of exogenous application are mentioned in Table 18.3.

The reducing and acidic properties of ascorbate are attributed to ene-diol group at C2/3 and hydroxyl group at C3 (Potters et al. 2002). Ascorbate oxidation synthesizes monodehydroascorbate (MDHA) which further disproportionate into ascorbate and dehydroascorbate (DHA). DHA is very unstable above pH 7, therefore, the total ascorbate pool is maintained in the reduced state. During normal growth conditions, 90% of the total ascorbate pool is maintained as a reduced state (Blokhina et al. 2003; Foyer and Noctor 2011). During stress conditions, ascorbate can directly scavenge ROS ($^1O_2$, $O_2^{\bullet-}$ and $OH^{\bullet}$) and helps to regenerate α-tocopherol from the tocopheroxyl radical thus protects the membrane. Ascorbate peroxidase (APX), MDHAR and DHAR are key enzymes required for ROS detoxification and ascorbate recycling (Winkler et al. 1994; Smirnoff 1996; Apel and Hirt 2004). Ascorbate also participates in the dissipation of excess excitation energy by being a cofactor of violaxanthin de-epoxidase (Gill and Tuteja 2010).

Exogenous application of AsA through foliar spraying, root feeding and seed pre-treatment effectively regulates antioxidative metabolism in plants (Athar et al. 2008). Several studies have reported that AsA mediates regulation of antioxidant defense metabolism in various crop plants grown under different stress condition such as salinity (okra, canola, barley) (Raza et al. 2013; Bybordi 2012; Agami 2014) and drought in canola (Shafiq et al. 2014). This shows the involvement of the activation of different antioxidant enzymes and the antioxidant defense system. Iron containing enzymes,

TABLE 18.3

Protective Effect of Ascorbic Acid under Various Stress Conditions with Different Methods of Exogenous Application

Plant species	AsA concentration	Method of application	Protective effects	References
Sunflower (*Helianthus annuus* L.)	100 and 200 mg L^{-1}	Pre-sowing treatment	Recorded significantly enhanced rate of germination, seedling biomass, plumule length under drought condition.	Ahmed et al. (2014 a)
	1 and 2 mM	Pre-sowing treatment	Enhanced germination rate and percentage, seed stamina index and fresh and dry weight under drought condition.	Fatemi (2014)
	100 and 200 mg L^{-1}	Foliar spray	Stearic acid, linoleic acid and palmitic acid concentration and oil yield under drought condition.	Ahmed et al. (2013)
	200 mgL-1	Foliar application	Enhanced salt tolerance	Noreen et al. (2017)
Barley (*Hordeum vulgare* L.)	1 mM	Pre-sowing treatment	Enhanced proline, RWC, chlorophyll and enzymatic antioxidants under salt stress.	Agami (2014)
	0.5 mM AA	Through rooting medium	Expression of salt-stress responsive genes including HvDRF2, HvWRKY12, HvBAS1, HvDHN3 and HvNHX1 in root and leaf of barley seedlings under salt stress.	Uçarlı (2019)
Wheat (*Triticum aestivum* L.)	200 mg L^{-1}	Foliar spray	Increased chlorophyll a and b, total soluble proteins, carbohydrates and carotenoids when exposed to drought condition.	Hussein et al. (2015)
	500 mg L^{-1}	Foliar spray	Recorded higher grain yield and yield related components under drought condition.	Mohamed (2013)
	100 µL L-1	Foliar application	Maintenance of chlorophyll and ascorbic acid content with higher APX activity under water deficit condition indicating better tolerance level.	Dwivedi et al. (2018)
Canola (*Brassica napus* L.)	50, 100 and 150 mg L^{-1}	Foliar spray	Improved shoot and root growth and weight under drought stress	Shafig et al. (2014)
Faba bean (*Vicia faba* L.)	200 and 400 mg L^{-1}	Foliar spray	Enhanced level of total carbohydrates, proteins, and solute concentration and enhanced level of Mg2+, Ca2+, P and K under salinity stress conditions.	Sadak et al. (2010)
Maize (*Zea mays* L.)	50, 100 and 150 mg L^{-1}	Foliar spray	Higher stem and leaf dry weights and leaf fresh weight as well as grain weight was recorded under drought conditions.	Dolatabadian et al. (2010)
	20 and 40 mg L^{-1}	Foliar and pre-sowing	Enhanced seedling growth and survival, chlorophyll b, membrane stability, leaf relative water content under cold stress conditions.	Ahmad et al. (2014 b)
	0.1, 0.3, and 0.5 mM	Foliar application	Improved photosynthetic processes, osmolytes and antioxidant defense systems alleviated the negative effects of Cadmium stress.	Zhang et al. (2019)

(Continued)

TABLE 18.3 (CONTINUED)
Protective Effect of Ascorbic Acid under Various Stress Conditions with Different Methods of Exogenous Application

Plant species	AsA concentration	Method of application	Protective effects	References
Sorghum (*Sorghum bicolor* L.)	50 and 100 mg L⁻¹	Foliar and pre-sowing	Improves germination percentage, thickness of xylem and phloem and leaf blade under salinity stress conditions.	Arafa et al. (2009)
Common bean (*Phaseolus vulgaris* L.)	100 mM	Foliar spray	Improved total chlorophyll contents and decreased ABA concentration during salinity stress.	Dolatabadian et al. (2009)
	200, or 400 mg L⁻¹AsA	Foliar spray	Enhanced water stress tolerance, increased enzymatic antioxidants (peroxidase), secondary metabolites (phenolic, flavonoids and tannins), malondialdehyde (MDA) and crop water productivity	Gaafar et al. (2020)
Safflower (*Carthamus tinctorius* L.)	55, 110 and 165µM	Pre-sowing Treatment	Enhanced germination percentage, seedling fresh and dry weights increased shoot and root lengths under salt stress condition.	Razaji et al. (2012)
	100 mg L⁻¹ and 150 mg L⁻¹	Foliar application	Enhanced shoot and root fresh and dry weights, plant height, chlorophyll and AsA contents as well as the activity of peroxidase (POD) enzyme under water deficit conditions.	Farooq et al. (2020)

APXs, use AsA as an electron source to dismutase H_2O_2 to H_2O and molecular oxygen (Mittler et al. 2004; van Doorn and Ketsa 2014). Expression of APX gene is affected by various abiotic stresses such as drought, light intensity and duration of salinity and abscisic acid (ABA) synthesis (Caverzan et al. 2012). This is highly modulated among different crop species depending upon the duration, nature and intensity of stress and plant type (Caverzan et al. 2012). Different isoforms of APX have been characterized for their specific function. Among three cytosolic APXs, APX1 and APX2 are well characterized for their enzymatic function (Suzuki et al. 2013). Expression of APX1 in *Arabidopsis* roots was found to catalyze H_2O_2 detoxification in the presence of electron source AsA (Fourcroy et al. 2004; Noctor et al. 2014). Recent studies have confirmed the involvement of APX1 in regulating oxidative stress (Maruta et al. 2012; Zhang et al. 2015). Further studies by Suzuki et al. (2013) confirm that APX1 is the most abundant form of APX involved in ROS regulation and in a protective function in *Arabidopsis thaliana*. The role of APX2 in photosynthetic electron chain mediated production of H_2O_2 was identified by Fryer et al. (2003). Moreover, the role of APX2 under various abiotic stress conditions such as heat, light and salinity has been recently reported in *Arabidopsis* (Ren et al. 2010; Jung et al. 2013; Suzuki et al. 2013). Recently, Chen et al. (2014) reported that the role of APX6 (the new form of APX) is in maintaining seedling vigor under oxidative stress.

Besides the antioxidative function, ascorbate also regulates cell division, the progression from Gap 1 (G1) to Synthesis (S) phase and cell elongation (Noctor and Foyer 1998). It is also essential for hydroxylation of proline during biosynthesis thus mediates root elongation, cell vascularization and cell wall expansion (Smirnoff 1996; Noctor and Foyer 1998). Therefore, the induced level of ascorbic acid is required to scavenge the harmful effects of ROS and regulate proper metabolic functioning. Exogenous application of ascorbic acid induces heat stress tolerance and chilling tolerance by modulating metabolism, osmolytes, antioxidants and transcriptional regulation of catalases and heat shock proteins in tomato (Alayafi 2020; Elkelish et al. 2020). Similar results were also observed for water stress tolerance in different cultivars of safflower (Farooq et al. 2020).

18.4.2 GLUTATHIONE (GSH)

Glutathione was first identified in 1888, and its structure was discovered after 1935. Glutathione is a universal tripeptide (γ-glutamyl-cysteinyl-glycine; γ-Glu-Cys-Gly; GSH) that contains long hydrophilic groups. The balance between two forms of glutathione, tripeptide glutathione (γ-glutamyl-cysteinyl-glycine, GSH) and glutathione disulfide (GSSG) is one of the key components to maintaining redox homeostasis in a plant cell (Noctor and Foyer 1998; Foyer and Noctor 2011). GSH is a crucial low molecular weight non-protein thiol that participates in ROS scavenging mechanisms (Potters et al. 2002). Being a reducing agent, GSH contributes to various biological processes, viz. cell division, sulfur metabolism, signal transduction, synthesis of protein and nucleic acid, the precursor of phytochelatins (which assist in regulating cellular heavy metal concentration), detoxification of xenobiotics and induction of stress-related genes (Potters et al. 2002; Blokhinaet al. 2003). It has been reported from almost all cellular organelles like chloroplast, mitochondria, endoplasmic reticulum, vacuole and cytosol (Gill and Tuteja 2010).

GSH synthesis in plants takes place in the chloroplast, cytosol and mitochondria (Mahmood et al. 2010; Zechmann and Müller 2010). However, a major two-step biochemical process occurs in the chloroplast. The synthesis of GSH is mediated through two enzymes, γ-glutamylcysteine synthetase (γ-ECS) and glutathione synthase, with the presence of three amino acids, namely glutamate, cysteine and glycine. In the first step of GSH biosynthesis, γ-carboxyl group of glutamate and α-amino group of cysteine forms an amide bond to provide γ-ECS, and the γ-ECS enzyme catalyzes this reaction. In the second step, GSH is formed in the presence of GSHS enzyme by amide bond formation between α-carboxyl group of the cysteinein γ-glutamylcysteine and the α-amino group of glycine to form GSH (Xiang and Oliver 1999). γ-ECS enzyme formed in the plastids were also transported into the cytosol and served as a precursor to GSH biosynthesis. Transported γ-ECS were used to synthesize GSH and transported back to mitochondria (Mahmood et al. 2010).

GSH is characterized for its high reductive potential due to the presence of central nucleophilic cysteine residue that detoxifies H_2O_2 and non-enzymatically scavenges $^1O_2$, $O_2^{•-}$ and $OH^•$. GSH can either makes adducts with reactive electrophiles via glutathionylation or converted into GSSG by reacting with ROS, thus protecting lipid, protein and nucleic acid, against oxidative stress (Foyer and Noctor 2011). It also supports the regeneration of other potential water-soluble antioxidants like ascorbic acid via the ascorbate-glutathione cycle (a part of Asada-Halliwell pathway) (Potters et al. 2002). DHA can also be recycled into ascorbate by GSH following a non-enzymatic pathway, but it works only at pH >7 and GSH concentration >1mM. Maintenance of GSH level either by *de novo* synthesis or by a glutathione reductase catalyzed reaction is a prerequisite for regulating the redox state of the cell (Foyer and Noctor 2011). Consequently, GSH/GSSG ratio is reported to be influenced by drought, salinity, chilling and heavy metal stress. The protective role of glutathione under various stress conditions is presented in Table 18.4.

GSH enhances various abiotic stress tolerance such as salinity, drought, high temperature (HT), low temperature and metal toxicity (Hasanuzzaman et al. 2011; Luo et al. 2011; Hasanuzzaman and Fujita 2013). Exogenous application of GSH improves various abiotic stress tolerances such as salinity tolerance in canola and onion (Khattab 2007; Aly-salama and Al-Mutawa 2009), cadmium tolerance in barley (Wang et al. 2011), drought tolerance in *Arabidopsis* (Chen et al. 2012) and drought, salt and aluminum tolerance (Nahar et al. 2016; 2015 a; 2015 b).

GSH reduces oxidative stress, protects the plasma membrane and prevents lipid peroxidation. Increased seedling establishment and growth was obtained by overexpressing glutathione-S-transferase (GST) and glutathione peroxidase (GPX) biosynthesis genes in tobacco (Roxas et al. 1997). In another study, El-Shabrawi et al. (2010) reported significantly higher GSH levels and both

TABLE 18.4

Protective Role of Glutathione under Various Stress Conditions

Plant Species	Glutathione Supplementation	Stress Level	Protective Effects	References
Salinity				
Brassica napus vs. Serw and Pactol	100 mg L^{-1} GSH, seed soaking for 24 h	100 and 200 mM NaCl	1. Improved shoot and root length 2. Increased level of SOD, CAT, POX, GPX and APX activities 3. Enhanced level of total soluble carbohydrates, soluble proteins, phenols and free amino acid 4. Improved level of total photosynthetic pigment contents	Khattab (2007)
Vigna radiata cv. Binamoog-1	1 mM GSH, 24 and 48 h	200 mM NaCl, 24 and 48 h	1. Reduction of MDA and H_2O_2 content, $O_2{}^-$ generation rate and LOX activity 2. Enhanced level of Chl b and total Chl contents in leaves 3. Enhanced SOD, APX, GR, GST and GPX enzyme activities	Nahar et al. (2015 a)
Oryza. sativa	2 mM GSH - foliar application	200 mM NaCl)	1. Increased grain filling 2. Increased spikelet fertility 3. Increased single plant yield	Hussain et al. (2016)
Drought				
Cicer arietinum and *C. reticulatum* Ladiz.	10 and 100 mM, foliar spray	withheld irrigation for 10 days	1. Decreased CAT, APX and GR activities 2. Increased AsA content	Çevik and Ünyayar (2015)
V. radiata cv. Binamoog-1	1 mM GSH, 24 and 48 h	Drought (25% PEG, −0.7 MPa)	1. Decreased MDA concentration, $O_2{}^-$ generation rate and LOX activity 2. Enhanced SOD, CAT, APX, MDHAR, DHAR, GR, GST and GPX enzyme activities	Nahar et al. (2015 b)
Heat				
Vigna radiata cv. Binamoog-1	0.5 mM GSH, 24 h seedling pretreatment	High temperature (42 °C, 24 h and 48 h)	1. Enhanced level of Chlorophyll b and total Chlorophyll concentration 2. Decreased MDA and H_2O_2 contents 3. Decreased CAT, GPX, DHAR, MDHAR and GR activities	Nahar et al. (2015 c)
Metal toxicity				
Oryza sativa lines 117 and 41	300 µM GSH	100 µM Cr^{6+}	1. Enhanced K, Mg, Fe and Mg contents in both shoot and root 2. Decreased concentration of Cr contents in both shoot and root 3. Enhanced level of GSH contents in leaf, stem and root	Qiu et al. (2013)
Hordeum vulgare cvs. Dong 17 and Weisuobuzhi	20 mg L^{-1} GSH	5.0 µM CdCl$_2$	1. Improved growth inhibition 2. Decreased level of Cd concentrations in leaves and roots 3. Enhanced net photosynthetic rate P(n), G(s), transpiration rate T(r) and stomatal conductance	Wang et al. (2011)
Gossypium hirsutum cvs. Coker312 and TM-1	50 µM GSH	50 µM Cd, 7 days	1. Decreased concentration of various Stress markers (TTC reduction, TSP, MDA, H_2O_2) 2. Decreased leaf Cd concentration 3. Increased SOD, POD, APX and GR enzymatic activities	Daud et al. (2016)

Glyoxalase I and Glyoxalase II activities in pokkali (salt tolerant genotype) than in IR64 (salt sensi-tive). Besides this, pokkali also exhibited a higher GSH/GSSG ratio and ascorbate (AsA)/dehydro-ascorbate (DHA) ratio along with higher activity for CAT, POX, SOD and GPX compared to IR64. These enhanced parameters in pokkali alleviate salt-induced damage to make the seedlings grow under salt conditions (El-Shabrawi et al. 2010).

In tobacco, increased GSH level, GSH redox state and activity of GPX, glutathione-S-transferase (GST) and Gly I were found to be correlated with salt tolerance (Hoque et al. 2008). External applica-tion of 0.5mM GSH improved the onion bulb epidermis damage, enhanced membrane absorptivity, and maintained cell integrity and viability when exposed to salt stress of 150 mM NaCl. In another study, GSH-treated maize plants showed reduced oxidative destruction, increased water retention and antioxidant enzyme, H+-adenosine triphosphatase (V-H+-ATPase) and H+-pyrophosphatase (V-H+-PPase) compared to control genotypes (Pei et al. 2019).

18.4.3 Tocopherol

Tocopherols are described as lipophilic antioxidants and are found as an integral part of biologi-cal membranes (Pandhair and Sekhon 2006). Tocopherols are of different types, among which α-tocopherol (or vitamin E) is attributed to confer the highest antioxidant activity, followed by β-, γ- and δ- tocopherols (Blokhina et al. 2003). Tocopherols protect lipid and other membrane compo-nents in chloroplast from ROS thus maintain the integrity of PS II (Shao et al. 2007). Tocopherols can also react with compounds like RO• and ROO•, preventing the propagation step of lipid peroxi-dation. It also acts as an ROS scavenger mainly for $^1O_2$ (Blokhinaet al. 2003). In addition to anti-oxidant functions, tocopherols also mediate the structure, fluidity and permeability of membranes (Wang and Quinn 2000).

Many previous reports concluded that alteration of tocopherol concentration directly or indi-rectly plays a significant role in abiotic stress tolerance. The alteration of tocopherol in plants dur-ing abiotic stress takes place in two different processes. The first process, during stress tocopherol synthesis, increases to get rid of ROS to avoid oxidative damage. In the second process, net tocoph-erol loss occurs due to a critical stress condition that degrades tocopherol, and synthesis fails to compensate for the amount of tocopherol. Because tocopherol scarcity cannot be reimbursed by rapid biosynthesis, this results in lipid peroxidation and, thus, leads to cell death. The first and second process is evident for stress tolerant and stress sensitive genotypes respectively (Munñe-Bosch 2005). These antioxidant properties are based on the quenching of singlet oxygen through the charge transfer mechanism.

α-tocopherol plays a significant role in reducing oxidative damage and increasing water and nutrients available to support growth, development and yield of the plants under various abiotic stress conditions (Shao et al. 2007). A single α-tocopherol molecule can neutralize up to 120 singlet oxygen molecules. Photosynthesis and metabolism serve as a potential source for ROS generated lipid peroxidation in plants. The amount of α-tocopherol in plants increases in response to various abiotic stresses (Noctor 2006). α-tocopherols can also scavenge lipid peroxide radicals to yield tocopheroxyl radical which is recycled back to the corresponding α-tocopherol by reacting with ascorbate and other antioxidants (Igamberdiev and Hill 2004).

Some previous studies revealed that foliar application of α-tocopherol played a crucial role in abiotic stress. The increased level of tocopherols in rice was associated with AsA and was con-firmed as one of the main responses of water stress (Chool Boo and Jung 1999). Oxidative stress in plants activates/alters the expression of genes responsible for tocopherol biosynthesis. Lipid-soluble antioxidant α-tocopherol is associated with the biological membrane of cells, mainly the membrane of the photosynthetic machinery (Mullineaux et al. 2006; Halliwell 2006). α-tocopherol concentra-tion in *Arabidopsis* showed a dramatic increase in response to light (Giacomelli et al. 2007). The α-tocopherol protects plants from photoinhibition and photooxidative stress (Abbasi et al. 2007; Havaux et al. 2005).

Overexpressing *tocopherol cyclase* (*IbTC*) in sweet potato revealed that α-tocopherol was the major form of tocopherol in sweet potato tissues. Compared to non-transgenic plants, transgenic plants showed increased tolerance to various environmental stresses including drought, salt and oxidative stresses. These findings suggest that *IbTC* is a key candidate gene for developing climate-resilient sweet potato varieties to combat climate change (Kim et al. 2019).

18.4.4 CAROTENOIDS

Synthesis and accumulation of carotenoids in plants influence the existence of oxygen, light and heat which affects accumulation, color degradation and oxidation. Carotenoids also represent a group of lipophilic antioxidants derived from the isoprenoid pathway. They are also known for their light-absorbing capacity between 400–550 nm of the visible spectrum (Sharma et al. 2012). Further, they pass the energy to chlorophyll molecules. Carotenoids are scavengers of $^1O_2$, an integral part of PS I system, and support the stabilization of light-harvesting complex and thylakoid membrane (Farré et al. 2010; Ramel et al. 2012). They also prevent the formation of $^1O_2$ from the triplet state of chlorophyll in the following manner (Pandhair and Sekhon 2006; Sharma et al. 2012):

$$3 \text{ chlorophyll} + 1\beta\text{-carotene} \rightarrow 1 \text{chlorophyll}$$

$$+ 3\beta\text{-carotene}*3\beta\text{-carotene}* \rightarrow 1\beta\text{-carotene} + \text{heat}$$

Zeaxanthin dissipates excess excitation energy into heat in a more efficient manner than carotenoids (Pandhair and Sekhon 2006). Carotenoids protect photosynthetic machinery against photooxidative stress in many ways. Light stress makes the light-harvesting complex (antenna system) over-excited; carotenoids that are present in β-carotene (reaction centers) and antenna system (lutein and neo-xanthin) protect the photosynthetic apparatus. Lycopene β-cyclase (LCYB) is a key candidate gene that converts the lycopene into α-carotene and β-carotene in the carotenoid biosynthesis pathway. The new allele IbLCYB2, isolated from roots of HVB-3 sweet potato overexpressed in Shangshu 19 sweet potato transgenic plants showed a significantly increased level of α-carotene, β-carotene, lutein, β-cryptoxanthin and zeaxanthin thus enhanced the level of tolerance to salt, drought and oxidative stresses (Kang et al. 2018).

18.4.5 RAFFINOSE FAMILY OF OLIGOSACCHARIDES (RFOS) AND SUGAR ALCOHOLS

The raffinose family of oligosaccharides (RFOs) is a group of soluble sugars derived from α-galactosyl of sucrose (Gangola et al. 2013; Raja et al. 2016). The most common trisaccharide raffinose is composed of raffinose, stachyose, verbascose and ajugose, and the tetra saccharide stachyose, which are actively accumulated during abiotic stress in crop plants (Raja et al. 2015; Ramadoss and Shunmugam 2014; Li et al. 2020). Carbohydrates including RFO and sugar alcohols also contribute to protect cells from oxidative damage and maintaining redox homeostasis (Couée et al. 2006; Nishizawa et al. 2008; Keunen et al. 2013). RFO has also been reported to be directly involved in ROS detoxification. During this detoxification process, RFO has been proposed to convert in their oxidized radical forms and are further regenerated by reacting with other antioxidants like ascorbic acid or flavonoids (Van den Ende and Valluru 2009). Raffinose is the first member of this family, followed by stachyose and verbascose (Gangola et al. 2013). Raffinose, as well as galactinol (precursor of RFO biosynthesis), plays an important role against oxidative stress in plants (Morsyet al. 2007; Nishizawa et al. 2008; Keunen et al. 2013) and seeds (Bailly et al. 2001; Lehner et al. 2006). Liu et al. (2007) found an enhanced tolerance to abiotic stress in *Arabidopsis* over-expressing *OsUGE-1* (*UDP-glucose 4-epimerase* from *Oryza sativa*) and associated it with an increased level of raffinose. The overexpression of galactinol synthase (*GS1*, *GS2* and *GS4*) and raffinose synthase (*RafS2*) in transgenic *Arabidopsis* increased the concentrations of galactinol and raffinose leading to increased tolerance to methyl viologen treatment and salinity or chilling stress

(Nishizawa et al. 2008). This tolerance is the result of improved ROS scavenging capacity due to the accumulation of galactinol and raffinose. Further, these transgenic plants exhibited significantly lower lipid peroxidation and higher PSII activities along with an increased level of other antioxidants (Nishizawa et al. 2008; Bolouri-Moghaddam et al. 2010).

During stress, soluble carbohydrates, sugar alcohols and RFOs help in maintaining osmotic adjustments and protein and membrane stabilization (Tarczynski et al. 1993; Amiard et al. 2003). Accumulation of RFOs was observed during drought stress and they function like osmolytes to stabilize cell wall protein and maintain cell turgor and/or act as antioxidants to reduce the accumulation of ROS (Nishizawa et al. 2008; Van den Ende and Valluru 2009; Stoyanova et al. 2011; Peshev et al. 2013). An increased level of RFOs was noticed during desiccation, thus helps in stabilization of phospholipids membrane (Farrant 2007). In the same way, mannitol and cyclitols have been reported to accumulate during salt and drought stress (Keller and Ludlow 1993; Wanek and Richter 1997). The sugar compounds and their role in various abiotic stress tolerance in crop plants are listed in Table 18.5. The sugar alcohols maintain the structural integrity of the membranes by substituting their hydroxyl group with OH group of water to maintain the hydrophilic interactions with the protein and membrane lipids. The stress accumulated solutes possess the most important property of not interfering with the normal metabolic activity of the cells.

Galactinol synthase is involved in the first step of biosynthesis of RFOs. Galactinol is synthesized from the transfer of a galactosyl moiety from UDP-galactose to myo-inositol. Further, galactinol also functions as a galactosyl donor to sucrose in the biosynthesis of stachyose and raffinose through stachyose and raffinose synthase, respectively. Higher chain RFOs (degree of polymerization DP >4) were produced independently by vacuolar galactosyl transferase (GGT). Earlier studies have shown that RFOs were found to be effective in protecting photophosphorylation and electron transport of chloroplast membrane against cold and high temperature stress (Santarius 1973). Overexpression of raffinose synthase and galactinol synthase (AtGolS1, AtGolS2, AtGolS4) in transgenic *Arabidopsis* plants showed a significantly increased amount of raffinose and galactinol concentration and resulted in oxidative stress tolerance and ROS scavenging tolerance (Nishizawa et al. 2008; Bolouri-Moghaddam et al. 2010). Recently, overexpression of *ZmRAFS* in *Arabidopsis* was found to be tolerant of drought conditions. Further, *ZmRAFS* mutant maize failed to produce raffinose and accumulate galactinol, and are more sensitive to drought stress. This indicated that *ZmRAFS* and its product raffinose were involved in drought tolerance (Li et al. 2020). As discussed above, galactinol synthase is a key enzyme for the synthesis of RFOs. Transcriptional analyses of *CsGolS1* under biotic and abiotic stress revealed that *CsGolS1* gene in *Camellia sinensis* is sensitive to low temperature, drought and abscisic acid, while the other two genes, *CsGolS2* and *CsGolS3*, are sensitive to biotic stress in particular to pest attack (Zhou et al. 2017). These results suggested that *CsGolS1* was mainly involved in abiotic stress and *CsGolS2* and *CsGolS3* were involved in biotic stress.

18.4.6 FLAVONOIDS

Flavonoids are secondary metabolites that consist of a large and common group of plant phenolics with more than 5000 differently described flavonoids in six major subclasses, including flavones, flavanols, flavanones, flavanols, anthocyanidins and isoflavones. Flavonoids have long been reported to demonstrate a wide range of responses to environment interactions (D'Auria and Gershenzon 2005; Agati and Tattini 2010). The biosynthesis of flavonoids in plants is upregulated when exposed to various abiotic stress conditions (Lillo et al. 2008; Olsen et al. 2009; Agati et al. 2011). Generally, different stress produces reactive oxygen species, and it has been postulated that flavonoids were also produced to combat stress-induced oxidative damage (Mittler et al. 2004; Mittler 2006). Some conditions in plants lead to inactivation of antioxidant enzymes that upregulate the biosynthesis of flavonoids, thus constitute a secondary ROS scavenging system in plants exposed to prolonged abiotic stress conditions.

TABLE 18.5

Sugar Compounds and Their Role in Various Abiotic Stress Tolerance in Crop Plants

Sugar	Transgene/Gene	Species	Enhanced Tolerance to	References
Disaccharides				
Trehalose	Trehalose-6-phosphate synthase	*L. esculentum*	Drought, oxidative stress (paraquat or H_2O_2), salinity	Cortina and Culiáñez-Macià (2005)
	Trehalose-6-phosphate Synthase and Trehalose-6-phosphate Phosphatase	*Zea mays*	Drought	Acosta-Pérez et al. (2020)
	Trehalose-6-phosphate synthase and phosphatase	*Arabidopsis*	Drought, salinity, temperature changes	Miranda et al. (2007)
	Trehalose-6-phosphate synthase/ phosphatase (TPSP)	*Oryza sativa*	Salinity and drought	Joshi et al. (2019)
	Trehalose phosphorylase	*N. tabacum*	Drought	Han et al. (2005)
	Trehalose-6-phosphate phosphatase	*Arabidopsis*	Drought	Lin et al. (2020)
	Trehalose synthase	*N. tabacum*	Drought, salinity	Zhang et al. (2005)
	ots A, otsB	*Oryza sativa*	Tolerance to salt and drought stress	Garg et al. (2002)
Raffinose Family Oligosaccharides (RFOs)				
Galactinol	Galactinol synthase	*Arabidopsis*	Oxidative stress (paraquat), chilling, drought, salinity	Nishizawa et al. (2008)
Raffinose	α-Galactosidase	Petunia x hybrida cv Mitchell	Freezing	Pennycooke et al. (2003)
	Raffinose synthase	*Zea mays*	Drought	Li et al. (2020)
	Galactinol Synthase	*Arabidopsis*	Salinity	Shen et al. (2020)
	UDP-glucose 4-epimerase (UDP, uridine diphosphate)	*Arabidopsis*	Drought, freezing, salinity	Liu et al. (2007)
Fructans				
	Levansucrase	*N. tabacum*	Freezing	Parvanova et al. (2004)
	SacB	Sugar beet	Drought stress	Pilon-Smits et al. (1999)
	Sucrose: sucrose 1-fructosyltransferase	*N. tabacum*	Freezing	Li et al. (2007)
	Sucrose: sucrose 1-fructosyltransferase and sucrose: fructan 6-fructosyltransferase	*O. sativa*	Chilling	Kawakami et al. (2008)
	Fructan 1-fructosyl-transferase from *Helianthus tuberosus*	Tobacco	Drought	Sun et al. (2020)
Sugar Alcohols				
Mannitol	Mannitol-1-phosphate dehydrogenase	Wheat	Water stress and salinity	Abebe et al. (2003)

(Continued)

TABLE 18.5 (CONTINUED)

Sugar Compounds and Their Role in Various Abiotic Stress Tolerance in Crop Plants

Sugar	Transgene/Gene	Species	Enhanced Tolerance to	References
	Mannitol-1-phosphate dehydrogenase	*O. sativa*	Drought, salinity	Pujni et al. (2007)
	Mannitol-1-phosphate dehydrogenase	Petunia x hybrida (Hook) Vilm. cv. Mitchell	Chilling	Chiang et al. (2005)
	Mannitol-1-phosphate dehydrogenase	*P. taeda*	Salinity	Tang et al. (2005)
	Mannose 6-phosphate reductase	*Arabidopsis*	Salinity	Zhifang and Loescher (2003)
Sorbitol				
Sorbitol	S6PDH	Japanese persimmon	Salt	Gao et al. (2001)
	Glucitol-6-phosphate dehydrogenase	P. taeda	Salinity	Tang et al. (2005)
	Sorbitol-6-phosphate dehydrogenase	*D. kaki*	Salinity	Deguchi et al. (2004)
Aldose / Aldehyde reductase	MsALR	Tobacco	Chemical and drought stress	Oberschall et al. (2000)
Invertase	Apoplastic Invertase			Fukushima et al. (2001)

Flavonoid synthesis is upregulated as an outcome of a wide range of abiotic stress tolerance and in response to UV-radiation (Lillo et al. 2008; Olsen et al. 2009; Agati et al. 2011). Because most of the abiotic stresses generate reactive oxygen species, it is proposed that flavonoids are accumulated/synthesized to combat the stress-induced oxidative damage in plants. When it comes to experimental evidence, only a few experiments have been carried out to understand the role of flavonoids in abiotic stress tolerance in crop plants. Walia et al. (2005) reported differential salinity tolerance in rice genotypes when exposed to NaCl. Also, the upregulation of F3'H gene that resulted in biosynthesis of ortho-dihydroxylationB-ring "antioxidant flavonoids" was observed to be higher in the salt susceptible genotype than in the salt tolerant rice genotype. In another study, enhanced levels of the flavonoid genes' (*CHS* and *GST*) expression was recorded during drought stress, which suggests the involvement of the role of flavonoids against water stress (Vasquez-Robinet et al. 2008).

The enhanced accumulation of two well-known antioxidants, myricetin and quercetin glycosides, was recorded in the salt sensitive *Myrtus communis* compared to the salt tolerant *Pistacia lentiscus* (Tattini et al. 2006) and are believed to be involved in the peroxidase-catalyzed enzymatic reduction of H_2O_2 in the palisade parenchyma cells. Induction of flavonoid biosynthesis in response to mild ozone stress suggested that counteracting O_3 induced oxidative damage. Flavonoid synthesis is affected by various abiotic stresses. Flavonoids or gene expression affected by treatment are presented in Table 18.6. A recent study regarding the functional characterization of flavonoid 3'-hydroxylase (CsF3'H) from *Crocus sativus* shows that it plays an important role in substrate specificity and overexpression in *Nicotiana benthamiana* and confers resistance to dehydration and UV-B stress through dihydroflavonol synthesis (Baba and Ashraf 2019).

TABLE 18.6

Changes in Flavonoid Metabolism in Response to Various Abiotic Stresses

Crop	Stress Treatment	Response	References
Lemnagibba	DMCU/DBMIB + cold + different light environments	Flavonoids accumulate in reaction to an electron transport chain redox signal, that can function independently from the ROS signal	Akhtar et al. (2010)
Ligustrum vulgare	High light + NaCl + UV-exclusion	Biosynthesis of quercetin 3-O-rutinoside, luteolin 7-O-glucoside in the same way induced by UV-radiation and root zone salinity. Accumulation of di-hydroxylated flavonoids upregulated by excess light even in absence of UV-radiation	Agati et al. (2011)
Ligustrum vulgare	High light + UV-exclusion	Enhanced accumulation of quercetin and luteolin in response to high light, even in the absence of UV-radiation	Agati et al. (2009)
Brassica oleracea	Cold	Cold enhances higher quercetin to kaempferol ratio in different cultivars, and builds up the tissue antioxidant activity	Schmidt et al. (2010) and Zietz et al. (2010)
Solanum lycopersicon	Nitrogen deprivation + cold + different light intensities	Enhanced quercetin biosynthesis was found in the low nitrogen + cold + high light treatment. Cold, nitrogen deprivation and light had a synergetic effect on PAL, CHS, F3'H, FLS upregulation	Løvdal et al. (2010)
Trifolium pratense	Mild O_3-exposure	Mild ozone stress increased total leaf phenolics	Saviranta et al. (2010)
Vitis vinifera	UV-exclusion + ABA or water control	Quercetin and kaempferol accumulation were reduced by UV-exclusion and enhanced by ABA	Berli et al. (2010)
Olea europaea	NaCl + high light	Enhanced flavonoid accumulation was detected by high light and salinity.	Remorini et al. (2009)
A. thaliana ecotype Colombia	Nitrogen deprivation + cold	Enhanced accumulation of quercetin than kaempferol	Olsen et al. (2009)
Arnica montana	Cold + UV-exclusion	Ortho-dihydroxy flavonoid biosynthesis was induced by low temperature.	Albert et al. (2009)

18.4.7 PROLINE

Several studies have shown that proline accumulation in plants in response to various abiotic stresses leads to abiotic stress tolerance (Shin et al. 2016; Bres et al. 2016; Mari et al. 2018; Nounjan et al. 2018; Abdel Latef et al. 2020 a, b). Proline accumulation in plants has been reported to be higher in stress-resistant/tolerant genotypes indicating the osmoregulatory mechanism of proline (De la Torre-González et al. 2018; Akbari et al. 2018). In rice, increased accumulation of proline was recorded in salt resistant/tolerant lines than sensitive genotypes (Nounjan et al. 2018). Osmotic adjustment by increasing the production of proline was recorded in salt tolerant groundnut cell lines (Jain et al. 2001). In the same way, overproduction of proline was recorded in pistachio (Akbari et al. 2018), *Anoectochilus* genus (Mei et al. 2018), pelargonium (Bres et al. 2016), tomato and lentil (De la Torre-González et al. 2018; Gaafar and Seyam 2018) against salt stress. An enhanced level of proline provides better protection for the plants against oxidative damage, but not osmotic adjustment, under salt stress conditions.

TABLE 18.7

Accumulation of Proline in Response to Abiotic Stresses

Crop	Stress	Accumulation and Mechanism of Proline	References
Wheat	Salinity stress	Increased accumulation of proline (osmotic adjustment)	Rady et al. (2019)
	Salinity stress	Increased accumulation of proline (increased in sensitive genotype than tolerant ones)	Tavakoli et al. (2016)
	Heat stress	Reduced accumulation of proline (proline accumulation was not correlated with heat tolerance, rather antioxidant enzymes)	Kumar et al. (2012)
	Heat stress	Enhanced level of proline (osmotic adjustment and proline production correlated with heat resistance)	Khan et al. (2015)
Rice	Salinity stress	Increased accumulation of proline (osmotic adjustment)	Nounjan et al. (2018)
	Heat stress	Increased accumulation of proline (although it further increased in heat-sensitive cultivar, it decreased in heat-tolerant ones)	Sánchez-Reinoso et al. (2014)
	Heat stress	Enhanced level of proline, correlated with thermotolerance	Ali et al. (2016)
	Cold stress	Enhanced level of proline was detected in cold-sensitive genotype than tolerant one	Aghaee et al. (2011)
Sorghum	Heat stress	Increased level of proline and osmotic adjustment	Gosavi et al. (2014)
	Cold stress	Enhanced level of proline (osmoprotectant, proline is higher in cold tolerant cultivar)	Vera-Hernández et al. (2018)
Soybean	Cold stress	Increased proline content and behaving like osmoprotectant	Borowski and Michałek (2014)
Chickpea	Cold stress	Increased level of proline in cold resistant cultivars	Tatar et al. (2013)
Barley	Drought stress	Increased proline content was not correlated with drought tolerance	Dbira et al. (2018)
Arabidopsis	Freezing and drought stress	Enhanced level of proline associated with stress tolerance	Ren et al. (2018)

Several other studies have also demonstrated the role of proline accumulation in various abiotic stress tolerance in plants. In wheat, water and temperature stress increased the production of proline, which regulated the physiological parameters such as leaf water potential and consequently improved the grain yield of wheat (Ahmed et al. 2017). Even though most of the studies related to the overproduction of proline with stress tolerance are specific to salt and drought stresses, proline accumulation was also recorded in cold tolerance (Table 18.7). For instance, cold tolerant/resistant sorghum and almond possess a higher proportion of proline accumulation than susceptible ones (Afshari and Parvane 2013; Vera-Hernández et al. 2018). In another study, chickpea cultivars exposed to cold temperature also produced an elevated level of proline, suggesting the role of proline in cold stress tolerance. Furthermore, rice genotypes tolerant to salt and cold stresses produce more proline than sensitive ones (Benitez et al. 2016). In this study, biosynthesis gene transcripts that are related to proline accumulation are significantly correlated with salinity tolerant genotypes under salinity stress, and this correlation was high with sensitive genotype during cold stress, thus implies the involvement of both glutamate and ornithine pathways for proline accumulation in tolerant genotypes both in salt and cold stress. On the other hand, rice genotypes sensitive to cold recorded a higher accumulation of proline (Aghaee et al. 2011). Exposure of mung bean seedlings to cold stress also reduced the endogenous proline content (Posmyk and Janas 2007). In another study, salt-sensitive wheat cultivar showed higher proline accumulation than tolerant genotypes, however, yield in salt tolerant genotypes was much better than sensitive ones (Tavakoli et al. 2016).

Different studies have shown that the accumulation of proline is related to different abiotic stress tolerance. This beneficial effect in enhanced proline accumulation is confirmed by means of the overexpressing of metabolic genes related to proline synthesis and exogenous application. Further, differential accumulation of proline with respect to stress as well as species/genotype was recorded in different studies. Also, this difference is mainly due to stress duration, stage of the stress, plant growth stage and other related environmental factors.

18.5 CONCLUSION

Due to climate change, current agriculture is facing the major problem of abiotic stress, which affects plant growth and development and results in significant yield loss. The identification and understanding of the role of non-enzymatic antioxidants and their physiological and molecular mechanisms of adaptation to various abiotic stresses in plants are important in order to cultivate varieties with improved agronomic and economic importance. During abiotic stress conditions, plants react in with different ways, in which non-enzymatic antioxidants are accumulated to provide better growth and development. Over the last few decades, several studies have pointed out the role of non-enzymatic antioxidants for abiotic stress tolerance. ROS at lower concentrations participate and regulate various physiological phenomena by mediating various signaling pathways. However, the higher concentration of ROS in plant cells leads to biomolecules' degradation, permanent cell death, apoptosis or hypersensitive response, collectively termed as the oxidative burst. Oxidative burst is the inevitable outcome of abiotic and biotic stresses. Therefore, different components of the antioxidant system have been targeted to improve plant performance under stress conditions. To date, this area has been explored a lot but some important aspects that need more attention are as follows: 1. How is ROS signaling generated and decoded as developmental cues? 2. How is the specificity of ROS signaling determined with non-enzymatic antioxidants? 3. What is the interconnection of ROS signaling with other signaling pathways? 4. What is the interaction among the various antioxidants? The next generation of analytical and molecular techniques can be utilized to fill these gaps in our understanding of ROS and plants' antioxidant systems. Besides this, new in vivo methodologies with optimized and widely applicable protocols for individual ROS will help us to deepen our knowledge in their role in abiotic stress tolerance. Implementation of new technologies, such as genome-editing technologies, high throughput genotyping, phenotyping using image analysis and biochemical methodologies, will accelerate the development of climate-resilient crop varieties based on non-enzymatic antioxidants.

REFERENCES

Abbasi, A. R., M. Hajirezaei, D. Hofius, U. Sonnewald, and L. M. Voll. "Specific roles of alpha- and gamma-tocopherol in abiotic stress responses of transgenic tobacco." *Plant Physiology* 143(4) (2007): 1720–1738.

Abdel Latef, A. A., A. Zaid, A. A. Abo-Baker, W. Salem, and M. F. Abu Alhmad. "Mitigation of copper stress in maize by inoculation with *Paenibacilluspolymyxa* and *Bacillus circulans.*" *Plants* 9(11) (2020b): 1513.

Abdel Latef, A. A., M. F. A. Abu Alhmad, M. Kordrostami, A. A. Abo–Baker, and A. Zakir. "Inoculation with Azospirillum lipoferum or Azotobacter chroococcum reinforces maize growth by improving physiological activities under saline conditions." *Journal of Plant Growth Regulation* 39(3) (2020a): 1293–1306.

Abebe, T., A. C. Guenzi, B. Martin, and J. C. Cushman. "Tolerance of mannitol-accumulating transgenic wheat to water stress and salinity." *Plant Physiology* 131(4) (2003): 1748–1755.

Acosta-Pérez, Phamela, BiankaDianey Camacho-Zamora, Edward A. Espinoza-Sánchez, Guadalupe Gutiérrez-Soto, Francisco Zavala-García, María Jazmín Abraham-Juárez, and Sugey Ramona Sinagawa-García. "Characterization of Trehalose-6-phosphate synthase and Trehalose-6-phosphate phosphatase genes and analysis of its differential expression in maize (*Zea mays*) seedlings under drought stress." *Plants* 9(3) (2020): 315.

Afshari, H., and T. Parvane. "Study the effect of cold treatments on some physiological parameters of 3 cold resistance almond cultivars." *Life Science Journal* 10(4) (2013): 4–16.

Agami, R. A. "Applications of ascorbic acid or proline increase resistance to salt stress in barley seedlings." *Biologia Plantarum* 58(2) (2014): 341–347.

Agati, Giovanni, Giovanni Stefano, Stefano Biricolti, and Massimiliano Tattini. "Mesophyll Distribution of 'antioxidant' flavonoid glycosides in Ligustrum vulgare leaves under contrasting sunlight irradiance." *Annals of Botany* 104(5) (2009): 853–861.

Agati, Giovanni, and Massimiliano Tattini. "Multiple functional roles of flavonoids in photoprotection." *The New Phytologist* 186(4) (2010): 786–793.

Agati, Giovanni, Stefano Biricolti, Lucia Guidi, Francesco Ferrini, Alessio Fini, and Massimiliano Tattini. "The biosynthesis of flavonoids is enhanced similarly by UV radiation and root zone salinity in *L. vulgare* leaves." *Journal of Plant Physiology* 168(3) (2011): 204–212.

Aghaee, A., F. Moradi, H. Zare-Maivan, F. Zarinkamar, H. Pour Irandoost, and P. Sharifi. "Physiological responses of two rice (*Oryza sativa* L.) genotypes to chilling stress at seedling stage." *African Journal of Biotechnology* 10(39) (2011): 7617–7621.

Ahmad, Ijaz, Shahzad Maqsood Ahmed Basra, and Abdul Wahid. "Exogenous application of ascorbic acid, salicylic acid and hydrogen peroxide improves the productivity of hybrid maize at low temperature stress." *International Journal of Agriculture and Biology* 16(4) (2014b): 825–830.

Ahmed, F., D. M. Baloch, M. J. Hassan, and N. Ahmed. "Role of plant growth regulators in improving oil quantity and quality of sunflower hybrids in drought stress." *Biologia* 59(2) (2013): 315–322.

Ahmed, F., D. M. Baloch, S. A. Sadiq, S. S. Ahmed, A. Hanan, S. A. Taran, N. Ahmed, and M. J. Hassan. "Plant growth regulators induced drought tolerance in sunflower (*Helianthus annuus* L.) hybrids." *Journal of Animal and Plant Sciences* 24(3) (2014a): 886–890.

Ahmed, Mukhtar, Qadir Ghulam, Farid Asif Shaheen, and Aslam Muhammad Aqeel. "Response of proline accumulation in bread wheat (*Triticum aestivum* L.) under rainfed conditions." *Journal of Agricultural Meteorology* (2017): D-14.

Akbari, Mohammad, Nasser Mahna, Katam Ramesh, Ali Bandehagh, and Silvia Mazzuca. "Ion homeostasis, osmoregulation, and physiological changes in the roots and leaves of pistachio rootstocks in response to salinity." *Protoplasma* 255(5) (2018): 1349–1362.

Akhtar, Tariq A., Hazel A. Lees, Mark A. Lampi, Daryl Enstone, Richard A. Brain, and Bruce M. Greenberg. "Photosynthetic redox imbalance influences flavonoid biosynthesis in *Lemnagibba*." *Plant, Cell and Environment* 33(7) (2010): 1205–1219.

Alayafi, Aisha Abdullah Mohammed. "Exogenous ascorbic acid induces systemic heat stress tolerance in tomato seedlings: Transcriptional regulation mechanism." *Environmental Science and Pollution Research International* 27(16) (2020): 19186–19199.

Albert, A., V. Sareedenchai, W. Heller, H. K. Seidlitz, and C. Zidorn. "Temperature is the key to altitudinal variation of phenolics in *Arnica montana* L. cv. ARBO." *Oecologia* 160(1) (2009): 1–8.

Ali, Muhammad Kazim, Abid Azhar, Haneef Ur Rehman, and Saddia Galani. "Antioxidant defence system and oxidative damages in rice seedlings under heat stress." *Pure and Applied Biology* 5(4) (2016): 1.

Aly-Salama, Karima Hamed, and M. M. Al-Mutawa. "Glutathione-triggered mitigation in salt-induced alterations in plasmalemma of onion epidermal cells." *International Journal of Agriculture and Biology (Pakistan)* 11(5) (2009): 639–642.

Amiard, Véronique, Annette Morvan-Bertrand, Jean-Pierre Billard, Claude Huault, Felix Keller, and Marie-Pascale Prud'homme. "Fructans, but not the sucrosyl-galactosides, raffinose and loliose, are affected by drought stress in perennial ryegrass." *Plant Physiology* 132(4) (2003): 2218–2229.

Anjum, Naser A., Sarvajeet S. Gill, Ritu Gill, Mirza Hasanuzzaman, Armando C. Duarte, Eduarda Pereira, Iqbal Ahmad, R. Tuteja, and Narendra Tuteja. "Metal/metalloid stress tolerance in plants: Role of ascorbate, its redox couple, and associated enzymes." *Protoplasma* 251(6) (2014): 1265–1283.

Anjum, Naser A., Shahid Umar, and Ming-Tsair Chan, eds. *Ascorbate-Glutathione Pathway and Stress Tolerance in Plants*. Springer Science & Business Media (2010).

Apel, Klaus, and Heribert Hirt. "Reactive oxygen species: Metabolism, oxidative stress, and signal transduction." *Annual Review of Plant Biology* 55 (2004): 373–399.

Arafa, A. A., M. A. Khafagy, and M. F. El-Banna. "The effect of glycinebetaine or ascorbic acid on grain germination and leaf structure of sorghum plants grown under salinity stress." *Australian Journal of Crop Science* 3(5) (2009): 294.

Athar, H., A. Khan, and M. Ashraf. "Exogenously applied ascorbic acid alleviates salt-induced oxidative stress in wheat." *Environmental and Experimental Botany* 63(1–3) (2008): 224–231.

Baba, Shoib Ahmad, and Nasheeman Ashraf. " Functional characterization of flavonoid 3'-hydroxylase, CsF3'H, from Crocus sativus L: Insights into substrate specificity and role in abiotic stress." *Archives of Biochemistry and Biophysics* 667 (2019): 70–78.

Bailly, Christophe, Catherine Audigier, Fabienne Ladonne, Marie Hélène Wagner, Françoise Coste, Françoise Corbineau, and Daniel Côme. "Changes in oligosaccharide content and antioxidant enzyme activities in developing bean seeds as related to acquisition of drying tolerance and seed quality." *Journal of Experimental Botany* 52(357) (2001): 701–708.

Banavath, J. N., T. Chakradhar, V. Pandit, S. Konduru, K. K. Guduru, C. S. Akila, S. Podha, and C. O. R. Puli. "Stress inducible overexpression of AtHDG11 leads to improved drought and salt stress tolerance in peanut (*Arachis hypogaea* L.)." *Frontiers in Chemistry* 6 (2018): 34.

Basu, U., A. G. Good, and G. J. Taylor. "Transgenic Brassica napus plants overexpressing aluminium-induced mitochondrial manganese superoxide dismutase cDNA are resistant to aluminium." *Plant, Cell and Environment* 24(12) (2001): 1278–1269.

Benitez, L. C., I. L. Vighi, P. A. Auler, M. N. do Amaral, G. P. Moraes, G. dos Santos Rodrigues, L. C. da Maia, A. M. de Magalhães Júnior, and E. J. B. Braga. "Correlation of proline content and gene expression involved in the metabolism of this amino acid under abiotic stress." *Acta Physiologiae Plantarum* 38(11) (2016): 267.

Berli, F. J., D. Moreno, P. Piccoli, L. Hespanhol-Viana, M. F. Silva, R. Bressan-Smith, J. B. Cavagnaro, and R. Bottini. "Abscisic acid is involved in the response of grape (*Vitis vinifera* L.) cv. Malbec leaf tissues to ultraviolet-B radiation by enhancing ultraviolet-absorbing compounds, antioxidant enzymes and membrane sterols." *Plant, Cell and Environment* 33(1) (2010): 1–10.

Blokhina, O., E. Virolainen, and K. V. Fagerstedt. "Antioxidants, oxidative damage and oxygen deprivation stress: A review." *Annals of Botany* 91(2) (2003): 179–194.

Bolouri-Moghaddam, Mohammad Reza, Katrien Le Roy, Li Xiang, Filip Rolland, and Wim Van den Ende. "Sugar signalling and antioxidant network connections in plant cells." *The FEBS Journal* 277(9) (2010): 2022–2037.

Borowski, Edward, and Slawomir Michalek. "The effect of chilling temperature on germination and early growth of domestic and Canadian soybean (*Glycine max* (L.) Merr.) cultivars." *Acta Scientiarum Polonorum: Hortorum Cultus* 13 (2014): 31–43.

Breś, Włodzimierz, Hanna Bandurska, Agnieszka Kupska, Justyna Niedziela, and Barbara Frąszczak. " Responses of pelargonium (Pelargonium × hortorum L.H. Bailey) to long-term salinity stress induced by treatment with different NaCl doses." *Acta Physiologiae Plantarum* 38(1) (2016): 26.

Broadbent, P., G. P. Creissen, B. Kular, A. R. Wellburn, and P. M. Mullineaux. "Oxidative stress responses in transgenic tobacco containing altered levels of glutathione reductase activity." *The Plant Journal* 8(2) (1995): 247–255.

Bybordi, Ahmad. "Effect of ascorbic acid and silicium on photosynthesis, antioxidant enzyme activity, and fatty acid contents in canola exposure to salt stress." *Journal of Integrative Agriculture* 11(10) (2012): 1610–1620.

Cao, S., X.-H. Du, L.-H. Li, Y.-D. Liu, L. Zhang, X. Pan, Y. Li, H. Li, and H. Lu. "Overexpression of Populus tomentosa cytosolic ascorbate peroxidase enhances abiotic stress tolerance in tobacco plants." *Russian Journal of Plant Physiology* 64(2) (2017): 224–234.

Caverzan, Andréia, Gisele Passaia, Silvia Barcellos Rosa, Carolina Werner Ribeiro, Fernanda Lazzarotto, and Márcia Margis-Pinheiro. "Plant responses to stresses: Role of ascorbate peroxidase in the antioxidant protection." *Genetics and Molecular Biology* 35(4) (2012): 1011–1019.

Çevik, Sertan, and Serpil Ünyayar. "The effects of exogenous application of ascorbate and glutathione on antioxidant system in cultivated Cicer arietinum and wild type C. reticulatum under drought stress." *Journal of Natural and Applied Sciences* 19(1) (2015).

Che, Y., N. Zhang, X. Zhu, S. Li, S. Wang, and H. Si. "Enhanced tolerance of the transgenic potato plants overexpressing Cu/Zn superoxide dismutase to low temperature." *Scientia Horticulturae* 261 (2020). http://www.ncbi.nlm.nih.gov/pubmed/108949.

Chen, C., I. Letnik, Y. Hacham, P. Dobrev, B. H. Ben-Daniel, R. Vanková, R. Amir, and G. Miller. "Ascorbate PEROXIDASE6 protects Arabidopsis desiccating and germinating seeds from stress and mediates cross talk between reactive oxygen species, abscisic acid, and auxin." *Plant Physiology* 166(1) (2014): 370–383.

Chen, J. H., H. W. Jiang, E. J. Hsieh, H. Y. Chen, C. T. Chien, H. L. Hsieh, and T. P. Lin. "Drought and salt stress tolerance of an Arabidopsis glutathione S-transferase U17 knockout mutant are attributed to the combined effect of glutathione and abscisic acid." *Plant Physiology* 158(1) (2012): 340–351.

Chiang, Yu-Jen, C. Stushnoff, A. E. McSay, M. L. Jones, and H. J. Bohnert. "Overexpression of mannitol-1-phosphate dehydrogenase increases mannitol accumulation and adds protection against chilling injury in petunia." *Journal of the American Society for Horticultural Science* 130(4) (2005): 605–610.

Choe, Yong-Hoe, Young-Saeng Kim, Il-Sup Kim, Mi-Jung Bae, Eun-Jin Lee, Yul-Ho Kim, Hyang-Mi Park, and Ho-Sung Yoon. "Homologous expression of γ-glutamylcysteine synthetase increases grain yield and tolerance of transgenic rice plants to environmental stresses." *Journal of Plant Physiology* 170(6) (2013): 610–618.

Chool Boo, Yong Chool, and Jin Jung. "Water deficit—Induced oxidative stress and antioxidative defenses in rice plants." *Journal of Plant Physiology* 155(2) (1999): 255–261.

Cortina, Carolina, and Francisco A. Culiáñez-Macià. "Tomato abiotic stress enhanced tolerance by trehalose biosynthesis." *Plant Science* 169(1) (2005): 75–82.

Couée, I., C. Sulmon, G. Gouesbet, and A. El Amrani. "Involvement of soluble sugars in reactive oxygen species balance and responses to oxidative stress in plants." *Journal of Experimental Botany* 57(3) (2006): 449–459.

D'Auria, John C., and Jonathan Gershenzon. "The secondary metabolism of *Arabidopsis thaliana*: Growing like a weed." *Current Opinion in Plant Biology* 8(3) (2005): 308–316.

Daud, M. K., Lei Mei, Muhammad Azizullah, Azizullah Dawood, Imran Ali, Qaisar Mahmood, Waheed Ullah, Muhammad Jamil, and S. J. Zhu. "Leaf-based physiological, metabolic, and ultrastructural changes in cultivated cotton cultivars under cadmium stress mediated by glutathione." *Environmental Science and Pollution Research* 23(15) (2016): 15551–15564.

Dbira, S., M. Al Hassan, P. Gramazio, A. Ferchichi, O. Vicente, J. Prohens, and M. Boscaiu. "Variable levels of tolerance to water stress (drought) and associated biochemical markers in Tunisian barley landraces." *Molecules* 23(3) (2018): 613.

De la Torre-González, A., D. Montesinos-Pereira, B. Blasco, and J. M. Ruiz. "Influence of the proline metabolism and glycine betaine on tolerance to salt stress in tomato (*Solanum lycopersicum* L.) commercial genotypes." *Journal of Plant Physiology* 231 (2018): 329–336.

Deguchi, Michihito, Yoshiko Koshita, Mei Gao, Ryutaro Tao, Takuya Tetsumura, Shohei Yamaki, and Yoshinori Kanayama. "Engineered sorbitol accumulation induces dwarfism in Japanese persimmon." *Journal of Plant Physiology* 161(10) (2004): 1177–1184.

Di, Hong, Yu Tian, Hongyue Zu, Xianyu Meng, Xing Zeng, and Zhenhua Wang. "Enhanced salinity tolerance in transgenic maize plants expressing a BADH gene from Atriplex micrantha." *Euphytica* 206(3) (2015): 775–783.

Dixit, Prachy, Prasun K. Mukherjee, V. Ramachandran, and Susan Eapen. "Glutathione transferase from Trichoderma virens enhances cadmium tolerance without enhancing its accumulation in transgenic Nicotiana tabacum." *PLOS ONE* 6(1) (2011).

Dolatabadian, A., S. A. M. Modarres Sanavy, and K. S. Asilan. "Effect of ascorbic acid foliar application on yield, yield component and several morphological traits of grain corn under water deficit stress conditions." *Notulae Scientia Biologicae* 2(3) (2010): 45–50.

Dolatabadian, A., S. A. M. Modarres Sanavy, and M. Sharifi. "Alleviation of water deficit stress effects by foliar application of ascorbic acid on *Zea mays* L.. " *Journal of Agronomy and Crop Science* 195(5) (2009): 347–355.

Dwivedi, S. K., G. Kumar, S. Basu, S. Kumar, K. K. Rao, and A. K. Choudhary. "Physiological and molecular aspects of heat tolerance in wheat." *SABRAO Journal of Breeding and Genetics* 50(2) (2018).

Elkelish, A., S. H. Qari, Y. S. A. Mazrou, K. A. A. Abdelaal, Y. M. Hafez, A. M. Abu-Elsaoud, G. E. Batiha, M. A. El-Esawi, and N. El Nahhas. "Exogenous ascorbic acid induced chilling tolerance in tomato plants Through modulating metabolism, osmolytes, antioxidants, and transcriptional regulation of catalase and heat shock proteins." *Plants* 9(4) (2020): 431.

El-Shabrawi, Hattem, Bhumesh Kumar, Tanushri Kaul, Malireddy K. Reddy, Sneh L. Singla-Pareek, and Sudhir K. Sopory. "Redox homeostasis, antioxidant defense, and methylglyoxal detoxification as markers for salt tolerance in Pokkali rice." *Protoplasma* 245(1–4) (2010): 85–96.

Eltayeb, Amin Elsadig, Naoyoshi Kawano, Ghazi Hamid Badawi, Hironori Kaminaka, Takeshi Sanekata, Isao Morishima, Toshiyuki Shibahara, S. Inanaga, and Kiyoshi Tanaka. "Enhanced tolerance to ozone and drought stresses in transgenic tobacco overexpressing dehydroascorbate reductase in cytosol." *Physiologia Plantarum* 127(1) (2006): 57–65.

Eltayeb, Amin Elsadig, Naoyoshi Kawano, Ghazi Hamid Badawi, Hironori Kaminaka, Takeshi Sanekata, Toshiyuki Shibahara, S. Inanaga, and Kiyoshi Tanaka. "Overexpression of monodehydroascorbate reductase in transgenic tobacco confers enhanced tolerance to ozone, salt and polyethylene glycol stresses." *Planta* 225(5) (2007): 1255–1264.

Eltayeb, Amin Elsadig, Shohei Yamamoto, Mohamed Elsadig Eltayeb Habora, Yui Matsukubo, Mitsuko Aono, Hisashi Tsujimoto, and Kiyoshi Tanaka. "Greater protection against oxidative damages imposed by various environmental stresses in transgenic potato with higher level of reduced glutathione." *Breeding Science* 60(2) (2010): 101–109.

Farooq, Ayesha, ShaziaAnwer Bukhari, Nudrat A. Akram, Muhammad Ashraf, Leonard Wijaya, Mohammed Nasser Alyemeni, and Parvaiz Ahmad. "Exogenously applied ascorbic acid-mediated changes in osmo-protection and oxidative defense system enhanced water stress tolerance in different cultivars of saf-flower (*Carthamus tinctorious* L.)." *Plants* 9(1) (2020): 104.

Farrant, Jill M. "Mechanisms of desiccation tolerance in angiosperm resurrection plants." In: Jenks M., Wood A. (ed.): *Plant Desiccation Tolerance* 51–90. Blackwell Publishing, Wallingford, UK.

Farré, Gemma, Georgina Sanahuja, Shaista Naqvi, Chao Bai, Teresa Capell, Changfu Zhu, and Paul Christou. "Travel advice on the road to carotenoids in plants." *Plant Science* 179(1–2) (2010): 28–48.

Fatemi, SeyedNabiladdin. "Ascorbic acid and it's effects on alleviation of salt stress in sunflower." *Annual Research and Review in Biology* 4 (2014): 3656–3665.

Fourcroy, Pierre, Gérard Vansuyt, Sergei Kushnir, Dirk Inzé, and Jean-François Briat. "Iron-regulated expres-sion of a cytosolic ascorbate peroxidase encoded by the APX1 gene in Arabidopsis seedlings." *Plant Physiology* 134(2) (2004): 605–613.

Foyer, Christine H., and Graham Noctor. "Ascorbate and glutathione: The heart of the redox hub." *Plant Physiology* 155(1) (2011): 2–18.

Fryer, Michael J., Louise Ball, Kevin Oxborough, Stanislaw Karpinski, Philip M. Mullineaux, and Neil R. Baker. "Control of ascorbate peroxidase 2 expression by hydrogen peroxide and leaf water status during excess light stress reveals a functional organisation of Arabidopsis leaves." *The Plant Journal: For Cell and Molecular Biology* 33(4) (2003): 691–705.

Fukushima, Eiichi, Yuuto Arata, Tsuyoshi Endo, Uwe Sonnewald, and Fumihiko Sato. "Improved salt toler-ance of transgenic tobacco expressing apoplastic yeast-derived invertase." *Plant and Cell Physiology* 42(2) (2001): 245–249.

Gaafar, Alaa A., Sami I. Ali, Mohamed A. El-Shawadfy, Zeinab A. Salama, Agnieszka Sekara, Christian Ulrichs, and Magdi T. Abdelhamid. "Ascorbic acid induces the increase of secondary metabolites, anti-oxidant activity, growth, and productivity of the common bean under water stress conditions." *Plants* 9(5) (2020): 627.

Gaafar, Reda M., and Maha M. Seyam. "Ascorbate–glutathione cycle confers salt tolerance in Egyptian lentil cultivars." *Physiology and Molecular Biology of Plants : An International Journal of Functional Plant Biology* 24(6) (2018): 1083–1092.

Gangola, Manu P., and Bharathi R. Ramadoss. "Sugars play a critical role in abiotic stress tolerance in plants." In: Shabir Wani (ed.), *Biochemical, Physiological and Molecular Avenues for Combating Abiotic Stress Tolerance in Plants*: 17–38. Academic Press (2018).

Gangola, Manu P., Yogendra P. Khedikar, Pooran M. Gaur, Monica Båga, and Ravindra N. Chibbar. "Genotype and growing environment interaction shows a positive correlation between substrates of raffinose family oligosaccharides (RFO) biosynthesis and their accumulation in chickpea (Cicer arietinum L.) seeds." *Journal of Agricultural and Food Chemistry* 61(20) (2013): 4943–4952.

Gao, Mei, Ryutaro Tao, Keisuke Miura, Abhaya M. Dandekar, and Akira Sugiura. "Transformation of Japanese persimmon (Diospyros kaki Thunb.) with apple cDNA encoding NADP-dependent sorbitol-6-phosphate dehydrogenase." *Plant Science : An International Journal of Experimental Plant Biology* 160(5) (2001): 837–845.

Garg, Ajay K., Ju-Kon Kim, Thomas G. Owens, Anil P. Ranwala, Yang Do Choi, Leon V. Kochian, and Ray J. Wu. "Trehalose accumulation in rice plants confers high tolerance levels to different abiotic stresses." *Proceedings of the National Academy of Sciences of the United States of America* 99(25) (2002): 15898–15903.

Giacomelli, Lisa, Antonio Masi, Daniel R. Ripoll, Mi Ja Lee, and Klaas J. van Wijk. "Arabidopsis thaliana deficient in two chloroplast ascorbate peroxidases shows accelerated light-induced necrosis when levels of cellular ascorbate are low." *Plant Molecular Biology* 65(5) (2007): 627–644.

Gill, Sarvajeet Singh, and Narendra Tuteja. "Reactive oxygen species and antioxidant machinery in abiotic stress tolerance in crop plants." *Plant Physiology and Biochemistry: PPB* 48(12) (2010): 909–930.

Gosavi, G. U., A. S. Jadhav, A. A. Kale, S. R. Gadaskh, B. D. Pawar, and V. P. Chimote. "Effect of heat-stress on proline, chlorophyll content, heat shock proteins and antioxidant enzyme activity insorghum (*Sorghum bicolor* L.) at seedling stage." *Indian Journal of Biotechnology* 13 (2014): 356–363.

Halliwell, Barry. "Reactive species and antioxidants. Redox biology is a fundamental theme of aerobic life." *Plant Physiology* 141(2) (2006): 312–322.

Han, Sang-Eun, Sang-Ryeol Park, Hawk-Bin Kwon, Bu-Young Yi, Gil-Bok Lee, and Myung-Ok Byun. "Genetic engineering of drought-resistant tobacco plants by introducing the trehalose phosphorylase (TP) gene from Pleurotussajor-caju." *Plant Cell, Tissue and Organ Culture* 82(2) (2005): 151–158.

Hasanuzzaman, Mirza, and Masayuki Fujita. "Exogenous sodium nitroprusside alleviates arsenic-induced oxidative stress in wheat (*Triticum aestivum* L.) seedlings by enhancing antioxidant defense and glyoxalase system." *Ecotoxicology* 22(3) (2013): 584–596.

Hasanuzzaman, Mirza, Mohammad Anwar Hossain, and Masayuki Fujita. "Nitric oxide modulates antioxidant defense and the methylglyoxal detoxification system and reduces salinity-induced damage of wheat seedlings." *Plant Biotechnology Reports* 5(4) (2011): 353.

Havaux, Michel, Françoise Eymery, Svetlana Porfirova, Pascal Rey, and Peter Dörmann. "Vitamin E protects against photoinhibition and photooxidative stress in *Arabidopsis thaliana*." *The Plant Cell* 17(12) (2005): 3451–3469.

Hoque, Md Anamul, Mst Nasrin Akhter Banu, Yoshimasa Nakamura, Yasuaki Shimoishi, and Yoshiyuki Murata. "Proline and glycinebetaine enhance antioxidant defense and methylglyoxal detoxification systems and reduce NaCl-induced damage in cultured tobacco cells." *Journal of Plant Physiology* 165(8) (2008): 813–824.

Hussain, BM Nahid, Sharif Ar Raffi, David J. Burritt, and Mohammad Anwar Hossain. "Exogenous glutathione improves salinity stress tolerance in rice (Oryza sativa L.)." *Plant Gene and Trait* 7(11) (2016): 1–17.

Hussein, Nehal M., M. I. Hussein, S. H. Gadel Hak, H. S. Shaalan, and M. A. Hammad. "Effect of two plant extracts and four aromatic oils on Tutaabsoluta population and productivity of tomato cultivar gold stone." *Journal of Plant Protection and Pathology* 6(6) (2015): 969–985.

Igamberdiev, Abir U., and Robert D. Hill. "Nitrate, NO and haemoglobin in plant adaptation to hypoxia: An alternative to classic fermentation pathways." *Journal of Experimental Botany* 55(408) (2004): 2473–2482.

Jain, M. J., G. M. Mathur, S. K. Koul, and N. S. Sarin. "Ameliorative effects of proline on salt stress-induced lipid peroxidation in cell lines of groundnut (Arachis hypogaea L.)." *Plant Cell Reports* 20(5) (2001): 463–468.

Joshi, Rohit, Khirod Kumar Sahoo, Anil Kumar Singh, Khalid Anwar, Raj Kumar Gautam, S. L. Krishnamurthy, S. K. Sopory, Ashwani Pareek, and Sneh Lata Singla-Pareek. "Engineering trehalose biosynthesis pathway improves yield potential in rice under drought, saline and sodic conditions." *Journal of Experimental Botany* 71(2) (2019): 653–668.

Jung, H. S., P. A. Crisp, G. M. Estavillo, B. Cole, F. Hong, T. C. Mockler, B. J. Pogson, and J. Chory. "Subset of heat-shock transcription factors required for the early response of Arabidopsis to excess light." *Proceedings of the National Academy of Sciences of the United States of America* 110(35) (2013): 14474–14479.

Kang, Chen, Hong Zhai, Luyao Xue, Ning Zhao, Shaozhen He, and Qingchang Liu. "A lycopene β-cyclase gene: IbLCYB2, enhances carotenoid contents and abiotic stress tolerance in transgenic sweetpotato." *Plant Science* 272 (2018): 243–254.

Kawakami, Akira, Yutaka Sato, and Midori Yoshida. "Genetic engineering of rice capable of synthesizing fructans and enhancing chilling tolerance." *Journal of Experimental Botany* 59(4) (2008): 793–802.

Keller, F., and M. M. Ludlow. "Carbohydrate metabolism in drought-stressed leaves of pigeonpea (*Cajanus cajan*)." *Journal of Experimental Botany* 44(8) (1993): 1351–1359.

Keunen, E. L. S., Darin Peshev, Jaco Vangronsveld, W. I. M. Van Den Ende, and A. N. N. Cuypers. "Plant sugars are crucial players in the oxidative challenge during abiotic stress: Extending the traditional concept." *Plant, Cell and Environment* 36(7) (2013): 1242–1255.

Khan, Sami U., Jalal U. Din, Abdul Qayyum, Noor E. Jan, and Matthew A. Jenks. "Heat tolerance indicators in Pakistani wheat (*Triticum aestivum* L.) genotypes." *Acta Botanica Croatica* 74(1) (2015): 109–121.

Khattab, H. "Role of glutathione and polyadenylic acid on the oxidative defense systems of two different cultivars of canola seedlings grown under saline conditions." *Australian Journal of Basic and Applied Sciences* 1(3) (2007): 323–334.

Kim, S. E., C. J. Lee, C. Y. Ji, H. S. Kim, S. U. Park, Y. H. Lim, W. S. Park et al. "Transgenic sweetpotato plants overexpressing tocopherol cyclase display enhanced α-tocopherol content and abiotic stress tolerance." *Plant Physiology and Biochemistry: PPB* 144 (2019): 436–444.

Kim, Yun-Hee, Ho Soo Kim, Sung-Chul Park, Chang Yoon Ji, JungWook Yang, Hyeong-Un Lee, and Sang-Soo Kwak. "Overexpression of swpa4 peroxidase enhances tolerance to hydrogen peroxide and high salinity-mediated oxidative stress in transgenic sweetpotato plants." *Plant Biotechnology Reports* 14 (2020): 1–7.

Kumar, Ranjeet R., Suneha Goswami, Sushil K. Sharma, Khushboo Singh, Kritika A. Gadpayle, Narender Kumar, Gyanendra K. Rai, Manorama Singh, and Raj D. Rai. "Protection against heat stress in wheat involves change in cell membrane stability, antioxidant enzymes, osmolyte, H2O2 and transcript of heat shock protein." *International Journal of Plant Physiology and Biochemistry* 4(4) (2012): 83–91.

Kwon, S. Y., S. M. Choi, Y. O. Ahn, H. S. Lee, H. B. Lee, Y. M. Park, and S. S. Kwak. "Enhanced stress-tolerance of transgenic tobacco plants expressing a human dehydroascorbate reductase gene." *Journal of Plant Physiology* 160(4) (2003): 347–353.

Lehner, Arnaud, Christophe Bailly, Brigitte Flechel, Pascal Poels, Daniel Côme, and Françoise Corbineau. "Changes in wheat seed germination ability, soluble carbohydrate and antioxidant enzyme activities in the embryo during the desiccation phase of maturation." *Journal of Cereal Science* 43(2) (2006): 175–182.

Li, H., A. Yang, X. Zhang, F. Gao, and J. Zhang. "Improving freezing tolerance of transgenic tobacco expressing sucrose: Sucrose 1-fructosyltransferase gene from Lactuca sativa." *Plant Cell, Tissue and Organ Culture* 89(1) (2007): 37–48.

Li, Tao, Yumin Zhang, Ying Liu, Xudong Li, Guanglong Hao, Qinghui Han, Lynnette MA Dirk et al. "Raffinose synthase enhances drought tolerance through raffinose synthesis or galactinol hydrolysis in maize and Arabidopsis plants." *Journal of Biological Chemistry* 295(23) (2020): 8064–8077.

Lillo, Cathrine, Unni S. Lea, and Peter Ruoff. "Nutrient depletion as a key factor for manipulating gene expression and product formation in different branches of the flavonoid pathway." *Plant, Cell and Environment* 31(5) (2008): 587–601.

Lim, Soon, Yun-Hee Kim, Sun-Hyung Kim, Suk-Yoon Kwon, Haeng-Soon Lee, Jin-Seog Kim, Kwang-Yun Cho, K. Paek, and Sang-Soo Kwak. "Enhanced tolerance of transgenic sweetpotato plants that express both CuZnSOD and APX in chloroplasts to methyl viologen-mediated oxidative stress and chilling." *Molecular Breeding* 19(3) (2007): 227–239.

Lin, Qingfang, Song Wang, Yihang Dao, Jianyong Wang, and Kai Wang. " Arabidopsis thaliana trehalose-6-phosphate phosphatase gene TPPI enhances drought tolerance by regulating stomatal apertures." *Journal of Experimental Botany* 71(14) (2020): 4285–4297.

Liu, H. L., X. Y. Dai, Y. Y. Xu, and K. Chong. "Over-expression of OsUGE-1 altered raffinose level and tolerance to abiotic stress but not morphology in Arabidopsis." *Journal of Plant Physiology* 164(10) (2007): 1384–1390.

Løvdal, Trond, Kristine M. Olsen, Rune Slimestad, Michel Verheul, and Cathrine Lillo. "Synergetic effects of nitrogen depletion, temperature, and light on the content of phenolic compounds and gene expression in leaves of tomato." *Phytochemistry* 71(5–6) (2010): 605–613.

Luo, Ya, Haoru Tang, and Yong Zhang. "Production of reactive oxygen species and antioxidant metabolism about strawberry leaves to low temperatures." *Journal of Agricultural Science* 3(2) (2011): 89.

Mahmood, Qaisar, Raza Ahmad, Sang-Soo Kwak, Audil Rashid, and Naser A. Anjum. "Ascorbate and glutathione: Protectors of plants in oxidative stress." In: Naser A. Anjum, Shahid Umar, and Ming-Tsair Chan (eds.), *Ascorbate-Glutathione Pathway and Stress Tolerance in Plants*: 209–229. Springer: Dordrecht (2010).

Mari, A. H., I. Rajpar, S. Tunio, and S. Ahmad. "Ions accumulation, proline content and juice quality of sugar beet genotypes as affected by water salinity." *Journal of Animal and Plant Sciences* 28(5) (2018): 1405–1412.

Maruta, Takanori, Takahiro Inoue, Masahiro Noshi, Masahiro Tamoi, Yukinori Yabuta, Kazuya Yoshimura, Takahiro Ishikawa, and Shigeru Shigeoka. " Cytosolic ascorbate peroxidase 1 protects organelles against oxidative stress by wounding- and jasmonate-induced H(2)O(2) in Arabidopsis plants." *Biochimica et Biophysica Acta (BBA)-General Subjects* 1820(12) (2012): 1901–1907.

Mei, Y., D. Qiu, S. Xiao, and D. Chen. "Evaluation of high temperature tolerance and physiological responses of different Anoectochilus germplasm resources." *Applied Ecology and Environmental Research* 16(5) (2018): 7017–7031.

Milligan, A. S., A. Daly, M. A. J. Parry, P. A. Lazzeri, and I. Jepson. "The expression of a maize glutathione S-transferase gene in transgenic wheat confers herbicide tolerance, both in planta and in vitro." *Molecular Breeding* 7(4) (2001): 301–315.

Miranda, José A., Nelson Avonce, Ramón Suárez, Johan M. Thevelein, Patrick Van Dijck, and Gabriel Iturriaga. "A bifunctional TPS–TPP enzyme from yeast confers tolerance to multiple and extreme abiotic-stress conditions in transgenic Arabidopsis." *Planta* 226(6) (2007): 1411–1421.

Mittler, Ron. "Abiotic stress, the field environment and stress combination." *Trends in Plant Science* 11(1) (2006): 15–19.

Mittler, Ron, Sandy Vanderauwera, Martin Gollery, and Frank Van Breusegem. "Reactive oxygen gene network of plants." *Trends in Plant Science* 9(10) (2004): 490–498.

Mohamed, Naheif Ebraheim Mohamed. "Behaviour of wheat Cv. Masr-1 plants to foliar application of some vitamins." *Nature and Science* 11(6) (2013): 1–5.

Molinari, Hugo Bruno Correa, Celso Jamil Marur, Edelclaiton Daros, Marília Kaphan Freitas De Campos, Jane Fiuza Rodrigues Portela De Carvalho, João Carlos Bespalhok Filho, Luiz Filipe Protasio Pereira, and Luiz Gonzaga Esteves Vieira. "Evaluation of the stress-inducible production of proline in transgenic sugarcane (Saccharum spp.): Osmotic adjustment, chlorophyll fluorescence and oxidative stress." *Physiologia Plantarum* 130(2) (2007): 218–229.

Morsy, Mustafa R., Laurent Jouve, Jean-François Hausman, Lucien Hoffmann, and James McD. Stewart. "Alteration of oxidative and carbohydrate metabolism under abiotic stress in two rice (*Oryza sativa* L.) genotypes contrasting in chilling tolerance." *Journal of Plant Physiology* 164(2) (2007): 157–167.

Mullineaux, Philip M., Stanislaw Karpinski, and Neil R. Baker. "Spatial dependence for hydrogen peroxide-directed signaling in light-stressed plants." *Plant Physiology* 141(2) (2006): 346–350.

Munné-Bosch, Sergi. "The role of α-tocopherol in plant stress tolerance." *Journal of Plant Physiology* 162(7) (2005): 743–748.

Nahar, K., M. Hasanuzzaman, M. M. Alam, and M. Fujita. "Roles of exogenous glutathione in antioxidant defense system and methylglyoxal detoxification during salt stress in mung bean." *Biologia Plantarum* 59(4) (2015a): 745–756.

Nahar, Kamrun, Mirza Hasanuzzaman, and Masayuki Fujita. "Physiological roles of glutathione in conferring abiotic stress tolerance to plants." In: Narendra Tuteja, and Sarvajeet S. Gill (eds.), *Abiotic Stress Response in Plants*: 155–184. John Wiley (2016).

Nahar, Kamrun, Mirza Hasanuzzaman, Md. Alam, and Masayuki Fujita. "Glutathione-induced drought stress tolerance in mung bean: Coordinated roles of the antioxidant defence and methylglyoxal detoxification systems." *AoB Plants* 7 (2015b): 1–18.

Nahar, Kamrun, Mirza Hasanuzzaman, Md. Alam, and Masayuki Fujita. "Exogenous glutathione confers high temperature stress tolerance in mung bean (*Vigna radiata* L.) by modulating antioxidant defense and methylglyoxal detoxification system." *Environmental and Experimental Botany* 112 (2015c): 44–54.

Nishizawa, Ayako, Yukinori Yabuta, and Shigeru Shigeoka. "Galactinol and raffinose constitute a novel function to protect plants from oxidative damage." *Plant Physiology* 147(3) (2008): 1251–1263.

Noctor, G., A. Mhamdi, and C. H. Foyer. "The roles of reactive oxygen metabolism in drought: Not so cut and dried." *Plant Physiology* 164(4) (2014): 1636–1648.

Noctor, Graham. "Metabolic signalling in defence and stress: The central roles of soluble redox couples." *Plant, Cell and Environment* 29(3) (2006): 409–425.

Noctor, Graham, and Christine H. Foyer. "Ascorbate and glutathione: Keeping active oxygen under control." *Annual Review of Plant Physiology and Plant Molecular Biology* 49(1) (1998): 249–279.

Noreen, Sibgha, Ayesha Siddiq, Kousar Hussain, Shakeel Ahmad, and Mirza Hasanuzzaman. "Foliar application of salicylic acid with salinity stress on physiological and biochemical attributes of sunflower (*Helianthus annuus* L.) CROP." *Acta Scientiarumpolonorum-Hortorum Cultus* 16(2) (2017): 57–74.

Nounjan, Noppawan, Pakkanan Chansongkrow, Varodom Charoensawan, Jonaliza L. Siangliw, Theerayut Toojinda, Supachitra Chadchawan, and Piyada Theerakulpisut. "High performance of photosynthesis and osmotic adjustment are associated with salt tolerance ability in rice carrying drought tolerance QTL: Physiological and co-expression network analysis.". *Frontiers in Plant Science* 9 (2018): 1135.

Oberschall, A., M. Deák, K. Török, L. Sass, I. Vass, I. Kovács, A. Fehér, D. Dudits, and G. V. Horváth. "A novel aldose/aldehyde reductase protects transgenic plants against lipid peroxidation under chemical and drought stresses." *The Plant Journal : For Cell and Molecular Biology* 24(4) (2000): 437–446.

Olsen, Kristine M., Rune Slimestad, Unni S. Lea, Cato Brede, T. Løvdal, Peter Ruoff, Michel Verheul, and Cathrine Lillo. "Temperature and nitrogen effects on regulators and products of the flavonoid pathway: Experimental and kinetic model studies." *Plant, Cell and Environment* 32(3) (2009): 286–299.

Pandhair, Varindra, and B. S. Sekhon. "Reactive oxygen species and antioxidants in plants: An overview." *Journal of Plant Biochemistry and Biotechnology* 15(2) (2006): 71–78.

Parvanova, D., A. Popova, I. Zaharieva, P. Lambrev, T. Konstantinova, S. Taneva, A. Atanassov, V. Goltsev, and D. Djilianov. "Low temperature tolerance of tobacco plants transformed to accumulate proline, fructans, or glycine betaine. Variable chlorophyll fluorescence evidence." *Photosynthetica* 42(2) (2004): 179–185.

Pei, L., R. Che, L. He, X. Gao, W. Li, and H. Li. "Role of exogenous glutathione in alleviating abiotic stress in maize (*Zea mays* L.)." *Journal of Plant Growth Regulation* 38(1) (2019): 199–215.

Pennycooke, Joyce C., Michelle L. Jones, and Cecil Stushnoff. "Down-regulating α-galactosidase enhances freezing tolerance in transgenic petunia." *Plant Physiology* 133(2) (2003): 901–909.

Peshev, Darin, Rudy Vergauwen, Andrea Moglia, E. Hideg, and Wim Van den Ende. "Towards understanding vacuolar antioxidant mechanisms: A role for fructans?." *Journal of Experimental Botany* 64(4) (2013): 1025–1038.

Pilon-Smits, Elizabeth A. H., Norman Terry, Tobin Sears, and Kees van Dun. "Enhanced drought resistance in fructan-producing sugar beet." *Plant Physiology and Biochemistry* 37(4) (1999): 313–317.

Posmyk, M. M., and K. M. Janas. "Effects of seed hydropriming in presence of exogenous proline on chilling injury limitation in Vigna radiata L. seedlings." *Acta Physiologiae Plantarum* 29(6) (2007): 509–517.

Potters, Geert, Laura De Gara, Han Asard, and Nele Horemans. "Ascorbate and glutathione: Guardians of the cell cycle, partners in crime?." *Plant Physiology and Biochemistry* 40(6–8) (2002): 537–548.

Prashanth, S. R., V. Sadhasivam, and Ajay Parida. "Over expression of cytosolic copper/zinc superoxide dismutase from a mangrove plant Avicennia marina in indica rice var Pusa Basmati-1 confers abiotic stress tolerance." *Transgenic Research* 17(2) (2008): 281–291.

Pujni, D., A. Chaudhary, and M. V. Rajam. "Increased tolerance to salinity and drought in transgenic indica rice by mannitol accumulation." *Journal of Plant Biochemistry and Biotechnology* 16(1) (2007): 1–7.

Qiu, Boyin, Fanrong Zeng, Shengguan Cai, Xiaojian Wu, Shamsi Imran Haider, Feibo Wu, and Guoping Zhang. "Alleviation of chromium toxicity in rice seedlings by applying exogenous glutathione." *Journal of Plant Physiology* 170(8) (2013): 772–779.

Rady, Mostafa M., Alpaslan Kuşvuran, Hesham F. Alharby, Yahya Alzahrani, and Sebnem Kuşvuran. "Pretreatment with proline or an organic bio-stimulant induces salt tolerance in wheat plants by improving antioxidant redox state and enzymatic activities and reducing the oxidative stress." *Journal of Plant Growth Regulation* 38(2) (2019): 449–462.

Raja, R. B., S. Agasimani, A. Varadharajan, and S. G. Ram. "Natural variability and effect of processing techniques on raffinose family oligosaccharides in pigeonpea cultivars." *Legume Research - an International Journal* 39(4) (2016): 528–532.

Raja, Ramadoss Bharathi, Ramya Balraj, Somanath Agasimani, Elango Dinakaran, Venkatesan Thiruvengadam, Jutti Rajendran Kannan Bapu, and Sundaram Ganesh Ram. "Determination of oligosaccharide fraction in a worldwide germplasm collection of chickpea ('Cicer arietinum'L.) using high performance liquid chromatography." *Australian Journal of Crop Science* 9(7) (2015): 605–613.

Ramadoss, Bharathi Raja, and Arun S. K. Shunmugam. "Anti-dietetic factors in legumes—Local methods to reduce them." *International Journal of Food and Nutritional Sciences* 3(3) (2014): 84–89.

Ramel, Fanny, Simona Birtic, S. Cuiné, C. Triantaphylidès, J. L. Ravanat, and M. Havaux. "Chemical quenching of singlet oxygen by carotenoids in plants.". *Plant Physiology* 158(3) (2012): 1267–1278.

Raza, Syed Hammad, Fahad Shafiq, Mahwish Chaudhary, and Imran Khan. "Seed invigoration with water, ascorbic and salicylic acid stimulates development and biochemical characters of okra (*Ablemoschus esculentus*) under normal and saline conditions." *International Journal of Agriculture and Biology* 15(3) (2013): 486–492.

Razaji, A., D. E. Asli, and M. Farzanian. "The effects of seed priming with ascorbic acid on drought tolerance and some morphological and physiological characteristics of safflower (Carthamus tinctorius L.)." *Annals of Biological Research* 3(8) (2012): 3984–3989.

Remorini, Damiano, Juan Carlos Melgar, Lucia Guidi, Elena Degl'Innocenti, Silvana Castelli, Maria Laura Traversi, R. Massai, and Massimiliano Tattini. "Interaction effects of root-zone salinity and solar irradiance on the physiology and biochemistry of *Olea europaea*." *Environmental and Experimental Botany* 65(2–3) (2009): 210–219.

Ren, Xiaozhi, Zhizhong Chen, Yue Liu, Hairong Zhang, Min Zhang, Qian Liu, Xuhui Hong, J. K. Zhu, and Zhizhong Gong. "ABO3, a WRKY transcription factor, mediates plant responses to abscisic acid and drought tolerance in Arabidopsis." *The Plant Journal : For Cell and Molecular Biology* 63(3) (2010): 417–429.

Ren, Yongbing, Min Miao, Yun Meng, Jiasheng Cao, Tingting Fan, Junyang Yue, Fangming Xiao, Yongsheng Liu, and Shuqing Cao. "DFR1-mediated inhibition of proline degradation pathway regulates drought and freezing tolerance in Arabidopsis." *Cell Reports* 23(13) (2018): 3960–3974.

Roxas, Virginia P., Roger K. Smith, Eric R. Allen, and Randy D. Allen. "Overexpression of glutathione S-transferase/glutathioneperoxidase enhances the growth of transgenic tobacco seedlings during stress." *Nature Biotechnology* 15(10) (1997): 988–991.

Sadak, M. Sh., M. M. Rady, N. M. Badr, and M. S. Gaballah. "Increasing sunflower salt tolerance using nicotinamide and α—Tocopherol." *International Journal of Academic Research* 2(4) (2010): 263–270.

Sánchez-Reinoso, Alefsi David, Gabriel Garcés-Varón, and Hermann Restrepo-Díaz. "Biochemical and physiological characterization of three rice cultivars under different daytime temperature conditions." *Chilean Journal of Agricultural Research* 74(4) (2014): 373–379.

Santarius, Kurt A. "The protective effect of sugars on chloroplast membranes during temperature and water stress and its relationship to frost, desiccation and heat resistance." *Planta* 113(2) (1973): 105–114.

Sarowar, Sujon, Eui Nam Kim, Young Jin Kim, Sung Han Ok, Ki Deok Kim, Byung Kook Hwang, and JeongSheop Shin. "Overexpression of a pepper ascorbate peroxidase-like 1 gene in tobacco plants enhances tolerance to oxidative stress and pathogens." *Plant Science* 169(1) (2005): 55–63.

Saviranta, Niina M. M., Riitta Julkunen-Tiitto, Elina Oksanen, and Reijo O. Karjalainen. "Leaf phenolic compounds in red clover (*Trifolium pratense* L.) induced by exposure to moderately elevated ozone." *Environmental Pollution* 158(2) (2010): 440–446.

Schmidt, Susanne, Michaela Zietz, Monika Schreiner, Sascha Rohn, Lothar W. Kroh, and Angelika Krumbein. "Genotypic and Climatic Influences on the Concentration and Composition of Flavonoids in Kale (Brassica oleraceavar. sabellica)." *Food Chemistry* 119(4) (2010): 1293–1299.

Shafiq, Sidra, Nudrat Aisha Akram, Muhammad Ashraf, and Amara Arshad. "Synergistic effects of drought and ascorbic acid on growth, mineral nutrients and oxidative defense system in canola (Brassica napus L.) plants." *Acta Physiologiae Plantarum* 36(6) (2014): 1539–1553.

Shao, H. B., L. Y. Chu, Z. H. Lu, and C. M. Kang. "Primary antioxidant free radical scavenging and redox signaling pathways in higher plant cells." *International Journal of Biological Sciences* 4(1) (2007): 8–14.

Sharma, Pallavi, Ambuj Bhushan Jha, Rama Shanker Dubey, and Mohammad Pessarakli. "Reactive oxygen species, oxidative damage, and antioxidative defense mechanism in plants under stressful conditions." *Journal of Botany* 2012 (2012): 1–26.

Shen, Yang, Bowei Jia, Jinyu Wang, Xiaoxi Cai, Bingshuang Hu, Yan Wang, Yue Chen, Mingzhe Sun, and Xiaoli Sun. "Functional analysis of Arabidopsis thaliana galactinol synthase AtGolS2 in response to abiotic stress." *Molecular Plant Breeding* 11(14) (2020): 1–14.

Shin, Hyunsuk, Sewon Oh, Rajeev Arora, and Daeil Kim. "Proline accumulation in response to high temperature in winter-acclimated shoots of Prunus persica: A response associated with growth resumption or heat stress?." *Canadian Journal of Plant Science* 96(4) (2016): 630–638.

Smirnoff, Nicholas. "Botanical briefing: The function and metabolism of ascorbic acid in plants." *Annals of Botany* 78(6) (1996): 661–669.

Song, J., R. Zhang, D. Yue, X. Chen, Z. Guo, C. Cheng, M. Hu, J. Zhang, and K. Zhang. "Co-expression of ApGSMT2g and ApDMT2g in cotton enhances salt tolerance and increases seed cotton yield in saline fields." *Plant Science: An International Journal of Experimental Plant Biology* 274 (2018): 369–382.

Stoyanova, Silviya, Jan Geuns, Eva Hideg, and Wim Van Den Ende. "The food additives inulin and stevioside counteract oxidative stress." *International Journal of Food Sciences and Nutrition* 62(3) (2011): 207–214.

Sun, Xuemei, Yuan Zong, Shipeng Yang, Lihui Wang, Jieming Gao, Ying Wang, Baolong Liu, and Huaigang Zhang. "A fructan: The fructan 1-fructosyl-transferase gene from *Helianthus tuberosus* increased the PEG-simulated drought stress tolerance of tobacco." *Hereditas* 157 (2020): 1–8.

Suzuki, Nobuhiro, Gad Miller, Carolina Salazar, Hossain A. Mondal, Elena Shulaev, Diego F. Cortes, Joel L. Shuman et al. "Temporal-spatial interaction between reactive oxygen species and abscisic acid regulates rapid systemic acclimation in plants." *The Plant Cell* 25(9) (2013): 3553–3569.

Tanaka, Y., T. Hibino, Y. Hayashi, A. Tanaka, S. Kishitani, T. Takabe, S. Yokota, and T. Takabe. "Salt tolerance of transgenic rice overexpressing yeast mitochondrial Mn-SOD in chloroplasts." *Plant Science* 148(2) (1999): 131–138.

Tang, Wei, Xuexian Peng, and Ronald J. Newton. "Enhanced tolerance to salt stress in transgenic loblolly pine simultaneously expressing two genes encoding mannitol-1-phosphate dehydrogenase and glucitol-6-phosphate dehydrogenase." *Plant Physiology and Biochemistry : PPB* 43(2) (2005): 139–146.

Tarczynski, Mitchell C., Richard G. Jensen, and Hans J. Bohnert. "Stress protection of transgenic tobacco by production of the osmolyte mannitol." *Science* 259(5094) (1993): 508–510.

Tatar, O., C. Ozalkan, and G. D. Atasoy. "Partitioning of dry matter, proline accumulation, chlorophyll content and antioxidant activity of chickpea (*Cicer arietinum* L.) plants under chilling stress." *Bulgarian Journal of Agricultural Science* 19(2) (2013): 260–265.

Tattini, Massimiliano, Damiano Remorini, P. Pinelli, G. Agati, E. Saracini, M. L. Traversi, and R. Massai. "Morpho-anatomical, physiological and biochemical adjustments in response to root zone salinity stress and high solar radiation in two Mediterranean evergreen shrubs, Myrtus communis and Pistacia lentiscus.". *New Phytologist* 170(4) (2006): 779–794.

Tavakoli, M., K. Poustini, and H. Alizadeh. "Proline accumulation and related genes in wheat leavesunder salinity stress." *Journal of Agricultural Science and Technology* 18 (2016): 707–716.

Tian, Fengxia, Wenqiang Wang, Chao Liang, Xin Wang, Guiping Wang, and Wei Wang. "Overaccumulation of glycine betaine makes the function of the thylakoid membrane better in wheat under salt stress." *The Crop Journal* 5(1) (2017): 73–82.

Tseng, MenqJiau, Cheng-Wei Liu, and Jinn-Chin Yiu. "Enhanced tolerance to sulfur dioxide and salt stress of transgenic Chinese cabbage plants expressing both superoxide dismutase and catalase in chloroplasts." *Plant Physiology and Biochemistry : PPB* 45(10–11) (2007): 822–833.

Uçarlı, Cüneyt. "Short-term effects of exogenous application of ascorbic acid on barley (Hordeum vulgare L.) seedlings under salinity stress." *International Journal of Environmental Research and Technology* 2(3) (2019): 37–43.

Upadhyaya, C. P., J. Venkatesh, M. A. Gururani, L. Asnin, K. Sharma, H. Ajappala, and S. W. Park. "Transgenic potato overproducing L-ascorbic acid resisted an increase in methylglyoxal under salinity stress via maintaining higher reduced glutathione level and glyoxalase enzyme activity." *Biotechnology Letters* 33(11) (2011): 2297.

Van Breusegem, Frank, L. Slooten, J. -M. Stassart, J. Botterman, T. Moens, M. Van Montagu, and D. Inze. "Effects of overproduction of tobacco MnSOD in maize chloroplasts on foliar tolerance to cold and oxidative stress.". *Journal of Experimental Botany* 50(330) (1999): 71–78.

Van Camp, Wim, K. Capiau, M. Van Montagu, D. Inzé, and L. Slooten. "Enhancement of oxidative stress tolerance in transgenic tobacco plants overproducing Fe-superoxide dismutase in chloroplasts.". *Plant Physiology* 112(4) (1996): 1703–1714.

Van den Ende, Wim, and Ravi Valluru. "Sucrose, sucrosyl oligosaccharides, and oxidative stress: Scavenging and salvaging?." *Journal of Experimental Botany* 60(1) (2009): 9–18.

Van Doorn, Wouter G., and Saichol Ketsa. "Cross reactivity between ascorbate peroxidase and phenol (guaiacol) peroxidase." *Postharvest Biology and Technology* 95 (2014): 64–69.

Vasquez-Robinet, Cecilia, Shrinivasrao P. Mane, Alexander V. Ulanov, Jonathan I. Watkinson, Verlyn K. Stromberg, David De Koeyer, Roland Schafleitner et al. "Physiological and molecular adaptations to drought in Andean potato genotypes." *Journal of Experimental Botany* 59(8) (2008): 2109–2123.

Venkatesh, J., C. P. Upadhyaya, J. Yu, A. Hemavathi, D. H. Kim, R. J. Strasser, and S. W. Park. "Chlorophyll a fluorescence transient analysis of transgenic potato overexpressing D-galacturonic acid reductase gene for salinity stress tolerance." *Horticulture, Environment, and Biotechnology* 53(4) (2012): 320–328.

Vera Hernandez, F. P., M. A. Ortega-Ramirez, M. Martinez Nunez, M. Ruiz-Rivas, and F. Rosas Cardenas. "Proline as a probable biomarker of cold stress tolerance in sorghum (*Sorghum bicolor*). Mex." *Journal of Biotechnology* 3 (2018): 77–86.

Walia, Harkamal, Clyde Wilson, Pascal Condamine, Xuan Liu, Abdelbagi M. Ismail, Linghe Zeng, Steve I. Wanamaker et al. "Comparative transcriptional profiling of two contrasting rice genotypes under salinity stress during the vegetative growth stage." *Plant Physiology* 139(2) (2005): 822–835.

Wanek, Wolfgang, and Andreas Richter. "Biosynthesis and accumulation of D-ononitol in *Vigna umbellata* in response to drought stress." *Physiologia Plantarum* 101(2) (1997): 416–424.

Wang, Fang, Fei Chen, Yue Cai, Guoping Zhang, and Feibo Wu. "Modulation of exogenous glutathione in ultrastructure and photosynthetic performance against Cd stress in the two barley genotypes differing in Cd tolerance." *Biological Trace Element Research* 144(1–3) (2011): 1275–1288.

Wang, Jiangying, Bin Wu, Hengfu Yin, Zhengqi Fan, Xinlei Li, Sui Ni, Libo He, and Jiyuan Li. "Overexpression of CaAPX induces orchestrated reactive oxygen scavenging and enhances cold and heat tolerances in tobacco." *BioMed Research International* 2017 (2017): 1–15.

Wang, Xiaoyuan, and Peter J. Quinn. "The location and function of vitamin E in membranes (review)." *Molecular Membrane Biology* 17(3) (2000): 143–156.

Wang, Yueju, Michael Wisniewski, Richard Meilan, Minggang Cui, Robert Webb, and Leslie Fuchigami. "Overexpression of cytosolic ascorbate peroxidase in tomato confers tolerance to chilling and salt stress." *Journal of the American Society for Horticultural Science* 130(2) (2005): 167–173.

Winkler, Barry S., Stephen M. Orselli, and Tonia S. Rex. "The redox couple between glutathione and ascorbic acid: A chemical and physiological perspective." *Free Radical Biology and Medicine* 17(4) (1994): 333–349.

Xiang, C. H. E. N. G., and D. Oliver. "Glutathione and its central role in mitigating plant stress." In: Pessarakli M (ed) *Handbook of Plant and Crop Stress*: 697–707. Marcel Dekker, Inc: New York (1999).

You, Lili, Qiping Song, Yuyong Wu, Shengchun Li, Chunmei Jiang, Ling Chang, Xinghong Yang, and Jiang Zhang. "Accumulation of glycine betaine in transplastomic potato plants expressing choline oxidase confers improved drought tolerance." *Planta* 249(6) (2019): 1963–1975.

Yu, T. A. O., Yang Sheng Li, Xue Feng Chen, Jing Hu, X. U. N. Chang, and Ying Guo Zhu. "Transgenic tobacco plants overexpressing cotton glutathione S-transferase (GST) show enhanced resistance to methyl viologen." *Journal of Plant Physiology* 160(11) (2003): 1305–1311.

Zechmann, Bernd, and Maria Müller. "Subcellular compartmentation of glutathione in dicotyledonous plants." *Protoplasma* 246(1–4) (2010): 15–24.

Zhang, Kangping, Guiyin Wang, Mingchen Bao, Longchang Wang, and X. Xie. "Exogenous application of ascorbic acid mitigates cadmium toxicity and uptake in Maize (*Zea mays* L.)." *Environmental Science and Pollution Research International* 26(19) (2019): 19261–19271.

Zhang, Mengru, Ming Gong, Yumei Yang, Xujuan Li, Haibo Wang, and Zhurong Zou. "Improvement on the thermal stability and activity of plant cytosolic ascorbate peroxidase 1 by tailing hyper-acidic fusion partners." *Biotechnology Letters* 37(4) (2015): 891–898.

Zhang, Shu-Zhen, Y. A. N. G. Ben-Peng, F. E. N. G. Cui-Lian, and T. A. N. G. Huo-Long. "Genetic Transformation of Tobacco with the Trehalose Synthase Gene from Grifolafrondosa Fr. enhances the resistance to drought and salt in tobacco." *Journal of Integrative Plant Biology* 47(5) (2005): 579–587.

Zhifang, G., and W. H. Loescher. "Expression of a celery mannose 6-phosphate reductase in Arabidopsis thaliana enhances salt tolerance and induces biosynthesis of both mannitol and a glucosyl-mannitol dimer." *Plant, Cell and Environment* 26(2) (2003): 275–283.

Zhou, Yu, Yan Liu, Shuangshuang Wang, Cong Shi, Ran Zhang, Jia Rao, Xu Wang et al. "Molecular cloning and characterization of galactinol synthases in *Camellia sinensis* with different responses to biotic and abiotic stressors." *Journal of Agricultural and Food Chemistry* 65(13) (2017): 2751–2759.

Zhu, B., J. Su, M. Chang, D. P. S. Verma, Y. Fan, and R. Wu. "Overexpression of a Δ1-pyrroline-5-carboxylate synthetase gene and analysis of tolerance to water- and salt-stress in transgenic rice." *Plant Science* 139(1) (1998): 41–48.

Zietz, Michaela, Annika Weckmüller, Susanne Schmidt, Sascha Rohn, Monika Schreiner, Angelika Krumbein, and Lothar W. Kroh. "Genotypic and climatic influence on the antioxidant activity of flavonoids in kale (*Brassica oleracea var. sabellica*)." *Journal of Agricultural and Food Chemistry* 58(4) (2010): 2123–2130.

Index

Printed in the United States
by Baker & Taylor Publisher Services